Rational Forms Involving $a + bu$

5. $\displaystyle \int \frac{u\,du}{a + bu} = \frac{1}{b^2}[a + bu - a\,\ln|a + bu|] + C$

6. $\displaystyle \int \frac{u^2\,du}{a + bu} = \frac{1}{b^2}[\tfrac{1}{2}(a + bu)^2 - 2a(a + bu) + a^2\,\ln|a + bu|] + C$

7. $\displaystyle \int \frac{u\,du}{(a + bu)^2} = \frac{1}{b^2}\left[\frac{a}{a + bu} + \ln|a + bu|\right] + C$

8. $\displaystyle \int \frac{u^2\,du}{(a + bu)^2} = \frac{1}{b^3}\left[a + bu - \frac{a^2}{a + bu} - 2a\,\ln|a + bu|\right] + C$

9. $\displaystyle \int \frac{du}{u(a + bu)} = -\frac{1}{a}\ln\left|\frac{a + bu}{u}\right| + C$

10. $\displaystyle \int \frac{du}{u^2(a + bu)} = -\frac{1}{au} + \frac{b}{a^2}\ln\left|\frac{a + bu}{u}\right| + C$

11. $\displaystyle \int \frac{du}{u(a + bu)^2} = \frac{1}{a(a + bu)} - \frac{1}{a^2}\ln\left|\frac{a + bu}{u}\right| + C$

Forms Involving $\sqrt{a + bu}$

12. $\displaystyle \int u\sqrt{a + bu}\,du = \frac{2(3bu - 2a)}{15b^2}(a + bu)^{3/2} + C$

13. $\displaystyle \int \frac{u\,du}{\sqrt{a + bu}} = \frac{2(bu - 2a)}{3b^2}\sqrt{a + bu} + C$

14a. $\displaystyle \int \frac{du}{u\sqrt{a + bu}} = \frac{1}{\sqrt{a}}\ln\left|\frac{\sqrt{a + bu} - \sqrt{a}}{\sqrt{a + bu} + \sqrt{a}}\right| + C \qquad \text{if } a > 0$

14b. $\displaystyle \qquad\qquad = \frac{2}{\sqrt{-a}}\tan^{-1}\sqrt{\frac{a + bu}{-a}} + C \qquad \text{if } a < 0$

15. $\displaystyle \int \frac{\sqrt{a + bu}}{u}\,du = 2\sqrt{a + bu} + a\int \frac{du}{u\sqrt{a + bu}}$

Forms Involving $a^2 \pm u^2$ and $u^2 - a^2$

16. $\displaystyle \int \frac{du}{a^2 + u^2} = \frac{1}{a}\tan^{-1}\frac{u}{a} + C$

17. $\displaystyle \int \frac{du}{a^2 - u^2} = \frac{1}{2a}\ln\left|\frac{a + u}{a - u}\right| + C$

18. $\displaystyle \int \frac{du}{u^2 - a^2} = \frac{1}{2a}\ln\left|\frac{u - a}{u + a}\right| + C$

(Continued on inside back cover)

THE CALCULUS TUTORING BOOK

Carol Ash
Robert B. Ash
Department of Mathematics
University of Illinois at Urbana-Champaign

IEEE
PRESS

The Institute of Electrical and Electronics Engineers, Inc., New York

Copyright © 1986 by
THE INSTITUTE OF ELECTRICAL
AND ELECTRONICS ENGINEERS, INC.
345 East 47th Street, New York, NY 10017-2394
All rights reserved.

PRINTED IN THE UNITED STATES OF AMERICA

ISBN 0-7803-1044-6(pbk)   IEEE Order Number: PP0177-6
ISBN 0-87942-183-5(case)   IEEE Order Number: PC0177-6

Library of Congress Cataloging-in-Publication Data
Ash, Carol, 1935–
   The calculus tutoring book.

   Includes index.
   "IEEE order number PC01776" — T.p. verso.
   1. Calculus.   I. Ash, Robert B.   II. Title.
QA303.A75 1986   515   85-23049

ISBN 0-87942-183-5

iv

CONTENTS

10/TOPICS IN THREE-DIMENSIONAL ANALYTIC GEOMETRY

11/PARTIAL DERIVATIVES

12/MULTIPLE INTEGRALS

APPENDIX

PREFACE

This is a text in calculus, written for students in mathematics and applied areas such as engineering, physics, chemistry, computer science, economics, biology, and psychology. The style is unlike that of the usual text that the student encounters when enrolling in a standard calculus sequence. We'll try to explain the reasoning behind our approach, which is based on more than 20 years of teaching experience.

Mathematicians and consumers of mathematics (such as engineers) seem to disagree as to what mathematics actually is. To a mathematician, it is important to distinguish between rigor and intuition. To an engineer, intuitive thinking, geometric reasoning, and physical deductions are all valid if they illuminate a problem, and a formal proof is often unnecessary or counterproductive.

Most calculus texts claim to be intuitive, informal, and even friendly, and in fact one can find many worked-out examples, as well as some geometric and physical reasoning. However, the dominant feature of these books is *formalism*. Definitions and theorems are stated precisely, and many results are proved at a level of rigor that is acceptable to a working mathematician. We admit to a twinge of embarrassment in arguing that this is bad. However, our calculus students have ranged from close to the best to be found anywhere, to far from the worst, and it seems entirely clear to us that most students are not ready for an abstract presentation, and they simply will not learn the formalism. The better students will succeed in reading around the abstractions, so that the textbook at least becomes useful as a source of examples.

Our approach uses informal language and emphasizes geometric and physical reasoning. The style is similar to that used in applied courses and, for this reason, students find the presentation very congenial. They do not regard calculus as a strange subject outside their normal experience. Invariably, a number of students are motivated toward further study of mathematics, and there is no better preparation than to learn to think intuitively, geometrically, and physically.

We expect that this text will be used for independent study, or as a supplement or reference for those who are having difficulty in a standard calculus course; for maximum benefit to the student, detailed solutions to all problems are supplied. (We have used the book as a classroom text, and have found the inclusion of detailed solutions to be a useful feature here as well.) The problems are limited in number so that it is feasible to work through all of them. They have been carefully chosen so that a student who does most of them will be well prepared for applications of calculus in later courses. The text and problems concentrate on basic material rather than fringe topics; as a result the book is of manageable size.

We believe that for a student encountering calculus for the first time, our approach is most appropriate. We hope that faculty who teach courses in which calculus is applied will, after seeing how well the approach works, try to influence departments of mathematics to change their style of teaching.

The close cooperation and teamwork of the staff at IEEE PRESS were invaluable. In particular, we would like to express our gratitude to David Boulanger, Associate Editor; W. Reed Crone, Managing Editor; and David L. Staiger, Staff Director.

We wanted the diagrams in the book to be freehand line drawings, similar to those sketched by an instructor at a blackboard or a student working at home. We thank our artist, Evan Polenghi, for carrying out our conception with skill and grace.

Above all, we thank Professor M. E. Van Valkenburg, Dean of the School of Engineering at the University of Illinois at Urbana-Champaign and Editor in Chief of IEEE PRESS, for making the publication of this text possible.

CAROL ASH
ROBERT ASH

1/FUNCTIONS

We begin calculus with a chapter on functions because virtually all problems in calculus involve functions. We discuss functions in general, and then concentrate on the special functions which will be used repeatedly throughout the course.

1.1 Introduction

A function may be thought of as an input-output machine. Given a particular input, there is a corresponding output. This process may be represented by various schemes, such as a table or a mapping diagram listing inputs and outputs (Fig. 1). Functions will usually be denoted by single letters, the most common being f and g. If the function g produces the output 3 when the input is 2, we write $g(2) = 3$.

TABLE

INPUT	OUTPUT
2	3
8	4
9	4
10	-1

MAPPING DIAGRAM

$2 \longrightarrow 3$
8
9 $ 4$
$10 \longrightarrow -1$

FIG. 1

Often functions are described with formulas. If $f(x) = x^2 + x$ then $f(3) = 9 + 3 = 12$, $f(a) = a^2 + a$, $f(a + b) = (a + b)^2 + (a + b) = a^2 + 2ab + b^2 + a + b$. We might refer to "the function $x^2 + x$" without using a special name such as f.

For example, if $f(x) = 2x - 9$ then

$$f(3) = 6 - 9 = -3$$
$$f(0) = -9$$
$$f(a) = 2a - 9$$
$$f(a + b) = 2(a + b) - 9 = 2a + 2b - 9$$
$$f(a) + f(b) = 2a - 9 + 2b - 9 = 2a + 2b - 18$$
$$f(3a) = 2(3a) - 9 = 6a - 9$$
$$3f(a) = 3(2a - 9) = 6a - 27$$
$$f(a^2) = 2a^2 - 9$$
$$(f(a))^2 = (2a - 9)^2 = 4a^2 - 36a + 81$$
$$f(-a) = 2(-a) - 9 = -2a - 9$$
$$-f(a) = -(2a - 9) = -2a + 9.$$

1

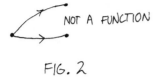

FIG. 2

The input of a function f is called the *independent variable,* while the output is the *dependent variable.* We say that the function f *maps x to $f(x)$,* and call $f(x)$ the *value* of the function at x. The set of inputs is called the *domain* of f, and the set of outputs is the *range.*

A function $f(x)$ is not allowed to send one input to more than one output. Figure 2 illustrates a correspondence that is *not* a function. For example, it is illegal to write $g(x) = \pm\sqrt{2x^2 + 3}$, since each value of x produces *two* outputs. It certainly is legal to write and use the *expression* $\pm\sqrt{2x^2 + 3}$, but it cannot be named $g(x)$ and called a function.

Functions often arise when a problem is translated into mathematical terms. The solution to the problem may then involve operating on the functions with calculus. Before continuing with functions in more detail we'll give an example of a function emerging in practice. Suppose a pigeon is flying from point A over water to point B on the beach (Fig. 3), and the energy required to fly is 60 calories per mile over water but only 40 calories per mile over land. (The effect of cold air dropping makes flying over water more taxing.) The problem is to find the path that requires minimum energy. The direct path from A to B is shortest, but it has the disadvantage of being entirely over water. The path ACB is longer, but it has the advantage of being mostly over land. In general, suppose the bird first flies from A to a point P on the beach x miles from C, and then travels the remaining $10 - x$ miles to B. The value $x = 0$ corresponds to the path ACB, and $x = 10$ corresponds to the path AB. The total energy E used in flight can be calculated as follows:

$$E = \text{energy expended over water} + \text{energy expended over land}$$
$$= \text{calories per water mile} \times \text{water miles}$$
$$+ \text{calories per land mile} \times \text{land miles}$$
$$= 60\,\overline{AP} + 40\,\overline{PB}$$

$$(1) \qquad = 60\sqrt{36 + x^2} + 40(10 - x), \qquad 0 \le x \le 10.$$

Thus the energy is a *function* of x. Calculus will be used in Section 4.2 to finish the problem and find the value of x that minimizes E.

In deriving (1), we restricted x so that $0 \le x \le 10$ since we assumed that to minimize energy the bird should fly to a point P *between C and B* as indicated in Fig. 3. Since problems often restrict the independent variable in a similar fashion, certain notation and terminology has become standard.

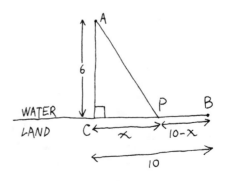

FIG. 3

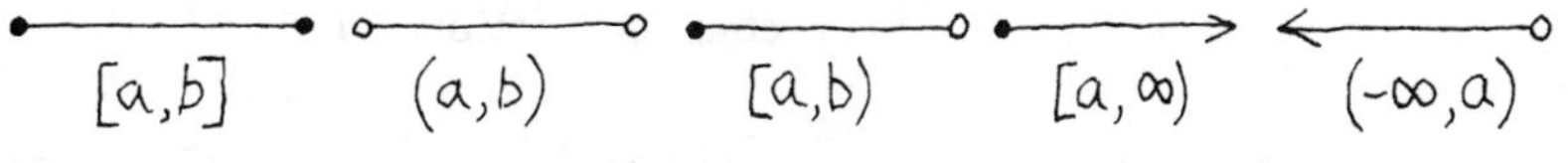

$$[a,b] \qquad (a,b) \qquad [a,b) \qquad [a,\infty) \qquad (-\infty,a)$$

FIG. 4

The set of all x such that $a \le x \le b$ is denoted by $[a,b]$ and called a *closed interval* (Fig. 4). With this notation, the variable x in (1) lies in the interval $[0,10]$. The set of all x such that $a < x < b$ is denoted by (a,b) and called an *open interval*. Similarly we use $[a,b)$ for the set of x where $a \le x < b$, $(a,b]$ for $a < x \le b$, $[a,\infty)$ for $x \ge a$, (a,∞) for $x > a$, $(-\infty,a]$ for $x \le a$, and $(-\infty,a)$ for $x < a$. In general, the square bracket, and the solid dot in Fig. 4, means that the endpoint belongs to the set; a parenthesis, and the small circle in Fig. 4, means that the endpoint does not belong to the set. The notation $(-\infty,\infty)$ refers to the set of all real numbers.

As another example of a function, consider the *greatest integer function*: Int x is defined as the largest integer that is less than or equal to x. Equivalently, Int x is the first integer at or to the left of x on the number line. For example, Int $5.3 = 5$, Int $5.4 = 5$, Int $7 = 7$, Int$(-6.3) = -7$. Note that for *positive* inputs, Int simply chops away the decimal part. The domain of Int is the set of all (real) numbers. (Elementary calculus uses only the real number system and excludes nonreal complex numbers such as $3i$ and $4 + 2i$.) The range of Int is the set of integers. Frequently, Int x is denoted by $[x]$. Many computers have an internal Int operation available. To illustrate one of its uses, suppose that a computer obtains a numerical result, such as $x = 2.1679843$, and is instructed to keep only the first 4 digits. The computer multiplies by 1000 to obtain 2167.9843, applies Int to get 2167, and then divides by 1000 to obtain the desired result 2.167 or, in our functional notation, $\frac{1}{1000}$ Int$(1000\,x)$.

Most work in calculus involves a few *basic* functions, which (amazingly) have proved sufficient to describe a large number of physical phenomena. As a preview, and for reference, we list these functions now, but it will take most of the chapter to discuss them carefully. The material is important preparation for the rest of the course, since the basic functions dominate calculus.

Table of Basic Functions

Type	Examples
Constant functions	$f(x) = 2 \qquad$ for all x, $g(x) = -\pi \qquad$ for all x
Power functions	$x^2,\ x^3,\ x,\ x^{1/2},\ x^{-1},\ x^{-99/5},\ x^{2.7}$
Trigonometric functions	sine, cosine, tangent, secant, cosecant, cotangent
Inverse trigonometric functions	$\sin^{-1}x,\ \cos^{-1}x,\ \tan^{-1}x$
Exponential functions	$2^x,\ 3^x,\ (\frac{1}{2})^x,\ 10^x$ and especially e^x, where $e = 2.71828\cdots$
Logarithm functions	$\log_2 x,\ \log_3 x,\ \log_{1/2} x,\ \log_{10} x$ and especially $\log_e x$, denoted $\ln x$

Problems for Section 1.1

1. Let $f(x) = 2 - x^2$ and $g(x) = (x - 3)^2$. Find

(a) $f(0)$ (d) $g(0)$ (g) $(g(b))^3$
(b) $f(1)$ (e) $g(1)$ (h) $f(2a + b)$
(c) $f(b^3)$ (f) $g(b^3)$ (i) the range of f and of g, if the domain is $(-\infty, \infty)$

2. Let $f(x) = |x|/x$.

(a) Find $f(-7)$ and $f(3)$.
(b) For what values of x is the function defined?
(c) With the domain from part (b), find the range of f.
(d) Does $f(2 + 3)$ equal $f(2) + f(3)$?
(e) Does $f(-2 + 6)$ equal $f(-2) + f(6)$?
(f) Does $f(a + b)$ ever equal $f(a) + f(b)$?

3. The number x_0 is called a fixed point of the function f if $f(x_0) = x_0$; i.e., a fixed point is a number that maps to itself. Find the fixed points of the following functions: (a) $|x|/x$ (b) Int x (c) x^2 (d) $x^2 + 4$.

4. Let $f(x) = 2x + 1$. Does $f(a^2)$ ever equal $(f(a))^2$?

5. If $f(x) = 2x + 3$ then $f(f(x)) = f(2x + 3) = 2(2x + 3) + 3 = 4x + 9$.

(a) Find $f(f(x))$ if $f(x) = x^3$.
(b) Find Int(Int x).
(c) If $f(x) = -x + 1$, find $f(f(x)), f(f(f(x)))$, and so on, until you see the pattern.

6. A charter aircraft has 350 seats and will not fly unless at least 200 of those seats are filled. When there are 200 passengers, a ticket costs $300, but each ticket is reduced by $1 for every passenger over 200. Express the total amount A collected by the charter company as a function of the number p of passengers.

1.2 The Graph of a Function

Information can usually be perceived more easily from a diagram than from a set of statistics or a formula. Similarly, the behavior of a function can often be better understood from its *graph*, which is drawn in a rectangular coordinate system by using the inputs as x-coordinates and the outputs as y-coordinates; i.e., the graph of f is the graph of the equation $y = f(x)$. In sketching a graph it may be useful to make a table of values of the input x and the corresponding output y.

The graph of the function $f(x) = -2x + 3$ is the line with equation $y = -2x + 3$ (Fig. 1). It has slope -2 and passes through the point $(0, 3)$.

The graph of Int x is shown in Fig. 2 along with a partial table of values used to help plot the graph. The graph shows for instance that as x increases from 2 toward 3, Int x, the y-coordinate in the picture, remains 2; when x reaches 3, Int x suddenly jumps to 3.

Example 1 The graph of a function g is given in Fig. 3. Various values of g can be read from the picture: since the point $(0, 6)$ is on the graph, we have $g(0) = 6$; similarly, $g(4) = 11$, $g(10) = 4$. Since P is lower than Q, we can tell that $g(2) < g(3)$. If $g(x)$ represents the final height of a tree when it is planted with x units of fertilizer, then using no fertilizer results in a 6-foot tree, using 10 units of fertilizer overdoses the tree and it grows to

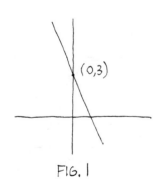

(0,3)

FIG. 1

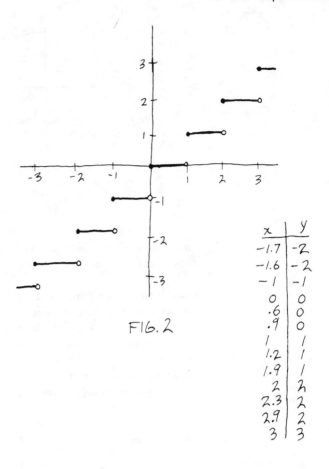

FIG. 2

x	y
-1.7	-2
-1.6	-2
-1	-1
0	0
$.6$	0
$.9$	0
1	1
1.2	1
1.9	1
2	2
2.3	2
2.9	2
3	3

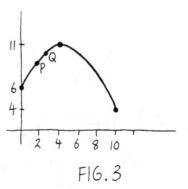

FIG. 3

only 4 feet, while 4 units of fertilizer produces an 11-foot tree, the maximum possible height according to the data.

The vertical line test   Not every curve can be the graph of a function. The curve in Fig. 4 is disqualified because one *x* is paired with several *y*'s, and a function cannot map one input to more than one output. In general, *a curve is the graph of a function if and only if no vertical line ever intersects the curve*

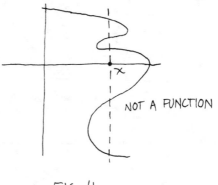

FIG.4

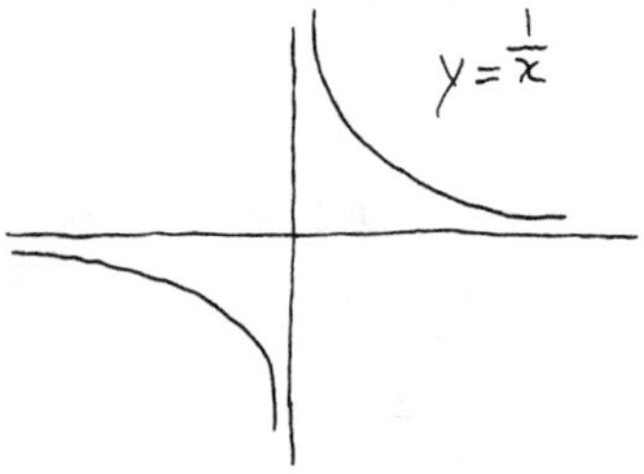

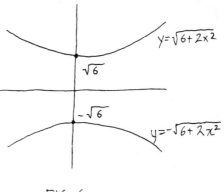

more than once. In other words, if a vertical line intersects the curve at all, it does so only once.

Equations versus functions The hyperbola in Fig. 5 is the graph of the equation $xy = 1$. It is also (solve for y) the graph of the function $f(x) = 1/x$. The hyperbola in Fig. 6 is the graph of the equation $y^2 - 2x^2 = 6$. It is *not* the graph of a function because it fails the vertical line test. However, the upper branch of the hyperbola is the graph of the function $\sqrt{2x^2 + 6}$ (solve for y and choose the positive square root since $y > 0$ on the upper branch), and the lower branch is the graph of the function $-\sqrt{2x^2 + 6}$.

FIG.5

FIG.6

Continuity If the graph of f breaks at $x = x_0$, so that you must lift the pencil off the paper before continuing, then f is said to be *discontinuous* at $x = x_0$. If the graph doesn't break at $x = x_0$, then f is *continuous* at x_0.

The function $-2x + 3$ (Fig. 1) is continuous (everywhere). On the other hand, Int x (Fig. 2) is discontinuous when x is an integer, and $1/x$ (Fig. 5) is discontinuous at $x = 0$.

Many physical quantities are continuous functions. If $h(t)$ is your height at time t, then h is continuous since your height cannot jump.

One-to-one functions, non-one-to-one functions and nonfunctions A function is not allowed to map one input to more than one output (Fig. 7).

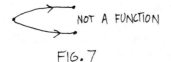

FIG. 7

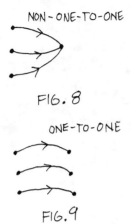

NON-ONE-TO-ONE

FIG. 8

ONE-TO-ONE

FIG. 9

But a function can map more than one input to the same output (Fig. 8), in which case the function is said to be *non-one-to-one*. A *one-to-one* function maps different inputs to different outputs (Fig. 9).

The function x^2 is not one-to-one because, for instance, inputs 2 and -2 both produce the output 4. The function x^3 is one-to-one since two different numbers always produce two different cubes.

A curve that passes the vertical line test, and thus is the graph of a function, will further be the graph of a one-to-one function if and only if no horizontal line intersects the curve more than once (horizontal line test). The function in Fig. 10 fails the horizontal line test and is not one-to-one because x_1 and x_2 produce the same value of y.

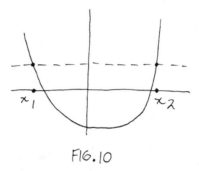

FIG. 10

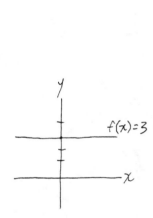

y

$f(x)=3$

x

FIG. 11

Constant functions If, for example, $f(x) = 3$ for all x, then f is called a *constant function*. The graph of a constant function is a horizontal line (Fig. 11). The constant functions are among the basic functions of calculus, listed in the table in Section 1.1.

Power functions Another group of basic functions consists of the *power functions* x^r, such as

$$x^2 = x \cdot x$$

$$x^{-1} = 1/x$$

$$x^{1/2} = \sqrt{x} \qquad \text{(the } positive \text{ square root of } x\text{)}$$

$$x^{-1/3} = \frac{1}{\sqrt[3]{x}}$$

$$x^{7/4} = \sqrt[4]{x^7} = (\sqrt[4]{x})^7$$

$$x^{2.6} = x^{26/10} = \sqrt[10]{x^{26}}.$$

To sketch the graph of x^3, we make a table of values and plot a few points. When the pattern seems clear, we connect the points to obtain the final graph (Fig. 12). The connecting process assumes that x^3 is continuous, something that seems reasonable and can be proved formally. In general, x^r is continuous wherever it is defined. If r is negative then x^r is not defined at $x = 0$ and is discontinuous there; the graph of $1/x$, that is, the graph of x^{-1}, is shown in Fig. 5 with a discontinuity at the origin. Figure 13 gives the graph of x^2 (a parabola) and of x^4. For $-1 < x < 1$, the graph of x^4 lies below the graph of x^2 since the fourth power of a number between -1 and

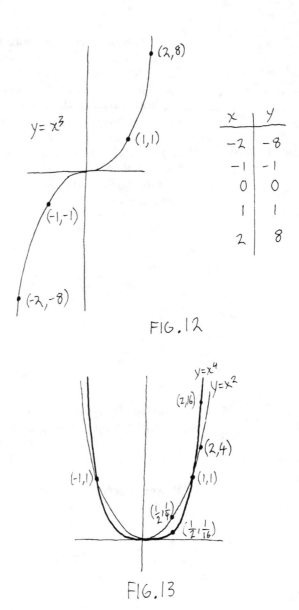

x	y
-2	-8
-1	-1
0	0
1	1
2	8

FIG. 12

FIG. 13

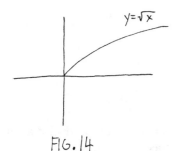

FIG. 14

1 is smaller than its square; otherwise x^4 lies above x^2. Figure 14 gives the graph of $y = \sqrt{x}$, the upper half of the parabola $x = y^2$.

Increasing and decreasing functions Suppose that whenever $a > b$, we have $f(a) > f(b)$; that is, as x increases, $f(x)$ increases also. In this case, f is said to be *increasing*. The graph of an increasing function rises to the right (Figs. 12 and 14).

Suppose that whenever $a > b$, we have $f(a) < f(b)$; that is, as x increases, $f(x)$ decreases. In this case, f is *decreasing*. The graph of a decreasing function falls to the right (Fig. 1).

The functions x^2 and x^4 (Fig. 13) decrease on the interval $(-\infty, 0]$ and increase on $[0, \infty)$; overall, on $(-\infty, \infty)$, they are neither increasing nor decreasing. The function $1/x$ (Fig. 5) decreases on the intervals $(-\infty, 0)$ and $(0, \infty)$ but is neither decreasing nor increasing on the interval $(-\infty, \infty)$.

Motion along a line   Suppose that at time t, the position x of a particle (such as a car) moving on a number line is given by the function $x = t^2$. Then at time $t = -3$, the particle is at position $x = 9$; at time $t = -1$, it is at position $x = 1$; at time $t = 0$, it is at position $x = 0$; at time $t = 4$, it is at position $x = 16$, and so on. Note that there is nothing mysterious about negative time. If time is measured in minutes, then t = 0 is a fixed time, such as 12:30 p.m. on Jan. 20, 1947, and negative values of t correspond to times before that moment. For example, $t = -3$ is 3 minutes earlier, that is, 12:27 p.m. Instead of drawing the graph of $x = t^2$ (a parabola in a t, x coordinate system), we might sketch the motion as in Fig. 15. Until time 0, the particle moves from right to left on the x-line and decelerates (look at the decrease in distance between consecutive times to see the deceleration). After time 0, the particle moves from left to right and accelerates. (For clarity, the right-to-left part of the motion is drawn above the left-to-right motion in Fig. 15, but, in reality, the particle is assumed to travel back and forth on the same road, not on a double-decker road.)

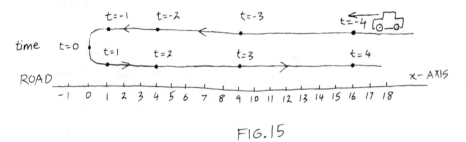

FIG. 15

One of the applications of calculus (Section 3.2) will be the computation of the speed and acceleration at any instant of time, given the position function.

Problems for Section 1.2

1. Sketch the graph. Is the function increasing? decreasing? one-to-one? continuous? (a) $2x$ (b) $x + |x|$ (c) $|x|/x$ (d) $f(x)$ is the larger of x and 3

2. Let $f(x)$ be 0 if x is an even integer, 1 if x is an odd integer, and undefined otherwise. Sketch the graph of f.

3. Figure 16 shows the graph of a function f.

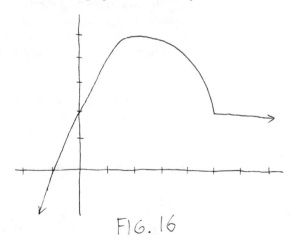

FIG. 16

(a) Find $f(-1)$, $f(0)$ and $f(6)$.
(b) Estimate x such that $f(x) = 4$.
(c) Find x such that $f(x) < 0$.

4. Suppose f is an increasing function. If x *decreases,* what does $f(x)$ do?
5. Are the following functions continuous?

(a) the cost $c(w)$ of mailing a package weighing w grams
(b) your weight $w(t)$ at time t

6. What can you conclude about the graph of f under the following conditions.

(a) $f(x) > 0$ for all x
(b) $f(x) > x$ for all x (for example, $f(5)$ is a number that must be larger than 5)

7. (a) Sketch the power functions x^{-3}, x^{-2}, $x^{-1/2}$ on the same set of axes. (b) Sketch the power functions x, x^5, x^7, x^8 on the same set of axes.
8. A function f is said to be *even* if $f(-x) = f(x)$ for all x; for example, $f(7) = 3$ and $f(-7) = 3$, $f(-4) = -2$ and $f(4) = -2$, and so on. A function is *odd* if $f(-x) = -f(x)$ for all x; for example, $f(3) = -12$ and $f(-3) = 12$, $f(-6) = -2$ and $f(6) = 2$, and so on. The functions $\cos x$ and x^2 are even, $\sin x$ and x^3 are odd, $2x + 3$ and $x^2 + x$ are neither.

(a) Figure 17 shows the graph of a function $f(x)$ for $x \geq 0$. If f is even, complete the graph for $x \leq 0$.
(b) Complete the graph in Fig. 17 if f is odd.

9. Find $f(x)$ if the graph of f is the line AB where $A = (1, 2)$ and $B = (2, 5)$.
10. Let $f(t)$ be the position of a particle on a number line at time t. Describe the motion if

(a) f is a constant function   (c) f is a decreasing function
(b) $f(t) = t - 2$ (d) $f(t) > 0$ for all t

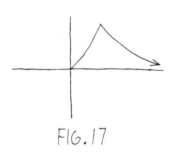

FIG. 17

1.3 The Trigonometric Functions

We continue with the development of the basic functions listed in Section 1.1 by considering the six trigonometric functions. The functions are entitled to be called basic because of their many applications, two of which (vibrations and electron flow) are described later in the section. We assume that you have studied trigonometry before starting calculus and therefore this section contains only a summary of the main results. A list of trigonometric identities and formulas is included at the end of the section for reference.

Definition of sine, cosine and tangent   Using Fig. 1, we define

(1) $$\sin \theta = \frac{y}{r}, \qquad \cos \theta = \frac{x}{r}, \qquad \tan \theta = \frac{y}{x} = \frac{\sin \theta}{\cos \theta}.$$

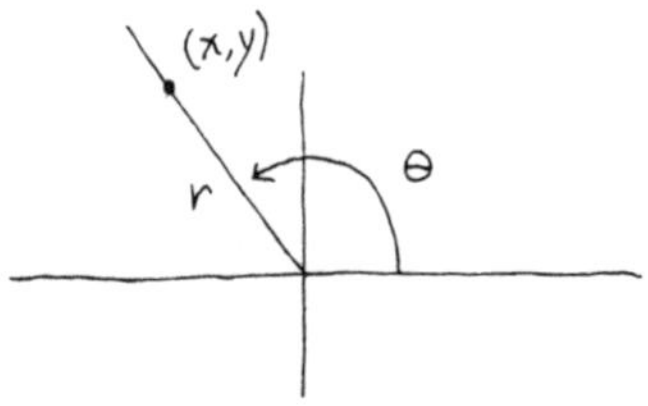

Figure 1 shows a positive θ corresponding to a counterclockwise rotation away from the positive x-axis. A negative θ corresponds to a clockwise rotation.

The distance r is always positive, but the signs of x and y depend on the quadrant. If $90° < \theta < 180°$, so that θ is a second quadrant angle, then

FIG. 1

x is negative and y is positive; thus $\sin \theta$ is positive, while $\cos \theta$ and $\tan \theta$ are negative. In general, Fig. 2 indicates the sign of $\sin \theta$, $\cos \theta$ and $\tan \theta$ for θ in the various quadrants.

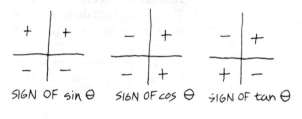

FIG. 2

Degrees versus radians   An angle of 180° is called π radians. More generally, to convert back and forth use

(2) $$\frac{\text{number of radians}}{\text{number of degrees}} = \frac{\pi}{180}.$$

Equivalently

(3) $$\text{number of degrees} = \frac{180}{\pi} \times \text{number of radians}$$

(4) $$\text{number of radians} = \frac{\pi}{180} \times \text{number of degrees}.$$

One radian is a bit more than 57°. Tables 1 and 2 list some important angles in both radians and degrees, and the corresponding functional values.

Table 1

Degrees	Radians	sin	cos	tan
0°	0	0	1	0
90°	$\pi/2$	1	0	none
180°	π	0	-1	0
270°	$3\pi/2$	-1	0	none
360°	2π	0	1	0

Table 2

Degrees	Radians	sin	cos	tan
30°	$\pi/6$	$\frac{1}{2}$	$\frac{1}{2}\sqrt{3}$	$1/\sqrt{3}$
45°	$\pi/4$	$\frac{1}{2}\sqrt{2}$	$\frac{1}{2}\sqrt{2}$	1
60°	$\pi/3$	$\frac{1}{2}\sqrt{3}$	$\frac{1}{2}$	$\sqrt{3}$

In most situations not involving calculus, it makes no difference whether we use radians or degrees, but it turns out (Section 3.3) that for the *calculus* of the trigonometric functions, it will be better to use radian measure.

One *geometric* instance where radians are preferable involves arc length on a circle. Suppose a central angle θ cuts off arc length s on a circle of radius r (Fig. 3). The entire circumference of the circle is $2\pi r$; the indicated arc length s is just a fraction of the entire circumference, namely, the fraction $\theta/360$ if θ is measured in degrees, and $\theta/2\pi$ if θ is measured in radians. Therefore, with θ in radian measure,

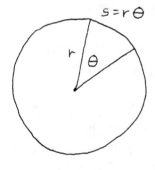

FIG. 3

(5) $$s = \frac{\theta}{2\pi} \cdot 2\pi r = r\theta.$$

If degrees are used, the formula is $s = \dfrac{\theta}{360} \cdot 2\pi r = \dfrac{\pi}{180} r\theta$, which is not as attractive as (5).

Reference angles   Trig tables list $\sin\theta$, $\cos\theta$ and $\tan\theta$ for $0 < \theta < 90°$. To find the functions for other angles, we use knowledge of the appropriate signs given in Fig. 2 plus reference angles, as illustrated in the following examples.

If θ is a second quadrant angle, its reference angle is $180° - \theta$, so $150°$ has reference angle $30°$ (Fig. 4), and

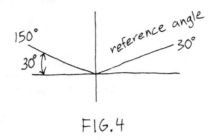

FIG.4

$$\sin 150° = \sin 30° = \tfrac{1}{2}, \qquad \cos 150° = -\cos 30° = -\tfrac{1}{2}\sqrt{3},$$
$$\tan 150° = -\tan 30° = -1/\sqrt{3}.$$

If θ is in the third quadrant, its reference angle is $\theta - 180°$, so $210°$ has reference angle $30°$ (Fig. 5), and

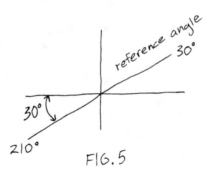

FIG.5

$$\sin 210° = -\sin 30° = -\tfrac{1}{2}, \qquad \cos 210° = -\cos 30° = -\tfrac{1}{2}\sqrt{3},$$
$$\tan 210° = \tan 30° = 1/\sqrt{3}.$$

If θ is in quadrant IV, its reference angle is $360° - \theta$, so $330°$ has reference angle $30°$ (Fig. 6), and

$$\sin 330° = -\sin 30° = -\tfrac{1}{2}, \qquad \cos 330° = \cos 30° = \tfrac{1}{2}\sqrt{3},$$
$$\tan 330° = -\tan 30° = -1/\sqrt{3}.$$

Right triangle trigonometry   In the right triangle in Fig. 7,

(6)
$$\sin\theta = \frac{\text{opposite leg}}{\text{hypotenuse}}, \qquad \cos\theta = \frac{\text{adjacent leg}}{\text{hypotenuse}},$$
$$\tan\theta = \frac{\text{opposite leg}}{\text{adjacent leg}}.$$

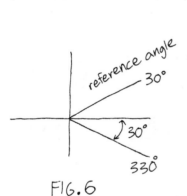

FIG.6

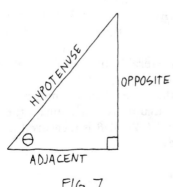

FIG.7

Graphs of sin x, cos x and tan x   Figures 8–10 give the graphs of the functions, with x measured in radians. The graphs show that sin x and cos x have *period* 2π (that is, they repeat every 2π units), while tan x has period π. Furthermore, $-1 \le \sin x \le 1$ and $-1 \le \cos x \le 1$, so that each function has *amplitude* 1. On the other hand, the tangent function assumes all values, that is, has range $(-\infty, \infty)$. Note that sin x and cos x are defined for all x, but tan x is not defined at $x = \pm\pi/2, \pm3\pi/2, \cdots$.

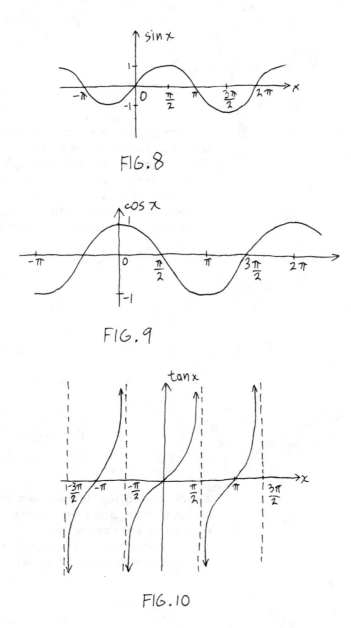

FIG. 8

FIG. 9

FIG. 10

The graph of $a \sin(bx + c)$ The function sin x has period 2π and amplitude 1. The function $3 \sin 2x$ has period π and amplitude 3 (Fig. 11). *In general, $a \sin bx$, for positive a and b, has amplitude a and period $2\pi/b$.* For example, $5 \sin \frac{1}{2}x$ has period 4π and amplitude 5.

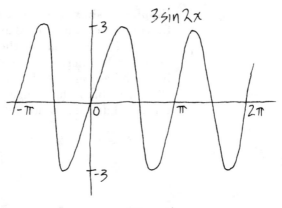

FIG. 11

The graph of $a \sin(bx + c)$ not only involves the same change of period and amplitude as $a \sin bx$ but is also *shifted*. As an example, consider $\sin(2x - \frac{1}{2}\pi)$. To sketch the graph, first plot a few points to get your bearings. For this purpose, the most convenient values of x are those which make the angle $2x - \frac{1}{2}\pi$ a multiple of $\pi/2$; the table in Fig. 12 chooses angles 0 and $\pi/4$ to produce points $(0, -1)$, $(\pi/4, 0)$ on the graph. Then continue on to make the amplitude 1 and the period π as shown in Fig. 12.

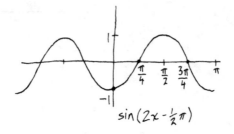

$$\sin\left(2x - \tfrac{1}{2}\pi\right)$$

x	y
0	$\sin\left(-\tfrac{1}{2}\pi\right) = -1$
$\tfrac{\pi}{4}$	$\sin 0 = 0$

FIG. 12

Application to simple harmonic motion   If a cork is pushed down in a bucket of water and then released (or, similarly, a spring is stretched and released), it bobs up and down. Experiments show that if a particular cork oscillates between 3 units above and 3 units below the water level with the timing indicated in Fig. 13, its height h at time t is given by $h(t) = 3 \sin \frac{1}{2}t$.

TIME $t = -\pi$ TIME $t = 0$ TIME $t = \pi$ TIME $t = 2\pi$ TIME $t = 3\pi$

FIG. 13

(Note that there is nothing strange about time π. It is approximately 3.14 minutes after time 0.) More generally, the amplitude, frequency and shift depend on the cork, the medium and the size and timing of the initial push down, but the oscillation, called *simple harmonic motion*, always has the form $a \sin(bt + c)$, or equivalently $a \cos(bt + c)$.

Another instance of simple harmonic motion involves the flow of the alternating current (a.c.) in a wire. Electrons flow back and forth, and if $i(t)$ is the current, that is, the amount of charge per second flowing in a given direction at time t, then $i(t)$ is of the form $a \sin(bt + c)$ or $a \cos(bt + c)$. If $i(t) = 10 \cos t$ then at time $t = 0$, 10 units of charge per second flow in the given direction; at time $t = \pi/2$, the flow momentarily stops; at time $t = \pi$, 10 units of charge per second flow opposite to the given direction.

The graph of $f(x) \sin x$ First consider two special cases. The graph of $y = 2 \sin x$ has amplitude 2 and lies between the pair of lines $y = \pm 2$ (Fig. 14), although usually we do not actually sketch the lines. The lines, which are reflections of one another in the x-axis, are called the *envelope* of $2 \sin x$. The graph of $y = -2 \sin x$ also lies between those lines; in addition, the effect of the negative factor -2 is to change the signs of y-coordinates, so the graph is the reflection in the x-axis of the graph of $2 \sin x$ (Fig. 14).

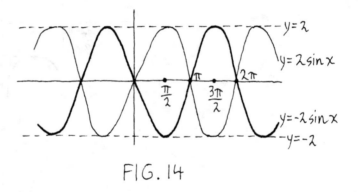

FIG. 14

Similarly, the graph of $x^3 \sin x$ is sandwiched between the curves $y = \pm x^3$ which we sketch as guides (Fig. 15). The curves, called the envelope of $x^3 \sin x$, are reflections of one another in the x-axis. Furthermore,

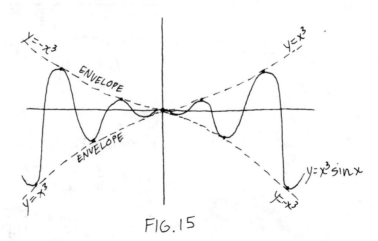

FIG. 15

whenever x^3 is negative (as it is to the left of the y-axis) we not only change the amplitude but also reflect sine in the x-axis to obtain $x^3 \sin x$. The result in Fig. 15 shows unbounded oscillations.

In general, *to sketch the graph of $f(x) \sin x$, first draw the curve $y = f(x)$ and the curve $y = -f(x)$, its reflection in the x-axis, to serve as the envelope. Then change the height of the sine curve so that it fits within the envelope, and in addition reflect the sine curve in the x-axis whenever $f(x)$ is negative.*

Secant, cosecant and cotangent By definition,

$$(7) \qquad \sec x = \frac{1}{\cos x}, \qquad \csc x = \frac{1}{\sin x}, \qquad \cot x = \frac{\cos x}{\sin x} = \frac{1}{\tan x}.$$

In each case, the function is defined for all values of x such that the denominator is nonzero. For example, $\csc x$ is not defined for $x = 0, \pm\pi, \pm 2\pi, \cdots$. The graphs are given in Figs. 16–18.

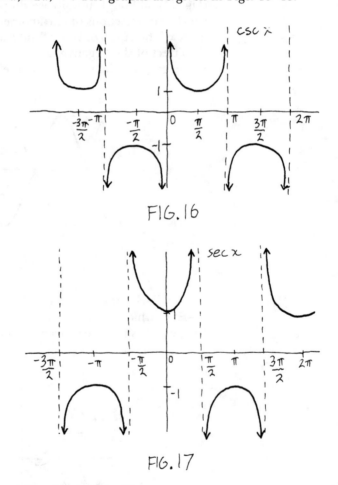

FIG. 16

FIG. 17

In a right triangle (Fig. 7),

$$(8) \qquad \sec \theta = \frac{\text{hypotenuse}}{\text{adjacent leg}}, \qquad \csc \theta = \frac{\text{hypotenuse}}{\text{opposite leg}},$$

$$\cot \theta = \frac{\text{adjacent leg}}{\text{opposite leg}}.$$

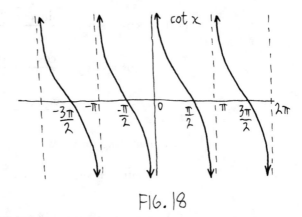

FIG. 18

Notation   It is standard practice to write $\sin^2 x$ for $(\sin x)^2$, and $\sin x^2$ to mean $\sin(x^2)$. Similar notation holds for the other trigonometric functions.

Standard trigonometric identities

Negative angle formulas

(9) $\sin(-x) = -\sin x, \qquad \cos(-x) = \cos x, \qquad \tan(-x) = -\tan x,$

$\csc(-x) = -\csc x, \qquad \sec(-x) = \sec x, \qquad \cot(-x) = -\cot x$

Addition formulas

$\sin(x + y) = \sin x \cos y + \cos x \sin y$

(10) $\sin(x - y) = \sin x \cos y - \cos x \sin y$

$\cos(x + y) = \cos x \cos y - \sin x \sin y$

$\cos(x - y) = \cos x \cos y + \sin x \sin y$

Double angle formulas

$\sin 2x = 2 \sin x \cos x$

(11) $\cos 2x = \cos^2 x - \sin^2 x = 1 - 2 \sin^2 x = 2 \cos^2 x - 1$

$\tan 2x = \dfrac{2 \tan x}{1 - \tan^2 x}$

Pythagorean identities

$\sin^2 x + \cos^2 x = 1$

(12) $1 + \tan^2 x = \sec^2 x$

$1 + \cot^2 x = \csc^2 x$

Half-angle formulas

(13) $\sin^2 \tfrac{1}{2} x = \dfrac{1 - \cos x}{2}$

$\cos^2 \tfrac{1}{2} x = \dfrac{1 + \cos x}{2}$

Product formulas

$\sin x \cos y = \dfrac{\sin(x + y) + \sin(x - y)}{2}$

(14) $\cos x \sin y = \dfrac{\sin(x + y) - \sin(x - y)}{2}$

$\cos x \cos y = \dfrac{\cos(x + y) + \cos(x - y)}{2}$

$\sin x \sin y = \dfrac{\cos(x - y) - \cos(x + y)}{2}$

Factoring formulas

$\sin x + \sin y = 2 \cos \dfrac{x - y}{2} \sin \dfrac{x + y}{2}$

(15) $\sin x - \sin y = 2 \cos \dfrac{x + y}{2} \sin \dfrac{x - y}{2}$

$\cos x + \cos y = 2 \cos \dfrac{x + y}{2} \cos \dfrac{x - y}{2}$

$\cos x - \cos y = 2 \sin \dfrac{x + y}{2} \sin \dfrac{y - x}{2}$

Reduction formulas

$\cos(\tfrac{1}{2}\pi - \theta) = \sin \theta$

(16) $\sin(\tfrac{1}{2}\pi - \theta) = \cos \theta$

$\cos(\pi - \theta) = -\cos \theta$

$\sin(\pi - \theta) = \sin \theta$

(17) *Law of Sines* (Fig. 19)

$\dfrac{\sin A}{a} = \dfrac{\sin B}{b} = \dfrac{\sin C}{c}$

(18) *Law of Cosines* (Fig. 19)

$c^2 = a^2b^2 - 2ab \cos C$

(19) *Area formula* (Fig. 19)

area of triangle $ABC = \tfrac{1}{2}ab \sin C$

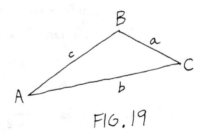

FIG. 19

Problems for Section 1.3

1. Convert from radians to degrees.
(a) $\pi/5$ (b) $5\pi/6$ (c) $-\pi/3$
2. Convert from degrees to radians.
(a) $12°$ (b) $-90°$ (c) $100°$

3. Evaluate without using a calculator.

(a) $\sin 210°$ (b) $\cos 3\pi$ (c) $\tan 5\pi/4$

4. Sketch the graph.

(a) $\sin \frac{1}{3}x$ (d) $5 \sin(\frac{1}{2}x + \pi)$
(b) $\tan 4x$ (e) $2 \cos(3x - \frac{1}{2}\pi)$
(c) $3 \cos \pi x$

5. Let $\sin x = a$, $\cos y = b$ and evaluate the expression in terms of a and b, if possible.

(a) $\sin(-x)$ (d) $-\cos y$
(b) $\cos(-y)$ (e) $\sin^2 x$
(c) $-\sin x$ (f) $\sin x^2$

6. In each of (a) and (b), use right triangle trigonometry to find an exact answer, rather than tables or a calculator which will give only approximations.

(a) Find $\cos \theta$ if θ is an acute angle and $\sin \theta = 2/3$.
(b) Find $\sin \theta$ if θ is acute and $\tan \theta = 7/4$.

7. Sketch the graph.

(a) $x \sin x$ (b) $x^2 \sin x$

1.4 Inverse Functions and the Inverse Trigonometric Functions

If a function maps a to b we may wish to switch the point of view and consider the *inverse function* which sends b to a. For example, the function defined by $F = \frac{9}{5}C + 32$ gives the fahrenheit temperature F as a function of the centigrade reading C. If we solve the equation for C to obtain $C = \frac{5}{9}(F - 32)$ we have the inverse function which produces C, given F. If the original function is useful, the inverse is probably also useful. In this section, we discuss inverses in general, and three inverse trigonometric functions in particular.

The inverse function Let f be a one-to-one function. The inverse of f, denoted by f^{-1}, is defined as follows: *if $f(a) = b$ then $f^{-1}(b) = a$.* In other words, the inverse maps "backwards" (Fig. 1). Only one-to-one functions have inverses because reversing a non-one-to-one function creates a pairing that is not a function (Fig. 2).

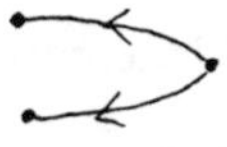

Given a table of values for f, a table of values for f^{-1} can be constructed by interchanging columns. A partial table for $f(x) = 3x$ and the corresponding partial table for its inverse are given below.

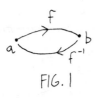

FIG. 1

NOT A FUNCTION

FIG. 2

x	$f(x)$		x	$f^{-1}(x)$
2	6		6	2
5	15		15	5
7	21		21	7

Clearly, $f^{-1}(x) = \frac{1}{3}x$. Note that we may also think of $\frac{1}{3}x$ as the "original" function with inverse $3x$. In general, *f and f^{-1} are inverses of each other.*

Figure 1 shows that if f and f^{-1} are applied successively (first f and then f^{-1}, or vice versa) the result is a "circular" trip which returns to the starting point. In other words,

(1) $$f^{-1}(f(x)) = x \quad \text{and} \quad f(f^{-1}(x)) = x.$$

For example, multiplying a number by 3 and then multiplying that result by 1/3 produces the original number.

Example 1 In functional notation, the centigrade/fahrenheit equations show that if $f(x) = \frac{9}{5}x + 32$ then $f^{-1}(x) = \frac{5}{9}(x - 32)$.

The graph of $f^{-1}(x)$ One of the advantages of an inverse function is that its properties, such as its graph, often follow easily from the properties of the original function. Comparing the graphs of f and f^{-1} amounts to comparing points such as $(2, 7)$ and $(7, 2)$ (Fig. 3). The points are reflections of one another in the line $y = x$. In general, *the graph of f^{-1} is the reflection of the graph of f in the line $y = x$*, so that the pair of graphs is symmetric with respect to the line. If $f(x) = x^2$, and $x \geq 0$ so that f is one-to-one, then $f^{-1}(x) = \sqrt{x}$. The symmetry of the two graphs is displayed in Fig. 4.

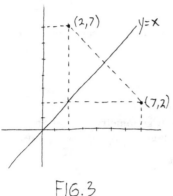

FIG. 3

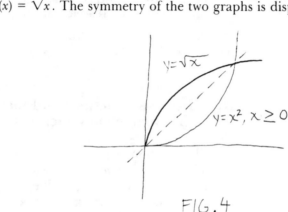

FIG. 4

The inverse sine function Unfortunately, the sine function as a whole doesn't have an inverse because it isn't one-to-one. But various pieces of the sine graph *are* one-to-one, in particular, any section between a low and a high point passes the horizontal line test and can be inverted. By convention, we use the part between $-\pi/2$ and $\pi/2$ and let $\sin^{-1}x$ be the inverse of this abbreviated sine function; that is, $\sin^{-1}x$ *is the angle between $-\pi/2$ and $\pi/2$ whose sine is x*. Equivalently,

(2) $\sin^{-1}a = b$ *if and only if* $\sin b = a$ *and* $-\pi/2 \leq b \leq \pi/2$.

The graph of $\sin^{-1}x$ is found by reflecting $\sin x$, $-\pi/2 \leq x \leq \pi/2$, in the line $y = x$ (Fig. 5). The domain of $\sin^{-1}x$ is $[-1, 1]$ and the range is $[-\pi/2, \pi/2]$.

The $\sin^{-1}$ function is also denoted by Sin^{-1} and arcsin. In computer programming, the abbreviation ASN of arcsin is often used.

Example 2 Find $\sin^{-1}\frac{1}{2}$.

Solution: Let $x = \sin^{-1}\frac{1}{2}$; then $\sin x = \frac{1}{2}$. We know that $\sin 30° = \frac{1}{2}$, $\sin(-330°) = \frac{1}{2}$, $\sin 150° = \frac{1}{2}, \cdots$. We must choose the angle between $-90°$ and $90°$; therefore $\sin^{-1}\frac{1}{2} = 30°$, or, in radians, $\sin^{-1}\frac{1}{2} = \pi/6$.

Example 3 Find $\sin^{-1}(-1)$.

Of all the angles whose sine is -1, the one in the interval $[-\pi/2, \pi/2]$ is $-\pi/2$. Therefore, $\sin^{-1}(-1) = -\pi/2$.

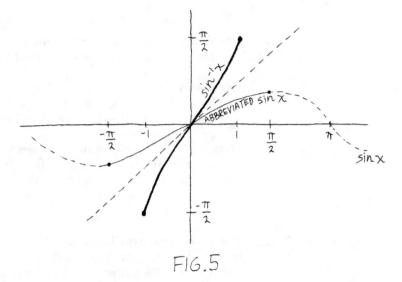

FIG.5

Warning   1. The angles $-\pi/2$ and $3\pi/2$ are coterminal angles; that is, as rotations from the positive x-axis, they terminate in the same place. However $-\pi/2$ and $3\pi/2$ are not the *same* angle or the same number, and arcsin(-1) is $-\pi/2$, *not* $3\pi/2$.

2. Although (1) states that $f^{-1}(f(x)) = x$, $\sin^{-1}(\sin 200°)$ is *not* 200°. This is because $\sin^{-1}$ is *not* the inverse of sine unless the angle is between $-90°$ and $90°$. The sine function maps 200°, along with many other angles, such as 560°, $-160°$, 340°, $-20°$, all to the same output. The $\sin^{-1}$ function maps in reverse to the particular angle between $-90°$ and $90°$. Therefore, $\sin^{-1}(\sin 200°) = -20°$.

The inverse cosine function   The cosine function, like the sine function, has no inverse, because it is not one-to-one. By convention, we consider the one-to-one piece between 0 and π, and let $\cos^{-1}x$ be the inverse of this abbreviated cosine function (Fig. 6). Thus, $\cos^{-1}x$ *is the angle between* 0 *and* π *whose cosine is* x. Equivalently,

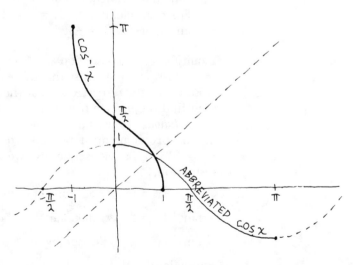

FIG.6

(3) $\cos^{-1}a = b$ *if and only if* $\cos b = a$ *and* $0 \le b \le \pi$.

The domain of $\cos^{-1}x$ is $[-1, 1]$ and the range is $[0, \pi]$.

The $\cos^{-1}$ function is also denoted by Cos^{-1}, arccos and ACN.

Example 4   Find $\cos^{-1}(-\frac{1}{2})$.

Solution: The angle between $0°$ and $180°$ whose cosine is $-\frac{1}{2}$ is $120°$. Therefore, $\cos^{-1}(-\frac{1}{2}) = 120°$, or in radians, $\cos^{-1}(-\frac{1}{2}) = 2\pi/3$.

Warning   The graphs of $\sin x$ and $\cos x$ wind forever along the x-axis, but the graphs of $\sin^{-1}x$ and $\cos^{-1}x$ (reflections of *portions* of $\sin x$ and $\cos x$) do *not* continue forever up and down the y-axis. They are shown *in entirety* in Figs. 5 and 6. (If either curve did continue winding, the result would be a *non*function.)

The inverse tangent function   The $\tan^{-1}$ function is the inverse of the branch of the tangent function through the origin (Fig. 7). In other words, $\tan^{-1}x$ *is the angle between* $-\pi/2$ *and* $\pi/2$ *whose tangent is* x. Equivalently,

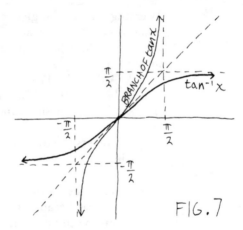

FIG. 7

(4) $\tan^{-1}a = b$ *if and only if* $\tan b = a$ *and* $-\pi/2 < b < \pi/2$.

The $\tan^{-1}$ function is also denoted Tan^{-1}, arctan and ATN.

For example, $\tan^{-1}(-1) = -\pi/4$ because $-\pi/4$ is between $-\pi/2$ and $\pi/2$ and $\tan(-\pi/4) = -1$.

Example 5   The equation $y = 2 \tan 3x$ does not have a unique solution for x. Restrict x suitably so that there *is* a unique solution and then solve for x. Equivalently, restrict x so that the function $2 \tan 3x$ is one-to-one, and then find the inverse function.

Solution: To use $\tan^{-1}$ as the inverse of tangent, the angle, which is $3x$ in this problem, must be restricted to the interval $(-\frac{1}{2}\pi, \frac{1}{2}\pi)$, that is, $-\frac{1}{2}\pi < 3x < \frac{1}{2}\pi$. Consequently, we choose $-\pi/6 < x < \pi/6$. With this restriction,

$\frac{1}{2}y = \tan 3x$ (divide both sides of the original equation by 2)

$\tan^{-1}\frac{1}{2}y = 3x$ (take $\tan^{-1}$ on both sides)

$\frac{1}{3}\tan^{-1}\frac{1}{2}y = x$ (divide by 3).

Equivalently, if $f(x) = 2 \tan 3x$ and $-\pi/6 < x < \pi/6$, then $f^{-1}(x) = \frac{1}{3}\tan^{-1}\frac{1}{2}x$.

Problems for Section 1.4

1. Suppose f is one-to-one so that it has an inverse. If $f(3) = 4$ and $f(5) = 2$, find, if possible, $f^{-1}(3), f^{-1}(4), f^{-1}(5), f^{-1}(2)$.

2. Find the inverse by inspection, if it exists.

(a) $x - 3$ (c) $1/x$
(b) Int x (d) $-x$

3. If $f(x) = 2x - 9$ find a formula for $f^{-1}(x)$.
4. Find $f^{-1}(f(17))$.
5. Show that an increasing function always has an inverse and then decide if the inverse is decreasing.
6. True or False? If f is continuous and invertible then f^{-1} is also continuous.
7. Are the following pairs of functions inverses of one another?
(a) x^2 and $\sqrt{x}$ (b) x^3 and $\sqrt[3]{x}$

8. Find the function value.

(a) $\cos^{-1}0$ (e) $\sin^{-1}(-\frac{1}{2}\sqrt{3})$
(b) $\sin^{-1}0$ (f) $\tan^{-1}1$
(c) $\sin^{-1}2$???? (g) $\tan^{-1}(-1)$
(d) $\cos^{-1}(-\frac{1}{2}\sqrt{3})$

9. Estimate $\tan^{-1}1000000$.
10. True or False? (a) If $\sin a = b$ then $\sin^{-1}b = a$ (b) If $\sin^{-1}c = d$ then $\sin d = c$.
11. Place restrictions on θ so that the equation has a unique solution for θ, and then solve. (a) $z = 3 + \frac{1}{2}\sin \pi\theta$ (b) $x = 5\cos(2\theta - \frac{1}{3}\pi)$
12. Odd and even functions were defined in Problem 8, Section 1.2. Do odd (resp. even) functions have inverses? If inverses exist, must they also be odd (resp. even)?

1.5 Exponential and Logarithm Functions

This section completes the discussion of the basic functions listed in Section 1.1 by considering the exponential functions and their inverses, the logarithm functions. As with the other basic functions, they have important physical applications, such as exponential growth, discussed in Section 4.9.

Exponential functions Functions such as 2^x, $(\frac{1}{4})^x$ and 7^x are called *exponential* functions, as opposed to *power* functions x^2, $x^{1/4}$ and x^7. In general, an exponential function has the form b^x, and is said to have *base b*.

Negative bases create a problem. If $f(x) = (-4)^x$ then $f(\frac{1}{2}) = \sqrt{-4}$ and $f(\frac{1}{4}) = \sqrt[4]{-4}$, which are *not* real. Similarly, there is no (real) $f(\frac{3}{2}), f(\frac{5}{2}), f(\frac{7}{2}), \cdots$; the domain of $(-4)^x$ is too riddled with gaps to be useful in calculus. (The power function $x^{1/2}$ also has a restricted domain, namely $[0,\infty)$, but at least the domain is an entire interval.) Because of this difficulty, *we do not consider exponential functions with negative bases.*

To sketch the graph of 2^x, we first make a table of values. (Remember that 2^{-7}, for example, is defined as $1/2^7$, and 2^0 is 1.)

x	-7	-3	-1	0	1	4	10
2^x	$\frac{1}{128}$	$\frac{1}{8}$	$\frac{1}{2}$	1	2	16	1024

For convenience, we used integer values of x in the table, but 2^x is also defined when x is not an integer. For example,

$$2^{2/3} = \sqrt[3]{2^2} = \sqrt[3]{4}, \qquad 2^{3.1} = 2^{31/10} = \sqrt[10]{2^{31}},$$

and the graph of 2^x also contains the points $(2/3, \sqrt[3]{4})$ and $(3.1, \sqrt[10]{2^{31}})$.

We plot the points from the table, and when the pattern seems clear, connect them to obtain the final graph (Fig. 1). The connecting process assumes that 2^x is continuous.† Figure 1 also contains the graphs of $(\frac{1}{2})^x$ and 3^x for comparison.

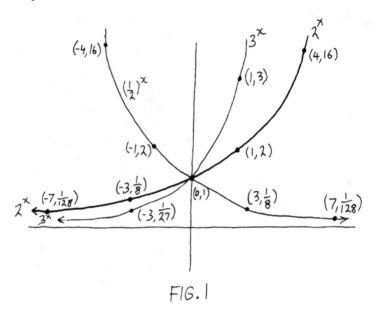

FIG. 1

The exponential function e^x In algebra, the most popular base is 10, while computer science often favors base 2. However, for reasons to be given in Section 3.3, calculus uses base e, a particular irrational number (that is, an infinite nonrepeating decimal) between 2.71 and 2.72; the official definition will be given in that section. Because calculus concentrates on base e, the function e^x is often referred to as *the* exponential function. It is sometimes written as $\exp x$; programming languages use $\text{EXP}(X)$.

Figure 2 shows the graph of e^x, along with 2^x and 3^x for comparison. Note that $2 < e < 3$, and correspondingly, the graph of e^x lies between the graphs of 2^x and 3^x. We continue to assume that exponential functions are continuous.

In practice, a value of e^x, such as e^2, may be approximated with tables or a calculator. Section 8.9 will indicate one method for evaluating e^x directly. A rough estimate of e^2 can be obtained by noting that since e is slightly less than 3, e^2 is somewhat less than 9.

---

†The connecting process also provides a definition of 2^x for irrational x, that is, when x is an infinite nonrepeating decimal, such as π. For example, $\pi = 3.14159\ldots$, and by connecting the points to make a continuous curve, we are defining 2^π by the following sequence of inequalities:

$$2^{3.14} < 2^\pi < 2^{3.141}$$

$$2^{3.141} < 2^\pi < 2^{3.1415}$$

$$2^{3.1415} < 2^\pi < 2^{3.14159}$$

$$\vdots$$

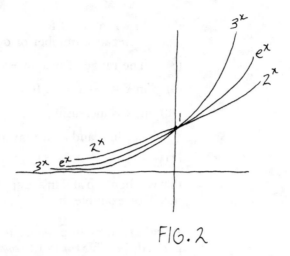

FIG. 2

The graph of e^x provides much information at a glance:

(1) e^x is defined for all x.

(2) $e^x > 0$; in fact, the range of e^x is $(0, \infty)$.

(3) e^x is increasing.

The function ln x Since e^x passes the horizontal line test and is one-to-one, it has an inverse, called the *natural logarithm* function and denoted by ln x. It is also written $\log_e x$ and called the logarithm with base e. In other words,

(4) $$\ln a = b \quad \text{if and only if} \quad e^b = a.$$

For example, if $e^{2p-q} = z$ then $\ln z = 2p - q$. As an important consequence of (4), since

(5) $$e^0 = 1 \quad \text{and} \quad e^1 = e,$$

we have

(6) $$\ln 1 = 0 \quad \text{and} \quad \ln e = 1.$$

The graph of ln x is the reflection of e^x in the line $y = x$ (Fig. 3). The graph reveals the following properties (7)–(10).

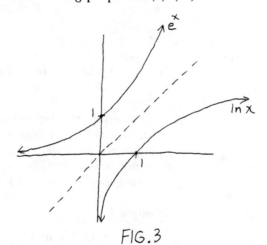

FIG. 3

(7)   ln x is defined for $x > 0$; we cannot take the logarithm of a negative number or of 0.

(8)   The range of ln x is $(-\infty, \infty)$.

(9)   ln x is negative if $0 < x < 1$, and positive if $x > 1$.

(10) ln x is increasing.

Since ln x and e^x are inverses,

(11) $$\ln e^x = x \quad \text{and} \quad e^{\ln x} = x;$$

that is, when exp and ln are applied successively to x, they "cancel each other out." For example, $\ln e^7 = 7$, $e^{\ln 8} = 8$, $\ln e^{a+b} = a + b$, $e^{\ln 6x} = 6x$.

Warning It is impossible to *take* ln of a negative number, but it is perfectly possible for ln x to *come out* negative. In fact, by (9), ln x is negative whenever $0 < x < 1$. For example, $\ln(-3)$ is impossible, but ln $x = -3$ is possible.

Laws of exponents and logarithms The familiar rules of exponents hold for e^x.

(12) $$e^x e^y = e^{x+y}$$

(13) $$e^x/e^y = e^{x-y}$$

(14) $$e^{-x} = 1/e^x$$

(15) $$(e^x)^y = e^{xy}.$$

We will derive the property of logarithms analogous to (12). Let $a = e^x$ and $b = e^y$ so that, by (4), $x = \ln a$ and $y = \ln b$. Then (12) becomes $ab = e^{\ln a + \ln b}$, which, by (4), may be rewritten as

(16) $$\ln ab = \ln a + \ln b.$$

Similarly, the other rules of exponents lead to the following laws of logarithms:

(17) $$\ln \frac{a}{b} = \ln a - \ln b$$

(18) $$\ln \frac{1}{a} = -\ln a$$

(this is a special case of (17) since $\ln \frac{1}{a} = \ln 1 - \ln a = 0 - \ln a = -\ln a$)

(19) $$\ln a^b = b \ln a.$$

We assume throughout that identities and equations involving the logarithm function never involve the logarithm of a negative number or 0. For example, we might use (19) to write $\ln x^2 = 2 \ln x$. It is understood that x must not be 0 or negative, so that $\ln x^2$ and $\ln x$ are both defined.

Note that $\ln x^2$ means $\ln(x^2)$, not $(\ln x)^2$.

Example 1

(a) $\ln 4 + \ln 3 = \ln 12$ (by (16))
(b) $\ln 81 = \ln 3^4 = 4 \ln 3$ (by (19))

(c) $\frac{1}{2}\ln 9 = \ln 9^{1/2} = \ln \sqrt{9} = \ln 3$ (by (19))
(d) $\ln e^3 = 3 \ln e = 3$ (by (19) and (6))
(e) $\ln 1/e = -\ln e = -1$ (by (18) and (6))

Warning **1.** $\ln 3x$ is not $3 \ln x$; instead, $\ln 3x = \ln 3 + \ln x$.
 2. $2 \ln 3x$ is neither $\ln 6x$ nor $6 \ln x$, nor $\ln 3x^2$; instead, $2 \ln 3x = \ln(3x)^2 = \ln 9x^2$.
 3. $\ln 2x + \ln 3x$ is not $\ln 5x$; instead, $\ln 2x + \ln 3x = \ln 6x^2$.

Example 2

(a) $\ln 3e^{4x} = \ln 3 + \ln e^{4x} = \ln 3 + 4x$ (by (16) and (11))
(b) $e^{2 \ln 3x} = e^{\ln(3x)^2} = e^{\ln 9x^2} = 9x^2$ (by (19) and (11))
(c) $2 \ln x + \ln x = \ln x^2 + \ln x = \ln x^3$ (by (19) and (16))

Logarithms with other bases There are logarithm functions with bases other than e, corresponding to exponential functions with bases other than e: $\log_2 x$ is the inverse of 2^x, $\log_3 x$ is the inverse of 3^x, $\log_{1/2} x$ is the inverse of $(\frac{1}{2})^x$, and so on. Since calculus uses the exponential function with base e, in this book we will consider only the logarithm function with base e, that is, $\ln x$.

The elementary functions We have now introduced all the basic functions listed in Section 1.1. However, applications often involve not only the basic functions, but combinations of them, such as the sum $x^2 + x$ or the product $x^2 \sin x$. Still another way of combining two functions f and g is to form the functions $f(g(x))$ and $g(f(x))$, called *compositions.* If $f(x) = \sin x$ and $g(x) = \sqrt{x}$ then $f(g(x)) = \sin \sqrt{x}$ and $g(f(x)) = \sqrt{\sin x}$. *The basic functions plus all combinations formed by addition, subtraction, multiplication, division and composition, a finite number of times, are referred to as the elementary functions.* For example, $\sin x$, $2x^3 + 4$, $\sin x^2$, $1/x$ and $x \cos 2x$ are elementary functions.

All the basic functions are continuous wherever they are defined, and it can be shown that the elementary functions also are continuous except where they are not defined, usually because of a zero in a denominator. For example, $e^{1/x}$ is continuous except at $x = 0$ where it is not defined, $(x^3 + \sin x)/(x - 1)$ is continuous except at $x = 1$ where it is not defined, $\sin x^2$ is continuous everywhere.

Solving equations involving e^x and $\ln x$ To solve the equation $e^x = 7$, take $\ln$ on both sides and use $\ln e^x = x$ to get $x = \ln 7$. To solve the equation $\ln x = -6$, take $\exp$ on both sides and use $e^{\ln x} = x$ to get $x = e^{-6}$.

Example 3 Solve $4 \ln(2x + 5) = 8$.

Solution:

$$\ln(2x + 5) = 2 \quad \text{(divide by 4)}$$

$$2x + 5 = e^2 \quad \text{(take exp)}$$

$$x = \tfrac{1}{2}(e^2 - 5) \quad \text{(algebra)}$$

Example 4 Solve $\ln 12x + \ln 3x = 4$.

First solution:

$$\ln 36x^2 = 4 \quad (\ln a + \ln b = \ln ab)$$

$$36x^2 = e^4 \qquad \text{(take exp)}$$

It looks as if the solution should be $x = \pm\frac{1}{6}e^2$, but if x is negative, then $12x$ and $3x$ are also negative, and there is no $\ln 12x$ or $\ln 3x$. Thus the only solution is $x = \frac{1}{6}e^2$.

Second solution:

$$e^{\ln 12x + \ln 3x} = e^4 \qquad \text{(take exp)}$$

$$e^{\ln 12x}e^{\ln 3x} = e^4 \qquad (e^{a+b} = e^a e^b)$$

$$(12x)(3x) = e^4 \qquad (e^{\ln a} = a)$$

$$36x^2 = e^4$$

$$x = \frac{1}{6}e^2 \qquad \text{(as in the first solution)}$$

Warning If $\ln 12x + \ln 3x = 4$, it is *not* correct to take exp *of each term* to get $12x + 3x = e^4$; if exp is used at all, it must be applied to *each entire side* of the equation, to obtain $e^{\ln 12x + \ln 3x} = e^4$. In general, if $p + q = 4$ then applying exp to both sides produces $e^{p+q} = e^4$, *not* $e^p + e^q = e^4$; and applying ln to both sides produces $\ln(p + q) = \ln 4$, *not* $\ln p + \ln q = \ln 4$.

Example 5 Solve $\ln(-x) = 3$. Note that writing $\ln(-x)$ does not violate the principle that it is impossible to take ln of a negative number. The function $\ln(-x)$ is defined for $-x > 0$, that is, for $x < 0$.
   Solution: Take exp on both sides to obtain $-x = e^3$, $x = -e^3$.

Solving inequalities involving e^x and ln x Consider the inequalities (a) $e^x < 5$ and (b) $\ln x > -\frac{1}{2}$. To solve (a), take ln on both sides to get the solution $x < \ln 5$. For (b), take exp on both sides to get $x > e^{-1/2}$.
   Note that, in general, we *can't* "do the same thing" to both sides of an inequality and expect another similar inequality to result. If $a > b$, we cannot conclude that $\sin a > \sin b$ (for example, $2\pi > 0$, but $\sin 2\pi = \sin 0$). If $a > b$, we cannot square both sides to conclude that $a^2 > b^2$ (for example, $2 > -3$, but $4 < 9$). However, if we operate on both sides of an inequality with an *increasing* function, the sense of the inequality *is* maintained. Since exp and ln are increasing functions (as opposed to the squaring function and the sine function which are not) it is true that if $a > b$ then $e^a > e^b$ and $\ln a > \ln b$, justifying the method for solving (a) and (b).

Problems for Section 1.5

   1. Arrange each set of numbers from smallest to largest without using tables or a calculator.

   (a) $e^{-10}, -e^{10}, e^{10}$
   (b) $e^{-1/2}, e^{1/3}, e^{-3}, e^{-5}, e^6$
   (c) $-e^6, -e^7$

   2. Simplify each expression.

   (a) $e^{\ln 7}$ (e) $e^{-\ln 1/2}$
   (b) $\ln e^4$ (f) $e^{1 + \ln 4}$
   (c) $e^{6\ln 2}$ (g) $\exp(\ln x + \ln y)$
   (d) $\ln \sqrt{e}$

3. Let $\ln 2 = a$, $\ln 3 = b$ and write each expression in terms of a and b.

(a) $\ln 6$ (g) $\ln 2 + \ln 3$
(b) $\ln 8$ (h) $(\ln 2)(\ln 3)$
(c) $\ln \sqrt{3}$ (i) $(\ln 2)/(\ln 3)$
(d) $\ln 81$ (j) $(\ln 2)^3$
(e) $\ln \frac{1}{2}$ (k) $\ln 2^3$
(f) $\ln \frac{3}{2}$

4. For which values of x is the function defined.

(a) $\ln(2x + 3)$ (d) $\ln \ln x$
(b) $\ln \sin \pi x$ (e) $\ln \ln \ln x$
(c) e^{3x-4} (f) $\ln \ln \ln \ln x$

5. Show that $-\ln(\sqrt{2} - 1)$ simplifies to $\ln(\sqrt{2} + 1)$.

6. True or False?

(a) If $\ln a = \ln b$, then $a = b$.
(b) If $e^a = e^b$, then $a = b$.
(c) If $\sin a = \sin b$, then $a = b$.

7. Show that $\exp\left(\dfrac{4 - 2\ln 3 - \ln 2}{3}\right)$ simplifies to $e\sqrt[3]{e/18}$

8. Show that $2^x = e^{x\ln 2}$. (In fact, some computers evaluate 2^3, not by finding $2 \cdot 2 \cdot 2$, but by converting 2^3 to $e^{3\ln 2}$ and evaluating that expression.)

9. Suppose a car travels on the number line so that its position at time t is e^t. Describe the car's motion during the time interval $(-\infty, \infty)$.

10. Solve

(a) $2e^{-x} - 3 = 0$ (k) $4\ln x + \ln 2x = 3$
(b) $\ln(2x + 7) = -1$ (l) $\ln(5x - 3) = \ln 2x$
(c) $e^x = -5$ (m) $\ln(5x + 3) = \ln 2x$
(d) $-2 < \ln x < 8$ (n) $\ln(x + 1) + \ln x = 2$
(e) $e^{2x+7} > 5$ (o) $e^x = e^{-x}$
(f) $-\ln x = 4$ (p) $x \ln x = 0$
(g) $\ln(-x) = 4$ (q) $xe^x + 2e^x = 0$
(h) $e^{5x+3} = e^{2x}$ (r) $e^x \ln x = 0$
(i) $\ln \ln x = -2$ (s) $\dfrac{25}{2 + \ln 3x} = 5$
(j) $\arcsin e^x = \pi/6$

11. Show that $\ln \frac{1}{2}\sqrt{2}$ simplifies to $-\frac{1}{2}\ln 2$.

12. A scientist observes the temperature T and the volume V in an experiment and finds that $\ln T$ always equals $-\frac{2}{3}\ln V$. Show that $TV^{2/3}$ must therefore be constant.

13. The equation $4\ln x + 2(\ln x)^2 = 0$ can be considered as a quadratic equation in the variable $\ln x$. Solve for $\ln x$, and then solve for x itself.

14. True or False? (a) If $a = b$, then $e^a = e^b$. (b) If $a + b = c$, then $e^a + e^b = e^c$.

15. Find the mistake in the following "proof" that $2 < 1$. We know that $(\frac{1}{2})^2 < \frac{1}{2}$, so $\ln(\frac{1}{2})^2 < \ln \frac{1}{2}$. Thus $2\ln \frac{1}{2} < \ln \frac{1}{2}$. Cancel $\ln \frac{1}{2}$ to get $2 < 1$.

1.6   Solving Inequalities Involving Elementary Functions

This section contains algebra needed in Chapters 3 and 4. A *simple* inequality such as $2x + 3 > 11$ is solved with the same maneuvers as the *equation* $2x + 3 = 11$ (the solution is $x > 4$), but, in general, inequalities are trickier than equations. For example, to solve $\dfrac{x^2 - 2x + 1}{x - 5} > 0$, we want to multiply on both sides by $x - 5$ to eliminate fractions. But if $x < 5$, then $x - 5$ is negative and multiplication by $x - 5$ reverses the inequality; if $x > 5$, then $x - 5$ is positive, and the inequality is not reversed. (For *equations*, this type of difficulty doesn't arise.) This section offers a straightforward method for solving inequalities of the form $f(x) > 0, f(x) < 0$, or equivalently for deciding where a function is positive and where it is negative.

In order for a function f to change from positive to negative, or vice versa, its graph must either cross or jump over the x-axis. Therefore, a nonzero continuous f cannot change signs; its graph must lie entirely on one side of the x-axis. Suppose f is 0 only at $x = -3$ and $x = 2$, and is discontinuous only at $x = 5$, so that *within* the open intervals $(-\infty, -3)$, $(-3, 2)$, $(2, 5)$ and $(5, \infty)$, f is nonzero and continuous. Then in each interval f cannot change signs and is either entirely positive or entirely negative. One possibility is shown in Fig. 1. In general, we have the following method for determining the sign of a function f, that is, for solving the inequalities $f(x) > 0, f(x) < 0$.

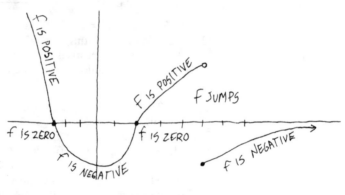

FIG. 1

Step 1 Find values of x where f is discontinuous. For an elementary function f, these occur where f is not defined, in practice because of a zero in a denominator.

Step 2 Find values of x where f is zero; that is, solve the equation $f(x) = 0$.

Step 3 Look at the open intervals in between. On each of the intervals, f maintains only one sign. To find the sign that f takes on each interval, test one number from each interval.

Example 1 Solve the inequalities

(1) $$\frac{x^2 - 2x + 1}{x - 5} > 0, \qquad \frac{x^2 - 2x + 1}{x - 5} < 0.$$

Equivalently, if $f(x) = \dfrac{x^2 - 2x + 1}{x - 5}$, decide where f is positive and where f is negative.

Solution: Step 1 The elementary function f is discontinuous only at $x = 5$, where it is not defined because of a zero in the denominator.

Step 2 Solve the equation $f(x) = 0$.

$$\frac{x^2 - 2x + 1}{x - 5} = 0$$

$x^2 - 2x + 1 = 0$ (multiply by $x - 5$; equivalently, a fraction is 0 if and only if its numerator is 0)

$(x - 1)^2 = 0$

$x = 1$

Step 3   Consider the intervals $(-\infty, 1)$, $(1, 5)$ and $(5, \infty)$. Test one value of x from each interval.

interval	a value of x in the interval	$f(x)$	sign of f in the interval
$(-\infty, 1)$	0	$-\frac{1}{5}$	negative
$(1, 5)$	2	$-\frac{1}{3}$	negative
$(5, \infty)$	6	25	positive

Therefore, $f(x)$ is positive for $x > 5$, and negative for $x < 1$ and for $1 < x < 5$. Equivalently, the solution to the first inequality in (1) is $x > 5$, and the solution to the second inequality is $x < 1$ or $1 < x < 5$.

Note that Steps 1 and 2 locate points where the function either jumps or touches the x-axis. These are places where f *might* (but doesn't have to) change sign by crossing or jumping over the x-axis. Indeed, in this example, f changes sign at $x = 5$ but not at $x = 1$. The graph in Fig. 2 shows what is happening. At $x = 1$, f touches the x-axis but does not cross, so there is no sign change. At $x = 5$, f happens to jump over the axis, so there is a sign change.

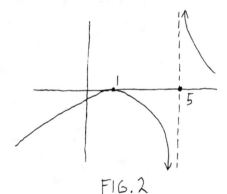

FIG. 2

Problems for Section 1.6

1. Decide where the function f is positive and where it is negative.

(a) $\dfrac{10 - 10x^2}{9(x - 3)^2}$ (d) $\dfrac{e^x}{x}$

(b) $\dfrac{x + 1}{x - 1}$ (e) $x^2 + x - 6$

(c) $x^2 - x + 2$

2. Solve

(a) $\dfrac{16}{x^2} + \dfrac{54}{x^3} > 0$ (b) $\dfrac{1}{2x} + \dfrac{9}{6x + 4} < 3$ (c) $\dfrac{1}{x^2 - 4} > 0$

1.7 Graphs of Translations, Reflections, Expansions and Sums

Considerable time is spent in mathematics finding graphs of functions because graphs can be extremely useful. It is possible to see from a graph where a function is positive, negative, increasing, decreasing, large, small, one-to-one, discontinuous, and so on, when it may be very hard to do this from a formula.

Suppose that the graph of $y = f(x)$ is known. We will develop efficient techniques for finding the graphs of certain variations of f. For example, in trigonometry it is shown that the graph of $\sin 2x$ can be obtained easily from the graph of $\sin x$ by changing the period to π. Similarly, the graph of $2 \sin x$ can be derived from the graph of $\sin x$ by changing the amplitude to 2. We will generalize these ideas to arbitrary graphs. In each case, the problem will be to find the graph of a variation of f, assuming that we have the graph of f. We are *not* concerned here with how the original graph was obtained. Perhaps it was found by plotting many points, possibly it was generated by a computer, it may be a standard curve such as $y = e^x$ or it may have been drawn using techniques of calculus, coming later.

We will first consider three variations in which an operation is performed on the variable x in the equation $y = f(x)$, resulting in horizontal changes in the graph. Then we examine three variations obtained by operating on the entire right-hand side of the equation $y = f(x)$, resulting in vertical changes in the graph. Results are summarized in Table 1. Finally we consider the graph of a sum of functions, given the individual graphs.

Horizontal translation The graph of $y = x^3 + 3x^2 - 1$ is given in Fig. 1. The problem is to draw the graph of the variation $y = (x - 7)^3 + 3(x - 7)^2 - 1$. First, look for a connection between the two tables of values.

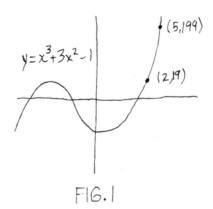

$y = x^3 + 3x^2 - 1$

(5,199)

(2,19)

FIG. 1

	OLD			NEW	
x	$y = x^3 + 3x^2 - 1$		x	$y = (x - 7)^3 + 3(x - 7)^2 - 1$	
2	$2^3 + 3(2^2) - 1 = 19$		9	$2^3 + 3(2^2) - 1 = 19$	
5	$5^3 + 3(5^2) - 1 = 199$		12	$5^3 + 3(5^2) - 1 = 199$	

Substituting $x = 9$ into the new equation involves the same arithmetic (because 7 is immediately subtracted away) as substituting $x = 2$ in the original equation. Similarly, $x = 12$ in the new equation produces the same calculation as $x = 5$ in the old equation. In general, if (a, b) is in the old table then $(a + 7, b)$ is in the new table. Now that we have a connection between the tables, how are the graphs related? The new point $(9, 19)$ is 7 units to the right of the old point $(2, 19)$. In general, given the (old) graph of $y = f(x)$, the (new) graph of $y = f(x - 7)$ is obtained by translating (i.e., shifting) the old graph to the *right* by 7 units (Fig. 2). This agrees with the familiar result that $x^2 + y^2 = r^2$ is a circle with center at the origin, while $(x - 7)^2 + y^2 = r^2$ is a circle centered at the point $(7, 0)$, that is, translated to the *right* by 7.

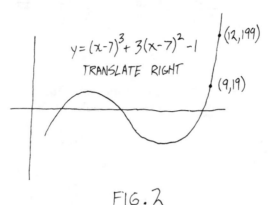

$$y = (x-7)^3 + 3(x-7)^2 - 1$$

TRANSLATE RIGHT

$(12, 199)$

$(9, 19)$

FIG. 2

Similarly, the graph of $y = f(x + 3)$ is found by translating $y = f(x)$ to the *left* by 3 units.

Horizontal expansion/contraction Consider the following two equations with their respective tables of values.

	OLD			NEW	
x	$y = x^3 + 3x^2 - 1$		x	$y = (5x)^3 + 3(5x)^2 - 1$	
2	$2^3 + 3(2^2) - 1 = 19$		2/5	$2^3 + 3(2^2) - 1 = 19$	
5	$5^3 + 3(5^2) - 1 = 199$		1	$5^3 + 3(5^2) - 1 = 199$	

Substituting $x = 2/5$ in the new equation produces the same calculation as $x = 2$ in the old equation (because each occurrence of 2/5 in the new equation is immediately multiplied by 5). If (a, b) is in the old table then $(a/5, b)$ is in the new table. In general, given the graph of $y = f(x)$ (Fig. 3a), the graph of $y = f(5x)$ is obtained by dividing x-coordinates by 5 so as to contract the graph horizontally (Fig. 3b). Similarly, the graph of $y = f(\frac{1}{3}x)$ is found by tripling x-coordinates so as to expand the graph of f horizontally (Fig. 3c). Note that in the expansion (resp. contraction), points *on* the y-axis do not move, but all other points move away from (resp. toward) the y-axis so as to triple widths (resp. divide widths by 5).

The expansion/contraction rule says that the graph of $y = \sin 2x$ is drawn by halving x-coordinates and contracting the graph of $y = \sin x$ horizontally. This agrees with the standard result from trigonometry that $y = \sin 2x$ is drawn by changing the period on the sine curve from 2π to π, a horizontal contraction.

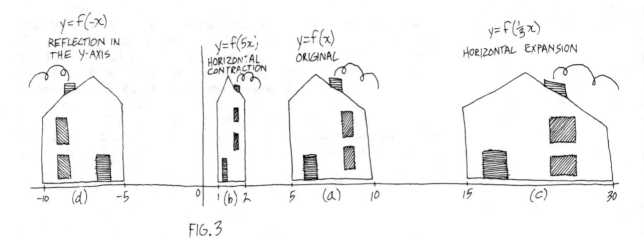

FIG. 3

Horizontal reflection   Consider the following two equations and their respective tables of values.

	OLD			NEW
x	$y = x^3 + 3x^2 - 1$		x	$y = (-x)^3 + 3(-x)^2 - 1$
2	$2^3 + 3(2^2) - 1 = 19$		-2	$2^3 + 3(2^2) - 1 = 19$
5	$5^3 + 3(5^2) - 1 = 199$		-5	$5^3 + 3(5^2) - 1 = 119$

Substituting $x = -2$ into the new equation results in the same calculation as $x = 2$ in the original. If (a, b) is in the old table then $(-a, b)$ is in the new table. In general, given the graph of $y = f(x)$ (Fig. 3a), the graph of $y = f(-x)$ is obtained by reflecting the old graph in the y-axis (Fig. 3d) so as to change the sign of each x-coordinate.

Vertical translation   Consider the equations

$$y = x^3 + 3x^2 - 1 \quad \text{and} \quad y = (x^3 + 3x^2 - 1) + 10.$$

For any fixed x, the y value for the second equation is 10 more than the first y. In general, given the graph of $y = f(x)$, the graph of $y = f(x) + 10$ is obtained by translating the original graph up by 10.† Similarly, the graph of $y = f(x) - 4$ is found by translating the graph of $y = f(x)$ down by 4.

Vertical expansion/contraction   Consider the equations

$$y = x^3 + 3x^2 - 1 \quad \text{and} \quad y = 2(x^3 + 3x^2 - 1).$$

For any fixed x, the y value for the second equation is twice the first y. In general, given the graph of $y = f(x)$, the graph of $y = 2f(x)$ is obtained by doubling the y-coordinates so as to expand the original graph vertically. Similarly, the graph of $y = \frac{2}{3}f(x)$ is found by multiplying heights by 2/3, so as to contract the graph of $f(x)$ vertically.

---

†The conclusion that $y = f(x) + 10$ is obtained by translating *up* by 10 may be compared with a corresponding result for circles, provided that we rewrite the equation as $(y - 10) = f(x)$. The circle $x^2 + y^2 = r^2$ has center at the origin, while $x^2 + (y - 10)^2 = r^2$ is centered at the point $(0, 10)$, that is, translated up by 10. Similarly, the graph of $(y - 10) = f(x)$ is obtained by translating $y = f(x)$ up by 10.

The familiar method for graphing $y = 2 \sin x$ (change the amplitude from 1 to 2) is a special case of the general method for $y = 2f(x)$ (double all heights).

Vertical reflection   Consider $y = f(x)$ versus $y = -f(x)$. The second y is always the negative of the first y. Thus, the graph of $y = -f(x)$ is obtained from the graph of $y = f(x)$ by reflecting in the x-axis. A special case appeared in Fig. 14 of Section 1.3 which showed the graphs of $y = 2 \sin x$ and $y = -2 \sin x$ as reflections of one another.

Table 1   Summary

Variation of $y = f(x)$	How to obtain the graph from the original $y = f(x)$
An operation is performed on the variable x	
$y = f(-x)$	Reflect the graph of $y = f(x)$ in the y-axis
$y = f(2x)$	Halve the x-coordinates of the graph of $y = f(x)$ so as to contract horizontally
$y = f(\tfrac{1}{3}x)$	Multiply the x-coordinates of the graph of $y = f(x)$ by 3 so as to expand horizontally
$y = f(x + 2)$	Translate the graph of $y = f(x)$ to the *left* by 2
$y = f(x - 3)$	Translate the graph of $y = f(x)$ to the *right* by 3
An operation is performed on $f(x)$, i.e., on the entire right-hand side	
$y = -f(x)$	Reflect the graph of $y = f(x)$ in the x-axis
$y = 2f(x)$	Double the y-coordinates of the graph of $y = f(x)$ so as to expand vertically
$y = \tfrac{1}{3}f(x)$	Multiply the y-coordinates of the graph of $y = f(x)$ by $\tfrac{1}{3}$ so as to contract vertically
$y = f(x) + 2$	Translate the graph of $y = f(x)$ *up* by 2
$y = f(x) - 3$	Translate the graph of $y = f(x)$ *down* by 3

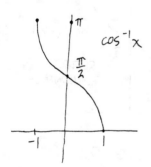

FIG. 4

Example 1   The graph of $\cos^{-1}x$ is shown in Fig. 4. Six variations are given in Figs. 5–10.

Warning   The graph of $f(x - 1)$ (note the *minus* sign) is obtained by translating $f(x)$ to the *right* (in the *positive* direction). The graph of $f(x) - 1$ (note the *minus* sign) is found by translating $f(x)$ *down* (in the *negative* direction).

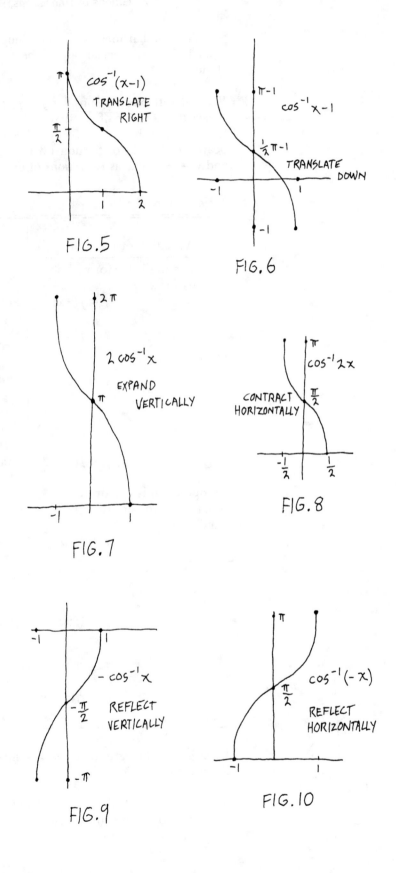

$\cos^{-1}(x-1)$
TRANSLATE RIGHT

FIG. 5

$\cos^{-1}x - 1$
TRANSLATE DOWN

FIG. 6

$2\cos^{-1}x$
EXPAND VERTICALLY

FIG. 7

$\cos^{-1}2x$
CONTRACT HORIZONTALLY

FIG. 8

$-\cos^{-1}x$
REFLECT VERTICALLY

FIG. 9

$\cos^{-1}(-x)$
REFLECT HORIZONTALLY

FIG. 10

The graph of $f(x) + g(x)$ Given the graphs of $f(x)$ and $g(x)$, to sketch $y = f(x) + g(x)$, add the heights from the separate graphs of f and g, as shown in Fig. 11. For example, the new point D is found by adding height $\overline{AB}$ to height $\overline{AC}$ to obtain the new height $\overline{AD}$. On the other hand, since point P has a negative y-coordinate, the new point R is found by subtracting length $\overline{PQ}$ from $\overline{QS}$ to get the new height $\overline{QR}$.

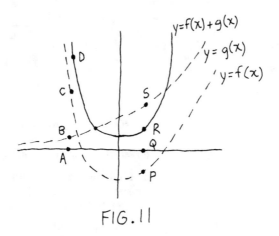

FIG. 11

To sketch $y = \cos x + \sin x$, draw $y = \cos x$ and $y = \sin x$ on the same set of axes, and then add heights (Fig. 12). For example, add height $\overline{AB}$ to height $\overline{AC}$ to obtain the new height $\overline{AD}$; at $x = \pi$, when the sine height is 0, the corresponding point on the sum graph is point E, lying *on* the cosine curve.

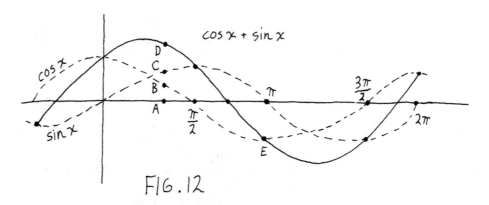

FIG. 12

Problems for Section 1.7

1. Sketch the graph and, in each case, include the graph of ln x for comparison

(a) $\ln(-x)$ (d) $\ln 2x$
(b) $-\ln x$ (e) $\ln(x + 2)$
(c) $2 \ln x$ (f) $2 + \ln x$

2. Figure 13 shows the graph of a function, which we denote by star x. Sketch the following variations given on the next page.

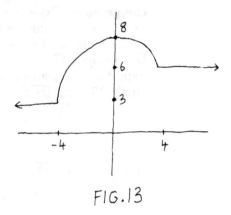

FIG.13

(a) star $\frac{1}{2}x$ (d) star $x - 2$
(b) $\frac{1}{2}$ star x (e) star$(-x)$
(c) star$(x - 2)$ (f) $-$star x

3. Find the new equation of the curve $y = 2x^7 + (2x + 3)^6$ if the curve is (a) translated left by 2   (b) translated down by 5.

4. Sketch the graph.

(a) $y = |\sin x|$ (d) $y = e^{|x|}$
(b) $y = |\ln x|$ (e) $y = \ln|x|$
(c) $y = |e^x|$

5. Sketch each trio of functions on the same set of axes.

(a) $x, \ln x, x + \ln x$
(b) $x, \ln x, x - \ln x$
(c) $x, \sin x, x + \sin x$

6. The variations $\sin^2 x$, $\sin^3 x$ and $\sqrt[3]{\sin x}$ were not discussed in the section. Sketch their graphs by graphically squaring heights, cubing heights and cube-rooting heights on the sine graph.

REVIEW PROBLEMS FOR CHAPTER 1

1. Let $f(x) = \sqrt{5 - x}$.

(a) Find $f(-4)$.
(b) For which values of x is f defined? With these values as the domain, find the range of f.
(c) Find $f(a^2)$ and $(f(a))^2$.
(d) Sketch the graph of f by plotting points. Then sketch the graph of f^{-1}, if it exists.

2. For this problem, we need the idea of the remainder in a division problem. If 8 is divided by 3, we say that the quotient is 2 and the remainder is 2. If 26.8 is divided by 3, the quotient is 8 and the remainder is 2.8. If 27 is divided by 3, the quotient is 9 and the remainder is 0.

If $x \geq 0$, let $f(x)$ be the remainder when x is divided by 3.

(a) Sketch the graph of f.
(b) Find the range of f.
(c) Find $f^{-1}(x)$ if it exists.
(d) Find $f(f(x))$.

3. Describe the graph of f under each of the following conditions.

(a) $f(a) = a$ for all a
(b) $f(a) \neq f(b)$ if $a \neq b$
(c) $f(a + 7) = f(a)$ for all a

4. If $\log_2 x$ is the inverse of 2^x, sketch the graphs of $\log_2 x$ and $\ln x$ on the same set of axes.

5. Find $\sin^{-1}(-\frac{1}{2}\sqrt{2})$.

6. Solve for x.

(a) $y = 2 \ln(3x + 4)$ (b) $y = 4 + e^{3x}$

7. Sketch the graph.

(a) $e^{-x} \sin x$ (e) $\sin^{-1} \frac{1}{2}x$
(b) $\sin^{-1}(x + 2)$ (f) $\sin 3\pi x$
(c) $\sin^{-1}x + \frac{1}{2}\pi$ (g) $2 \cos(4x - \pi)$
(d) $\frac{1}{2} \sin^{-1}x$

8. The functions $\sinh x = \frac{1}{2}(e^x - e^{-x})$ and $\cosh x = \frac{1}{2}(e^x + e^{-x})$ are called the *hyperbolic sine* and *hyperbolic cosine*, respectively.

(a) Sketch their graphs by first drawing $\frac{1}{2}e^x$ and $\frac{1}{2}e^{-x}$.
(b) Show that $\cosh^2 x - \sinh^2 x = 1$ for all x.

9. Solve the equation or inequality.

(a) $\ln x - \ln(2x - 3) = 4$ (c) $2e^x + 8 < 0$

(b) $\ln x < -8$ (d) $\dfrac{1}{x - 3} > \dfrac{1}{4x}$

10. Simplify $5e^{2\ln 3}$.

11. Show that $\ln x - \ln 5x$ simplifies to $-\ln 5$.

2/LIMITS

2.1 Introduction

We begin the discussion of limits with some examples. As you read them, you will become accustomed to the new language and, in particular, see how limit statements about a function correlate with the graph of the function. The examples will show how limits are used to describe discontinuities, the "ends" of the graph where $x \to \infty$ or $x \to -\infty$, and asymptotes. (An asymptote is a line, or, more generally, a curve, that is approached by the graph of f.) Limits will further be used in Sections 3.2 and 5.2 where they are fundamental for the definitions of the derivative and the integral, the two major concepts of calculus.

A limit definition The graph of a function f is given in Fig. 1. Note that as x gets closer to 2, but not equal to 2, $f(x)$ gets closer to 5. We write $\lim_{x \to 2} f(x) = 5$ and say that as x approaches 2, $f(x)$ approaches 5. Equivalently, if $x \to 2$ then $f(x) \to 5$. This contrasts with $f(2)$ itself which is 3.

If point A in Fig. 1 is moved vertically or removed entirely, the limit of $f(x)$ as $x \to 2$ remains 5. In other words, if the value of f at $x = 2$ is changed from 3 to anything else, including 5, or if no value is assigned at all to $f(2)$, we still have $\lim_{x \to 2} f(x) = 5$.

In general, we write

$$\lim_{x \to a} f(x) = L$$

if, for all x sufficiently close, *but not equal*, to a, $f(x)$ is forced to stay as close as we like, and possibly equal, to L.

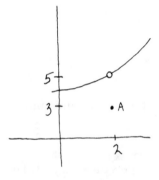

FIG. 1

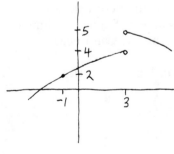

FIG. 2

One-sided limits In Fig. 2, there is no $f(3)$, but we write

(1)
$$\lim_{x \to 3-} f(x) = 4 ,$$

meaning that if x approaches 3 *from the left*, that is, through values *less* than 3 such as $2.9, 2.99, \cdots$, then $f(x)$ approaches 4; and

(2)
$$\lim_{x \to 3+} f(x) = 5 ,$$

meaning that if x approaches 3 *from the right*, that is, through values *greater* than 3 such as $3.1, 3.01, \cdots$, then $f(x)$ approaches 5.

We call (1) a left-hand limit and (2) a right-hand limit. The symbols $3-$ and $3+$ are not new numbers; they are symbols that are used only in the context of a limit statement to indicate from which direction 3 is approached.

In this example, if we are asked simply to find $\lim_{x \to 3} f(x)$, we have to conclude that the limit does not exist. Since the left-hand and right-hand limits disagree, there is no single limit to settle on.

Infinite limits Let

$$f(x) = \frac{1}{x-3}.$$

A table of values and the graph are given in Fig. 3. There is no $f(3)$, but

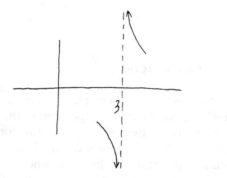

x	$f(x)$
2.8	-5
2.9	-10
2.95	-20
2.99	-100

x	$f(x)$
3.2	5
3.1	10
3.05	20
3.01	100

FIG. 3

we write

$$\lim_{x\to 3+} f(x) = \infty$$

meaning that as x approaches 3 from the right, $f(x)$ becomes unboundedly large; and we write

$$\lim_{x\to 3-} f(x) = -\infty$$

to convey that as x approaches 3 from the left, $f(x)$ gets unboundedly large and negative.

There is no value for $\lim_{x\to 3} f(x)$, since the left-hand and right-hand limits do not agree. We do *not* write $\lim_{x\to 3} f(x) = \pm\infty$.

In general, $\lim_{x\to a} f(x) = \infty$ means that for all x sufficiently close, but not equal, to a, $f(x)$ can be forced to stay as large as we like. Similarly, a limit of $-\infty$ means that $f(x)$ can be made to stay arbitrarily large and negative.

Limits as $x \to \infty$, $x \to -\infty$ For the function in Fig. 4, we write

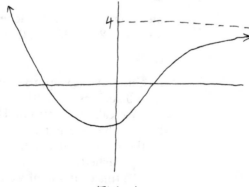

FIG. 4

(3)
$$\lim_{x \to \infty} f(x) = 4$$

to indicate that as x becomes unboundedly large, far out to the right on the graph, the values of y get closer to 4. More precisely,

(3')
$$\lim_{x \to \infty} f(x) = 4-$$

because the values of y are always less than 4 as they approach 4. Both (3) and (3') are correct, but (3') supplies more information since it indicates that the graph of $f(x)$ approaches its asymptote, the line $y = 4$, from *below*.

For the same function, $\lim_{x \to -\infty} f(x) = \infty$ because the graph rises unboundedly to the left.

If a function $f(t)$ represents height, voltage, speed, etc., at time t, then $\lim_{t \to \infty} f(t)$ is called the *steady state* height, voltage, speed, and is sometimes denoted by $f(\infty)$. It is often interpreted as the eventual height, voltage, speed reached after some transient disturbances have died out.

Example 1 There is no limit of $\sin x$ as $x \to \infty$ because as x increases without bound, $\sin x$ just bounces up and down between -1 and 1.

Example 2 The graph of e^x (Section 1.5, Fig. 2) rises unboundedly to the right, so

(4)
$$\lim_{x \to \infty} e^x = \infty.$$

Alternatively, consider the values $e^{100}, e^{1000}, e^{10000}, \cdots$ to see that the limit is ∞. We sometimes abbreviate (4) by writing $e^\infty = \infty$.

The left side of the graph of e^x approaches the x-axis asymptotically (from above), so

(5)
$$\lim_{x \to -\infty} e^x = 0.$$

Alternatively, consider $e^{-100} = 1/e^{100}, e^{-1000} = 1/e^{1000}, \cdots$ to see that the limit is 0 (more precisely, 0+). The result in (5) may be abbreviated by $e^{-\infty} = 0$.

Warning The limit of a function may be L even though f never reaches L. The limit must be *approached,* but *not necessarily attained.* We have $\lim_{x \to -\infty} e^x = 0$ although e^x never reaches 0; for the function f in Fig. 1, $\lim_{x \to 2} f(x) = 5$ although $f(x)$ never attains 5.

Example 3 The graph of $\ln x$ (Section 1.5, Fig. 3) rises unboundedly to the right, so

(6)
$$\lim_{x \to \infty} \ln x = \infty.$$

The graph of $\ln x$ drops asymptotically toward the y-axis, so

(7)
$$\lim_{x \to 0+} \ln x = -\infty.$$

Limits of continuous functions *If f is continuous at $x = a$ so that its graph does not break, then* $\lim_{x \to a} f(x)$ *is simply $f(a)$.* For example, in Fig. 2, $\lim_{x \to -1} f(x) = f(-1) = 2$. If there is a *dis*continuity at $x = a$, then either $\lim_{x \to a} f(x)$ and $f(a)$ disagree, or one or both will not exist.

Example 4 The function $x^3 - 2x$ is continuous (the elementary functions are continuous except where they are not defined) so to find the limit as x approaches 2, we can merely substitute $x = 2$ to get $\lim_{x \to 2}(x^3 - 2x) = 8 - 4 = 4$.

Some types of discontinuities Figure 1 shows a *point discontinuity* at $x = 2$, Fig. 2 shows a *jump discontinuity* at $x = 3$ and Fig. 3 shows an *infinite discontinuity* at $x = 3$. In general, a function f has a point discontinuity at $x = a$ if $\lim_{x \to a} f(x)$ is finite but not equal to $f(a)$, either because the two values are different or because $f(a)$ is not defined. The function has a jump discontinuity at $x = a$ if the left-hand and right-hand limits are finite but unequal. Finally, f has an infinite discontinuity at $x = a$ if at least one of the left-hand and right-hand limits is ∞ or $-\infty$. A function with an infinite discontinuity at $x = a$ is said to *blow up* at $x = a$.

Problems for Section 2.1

1. Find the limit

(a) $\lim_{x \to 3} x^2$ (e) $\lim_{x \to \infty}(\frac{1}{3})^x$
(b) $\lim_{x \to \infty} \sqrt{x}$ (f) $\lim_{x \to \pi/2} \tan x$
(c) $\lim_{x \to 0} \cos x$ (g) $\lim_{x \to 2}(x^2 + 3x - 1)$
(d) $\lim_{x \to -\infty} \tan^{-1}x$

2. Find $\lim \text{Int } x$ as (a) $x \to 3-$ (b) $x \to 3+$
3. Find $\lim |x|/x$ as (a) $x \to 0-$ (b) $x \to 0+$
4. Find $\lim \tan x$ as (a) $x \to \frac{1}{2}\pi-$ (b) $x \to -\frac{1}{2}\pi$
5. (a) Draw the graph of a function f such that f is increasing, but $\lim_{x \to \infty} f(x)$ is *not* ∞. (b) Draw the graph of a function f such that $\lim_{x \to \infty} f(x) = \infty$, but f is *not* an increasing function.
6. Identify the type of discontinuity and sketch a picture.

(a) $\lim_{x \to 3} f(x) = 2$ and $f(3) = 6$
(b) $\lim_{x \to 3} f(x) = \infty$
(c) $\lim_{x \to 2+} f(x) = 4$ and $\lim_{x \to 2-} f(x) = 7$
(d) $\lim_{x \to 3+} f(x) = -\infty$ and $\lim_{x \to 3-} f(x) = 5$

7. Does $\lim_{a \to 0} f(2 + a)$ necessarily equal $f(2)$?
8. Use limits to describe the asymptotic behavior of the function in Fig. 5.

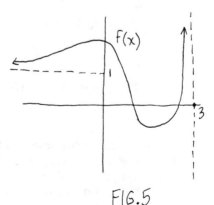

FIG.5

9. Let $f(x) = 0$ if x is a power of 10, and let $f(x) = 1$ otherwise. For example, $f(100) = 0, f(1000) = 0, f(983) = 1$. Find

(a) $\lim_{x \to 65} f(x)$

(b) $\lim_{x \to 100} f(x)$

(c) $\lim_{x \to \infty} f(x)$

10. Use the graph of $f(x)$ to find $\lim_{x \to \infty} f(x)$ if

(a) $f(x) = x \sin x$ (b) $f(x) = \dfrac{\sin x}{x}$

2.2 Finding Limits of Combinations of Functions

The preceding section considered problems involving individual basic functions, such as e^x, $\sin x$ and $\ln x$. We now examine limits of combinations of basic functions, that is, limits of elementary functions in general, and continue to apply limits to curve sketching.

Limits of combinations   To find the limit of a combination of functions we find all the "sublimits" and put the results together sensibly, as illustrated by the following example.

Consider

$$\lim_{x \to 0+} \frac{x^2 + 5 + \ln x}{2e^x}.$$

We can't conveniently find the limit simply by looking at the graph of the function because we don't have the graph on hand. In fact, finding the limit will help *get* the graph. The graph exists only for $x > 0$ because of the term $\ln x$, and finding the limit as $x \to 0+$ will give information about how the graph "begins." We find the limit by combining sublimits. If $x \to 0+$ then $x^2 \to 0$, 5 remains 5 and $\ln x \to -\infty$. The sum of three numbers, the first near 0, the second 5 and the third large and negative, is itself large and negative. Therefore, the numerator approaches $-\infty$. In the denominator, $e^x \to 1$ so $2e^x \to 2$. A quotient with a large negative numerator and a denominator near 2 is still large and negative. Thus, the final answer is $-\infty$. We abbreviate all this by writing

$$\lim_{x \to 0^+} \frac{x^2 + 5 + \ln x}{2e^x} = \frac{0 + 5 + (-\infty)}{2} = \frac{-\infty}{2} = -\infty \quad \text{(Fig. 1)}.$$

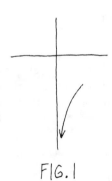

FIG. 1

In each limit problem involving combinations of functions, find the individual limits and then put them together. The last section emphasized the former so now we concentrate on the latter, especially for the more interesting and challenging cases where the individual limits to be combined involve the number 0 and/or the symbol ∞.

Consider $\infty/0-$, an abbreviation for a limit problem where the numerator grows unboundedly large and the denominator approaches 0 from the left. To put the pieces together, examine say

$$\frac{100}{-1/2} = -200, \qquad \frac{1000}{-1/7} = -7000, \qquad \cdots,$$

which leads to the answer $-\infty$. In abbreviated notation, $\infty/0- = -\infty$.

Consider $2/\infty$, an abbreviation for a limit problem in which the numerator approaches 2 and the denominator grows unboundedly large. Compute fractions like

$$\frac{1.9}{100} = .019, \qquad \frac{2.001}{1000} = .002001, \qquad \cdots$$

to see that the limit is 0. In abbreviated notation, $2/\infty = 0$ or, more precisely, $2/\infty = 0+$.

To provide further practice, we list more limit results in abbreviated form. If you understood the preceding examples you will be able to do the following similar problems when they occur (without resorting to memorizing the list).

$0 \times 0 = 0$	$\infty - 4 = \infty$	$\dfrac{0}{\infty} = 0$
$0 + 0 = 0$	$\dfrac{\infty}{8} = \infty$	
$\dfrac{0}{3} = 0$		$\dfrac{0}{-\infty} = 0$
	$-2 \times \infty = -\infty$	
$4^0 = 1$	$\infty^3 = \infty$	$\dfrac{\infty}{0+} = \infty$
$\dfrac{5}{0+} = \infty$	$\infty + \infty = \infty$	
	$\infty \times \infty = \infty$	$\dfrac{\infty}{0-} = -\infty$
$\dfrac{5}{0-} = -\infty$	$\infty \times -\infty = -\infty$	$1^0 = 1$
$3^\infty = \infty$	$(6-) \times \infty = \infty$	$(0+)^\infty = 0$
$\left(\dfrac{1}{2}\right)^\infty = 0$		$\infty^1 = \infty$
		$\infty^{1/2} = \infty$
$\dfrac{2}{-\infty} = 0$		$(0+)^1 = 0$

Example 1 $\lim_{x \to \infty} e^x \ln x = \infty \times \infty = \infty$, $\lim_{x \to 0+} e^x \ln x = 1 \times -\infty = -\infty$.

The graph of $a + be^{cx}$ Consider the function $f(x) = 2 - e^{3x}$. From Section 1.7 we know that the graph can be obtained from the graph of e^x by reflection, contraction and translation. The result is a curve fairly similar to the graph of e^x, but in a different location. The fastest way to determine the new location is to take limits as $x \to \infty$ and $x \to -\infty$, and perhaps plot one convenient additional point as a check:

$$f(\infty) = 2 - e^\infty = 2 - \infty = -\infty$$
$$f(-\infty) = 2 - e^{-\infty} = 2 - 0 = 2$$

and, as a check,

$$f(0) = 2 - 1 = 1.$$

The three computations lead to the graph in Fig. 2.

Example 2 Let

$$f(x) = \frac{2}{5 - x}.$$

Then f is not defined at $x = 5$. Find $\lim_{x \to 5} f(x)$ and sketch the graph of f in the vicinity of $x = 5$.

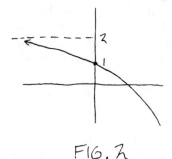

FIG. 2

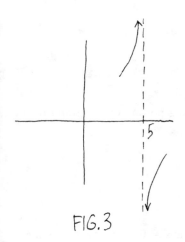

FIG. 3

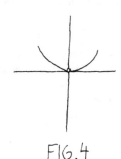

FIG. 4

Solution: We have $\lim_{x \to 5} \dfrac{2}{5-x} = \dfrac{2}{0}$. On closer examination, if x remains larger than 5 as it approaches 5, then $5 - x$ remains less than 0 as it approaches 0. Thus

$$\lim_{x \to 5+} \frac{2}{5-x} = \frac{2}{0-} = -\infty \quad \text{and (similarly)} \quad \lim_{x \to 5-} \frac{2}{5-x} = \frac{2}{0+} = \infty.$$

Since the left-hand and right-hand limits disagree, $\lim_{x \to 5} f(x)$ does not exist. However, the one-sided limits are valuable for revealing that f has an infinite discontinuity at $x = 5$ with the asymptotic behavior indicated in Fig. 3.

Warning A limit problem of the form $2/0$ does not necessarily have the answer ∞. Rather, $2/0+ = \infty$ while $2/0- = -\infty$. In general, in a problem which is of the form $(\text{non } 0)/0$, it is important to examine the denominator carefully.

Example 3 Let $f(x) = e^{-1/x^2}$. Determine the type of discontinuity at $x = 0$ where f is not defined.
  Solution:

$$\lim_{x \to 0} e^{-1/x^2} = e^{-1/0+} = e^{-\infty} = 0+ \qquad (\text{Fig. 4}).$$

Therefore f has a point discontinuity at $x = 0$. If we choose the natural definition $f(0) = 0$, we can remove the discontinuity and make f continuous. In other words, for all practical purposes, e^{-1/x^2} is 0 when $x = 0$.
  In general, if a function g has a point discontinuity at $x = a$, the discontinuity is called *removable* in the sense that we can define or redefine $g(a)$ to make the function continuous. On the other hand, jump discontinuities and infinite discontinuities are not removable. There is no way to define $f(5)$ in Example 2 (Fig. 3) so as to remove the infinite discontinuity and make f continuous.

Problems for Section 2.2

1. Find

(a) $\dfrac{-3}{\infty}$

(b) $\dfrac{\infty}{-4}$

(c) $\infty - 4$

(d) $\dfrac{-19}{0-}$

(e) $\dfrac{1}{-\infty}$

(f) $e^{1/\infty}$

(g) $1/e^x$

(h) 3^x

(i) $\left(\dfrac{1}{4}\right)^x$

(j) $(-\infty)^3$

2. Find

(a) $\lim_{x \to 0+} (\ln x)^2$

(b) $\lim_{x \to \infty} \dfrac{1}{\ln x}$

(c) $\lim_{x \to 0+} (x - \ln x)$

(d) $\lim_{x \to 4} e^{x-4}$

(e) $\lim_{x \to 2} \ln(3x - 5)$

(f) $\lim_{x \to -4} \dfrac{x+5}{x+4}$

(g) $\lim\limits_{x \to 2} x(x + 4)$ (j) $\lim\limits_{x \to \infty} x \cos \dfrac{1}{x}$

(h) $\lim\limits_{x \to -\infty} e^{x-4}$

(i) $\lim\limits_{x \to \pi/2} \dfrac{3}{\sin x - 1}$ (k) $\lim\limits_{x \to 0+} \dfrac{e^x}{\ln x}$

3. Find the limit and sketch the corresponding portion of the graph of the function:

(a) $\lim\limits_{x \to 0} \dfrac{1}{x^2}$ (b) $\lim\limits_{x \to 0} \dfrac{1}{\sin x}$ (c) $\lim\limits_{x \to 1} \dfrac{2}{x - x^3}$

4. Use limits to sketch the graph:

(a) $e^{-3x} - 2$ (b) $3 + 2e^{5x}$

5. The function $f(x) = e^{1/x}$ has a discontinuity at $x = 0$ where it is not defined. Decide if the discontinuity is removable and, if so, remove it with an appropriate definition of $f(0)$.

6. Let $f(x) = \sin 1/x$.

(a) Try to find the limit as $x \to 0+$. In this case, f has a discontinuity which is neither point nor jump nor infinite. The discontinuity is called *oscillatory*.

(b) Find the limit as $x \to \infty$.

(c) Use (a) and (b) to help sketch the graph of f for $x > 0$.

2.3 Indeterminate Limits

The preceding section considered many limit problems, but deliberately avoided the forms $0/0$, $0 \times \infty$, $\infty - \infty$ and a few others. This section discusses these forms and explains why they must be evaluated with caution.

Consider $0/0$, an abbreviation for

$$\lim_{x \to a} \frac{\text{function } f(x) \text{ which approaches } 0 \text{ as } x \to a}{\text{function } g(x) \text{ which approaches } 0 \text{ as } x \to a}.$$

Unlike problems say of the form $0/3$, which *all* have the answer 0, $0/0$ problems can produce a variety of answers. Suppose that as $x \to a$, we have the following table of values:

numerator	.1	.01	.001	.0001	$\cdots$
denominator	.1	.01	.001	.0001	$\cdots$

Then the quotient approaches 1. But consider a second possible table of values:

numerator	2/3	2/4	2/5	2/6	$\cdots$
denominator	1/3	1/4	1/5	1/6	$\cdots$

In this case the quotient approaches 2. Or consider still another possible table of values:

numerator	1/2	1/3	1/4	1/5	$\cdots$
denominator	.1	.01	.001	.0001	$\cdots$

Then the quotient approaches ∞. Because of this unpredictability, the limit form $0/0$ is called indeterminate. In general, *a limit form is indeterminate when different problems of that form can have different answers.* The characteristic of an indeterminate form is a conflict between one function pulling one way and a second function pulling another way.

In a 0/0 problem, the small numerator is pulling the quotient toward 0, while the small denominator is trying to make the quotient ∞ or −∞. The result depends on how "fast" the numerator and denominator each approach 0.

In a problem of the form ∞/∞, the large numerator is pulling the quotient toward ∞, while the large denominator is pulling the quotient toward 0. The limit depends on how fast the numerator and denominator each approach ∞.

In a problem of the form $(0+)^0$, the base, which is positive and nearing 0, is pulling the answer toward 0, while the exponent, which is nearing 0, is pulling the answer toward 1. The final answer depends on the particular base and exponent, and on how "hard" they pull.

In a problem of the form $0 \times \infty$, the factor approaching 0 is trying to make the product small, while the factor growing unboundedly large is trying to make the product unbounded. In an ∞^0 problem, the base tugs the answer toward ∞ while the exponent, which is nearing 0, pulls toward 1. In a 1^x problem, the base, which is nearing 1, pulls the answer toward 1, while the exponent wants the answer to be ∞ if the base is larger than 1, or 0 if the base is less than 1. In a problem of the form $\infty - \infty$, the first term pulls toward ∞ while the second term pulls toward −∞. Thus, $0 \times \infty$, ∞^0, 1^x and $\infty - \infty$ are also indeterminate.

Here is a list of indeterminate forms:

(1)
$$\frac{0}{0}, \frac{\infty}{\infty}, \frac{-\infty}{\infty}, \frac{-\infty}{-\infty}, \frac{\infty}{-\infty}, 0 \times \infty, 0 \times -\infty, \infty - \infty, (-\infty) - (-\infty), (0+)^0, 1^x, \infty^0.$$

Every indeterminate limit problem *can* be done; we do not accept "indeterminate" as a final answer. For example, if a problem is of the form 0/0, there *is* an answer (perhaps 0, or 1, or −2, or ∞, or −∞, or "no limit"), but it usually requires a special method. We discuss one method in this section, but most indeterminate problems require techniques from differential calculus. Further discussion appears in Section 4.3.

Highest power rule The problem $\lim_{x \to \infty}(2x^3 - x^2)$ is of the indeterminate form $\infty - \infty$, but by factoring out the highest power we have

$$\lim_{x \to \infty} 2x^3 \left(1 - \frac{1}{2x}\right) = \infty \times \left(1 - \frac{1}{\infty}\right) = \infty \times (1 - 0) = \infty \times 1 = \infty.$$

The final limit depends entirely on $2x^3$ since the second factor approaches 1. This illustrates the proof of the following general principle:

(2)

As $x \to \infty$ or $x \to -\infty$, a polynomial has the same limit as its term of highest degree.

For example, $\lim_{x \to -\infty}(x^4 + 2x^2 + 3x - 2) = \lim_{x \to -\infty} x^4 = \infty$.

Similarly the problem $\lim_{x \to -\infty} \dfrac{x^3 - x^2 - 1}{5x^3 + 7x + 2}$ is of the indeterminate form ∞/∞, but by factoring out the highest power in the numerator and denominator we have

$$\lim_{x \to -\infty} \frac{x^3 - x^2 - 1}{5x^3 + 7x + 2} = \lim_{x \to -\infty} \frac{x^3}{5x^3} \cdot \frac{1 - \dfrac{1}{x} - \dfrac{1}{x^3}}{1 + \dfrac{7}{5x^2} + \dfrac{2}{5x^3}}.$$

The second factor is of the form $\dfrac{1 - 0 - 0}{1 + 0 + 0}$, and approaches 1. Therefore,

$$\lim_{x \to -\infty} \frac{x^3 - x^2 - 1}{5x^3 + 7x + 2} = \lim_{x \to -\infty} \frac{x^3}{5x^3} = \lim_{x \to -\infty} \frac{1}{5} \text{ (by canceling)} = \frac{1}{5}.$$

In general, we have the following principle:

(3)

> As $x \to \infty$ or $x \to -\infty$, a quotient of polynomials has the same limit as the quotient
>
> $$\frac{\text{term of highest degree in numerator}}{\text{term of highest degree in denominator}}$$
>
> which cancels to an expression whose limit is easy to evaluate.

Example 1 Describe the left end of the graph of $\dfrac{x^5 + x^3 + 1}{6x^3 - 7x^2 + x + 4}$.

Solution: By the highest power rule,

$$\lim_{x \to -\infty} \frac{x^5 + x^3 + 1}{6x^3 - 7x^2 + x + 4} = \lim_{x \to -\infty} \frac{x^5}{6x^3} = \lim_{x \to -\infty} \frac{x^2}{6} = \infty.$$

Therefore at the left end, the curve rises unboundedly.

Warning The highest power rule for polynomials and quotients of polynomials is designed only for problems in which $x \to \infty$ or $x \to -\infty$. The highest powers do *not* dominate if $x \to 6$ or $x \to -10$ or $x \to 0$. In fact if $x \to 0$ then the *lowest* powers dominate because the higher powers of a *small* x are much smaller than the lower powers.

Summary To find the limit of a combination of functions, find all the sublimits.

If you are fortunate, the result will be in a form that can be evaluated immediately; for example, 8/4, which is 2, or $3 \times \infty$, which is ∞.

If the sublimits produce a result of the form 6/0, then the denominator must be examined more carefully. If it is 0+, then the answer is ∞; if it is 0−, then the answer is $-\infty$; and if the denominator is neither (perhaps it is sometimes 0+ and sometimes 0−) then no limit exists.

If the sublimits produce an indeterminate form, perhaps the highest power rule will help; if not, wait for methods coming later.

Problems for Section 2.3

1. $\lim_{x \to \infty}(x^3 - x^4)$

2. $\lim \dfrac{2x^{99} + x^{88} - 7}{x^{34} + 2}$ as (a) $x \to \infty$ (b) $x \to 0$ (c) $x \to 1$

3. $\lim \dfrac{x}{1 - x}$ as (a) $x \to \infty$ (b) $x \to 1+$ (c) $x \to 1-$ (d) $x \to -\infty$

4. $\lim \dfrac{3x^4 - 2x + 4}{x^4 - x}$ as (a) $x \to \infty$ (b) $x \to 1-$

5. $\lim_{x \to \infty} \dfrac{(x + 3)(2 - x)}{(2x + 3)(x - 5)}$

6. $\lim_{x \to 2} \dfrac{x^2 - 4}{x - 2}$

7. $\lim_{x \to \infty} \dfrac{2x - 5}{3x^2 + 4x}$

REVIEW PROBLEMS FOR CHAPTER 2

1. Find

(a) $\lim_{x \to 0} x \cos x$ (c) $\lim_{x \to \infty} e^{-x} \cos x$

(b) $\lim_{x \to 0}(x + \cos x)$ (d) $\lim_{x \to -\infty} \dfrac{2x + 4}{3x + 5}$

2. Find $\lim \dfrac{2x^4 + 3x}{x^2 + 5}$ as (a) $x \to -\infty$ (b) $x \to 2$

3. Find $\lim \dfrac{2}{x^3 - x^2}$ as (a) $x \to \infty$ (b) $x \to 0$ (c) $x \to 1$

4. Find $\lim(2x - 4x^3)$ as (a) $x \to \infty$ (b) $x \to 2$

5. (a) Show that $\ln \sin x$ is not defined at $x = 0$ or $x = \pi$, or as $x \to 0-$ or $x \to \pi+$. (b) Find $\lim_{x \to 0+} \ln \sin x$. (c) Find $\lim_{x \to \pi-} \ln \sin x$.

6. Use limits to help sketch the graph of $1 - e^{2x}$.

7. Suppose f is not defined at $x = 3$. Identify the type of discontinuity and decide if it is removable if

(a) $\lim_{x \to 3} f(x) = 5$

(b) $\lim_{x \to 3-} f(x) = 6$ and $\lim_{x \to 3+} f(x) = \infty$

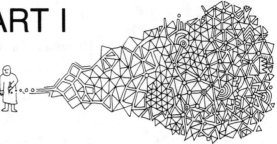

3.1 Preview

This section considers two problems which introduce one of the fundamental ideas of calculus. Subsequent sections continue the development systematically.

Velocity   Suppose that the position of a car on a road at time t is $f(t) = 12t - t^3$. Assume time is in hours and distance is in miles. Then $f(0) = 0$, $f(1) = 11$, $f(2) = 16$, so the car is at position 0 at time 0, at position 11 at time 1, and so on (Fig. 1). The problem is to find the speedometer reading at any instant of time.

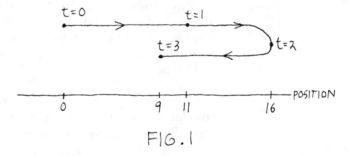

FIG. 1

It is easy to find *average* speeds. For example, in the two hours between times $t = 0$ and $t = 2$, 16 miles are covered so the average speed is 8 mph. An average speed over a period of time is not the same as the instantaneous speedometer readings at each moment in time, but we can use averages to find the instantaneous speed for an arbitrary time t.

First consider the period between times t and $t + \Delta t$. (The symbol Δt is considered a single letter, like h or k, and is commonly used in calculus to represent a small change in t.) The quotient

$$(1) \qquad \frac{\text{change in position}}{\text{change in time}} = \frac{\text{later position} - \text{earlier position}}{\Delta t}$$

$$= \frac{f(t + \Delta t) - f(t)}{\Delta t}$$

is called the *average velocity*. It will be positive if the car is moving to the right, and negative if the car is moving to the left (when the later position is a smaller number than the earlier position). The average speed is the absolute value of the average velocity.

To find the *instantaneous velocity* at time t consider average velocities, but for smaller and smaller time periods, that is, for smaller and smaller values of Δt. In particular, we take the instantaneous velocity at time t to be

(2) $$\lim_{\Delta t \to 0} \frac{f(t + \Delta t) - f(t)}{\Delta t}.$$

Therefore, for our specific function $f(t) = 12t - t^3$,

instantaneous velocity at time t

$$= \lim_{\Delta t \to 0} \frac{12(t + \Delta t) - (t + \Delta t)^3 - (12t - t^3)}{\Delta t}$$

$$= \lim_{\Delta t \to 0} \frac{12t + 12\,\Delta t - t^3 - 3t^2\,\Delta t - 3t(\Delta t)^2 - (\Delta t)^3 - 12t + t^3}{\Delta t}$$

$$= \lim_{\Delta t \to 0} (12 - 3t^2 - 3t\,\Delta t - (\Delta t)^2)$$

$$= 12 - 3t^2.$$

We began with $f(t) = 12t - t^3$ representing position. The function $12 - 3t^2$ just obtained is called the *derivative* of f and is denoted by $f'(t)$. It represents the car's instantaneous velocity. If the derivative is positive then the car is traveling to the right, and if the derivative is negative the car is traveling to the left; the absolute value of the derivative is the speedometer reading.† Velocity is even more useful than speed because the sign of the velocity provides extra information about the direction of travel. For example, $f'(0) = 12$, indicating that at time $t = 0$, the car is traveling to the right at speed 12 mph. Similarly, $f'(2) = 0$, so at time 2 the car has temporarily stopped; $f'(3) = -15$, so at time 3 the car is traveling to the left at 15 mph.

Slope   The slope of a line is used to describe how a line slants and, as a corollary, to identify parallel and perpendicular lines. The problem is to assign slopes to curves in general.

A curve that is not a line will not have a unique slope; instead the slope will change along the curve. It will be positive and large when the curve is rising steeply, positive and small when the curve is rising slowly, and negative when the curve is falling (Fig. 2).

To compute the slope at a particular point A on a curve, we draw a line tangent to the curve at the point (Fig. 2) and take the slope of the tangent line to be the slope of the curve. If the curve is the graph of a function $f(x)$, then the problem is to find the slope of the tangent line at a typical point A with coordinates $(x, f(x))$. We can't determine the slope immediately because we have only one point on the tangent, and we need two points to find the slope of a line. However, we can get the slope of the tangent by a limiting process. Consider a point B on the curve *near* A with coordinates $(x + \Delta x, f(x + \Delta x))$. (Figure 2 shows Δx positive since B is to the right of A; Δx can also be negative, in which case B is to the left of A.) The line AB is called a secant and has slope

(1′) $$\frac{\text{change in } y\text{-coordinate}}{\text{change in } x\text{-coordinate}} = \frac{\Delta y}{\Delta x} = \frac{f(x + \Delta x) - f(x)}{\Delta x}$$

†Initially, in (1), we assumed that $\Delta t > 0$ so that $t + \Delta t$ is a later time than t. However, the limit in (2) allows Δt to be negative as well. In that case, a similar argument will show that the derivative obtained still represents an instantaneous velocity with these properties.

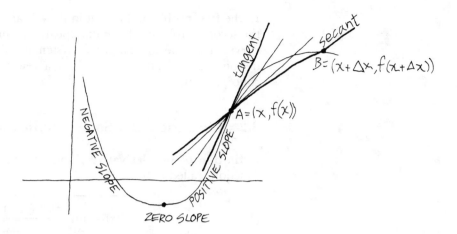

FIG. 2

which is equivalent to (1), but with the independent variable named x instead of t. If we slide point B along the curve toward point A, the secant begins to resemble the tangent at point A. Figure 2 shows some of the in-between positions as the original secant AB approaches the tangent line. This sliding is done mathematically by allowing Δx to approach 0 in (1′). Therefore we choose

$$(2′) \qquad \lim_{\Delta x \to 0} \frac{f(x + \Delta x) - f(x)}{\Delta x}$$

as the slope of the tangent, and hence as the slope of the curve at point A.

From the calculations in the velocity problem we know that if $f(x) = 12x - x^3$ then the limit in (2′) is $12 - 3x^2$, denoted $f'(x)$. Since $f(1) = 11$ and $f'(1) = 9$, the point $(1, 11)$ is on the graph of f, and at that point the slope is 9. Similarly, $f(2) = 16$ and $f'(2) = 0$, so the slope on the graph at the point $(2, 16)$ is 0; $f(3) = 9$ and $f'(3) = -15$ so the slope at the point $(3, 9)$ is -15. Figure 3 shows a partial graph of f.

In the first problem, (1) appeared as an average velocity; in the second problem, the same quotient, eq. (1′), represented the slope of a secant line.

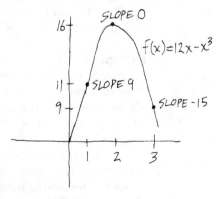

FIG. 3

In the first problem, the limit in (2) was an instantaneous velocity $f'(t)$; in the second problem, the limit appeared again in (2′) as the slope $f'(x)$ of a curve. It is time to examine f' systematically. In the next section we will define the derivative and look at a few applications to help make the concept clear.

3.2   Definition and Some Applications of the Derivative

Definition of the derivative   The derivative of a function f is another function, called f', defined by

(1)
$$f'(x) = \lim_{\Delta x \to 0} \frac{f(x + \Delta x) - f(x)}{\Delta x}.$$

(We will assume for the present that the limit exists. Section 3.3 discusses instances when it does not exist.) Equivalently, if y is a function of x, the derivative y' is defined by

(2)
$$y' = \lim_{\Delta x \to 0} \frac{\text{change in } y}{\text{change in } x} = \lim_{\Delta x \to 0} \frac{\Delta y}{\Delta x}.$$

The process of finding the derivative is called *differentiation*. The branch of calculus dealing with the derivative is called *differential calculus*.

Speed and velocity   Section 3.1 showed that *if $f(t)$ is the position of a particle on a number line at time t* then $f'(t)$ is the velocity of the particle. If the velocity is positive, the particle is traveling to the right; if the velocity is negative, the particle is traveling to the left. *The speed of the particle is the absolute value of the velocity,* that is, *the speed is $|f'(t)|$*. If $f(3) = 12$ and $f'(3) = -4$ then at time 3 the particle is at position 12 with velocity -4, so it is traveling to the left at speed 4.

Slope   Section 3.1 showed that $f'(x)$ is the slope of the tangent line at the point $(x, f(x))$ on the graph of f. Thus $f'(x)$ is taken to be the slope of the graph of f at the point $(x, f(x))$. If the slope is positive, then the curve is rising to the right; if the slope is negative, the curve is falling to the right. If $f(3) = 12$ and $f'(3) = -4$ then the point $(3, 12)$ is on the graph of f, and at that point the slope is -4.

Example 1   Figure 1 gives the graph of a function f. Values of f' may be estimated from the slopes on the graph of f. It looks as if $f'(-3)$ is a large positive number since the curve is rising steeply at $x = -3$. The curve levels off and has a horizontal tangent line at $x = -2$, so $f'(-2) = 0$. Similarly, $f'(-1)$ is large and negative, while $f'(100)$ is a small negative number. We can plot a rough graph of the function f' (Fig. 2) by plotting points such as $A = (-3, \text{large positive})$, $B = (-2, 0)$, $C = (-1, \text{large negative})$, $D = (100, \text{small negative})$. Note that on the graph of f' we treat values of f' as y-coordinates, just as we do for *any* function, although the values of f' were obtained originally as slopes on the graph of f.

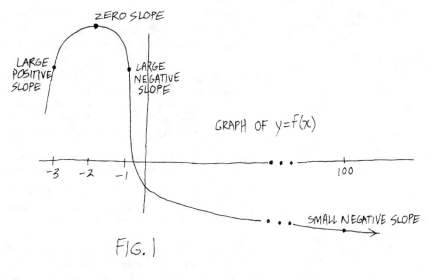

ZERO SLOPE

LARGE POSITIVE SLOPE

LARGE NEGATIVE SLOPE

GRAPH OF y=f(x)

-3   -2   -1

100

SMALL NEGATIVE SLOPE

FIG. 1

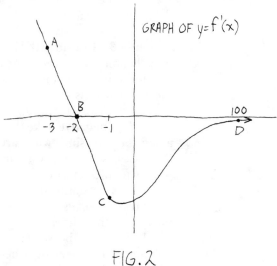

GRAPH OF y=f'(x)

A

B

-3   -2   -1

100

D

C

FIG. 2

Notation   If $y = f(x)$ there are many symbols for the derivative of f. Some of them are

$$f', \quad f'(x), \quad \frac{df}{dx}, \quad \frac{d}{dx} f(x), \quad D_x f, \quad Df, \quad y', \quad \frac{dy}{dx}.$$

The notation dy/dx looks like a fraction but is intended to be a single inviolate symbol.

More general physical interpretation of the derivative   So far, the derivative is a velocity if f represents position, and is a slope on the graph of f. More abstractly, the quotient

$$\frac{\text{change in } f}{\text{change in } x} = \frac{f(x + \Delta x) - f(x)}{\Delta x}$$

is the average rate of change of f with respect to x on the interval between

x and $x + \Delta x$. Thus $f'(x)$ *is the instantaneous rate of change of f with respect to* x. Suppose $f(3) = 13$ and $f'(3) = -4$. If x increases, y (that is, $f(x)$) changes also, and when x reaches 3, y is 13. At that moment y is decreasing instantaneously by 4 units for each unit increase in x.

In general, we have the following connection between the sign of the derivative and the behavior of f.

(3) | *If $f'(x)$ is positive on an interval then f increases on that interval.* In particular, a graph with positive slope is rising to the right.

(4) | *If $f'(x)$ is negative on an interval then f decreases on that interval.* In particular, a graph with negative slope is falling to the right.

(5) | *If $f'(x)$ is zero on an interval then f is constant on that interval.* In particular, a graph with zero slope is a horizontal line.

Example 2 Let $f(t)$ be the temperature at time t (measured in hours). Then $f'(t)$ is the rate at which the temperature is changing per hour. If $f(2) = 40$ and $f'(2) = -5$ then at time 2 the temperature is 40° and is dropping at that moment by 5° per hour.

Example 3 Consider the steering wheel of your car with the front wheels initially pointing straight ahead. Let θ be the angle through which you turn the steering wheel, and let $f(\theta)$ be the corresponding angle through which the front wheels turn (Fig. 3). As in trigonometry, positive angles mean counterclockwise turning.

If f' is negative, take the car back to the dealer, driving very cautiously along the way, since wires are crossed somewhere. When θ increases, $f(\theta)$ decreases, so when you turn the steering wheel counterclockwise, the wheels turn clockwise.

If f' is constantly 0, again take the car back to the dealer, but you'll need a tow truck, because no matter how the steering wheel is turned there is no turning in the wheels.

If f' is 10, the steering is overly sensitive, since for each degree of turning of the steering wheel there is 10 times as much turning of the wheels (in the same direction at least, since 10 is positive). Even $f' = 1$ is probably too large; $f' = \frac{1}{4}$ is more reasonable. In this case, as you turn the steering wheel in a particular direction, the wheels also turn in that direction (because $\frac{1}{4}$ is positive), but each degree of turning in the steering wheel produces only $\frac{1}{4}°$ of turning in the front wheels.

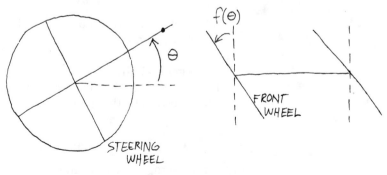

FIG.3

Higher derivatives The function f' is the derivative of f. The derivative of f' is yet another function, called the *second derivative* of f and denoted by f''. A second derivative may sound twice as complicated as a first derivative, but if f'' is regarded as the *first* derivative of f' it isn't a new idea at all: f'' *is the instantaneous rate of change of f' with respect to x*. If $f''(6) = 7$ then, when $x = 6, f'$ is in the process of increasing by 7 units for every unit increase in x. There are many notations for the second derivative, such as

$$f'', \qquad f^{(2)}, \qquad y'', \qquad \frac{d^2y}{dx^2}, \qquad \frac{d^2f}{dx^2}, \qquad \frac{d^2}{dx^2}f(x).$$

Similarly, f''', the derivative of f'', is called the third derivative of f, and so on.

Example 4 Let C be the cost (in dollars) of a standard shopping cart of groceries at time t (measured in days). Suppose that at a certain time, $dC/dt = 2$ and $d^2C/dt^2 = -.03$. Then at this instant, C is going up by \$2 per day (inflation), but the \$2/day figure is in the process of going down by 3¢/day per day (the rate of inflation is tapering off slightly). If the second derivative remains $-.03$ for a while then in another day, the first derivative will decrease to 1.97, and C will be rising by only \$1.97 per day. If the second derivative remains $-.03$ long enough, the first derivative will eventually become zero and then negative, and C will start to fall.

Acceleration Let $f(t)$ be the position of a particle on a number line at time t (use miles and hours) so that $f'(t)$ is the velocity of the particle. The problem is to interpret $f''(t)$ from this point of view.

Suppose $f'(3) = -7$ and $f''(3) = 2$. Then, at time 3, the particle is moving to the left at 7 mph. Since f'' is the rate of change of f', the velocity, which is -7 at this instant, is in the process of increasing by 2 mph per hour, changing from -7 toward -6 and upwards. The absolute value of the velocity is getting smaller so the speed is decreasing. Thus the car is slowing down (decelerating) by 2 mph at this instant.

Unfortunately, the word acceleration has two meanings. *Physicists and mathematicians call f'' the acceleration; their acceleration is the rate of change of VELOCITY. But drivers use acceleration to mean the rate of change of SPEED,* that is, an indication that the car is speeding up or slowing down. The (mathematician's) acceleration $f''(x)$ does not, by itself, determine whether a *driver* is accelerating or decelerating; both f'' and f' must be considered. If $f'(3) = 7$ and $f''(3) = 2$ then, at time 3, the particle is traveling to the right at 7 mph, and the velocity, which is 7 at this instant, is in the process of increasing by 2 mph per hour. Its absolute value is increasing and the car is speeding up by 2 mph per hour. Further examination of the four possible combinations of signs gives the following general result:

(6) *If the velocity f' and the acceleration f'' have the same sign then the particle is speeding up (accelerating). If they have opposite signs then the particle is slowing down (decelerating).* For example, suppose $f''(4) = -5$. If $f'(4)$ is also negative, then at time 4 the particle is accelerating by 5 mph per hour. If $f'(4)$ is positive, then at time 4 the particle is decelerating by 5 mph per hour.

Warning If the acceleration f'' is positive, it is *not* necessarily true that the particle is speeding up. If the acceleration f'' is negative, it is *not* necessarily true that the particle is slowing down. The conclusions are true if the particle is traveling to the right, but the conclusions are false if the particle is traveling to the left.

Units   If $f(t)$ is the temperature at time t (measured in hours) then the units of f' are degrees/hour, and the units of f'' are degrees/hour per hour, that is, degrees/hour². If $f(t)$ is position at time t (miles and hours) then the units of the velocity f' are miles/hour, and the units of the acceleration f'' are miles/hour per hour, or miles/hour². In general, if f is a function of x then the units of f' are (units of f)/(unit of x), and the units of f'' are (units of f)/(unit of x)².

Concavity   The derivative $f'(x)$ is the slope of the graph of $f(x)$ at the point $(x, f(x))$. The problem is to interpret the second derivative $f''(x)$ from a geometric point of view.

   If f' is positive then the graph of f is rising to the right, but this still allows some leeway. The graph can "bend" in two possible ways as it goes up. The two types of bending are called *concave up* and *concave down* (Fig. 4). Similarly, when f' is negative, the graph of f has negative slope but the graph can be either concave up or concave down (Fig. 5).

   We can use the second derivative to detect the concavity. If f'' is positive on an interval then f' is increasing, so the graph of f has increasing slope, as in Figs. 4(a) and 5(a). If f'' is negative on an interval then f' is decreasing,

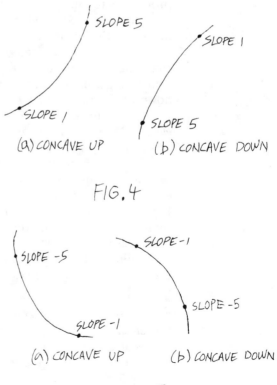

so the graph of f has decreasing slope, as in Figs. 4(b) and 5(b). If f'' is zero on an interval then the slope f' is constant, and the graph of f is a line. We summarize as follows.

(7)

f'' on an interval	graph of f in that interval
positive	concave up
negative	concave down
zero	a line

A point on the graph of f at which the concavity changes is called a *point of inflection.*

Example 6 Suppose $f'(x) > 0$ for $2 \le x \le 7$, $f''(x) < 0$ for $2 \le x < 5$, $f''(5) = 0$, and $f''(x) > 0$ for $5 < x \le 7$. Sketch a graph consistent with the data.

Solution: The graph of f rises on $[2, 7]$, is concave down until $x = 5$ and then switches to concave up. The point $(5, f(5))$ is a point of inflection (Fig. 6).

The sketch deliberately omits the axes (but assumes, as usual, that they are horizontal and vertical). Since we have no information about the values of f, we don't know any specific heights on the graph. The curve can intersect the x-axis, or lie entirely above or below it.

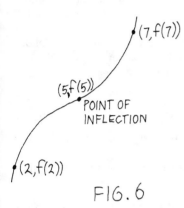

FIG. 6

Problems for Section 3.2

1. If the curve in Fig. 7 is the graph of f, estimate $f'(0)$, $f'(-100)$ and $f'(100)$. Sketch the graph of $f'(x)$.

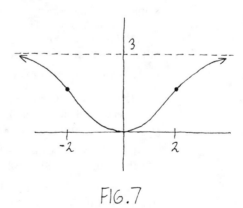

FIG. 7

2. Let p be the price of a camera and S the number of sales. Find the probable sign of dS/dp.

3. Let y be the distance (in feet) from a submerged water bucket up to the top of the well at time t (in seconds). Suppose $dy/dt = -2$ at a particular instant. Which way is the bucket moving, and how fast is it going?

4. If dy/dx is positive, how does y change if x *decreases?*

5. Let $f(x)$ be your height in inches at age x, and let $f'(13.7) = 2$.

(a) By about how much will you grow between age 13.7 and age 14?

(b) Why is your answer to (a) only approximate?

6. A street (number line) is lined with houses. Let $f(x)$ be the number of people living in the interval $[0,x]$. For example, if $f(8) = 100$ then 100 people live in the interval $[0,8]$.

(a) What does $f(x + \Delta x) - f(x)$ represent in this context?
(b) What does the quotient $\dfrac{f(x + \Delta x) - f(x)}{\Delta x}$ represent?
(c) What does $f'(x)$ represent?
(d) What values of $f'(x)$ are impossible?

7. Suppose Smith's salary is x dollars and Brown's salary is y dollars. If Smith's salary increases, how will Brown fare in comparison if dy/dx is   (a) 2   (b) 1/2   (c) -1 (d) 0?

8. Let x be the odometer reading of a vehicle and $f(x)$ the number of gallons of gasoline it has consumed since purchase. Describe $f'(x)$ for a van and for a motorcycle (what units? positive? negative? which is larger?).

9. True or False?

(a) If $f(2) = g(2)$, then $f'(2) = g'(2)$.
(b) If f is increasing, then f' is increasing.
(c) If f is a periodic function, that is, f repeats every b units, then f' is also periodic.
(d) If f is even, then f' is even (even functions were defined and their graphs discussed in Problem 8 of Section 1.2).

10. The posted speed limit at position x on a straight road is $L(x)$, and a car travels so that at time t its position on the road is $f(t)$. For example, if $f(2) = 3$ and $L(3) = 50$ then, at time 2, the car is at position 3 on the road and the posted speed limit is 50 mph. Suppose that at time 6 the car breaks the law and exceeds the speed limit. Express this fact mathematically using a derivative and an inequality.

11. Let $f(x) = x$ for all x. Find $f'(x)$ (a) using the definition in (1)   (b) using slope   (c) using velocity.

12. If the curve in Fig. 7 is the graph of g', sketch a possible graph of g.

13. Let $f(t)$ be the temperature in your city at time t. If it is uncomfortably hot at time $t = 2$, are you pleased or displeased with the indicated data?

(a) $f'(2) = 6, f''(2) = -4$ (b) $f'(2) = -6, f''(2) = -4$ (c) $f'(2) = 0$

14. Let $s(t)$ be the position of a particle on a line at time t (miles and hours). Find the direction of motion and the speed at time 3. Is the particle speeding up or slowing down, and at what rate?

(a) $s'(3) = -4, s''(3) = -1$ (c) $s'(3) = 0, s''(3) = 2$
(b) $s'(3) = 5, s''(3) = -2$ (d) $s'(3) = 2, s''(3) = 0$

15. Suppose $f(2) = 3$, $f(10) = 4$; $f'(x)$ is positive on $[2,8)$, zero at $x = 8$, and negative on $(8,10]$; f'' is positive on $[2,6)$, zero at $x = 6$, and negative on $(6,10]$. Sketch a rough graph of f on $[2,10]$.

16. What kind of second derivative (positive? negative? large? small?) would the car owner prefer in Example 3?

17. If $f'(x)$ decreases from 5 to 1 as x increases from 3 to 4, what can you conclude about $f(x)$ and $f''(x)$ for $3 \le x \le 4$?

18. Let $f(x)$ be the cost to a refinery of starting up production and turning out x barrels of oil.

(a) What does it mean if $f(60) = 400$?
(b) $f'(x)$ is called the marginal cost. What does it represent to the refinery? In particular, what does it mean if $f'(60) = 21$ and $f'(100) = 10$?
(c) Suppose $f(10) = 200$ and $f'(10) = 3$. Interpret physically.

3.3 Derivatives of the Basic Functions

We now begin computing derivatives. In this section we find the derivatives of (almost) all the basic functions; a summary appears at the end of the section. Sections 3.5 and 3.6 will develop rules for differentiating combinations (sums, products, quotients, compositions) of the basic functions. Then you will be able to differentiate any elementary function. (If the derivatives of the basic functions x^2 and $\sin x$ are known, along with the rules for differentiating compositions and products, then such elementary functions as $\sin x^2$ and $x^2 \sin x$ can be differentiated.)

Derivative of a constant function   If $f(x)$ is a constant function then the graph of f is a horizontal line and has slope 0. Thus $f'(x) = 0$. In other words, $D_x c = 0$ for any constant c.

Derivative of the function x   The graph of $f(x) = x$ is the line $y = x$. The line has slope 1 so $f'(x) = 1$. In other words $D_x x = 1$. (See also Problem 11 in the preceding section.)

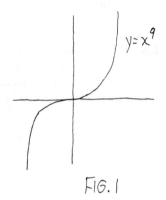

$y = x^9$

FIG. 1

Derivative of the function x^9   It is easy to find $D_x c$ and $D_x x$ using slopes. However, the graph of x^9 (Fig. 1) has varying slope, so $D_x x^9$ is not easy to predict. To get the precise formula for $f'(x)$, we use the definition of the derivative:

$$f'(x) = \lim_{\Delta x \to 0} \frac{f(x + \Delta x) - f(x)}{\Delta x} = \lim_{\Delta x \to 0} \frac{(x + \Delta x)^9 - x^9}{\Delta x}.$$

Now, expand $(x + \Delta x)^9$ by the binomial theorem (Appendix A4) to get

$$f'(x) = \lim_{\Delta x \to 0} \frac{x^9 + 9x^8 \Delta x + a_2 x^7 (\Delta x)^2 + \cdots + a_8 x (\Delta x)^8 + (\Delta x)^9 - x^9}{\Delta x}.$$

(The values of the coefficients $a_2, \cdots, a_8$ will turn out to be unimportant, so we don't bother computing them.) Then

$$f'(x) = \lim_{\Delta x \to 0}[9x^8 + a_2 x^7 \Delta x + \cdots + a_8 x (\Delta x)^7 + (\Delta x)^8] = 9x^8.$$

Thus $D_x x^9 = 9x^8$. Note that the slope $9x^8$ is a large positive number when $x = \pm 4$ for example, corresponding to the steep rise in the graph of x^9 at $x = \pm 4$, and $9x^8$ is a small positive number when x is $\pm\frac{1}{2}$, corresponding to the gentle rise in the graph at $x = \pm\frac{1}{2}$.

Derivative of x^r   The formula $D_x x^9 = 9x^8$ is a special case of the more general pattern $D_x x^r = rx^{r-1}$. This pattern, called the *power rule*, also works for every other power function: to differentiate x^r, lower the exponent by 1 and drop the old exponent down to become a multiplier. For example, $D_x x^2 = 2x$, $D_x x^3 = 3x^2$, and similarly

$$\frac{d(1/x^3)}{dx} = \frac{d(x^{-3})}{dx} = -3x^{-4} = -\frac{3}{x^4} \quad \text{(the exponent } -3 \text{ goes } down \text{ to } -4\text{),}$$

$$\frac{d(\sqrt{x})}{dx} = \frac{d(x^{1/2})}{dx} = \frac{1}{2}x^{-1/2} = \frac{1}{2\sqrt{x}}.$$

The proof of the power rule for $x^2, x^3, x^4, \cdots$ is similar to the proof for x^9. The rule holds for $r = 1$ since the desired formula $D_x x^1 = 1x^0$ amounts to the formula $D_x x = 1$, already proved. Section 3.5 will prove the power rule for r a negative integer and Section 3.7 will give the proof for fractional r.

Warning   There are many ways to indicate that the derivative of x^3 is $3x^2$. For example, you may write $D_x x^3 = 3x^2$, $d(x^3)/dx = 3x^2$, if $f(x) = x^3$ then $f'(x) = 3x^2$. But do *not* write $f'(x^3) = 3x^2$ and do *not* write $x^3 = 3x^2$.

Letters other than x and y may be used. If $z = t^2$ then $dz/dt = 2t$; if $f(u) = u^4$ then $f'(u) = 4u^3$.

Example 1   Find the slope at the point $(2, 8)$ on the graph of $y = x^3$ and find the equation of the tangent line at the point.

Solution: If $f(x) = x^3$ then $f'(x) = 3x^2$ and $f'(2) = 12$. So the slope at $(2, 8)$ is 12. The tangent line has slope 12 and contains the point $(2, 8)$ so its equation is $y - 8 = 12(x - 2)$.

Derivative of sin x   We can make an educated guess for the derivative of $\sin x$, based on slopes on the sine curve (Fig. 8 of Section 1.3). It looks as if the slope of $\sin x$ at $x = 0$ is about 1, the slope at $x = \pi/2$ is 0, the slope at $x = \pi$ is -1, the slope at $x = 3\pi/2$ is 0, and so on. Thus, the derivative of $\sin x$ is a function with the following table of values:

x	derivative of $\sin x$
0	1
$\pi/2$	0
π	-1
$3\pi/2$	0
2π	1

A well-known function that has these values is $\cos x$; and we guess that $D_x \sin x = \cos x$.

We will continue with the proof to confirm the guess, but must admit that students who find it too lengthy to read can grow up to lead rich full happy lives anyway. For the proof we use the definition of the derivative.

$$D_x \sin x = \lim_{\Delta x \to 0} \frac{\sin(x + \Delta x) - \sin x}{\Delta x}$$

$$= \lim_{\Delta x \to 0} \frac{2 \cos \frac{1}{2}(2x + \Delta x) \sin \frac{1}{2}\Delta x}{\Delta x} \quad \text{(by the identity in (15) of Section 1.3)}$$

(1)
$$= \lim_{\Delta x \to 0} \cos \tfrac{1}{2}(2x + \Delta x) \frac{\sin \frac{1}{2}\Delta x}{\frac{1}{2}\Delta x} \quad \text{(rearrange)}.$$

As $\Delta x \to 0$, the first factor in (1) approaches $\cos x$. If we let $\theta = \frac{1}{2}\Delta x$ for convenience, the second factor is $(\sin \theta)/\theta$ where $\theta \to 0$. Therefore, to complete the proof we must show that

(2)
$$\lim_{\theta \to 0} \frac{\sin \theta}{\theta} = 1 .$$

First consider the special case where $\theta \to 0+$ so that we may use a picture with a positive angle θ. Consider a circle of radius 1 and a sector with angle θ (Fig. 2). Then

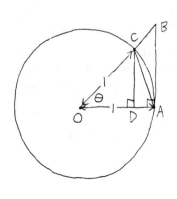

FIG. 2

(3)   area of triangle $OAB = \frac{1}{2}bh = \frac{1}{2}\overline{OA} \cdot \overline{AB} = \frac{1}{2}\overline{AB}$

and

(4)   area of triangle $OAC = \frac{1}{2}bh = \frac{1}{2}\overline{OA} \cdot \overline{DC} = \frac{1}{2}\overline{DC}$.

By trigonometry,

(5) $\tan \theta = \dfrac{\overline{AB}}{\overline{OA}} = \overline{AB}$ and $\sin \theta = \dfrac{\overline{DC}}{\overline{OC}} = \overline{DC}$.

Therefore, by (3), (4) and (5),

area of triangle $OAB = \frac{1}{2}\tan \theta$ and   area of triangle $OAC = \frac{1}{2}\sin \theta$.

(6)

The area of the entire circle with radius 1 is π, and the sector OAC is a fraction of the circle, namely, the fraction $\theta/2\pi$ if θ is measured in radians. Therefore

(7)   area of sector $OAC = \dfrac{\theta}{2\pi} \cdot \pi = \frac{1}{2}\theta$.

Now we are ready to put the ingredients together to prove (2). Since area of triangle $OAC <$ area of sector $OAC <$ area of triangle OAB, we have, by (6) and (7), $\frac{1}{2}\sin \theta < \frac{1}{2}\theta < \frac{1}{2}\tan \theta$. Divide each term by $\frac{1}{2}\sin \theta$ (which is positive since $\theta \to 0+$) to get

$$1 < \frac{\theta}{\sin \theta} < \frac{1}{\cos \theta},$$

and take reciprocals to obtain

$$\cos \theta < \frac{\sin \theta}{\theta} < 1 .$$

We know that $\lim_{\theta \to 0+} \cos \theta = 1$, so as $\theta \to 0+$, $(\sin \theta)/\theta$ is squeezed between 1 and a quantity approaching 1. Therefore $\lim_{\theta \to 0+} \dfrac{\sin \theta}{\theta} = 1$. For the case where $\theta \to 0-$, note that $(\sin \theta)/\theta$ takes on the same values when θ approaches 0 from the left as from the right; that is, $(\sin \theta)/\theta$ is the same whether θ equals b or $-b$ since

$$\frac{\sin(-b)}{-b} = \frac{-\sin b}{-b} = \frac{\sin b}{b} .$$

Therefore, more generally, we have the two-sided limit in (2). This in turn concludes the proof that $D_x \sin x = \cos x$.

Derivative of cos x   To find $D_x \cos x$ note that the cosine and sine graphs (Figs. 8 and 9 of Section 1.3) are translations of one another. The slope at x on the cosine graph is the same as the slope at $x + \frac{1}{2}\pi$ on the sine graph. In other words, $\cos'x = \sin'(x + \frac{1}{2}\pi)$. But $\sin'$ is cos, so $D_x \cos x = \cos(x + \frac{1}{2}\pi)$. Furthermore, $\cos(x + \frac{1}{2}\pi) = -\sin x$. To see this, either use the trig identity for $\cos(x + y)$ or note that the cosine curve translated to the left by $\frac{1}{2}\pi$ is the same as a reflected sine curve (Fig. 3). Therefore we have the final result $D_x \cos x = -\sin x$. (Equivalently, $D_\theta \cos \theta = -\sin \theta$, $D_y \cos y = -\sin y$, $D_u \cos u = -\sin u$, and so on.)

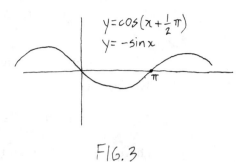

$$y = \cos\left(x + \tfrac{1}{2}\pi\right)$$
$$y = -\sin x$$

FIG. 3

Derivatives of the other trigonometric functions   The functions tan x, cot x, sec x and csc x are various quotients of sin x and cos x. We will find their derivatives in Section 3.5 using a quotient rule, but for completeness we include them in the table of basic derivatives in this section.

Notation   If $f(x) = \sin x$ then $f'(\pi)$ means the value of the derivative when $x = \pi$. Thus, $f'(\pi) = \cos \pi = -1$. We might also let $y = \sin x$ and use the notation $y'|_{x=\pi} = \cos x|_{x=\pi} = \cos \pi = -1$.

Radians versus degrees   Radian measure is used in calculus rather than degrees because the derivative formula for sin x (and hence all the other trigonometric functions) is simpler in radians. We will explain why in this paragraph but if you find it difficult, as many students do, consider it optional.

The rate of change of sin x is different when x is measured in radians than when x is measured in degrees. In particular, sin x changes more rapidly with respect to x when x represents radians. A change of 1 radian has more effect on sin x than a change of 1 degree. In fact, 1 radian has the same effect as approximately 57°. Equivalently, if the rate of change of sin x *per radian* is q then the rate of change of sin x *per degree* is approximately $\frac{1}{57}q$, actually $\frac{\pi}{180}q$. Therefore the formula $D_x \sin x = \cos x$, which holds when radian measure is used, becomes $D_x \sin x = \frac{\pi}{180} \cos x$ when degree measure is used.

Both the guess and the proof of the derivative formula $D_x \sin x = \cos x$ were based on radian measure. In the proof, formula (7) assumed radian measure. Similarly, the guess was based on a graph of sin x using radian measure on the x-axis. If degrees are used (Fig. 4) then the graph of sin x has a different appearance. The slopes are smaller, ranging between $-1/57$ and $1/57$ approximately (actually between $-\pi/180$ and $\pi/180$) rather than between -1 and 1. Slopes read from Fig. 4 lead to $D_x \sin x = \frac{\pi}{180} \cos x$, x in degrees.

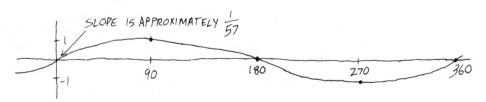

SLOPE IS APPROXIMATELY $\frac{1}{57}$

FIG. 4

The formula $D_x \sin x = \cos x$, x in radians, is simpler than $D_x \sin x = \frac{\pi}{180} \cos x$, x in degrees. Therefore radian measure is used in calculus.

Derivative of e^x and a definition of the number e   Finding $D_x e^x$ is a substantial and difficult problem, especially since it is at this stage that we must define the number e. We'll start by assuming that we have not yet singled out a favorite base, and try to find the derivative of b^x, where b is a fixed positive number. We have

$$(8) \quad \begin{aligned} D_x b^x &= \lim_{\Delta x \to 0} \frac{b^{x+\Delta x} - b^x}{\Delta x} \quad \text{(definition of the derivative)} \\ &= \lim_{\Delta x \to 0} b^x \left[\frac{b^{\Delta x} - 1}{\Delta x} \right] \quad \text{(factor)}. \end{aligned}$$

Now look at sublimits. The factor b^x does not change since it does not contain Δx. Thus we concentrate on finding the limit of the second factor,

$$(9) \quad \frac{b^{\Delta x} - 1}{\Delta x},$$

which is of the indeterminate form $0/0$. The quotient in (9) happens to be the slope of the line through the points $(0, 1)$ and $(\Delta x, b^{\Delta x})$, a secant line on the graph of b^x (Fig. 5). If $\Delta x \to 0$, then the point $(\Delta x, b^{\Delta x})$ slides along the graph toward the point $(0, 1)$ and the secant approaches the tangent line. Therefore, the limit of (9) is the slope of the tangent line, or equivalently, the slope on the graph of b^x at $(0, 1)$. Consequently (8) becomes

$$(8') \quad D_x b^x = m b^x \quad \text{where } m \text{ is the slope at } (0, 1) \text{ on the graph of } b^x.$$

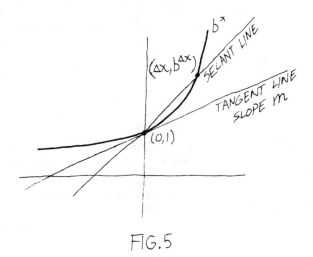

FIG.5

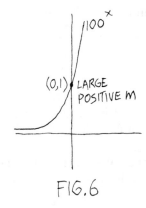

FIG.6

The value of m depends on the value of b. The slope m at the point $(0, 1)$ on the graph of 100^x (Fig. 6) is a large positive number; thus $D_x 100^x = m\,100^x$ where m is a specific large positive number. On the other hand, the slope m at $(0, 1)$ on the graph of 1.01^x (Fig. 7) is a very small positive number. We have the most convenient version of $(8')$ when the slope at $(0, 1)$ on the graph of b^x is 1. Somewhere between the extremes of 100^x and 1.01^x, there is such a b^x (Fig. 8). That particular b is named e. Thus we arrive

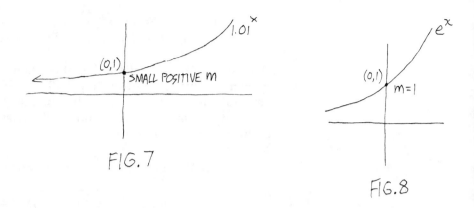

FIG.7

FIG.8

at the following definition of e: e *is the base such that the graph of b^x has slope* 1 *at the point* $(0, 1)$. This definition of e is not yet of computational value; in fact we cannot tell immediately from the definition that e is between 2.71 and 2.72. (One of the ways of computing e will be demonstrated later in Section 8.9.) However, with the definition of e we do immediately have the derivative of e^x. Set $m = 1$, $b = e$ in (8′) to get $D_x e^x = e^x$.

The derivative of the inverse function   If we find the general connection between the derivatives of inverse functions, we can use it to *easily* find the derivatives of $\ln x$, $\sin^{-1} x$ and $\cos^{-1} x$, now that we have derivatives for e^x, $\sin x$ and $\cos x$.

Suppose y is an invertible function of x. Then x is a function of y, and we want the connection between the original derivative dy/dx and the inverse derivative dx/dy. Suppose $dy/dx = 3$, meaning that if x increases, then y increases 3 times as much. If the perspective is changed, and y is viewed as the independent variable, then if y increases, x also increases, but only $1/3$ as much; that is, $dx/dy = \frac{1}{3}$. In general,

$$(10) \qquad \boxed{\frac{dx}{dy} = \frac{1}{\dfrac{dy}{dx}}.}$$

The inverse formula is easy to remember, because if we pretend that dy/dx and dx/dy are fractions, the formula looks like standard algebra.

Derivative of $\ln x$ Let $y = \ln x$. Then $x = e^y$, and

$$\frac{d(\ln x)}{dx} = \frac{dy}{dx} = \frac{1}{\dfrac{dx}{dy}} = \frac{1}{e^y}.$$

We don't stop here because when y is a function of x we expect the derivative to be a function of x also. Thus we must express $1/e^y$ in terms of x, which is easy because $e^y = x$. Therefore, $dy/dx = 1/x$, that is, $D_x \ln x = 1/x$.

Derivatives of the inverse trigonometric functions   We continue to take advantage of (10). To find the derivative of $\sin^{-1} x$, let $y = \sin^{-1} x$, so that $x = \sin y$ where $-\frac{1}{2}\pi \le y \le \frac{1}{2}\pi$. Then

(11)
$$\frac{d(\sin^{-1}x)}{dx} = \frac{dy}{dx} = \frac{1}{\frac{dx}{dy}} = \frac{1}{\cos y}.$$

We want to express the answer in terms of x since y is a function of x. We know that $\sin y = x$, and $\cos^2 y = 1 - \sin^2 y$ by a trig identity, so $\cos^2 y = 1 - x^2$. Thus $\cos y$ is either $\sqrt{1 - x^2}$ or $-\sqrt{1 - x^2}$. In this case, y is an angle between $-\frac{1}{2}\pi$ and $\frac{1}{2}\pi$, so its cosine is positive. Therefore $\cos y = \sqrt{1 - x^2}$ and $D_x \sin^{-1}x = 1/\sqrt{1 - x^2}$.

Derivatives of $\cos^{-1}x$ and $\tan^{-1}x$ may be obtained similarly and are listed in the table of basic derivatives.

Table of basic derivatives

$D_x c = 0$	$D_x \sin x = \cos x$	$D_x \sin^{-1}x = \dfrac{1}{\sqrt{1 - x^2}}$
$D_x x = 1$	$D_x \cos x = -\sin x$	
$D_x x^r = rx^{r-1}$	$D_x \tan x = \sec^2 x$	$D_x \cos^{-1}x = -\dfrac{1}{\sqrt{1 - x^2}}$
(power rule)	$D_x \cot x = -\csc^2 x$	
$D_x \ln x = \dfrac{1}{x}$	$D_x \sec x = \sec x \tan x$	$D_x \tan^{-1}x = \dfrac{1}{1 + x^2}$
	$D_x \csc x = -\csc x \cot x$	
$D_x e^x = e^x$		

Problems for Section 3.3

1. Find

(a) $D_x x^6$
(b) $D_x 1/x^6$
(c) $D_x x^{8/7}$
(d) $D_u \sqrt[3]{u}$
(e) $\dfrac{d}{dx}\left(\dfrac{1}{\sqrt{x}}\right)$
(f) $\dfrac{d(x^{2/3})}{dx}$
(g) $D_x 0$
(h) $\dfrac{d(e^t)}{dt}$
(i) $D_x 4$

2. If $f(z) = \ln z$, find $f'(z)$.
3. If $y = x$, find y'.
4. If $f(x) = 7$ for all x, find $f'(x)$.
5. If $u = \tan t$, find du/dt.
6. Find y' and y'' if (a) $y = \ln x$ (b) $y = \sin x$ (c) $y = e^x$.
7. If $f(x) = 1/\sqrt{x}$ find $f'(17)$.
8. If $f(x) = \sin x$ find $f(\pi)$ and $f'(\pi)$.
9. Differentiate the function.

(a) x^{-3}
(b) x^{14}
(c) $\sqrt{x^5}$
(d) $1/x^5$
(e) x
(f) $\ln x$
(g) $x^{-1/3}$
(h) x^4
(i) $1/x^4$
(j) $\dfrac{1}{x}$
(k) $\dfrac{1}{x^2}$

10. Examine the graph of $\ln x$ and convince yourself that the slopes do look like $1/x$.

11. Use (10) together with the derivative formula for tan x to prove the derivative formula for $\tan^{-1}x$.

12. The $\sin^{-1}$ function is the inverse of sin x when x is restricted to $[-\frac{1}{2}\pi, \frac{1}{2}\pi]$. Consider a second $\sin^{-1}$ function, called II $\sin^{-1}$, defined as the inverse of sin x when x is restricted to $[\pi/2, 3\pi/2]$.

(a) Sketch the graph of $y = $ II $\sin^{-1}x$.
(b) Does the derivative of II $\sin^{-1}x$ equal $1/\sqrt{1-x^2}$? If not, find its derivative.

13. If $a = b^{-4}$, find da/db and db/da directly and verify that $\dfrac{da}{db} = \dfrac{1}{db/da}$.

14. A block bounces up and down on a spring so that at time t, its height is sin t (use meters and seconds).

(a) Find the speed of the block at time $t = 2\pi/3$.
(b) Is the block speeding up or slowing down at time $t = 2\pi/3$, and by how much?
(c) When is the speed of the block maximum? minimum?

15. Find the slope at $(-2, 16)$ on the graph of $y = x^4$ and find the equations of the lines tangent and perpendicular to the graph at the point.

3.4 Nondifferentiable Functions

It is possible for a function *not* to have a derivative for some value of x. We mention this possibility not because it will happen frequently and hinder you in later work, but because you will understand the derivative better if you see examples where one doesn't exist. A function that doesn't have a derivative at $x = x_0$ must correspondingly have a graph with no slope at the point $(x_0, f(x_0))$. We will illustrate a few (but not all) of the ways in which this can happen.

Discontinuities   Imagine traveling from left to right along the graph of f in Fig. 1. It is a vertical step up to point A and then a vertical step back down again, so we say that the left-hand slope at A is ∞ and the right-hand slope is $-\infty$. But even if we are willing to accept infinite derivatives, the left-hand and right-hand slopes don't agree. Thus f is not differentiable at $x = 2$; that is, there is no $f'(2)$.

Continuing from left to right in Fig. 1, it is a vertical step up to the point B and then a slope of approximately 1 leaving point B. Thus, the

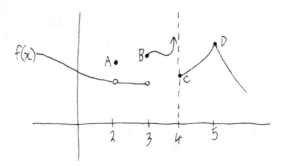

FIG. 1

left-hand slope is ∞, and the right-hand slope is about 1. The disagreement means that there is no $f'(3)$.

Similarly, f is not differentiable at $x = 4$, and in general, *if f is discontinuous at $x = x_0$, then f is not differentiable at $x = x_0$. (Equivalently, if f is differentiable then f is continuous.)*

Cusps Continuing from left to right in Fig. 1, the slope coming into point D, the left-hand slope, is about 1, while the slope leaving the point, the right-hand slope, is about -2. Since the two values disagree, there is no slope assigned to D and there is no $f'(5)$. We call point D a *cusp*. In general, *a cusp arises when the graph is continuous but suddenly changes direction* (so that the curve is not "smooth"), *and in this case f is not differentiable.*

Note that differentiability is a more exclusive property than continuity: a differentiable function must be continuous, but a continuous function need not be differentiable (at the cusp in Fig. 1, f is continuous but not differentiable). In other words, the collection of differentiable functions is a subset of the collection of continuous functions.

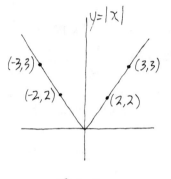

FIG. 2

Example 1 Let $f(x) = |x|$. The graph of f (Fig. 2) has a cusp at $x = 0$, so there is no $f'(0)$. In particular, the figure shows that the left-hand slope is -1 and the right-hand slope is 1. Let's try to find $f'(0)$ using the definition of the derivative to see what happens:

$$f'(0) = \lim_{\Delta x \to 0} \frac{f(0 + \Delta x) - f(0)}{\Delta x} = \lim_{\Delta x \to 0} \frac{|0 + \Delta x| - |0|}{\Delta x} = \lim_{\Delta x \to 0} \frac{|\Delta x|}{\Delta x}.$$

The limit doesn't exist because the left-hand limit is -1 and the right-hand limit is 1 (see Problem 3, Section 2.1). Again we conclude that the left-hand slope is -1, the right-hand slope is 1, and there is no $f'(0)$.

3.5 Derivatives of Constant Multiples, Sums, Products and Quotients

Now that we have derivatives for the basic functions, we'll continue by looking at combinations of functions. All our combination rules assume that we are working with differentiable functions.

The constant multiple rule for the derivative of $cf(x)$ The graph of $2f(x)$ is a vertical expansion of the graph of $f(x)$, which makes it twice as steep (for example, see Figs. 4 and 7 in Section 1.7). Thus $D_x 2f(x) = 2D_x f(x)$ and, in general, for any constant c,

(1)
$$\boxed{D_x cf(x) = cD_x f(x).}$$

The constant factor c can be "pulled out" of the differentiation problem. In other words, *slide past the constant and then start differentiating.* If $f(x) = 3 \sin x$ then $f'(x) = 3 \cos x$. If $f(x) = -\tan x$ then $f'(x) = -\sec^2 x$.

Combining the power rule with (1), we have $D_x 4x^3 = 4 \cdot 3x^2 = 12x^2$. Similarly, $D_x 8x^2 = 16x$, and $D_x(-\frac{1}{2}x^8) = -4x^7$.

Combining the formula $D_x x = 1$ with (1), we have $D_x 8x = 8 \cdot 1 = 8$. Similarly, $D_x \frac{1}{2}x = \frac{1}{2}$, $D_x 7x = 7$, $D_x(-x) = -1$ and so on.

Note that (1) includes the case of a constant *divisor*. For example,

$$D_t \frac{\ln t}{7} = D_t \frac{1}{7} \ln t = \frac{1}{7} \cdot \frac{1}{t} = \frac{1}{7t}$$

and

$$\frac{d}{dx}\left(\frac{1}{2x^4}\right) = \frac{d}{dx}(\tfrac{1}{2}x^{-4}) = -2x^{-5} = -\frac{2}{x^5}.$$

The sum rule for the derivative of $f(x) + g(x)$ By definition of the derivative,

$$D_x(f(x) + g(x)) = \lim_{\Delta x \to 0} \frac{f(x + \Delta x) + g(x + \Delta x) - (f(x) + g(x))}{\Delta x}.$$

To evaluate this limit, first rearrange to separate the f and g parts.

$$D_x(f(x) + g(x)) = \lim_{\Delta x \to 0}\left(\frac{f(x + \Delta x) - f(x)}{\Delta x} + \frac{g(x + \Delta x) - g(x)}{\Delta x}\right).$$

Further separation is possible since the limit of a combination of functions is computed by finding the individual limits; in this case, the limit of the sum is the sum of the limits. Therefore

$$D_x(f(x) + g(x)) = \lim_{\Delta x \to 0} \frac{f(x + \Delta x) - f(x)}{\Delta x} + \lim_{\Delta x \to 0} \frac{g(x + \Delta x) - g(x)}{\Delta x}.$$

But the first limit on the right-hand side is $f'(x)$, by definition of the derivative, and the second limit is $g'(x)$. Thus the sum rule is

(2)
$$\boxed{D_x(f + g) = D_x f + D_x g.}$$

The derivative of the sum is the sum of the derivatives. In other words, *differentiate f and g separately, and then add.* For example, $D_x(2x^3 + 7x^2 - 3x + 4) = 6x^2 + 14x - 3$.

The product rule for the derivative of $f(x)g(x)$ Again we'll use the definition of the derivative:

$$D_x f(x)g(x) = \lim_{\Delta x \to 0} \frac{f(x + \Delta x)g(x + \Delta x) - f(x)g(x)}{\Delta x}.$$

Now add *and* subtract $f(x + \Delta x)g(x)$ in the numerator, which is strange but legal, to get

$$D_x f(x)g(x) =$$
$$\lim_{\Delta x \to 0} \frac{f(x + \Delta x)g(x + \Delta x) - f(x + \Delta x)g(x) + f(x + \Delta x)g(x) - f(x)g(x)}{\Delta x}.$$

Then factor and rearrange:

$$D_x f(x)g(x) = \lim_{\Delta x \to 0}\left(f(x + \Delta x)\frac{g(x + \Delta x) - g(x)}{\Delta x} + \frac{f(x + \Delta x) - f(x)}{\Delta x}g(x)\right).$$

Now there are four sublimits to examine. To find $\lim_{\Delta x \to 0} f(x + \Delta x)$, we simply substitute $\Delta x = 0$ because f is assumed differentiable, hence continuous. Thus the limit is $f(x)$. For the next two sublimits, we have, by definition of the derivative,

$$\lim_{\Delta x \to 0} \frac{g(x + \Delta x) - g(x)}{\Delta x} = g'(x) \quad \text{and} \quad \lim_{\Delta x \to 0} \frac{f(x + \Delta x) - f(x)}{\Delta x} = f'(x).$$

Finally, $\lim_{\Delta x \to 0} g(x) = g(x)$ because Δx does not appear in the expression $g(x)$. Thus the final limit is $f(x)g'(x) + f'(x)g(x)$; that is, the product rule is

(3) $$\boxed{(fg)' = fg' + f'g.}$$

The derivative of a product is the first factor times the derivative of the second plus the second times the derivative of the first. If $f(x) = x^3 \sin x$ then $f'(x) = x^3 \cos x + 3x^2 \sin x$.

Warning The derivative of $x^3 \sin x$ is *not* $3x^2 \cos x$. The derivative of a product fg is *not* found by differentiating f and g separately and multiplying.

Example 1

$$\frac{d(x^3 \ln x)}{dx} = x^3 \cdot \frac{1}{x} + 3x^2 \ln x = x^2 + 3x^2 \ln x.$$

The product rule for more than two factors If $y = fg$ then $y' = fg' + f'g$. Suppose $y = fgh$, a product of three functions. By grouping, we can rewrite y as $f(gh)$ which represents y as a product of two factors, although one of the two factors is itself a product. Then

$y' = f(gh)' + f'(gh)$ (product rule for two factors)

$= f(gh' + g'h) + f'(gh)$ (product rule for two factors again)

$= fgh' + fg'h + f'gh.$

Therefore the product rule for three factors is

(4) $$(fgh)' = fgh' + fg'h + f'gh.$$

If $f(x) = x^2 \sin x \cos x$ then

$f'(x) = (x^2 \sin x)(-\sin x) + x^2 \cos x \cos x + 2x \sin x \cos x$

$= -x^2 \sin^2 x + x^2 \cos^2 x + 2x \sin x \cos x.$

Similar results hold for products of four or more factors.

Warning Certain possibly ambiguous notations have standard interpretations in mathematics. The notation $\tan xe^x$ is assumed to mean $\tan(xe^x)$. If you intend $(\tan x)(e^x)$ then you must insert the appropriate parentheses, or better still write $e^x \tan x$ which is unambiguous. Similarly, $\sin x \cos x$ means $(\sin x)(\cos x)$, $\sin x^2$ means $\sin(x^2)$ and $\sin^2 x$ means $(\sin x)^2$. Be careful to have your notation match your intention.

The quotient rule for the derivative of $f(x)/g(x)$ By the definition of the derivative,

$$D_x \frac{f(x)}{g(x)} = \lim_{\Delta x \to 0} \frac{\dfrac{f(x + \Delta x)}{g(x + \Delta x)} - \dfrac{f(x)}{g(x)}}{\Delta x}.$$

Simplify the fraction on the right-hand side by multiplying numerator and denominator by $g(x)g(x + \Delta x)$ to get

$$D_x \frac{f(x)}{g(x)} = \lim_{\Delta x \to 0} \frac{g(x)f(x + \Delta x) - f(x)g(x + \Delta x)}{\Delta x\, g(x)g(x + \Delta x)}.$$

Add and subtract $f(x)g(x)$ in the numerator to obtain

$$D_x \frac{f(x)}{g(x)} = \frac{g(x)f(x + \Delta x) - f(x)g(x) - f(x)g(x + \Delta x) + f(x)g(x)}{\Delta x\, g(x)g(x + \Delta x)}.$$

Factor and rearrange to get

$$D_x \frac{f(x)}{g(x)} = \lim_{\Delta x \to 0} \frac{g(x)\left(\dfrac{f(x + \Delta x) - f(x)}{\Delta x}\right) - f(x)\left(\dfrac{g(x + \Delta x) - g(x)}{\Delta x}\right)}{g(x)g(x + \Delta x)}.$$

Finally, find the separate sublimits as in the proof of the product rule, to produce the quotient rule

(5)
$$\boxed{\left(\frac{f}{g}\right)' = \frac{gf' - fg'}{g^2}.}$$

The derivative of a quotient is the denominator times the derivative of the numerator, minus the numerator times the derivative of the denominator, all divided by the square of the denominator.

Example 2   By the quotient rule,

$$D_x \frac{4x}{3x + 5} = \frac{(3x + 5)\cdot 4 - 4x \cdot 3}{(3x + 5)^2} = \frac{20}{(3x + 5)^2}.$$

Warning   It is correct but *silly* to use the quotient rule to write

$$D_x \frac{x^2 + 3x}{6} = \frac{6(2x + 3) - (x^2 + 3x)\cdot 0}{36} = \frac{2x + 3}{6}.$$

Instead, write the function as $\frac{1}{6}(x^2 + 3x)$ and use the constant multiple rule to get the derivative $\frac{1}{6}(2x + 3)$ immediately.

Delayed proof of the tangent derivative formula   The formula $D_x \tan x = \sec^2 x$, stated in Section 3.3, can now be justified by the quotient rule

$$D_x \tan x = D_x \frac{\sin x}{\cos x}$$

$$= \frac{\cos x \cos x - \sin x(-\sin x)}{\cos^2 x}$$

$$= \frac{\cos^2 x + \sin^2 x}{\cos^2 x}$$

$$= \frac{1}{\cos^2 x} \qquad \text{(by a trigonometric identity)}$$

$$= \sec^2 x.$$

The derivatives of cot x, sec x and csc x can be found in a similar manner.

Delayed proof of the power rule $D_x x^r = r x^{r-1}$ when r is a negative integer Consider $D_x x^{-9}$ for example. By the quotient rule and the *previously proved* case of the power rule for r a positive integer (Section 3.3) we have

$$D_x x^{-9} = D_x \frac{1}{x^9} = \frac{x^9 \cdot 0 - 1 \cdot 9x^8}{(x^9)^2} = \frac{-9x^8}{x^{18}} = -9x^{-10}.$$

The proof in the general case is handled in the same way, with -9 replaced by an arbitrary negative integer r.

The derivative of a function "with two formulas" Suppose $f(x) = |\ln x|$. Then $f(x) = \ln x$ when $\ln x \geq 0$ but $f(x) = -\ln x$ when $\ln x < 0$. Thus

$$f(x) = \begin{cases} -\ln x & \text{if } 0 < x < 1 \\ \ln x & \text{if } x \geq 1 \end{cases} \quad \text{so} \quad f'(x) = \begin{cases} -1/x & \text{if } 0 < x < 1 \\ 1/x & \text{if } x > 1. \end{cases}$$

(The graph of f (see Problem 4b of Section 1.7) has a cusp at $x = 1$ and f is not differentiable there. In fact, set $x = 1$ in the formula $-1/x$ to obtain the left-hand slope -1 at the cusp, and set $x = 1$ in the formula $1/x$ to obtain the right-hand slope 1, a different value.)

In general, if $f(x)$ is defined by different formulas on various intervals then $f'(x)$ is found by differentiating each formula separately.

Example 3 We discussed velocity and acceleration in Section 3.2 but did not actually compute them in that section since efficient techniques of differentiation had not yet been developed. If $f(t) = t^3 - 3t^2 - 45t$ is the position of a particle at time t, we are now prepared to describe its motion using derivatives.

The velocity is $f'(t) = 3t^2 - 6t - 45$. To determine when the particle travels left and when it travels right, we will determine the sign of $f(t)$ using the method of Section 1.6. The function $f'(t)$ has no discontinuities, and is 0 when

$$3t^2 - 6t - 45 = 0$$
$$t^2 - 2t - 15 = 0$$
$$(t + 3)(t - 5) = 0$$
$$t = -3, 5.$$

To find the sign of $f'(t)$ in the intervals $(-\infty, -3)$, $(-3, 5)$ and $(5, \infty)$, test a value of $f(t)$ for t in each interval. For example, $f'(-100)$ is positive so $f'(t)$ is positive in $(-\infty, -3)$. The results are shown in Table 1.

Table 1

Time interval	Sign of f'	Particle
$(-\infty, -3)$	positive	moves right
$(-3, 5)$	negative	moves left
$(5, \infty)$	positive	moves right

We continue further to determine the sign of the acceleration $f''(t) = 6t - 6$. The function $f''(t)$ is continuous, and is 0 when $6t - 6 = 0$, $t = 1$. Table 2 shows the sign of $f''(t)$.

Table 2

Time interval	Sign of f''
$(-\infty, 1)$	negative
$(1, \infty)$	positive

By (6) of Section 3.2, the particle accelerates when f' and f'' have the same sign, and decelerates when f' and f'' have opposite signs. Table 3 combines Tables 1 and 2 to display the sign pattern.

Table 3

Time interval	Sign of f'	Sign of f''	Particle
$(-\infty, -3)$	positive	negative	moves right, slows down
$(-3, 1)$	negative	negative	moves left, speeds up
$(1, 5)$	negative	positive	moves left, slows down
$(5, \infty)$	positive	positive	moves right, speeds up

It is helpful to locate a few positions precisely before plotting the motion. Some key values of $f(t)$ are $f(-\infty) = -\infty, f(-3) = 81, f(1) = -47,$ $f(5) = -175, f(\infty) = \infty$. Figure 1 shows the final result.

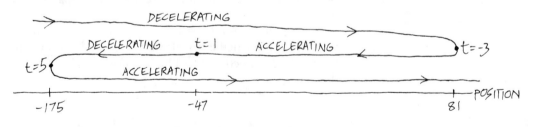

FIG. 1

Example 4 Section 3.2 discussed slopes, and now we are ready to actually compute some. Use the derivative to find the vertex of the parabola $y = 2x^2 + 8x + 9$, and sketch its graph.

Solution: At the vertex of a parabola the slope is 0. We have $y' = 4x + 8$, which is 0 when $x = -2$. If $x = -2$ then $y = 1$, so the vertex is $(-2, 1)$.

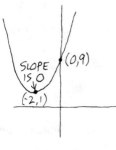

SLOPE
IS 0

(0,9)

(-2,1)

FIG. 2

We know that the parabola opens upward since the coefficient of x^2 is positive. Alternatively, $y'' = 4$, and a positive second derivative implies that the curve is concave up. Figure 2 gives the graph.

Problems for Section 3.5

1. Find $f'(x)$ if (a) $f(x) = 3x^6 + \cos x$ (b) $f(x) = 2x^5 - 6x^3 - 4x + 5$.

2. Find $y^{(4)}$, the fourth derivative of y.

(a) $y = 1/x$ (b) $y = \sin x$ (c) $y = x$

3. Differentiate

(a) $\dfrac{x^3}{2}$　　(h) $\sec x \tan x$

(b) $2x^3$　　(i) $2e^x \ln x + 5x^2$

(c) $\dfrac{2}{x^3}$　　(j) $2e^x + \ln x$

(d) $\dfrac{1}{2x^3}$　　(k) $4x^2 \tan^{-1}x$

(e) $\dfrac{x^3 + 2x}{3}$　　(l) $x^3 \sin x \tan x$

(f) $2x^3 \cos x$ (m) $\dfrac{3}{x}$

(g) $\sqrt{x}\, \ln x$ (n) $\dfrac{1}{3x}$

4. Find $f^{(5)}(r)$, the fifth-order derivative, if (a) $f(r) = r^5$ (b) $f(r) = r^4$ (c) $f(r) = r^4 \ln r$.

5. If $f(x) = 3x^4 - 2x$, find $f(-2)$, $f'(-2)$ and $f''(-2)$.

6. Find

(a) $\dfrac{d(xe^x)}{dx}$ (c) $\dfrac{d^3(xe^x)}{dx^3}$

(b) $\dfrac{d^2(xe^x)}{dx^2}$ (d) $\dfrac{d^n(xe^x)}{dx^n}$

7. Differentiate the function (a) $\dfrac{1 + 3x}{6x + x^2}$ (b) $\dfrac{\sin x}{x}$ (c) $\dfrac{xe^x}{1 + 3e^x}$.

8. Prove that the derivative of $\sec x$ really is $\sec x \tan x$.

9. Find $f'(x)$ if (a) $f(x) = |\sin x|$ (b) $f(x) = \begin{cases} 2x^3 - 4 & \text{if } x \le 2 \\ 9 & \text{if } x > 2 \end{cases}$.

10. Find the slope on the graph of $y = 2x^3 + 6x$ at the point $(1,8)$. Then find the equation of the tangent line at the point.

11. Find the equation of the line perpendicular to the graph of $y = 5 - x^4$ at the point $(2, -11)$.

12. Use the second derivative to find the concavity of (a) $y = \sin x$ and (b) $y = x^3$ and verify the accuracy of the graphs drawn in Sections 1.2 and 1.3.

13. Suppose the position of a particle at time t is $t^2 - 3t^3$. Find its speed at time $t = 2$. Is the car speeding up or slowing down at time $t = 2$, and by how much?

14. If $f(x) = x^2 + ax + b$, and the line $y = 2x - 2$ is tangent to the graph of f at the point $(3,4)$, find a and b.

15. Find the vertex of the parabola $y = -3x^2 - 4x + 2$ and sketch its graph.

16. Let $f(x) = \begin{cases} x^2 & \text{if } x \le 4 \\ ax + b & \text{if } x > 4 \end{cases}$. Find a and b so that the graph of f has neither a discontinuity nor a cusp at $x = 4$.

17. If $y = x \sin x$, show that $y'' + y = 2 \cos x$.

18. Suppose the temperature T at hour t is $t^3 - 15t$. Use T, T' and T'' to describe the weather at time 3.

19. Use calculus to help sketch the graph of the function if

$$y = \begin{cases} -x^2 + 8x & \text{for } x \le 4 \\ 16 & \text{for } 4 < x < 6 \\ x^2 - 20x + 100 & \text{for } x \ge 6 \end{cases}$$

20. If the position of a particle at time t is $12t - t^3$, sketch its motion, showing the direction of travel and when it speeds up and slows down.

3.6   The Derivative of a Composition

In this section we continue to find derivatives of combinations of functions so that you may differentiate all the elementary functions.

The chain rule for the derivative of a composition   Compositions of the basic functions, such as e^{2x} and $\sin x^2$, occur frequently, and the chain rule we are about to derive is very important.

The composition $y = \sin x^2$ can be written as $y = \sin u$ where $u = x^2$. In general, a composition can be denoted by $y = y(u)$ where $u = u(x)$, meaning that y is a function of u, and u in turn is a function of x. We want to express the composition derivative dy/dx in terms of the individual derivatives dy/du and du/dx. Suppose $dy/du = 3$ and $du/dx = 2$. Then, if x increases, u increases twice as fast, and in turn, y increases 3 times as fast as u. Overall, y is increasing 6 times as fast as x; that is, $dy/dx = 3 \cdot 2 = 6$.

In general, we have the following chain rule:

(1)
$$\boxed{\dfrac{dy}{dx} = \dfrac{dy}{du}\dfrac{du}{dx}.}$$

This form of the chain rule is easy to remember because if we pretend that dy/dx is a fraction with numerator dy and denominator dx, and similarly that dy/du and du/dx are fractions, then the right side "cancels" to the left side.

For example, let $y = \sin x^2$. Then $y = \sin u$ where $u = x^2$ and, by the chain rule,

$$\frac{dy}{dx} = \frac{dy}{du}\frac{du}{dx} = \cos u \cdot 2x = \cos x^2 \cdot 2x = 2x \cos x^2.$$

Before continuing with more examples, we will restate the chain rule in a form that is more useful for rapid computation. The last example shows that the basic derivative formula $D_x \sin x = \cos x$ leads to the result

$$D_x \sin x^2 = \cos x^2 \cdot 2x \qquad (\textit{insert the extra factor } 2x).$$

More generally, from any known derivative formula $D_x f(x) = f'(x)$, we get

(2)
$$\boxed{D_x f(u(x)) = f'(u)u'(x) \qquad (\textit{insert the extra factor } u'(x)).}$$

The result in (2) is a restatement of the chain rule from (1). It says that if $D_x f(x)$ is known, probably from the list of basic derivatives, and x is replaced by something else so that a composition is created, then

$$\boxed{D_x f(\text{thing}) = f'(\text{thing}) \cdot D_x \text{ thing}.}$$

In other words, differentiate "as usual," and then multiply by D_x thing. The table of basic derivatives can be rewritten to incorporate the chain rule.

$$
\begin{array}{ll}
D_x u^r = r u^{r-1} u'(x) & D_x \sec u = \sec u \tan u \cdot u'(x) \\[2mm]
D_x \ln u = \dfrac{1}{u} u'(x) & D_x \csc u = -\csc u \cot u \cdot u'(x) \\[2mm]
D_x e^u = e^u u'(x) & D_x \sin^{-1} u = \dfrac{1}{\sqrt{1-u^2}} u'(x) \\[2mm]
D_x \sin u = \cos u \cdot u'(x) & \\[2mm]
D_x \cos u = -\sin u \cdot u'(x) & D_x \cos^{-1} u = -\dfrac{1}{\sqrt{1-u^2}} u'(x) \\[2mm]
D_x \tan u = \sec^2 u \cdot u'(x) & \\[2mm]
D_x \cot u = -\csc^2 u \cdot u'(x) & D_x \tan^{-1} u = \dfrac{1}{1+u^2} u'(x)
\end{array}
$$

Example 1   If $f(x) = \ln 3x$ then f is of the form $\ln u$, so by the chain rule for $D_x \ln u$,

$$f'(x) = \frac{1}{3x} \cdot D_x 3x = \frac{1}{3x} \cdot 3 = \frac{1}{x}.$$

Example 2   If $y = (3x^2 - 4x)^{25}$ then y is of the form u^{25} so, by the chain rule for $D_x u^r$,

$$y' = 25(3x^2 - 4x)^{24} D_x(3x^2 - 4x) = 25(3x^2 - 4x)^{24}(6x - 4).$$

Warning   The most common mistake made in computing derivatives is the omission of the extra step demanded by the chain rule. For example, $D_x \sin x = \cos x$ but $D_x \sin x^2$ is *not* $\cos x^2$; rather, $D_x \sin x^2 = 2x \cos x^2$. Similarly, $D_x e^x = e^x$ but $D_x e^{3x}$ is *not* e^{3x}; rather, $D_x e^{3x} = 3 e^{3x}$.

Example 3   If $y = \sec 2x$, find y' and y''.
   Solution: By the chain rule,

$$y' = \sec 2x \tan 2x \cdot D_x 2x = 2 \sec 2x \tan 2x.$$

Then

$$y'' = 2 D_x(\sec 2x \tan 2x) \qquad (\text{rule for } D_x cf)$$

$$= 2(\sec 2x \cdot D_x \tan 2x + \tan 2x \cdot D_x \sec 2x) \qquad (\text{product rule}).$$

Now use the chain rule to differentiate $\tan 2x$ and $\sec 2x$ and obtain

$$y'' = 2(\sec 2x \cdot \sec^2 2x \cdot 2 + \tan 2x \cdot \sec 2x \tan 2x \cdot 2)$$

$$= 4 \sec^3 2x + 4 \sec 2x \tan^2 2x.$$

Example 4   Let $z = \cos^3 5\theta$. The notation means $(\cos 5\theta)^3$ so z is of the form u^3. Then

$$z'(\theta) = 3(\cos 5\theta)^2 D_\theta \cos 5\theta \qquad (\text{by the chain rule})$$

$$= 3(\cos 5\theta)^2 \cdot -\sin 5\theta \cdot 5 \qquad (\text{by the chain rule again})$$

$$= -15 \cos^2 5\theta \sin 5\theta.$$

Note that $(\cos 5\theta)^3$ is a composition of *three* functions, and the chain rule is used twice to find its derivative.

Example 5   Find dy/dx if $y = 1/(3x^2 + 4)$.
First solution: Write y as $(3x^2 + 4)^{-1}$ and use the chain rule to obtain

$$\frac{dy}{dx} = -(3x^2 + 4)^{-2} \cdot 6x = -\frac{6x}{(3x^2 + 4)^2}.$$

Second solution: By the quotient rule,

$$\frac{dy}{dx} = \frac{(3x^2 + 4) \cdot 0 - 1 \cdot 6x}{(3x^2 + 4)^2} = -\frac{6x}{(3x^2 + 4)^2}.$$

Problems for Section 3.6

In Problems 1–56, find the derivative of the function.

1. e^{6x}

2. $\sin 2x$

3. e^{-x}

4. $-e^x$

5. $\sin^{-1}(3 - x)$

6. $2 \cos 5x$

7. $x^2 \sin 5x$

8. $5xe^{2x}$

9. $\dfrac{1}{2 + \sin x}$

10. $\sin e^x$

11. $e^{-x} \cos 4x$

12. $x^3(2x + 5)^6$

13. $2 \cos 5x$

14. $\ln(5 - x)$

15. $\ln \cos x$

16. e^{5+2x}

17. $\sqrt{3 + x^2}$

18. $\tan^{-1} \frac{1}{2}x$

19. $\dfrac{4}{\cos 5x}$

20. $\sin \pi x$

21. $\cos^3 x$

22. $\sin \dfrac{1}{x}$

23. $e^{\sqrt{x}}$

24. $e^{1/x}$

25. $(\tan^{-1}x)^3$

26. $(x^2 + 4)^3$

27. $\sin x^4$

28. $\cos^4 x$

29. $\dfrac{1}{\sqrt{x^2 + 4x}}$

30. $\ln x^3$

31. $(\ln x)^3$

32. $\dfrac{1}{\ln x}$

33. $\sin^2 x$

34. $x \cos 2x$

35. $\cos(3 - x)$

36. $\cot e^x$

37. $x^3 e^{8x} \sin 4x$

38. $x \ln(2x + 1)$

39. $(3x + 4)^6$

40. $\sec^3 3x^4$

41. $(4 - x)^6$

42. $\dfrac{2 + 7x}{2}$

43. $3 \sin^{-1} \frac{1}{2}x$

44. $\ln \sin e^x$

45. $\cos^3 4x$

46. $e^x \ln x$

47. $\dfrac{1}{e^x + 1}$

48. $\csc 4x$

49. $5 + 4 \ln \ln x$

50. $\sqrt{\ln x}$

51. $\ln \sqrt{x}$

52. $x^2 \ln 3$

53. $\ln|x|$

54. $\dfrac{4x}{\sqrt{2x + 3}}$

55. $\sin \dfrac{x^2 + 2}{x + 1}$

56. $\sqrt{\dfrac{2 - x}{3x + 4}}$

57. The kinetic energy of an object with mass m and speed v is $\frac{1}{2}mv^2$. More specifically, if m and v are functions of time t then the kinetic energy is $\frac{1}{2}m(t)v^2(t)$. Suppose at a certain time, the mass is 5 grams, the speed is 3 meters per second, the mass is increasing by 2 grams per second and the speed is decreasing by 1 meter per second. Is the kinetic energy increasing or decreasing at this moment and by how much?

58. Find $D_x \ln \ln \ln \ln \cdots \ln \ln \ln 2x$, where there are 639 logarithm functions in the composition.

59. Let $f(x)$ be an arbitrary differentiable function. Differentiate the indicated combinations

(a) $\cot f(x)$ (d) $\ln f(x)$
(b) $xf(x)$ (e) $e^{f(x)}$
(c) $(f(x))^3$

60. Suppose star x is a function whose derivative is $e^x(x^3 + 3)$. Find D_x star $3x$.

61. Let $w = 3e^{\sec 2\theta}$. Find $w''(\theta)$.

62. Find the equations of the lines tangent and perpendicular to the graph of $y = (2 - x)^4$ at the point $(3, 1)$.

63. Find the 99th and 100th derivative of $1/(2 + 3x)$.

64. A 10-foot ladder leans up against a wall. Let x be the distance from the foot of the ladder to the base of the wall, and let y be the distance from the top of the ladder to the ground below. If the ladder slides down the wall then x increases while y decreases. Find the rate of change of y with respect to x in general. Then find the rate of change in particular when $x = 1$ and again when $x = 9$.

3.7 Implicit Differentiation and Logarithmic Differentiation

Implicit differentiation   Suppose we want the slope on the graph of

$$(1) \qquad\qquad y^3 - 6x^2 = 3$$

at the point $(-2, 3)$. The equation defines y *implicitly* as a function of x. When the equation is solved for y to obtain

$$(2) \qquad\qquad y = (6x^2 + 3)^{1/3},$$

then y is expressed *explicitly* as a function of x. From the explicit description in (2),

$$y' = \frac{1}{3}(6x^2 + 3)^{-2/3} \cdot 12x = \frac{4x}{(6x^2 + 3)^{2/3}},$$

so $y'|_{x=-2} = -\frac{8}{9}$. Therefore the slope at the point $(-2, 3)$ is $-\frac{8}{9}$.

It is possible to find the derivative y' *without* having the explicit expression for y. This is particularly useful for equations that are too difficult to solve for y. To find y' from the implicit description in (1), differentiate with respect to x on both sides. In this procedure y *is treated as a function of x*, so that the derivative of y^3 with respect to x is $3y^2y'$ by the chain rule. Then

$$3y^2y' - 12x = 0$$

$$y' = \frac{4x}{y^2}$$

$$y'|_{x=-2, y=3} = -\frac{8}{9}.$$

Therefore, the slope at the point $(-2, 3)$ is $-8/9$, as before.

The process of finding y' without first solving for y is called *implicit differentiation*.

Note that the derivative of x^4 with respect to x is $4x^3$, but if y is a function of x then the derivative of y^4 *with respect to x* is $4y^3y'$, by the chain rule. Similarly, if the differentiation is *with respect to x*, then the derivative of e^x is e^x but the derivative of e^y is e^yy'; the derivative of $\sin x$ is $\cos x$ but the derivative of $\sin y$ is $y'\cos y$.

The derivative of a term such as x^3y^5 with respect to x requires the product rule *and* the chain rule:

$$D_xx^3y^5 = x^3D_xy^5 + y^5D_xx^3 = x^3\cdot 5y^4y' + y^5\cdot 3x^2 = 5x^3y^4y' + 3x^2y^5.$$

Warning   Don't omit the extra occurrences of y' demanded by the chain rule.

Example 1   The equation $y^3 + x^2y + x^2 - 3y^2 = 0$ is not easy to solve for y, and as a matter of fact it does not have a unique solution for y since a cubic equation has *three* solutions. The equation implicitly defines three functions, corresponding to the indicated three sections of the graph in Fig. 1. By a single implicit differentiation we can find the derivative of each function.

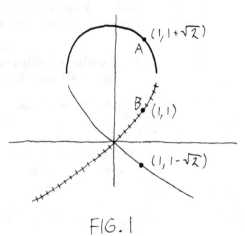

FIG. 1

Differentiate on both sides of the equation with respect to x (use the product rule on x^2y) to obtain

$$3y^2y' + x^2y' + 2xy + 2x - 6yy' = 0.$$

Although it is difficult to solve the original equation for y, it is easy to solve the differentiated equation for y':

$$(3y^2 + x^2 - 6y)y' = -2xy - 2x,$$

$$y' = \frac{-2xy - 2x}{3y^2 + x^2 - 6y}.$$

The derivative formula holds for each of the implicitly defined functions. To find the slope at the point B, substitute $x = 1$, $y = 1$ to get $y' = 2$. Similarly, substitute $x = 1, y = 1 + \sqrt{2}$ to find that the slope at point A is $-1 - \frac{1}{2}\sqrt{2}$ (appropriately negative, since the curve is falling at A).

Delayed proof of the power rule $D_x x^r = r x^{r-1}$ for fractional r Consider $y = x^{4/3}$ for example. Assuming that the function is differentiable, we are now ready to use implicit differentiation to show that y' really is $\frac{4}{3} x^{1/3}$ as claimed in Section 3.3. Cube both sides of $y = x^{4/3}$ to obtain $y^3 = x^4$, an implicit description of y. This appears to be a step backwards when we began with the explicit function $y = x^{4/3}$ but the implicit version has the advantage of involving only integer exponents. Then, by the *previously proved* cases of the power rule for r an integer (Sections 3.3–3.5), we have $3y^2 y' = 4x^3$, so

$$y' = \frac{4x^3}{3y^2} = \frac{4x^3}{3(x^{4/3})^2} = \frac{4x^3}{3x^{8/3}} = \frac{4}{3} x^{1/3}.$$

as desired. The proof in the general case is handled in the same way, but with 4/3 replaced by p/q where p and q are arbitrary integers.

Logarithmic differentiation There are three kinds of functions involving exponents.

1. The base contains the variable x and the exponent is a constant, such as $(3x + 4)^5$ and $\sin^3 x$.
2. The base is e and the exponent contains the variable x, such as e^x and e^{3x}.
3. The base is *not* e and the exponent contains the variable x, such as 2^x, $(x^2 + 2x)^{x^3}$ and $(\sin x)^x$. (As usual, for this type we consider only positive bases. The domain of the function $(\sin x)^x$ is taken to be the set of x for which $\sin x$ is positive.)

Derivatives of the first two types have already been discussed. To differentiate the first type, use $D_x u^r = r u^{r-1} D_x u$. For example, $D_x(3x + 4)^5 = 5(3x + 4)^4 \cdot 3 = 15(3x + 4)^4$. To differentiate the second type, use $D_x e^u = e^u D_x u$. For example, $D_x e^{3x} = 3e^{3x}$.

Consider $y = (\sin x)^x$, a function of the third type. To find its derivative, first take logarithms on both sides and use $\ln a^b = b \ln a$ to obtain

$$(3) \qquad\qquad \ln y = x \ln \sin x.$$

This redescribes y implicitly (a step backwards) but it has the advantage of avoiding exponents. Differentiate implicitly in (3) and use the product rule on $x \ln \sin x$ to get

$$\frac{1}{y} y' = x D_x(\ln \sin x) + \ln \sin x \cdot D_x x = x \frac{1}{\sin x} \cdot \cos x + \ln \sin x.$$

Therefore

$$(4) \qquad\qquad y' = y(x \cot x + \ln \sin x).$$

When y' is obtained by implicit differentiation, it is expressed in terms of x *and* y, as in (4). However, in this case we may replace y by the explicit expression $(\sin x)^x$ to obtain the final answer $y' = (\sin x)^x (x \cot x + \ln \sin x)$.

The process of taking logarithms on both sides of $y = f(x)$ and then finding y' by implicit differentiation is called *logarithmic differentiation*. It is used to differentiate functions of the third kind and, in general, may be used in any problem in which $\ln f(x)$ is easier to differentiate than $f(x)$.

Warning $D_x(\sin x)^x$ is *not* $x(\sin x)^{x-1}$.

Example 2   Find $D_x 8^x$.

Solution: If $y = 8^x$ then $\ln y = x \ln 8$, which we may write more suggestively as $\ln y = (\ln 8)x$. Note that $\ln 8$ is a *constant*. Just as the derivative of $5x$ is 5, so the derivative of $(\ln 8)x$ is simply the number $\ln 8$. Thus by implicit differentiation we have

$$\frac{1}{y}y' = \ln 8$$

$$y' = y \ln 8.$$

Replace y by the explicit expression 8^x to get the final answer $D_x 8^x = 8^x \ln 8$.

Warning $D_x 8^x$ is *not* $x 8^{x-1}$.

Problems for Section 3.7

1. Find dy/dx if (a) $y = x \sin y$ (b) $x + y = y \tan y + x \tan x$.
2. Find dy/dx and dx/dy if $y = \cos(x^2 + y^2)$.
3. Find the line tangent to the graph of the equation at the indicated point, first by solving for y, and then again by implicit differentiation.

(a) $x^2 + y^2 = 1$, point $(\frac{1}{2}, -\frac{1}{2}\sqrt{3})$ (b) $\sqrt{x} + \sqrt{y} = 3$, point $(1,4)$

4. If $\ln y = 1 - xy$ defines $y = f(x)$, find $f'(0)$.
5. Show that the ellipse $4x^2 + 9y^2 = 72$ and the hyperbola $x^2 - y^2 = 5$ intersect perpendicularly, that is, at the point of intersection, the product of the slopes is -1.
6. If $y(x)$ is defined implicitly by $e^{xy} = y$, show that y satisfies the equation $(1 - xy)y' = y^2$.
7. Let $y = \dfrac{x^3 \sin x}{x^2 + 4}$. Find y' with (a) the product rule for three factors and
(b) logarithmic differentiation.
8. Differentiate the function:

(a) 2^x (e) $(2x + 3)^4$
(b) x^x (f) 4^{2x+3}
(c) $x^{\sin x}$ (g) e^x
(d) x^3 (h) $(2x + 3)^{4x}$

3.8   Antidifferentiation

So far we have concentrated on finding f', given f. We now turn to the problem of finding f, given f'. This process is called *antidifferentiation*. One important application occurs at the end of the section and more applications will appear later.

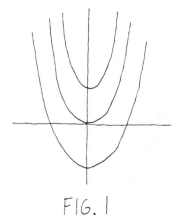

FIG. 1

The set of antiderivatives of a function   We say that $\frac{1}{4}x^4$ is an *antiderivative* of x^3 because $D_x\frac{1}{4}x^4 = x^3$. Also, $D_x(\frac{1}{4}x^4 + 7) = x^3$, $D_x(\frac{1}{4}x^4 - 2) = x^3$ and, in general, $D_x(\frac{1}{4}x^4 + C) = x^3$ where C is an arbitrary constant. Therefore all functions of the form $\frac{1}{4}x^4 + C$ are antiderivatives of x^3. All of the antiderivatives of x^3 have "parallel" graphs (Fig. 1) in the sense that they all have slope x^3. There are no antiderivatives of x^3 *except* the functions $\frac{1}{4}x^4 + C$ since the *only* way to produce the slope x^3 is to translate $\frac{1}{4}x^4$ up or down.

The notation $\int f(x)\,dx$ stands for the entire collection of antiderivatives of $f(x)$, and we write

$$\int x^3\,dx = \frac{x^4}{4} + C.$$

Some antiderivative formulas Antiderivatives for some of the basic functions can be obtained by reversing derivative formulas. We have $D_x \sin x = \cos x$, so $\int \cos x\,dx = \sin x + C$. Similarly, $D_x \cos x = -\sin x$, so $\int(-\sin x)\,dx = \cos x + C$. However, it is more useful to have a formula for $\int \sin x\,dx$, since it is $\sin x$ and not $-\sin x$ that is considered the basic function. Therefore, we use $D_x(-\cos x) = \sin x$ to obtain $\int \sin x\,dx = -\cos x + C$. Proceeding in this way, we assemble the following list.

(1) $\quad \int k\,dx = kx + C \qquad$ (where k stands for a constant)

(2) $\quad \int \sin x\,dx = -\cos x + C$

(3) $\quad \int \cos x\,dx = \sin x + C$

(4) $\quad \int e^x\,dx = e^x + C$

(5) $\quad \int x^r\,dx = \dfrac{x^{r+1}}{r+1} + C, \qquad r \neq -1$

(6) $\quad \int \dfrac{1}{x}\,dx = \ln x + C, \qquad x > 0$

In (6), the function $1/x$ is defined for $x \neq 0$ but the antiderivative $\ln x$ is defined only for $x > 0$. We can do better if we observe that by Problem 53 in Section 3.6, $D_x \ln|x| = 1/x$. Therefore we can extend (6) to

(6′) $\qquad \int \dfrac{1}{x}\,dx = \ln|x| + C, \qquad x \neq 0.$

Both $\ln x$ and $\ln|x|$ differentiate to $1/x$, but $\ln|x|$ has the advantage of being defined for all $x \neq 0$, while $\ln x$ is defined only for $x > 0$.

If we reverse the formula $D_x \tan x = \sec^2 x$, we have

(7) $\qquad \int \sec^2 x\,dx = \tan x + C.$

This is not as "basic" as (1)–(6), but we'll take what we can get. Similarly,

(8) $\qquad \int \csc^2 x\,dx = -\cot x + C$

(9)
$$\int \sec x \, \tan x \, dx = \sec x + C$$

(10)
$$\int \csc x \, \cot x \, dx = -\csc x + C$$

(11)
$$\int \frac{1}{\sqrt{1 - x^2}} \, dx = \sin^{-1}x + C$$

(12)
$$\int \frac{1}{1 + x^2} \, dx = \tan^{-1}x + C$$

We do not yet have antiderivatives for ln x, the basic trigonometric functions other than cos x and sin x, or the inverse trig functions, because there is no well-known derivative formula whose *answer* is any of these functions.

Example 1 $\displaystyle\int x^5 \, dx = \frac{x^6}{6} + C.$

Example 2 $\displaystyle\int \frac{1}{t^5} \, dt = \int t^{-5} \, dt = \frac{t^{-4}}{-4} + C = -\frac{1}{4t^4} + C.$

Selecting a particular antiderivative   Consider the function f such that $f'(x) = x^3$ *and* $f(2) = 3$. To find f we must select from all parallel curves with slope x^3, the particular one through the point $(2, 3)$. (Just as a line is determined by a point and a slope number, a curve, more generally, is determined by a point and a slope function.)

If $y' = x^3$ then $y = \frac{1}{4}x^4 + C$. To find C, set $x = 2$, $y = 3$ to obtain $3 = 4 + C$, $C = -1$. Therefore $f(x) = \frac{1}{4}x^4 - 1$.

Antiderivatives of the elementary functions   We would like to follow the same strategy for antidifferentiation that we used for differentiation, that is, find antiderivatives for all the basic functions and then use combination rules to find antiderivatives for all the elementary functions.

It's easy to find rules for constant multiples and sums. For example, $\int 6 \cos x \, dx = 6 \sin x + C$ because $D_x 6 \sin x = 6 \cos x$. Similarly, $\int (x^3 + \cos x) \, dx = \frac{1}{4}x^4 + \sin x + C$ because $D_x(\frac{1}{4}x^4 + \sin x + C) = x^3 + \cos x$. In general,

(13)
$$\boxed{\int cf(x) \, dx = c \int f(x) \, dx}$$

and

(14)
$$\boxed{\int [f(x) + g(x)] \, dx = \int f(x) \, dx + \int g(x) \, dx.}$$

For example, $\int (2x^4 + 3x - 4) \, dx = \frac{2}{5}x^5 + \frac{3}{2}x^2 - 4x + C$.

But there are no other easy rules. We are collecting information about antidifferentiation by reversing differentiation formulas, and a reversed formula is often not of the same character as the original. The reverse of the *basic* derivative formula $D_x \tan x = \sec^2 x$ becomes an antiderivative formula for the *nonbasic* function $\sec^2 x$. Similarly, the reverse of the product rule $(fg)' = fg' + f'g$ is $\int (fg' + f'g)\,dx = fg$, which is no longer a product rule.

Since we are missing some of the basic antiderivative formulas and combination rules, we are thwarted, at least temporarily, in the effort to antidifferentiate all the elementary functions. It will turn out that there simply are no product, quotient, or composition rules and, in fact, the antiderivatives of some elementary functions don't have nice formulas at all. All of Chapter 7 will be devoted to overcoming these difficulties. In the meantime, the scope of (1)–(12) can be widened sufficiently so that even before Chapter 7, some significant applications can be discussed.

Extending known antiderivative formulas If we know an antiderivative for $f(x)$, we can also find an antiderivative for $f(ax + b)$. For example, consider $\int \cos(\pi x + 7)\,dx$. We might guess that the answer is $\sin(\pi x + 7) + C$, but differentiate back to see that this is not quite right, since, by the chain rule, $D_x \sin(\pi x + 7) = \cos(\pi x + 7) \cdot \pi$. We don't want the extra factor π, so we refine our guess to

$$\int (\cos \pi x + 7)\,dx = \frac{1}{\pi} \sin(\pi x + 7) + C.$$

This is correct because

$$D_x \frac{1}{\pi} \sin(\pi x + 7) = \frac{1}{\pi} \cos(\pi x + 7) \cdot \pi = \cos(\pi x + 7).$$

In general,

(15)

> *if $F(x)$ is an antiderivative of $f(x)$ then*
>
> $$\int f(ax + b)\,dx = \frac{1}{a} F(ax + b) + C.$$
>
> In other words, *if x is replaced by $ax + b$ in (1)–(12), antidifferentiate "as usual" but insert the extra factor $1/a$.*

Example 3 $\displaystyle \int e^{3x}\,dx = \frac{1}{3} e^{3x} + C.$

Example 4 $\displaystyle \int e^{x/2}\,dx = 2e^{x/2} + C.$

Example 5 $\displaystyle \int \frac{1}{5x - 8}\,dx = \frac{1}{5} \ln|5x - 8| + C.$

Example 6 $\displaystyle \int \frac{1}{(4 - x)^3}\,dx = \int (4 - x)^{-3}\,dx = -1 \cdot \frac{(4 - x)^{-2}}{-2} + C$

$$= \frac{1}{2(4 - x)^2} + C.$$

Warning   1. The answer to Example 6 is *not* $\ln(4-x)^3$ because the derivative of $\ln(4-x)^3$ is $\dfrac{1}{(4-x)^3}$ *times* $3(4-x)^2 \cdot -1$ by the chain rule.

2. Any antidifferentiation problem can be checked by differentiating the answer. (The catch is that you must be able to differentiate correctly to catch mistakes in the antidifferentiation.)

3. Within the context of this section, the *only* functions $f(x)$ which you are prepared to antidifferentiate are those in (1)–(12), along with their constant multiples, sums and variations of the form $f(ax+b)$ where a and b are constants.

Example 7   Assume $x > 0$ so that (6) can be used instead of (6'), and find $\int \dfrac{1}{4x}\, dx$.

First solution: $\int \dfrac{1}{4x}\, dx = \dfrac{1}{4} \int \dfrac{1}{x}\, dx = \dfrac{1}{4} \ln x + C$.

Second solution: $\int \dfrac{1}{4x} = \dfrac{1}{4} \ln 4x + C$ (by (15)).

We seem to have two different answers, $\frac{1}{4} \ln x + C$ and $\frac{1}{4} \ln 4x + C$. But

$$\dfrac{1}{4} \ln 4x + C$$

$$= \dfrac{1}{4}(\ln 4 + \ln x) + C = \dfrac{1}{4}\ln x + \dfrac{1}{4}\ln 4 + C = \dfrac{1}{4}\ln x + D.$$

The arbitrary constant C plus the particular constant $\frac{1}{4}\ln 4$ is another arbitrary constant D. Therefore the two solutions do agree.

An application of antidifferentiation and an introduction to parametric equations   Suppose that a gun has a muzzle velocity of 60 feet per second, and is fired from a 40 foot hill at an angle of 30° with the horizontal. What is the path of the bullet? Where does it land? For how long is it in flight? How high does it get?

Establish a coordinate system so that the gun is at the point $(0, 40)$ (Fig. 2). Physicists do the problem in two parts, worrying separately about the x-coordinate $x(t)$ and the y-coordinate $y(t)$ of the bullet at time t. They separate the muzzle velocity into a horizontal speed and a vertical speed as follows. The muzzle velocity 60 together with the 30° angle is represented by an arrow 60 units long at angle 30° with the horizontal. By trigonometry, the horizontal arrow in Fig. 2 has length $30\sqrt{3}$, and the vertical arrow

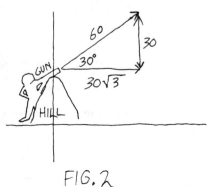

FIG. 2

has length 30. Physicists conclude that the bullet can be considered to have horizontal speed $30\sqrt{3}$ feet per second and vertical speed 30 feet per second.

Let's continue with the vertical part of the problem. Let $t = 0$ be the time at which the bullet is initially fired (any other choice would be all right, too). Since the bullet is fired at time 0 from the point $(0, 40)$, we have $y(0) = 40$. Also, the bullet is initially moving upward with vertical speed 30, so $y'(0) = 30$. Furthermore, from basic physics, the gravitational field of the earth causes any vertical velocity to decrease by 32 feet/second per second, so

$$y''(t) = -32 \quad \text{for all } t.$$

Now, work backwards to find $y'(t)$ and then $y(t)$. We have $y'(t) = -32t + C$. To determine C, use $y'(0) = 30$, and set $t = 0$, $y' = 30$ to get $30 = -32 \cdot 0 + C$, $C = 30$. Therefore,

(16) $$y'(t) = -32t + 30.$$

Antidifferentiate again to get $y(t) = -16t^2 + 30t + K$. To determine K, use $y(0) = 40$, and set $t = 0$, $y = 40$ to get $40 = -16 \cdot 0 + 30 \cdot 0 + K$, $K = 40$. Thus,

(17) $$y(t) = -16t^2 + 30t + 40.$$

Consider the horizontal part of the problem. By Newton's laws of motion, an object will maintain its initial horizontal velocity (until the vertical component of velocity causes a crash), so

$$x'(t) = 30\sqrt{3} \quad \text{for all } t.$$

Therefore $x(t) = 30\sqrt{3}\, t + Q$. Since $x(0) = 0$ we have $0 = 30\sqrt{3} \cdot 0 + Q$, $Q = 0$. Thus

(18) $$x(t) = 30\sqrt{3}\, t.$$

Now we can answer all of the questions about the bullet. It lands when $y = 0$, so set $y = 0$ in (17) and solve for t to get $t = \dfrac{15 \pm \sqrt{865}}{16}$, $t = -.9$ or 2.775 approximately. Ignore the negative solution, since the experiment starts at time $t = 0$. Thus the bullet lands about 2.775 seconds after being fired. From (18), if $t = 2.775$ then $x = 144$ approximately. Therefore the bullet travels about 144 feet horizontally before landing (Fig. 3).

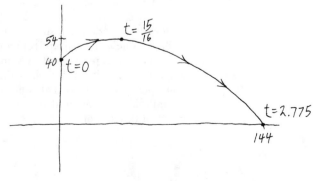

FIG.3

To find its maximum height, note that the bullet has positive velocity as it rises, negative velocity as it falls, and reaches a peak at the instant its velocity is 0. From (16), $y' = 0$ when $t = 15/16$, and, from (17), at this moment $y = 54$ approximately. So the bullet rises to a maximum height of about 54 feet.

In general, a curve in the plane may be described with one equation in x and y, or by a pair of equations, such as (17) and (18), which give x and y in terms of a third variable, t in this case. The two equations $x = x(t)$, $y = y(t)$ are called *parametric equations,* and t is called a *parameter.* If (18) is solved for t and substituted into (17), we have

$$(19) \qquad y = -16\left(\frac{x}{30\sqrt{3}}\right)^2 + 30\left(\frac{x}{30\sqrt{3}}\right) + 40 ,$$

a nonparametric description of the bullet's path. Equation (19) is of the form $y = ax^2 + bx + c$, and therefore the path is a parabola.

Problems for Section 3.8

1. Find

(a) $\int 3 \sin x\, dx$ (g) $\int \frac{1}{x^5}\, dx$

(b) $\int \sin 3x\, dx$ (h) $\int \sqrt{x}\, dx$

(c) $\int u^4\, du$ (i) $\int \frac{1}{\sqrt{x}}\, dx$

(d) $\int \sec \frac{x}{\pi} \tan \frac{x}{\pi}\, dx$ (j) $\int x^8\, dx$

(e) $\int \frac{1}{t^3}\, dt$ (k) $\int \frac{1}{2x^2}\, dx$

(f) $\int x^{-1}\, dx$ (l) $\int \frac{4}{x^2}\, dx$

2. Find $f(x)$ if $f'(x) = \sin x + x^2$ and $f(0) = 10$.
3. Find all functions $f(x)$ such that $f'''(x) = 5$.
4. A particle traveling on a number line has velocity $7 - t^2$ at time t. If it is at position 4 at time 3, where is it at time 6?
5. Find y if $y' = 2x + 3$ and $y = -2$ when $x = 1$.

6. We know that $\int \frac{1}{x}\, dx = \ln x + C$. Does $\int \frac{1}{\sin x}\, dx$ equal $\ln \sin x + C$?

7. We know that $\int \cos x\, dx = \sin x + C$. (a) Does $\int \cos^2 x\, dx$ equal $\sin^2 x + C$? (b) Does $\int \cos 2x\, dx$ equal $\sin 2x + C$?   (c) Does $\int 3 \cos x\, dx$ equal $3 \sin x + C$? (d) Does $\int \cos x^2\, dx$ equal $\sin x^2 + C$?

8. A stone is thrown up from a point 24 feet above the ground with an initial velocity of 40 feet per second. Assume that the only force acting on the stone is the force due to gravity which gives the stone a constant acceleration of -32 feet/second per second. How high will the stone rise and when will it hit the ground?

In Problems 9–35, find an antiderivative for the function, if possible within the context of this section.

9. $\dfrac{3}{3 - x}$

10. $\dfrac{1}{2x + 5}$

11. $\sqrt{x^2 + 5}$

12. $\dfrac{5}{x}$

13. $\dfrac{1}{5x}$

14. $\dfrac{1}{2 + x}$

15. $\dfrac{1}{2x^3 + 3}$

16. $7 \cos \pi x$

17. $\cos x^3$

18. $\dfrac{x^2 + 6x}{5}$

19. $\sec x$

20. $\dfrac{5}{(3x + 6)^2}$

21. $\sqrt{2 + \dfrac{1}{4}x}$

22. $\dfrac{2}{1 + x^2}$

23. $\dfrac{2}{1 - x^2}$

24. $\sin \dfrac{x}{6}$

25. $\cos \dfrac{2\pi x}{3}$

26. $\dfrac{x^6}{6}$

27. $3e^{-x}$

28. $e^{\sin x}$

29. $x^2 + x + \dfrac{1}{x} + \dfrac{1}{x^2} + \dfrac{1}{x^3}$

30. e^{2x}

31. π

32. $(3x + 4)^4$

33. $\dfrac{2}{x^3}$

34. $\dfrac{x^3}{2}$

35. $\dfrac{1}{2x^3}$

In Problems 36–59, perform the indicated antidifferentiation, if possible within the context of this section.

36. $\displaystyle\int \dfrac{1}{5x^3}\, dx$

37. $\displaystyle\int \dfrac{1}{\sqrt{t}}\, dt$

38. $\displaystyle\int 3x^3\, dx$

39. $\displaystyle\int \dfrac{1}{3x^3}\, dx$

40. $\displaystyle\int \dfrac{1}{x}\, dx$

41. $\displaystyle\int \dfrac{1}{2 - 3x^3}\, dx$

42. $\displaystyle\int (2 - 3x^2)\, dx$

43. $\displaystyle\int \dfrac{1}{(2 - 3x)^3}\, dx$

44. $\displaystyle\int (x^4 + 5)\, dx$

45. $\displaystyle\int dx$

46. $\displaystyle\int \sin 3u\, du$

47. $\displaystyle\int \sin^3 x\, dx$

48. $\displaystyle\int e^{-2x}\, dx$

49. $\displaystyle\int \ln x\, dx$

50. $\displaystyle\int \dfrac{3}{x^4}\, dx$

51. $\displaystyle\int \dfrac{1}{1 - v}\, dv$

52. $\displaystyle\int \dfrac{2}{3 + 4x}\, dx$

53. $\displaystyle\int \dfrac{1}{\sin x}\, dx$

54. $\displaystyle\int 4e^{5x}\, dx$

55. $\displaystyle\int \sqrt{3 - x}\, dx$

56. $\displaystyle\int \dfrac{5t + 3}{2}\, dt$

57. $\displaystyle\int \dfrac{5x^3 + 3}{2}\, dx$

58. $\displaystyle\int \dfrac{2}{5x^3 + 3}$

59. $\displaystyle\int (2x + 3)^5\, dx$

REVIEW PROBLEMS FOR CHAPTER 3

1. Let $f(t)$ be the number of gallons of water that has spurted through a hole in the dike during the t hours since the leak started. For example if $f(3) = 100$ then 100 gallons flowed in during the first 3 hours of the leak.

(a) What does the derivative $f'(t)$ represent? If $f'(3) = 20$ and $f''(3) = -1$, are the residents of the flood plain happy or unhappy?
(b) What value(s) of $f'(t)$ is the flood plain rooting for?

In Problems 2–36, differentiate the function.

2. $\sin(2x + 3\pi)$

3. $x \sin x$

4. $\tan^{-1}x^2$

5. $\dfrac{1}{2 - x}$

6. $\ln(2 - x)$

7. $\dfrac{1}{x}$

8. $\dfrac{1}{4x^2}$

9. e^{-2x}

10. $-e^x$

11. $\tan 3x$

12. $\dfrac{3}{x}$

13. $x^2(2 - 3x)^7$

14. $x \sin^{-1}x$

15. $\dfrac{2 + \sin 4x}{5}$

16. $3xe^x \sec x$

17. $\dfrac{\cos x}{6}$

18. 4^x

19. x^4

20. e^{8-x}

21. $(8 - x)^3$

22. $(8 - x)^x$

23. $\left(\dfrac{2x + 3}{5}\right)^4$

24. $\sqrt{2x + 5}$

25. $\dfrac{2}{3 + 2x}$

26. $e^{\sqrt{x}}$

27. $\dfrac{4 - 2x}{7}$

28. $\dfrac{x}{2x + 3}$

29. $x \sin \dfrac{1}{x}$

30. $\dfrac{e^x}{x}$

31. $\dfrac{(x + 2)}{4}$

32. $\dfrac{2}{3x}$

33. $\cos^3 2x$

34. $3 \sin e^{2x}$

35. $\dfrac{1}{7x^3 + 2x - 5}$

36. $\dfrac{2x + 3}{5x - 4}$

37. A car particle's positions on a number line at time t is $t^2 - 2t^3 + 1$. Find the particle's position, speed, velocity and direction of motion at time $t = 2$. Is it speeding up or slowing down at time $t = 2$, and by how much?

38. Sketch a possible graph of f if $f'(x)$ is positive in the intervals $(-\infty, 3)$ and $(5, \infty)$, negative in the interval $(3, 5)$ and zero at $x = 3, 5$; and $f''(x)$ is negative for x in $(-\infty, 4)$, zero at $x = 4$ and positive for x in $(4, \infty)$.

39. If $f'(x) = x^3 - 2x$ and the graph contains the point $(2, -2)$, find $f(x)$.

40. If $f(t) = t^3 + 3t^2 + 1$ is the position of a particle at time t, sketch a picture of its motion, indicating its direction, when it speeds up, and when it slows down.

41. Find $D_x \sin \sin \sin \sin \sin \cdots \sin \sin 2x$, where the composition contains 825 sine functions.

42. Let $y = \frac{1}{3}x^3 + \frac{1}{2}x$. (a) Show that y is an increasing function of x. (b) Suppose x increases and has just reached $\frac{1}{2}$. At this instant is y increasing faster or slower than x?

43. Use derivatives to see if the graph of e^x really has the concavity indicated in Fig. 2 of Section 1.5.

44. Find dy/dx (a) $xy + 3xy^2 = 62 - x$ (b) $\sin x + \sin y = 6$.

45. Find

(a) $D_x \dfrac{5x + 2}{x^2 + 3x}$ (e) $D_x(x + e^x)$

(b) $\dfrac{d(x^3 \ln x)}{dx}$ (f) $d(te^t)/dt$

(c) $D_t(\ln t)^{2t}$ (g) $D_x \dfrac{2x}{\sqrt{3x + 4}}$

(d) $\dfrac{d|3x - 6|}{dx}$ (h) $D_x e^{|x|}$

46. Find y' and y''.

(a) $y = 3x \sin x$ (c) $y = x^4 \cos x^2$
(b) $y = |1 - \ln x|$ (d) $y = 5^x$

47. Find the 19th and 20th derivatives of $1/\sqrt{2 + 5x}$.

48. Show that the lines tangent to the graph of $xy = 1$ in the first quadrant form triangles all of which have the same area (Fig. 1 shows two such triangles).

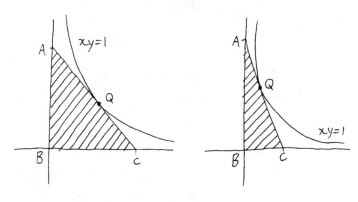

FIG. 1

49. The product rule states that $(fg)' = fg' + f'g$. Differentiate again to get a product rule for $(fg)''$, and again for $(fg)'''$ and again for $(fg)''''$. Look at the pattern and invent a product rule for $(fg)^{(n)}$, the nth derivative of fg.

50. Suppose that a one-dimensional object placed on a slide (number line) is projected onto a screen (another number line) so that the point x on the slide projects to the point x^2 on the screen. If a 2-foot object AB is placed with A at $x = -2$ and B at $x = -4$ (Fig. 2) then its image is *magnified* (to 12 feet), *distorted* (the magnification is "uneven" — for example, the right half of the object has a 5-foot image while the left half has a 7-foot image), and *reversed* (A is to the right of B but the image of A is to the left of the image of B).

Consider a projector which sends x to $f(x)$, where f is an arbitrary function instead of x^2 in particular. What type of derivative (positive? negative? large? small? etc.) is to be expected when the image is (a) reversed (a′) unreversed (b) magnified (b′) reduced (c) distorted (c′) undistorted.

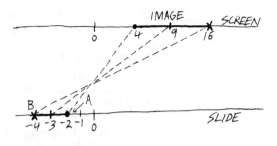

FIG. 2

Do Problems 51–62 if possible within the context of this section.

51. $\displaystyle\int \frac{1}{7x}\, dx$

52. $\displaystyle\int \frac{1}{7x^2}\, dx$

53. $\displaystyle\int \frac{1}{(4x-2)^3}\, dx$

54. $\displaystyle\int (4x+2)\, dx$

55. $\displaystyle\int e^{5x}\, dx$

56. $\displaystyle\int \sin \frac{1}{2}\,\pi x\, dx$

57. $\displaystyle\int \sin^3 \frac{1}{2}\,\pi x\, dx$

58. $\displaystyle\int \frac{1}{x^2+x}\, dx$

59. $\displaystyle\int \frac{1}{3-t}\, dt$

60. $\displaystyle\int \frac{1}{\sqrt{3-t}}\, dt$

61. $\displaystyle\int \sqrt{1+2x}\, dx$

62. $\displaystyle\int \sqrt{1+2x^2}\, dx$

63. Find (a) $D_x x^5$ (b) $\displaystyle\int x^5\, dx$ (c) $D_x \dfrac{1}{x^4}$ (d) $\displaystyle\int \dfrac{1}{x^4}\, dx$.

4/THE DERIVATIVE   PART II

4.1   Relative Maxima and Minima

It is useful to be able to locate the peaks and valleys, called relative extrema, on the graph of a function f. They help in making an accurate sketch, and can also be used to find the overall highest and lowest values of f, called absolute extrema, for such purposes as maximizing profit and minimizing cost. This section shows how to find relative extrema and later sections continue the applications to graphs and absolute extrema.

Definition of relative extrema   A function f has a *relative maximum* at x_0 if $f(x_0) \geq f(x)$ for all x near x_0. Similarly, f has a *relative minimum* at x_0 if $f(x_0) \leq f(x)$ for all x near x_0. Figure 1 shows relative maxima at x_2 and x_4, and relative minima at x_3 and x_5.

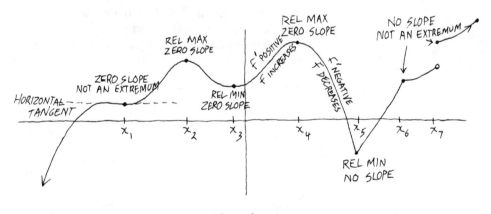

FIG. 1

Critical numbers   Consider the graph of the function in Fig. 1. At the relative extrema where a slope exists (at x_2, x_3 and x_4), that slope is 0. For example, the relative maximum at $x = x_2$ occurs when the function increases and then decreases. The slope changes from positive to negative, and is 0 *at* the maximum point. In general, *if f is differentiable and f has a relative extreme value at x_0 then $f'(x_0) = 0$.* Equivalently, *if $f'(x_0)$ is a nonzero number then f cannot have a relative extreme value at x_0.*

On the other hand, *if $f'(x_0) = 0$ then a relative extreme value may* (see x_2, x_3, x_4) *but need not* (see x_1) *occur.*

Similarly, *if f is not differentiable at a point then a relative extreme value may* (see the cusp at x_5) *but need not* (see the cusp at x_6 and the jump at x_7) *exist.*

If $f'(x_0) = 0$ or $f'(x_0)$ does not exist then x_0 is called a critical number. The preceding discussion shows that the list of critical numbers includes all the relative maxima, all the relative minima, and possible nonextrema as well. In other words, *critical numbers do not necessarily produce maxima or minima, but they are the only candidates.* In Fig. 1, x_1 through x_7 are critical numbers, but the function does not have a relative extreme value at x_1, x_6 or x_7.

There are two standard methods for classifying critical numbers.

First derivative test   Let f be continuous. To identify a critical number x_0 as a relative maximum, relative minimum or neither, examine the sign of the first derivative to the left and right of x_0. If the derivative changes from positive to negative, so that f increases and then decreases, f has a relative maximum at x_0 (see x_4 in Fig. 1). If the derivative changes from negative to positive then f has a relative minimum at x_0 (see x_3 in Fig. 1). Otherwise, f has neither.

Example 1   Let $f(x) = 4x^5 - 5x^4 - 40x^3$. Find the relative extrema of f and sketch the graph.

Solution: Solve $f'(x) = 0$ to find some critical numbers.

$$20x^4 - 20x^3 - 120x^2 = 0$$
$$20x^2(x^2 - x - 6) = 0$$
$$x^2(x - 3)(x + 2) = 0$$
$$x = 0, 3, -2.$$

The function is differentiable everywhere, so there are no critical numbers other than 0, 3 and -2.

Determine the sign of $f'(x)$ in the intervals between the critical numbers by testing one value from each interval, as described in Section 1.6.

Interval	Sign of f'	Behavior of f	Relative Extrema
$(-\infty, -2)$	positive	increases	rel max at $x = -2$
$(-2, 0)$	negative	decreases	
			no extremum at $x = 0$ (but the graph is instantaneously horizontal as it falls through $x = 0$)
$(0, 3)$	negative	decreases	rel min at $x = 3$
$(3, \infty)$	positive	increases	

Finally, we find the y-coordinates corresponding to the critical numbers, namely, $f(-2) = 112$, $f(0) = 0$ and $f(3) = -513$, and use them to plot the graph in Fig. 2.

Second derivative test   This test is applicable to the type of critical point at which $f'(x_0) = 0$. In this case, if $f''(x_0) < 0$ then in addition to zero slope we visualize downward concavity at $x = x_0$ (see x_4 in Fig. 1) and expect a relative maximum. If $f''(x_0) > 0$ then in addition to zero slope we picture upward concavity at $x = x_0$ (see x_3 in Fig. 1) and expect a relative minimum. In general we have the following conclusions.

(1)   If $f'(x_0) = 0$ and $f''(x_0) < 0$ then f has a *relative maximum* at x_0.

(2)   If $f'(x_0) = 0$ and $f''(x_0) > 0$ then f has a *relative minimum* at x_0.

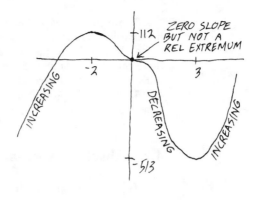

FIG. 2

(3) If $f'(x_0) = 0$ and $f''(x_0) = 0$ then *no conclusion can be drawn.* As problems will demonstrate, it is possible for there to be a relative maximum, or a relative minimum, or neither at x_0. Another method must be used in this case, such as the first derivative test.

With the second derivative test, a decision about a critical number x_0 is made by examining f'' only *at* x_0; with the first derivative test, the decision is made by examining f' *to the left and right of x_0.* The second derivative test is perhaps more elegant; on the other hand, the first derivative test never fails to produce a conclusion, whereas the second derivative test is inconclusive in case (3).

Example 2 Find the relative extrema in Example 1, using the second derivative test this time.

Solution: Again find the critical numbers $x = 0, 3, -2$. We have $f''(x) = 80x^3 - 60x^2 - 240x$, so $f''(-2) = -400, f''(0) = 0$ and $f''(3) = 900$. Therefore f has a relative maximum at $x = -2$ and a relative minimum at $x = 3$. The second derivative test is inconclusive for $x = 0$. We must resort to the first derivative test for the intervals $(-2, 0)$ and $(0, 3)$ as in Example 1 to show that f does not have a relative extremum at $x = 0$.

Problems for Section 4.1

1. Use (i) the first derivative test and (ii) the second derivative test to locate relative maxima and minima.

(a) $f(x) = x^3 - 3x^2 - 24x$ (d) $\dfrac{e^x}{x}$

(b) $x^4 - x^2$ (e) $x \ln x$

(c) $x^5 + x$

2. Locate relative maxima and minima, if possible, with the given information.

(a) $f'(2) = 0, f'(x) < 0$ for $1.9 < x < 2,$ (e) $f'(2) = 0, f''(2) = 6$
$\quad\ f'(x) > 0$ for $2 < x < 2.001$ (f) $f'(2) = 0, f''(2) = 0$

(b) $f'(2) = 0$ (g) $f'(6) < 0, f'(7) = 0,$
(c) $f''(2) = 0$ $\qquad f'(8) > 0$
(d) $f'(2) = 3$

3. Suppose f has a relative minimum at x_0 and a relative maximum at x_1. Is it necessarily true that $f(x_0) < f(x_1)$?

4. Use the functions x^3, x^4 and $-x^4$ to show that when $f'(x_0) = 0$ and $f''(x_0) = 0$, there may be a relative maximum, a relative minimum or neither at x_0, thus verifying part (3) of the second derivative test.

5. Sketch the graph of a function f so that $f'(3) = f'(4) = 0$ and $f'(x) > 0$ otherwise.

4.2   Absolute Maxima and Minima

If $f(x)$ is the profit when a factory hires x workers then, instead of puny *relative* maximum values, we want to find *the* maximum, often referred to as the *absolute* maximum. This section shows how to find the (absolute) extrema for a function $f(x)$. Furthermore, the extrema are usually to be found for x restricted to a particular interval; in the factory example we must have $x \geq 0$ since the number of workers can't be negative, and (say) $x \leq 500$ by Fire Department safety regulations.

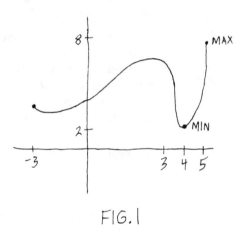

FIG.1

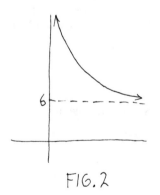

FIG.2

To see extrema graphically, consider Fig. 1, showing a function defined on the interval $[-3, 5]$. Its highest value is 8, when $x = 5$, and its lowest value is 2, when $x = 4$. The function has a relative maximum at $x = 3$, but *the* maximum is at $x = 5$. The function has a relative minimum at $x = 4$, and *the* minimum also occurs here. As another example, the function in Fig. 2, defined on $(0, \infty)$, has no maximum value because $f(x)$ can be made as large as we like by letting x approach 0 from the right. In this case, we will adopt the convention that the maximum is ∞ when $x = 0+$. Similarly, the function has no minimum because $f(x)$ gets closer and closer to 6 without reaching it. As a convenient shorthand in this case (albeit an abuse of terminology) we will say that the minimum is 6 when $x = \infty$.†

Finding maxima and minima   The extrema of a function occur either at the end of the graph (see the maximum at $x = 5$ in Fig. 1), or at one of the relative extrema (see the minimum at $x = 4$ in Fig. 1), or at an infinite

---

†More precisely, 6 is called the *infimum* of f rather than the minimum because f never *reaches* 6. Similarly, a "maximum that is not attained," such as $\pi/2$ for the arctangent function, is called a *supremum*.

discontinuity (see the maximum at $x = 0+$ in Fig. 2). To locate the maximum and the minimum, first find the following candidates.

(A) Critical values of f Find critical numbers by solving $f'(x) = 0$, and by finding places where the derivative does not exist, a less likely source. For each critical number x_0, find $f(x_0)$, called a *critical value* of f. This list contains all the relative maxima and relative minima, and possibly some values of f with no particular max/min significance. It is not necessary to decide which critical value of f serves which purpose. Include them all in the candidate list without classifying them.

(B) End values of f If a function f is defined for $a \le x \le b$ then the end values of f are $f(a)$ and $f(b)$. If f is defined on $[a, \infty)$ then the end values are $f(a)$ and $f(\infty)$, that is, $\lim_{x \to \infty} f(x)$.

(C) Infinite values of f In practice, f may become infinite at the ends where $x \to \infty$ or $x \to -\infty$ (overlapping with candidates from (B)), or at a place where a denominator is 0.

The largest of the candidates from (A)–(C) is the maximum value of f and the smallest is the minimum value. (Candidates from (C) are immediate winners.)

Example 1   Find the maximum value of $f(x) = x^4 + 4x^3 - 6x^2 - 8$ for $0 \le x \le 1$.
Solution: We have $f'(x) = 4x^3 + 12x^2 - 12x$. Find the critical numbers by solving $f'(x) = 0$ to get $4x(x^2 + 3x - 3) = 0$, $x = 0$, $\dfrac{-3 \pm \sqrt{21}}{2}$. But $\frac{1}{2}(-3 - \sqrt{21})$ is negative, and hence not in $[0, 1]$, so ignore it. Count $\frac{1}{2}(-3 + \sqrt{21})$ since it is about .79 and *is* in $[0, 1]$.
The candidates are $f(0) = -8$ which is both a critical value of f and an end value, the critical value $f(\frac{1}{2}[-3 + \sqrt{21}])$ which is approximately $f(.79)$, or -9.4, and the end value $f(1) = -9$. The largest of these, -8, is the maximum.

Warning   The preceding example asked for the maximum *value* of f, so the answer is -8, not $x = 0$. If the problem had asked *where* f has its maximum, then the answer would be $x = 0$. Make your answer fit the question.

Example 2   We don't always have to rely on calculus to produce maxima and minima. Consider $f(x) = \dfrac{4}{1 + x^2}$. By inspection, the largest value of f is 4, when $x = 0$; any other value of x would increase the denominator and therefore decrease $f(x)$. The smallest value of f is 0 when $x = \pm\infty$, since this maximizes the denominator and therefore minimizes f.

Example 3   Let $f(x) = \dfrac{x}{4 - x}$. Find the maximum and minimum values of $f(x)$ on (a) $(-\infty, \infty)$ and (b) $[6, \infty)$.
Solution: (a) The function has an infinite discontinuity at $x = 4$ since

$$f(4-) = \lim_{x \to 4-} \frac{x}{4 - x} = \frac{4}{0+} = \infty \quad \text{and} \quad f(4+) = \lim_{x \to 4+} \frac{x}{4 - x} = \frac{4}{0-} = -\infty.$$

There is no need to search for other candidates. We say that the maximum is ∞ and the minimum is $-\infty$.

(b) Since 4 is not in $[6, \infty)$, we ignore the infinite discontinuity now. There are no critical numbers since, by the quotient rule,

$$f'(x) = \frac{(4 - x) - x(-1)}{(4 - x)^2} = \frac{4}{(4 - x)^2}$$

which is never 0. The only candidates are the end values $f(6) = -3$ and $f(\infty)$. By the highest power rule (Section 2.3),

$$f(\infty) = \lim_{x \to \infty} \frac{x}{4 - x} = \lim_{x \to \infty} \frac{x}{-x} = \lim_{x \to \infty}(-1) = -1.$$

Therefore, the minimum value of f is -3 (when $x = 6$) and its maximum is -1 (when $x = \infty$).

Example 4   In (1) of Section 1.1 we found that the energy E used by a pigeon flying on the route APB (Fig. 3 of Section 1.1) is

$$E(x) = 60\sqrt{36 + x^2} + 40(10 - x) \qquad \text{for } 0 \le x \le 10.$$

We are now ready to finish the problem and find the value of x that minimizes E.

Solve $E'(x) = 0$ to find critical numbers.

$$\frac{60x}{\sqrt{36 + x^2}} - 40 = 0$$

$$\frac{60x}{\sqrt{36 + x^2}} = 40$$

$$3x = 2\sqrt{36 + x^2}$$

$$9x^2 = 4(36 + x^2) \qquad \text{(square both sides)}$$

$$5x^2 = 4 \cdot 36$$

$$x^2 = \frac{4 \cdot 36}{5}$$

$$x = \frac{2 \cdot 6}{\sqrt{5}} = \frac{12}{\sqrt{5}} = 5.4 \qquad \text{(approximately)}.$$

Therefore, the only critical value of E is $E(12/\sqrt{5})$ which is approximately $E(5.4)$, or 670. The end values are $E(0) = 760$ and $E(10) = 700$ (approximately). The smallest of the three candidates is 670. Therefore, in Fig. 3 of Section 1.1, the best the pigeon can do is to fly across the water to a point P about 5.4 miles from C and then fly the remaining 4.6 miles to town along the beach.

Example 5   Find the point on the graph of $y = \sqrt{x}$ which is nearest the point $(2, 0)$.

Solution: A typical point on the curve is $(x, \sqrt{x})$ (Fig. 3 on next page). By the distance formula, the distance from this point to $(2, 0)$, that is, the function to be minimized, is $d(x) = \sqrt{(x - 2)^2 + x}$ for $x \ge 0$. As a shortcut, *to find a value of x that minimizes (maximizes) an entire square root, it is sufficient to find*

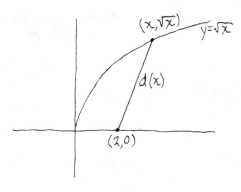

FIG. 3

a value of x that minimizes (maximizes) the expression under the square root sign; that is, $\sqrt{R(x)}$ is smallest (largest) when $R(x)$ is smallest (largest). Therefore we can work with $R(x) = (x - 2)^2 + x$, a slight advantage, since $R(x)$ is simpler than $d(x)$. We have $R'(x) = 2(x - 2) + 1$, which is 0 when $x = 3/2$. Therefore, the candidates are the critical number $x = \frac{3}{2}$ and the ends where $x = 0$, $x = \infty$. The closest point must be chosen from $(0, 0)$, $(\frac{3}{2}, \sqrt{\frac{3}{2}})$ and points far out to the right on the curve. Clearly, points far out to the right make the distance approach ∞ so we will not find a minimum from that source. The distance from $(0, 0)$ to $(2, 0)$ is 2. The distance from $(\frac{3}{2}, \sqrt{\frac{3}{2}})$ to $(2, 0)$ is $\sqrt{\frac{1}{4} + \frac{3}{2}} = \sqrt{\frac{7}{4}}$, which is less than 2. Consequently the closest point is $(\frac{3}{2}, \sqrt{\frac{3}{2}})$.

Example 6 A tin can is to be manufactured with volume V (V is a fixed constant throughout the problem). To save money, the manufacturer wants to minimize the amount of material, that is, minimize the surface area A. What dimensions should the can have?

Solution: The relevant geometry formulas for a circular cylinder with radius r and height h are

(1) $V = \pi r^2 h$

(2) lateral surface area $= 2\pi r h$

(3) top circular surface area $=$ bottom circular surface area $= \pi r^2$.

From (2) and (3), the function A to be minimized is given by

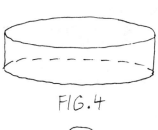

FIG. 4

(4) $$A = 2\pi r h + 2\pi r^2.$$

Before using any calculus, we can see that if r is very large and h very small (Fig. 4), but still satisfying (1) as required, then A will be huge because of the top and bottom pieces. On the other hand, if r is very small and h very large (Fig. 5), then A will be huge because of the lateral surface area, since

$$\text{lateral surface area} = 2\pi r h = 2\pi r \cdot \frac{V}{\pi r^2} = \frac{2V}{r},$$

which blows up as $r \to 0+$. Thus, extreme shapes require large A, and a tin can in between will use the least material. In other words, if A is considered as a function of r for $r \geq 0$, then A has a maximum value of ∞ at the ends where $r = 0$, ∞ and the minimum will occur at a critical number within the interval $(0, \infty)$.

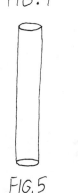

FIG. 5

Although A depends on both r and h, we can eliminate h by solving (1) for h and substituting in (4) to obtain

$$A = 2\pi r \cdot \frac{V}{\pi r^2} + 2\pi r^2 = \frac{2V}{r} + 2\pi r^2.$$

Then

$$A'(r) = -\frac{2V}{r^2} + 4\pi r.$$

Solve $A'(r) = 0$ to obtain

$$(5) \qquad r^3 = \frac{V}{2\pi}, \qquad r = \sqrt[3]{\frac{V}{2\pi}}.$$

The corresponding value of h can be found by using $h = V/\pi r^2$. Better still, for a more attractive answer, go back to $r^3 = V/2\pi$ in (5) and replace V by $\pi r^2 h$ to obtain $h = 2r$. Therefore, if the volume is fixed, the tin can with minimum surface area has a height which is twice its radius.

As another method, leave A in terms of r and h, and consider that h is a function of r defined implicitly by (1) (alternatively, r may be considered a function of h). Differentiate with respect to r in (4) to obtain $A' = 2\pi h + 2\pi r h' + 4\pi r$, and set $A' = 0$ to get

$$(6) \qquad h + r h' + 2r = 0.$$

Differentiate implicitly with respect to r in (1) to obtain $0 = \pi r^2 h' + 2\pi r h$, $h' = -2h/r$. Substituting this into (6) gives $h + r \cdot (-\frac{2h}{r}) + 2r = 0$, or $h = 2r$ as in the first method.

Example 7 Points A and B are a and b feet from a wall, respectively (Fig. 6). How can we leave A and bounce off the wall to B so as to minimize the total distance from A to the wall to B?

Solution: The total distance is very large if the ricochet point P is either far above A or far below B. We expect that somewhere on the wall *between* A and B is a point at which the distance is a minimum.

Let c be the fixed distance and x the variable distance indicated in Fig. 6, and let $f(x)$ be the distance $\overline{APB}$ to be minimized. Then

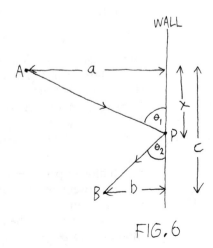

FIG. 6

$$f(x) = \overline{AP} + \overline{PB} = \sqrt{x^2 + a^2} + \sqrt{(c - x)^2 + b^2} \qquad \text{for } 0 \le x \le c.$$

Hence

$$f'(x) = \frac{x}{\sqrt{x^2 + a^2}} - \frac{c - x}{\sqrt{(c - x)^2 + b^2}} = \cos \theta_1 - \cos \theta_2.$$

We switch from the variable x to the angles θ_1 and θ_2 to simplify the algebra. The derivative is 0 if $\cos \theta_1 = \cos \theta_2$ which, for acute angles, means $\theta_1 = \theta_2$. Thus the only candidate is the point at which $\theta_1 = \theta_2$, and hence the condition for minimum distance is simply that $\theta_1 = \theta_2$.

By a law of physics (Fermat's principle), if light is reflected off a surface from A to B, the total time, hence distance, is minimized. Therefore light travels so that the angle of incidence equals the angle of reflection.

Example 8 Two corridors of widths 8 and 27 meet at right angles. What is the longest steel girder that can slide around the corner without getting stuck?

Solution: Consider all line segments of the type shown in Fig. 7. As the girder is maneuvered most efficiently around the corner, at each instant it hugs the corner as these segments do. If the girder is longer than any of the segments, it will not fit (we assume the thickness of the girder is negligible). Equivalently, *if the girder is longer than the smallest segment, it will get stuck;* we have therefore turned the problem into a minimization. *The longest girder that will survive has the same length as the shortest segment.*

Let θ be the angle in Fig. 8 and let L be the length of the indicated segment AC. Then

$$L(\theta) = \overline{AB} + \overline{BC} = \frac{8}{\sin \theta} + \frac{27}{\cos \theta} = 8 \csc \theta + 27 \sec \theta,$$

$$\text{where } 0° \le \theta \le 90°.$$

Figure 7 shows that values of θ near $0°$ and $90°$ correspond to very long segments, so the minimum length will occur at a critical angle in between. We have

$$L'(\theta) = -8 \csc \theta \cot \theta + 27 \sec \theta \tan \theta.$$

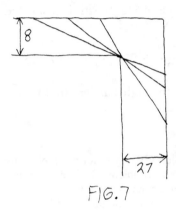

FIG. 7

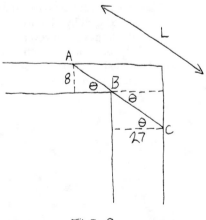

FIG. 8

Solve $L'(\theta) = 0$ to find the critical angles:

$$27 \sec \theta \tan \theta = 8 \csc \theta \cot \theta$$

$$\tan^3\theta = \frac{8}{27}$$

$$\tan \theta = \frac{2}{3}.$$

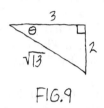

FIG.9

An approximate value of θ can be found from tables or a calculator, but the problem asks for the minimum L, and not the value of θ that produces it. To compute L efficiently, use the right triangle of Fig. 9 with legs labeled so that $\tan \theta = 2/3$. Then the hypotenuse is $\sqrt{13}$ and

$$\text{minimum } L = 8 \csc \theta + 27 \sec \theta = 8 \cdot \frac{\sqrt{13}}{2} + 27 \cdot \frac{\sqrt{13}}{3} = 13\sqrt{13}.$$

Thus the longest girder that can be carried through has length $13\sqrt{13}$.

Problems for Section 4.2

(If you have difficulty setting up verbal problems, you are not unique. Many students find the computational aspects of extremal problems fairly routine but (understandably) don't know how to begin problems such as Example 8.)

1. Find the maximum and minimum values of $f(x)$ on the indicated intervals.

(a) $f(x) = x^3 + x^2 - 5x - 5$ (i) $(-\infty, \infty)$ (ii) $[0, 2]$ (iii) $[-1, 0]$

(b) $f(x) = \dfrac{e^x}{x}$ (i) $[-2, 2]$ (ii) $[0, 2]$ (iii) $(-\infty, 0]$

(c) $f(x) = \dfrac{x - 2}{x^2 - 3}$ (i) $[0, 5]$ (ii) $[2, 5]$

(d) $f(x) = x^3 + x^2 - x + 3$, $[0, 4]$

2. Suppose $f'(x)$ is always negative. Find the largest and smallest values of f on $[3, 4]$.

3. Without using any calculus at all, find the largest and smallest values of $\sqrt{2 + x^2}$ for x in $(-\infty, \infty)$.

4. A charter aircraft has 350 seats and will not fly unless at least 200 of those seats are filled. When there are 200 passengers, a ticket costs \$300, but each ticket is reduced by \$1 for every passenger over 200. What number of passengers yields the largest total revenue? smallest total revenue?

5. A builder with 200 feet of wire wants to fence off a rectangular garden using an existing 100-foot stone wall as part of the boundary (Fig. 10). How should it be done to get maximum area? minimum area?

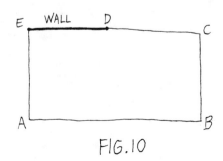

FIG.10

6. A rectangular house is built on the corner of a right triangular lot with legs 100 and 150 (Fig. 11). What dimensions for the house will produce maximum floor space?

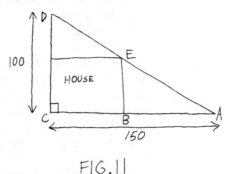

FIG. 11

7. A farmer has calves which weight 100 pounds each and are gaining weight at the rate of 1.2 pounds per day. If she sells them now she can realize a profit of 12 cents per pound. But since the price of cattle feed is rising, her profit per pound is falling by 1/40 of a cent per day. If she sells right now she gets the higher profit per pound but is selling skinny cows. If she waits to sell fat cows she makes less per pound. When should she sell?

8. Let $f(x) = -x^3 - 5x^2 - 13x + 4$; find the maximum and minimum slope on the graph of f for $0 \le x \le 1$.

9. At midnight, car B is 100 miles due south of car A. Then A moves east at 15 mph and B moves north at 20 mph. At what time are they closest together?

10. Given the ellipse $4x^2 + 9y^2 = 36$ and the point $Q = (1, 0)$, find the points on the ellipse nearest and furthest from Q.

11. Of all the rectangles inscribed in a semicircle with fixed radius r, which one has maximum area? minimum area?

12. A truck is to travel at constant speed s for 600 miles down a highway where the maximum speed allowed is 80 mph and the minimum speed is 30 mph. When the speed is s, the gas and oil cost $(5 + \frac{1}{10}s)$ cents per mile, so the slower the truck the less the transport company pays for gas and oil. The truck driver's salary is \$3.60 per hour (use 360 cents per hour so that all money is measured in cents). Thus, the faster the truck the less time it takes and the less the company must pay the driver. Find the most economical speed and least economical speed for the trip.

13. A wire 16 feet long is cut into two pieces, one of which is bent to form a square and the other to form a circle. How should the wire be cut so as to maximize the total area of square plus circle?

14. Suppose you wish to use the least amount of fencing to fence off a rectangular garden with fixed area A. What is the best you can do?

15. A motel with 100 rooms sells out each night at a price of \$50 per room. For each \$2 increase in price it is anticipated that an additional room will be vacant. What price should be charged in order to maximize income?

4.3 L'Hôpital's Rule and Orders of Magnitude

Section 2.3 identified a group of indeterminate limit forms, and we are now prepared to evaluate indeterminate limits, beginning in this section with quotients.

Consider

(1)
$$\lim_{x \to 3} \frac{x^3 - 3x - 18}{x^2 - 9}$$

which is of the indeterminate form 0/0. We will find the limit by working with the graphs of the numerator and denominator separately, and then extract a method for problems of this form in general. Each graph crosses the x-axis at $x = 3$ (which is why the problem is of the form 0/0). The graph of the numerator has slope 24 as it crosses, because the derivative of the numerator is $3x^2 - 3$, which is 24 when $x = 3$. The graph of the denominator has slope 6 when it crosses, because the derivative of the denominator is $2x$, which is 6 when $x = 3$ (Fig. 1 on next page). The limit in (1) depends on the ratio of the heights *near* $x = 3$ (at $x = 3$ we have the meaningless ratio 0/0). The two functions start "even" on the x-axis, the "starting line," at position $x = 3$, but the graph of the numerator is rising above the x-axis 4 times as steeply as the graph of the denominator. Thus, near $x = 3$, the graph of the numerator is about 4 times as high above the x-axis as the graph of the denominator. It follows that the ratio of their heights *near* $x = 3$ is *near* 4, and the *limit* in (1) *is* 4. The number 4 came from the computation 24/6 which in turn came from examining the quotient

$$\frac{\text{numerator derivative } 3x^2 - 3}{\text{denominator derivative } 2x}$$

at $x = 3$. This suggests that if $\lim_{x \to a} \dfrac{f(x)}{g(x)}$ is of the indeterminate form 0/0, it can be found by switching to $\lim_{x \to a} \dfrac{f'(x)}{g'(x)}$. This result holds not only for 0/0, but can be shown (with a different argument) to hold for the other indeterminate quotients as well. The following rule contains the details.

L'Hôpital's rule   Suppose $\lim_{x \to a} \dfrac{f(x)}{g(x)}$ is one of the indeterminate forms

$$\frac{0}{0}, \quad \frac{\infty}{\infty}, \quad \frac{-\infty}{\infty}, \quad \frac{\infty}{-\infty}, \quad \frac{-\infty}{-\infty}.$$

Switch to $\lim_{x \to a} \dfrac{f'(x)}{g'(x)}$.

If the new limit is L, ∞ or $-\infty$ then the original limit is L, ∞ or $-\infty$, respectively.

If the new limit does not exist because $f'(x)/g'(x)$ oscillates badly then we have no information about the original quotient (which does *not* necessarily oscillate also); L'Hôpital's rule does not help in this situation.

If the new limit is still an indeterminate quotient, L'Hôpital's rule may be used again.

The rule is also valid for limit problems in which $x \to a+$, $x \to a-$, $x \to \infty$ and $x \to -\infty$.

Example 1   Find $\lim_{x \to \infty} \dfrac{3x^3 + 6x^2 - 5}{2x^3 + 5x^2 - 3x}$, which is of the indeterminate form ∞/∞.

Solution: In this particular problem two methods are available, the highest power rule from Section 2.3 and L'Hôpital's rule. With the first method

$$\lim_{x \to \infty} \frac{3x^3 + 6x^2 - 5}{2x^3 + 5x^2 - 3x} = \lim_{x \to \infty} \frac{3x^3}{2x^3} = \lim_{x \to \infty} \frac{3}{2} = \frac{3}{2}.$$

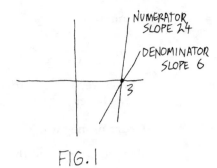

FIG. 1

With the second method,

$$(2) \quad \lim_{x \to \infty} \frac{3x^3 + 6x^2 - 5}{2x^3 + 5x^2 - 3x} = \frac{\infty}{\infty} = \lim_{x \to \infty} \frac{9x^2 + 12x}{6x^2 + 10x} = \frac{\infty}{\infty} = \lim_{x \to \infty} \frac{18x + 12}{12x + 10}$$

$$= \frac{\infty}{\infty} = \lim_{x \to \infty} \frac{18}{12} = \frac{3}{2}.^{\dagger}$$

As L'Hôpital's rule is applied repeatedly in this example, the lower powers differentiate away first, showing that the highest powers dominate as $x \to \infty$, in agreement with the highest power rule.

Example 2

$$(3) \qquad \lim_{x \to 0} \frac{\sin x}{x} = \frac{0}{0} = \lim_{x \to 0} \frac{\cos x}{1} (\text{L'Hôpital's rule}) = 1.^{\ddagger}$$

The result in (3) shows that if an angle θ is small, and is measured in radians (so that the derivative of $\sin \theta$ is $\cos \theta$) then $\sin \theta$ and θ are about the same size since their ratio is near 1. This is important in physics and engineering where many calculations may be simplified by replacing $\sin \theta$ by θ for small θ.

Warning L'Hôpital's rule applies only to indeterminate quotients. It should not be used (nor is it necessary) for limits of the form $2/\infty$ (the answer is immediately 0) or $3/0-$ (the anwer is $-\infty$) or $6/2$ (the answer is 3) and so on.

Example 3 By L'Hôpital's rule,

$$\lim_{x \to \infty} \frac{x^2}{e^x} = \frac{\infty}{\infty} = \lim_{x \to \infty} \frac{2x}{e^x} = \frac{\infty}{\infty} = \lim_{x \to \infty} \frac{2}{e^x} = \frac{2}{\infty} = 0.$$

The result indicates that while both x^2 and e^x grow unboundedly large as $x \to \infty$, e^x grows faster.

---

†We should not equate the original limit with the new limit at line (2) until *after* we have determined that the latter limit is either a number L, or ∞ or $-\infty$. However, it is customary to anticipate the situation and write the solution in the more compact form indicated.

‡As part of the proof in Section 3.3 that $D \sin x = \cos x$, we used geometry to show that $\lim_{x \to 0}(\sin x)/x = 1$. Since L'Hôpital's method is so much simpler than the geometric proof, you may wonder why we used geometry in the first place. We needed the limit in order to derive $D \sin x = \cos x$. But *before* the derivative formula is available we cannot do the differentiation necessary to apply L'Hôpital's rule. Thus we resorted to the geometric argument. The use of L'Hôpital's rule in Example 2 must be regarded as a check on previous work, rather than as an independent derivation.

Example 4

$$\lim_{x \to \infty} \frac{x}{\ln x} = \frac{\infty}{\infty} = \lim_{x \to \infty} \frac{1}{1/x} \text{ (L'Hôpital's rule)} = \lim_{x \to \infty} x \text{ (algebra)} = \infty.$$

Therefore, while both x and $\ln x$ grow unboundedly large as $x \to \infty$, x grows faster.

Order of magnitude   Suppose $f(x)$ and $g(x)$ both approach ∞ as $x \to \infty$ so that $\lim_{x \to \infty} \frac{f(x)}{g(x)}$ is of the form ∞/∞. If the limit is ∞ then $f(x)$ is said to be of a *higher order of magnitude* than $g(x)$; that is, f grows faster than g. If the limit is 0 then $f(x)$ has a *lower order of magnitude* than $g(x)$. If the limit is a positive number L then $f(x)$ and $g(x)$ have the *same order of magnitude*.

Examples 3 and 4 show that e^x is of a higher order of magnitude than x^2, and x is of a higher order of magnitude than $\ln x$. Similarly it can be shown that for any positive r, e^x grows faster than the power function x^r, and x^r grows faster than $\ln x$. (When r is negative, x^r doesn't grow at all as $x \to \infty$.)

The pecking order below in (4) contains some well-known functions which approach ∞ as $x \to \infty$, and lists them in increasing order of magnitude, from slower to faster.

(4) $$\ln x, (\ln x)^2, (\ln x)^3, \cdots, \sqrt{x}, x, x^{3/2}, x^2, x^3, \cdots, e^x.$$

Examples 3 and 4 illustrate how the order of the functions in (4) is justified. Functions which remain bounded as $x \to \infty$, such as $\sin x$, $\tan^{-1} x$ or constant functions, may be considered to have a lower order of magnitude than any of the functions in (4). Many indeterminate limit problems of the form ∞/∞ can be handled by inspection of the ordering in (4). For example, $\lim_{x \to \infty} e^x/x^4$ is of the indeterminate form ∞/∞; the function e^x is of a higher order of magnitude than x^4 and the answer is ∞.

Note that the list in (4) is not intended to be, and indeed can never be made, complete. There are functions slower than $\ln x$, faster than e^x, in between $\sqrt{x}$ and x, and so on.

The concept of order of magnitude is useful in many applications. Suppose $f(x)$ is the running time of a computer program which solves a problem of "size" x. Programs involving a "graph with x vertices" might require a running time of x^3 seconds, or x^4 seconds (worse), or e^x seconds (much worse, for large x), depending on the type of problem. If $f(x)$ is a power function, then the problem is said to run in polynomial time and is called *tractable;* if $f(x) = e^x$, the problem is said to require exponential time and is called *intractable.* Tractability depends on the order of magnitude of $f(x)$, and computer scientists draw the line between power functions and e^x. A major branch of computer science is devoted to determining whether a program runs in polynomial or exponential time. If it takes exponential time to find the "best" solution (such as the sales route with a minimum amount of driving time) then we often must settle for a less than optimal solution (a sales route with slightly more than the minimum driving time) that can be found in polynomial time.

Order of magnitude of a constant multiple   Consider $4x^2$ versus x^2. We have $\lim_{x \to \infty} 4x^2/x^2 = \lim_{x \to \infty} 4 = 4$. Since the limit is a positive number, not 0 or ∞, $4x^2$ and x^2 have the same order of magnitude, even though one is

4 times the other. In general, $f(x)$ *and* $cf(x)$ *have the same order of magnitude for any positive constant c.*

Highest order of magnitude rule   We can extend the highest power rule from Section 2.3: the proofs involve similar factoring arguments which we omit. As $x \to \infty$, a sum of functions on the list in (4) has the same limit as the term with the highest order of magnitude and, in fact, the sum has the same order of magnitude as that term. For example, $e^{2x} - x^4$ has the same order of magnitude as e^{2x} and

$$\lim_{x \to \infty}(e^{2x} - x^4) = \infty - \infty = \lim_{x \to \infty} e^{2x} = \infty.$$

As $x \to \infty$, a quotient involving functions on the list in (4) has the same limit as

$$\frac{\text{term with highest order of magnitude in the numerator}}{\text{term with highest order of magnitude in the denominator}}$$

and the final answer depends on which of the remaining terms has higher order of magnitude. For example,

$$\lim_{x \to \infty} \frac{3 - e^x}{x^3 + 2x} = \frac{-\infty}{\infty} = \lim_{x \to \infty} \frac{-e^x}{x^3} = -\lim_{x \to \infty} \frac{e^x}{x^3} = -\infty$$

since e^x has a higher order of magnitude than x^3.

Warning   The highest *power* rule is only valid for problems where $x \to \pm\infty$. The highest *order of magnitude* rule is even more restrictive. It applies only when $x \to \infty$ since the increasing orders of magnitude in (4) hold only in that case.

Problems for Section 4.3

1. Find $\lim \dfrac{x^3 - 5x + 4}{x^2 - 3x + 2}$ as (a) $x \to 1$ (b) $x \to 0$ (c) $x \to \infty$.

2. Find

(a) $\displaystyle\lim_{x \to \infty} \frac{x^2}{\ln x}$ (f) $\displaystyle\lim_{x \to \infty} \frac{e^{-x}}{1 + e^{-x}}$

(b) $\displaystyle\lim_{x \to 2} \frac{\ln(x - 1)}{x - 2}$ (g) $\displaystyle\lim_{x \to 0+} \frac{\ln x}{e^{1/x}}$

(c) $\displaystyle\lim_{x \to 0+} \frac{\ln x}{x}$ (h) $\displaystyle\lim_{x \to 0+} \frac{\ln 2x}{3x}$

(d) $\displaystyle\lim_{x \to \infty} \frac{x^4 + x^3}{e^x}$ (i) $\displaystyle\lim_{x \to \infty} \frac{\ln 2x}{3x}$

(e) $\displaystyle\lim_{x \to 0} \frac{\sin x - x}{\cos x - 1}$

3. Use L'Hôpital's rule to verify that $(\ln x)^{27}$ has a lower order of magnitude than x.

4. Both (a) $\lim_{x \to 0} \dfrac{\sin 3x}{2x}$ and (b) $\lim_{x \to 0} \dfrac{\sin^2 x}{x}$ are of the form 0/0 and can be done using L'Hôpital's rule. But they can also be cleverly done using the fact (Example 2) that $\lim_{x \to 0} \dfrac{\sin x}{x} = 1$. Do them both ways.

5. What is wrong with the following double application of L'Hôpital's rule?
$$\lim_{x \to 1} \frac{4x^2 - 2x - 2}{3x^2 - 4x + 1} = \lim_{x \to 1} \frac{8x - 2}{6x - 4} = \lim_{x \to 1} \frac{8}{6} = \frac{4}{3}$$

6. For each pair of functions, decide which has a higher order of magnitude.
(a) $3e^x, 4e^x$ (b) e^{3x}, e^{5x} (c) $\ln 3x, \ln 4x$

7. The graph of $(\sin x)/x$ can be drawn using the procedure of Section 1.3 for $f(x) \sin x$ where $f(x) = 1/x$. The tricky part is handling the graph near $x = 0$ when $\sin x$ approaches 0 and the envelope $1/x$ blows up. Sketch the entire graph.

4.4   Indeterminate Products, Differences and Exponential Forms

The preceding section discussed indeterminate quotients. We conclude the discussion of indeterminate limits in this section with methods for the remaining forms.

The forms $0 \times \infty$ and $0 \times -\infty$ L'Hôpital's rule applies only to indeterminate *quotients*. To do an indeterminate *product*, use algebra or a substitution to transform the product into a quotient to which L'Hôpital's rule does apply. For example, consider $\lim_{x \to 0+} x \ln x$ which is of the form $0 \times -\infty$. Use algebra to change the numerator x to a denominator of $1/x$ to get

(1) $\quad \lim_{x \to 0+} x \ln x = 0 \times -\infty = \lim_{x \to 0+} \frac{\ln x}{1/x} = \frac{-\infty}{\infty}$

(2) $\quad\quad\quad\quad = \lim_{x \to 0+} \frac{1/x}{-1/x^2}$ (use L'Hôpital's rule on the quotient)

(3) $\quad\quad\quad\quad = \lim_{x \to 0+}(-x)$ (by algebra) $= 0$.

In general, *for indeterminate products, try flipping one factor (preferably the simpler one) and putting it in the denominator to obtain an indeterminate quotient. Then continue with L'Hôpital's rule.*

As a second method in this example, let $u = 1/x$. Then $x = 1/u$ and as $x \to 0+$ we have $u \to \infty$, so

$$\lim_{x \to 0+} x \ln x = \lim_{u \to \infty} \frac{\ln 1/u}{u} = \lim_{u \to \infty} \frac{-\ln u}{u} \quad \text{(law of logarithms)}$$

which is of the form $-\infty/\infty$. Since u has a higher order of magnitude than $\ln u$, the answer is 0. In general, *as a second method for indeterminate products, try letting u be the reciprocal of one of the factors,* preferably the simpler one.

The function $x \ln x$ is defined only for $x > 0$, but this limit problem shows that for all practical pruposes $x \ln x$ is 0 when $x = 0$, and the graph can be considered to begin at the origin. In applied areas where the limit occurs frequently, the result is abbreviated by writing $0 \ln 0 = 0$.

Warning   1. Don't use L'Hôpital's rule indiscriminately. It applies *only* to indeterminate quotients and not to other indeterminate forms, and not to nonindeterminate problems, which can always be done directly.

2. Simplify algebraically whenever possible. If (2) is left unsimplified it is of the indeterminate form $\infty/-\infty$, but canceling produces (3) which is not indeterminate and gives the immediate answer 0.

The forms $\infty - \infty$ and $(-\infty) - (-\infty)$ L'Hôpital's rule applies only to indeterminate quotients, so other methods must be used for indeterminate differences. We will describe two possibilities.

If $x \to \infty$, a limit involving functions from the pecking order in (4) of Section 4.3 may be found using the highest order of magnitude rule. For example, $\lim_{x \to \infty}(x - \ln x)$ is of the form $\infty - \infty$; the answer is ∞ since x has a higher order of magnitude than $\ln x$.

If a problem involves the difference of two fractions, they can be combined algebraically into a single quotient, to which L'Hôpital's rule may be applied, if necessary. For example, consider $\lim_{x \to 0}\left(\frac{1}{x} - \frac{1}{x^2}\right)$. If $x \to 0-$, the limit is of the form $(-\infty) - \infty$, so the left-hand limit is $-\infty$. But the right-hand limit is of the indeterminate form $\infty - \infty$. In either case, we can use algebra to combine the fractions and obtain

$$\lim_{x \to 0}\left(\frac{1}{x} - \frac{1}{x^2}\right) = \lim_{x \to 0}\frac{x - 1}{x^2} = \frac{-1}{0+} = -\infty.$$

The forms $(0+)^0$, 1^∞ and ∞^0 We will illustrate with an example how to use logarithms to change exponential problems into products. Consider $\lim_{x \to \infty}(1 + \frac{.06}{x})^x$ which is of the indeterminate form 1^∞. Let $y = (1 + \frac{.06}{x})^x$. Take ln on both sides, and use $\ln a^b = b \ln a$, to obtain $\ln y = x \ln(1 + \frac{.06}{x})$. Then

$$\lim_{x \to \infty} \ln y = \lim_{x \to \infty} x \ln\left(1 + \frac{.06}{x}\right) = \infty \times \ln 1 = \infty \times 0.$$

To turn the indeterminate product into a quotient, one method is to let $u = 1/x$. Then $x = 1/u$, and as $x \to \infty$ we have $u \to 0+$, so

$$\lim_{x \to \infty} \ln y = \lim_{u \to 0+} \frac{\ln(1 + .06u)}{u} = \frac{0}{0}$$

$$= \lim_{u \to 0+} \frac{\dfrac{1}{1 + .06u} \cdot .06}{1} \qquad \text{(apply L'Hôpital's rule to the quotient)}$$

$$= .06.$$

If $\ln y$ approaches $.06$ then y itself approaches $e^{.06}$. So as a final answer,

$$(4) \qquad \lim_{x \to \infty}\left(1 + \frac{.06}{x}\right)^x = e^{.06}.$$

In general, *if $\lim f(x)$ is an indeterminate exponential form, let $y = f(x)$ and compute $\ln y$, which will no longer involve exponents. Find $\lim \ln y$, and if that answer is L, then the answer to the original problem is e^L.*

Warning In the preceding problem, the answer is $e^{.06}$, not $.06$. Don't forget this last step.

An application to compound interest Suppose an amount A (dollars) is deposited in a bank which pays 6% annual interest compounded three times a year. The bank divides the 6% figure into three 2% increments, and after four months pays 2% on amount A. Thus the four month balance is $A + .02A = A(1 + .02) = 1.02A$. In other words, the balance has been multiplied by 1.02. After eight months, the depositor receives 2% interest on amount $1.02A$, so the money is again multiplied by 1.02. Similarly, after twelve months, the bank pays a final 2% which again multiplies the balance by 1.02. Therefore after one year, amount A, compounded at 6% three times a year, accumulates to $(1.02)^3A$, that is, to $(1 + \frac{.06}{3})^3A$. More generally, if the bank pays $r\%$ interest compounded x times a year, then A grows to $A(1 + \frac{r}{x})^x$ at the end of the year. If the bank generously compounds your money not just x times a year but "continually" then A grows to $\lim_{x\to\infty} A(1 + \frac{r}{x})^x$. As a generalization of (4) we have

$$(5) \qquad \lim_{x\to\infty}\left(1 + \frac{r}{x}\right)^x = e^r,$$

so A grows to Ae^r. For example, \$1 compounded continually at 6% will grow to $e^{.06}$ dollars in a year, or approximately \$1.062, compared with \$1.06 obtained with simple interest.

A formula for the number e   We defined e in Section 3.3, but otherwise have given no indication of how to compute e to any desired number of decimal places. If r is set equal to 1 in (5), we have

$$e = \lim_{x\to\infty}\left(1 + \frac{1}{x}\right)^x.$$

(In banking circles, this means that \$1 compounded continually at 100% interest grows to \$$e$ after a year.) The accompanying computer program prints out values of $(1 + \frac{1}{x})^x$ for larger and larger x, and therefore the values are approaching e. But if we pick out a value far down on the list and call it "approximately e", we have no way of knowing how close this is to e. (For example, is the approximation accurate in the first three decimal places, or would even these places change as we continue computing?) An approximation *with* an error estimate would be much more useful, and we'll have such an estimate for e in Section 8.9

```
0020 PRINT "X", "(1 + 1/X)⌐X"
0030 FOR N=2000 TO 8000 STEP 1000
0040    PRINT N,(1+1/N)⌐N
0050 NEXT N

*RUN
X                (1 + 1/X)⌐X
   2000          2.7176026
   3000          2.7178289
   4000          2.7179421
   5000          2.7180101
   6000          2.7180553
   7000          2.7180877
   8000          2.718112

END AT 0050
```

Problems for Section 4.4

1. Find $\lim xe^{-x}$ as (a) $x \to \infty$ (b) $x \to 0$ (c) $x \to -\infty$.
2. Find $\lim(x^2 - \ln x)$ as (a) $x \to 1$ (b) $x \to 0+$ (c) $x \to \infty$.
3. Find $\lim(x - e^x)$ as (a) $x \to \infty$ (b) $x \to -\infty$.

4. Sketch the graph of $xe^{1/x}$ near $x = 0$ after finding limits as $x \to 0+$ and $x \to 0-$.

5. Find

(a) $\lim\limits_{x \to 0+} (\tan x)(\ln x)$ (f) $\lim\limits_{x \to 0+} (1 + x)^{1/x}$

(b) $\lim\limits_{x \to 0+} e^x \ln x$ (g) $\lim\limits_{x \to \infty} x^x$

(c) $\lim\limits_{x \to \infty} x^2 \sin \dfrac{1}{x}$ (h) $\lim\limits_{x \to \infty} x(e^{1/x} - 1)$

(d) $\lim\limits_{x \to \infty} x^{1/x}$ (i) $\lim\limits_{x \to 2+} (x - 2)^x$

(e) $\lim\limits_{x \to 0+} x^{1/x}$ (j) $\lim\limits_{x \to 0+} (e^x + 4x)^{2/x}$

4.5 Drawing Graphs of Functions

In this section we'll list some of the aids already discussed for sketching graphs, and add new ones involving the derivative. For any particular function you may find some, but not necessarily all, items on the list useful in producing a graph.

1. Ends If f is defined on $(-\infty, \infty)$, find $\lim_{x \to \infty} f(x)$ and $\lim_{x \to -\infty} f(x)$ to determine the ends of the graph. If f is defined only on $(a, b]$ for instance, find $f(b)$ and $\lim_{x \to a+} f(x)$ to determine the ends.

2. Gaps If f is defined around but not at $x = x_0$ (in practice, because of a zero in a denominator), find $\lim_{x \to x_0} f(x)$, or if necessary find the right-hand and left-hand limits separately, to discover the nature of the gap.

3. Relative extrema Find the critical numbers and classify them as relative maxima, relative minima or neither, using the first or second derivative test. This identifies the rise and fall of the graph. Furthermore, find the values of y corresponding to the critical numbers so that a few significant points can be plotted accurately.

4. Concavity Determine the sign of f'', with the method of Section 1.6, and use it to decide where f is concave up (f'' positive) and concave down (f'' negative). Often, approximately correct concavity is created automatically as you employ other graphing aids, so you may decide that using f'' to determine precise concavity is not worth it.

5. Familiar graphs If the new graph is related to a familiar graph then you have a head start, as the following examples illustrate.

The graph of $y = 2 + (x - 3)^2$ is the parabola $y = x^2$ translated to the right by 3 and up by 2 (Section 1.7).

The graphs of $y = a \sin(bx + c)$ and $y = a \sin b(x + c)$ are sinusoidal. Each has amplitude a and period $2\pi/b$, and the translation is best identified by plotting a few points (Section 1.3).

The graph of $y = f(x) \sin x$ is drawn by changing the heights on the sine curve so that it fits within the envelope $y = \pm f(x)$ (Section 1.3).

The graph of $y = a + be^{\alpha}$ has the shape of an exponential curve. It is located on the axes by plotting a point and finding limits as $x \to \pm\infty$ (Section 2.2).

Example 1 Sketch the graph of $f(x) = 1 - \dfrac{6}{x} + \dfrac{9}{x^2}$.

Solution: Find $\lim_{x \to \infty} f(x) = 1 - 0 + 0 = 1$ and $\lim_{x \to -\infty} f(x) = 1 - 0 + 0 = 1$, which indicates that the line $y = 1$ is an asymptote at each end of the graph.

The function is not defined at $x = 0$, so consider the limit as $x \to 0$. It is an advantage to let $u = 1/x$ so that the problem becomes $\lim(1 - 6u + 9u^2)$ as $u \to \infty$ (if $x \to 0+$) or $u \to -\infty$ (if $x \to 0-$). By the highest power rule, $9u^2$ dominates in each case and the limit is ∞. Therefore $\lim_{x \to 0} f(x) = \infty$, and the graph approaches the positive y-axis asymptotically from each side. (Intuitively, the term $9/x^2$ is so large as $x \to 0$ that it dominates $f(x)$.)

To find relative extrema, first find $f'(x) = \dfrac{6}{x^2} - \dfrac{18}{x^3}$. The derivative is 0 when $6x^3 = 18x^2$, $x = 3$. The derivative doesn't exist when $x = 0$, but neither does f; we have already found that f blows up at $x = 0$. The following table displays the pertinent information about the sign of the derivative and the behavior of f.

Interval	Sign of f'	Graph of f
$(-\infty, 0)$	positive	rises
$(0, 3)$	negative	falls
$(3, \infty)$	positive	rises

Therefore, f has a relative minimum at $x = 3$. (Alternatively, $f''(x) = -\dfrac{12}{x^3} + \dfrac{54}{x^4}$ so $f''(3)$ is positive. Therefore, by the second derivative test, f has a relative minimum at $x = 3$.) When $x = 3$, we have $y = 0$ so the relative minimum occurs at the point $(3, 0)$.

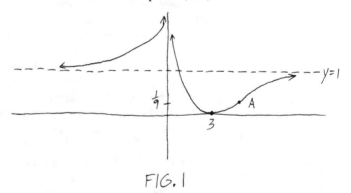

FIG. 1

So far we have the curve in Fig. 1, with the concavity tentatively suggested by the rise, fall, and asymptotic behavior of f. In this example, we'll check the concavity with the second derivative which has already been computed above. It is discontinuous at $x = 0$, and is 0 when $-12x + 54 = 0$, $x = 4\frac{1}{2}$. We collect the relevant information about the sign of f'' and the behavior of f.

Interval	Sign of f''	Graph of f
$(-\infty, 0)$	positive	concave up
$(0, 4\frac{1}{2})$	positive	concave up
$(4\frac{1}{2}, \infty)$	negative	concave down

This confirms the concavity in Fig. 1. Since $f(4\frac{1}{2}) = \frac{1}{9}$, the point of inflection at A is $(4\frac{1}{2}, \frac{1}{9})$.

Example 2 Sketch the graph of $y = \ln(x^3 + 8)$.

Solution: It is not always necessary to use all of the five aids described. If f is a variation of a familiar function g (the logarithm in this case), it may be possible to sketch the graph of f quickly by plotting a few points and using known properties of g.

The function f is defined only if $x^3 + 8 > 0$, $x > -2$. Then, as x increases, $x^3 + 8$ increases, and in turn, so does $\ln(x^3 + 8)$. Thus the graph always rises. For the right end, $\lim_{x \to \infty} \ln(x^3 + 8) = \ln \infty = \infty$. For the left end, $\lim_{x \to (-2)+} \ln(x^3 + 8) = \ln 0+ = -\infty$. Therefore, the usual asymptotic behavior of the logarithm function at $x = 0$ now takes place at $x = -2$. Also, the graph crosses the x-axis, not at $x = 1$, but when $x^3 + 8 = 1$, $x = \sqrt[3]{-7}$.

For large x, the highest power rule suggests that $f(x)$ behaves like $\ln x^3$, which is $3 \ln x$. Therefore, far out to the right, the graph of f is approximately 3 times the height of the graph of $\ln x$. A rough sketch is given in Fig. 2.

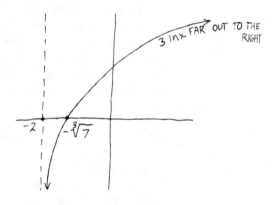

FIG. 2

Problems for Section 4.5

In Problems 1–22, sketch the graph of the function $f(x)$.

1. $-x^2 + 4x + 5$

2. $x^4 + 2x^3$

3. $x^{3/2}$

4. $x^{2/3}$

5. $x^4 + x^3 + 5x^2$

6. $2e^{-3x}$

7. $\sin\left(2x - \dfrac{\pi}{6}\right)$

8. $x\sqrt{2 - x^2}$

9. $\dfrac{\cos x}{x}$

10. $e^{-1/x}$

11. xe^x

12. $x^2 e^{-x}$

13. $x \ln x$

14. $x - \ln x$

15. $\dfrac{x - 1}{x + 1}$

16. $e^{-x} \sin x$

17. $-e^{-2x} - 4$

18. $3 \cos(\frac{1}{2}\pi x + \frac{1}{2}\pi)$

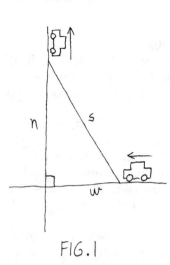

FIG. 1

19. e^x/x^5

20. e^{-x^2}

21. $x + \dfrac{1}{x}$

22. $\dfrac{4}{1 + x^2}$

23. (a) Sketch the graph of $\dfrac{\ln x}{x}$.   (b) Use part (a) to help sketch the graph of $\dfrac{\ln |x|}{x}$.

4.6 Related Rates

Suppose two (or more) quantities are related to one another. If one quantity is changing instantaneously with time, we can use differential calculus to determine how the other changes.

Example 1   Two cars travel west and north on perpendicular highways as indicated in Fig. 1. The problem is to decide if the cars are separating or getting closer. (Picture an elastic string between the two cars. Is the string getting shorter or longer?)

We do not have enough information to solve the problem at this stage. The westbound car is trying to close the gap while the northbound car is trying to increase it. What actually happens will be determined by the speeds of the cars, and also (although this is less obvious) by their distances from the intersection of the roads. Thus we continue stating the problem by asking if the cars are separating or getting closer at the *particular instant* when the westbound car is traveling at 25 mph, the northbound car is traveling at 10 mph, and they are respectively 5 miles and 12 miles from the intersection.

Now let's set up the problem so that we can use derivatives.

Step 1   Identify the functions involved.

In our problem, with t standing for time, one of the functions is the distance $n(t)$ from the northbound car to the intersection (Fig. 1). (The 10 mph is a specific value of dn/dt and the 12 miles is a value of n.) Similarly, the other functions needed are $w(t)$, the distance from the westbound car to the intersection, and $s(t)$, the distance between the two cars.

Step 2   Find a *general* connection among the functions.

In our problem, $s^2 = n^2 + w^2$ by the Pythagorean theorem. More precisely, $s^2(t) = n^2(t) + w^2(t)$ since s, n and w are functions of t.

Step 3   Differentiate with respect to t on both sides of the equation from Step 2 to get a *general* connection among the derivatives of the functions involved.

In our problem

(1)
$$2s\frac{ds}{dt} = 2n\frac{dn}{dt} + 2w\frac{dw}{dt}.$$

Note that the derivative of s^2 with respect to s is $2s$, but the derivative of $s^2(t)$ with respect to t is $2s \cdot ds/dt$ by the chain rule. Don't forget the factor ds/dt, and similarly the factors dn/dt and dw/dt, in (1).

Step 4   Substitute the specific data for the particular instant of interest.

In our problem, the instant occurs when $w = 5$ and $n = 12$, so $s = 13$ by the Pythagorean theorem. Also $dn/dt = 10$ (*positive* because when the car moves north at 10 mph the distance n is *increasing*) and $dw/dt = -25$ (*negative* because when the car moves west at 25 mph, the distance w is *decreasing*). Substitute these values into (1) and solve for ds/dt to obtain

(2)
$$\frac{ds}{dt} = \frac{n\dfrac{dn}{dt} + w\dfrac{dw}{dt}}{s} = \frac{(12)(10) + (5)(-25)}{13} = -\frac{5}{13}.$$

Therefore, at this moment, the distance s is *decreasing*, so the cars are getting closer by 5/13 miles per hour.

Note from (2) that the change in the gap between the cars depends not only, as expected, on their speeds and directions (because the formula for ds/dt involves the velocities dn/dt and dw/dt) but also on their distances to the intersection (because the formula contains n and w). For example, suppose the westbound and northbound cars travel at 25 mph and 10 mph again, but this time are respectively 2 miles and 6 miles from the inter-section, so that $w = 2, n = 6, s = \sqrt{40}$. Then ds/dt in (2) is *positive*, namely $10/\sqrt{40}$, and the cars are moving further apart at this instant.

Warning Be careful about signs when assigning values to derivatives. Suppose a bucket is being hauled *up* a well at 2 ft/sec. If $x(t)$ is the distance from the bucket to the top of the well, and $y(t)$ is the distance from the bucket to the bottom of the well, then x is *decreasing* by 2 ft/sec, while y is *increasing* by 2 ft/sec. Thus $dx/dt = -2$ and $dy/dt = 2$.

Example 2 A TV camera 10 meters across from the finish line is turning to stay trained on a runner heading toward the line (Fig. 2). When the runner is 9 meters from the finish line, the camera is turning at .1 radians per second. How fast is the runner going at this moment?

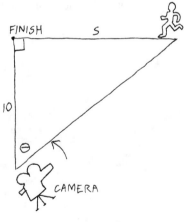

FIG. 2

Solution:

Step 1 Let t stand for time. Let $\theta(t)$ be the angle indicated in Fig. 2 and let $s(t)$ be the distance from the runner to the finish line.

Step 2 The general connection between the functions is $s = 10 \tan \theta$, or more precisely $s(t) = 10 \tan \theta(t)$.

Step 3 Differentiate with respect to t to obtain $ds/dt = 10 \sec^2\theta(d\theta/dt)$.

Step 4   At the moment of interest, $d\theta/dt = -.1$ (*negative because θ is decreasing*) and $s = 9$. Therefore the hypotenuse of the triangle is $\sqrt{181}$ and $\sec\theta = \sqrt{181}/10$. Thus

$$\frac{ds}{dt} = 10\left(\frac{181}{100}\right)(-.1) = -1.81 .$$

The negative sign is well deserved as an indication that s is decreasing. Since the problem asked only for the *speed* of the runner, the answer is 1.81 meters per second.

Problems for Section 4.6

(As with the section on maximum/minimum problems, this section contains verbal problems that students sometimes find difficult to set up.)

1. A snowball is melting at the rate of 10 cubic feet per minute. At what rate is the radius changing when the snowball is 2 feet in radius?

2. At a fixed instant of time, the base of a rectangle is 6, its height is 8, the base is growing by 4 ft/sec, and the height is shrinking by 3 ft/sec. How fast is the area of the rectangle changing at this instant?

3. A baseball diamond is 90 feet square. A runner runs from first base to second base at 25 ft/sec. How fast is he moving away from home plate when he is 30 feet from first base?

4. Water flows at 8 cubic feet per minute into a cylinder with radius 4. How fast is the water level rising?

5. An equilateral triangle is inscribed in a circle. Suppose the radius of the circle increases at 3 ft/sec. How fast is the area of the triangle increasing when the radius is 4?

6. A light 5 miles offshore revolves at 1 revolution per minute, that is, at 2π radians per minute (Fig. 3). When the light is directed toward the beach, the spot of light moves up the beach as the source revolves. How fast is the spot moving when it is 12 miles from the foot A of the source?

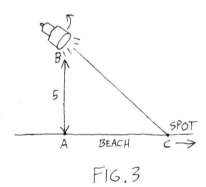

FIG. 3

7. A cone with height 20 and radius 5 is filled with a hose which pumps in water at the rate of 3 cubic meters per minute. When the water level is 2 meters, how fast is the level rising?

8. As you walk away from a light source at a constant speed of 3 ft/sec, your shadow gets longer (Fig. 4). The shadow's feet move at 3 ft/sec and it follows that the head of the shadow must move faster than 3 ft/sec to account for the lengthening. How fast does the head move if you are 6 feet tall and the source is 15 feet high?

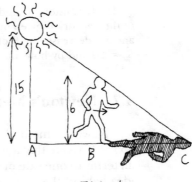

FIG. 4

9. Consider a cone with radius 6 and height 12 (centimeters).

(a) If water is leaking out at the rate of 10 cubic centimeters per minute, how fast is the water level dropping at the moment when the level is 3 centimeters?

(b) Suppose water leaks from the cone. When the water level is 6 centimeters, it is observed to be dropping at the rate of 2 centimeters per minute. How fast is the leak at this instant?

(c) Suppose the cone is not leaking, but the water is evaporating at a rate equal to the square root of the exposed circular area of the cone of water. How fast is the water level dropping when the level is 2 centimeters?

10. A stone is dropped into a lake, causing circular ripples whose radii increase by 2 m/sec. How fast is the disturbed area growing when the outer ripple has radius 5?

11. Consider the region between two concentric circles, a washer, where the inner radius increases by 4 m/sec and the outer radius increases at 2 m/sec. Is the area of the region increasing or decreasing, and by how much, at the moment the two radii are 5 meters and 9 meters?

12. Let triangle ABC have a right angle at C. Point A moves away from C at 6 m/sec while point B moves toward C at 4 m/sec. At the instant when $\overline{AC} = 12$, $\overline{BC} = 10$, is the area increasing or decreasing, and by how much?

13. A sphere is coated with a thick layer of ice. The ice is melting at a rate proportional to its surface area. Show that the thickness of the ice is decreasing at a constant rate.

14. A fish is being reeled in at a rate of 2 m/sec (that is, the fishing line is being shortened by 2 m/sec) by a person sitting 30 meters above the water (Fig. 5). How fast is the fish moving through the water when the line is 50 meters? when the line is only 31 meters?

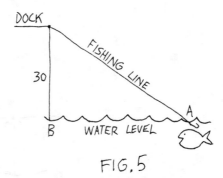

FIG. 5

15. If resistors R_1 and R_2 are connected in parallel, then the total resistance R of the network is given by $1/R = 1/R_1 + 1/R_2$. If R_1 is increasing by 2 ohms/min, and R_2 decreases by 3 ohms/min, is R increasing or decreasing when $R_1 = 10$, $R_2 = 20$ and by how much?

4.7 Newton's Method

Newton's method uses calculus to try to solve equations of the form $f(x) = 0$. (Note that any equation can be written in this form by transferring all terms to one side of the equation.) First we'll demonstrate the geometric idea behind the method.

Solving $f(x) = 0$ is equivalent to finding where the graph of the function f crosses the x-axis. Begin by guessing the root, and call the first guess x_1 (Fig. 1). Draw the tangent line to the graph of f at the point $(x_1, f(x_1))$. Let x_2 be the x-coordinate of the point where the tangent line crosses the x-axis. Now start again with x_2. Draw the line tangent to the graph of f at the point $(x_2, f(x_2))$ and let x_3 be the x-coordinate of the point where the tangent line crosses the x-axis. In Fig. 1, the numbers $x_1, x_2, x_3, \cdots$ approach the root; in Fig. 2, $x_1, x_2, x_3, \cdots$ do not approach the root (a change in concavity near the root is dangerous). However, more often than not, the situation in Fig. 1 prevails and Newton's method does work. It is certainly worth a try, especially if a computer or calculator is available to do most of the work.

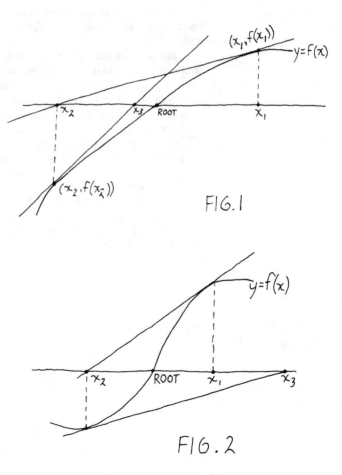

FIG. 1

FIG. 2

Now let's translate the geometry into a computational procedure. The line through the point $(x_1, f(x_1))$ and tangent to the graph of f must have slope $f'(x_1)$. By the point–slope formula, the equation of the tangent line is $y - f(x_1) = f'(x_1)(x - x_1)$. Set $y = 0$ and solve for x to find that the line crosses the x-axis when $x = x_1 - \dfrac{f(x_1)}{f'(x_1)}$. This value of x is taken to be x_2. In general, each new value of x is generated from the preceding one as follows:

$$(1) \quad \text{new } x = \text{last } x - \frac{f(\text{last } x)}{f'(\text{last } x)} \quad \text{or, equivalently,} \quad x_{n+1} = x_n - \frac{f(x_n)}{f'(x_n)}.$$

To see the method in operation, consider the computer program in (2) for solving $f(x) = x^3 - 10x^2 + 22x + 6 = 0$. When the program is run, it requests (with a question mark) a first guess at a root. After receiving the guess, it calculates successive values of x from (1), along with the corresponding values of $f(x)$. When two successive values of x differ by less than .00005, line 60 instructs the program to stop. If the values of $f(x)$ approach 0, then the values of x are approaching a root, and the last value of x can be taken to approximate the root.

To choose a first guess, note that $f(-1) < 0$, $f(2) > 0$. Since f is continuous, the graph of f must cross the x-axis between $x = -1$ and $x = 2$. Therefore, we began by running the program with the guess $x = 2$.

```
0010 INPUT X
0020 DEF FNF(X)=X*X*X-10*X*X+22*X+6
0030 DEF FND(X)=3*X*X-20*X+22
0040 PRINT "X","F(X)"
0050 LET Y=X-FNF(X)/FND(X)
0055 PRINT Y,FNF(Y)
0060 IF ABS(X-Y)<.00005 THEN GO TO 0080
0065 LET X=Y
0070 GO TO 0050
0080 END
```
(2)

```
*RUN
? 2
```

X	F(X)
5	−9
2	18
5	−9
2	18
5	−9
2	18

STOP AT 0055

The printout shows values of f which do *not* approach 0, so the values of x do *not* approach a root. The first tangent line at $x = 2$ leads to $x = 5$, but the second tangent line leads back to $x = 2$, the third tangent line is the same as the first and leads back to $x = 5$, and so on. We had to hit the escape button and stop the program manually, or it would have run forever, producing useless and repetitive results.

We ran the program again, this time with first guess $x = 1$. The printout shows values of $f(x)$ approaching 0. (The computer notation $E-15$

indicates a factor of 10^{-15}. Thus the last value of f, $-2.6645353E-15$, is $-2.6645353 \cdot 10^{-15}$, a very small number.)

```
*RUN
 ?  1
X                    F(X)
-2.8                 -155.952
-1.2638298           -39.795585
-.49953532           -7.6097842
-.26709965           -.60866996
-.24501119           -5.2591762E-03
-.24481698           -4.0487743E-07
-.24481697           -2.6645353E-15
```

END AT 0080

Therefore $x = -.24481697$ is an approximate root, but we do not know how many accurate decimal places we have. (One way to determine accuracy is to increase x until $f(x)$ changes from negative to positive. For example, $f(-.24481690) = .000002$, so there must be a root between $-.24481697$ and $-.24481690$, and the decimal places $-.2448169$ are correct.) Since the last two entries in the x column agree through 7 digits it is common practice to use the first 6 rounded digits, namely $-.244817$. This does not guarantee six place accuracy but merely provides a convenient stopping place for the procedure.

Problems for Section 4.7

Use Newton's method and continue until two successive approximations agree to the indicated number of decimal places. Then check the accuracy by searching for a sign change in $f(x)$ as above.

1. Find $\sqrt{39}$ by solving $x^2 = 39$ for the positive value of x. Use $x = 6$ as the initial guess and stop after agreement in two decimal places.

2. Find the cube root of 173; at least 3 decimal places.

3. Solve $e^x = 3 - x^2$; 3 decimal places. Begin by sketching the graphs of e^x and $3 - x^2$ on the same set of axes. Examine their intersections to determine the number and approximate values of solutions.

4. Find a solution of $\tan x = x$ (if possible) in interval $(0, \pi/2)$ and then again in $(\pi/2, 3\pi/2)$; 3 decimal places.

4.8   Differentials

As a by-product of the derivative of $f(x)$, which measures the rate of change of $f(x)$ with respect to x, we will develop the differential of $f(x)$ to describe the effect on $f(x)$ of a small change in x. The immediate results may not seem exciting, but in Section 5.3 the result in $(1')$ below will be used to explain the Fundamental Theorem of Calculus, in Section 6.1 the shell volume formulas developed here will be used to find moments of inertia of spheres and cylinders, and in Chapter 7, the new differential notation of this section will be used throughout.

Approximating a change in y   Suppose $y = f(x)$, and we start with a particular value of x and change it slightly by Δx so that there is a corresponding change Δy in y. The precise connection between Δx, Δy and f' is given by

$$f'(x) = \lim_{\Delta x \to 0} \frac{\Delta y}{\Delta x}.$$

If the limit is removed so that we are no longer entitled to claim equality, we have Δy approximately equal to $f'(x)\,\Delta x$; i.e., $\Delta y \sim f'(x)\,\Delta x$. The symbols dx and dy, called *differentials,* are defined as follows: $dx = \Delta x$, $dy = f'(x)\,dx$. With this notation we have

(1) $$\text{change } \Delta y \text{ in } y \sim f'(x)\,\overbrace{\Delta x}^{dx}. $$
$$\underbrace{}_{dy}$$

In other words, *dx is simply Δx, a change in x. The corresponding change Δy in y is approximated by $f'(x)\,dx$, denoted by dy.*

To see the geometric interpretation of approximating the change in y by $f'(x)\,dx$, consider the graph of $y = f(x)$. If the value of x is changed by dx, then the corresponding change in y is the change in the height on the graph of f (Fig. 1). On the other hand, consider the tangent line at the point $(x, f(x))$; its slope is $f'(x)$. As x changes by dx,

$$\frac{\text{change in } y \text{ on the tangent line}}{\text{change } dx \text{ in } x} = \text{slope of the tangent line},$$

so

$$\text{change in } y \text{ on the tangent line} = f'(x)\,dx.$$

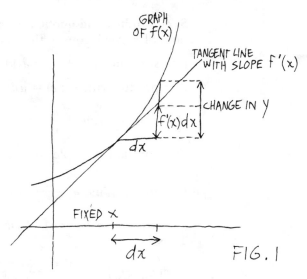

FIG. 1

Therefore, $f'(x)\,dx$ is the change in the height of the tangent line (Fig. 1). We call $f'(x)\,dx$ the *linear approximation* to the change in y; it approximates the rise or fall of the graph of f by the rise or fall of the tangent line. The error in the approximation is the difference between the height of the tangent line and the height of the graph of f, and approaches 0 as dx approaches 0. In fact, it can be shown that the error approaches 0 faster than dx.

The symbols Δx and dx both represent a change in x.† Mathematicians use the notation Δy for the change in y, and use dy for $f'(x)\,dx$ which

†The symbol dx in the antiderivative notation $\int f(x)\,dx$ is another story. It is not a small change in x; rather, it indicates that the antidifferentiation is to be done with respect to the variable x.

approximates the change in y (see (1)). In applied fields, and in this text, the distinction between $f'(x)\,dx$ and the change in y is often blurred, and both are referred to as dy; i.e., we often take the liberty of claiming that

(1') $\boxed{dy = f'(x)\,dx = \text{change in } y \text{ when } x \text{ changes by } dx.}$

Example 1   Let $y = x^3$. As usual, we write $\dfrac{dy}{dx} = 3x^2$ to mean that the derivative of y is $3x^2$. The differential version is $dy = 3x^2\,dx$, interpreted to mean that if x changes by dx there is a corresponding change in y given approximately by $3x^2\,dx$.

Example 2   We have $d(\sin x) = \cos x\,dx$; that is, the differential of $\sin x$ is $\cos x\,dx$. If x changes by dx then $\sin x$ changes by approximately $\cos x\,dx$.

Warning   Don't omit the dx and write $d(\sin x) = \cos x$ when you really mean either $d(\sin x) = \cos x\,dx$, $D\sin x = \cos x$ or $d(\sin x)/dx = \cos x$.

Example 3   Find the linear approximation to the change in x^5 when x changes from 2 to 1.999.
   Solution: We have $f(x) = x^5$, so $f'(x) = 5x^4$. When x changes from the value 2 by $dx = -.001$, the linear approximation to the change in x^5 is $f'(2)\,dx$, which is $(80)(-.001)$ or $-.08$.

Sum, product and quotient rules for differentials   Let u and v be functions of x. Analogous to the rules for derivatives, we have

(2)   sum rule $d(u + v) = d(u) + d(v)$

(3)   product rule $d(uv) = u\,d(v) + v\,d(u)$

(4)   quotient rule $d\left(\dfrac{u}{v}\right) = \dfrac{v\,d(u) - u\,d(v)}{v^2}$

(5)   constant multiple rule $d(cu) = c\,d(u)$, where c is a constant.

Example 4   Find $d\left(\dfrac{x^2}{2x + 3}\right)$.

   First solution (directly): As in Examples 1 and 2, we simply find $f'(x)\,dx$. Thus

$$d\left(\frac{x^2}{2x + 3}\right) = \frac{2x(2x + 3) - x^2 \cdot 2}{(2x + 3)^2}\,dx \qquad \text{(derivative quotient rule)}$$

$$= \frac{2x^2 + 6x}{(2x + 3)^2}\,dx.$$

   Second solution (differential quotient rule): By (5),

$$d\left(\frac{x^2}{2x + 3}\right) = \frac{(2x + 3)\,d(x^2) - x^2\,d(2x + 3)}{(2x + 3)^2}$$

$$= \frac{(2x + 3) \cdot 2x\,dx - x^2 \cdot 2dx}{(2x + 3)^2}$$

$$= \frac{2x^2\,dx + 6x\,dx}{(2x + 3)^2}.$$

Volume of a spherical shell Consider a hollow rubber ball with inner radius r and thickness dr (Fig. 2). The problem is to find a formula for the volume of this spherical shell, in other words, the volume of the rubber in the ball and not the volume of the air it holds. We can get an exact but ugly formula, and then an approximate but simpler one.

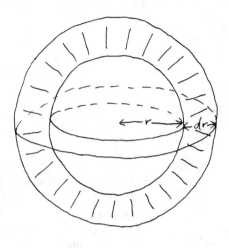

FIG. 2

To find a *precise* formula, think of the volume of the rubber material as the difference between the overall sphere of radius $r + dr$ and the inner sphere of air with radius r. The volume of a sphere of radius r is $V = \frac{4}{3}\pi r^3$, so

$$\text{shell volume} = \text{outer sphere} - \text{inner sphere}$$

(6)
$$= \frac{4}{3}\pi(r + dr)^3 - \frac{4}{3}\pi r^3$$

$$= 4\pi r^2\, dr + 4\pi r (dr)^2 + \frac{4}{3}\pi (dr)^3.$$

To find an *approximate* formula, think of the volume of the rubber material as the change in the volume V of the inner sphere when its radius r is increased by dr. If the change is referred to as dV and we use $dV = V'(r)\,dr$ then we have the (approximate) shell volume formula

(7)
$$dV = 4\pi r^2\, dr.$$

Note that the difference between (6) and (7) is $4\pi r (dr)^2 + \frac{4}{3}\pi (dr)^3$ which is *very* small if dr is small. When the shell formulas of this section are used in Section 6.1, it will be in situations where $dr \to 0$, which justifies the use of (7) as *the* volume formula of the spherical shell.

Area of a circular shell The circular shell (washer) of Fig. 3 has inner radius r and thickness dr. We want a formula for its area, comparable to (7). The inner circle has area $A = \pi r^2$ and the area of the shell is the change in A when r increases by dr. If the change in A is called dA, and we use $dA = A'(r)\,dr$, we have the shell area formula

(8)
$$dA = 2\pi r\, dr.$$

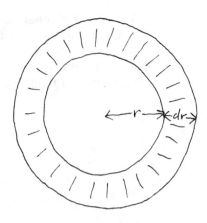

FIG. 3

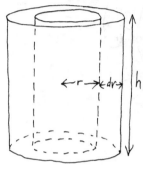

FIG. 4

Volume of a cylindrical shell   Consider a piece of glass tubing with inner radius r, thickness dr, and height h (Fig. 4). We want a nice formula for the volume of the cylindrical shell, that is, the volume of the glass material alone, and not the air inside. The inner cylinder has volume $V = \pi r^2 h$, and the shell volume is the change in V when r changes by dr and h stays fixed. If the change in V is called dV, and we use $dV = V'(r)\,dr$, where h is regarded as a constant in the differentiation process, then we have the shell volume formula

$$(9) \qquad\qquad dV = 2\pi r h\,dr.$$

The notation dy/dx   When dx and dy are used to represent small changes in x and y in the notation of (1'), the symbol dy/dx has two meanings. It can represent the actual fraction

$$(10) \qquad\qquad \frac{\text{small change in } y}{\text{small change in } x}$$

or it can mean the derivative of y with respect to x, that is, $f'(x)$. More precisely, the fraction approaches the derivative as $dx \to 0$. Until now, it has been illegal to consider the derivative symbol dy/dx as a fraction, except as a mnemonic device. Now it is acceptable to think of dy/dx as the fraction in (10). Many practitioners take the convenient liberty of sliding back and forth between the fraction and derivative interpretations of dy/dx (under the baleful glare of the mathematician). We will give an illustration.

Suppose a researcher is interested in the connection between stimulus (what is actually done to a person) and sensation (what the person feels). If salt is put in food, is the salt actually *tasted*? Suppose x is the number of milligrams of salt injected into a doughnut, and T is the salty taste reported by the doughnut eater on a taste scale where 0 indicates no salt taste and higher values indicate a very salty taste. How does x affect T? In particular, if x is increased by a small amount $dx = .1$, does T go up by a correspondingly small amount $dT = .1$? Experimenters have found that the answer is no; a change in x does not necessarily produce a change in T of similar, or even proportional, size; that is, dT is not $k\,dx$. Rather, if the doughnut is not very salty to begin with then a small change in the amount x of salt produces a large change in the perception T. If the doughnut is very salty, then the same small change in x goes virtually unnoticed so that T is practically unchanged. A similar phenomenon occurs in weightlifting. If you are lifting 10 pounds, you will notice an extra half pound, but if you are lifting 1000 pounds, you will barely feel an extra half pound. The experimenter's hypothesis for the connection between dx and dT is $dT = \dfrac{k\,dx}{x}$ where k is a fixed constant depending on the particular stimulus; this hypothesizes that the larger the value of x (that is, the saltier the doughnut), the less the effect of dx on T. The hypothesis may be written as $dT/dx = k/x$, and switching from the fraction interpretation of dT/dx to the derivative interpretation we have $T'(x) = k/x$. Antidifferentiate to get $T = k \ln x + C$. Therefore, one hypothesis proposes a logarithmic connection between stimulus x and sensation T.

Problems for Section 4.8

1. Find the differential.

(a) $d(\sqrt{x}\,)$ (d) $d\left(\dfrac{\sin x}{x}\right)$

(b) $d(\cos x)$ (e) $d(\sin x^5)$

(c) $d(x^5 \sin x)$ (f) $d(5)$

2. Find dy if $y = 2x^3 + 3$.

3. Find df if $f(x) = x + 3$.

4. Use linear approximations to make the following estimates. (a) Estimate the change in $x^3 + x^2$ as x changes from 3 to 2.9999.   (b) Estimate the change in $\sqrt[4]{x}$ when x changes from 16 to 16.1.

5. Use the methods which produced the shell formulas in (7)–(9) to find (a) the area dA of the equilateral triangular shell (Fig. 5) with "radius" r and thickness dr, and   (b) the volume dV of the conical shell (Fig. 6) with height h, radius r and "thickness" dr (that is, the volume of the sugar wafer and not of the ice cream inside).

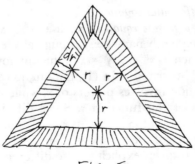

FIG.5

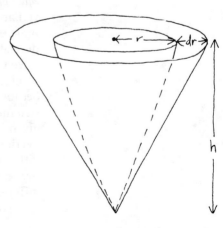

FIG.6

4.9   Separable Differential Equations

Differential equations constitute a vast topic, an entire branch of mathematics, and this section is only a bare introduction. We will use simple calculus to solve one type of differential equation.

To see how differential equations arise, consider a 10-liter punch bowl, initially filled with cider, being drunk at the rate of 2 liters per minute. As the punch is drunk, the bowl is simultaneously refilled, but with whiskey, not cider. Initially, there is no whiskey in the bowl, but gradually the whiskey content increases, until at "time ∞", the bowl is entirely filled with whiskey. The problem is to find a function $w(t)$ to give the number of liters of whiskey in the bowl at time t.

So far, the only known value of w is $w(0) = 0$. But we have information about the rate of change of w, that is, about $w'(t)$, the net liters of whiskey coming into the bowl per minute:

$w'(t)$ = IN − OUT

  = whiskey poured in per minute − whiskey drunk per minute.

The whiskey is poured in at the constant rate of 2 liters/min, so IN = 2, but the OUT rate is harder. The *punch* is drunk at the rate of 2 liters/min, but since the whiskey content of the punch varies from minute to minute, the OUT rate for *whiskey* is not 2 liters/minute; instead it is 2 times the fraction of the bowl which is whiskey at the moment under consideration. That fraction is

$$\frac{\text{liters of whiskey in bowl at time } t}{10},$$

that is, $\frac{1}{10}w(t)$ where $w(t)$ is the unknown function. Therefore $w'(t) = 2 - 2 \cdot \frac{1}{10}w(t)$. So instead of finding $w(t)$ immediately, we have

(1) $$w'(t) = 2 - \frac{1}{5}w(t),$$

called a *differential equation.*

In an *algebraic equation,* such as $x^3 - x^2 = 2x + 3$, the unknown is a *number,* frequently named x, although any letter can be used. In a *differential equation,* such as $y'' + 2xy = xy'$, the unknown is a *function,* usually named $y(x)$ and abbreviated y. In (1), the unknown is the function $w(t)$. An algebraic equation involves powers of x, while a differential equation involves derivatives of the function y. Some differential equations can be easily solved. A solution to $y' = 3x^2$ is $y = x^3$, and the complete solution is the set of all functions of the form $y = x^3 + C$. This is an easy differential equation because y' is given explicitly. The differential equation $y' = 2 - \frac{1}{5}y$ (a restatement of (1) with $w(t)$ replaced by y) is harder. It may look as if y' is given, but since the right side involves y, the equation only reveals a connection between y and y', and the solution is not obtained by antidifferentiating the right-hand side with respect to x. We will develop a procedure for "separating the variables" (if possible) before antidifferentiating, and then return to (1).

To illustrate the method, we will consider the differential equation

(2) $$y' = \frac{x}{y^2}.$$

Rewrite the equation as

(3) $$y^2(x)y'(x) = x,$$

and antidifferentiate on both sides with respect to x to obtain

(4) $$\int y^2(x)y'(x)\,dx = \int x\,dx.$$

To compute the left-hand side, note that the derivative of $\frac{1}{3}y^3$ *with respect to* x is $y^2 y'$, so we have

(5) $$\frac{1}{3}y^3 = \frac{1}{2}x^2 + C.$$

An arbitrary constant is inserted on one side only, as explained below. The procedure in (2)–(5) is usually written in a second notation, which might be considered an abuse of language, but which is easier to use and produces the same result. In this second notation, we have

(2') $$\frac{dy}{dx} = \frac{x}{y^2}$$

(3') $$y^2\,dy = x\,dx \qquad \text{(multiply by } y^2\,dx \text{ on both sides)}$$

(4') $$\int y^2\,dy = \int x\,dx$$

(5') $$\frac{1}{3}y^3 = \frac{1}{2}x^2 + C.$$

In future examples, we'll follow standard procedure and use the second notation.

So far, the function y has been found implicitly in (5'). The *explicit* solution is

(6) $$y = \sqrt[3]{\frac{3}{2}x^2 + 3C} \quad \text{or, equivalently,} \quad y = \sqrt[3]{\frac{3}{2}x^2 + D}.$$

More generally, *if it is possible to separate the variables so that the differential equation has the form*

$$(\text{expression in } x)\,dx = (\text{expression in } y)\,dy,$$

(as in (3') for example), *then the equation is called separable, and is solved by antidifferentiating on both sides.* (Only *first order equations*, that is, equations involving y' but not $y'', y''', \cdots$, may be separated.) The process usually leads to an implicit description of y. If it is feasible to solve for y explicitly, we do so, but otherwise we settle for an implicit version.

The algebra of arbitrary constants The algebraic rules for combining arbitrary constants are quite enjoyable. If A and B are arbitrary constants then so are $A + B$, $3A$, $A - B$, AB, etc., and may be named renamed C_1, C_2, C_3, C_4, etc. In (6), $3C$ became D because $3C$ and D are equally arbitrary. Similarly, in (5'), we did not write $\frac{1}{3}y^3 + K = \frac{1}{2}x^2 + C$, because $C - K$ would combine to one constant anyway.

Warning 1. Don't turn $C + x$ or Cx into D. A constant cannot swallow a variable. The curves of the form $y = Ax^2$ form a family of parabolas, containing $y = 3x^2$, $y = -5x^2$ and so on, but if Ax^2 is incorrectly combined to K, then the family becomes $y = K$, which is a set of horizontal lines.

2. Don't wait until the end of the problem to insert an arbitrary constant. At line (5′), don't write $\frac{1}{3}y^3 = \frac{1}{2}x^2$, $y = \sqrt[3]{\frac{3}{2}x^2}$ and *then* add the neglected constant to get the *wrong* answer $y = \sqrt[3]{\frac{3}{2}x^2} + C$. The constant must be inserted *at the antidifferentiation step, not later.*

Nonseparable example   If $y' = x + y$ so that $dy = (x + y)\,dx$, there is no way to continue and separate the variables. If both sides are divided by $x + y$, then x turns up on the same side as dy. The method of this section simply doesn't apply.

Antiderivatives for $1/x$ The usual rule is $\int (1/x)\,dx = \ln x + C$, but it is also true that

(7) $$\int \frac{1}{x}\,dx = \ln Kx\,,$$

since $\ln Kx = \ln K + \ln x = C + \ln x$. The version in (7) is often more useful. It will also be convenient to ignore absolute value signs and use $\ln x$ and $\ln Kx$ instead of $\ln|x|$ and $\ln|Kx|$. In physical applications of differential equations, it is likely that variables and arbitrary constants will be positive, and even if they are not, it is fortunately the case that omitting the absolute values in intermediate steps usually leads to the same *final* solution as including them. In general, it is often easier to relax our standards in solving a differential equation (such as omitting absolute values in (7)) and, if in doubt, substitute the proposed solution into the equation. If the equation is satisfied then the proposed solution must be correct.

Example 1   We will continue the punch bowl problem by solving (1).

$$\frac{dw}{dt} = 2 - \frac{1}{5}w$$

$$\frac{dw}{2 - \frac{1}{5}w} = dt \qquad \text{(multiply by } dt \text{ and divide by } 2 - \frac{1}{5}w \text{ to separate the variables)}$$

$$-5 \ln K\left(2 - \frac{1}{5}w\right) = t \qquad \text{(antidifferentiate)}$$

$$\ln K\left(2 - \frac{1}{5}w\right) = -\frac{1}{5}t \qquad \text{(divide by } -5)$$

$$K\left(2 - \frac{1}{5}w\right) = e^{-t/5} \qquad \text{(take exp on both sides)}$$

$$2 - \frac{1}{5}w = Ae^{-t/5} \qquad \text{(Let } 1/K \text{ be named } A)$$

(8) $\quad w = 10 - Be^{-t/5} \qquad \text{(Let } 5A \text{ be named } B).$

Equation (8) describes many solutions and is called the *general solution*. In this problem we want the *particular solution* satisfying the condition $w(0) = 0$ (the punch bowl contains no whiskey at time 0). Substitute $t = 0$,

$w = 0$ to get $0 = 10 - Be^0$, $B = 10$. Therefore, the final solution is $w = 10 - 10e^{-t/5}$. Note that, as expected, the steady state solution is $w(\infty) = 10$; after a long time, the punch is essentially all whiskey.

Exponential growth and decay   If you have ever waited for a cup of hot coffee to cool down, you have probably noticed that liquids do not cool at a constant rate. If the net temperature of a particular liquid (that is, degrees above room temperature) is 150° at time $t = 0$, and the liquid is cooling at that instant by 50° per minute, then it does not continue to cool at 50° per minute. Rather, by experimentation and physical law, when its temperature has decreased to 99°, it will be cooling at only 33° per minute; for this particular liquid, the cooling rate is 1/3 of the net temperature. The problem is to find a formula for $y(t)$, the net temperature of the liquid at time t.

Since the cooling rate for this liquid is 1/3 its net temperature, $y' = -\frac{1}{3}y$. The negative sign is designed to make y' *negative* since the liquid's temperature is *decreasing*. Then

$$\frac{dy}{dt} = -\frac{1}{3}y$$

(9)
$$\frac{dy}{y} = -\frac{1}{3}\,dt$$

$$\ln Ky = -\frac{1}{3}t$$

$$Ky = e^{-t/3}$$

$$y = \frac{1}{K}e^{-t/3}$$

(10)
$$y = Ce^{-t/3}.$$

(Instead of line (9) we could just as well have used $\frac{3}{y}\,dy = -dt$, or $-\frac{3}{y}\,dy = dt$, etc. All ultimately lead to $y = Ce^{-t/3}$.)

To determine the particular solution satisfying the initial condition $y = 150$ when $t = 0$, substitute in (10) to get $150 = Ce^0$, $C = 150$. Therefore the final solution is $y = 150e^{-t/3}$. The graph of the solution is an exponential curve with $y(0) = 150$ and $y(\infty) = 0$ (Fig. 1). Theoretically, the liquid never reaches room temperature (that is, zero net temperature), but approaches room temperature as $t \to \infty$. For example, to find how long it

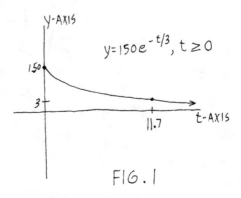

FIG. 1

takes for the liquid to cool from 150° to 3° (net temperature), set $y = 3$ and solve for t to get $\frac{1}{50} = e^{-t/3}$, $-\frac{1}{3}t = \ln\frac{1}{50} = -\ln 50$, $t = 3\ln 50$, or approximately 11.7 minutes.

Net temperature is not the only quantity that changes in such a way that the rate of change is proportional to "how much is there." If a particular cell has a mass of 99 milligrams and is growing at 33 milligrams per minute, then it does not continue to grow at 33 mg/min. Instead, when the cell grows to 150 mg, it will be growing faster, namely, at the rate of 50 mg/min. In general, the rate of growth of a cell is proportional to its mass (until the cell reaches a certain size and the rate of growth satisfies a different law, since cells do not grow arbitrarily large). Radioactive decay is another example; the rate of decay of material is proportional to the amount of material. Similarly, population growth is proportional to the size of the population. In general, the net temperature, population size, cell mass and amount of a radioactive substance at time t all satisfy a differential equation of the form $y' = by$. The value of the constant b (which was $-1/3$ in the liquid cooling example above) depends on the particular liquid, population, cell or substance; it is positive if the quantity is growing and negative if it is decaying. The solution is of the form $y = Ce^{bt}$. This type of growth or decay is called *exponential*.

Orthogonal trajectories   An *orthogonal trajectory* for a family (collection) of curves is a curve which intersects each member of the family at right angles. The equation $x^2 + y^2 = K$, $K \geq 0$, describes a family of circles (for example, $K = 9$ corresponds to the circle with radius 3 and center at the origin). The orthogonal trajectories for the family are lines through the origin (Fig. 2). The lines and circles constitute a pair of orthogonal families. The physical significance of the orthogonal trajectories depends on the

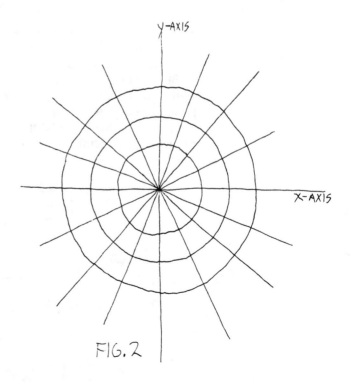

FIG. 2

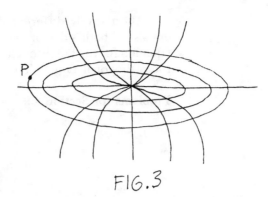

FIG.3

purpose of the original family. If the given curves are isotherms, that is, curves of constant temperature, then the orthogonal trajectories are heat flow lines (Section 11.6).

Consider the family of ellipses

(11) $$x^2 + 4y^2 = K, \qquad K \geq 0 \qquad \text{(Fig. 3)}.$$

The orthogonal trajectories are not geometrically obvious, but they can be found using differential equations.

Step 1 Find a differential equation for the given family. In (11), treat y as a function of x and differentiate implicitly to get $2x + 8yy' = 0$. Therefore the family has the differential equation

(12) $$y' = -\frac{x}{4y}.$$

At every point (x, y) on an ellipse in the family, the slope is $-x/4y$. For example, at point P in Fig. 3, x is negative and large, y is positive and small, $-x/4y$ is a large positive number, and correspondingly the slope on the ellipse at P is a large positive number.

Step 1 goes backwards from the family of curves in (11), usually considered to be the "solution", to the differential equation in (12), usually regarded as the "problem."

Step 2 Find a differential equation for the *orthogonal* family. Perpendicular curves have slopes which are negative reciprocals, so the orthogonal family has the differential equation $y' = 4y/x$. In other words, at every point (x, y) on an orthogonal trajectory, the slope is $4y/x$.

Step 3 Solve the differential equation from Step 2 to obtain the orthogonal family.

$$\frac{dy}{dx} = \frac{4y}{x}$$

$$\frac{dy}{y} = \frac{4}{x} \, dx$$

$$\ln Ky = 4 \ln x = \ln x^4$$

$$Ky = x^4$$

$$y = Ax^4.$$

Thus the orthogonal trajectories are the curves of the form $y = Ax^4$ (Fig. 3).

Alternatively, differential notation may be used. In Step 1, take differentials on both sides of (11) to obtain $2x\,dx + 8y\,dy = 0$, the differential equation for the family of ellipses. In Step 2, switch to $2x\,dy - 8y\,dx = 0$ for the orthogonal family. The solution then continues as before in Step 3.

Problems for Section 4.9

1. Solve

(a) $y' = -x \sec y$ (d) $y' = \dfrac{y}{2x + 3}$

(b) $dx + x^3 y\,dy = 0$ (e) $x^2\,dy = e^y\,dx$

(c) $x^2 + y^4\dfrac{dy}{dx} = 0$ (f) $y' = \dfrac{5x + 3}{y}$

2. Find the particular solution satisfying the given condition.

(a) $y' = xy,\ y(1) = 3$ (c) $y'e^y/x = 3,\ y(0) = 2$
(b) $yy' + 5x = 3,\ y(2) = 4$ (d) $y' = y^4 \cos x,\ y(0) = 2$

3. (a) Solve $xy' = 2y$ and sketch the family of solutions.   (b) Find the particular solution in the family through the point $(2, 3)$.

4. Find the orthogonal trajectories for the given family and sketch both families
(a) $x^2 + 2y^2 = C$ (b) $y = Ce^{-3x}$ (c) $2x^2 - y^2 = K$.

5. Suppose a substance decays at a rate equal to $1/10$ the amount of the substance. (a) Find a general solution for the amount $y(t)$ at time t.   (b) Find $y(t)$ if the initial amount is 75 grams.   (c) Find the half-life of the substance, that is, the length of time it takes for the substance to decay to half its original amount, and verify that the answer is independent of the initial amount.

6. Suppose the rate of growth of a cell is equal to $\frac{1}{2}$ its mass. Find the mass of the cell at time 3 if its initial mass is 2.

7. The velocity $v(t)$ of a falling object with mass m satisfies the differential equation $mv' = mg - cv$, where g and c in addition to m are constants. (The equation is derived from physical principles. The object experiences a downward force mg, due to gravity, and a retarding force cv proportional to its velocity, due to air resistance. Their sum, that is, the total force, is mv' since force equals mass times acceleration.) Find $v(t)$ if the initial velocity is 0, and then find the steady state velocity $v(\infty)$.

REVIEW PROBLEMS FOR CHAPTER 4

1. If P is the pressure of a gas, V its volume and T its temperature, then $PV = kT$ where k is a positive constant depending on the particular gas. Suppose at a fixed instant of time, $T = 20$, $V = 10$, P is decreasing by 2 pressure units per second and T is increasing by 3 temperature units per second. Is V increasing or decreasing at this moment, and by how much?

2. Find $\lim \dfrac{\ln \ln x}{\ln x}$ as (a) $x \to \infty$ (b) $x \to 1+$.

3. Sketch the graph of xe^{-x}.

4. Of all pairs of numbers whose sum is 10, which pair has the maximum product?

5. Find $d(xe^{2x})$.

6. Which of each pair has a higher order of magnitude? (a) $\ln x$, $\ln x^2$ (b) e^x, e^{x^2}.

7. At one instant, the edge of a cube is 3 meters and is growing by 2 m/sec. How fast is the volume growing at this moment?

8. Sketch the graph. (a) $3 \sin 2(x - \pi/3)$ (b) $2 + 5e^{-3x}$.

9. Find (a) $\lim_{x \to 0+} e^x \ln x$ (b) $\lim_{x \to 0+} x^{\tan x}$.

10. Show that of all rectangles with a given diagonal, the square has the largest area.

11. Sketch the graph of $y = \dfrac{2x}{x^2 + 1}$.

12. Find the relative extrema of each function three ways: with the first derivative test, with the second derivative test and with no derivatives at all. (a) $\sin^4 x$ (b) $(x + 2)^2 + 1$.

13. Let y be a function of t. Solve $t^2 y' = y$ with the condition that the steady state solution is $y = 2$, i.e., if $t = \infty$ then $y = 2$.

14. A gardener with 100 feet of wire wants to fence in a rectangular plot and further fence it into four smaller rectangles (not necessarily of equal width), as indicated in Fig. 1. How should it be done so as to maximize the total area.

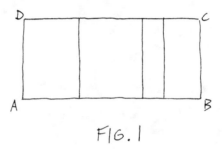

FIG. 1

15. Find the maximum and minimum values of $x \ln x + (1 - x) \ln(1 - x)$.

16. Let $f(x) = x^3 - 2x^2 + 3x - 4$. (a) Show that f is an increasing function. (b) Use part (a) to show that the equation $f(x) = 0$ has exactly one root.   (c) Choose a reasonable initial value of x for Newton's method.   (d) Continue with Newton's method until successive approximations agree in 3 decimal places and check the accuracy of those places.

5/THE INTEGRAL PART I

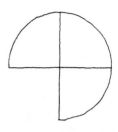

5.1 Preview

This section considers two problems to introduce the idea behind integral calculus.

Averages If your grades are 70%, 80% and 95% then your average grade is $\dfrac{70 + 80 + 95}{3}$ or 81.7%. Carrying this a step further, suppose the 70% was earned in an exam which covered three weeks of work, the 80% exam grade covered four weeks of work, and the 95% covered six weeks of material (Fig. 1). For an appropriate average, each grade is *weighted* by the corresponding number of weeks:

$$\text{weighted average} = \frac{(70)(3) + (80)(4) + (95)(6)}{13} = 84.6\%.$$

Note that we divide by 13, the sum of the weights, that is, the length of the school term, rather than by 3, the number of grades.

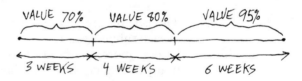

FIG. 1

For the most general situation, let f be a function defined on an interval $[a, b]$. The problem is to compute an average value for f. To simulate the situation in Fig. 1, begin by dividing $[a, b]$ into many subintervals, say 100 of them (Fig. 2). The subintervals do not have to be of the same length, but they should all be small. Let dx_1 denote the length of the first subinterval, let dx_2 be the length of the second subinterval, and so on. Pick a number in

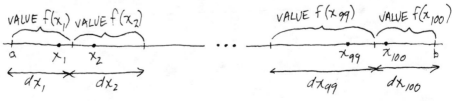

FIG. 2

137

each subinterval; let x_1 be the number chosen from the first subinterval, x_2 the number chosen from the second subinterval, and so on. Pretend that f is constant in each subinterval, and in particular has the value $f(x_1)$ throughout the first subinterval, the value $f(x_2)$ throughout the second subinterval, and so on. With this pretense we may find an average value in Fig. 2 as we did in Fig. 1:

$$\text{average value of } f = \frac{f(x_1)\,dx_1 + f(x_2)\,dx_2 + \cdots + f(x_{100})\,dx_{100}}{dx_1 + dx_2 + \cdots + dx_{100}}$$

(approximately).

The length of each subinterval is used as a weight, and the sum of the weights $dx_1 + \cdots + dx_{100}$ in the denominator is the length $b - a$ of the interval itself.

We use some abbreviations to avoid writing subscripts and long sums. First of all, the sum

$$f(x_1)\,dx_1 + f(x_2)\,dx_2 + \cdots + f(x_{100})\,dx_{100}$$

is abbreviated

$$\sum_{i=1}^{100} f(x_i)\,dx_i\,.$$

The letter Σ is called a summation symbol. If we take the liberty of allowing an unsubscripted dx to stand for the length of a typical subinterval, and an unsubscripted x to stand for the number chosen in that subinterval (Fig. 3), we can further abbreviate the sum by $\Sigma f(x)\,dx$. Thus we write

$$\text{average } f = \frac{\Sigma f(x)\,dx}{b - a} \qquad \text{(approximately)}\,.$$

This isn't the *precise* average value of f because it pretends that f is constant in each subinterval. If the subintervals are very small, which forces them to become more numerous, then (a continuous) f doesn't have much opportunity to change within a subinterval, and the pretense is not far from the truth. Therefore to get closer to the precise average, use 100 small subintervals, then repeat with 200 even smaller subintervals, and continue in this fashion. In general,

$$(1) \qquad\qquad \text{average value of } f = \lim_{dx \to 0} \frac{\Sigma f(x)\,dx}{b - a}\,.$$

We don't intend to find any averages yet because computing $\Sigma f(x)\,dx$ is too tedious to do directly. Much of this chapter is designed to bypass direct computation and obtain numerical answers easily.

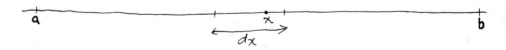

FIG. 3

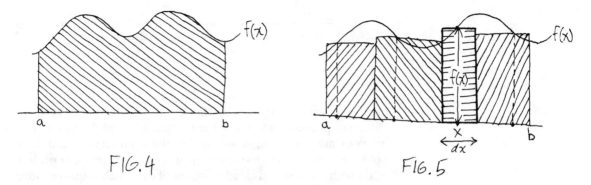

FIG.4

FIG.5

Area under a curve   Areas of rectangles are familiar, but consider the region under the graph of the function f between $x = a$ and $x = b$ (Fig. 4). The problem is to find its area. Begin by dividing the interval $[a, b]$ into many small pieces. Let dx be the length of a typical subinterval, and let x be a number in this subinterval. Build a thin rectangle with a base dx and height $f(x)$. (Figure 5 shows $[a, b]$ divided into four subintervals with four corresponding rectangles.) The area of the typical rectangle is $f(x) dx$. The entire region can be filled with such rectangles, and therefore the area under the graph is approximately the sum of rectangular areas, or $\sum f(x) dx$. The area is not necessarily $\sum f(x) dx$ *precisely* because the rectangles underlap and overlap the original region. However, there will be less underlap and overlap if the values of dx are small, so it appears sensible to claim that

(2) $$\text{area under the graph of } f = \lim_{dx \to 0} \sum f(x)\, dx.$$

Although averages and areas seem to be very different concepts, the new idea of $\lim_{dx \to 0} \sum f(x) dx$ appears in both (1) and (2). Beginning in the next section we will give the limit an official name, find ways to compute it, and present many more applications.

5.2   Definition and Some Applications of the Integral

Definition of the integral   Let f be a function defined on the interval $[a, b]$. Begin by dividing the interval into (say) 100 subintervals of lengths $dx_1, dx_2, \cdots, dx_{100}$, and choosing numbers $x_1, x_2, \cdots, x_{100}$ in the subintervals (Fig. 1). Find

$$\sum_{i=1}^{100} f(x_i)\, dx_i = f(x_1)\, dx_1 + f(x_2)\, dx_2 + \cdots + f(x_{100})\, dx_{100},$$

which we abbreviate by $\sum f(x) dx$. Figure 2 shows the correspondingly abbreviated picture. The sum is a weighted sum of 100 "representative" values

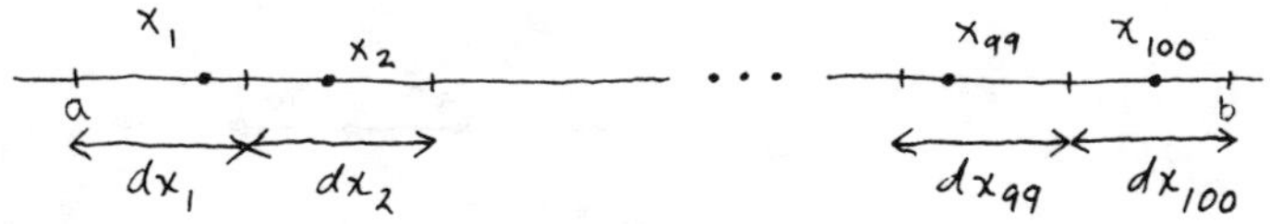

FIG. 1

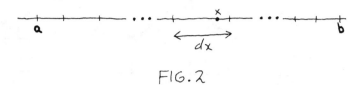

FIG. 2

of f, each value weighted by the length of the subinterval it represents. Different people performing the computation might choose different subintervals and different values within the subintervals, and their sums will not necessarily agree. However, suppose the process is repeated again and again with smaller and smaller values of dx, which requires more and more subintervals. It is likely that the resulting sums will be close to one particular number eventually, that is, the sums will approach a limit. The limit is called the *integral* of f on $[a,b]$ and is denoted by $\int_a^b f(x)\,dx$.

That is, the integral is defined by

(1)
$$\int_a^b f(x)\,dx = \lim_{dx \to 0} \sum f(x)\,dx.$$

For a simplistic but useful viewpoint, we can ignore the limit and consider $\int_a^b f(x)\,dx$ as merely $\sum f(x)\,dx$, found using many subintervals of $[a,b]$. In other words, think of the integral as adding many representative values of f, each value weighted by the length of the subinterval it represents.

The process of computing an integral is called *integration*. The integral symbol $\int$ is an elongated S for "sum" (the same symbol was used in a different context for antidifferentiation) and the symbols a and b attached to it indicate the interval of integration. The numbers a and b are called the *limits of integration*, and f is called the *integrand*. The sums of the form $\sum f(x)\,dx$ are called *Riemann sums*.

Example 1 To illustrate the definition we will try to find $\int_1^2 \frac{1}{x^2}\,dx$. The computer program in (2) finds some Riemann sums using n subintervals, for $n = 100, 300, 500, 700, 900$ and 1100. For convenience in writing the program we chose subintervals of equal length, and numbers $x_1, \cdots, x_n$ at the left ends of the subintervals. For example, in its third run, with $n = 500$, the computer divides $[1,2]$ into 500 subintervals of length

$$dx = \frac{b-a}{n} = \frac{2-1}{500} = .002 \qquad \text{(Fig. 3)}$$

and chooses $x_1 = 1, x_2 = 1.002, x_3 = 1.004, \cdots, x_{500} = 1.998$. Then the computer evaluates the Riemann sum

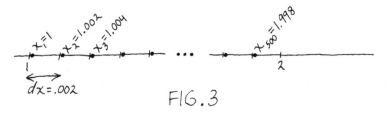

FIG. 3

$$\sum \frac{1}{x^2}\, dx = \frac{1}{(1)^2}(.002) + \frac{1}{(1.002)^2}(.002)$$

$$+ \frac{1}{(1.004)^2}(.002) + \cdots + \frac{1}{(1.998)^2}(.002)$$

to get .500751.

```
10  DEF FNF (X)= 1/(X*X)
20  A=1
30  B=2
35  PRINT "N", "RIEMANN SUM"
40  FOR N = 100 TO 1200 STEP 200
50  D = (B−A)/N
60  L= FNF(A)
70  FOR I = 1 TO N−1
80  L = L + FNF(A + I*D)
90  NEXT I
100 L = L*D
130 PRINT N,L
140 NEXT N
150 END
READY.
RNH
```

(2)

N	RIEMANN SUM
100	.503765
300	.501252
500	.500751
700	.500536
900	.500417
1100	.500341

This printout suggests that the Riemann sums approach a limit. It can be shown that for still larger values of n and smaller values of dx, the Riemann sums continue to approach a limit, even if the subintervals are not of the same length, and no matter how $x_1, \cdots, x_n$ are chosen in the subintervals. Although the computer program alone is not sufficient to determine the limit (that is, the integral), it suggests that $\int_1^2 \frac{1}{x^2}\, dx$ might be .5. In Section 5.3 we will bypass this attempt at direct computation and find the integral easily.

Integrals and average values   As one of the applications of the integral, (1) of the preceding section showed that

(3)
$$\text{average value of } f \text{ in } [a, b] = \frac{\displaystyle\int_a^b f(x)\, dx}{b - a}.$$

Think of the numerator as a weighted sum of "grades" and the denominator as the sum of the weights.

Integrals and area   The preceding section indicated a relation between the area under the graph of a function f and $\int_a^b f(x)\,dx$. We'll examine this more carefully now. It will seem as if there are several different connections between integrals and areas, but they will be summarized into one general conclusion in (8).

Case 1   The graph of f lies above the x-axis.

Figure 4 shows the area under the graph, and a typical rectangle with area $f(x)\,dx$. The integral adds the terms $f(x)\,dx$ and takes a limit as dx approaches 0, so $\int_a^b f(x)\,dx$ adds an increasing number of thinning rectangles. The limit process is considered to alleviate the underlap and overlap and, therefore,

(4)   area between the graph of f and the interval $[a, b]$ on the x-axis

$$= \int_a^b f(x)\,dx.$$

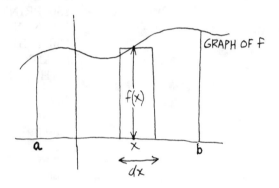

FIG.4

Case 2   The graph of f lies below the x-axis.

Figure 5 shows the region between the x-axis and the graph of f. The area is positive (*all areas are positive*), but the terms $f(x)\,dx$ are negative because $f(x)$ is negative. Hence the area of the indicated rectangle is $-f(x)\,dx$, not $f(x)\,dx$. The integral adds the terms $f(x)\,dx$ so the integral is a negative number, and

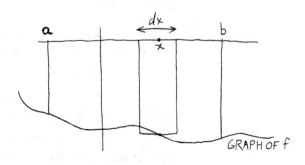

FIG.5

(5)   area between the graph of f and the interval $[a, b]$ on the x-axis

$$= -\int_a^b f(x)\, dx$$

or, equivalently,

(6) $$\int_a^b f(x)\, dx$$

$= -($area between the graph of f and the interval $[a, b]$ on the x-axis$)$.

Case 3   The graph of f crosses the x-axis.

Figure 6 shows the area between the graph and the x-axis, while Fig. 7 shows six subintervals of $[a, b]$ with corresponding rectangles. Then

$$\sum f(x)\, dx = f(x_1)\, dx_1 + f(x_2)\, dx_2 + f(x_3)\, dx_3 + f(x_4)\, dx_4$$

$$+ f(x_5)\, dx_5 + f(x_6)\, dx_6$$

$$= A_1 + A_2 - A_3 - A_4 + A_5 + A_6$$

(because $f(x_3)$ and $f(x_4)$ are negative)

$$= \text{I} - \text{II} + \text{III} \quad \text{(approximately)}.$$

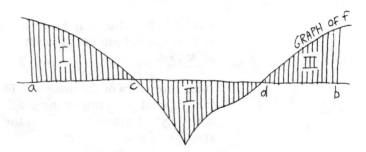

FIG. 6

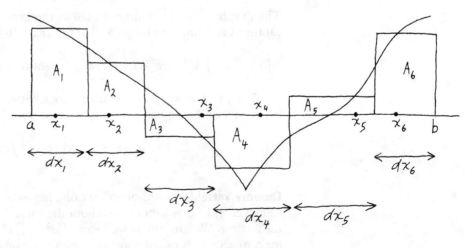

FIG. 7

On passing to the limit, we have

(7) $$\int_a^b f(x)\,dx = \text{area I} - \text{area II} + \text{area III} \qquad (\text{exactly}).$$

In all cases, remember that areas are positive but integrals can be negative if more area is captured below the x-axis than above the x-axis. The single rule covering all cases is

(8) $$\boxed{\int_a^b f(x)\,dx = \text{area above the } x\text{-axis} - \text{area below the } x\text{-axis}.}$$

Example 2   Suppose the problem is to compute the area of the shaded region in Fig. 6. The answer is *not* $\int_a^b f(x)\,dx$ since the integral is I − II + III and we want I + II + III. Instead, find the points c and d where the graph of f crosses the x-axis. Then

$$\text{I} + \text{II} + \text{III} = \int_a^c f(x)\,dx - \int_c^d f(x)\,dx + \int_d^b f(x)\,dx.$$

Warning   Area II in Fig. 6 is *not* negative (areas are never negative). It is the integral $\int_c^d f(x)\,dx$ that is negative, not the area.

Example 3   The graph of $\sin x$ on the interval $[0, 2\pi]$ (Fig. 8 of Section 1.3) determines as much area above the x-axis as below, so by (8), $\int_0^{2\pi} \sin x\,dx = 0$.

Some properties of the integral   The graph of $f + g$ is found by building the graph of g on top of the graph of f (Section 1.7), so the area determined by the graph of $f + g$ is the sum of the areas determined by the graphs of f and g. Therefore

(9) $$\int_a^b [f(x) + g(x)]\,dx = \int_a^b f(x)\,dx + \int_a^b g(x)\,dx.$$

The graph of $6f(x)$ is 6 times as tall as the graph of f. Therefore the area captured is 6 times as large, and $\int_a^b 6f(x)\,dx = 6\int_a^b f(x)\,dx$. In general,

(10) $$\int_a^b kf(x)\,dx = k\int_a^b f(x)\,dx \qquad \text{where } k \text{ is a constant}.$$

Finally, if $a < b < c$, then the area between a and b plus the area between b and c equals the area between a and c, so

(11) $$\int_a^b f(x)\,dx + \int_b^c f(x)\,dx = \int_a^c f(x)\,dx.$$

Dummy variables   Although we don't have the techniques to compute its value yet, $\int_0^2 x^3\,dx$ is a *number,* without the variable x appearing anywhere in the answer. We can just as well write $\int_0^2 t^3\,dt$, $\int_0^2 z^3\,dz$ or $\int_0^2 a^3\,da$. The letter x (or t, or z or a) is called a *dummy variable* because it is entirely arbitrary. If $\int_0^2 x^3\,dx$ were 4, then $\int_0^2 b^3\,db$ would also be 4. In general, $\int_a^b f(x)\,dx = \int_a^b f(t)\,dt = \int_a^b f(u)\,du$, and so on. (Equivalently, the horizontal axis may be named an x-axis or a t-axis or a u-axis.)

Mathematical models How do we know that $\int_a^b f(x)\,dx$ computes the area in Fig. 4 *exactly*? We don't! There is a philosophical point involved here. Most non-mathematicians agree that area is a measure of how spacious a region is, but do not give a precise definition of area. They believe that the integral can be used to compute area because they visualize adding many rectangular areas, with the limit process wiping out overlap and underlap. Most mathematicians on the other hand *define* the area in Fig. 4 to be $\int_a^b f(x)\,dx$. In a sense, this just begs the question because it is still up to the non-mathematician to decide whether the definition really captures physical spaciousness.

In general, mathematics is used to make *models*. The integral $\int_a^b f(x)\,dx$ is the mathematical model for the area in Fig. 4, just as $|f'(x)|$ is the model for the speed of a car traveling to position $f(x)$ at time x. It can never be *proved* that the mathematical model completely mirrors the physical idea, and neither can the connection be *defined* into existence. It is ultimately the responsibility of those who work with physical concepts to decide whether they approve of the mathematical models offered them. The models in this text (for area, volume, slope, speed, average value, tangent line and so on) have endured for centuries. Their "exactness" cannot be proved. The best we can do is demonstrate their reasonableness and cite their wide acceptance.

Problems for Section 5.2

1. Use areas to compute the integral.

(a) $\displaystyle\int_{-1}^{4} 6\,dx$ (b) $\displaystyle\int_{-1}^{3} x\,dx$ (c) $\displaystyle\int_{-2}^{2} x^3\,dx$

2. Use integrals to express the area between the graph of $y = \ln x$ and the x-axis for

(a) $1 \le x \le 5$ (b) $\dfrac{1}{2} \le x \le 1$ (c) $\dfrac{1}{3} \le x \le 7$

3. Decide which is the larger of each pair of integrals.

(a) $\displaystyle\int_{0}^{3} x^2\,dx,\ \int_{-1}^{3} x^2\,dx$ (b) $\displaystyle\int_{0}^{3} x^3\,dx,\ \int_{-1}^{3} x^3\,dx$ (c) $\displaystyle\int_{-1}^{0} x^3\,dx,\ \int_{-2}^{0} x^3\,dx$

4. Decide if the integral is positive, negative or zero.

(a) $\displaystyle\int_{0}^{3\pi/2} \cos x\,dx$ (b) $\displaystyle\int_{0}^{2\pi} \cos^2 x\,dx$

5. True or false?

(a) If $f(x) < 0$ for all x in $[a, b]$ then $\int_a^b f(x)\,dx < 0$.
(b) If $\int_a^b f(x)\,dx < 0$ then $f(x) < 0$ for all x in $[a, b]$.
(c) If $f(x) \le g(x)$ for x in $[a, b]$ then $\int_a^b f(x)\,dx \le \int_a^b g(x)\,dx$.

6. (a) Use area to show that $\int_0^{2\pi} \sin^2 x\,dx = \int_0^{2\pi} \cos^2 x\,dx$. (b) Use part (a) and the identity $\sin^2 x + \cos^2 x = 1$ to show that $\int_0^{2\pi} \sin^2 x\,dx = \pi$.

7. Let $A_1 = \int_a^b f(x)\,dx$.

(a) Consider area and translation to decide which of the following is equal to A_1: $A_2 = \int_{a+3}^{b+3} f(x)\,dx$, $A_3 = \int_{a+3}^{b+3} f(x + 3)\,dx$, $A_4 = \int_{a+3}^{b+3} f(x - 3)\,dx$.
(b) Let $A_5 = \int_{a/2}^{b/2} f(2x)\,dx$. Use area and expansion/contraction to find the connection between A_1 and A_5.

8. If $\int_a^b 4x^3\, dx = 10$, find (a) $\int_a^b 4t^3\, dt$ (b) $\int_a^b x^3\, dx$.

9. Express with an integral the area of a circle of radious R (begin with a semicircle and then double).

5.3   The Fundamental Theorem of Calculus

So far, we have no general method for evaluating an arbitrary integral $\int_a^b f(x)\, dx$. The Fundamental Theorem will provide a nice way to compute the integral, provided that f can be antidifferentiated. The theorem says that to find $\int_a^b f(x)\, dx$, first find an antiderivative F of f. Then evaluate F at $x = b$ and at $x = a$, and subtract $F(a)$ from $F(b)$. The result is the value of the integral. We will first state the theorem formally, do some examples, and then discuss informally why the method works.

Fundamental Theorem *If f is continuous on $[a, b]$ and F is an antiderivative of f then*

(1)
$$\boxed{\int_a^b f(x)\, dx = F(b) - F(a).}$$

For example, the function $\ln x$ is an antiderivative of $1/x$, so

$$\int_1^2 \frac{1}{x}\, dx = \ln 2 - \ln 1 = \ln 2 - 0 = \ln 2\,.$$

The computation $F(b) - F(a)$ is often denoted by $F(x)\big|_a^b$; the symbol $\big|_a^b$ declares the intention of substituting b and a, and subtracting.

Example 1   We expect $\int_0^3 x\, dx$ to be the area in Fig. 1, namely $\frac{1}{2} \cdot 3 \cdot 3 = 9/2$; indeed,

$$\int_0^3 x\, dx = \frac{1}{2} x^2 \Big|_0^3 = \frac{9}{2} - 0 = \frac{9}{2}\,.$$

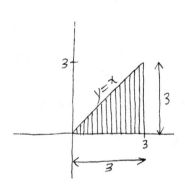

FIG. 1

Using a different antiderivative   Suppose we use $\frac{1}{2} x^2 + 7$ as an antiderivative of x in Example 1, instead of $\frac{1}{2} x^2$. Then we find that

$$\int_0^3 x\, dx = \left(\frac{1}{2} x^2 + 7 \right) \Big|_0^3 = \left(\frac{9}{2} + 7 \right) - (0 + 7) = \frac{9}{2}\,.$$

Notice that the 7 eventually canceled out. Any antiderivative of x is acceptable, and all produce the same final value for the integral. Thus, we might as well use the simplest possible antiderivative, $\frac{1}{2} x^2$.

Example 2   Example 1 of the preceding section used Riemann sums for $\int_1^2 1/x^2\, dx$ to estimate that the integral is near .5. An antiderivative of $1/x^2$ is $-1/x$, so by the Fundamental Theorem,

$$\int_1^2 \frac{1}{x^2}\, dx = -\frac{1}{x} \Big|_1^2 = -\frac{1}{2} - (-1) = \frac{1}{2}\,.$$

Example 3 Find $\int_{-3}^{-2} \dfrac{1}{x}\, dx$.

Solution: $\int_{-3}^{-2} \dfrac{1}{x}\, dx = \ln|x| \big|_{-3}^{-2} = \ln 2 - \ln 3 = \ln \frac{2}{3}$. Note that while $\ln x$ is an antiderivative for $1/x$ if $x > 0$, it is useless in a situation in which $x < 0$. To integrate $1/x$ on $[-3, -2]$, use the antiderivative $\ln|x|$.

The integral of a constant function Consider $\int_a^b 6\, dx$. The integral computes the area of a rectangle with base $b - a$ and height 6 (Fig. 2) so the integral is $6(b - a)$. As another approach, $\int_a^b 6\, dx = 6x \big|_a^b = 6b - 6a = 6(b - a)$. In general, if k is a constant then

(2)
$$\int_a^b k\, dx = k(b - a).$$

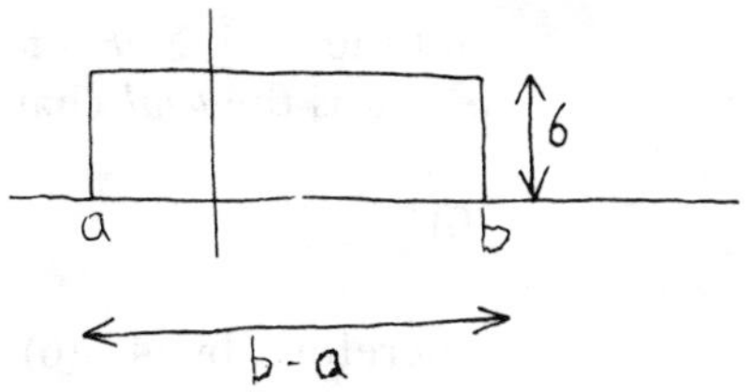

FIG. 2

The integral of the zero function If $f(x) = 0$, then the area between the graph of f and the x-axis is 0, since the graph of f *is* the x-axis. Thus

(3)
$$\int_a^b 0\, dx = 0.$$

As another approach, every Riemann sum $\int f(x)\, dx$ is 0 because each value of f is 0, so the integral must be 0. As still another approach, any constant function C is an antiderivative of the zero function, so $\int_a^b 0\, dx = C \big|_a^b$. Since the constant function C remains C no matter what value, a or b, is substituted for the absent x, the integral is $C - C$, or 0.

Informal proof of the Fundamental Theorem Since F is an antiderivative of f, we may rewrite (1) as

(1′)
$$\int_a^b F'(x)\, dx = F(b) - F(a).$$

We wish to show why (1′) holds. To evaluate the integral, divide $[a, b]$ into many subintervals. Figure 3 shows a typical subinterval with length dx, containing point x, where we assume $dx \to 0$. Then, by definition,

(4)
$$\int_a^b F'(x)\, dx = \sum F'(x)\, dx.$$

From (1′) of Section 4.8, each $F'(x)\, dx$ is the change dF in the function F as

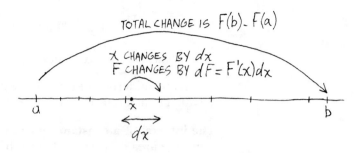

FIG.3

x changes by dx. Therefore

(5) $$\sum F'(x)\,dx = \sum dF.$$

But the sum, $\sum dF$, of all the changes in F as x changes little by little from a to b is the *total* change $F(b) - F(a)$ (Fig. 3 again); that is,

(6) $$\sum dF = F(b) - F(a).$$

Therefore, by (4)–(6), $\int_a^b F'(x)\,dx = F(b) - F(a)$, as desired.

Example 4   Find the average value of x^3 on the interval $[0, 3]$.
   Solution:

$$\text{average } x^3 = \dfrac{\displaystyle\int_0^3 x^3\,dx}{3 - 0} = \dfrac{\left.\dfrac{x^4}{4}\right|_0^3}{3} = \dfrac{27}{4}.$$

Example 5   Find the area indicated in Fig. 4.

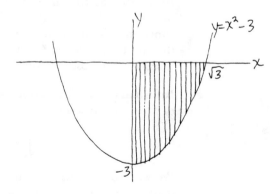

FIG.4

   First solution: The curve crosses the x-axis at $\sqrt{3}$. The region is below the x-axis, so

$$\text{area} = -\int_0^{\sqrt{3}} (x^2 - 3)\,dx = -\left(\frac{x^3}{3} - 3x\right)\Bigg|_0^{\sqrt{3}} = -(\sqrt{3} - 3\sqrt{3}) = 2\sqrt{3}.$$

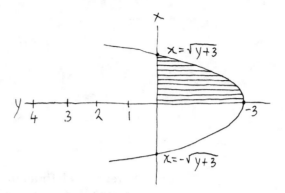

FIG. 5

Second solution: Turn Fig. 4 sideways to get Fig. 5, and consider the vertical axis to be the *x*-axis and the horizontal axis to be the *y*-axis. From this point of view, the region is above the horizontal axis, between $y = -3$ and $y = 0$, and under the graph of the function $x = \sqrt{y + 3}$. (The lower, irrelevant, portion of the parabola is $x = -\sqrt{y + 3}$.) Thus

$$\text{area} = \int_{-3}^{0} \sqrt{y + 3}\, dy = \frac{2}{3}(y + 3)^{3/2}\Big|_{-3}^{0} = \frac{2}{3}(3)^{3/2} - 0 = \frac{2}{3}3\sqrt{3} = 2\sqrt{3}.$$

The interval of integration is still named $[-3, 0]$ even though the *y*-axis is drawn so that *y* increases from right to left, and we use $\int_{-3}^{0}$ as usual, not $\int_{0}^{-3}$. (In fact if you view Fig. 5 from *behind* the page, the horizontal axis is still the *y*-axis, but now *y* increases in the usual manner from left to right.)

The integral of a function with several formulas   Suppose

$$f(x) = \begin{cases} x^2 & \text{if } x \leq 3 \\ 2x + 3 & \text{if } 3 < x < 7 \\ 17 - x & \text{if } x \geq 7. \end{cases}$$

To find say $\int_{0}^{10} f(x)\, dx$, use (11) of the preceding section:

$$\int_{0}^{10} f(x)\, dx = \int_{0}^{3} x^2\, dx + \int_{3}^{7} (2x + 3)\, dx + \int_{7}^{10} (17 - x)\, dx$$

$$= \frac{x^3}{3}\Big|_{0}^{3} + (x^2 + 3x)\Big|_{3}^{7} + \left(17x - \frac{x^2}{2}\right)\Big|_{7}^{10}$$

$$= 9 + 52 + 25.5$$

$$= 86.5 .$$

Example 6   Find $\int_{-2}^{3} e^{|x|}\, dx$.
   Solution: Since

$$e^{|x|} = \begin{cases} e^{x} & \text{for } x \geq 0 \\ e^{-x} & \text{for } x < 0, \end{cases}$$

we have

$$\int_{-2}^{3} e^{|x|}\, dx = \int_{-2}^{0} e^{-x}\, dx + \int_{0}^{3} e^{x}\, dx$$

$$= -e^{-x}\Big|_{-2}^{0} + e^{x}\Big|_{0}^{3}$$

$$= -1 + e^{2} + e^{3} - 1$$

$$= -2 + e^{2} + e^{3}.$$

Definite versus indefinite integrals   So far the symbol $\int$ has been used in two ways. First, $\int_{a}^{b} f(x)\, dx$ is an integral, defined as the limit of the Riemann sums $\Sigma f(x)\, dx$. In this context, dx stands for the length of a typical subinterval of $[a, b]$. Second, $\int f(x)\, dx$ is the collection of all antiderivatives of $f(x)$. In this context, the symbol dx is an instruction to antidifferentiate with respect to the variable x. The symbol $\int$ is used in $\int_{a}^{b} f(x)\, dx$ because it signifies summation. The same symbol is used for antidifferentiation because one of the methods of computing an integral (using the Fundamental Theorem) begins with antidifferentiation.

Frequently, both $\int_{a}^{b} f(x)\, dx$ and $\int f(x)\, dx$ are referred to as integrals; in particular, $\int_{a}^{b} f(x)\, dx$ is called a *definite integral* and $\int f(x)\, dx$ an *indefinite integral*. We will usually continue to call the former an integral and the latter an antiderivative. No matter which terminology you encounter, it will always be true, for example, that $\int 3x^{2}\, dx = x^{3} + C$ while $\int_{2}^{3} 3x^{2}\, dx = 19$.

Problems for Section 5.3

In Problems 1–21, evaluate the integral.

1. $\displaystyle\int_{-1}^{2} (6x^{2} - 3x + 2)\, dx$

2. $\displaystyle\int_{1}^{3} (3 - t)\, dt$

3. $\displaystyle\int_{0}^{2} (3x^{5} - 2x^{2})\, dx$

4. $\displaystyle\int_{\pi/3}^{\pi/2} \sin 2x\, dx$

5. $\displaystyle\int_{0}^{1} \frac{1}{1 + x^{2}}\, dx$

6. $\displaystyle\int_{0}^{1/2} \sin \pi x\, dx$

7. $\displaystyle\int_{1}^{5} \frac{1}{x}\, dx$

8. $\displaystyle\int_{2}^{3} \frac{1}{6x^{3}}\, dx$

9. $\displaystyle\int_{1}^{5} 3\sqrt{x}\, dx$

10. $\displaystyle\int_{1}^{9} \sqrt{10 - x}\, dx$

11. $\displaystyle\int_{3}^{4} \frac{1}{2x + 1}\, dx$

12. $\displaystyle\int_{-2}^{5} 4\, dx$

13. $\displaystyle\int_{0}^{\pi/4} \sec^{2}x\, dx$

14. $\displaystyle\int_{2}^{5} dx$

15. $\displaystyle\int_{-1}^{2} (x^{3} + 2)^{2}\, dx$

16. $\displaystyle\int_{2}^{4} \frac{\left(\frac{1}{2}x + 7\right)^{3}}{4}\, dx$

17. $\displaystyle\int_{-1}^{1} \left(\frac{x + 3}{5}\right)^{7} dx$

18. $\displaystyle\int_{-5}^{-4} \frac{1}{3x}\, dx$

19. $\displaystyle\int_{-3}^{0} 5 \cos \frac{1}{2} \pi x\, dx$

20. $\displaystyle\int_{2}^{3} \frac{1}{(2x - 9)^{3}}\, dx$

21. $\displaystyle\int_{0}^{1} e^{3x}\, dx$

22. Find the area of the triangle with vertices $A = (0,0)$, $B = (4,2)$, $C = (6,0)$ (a) using a geometric formula and (b) using an integral.

23. Find the average value of $\sin x$ on $[0, \pi]$.

24. Find (a) $\int x^3\, dx$ and (b) $\int_1^2 x^3\, dx$.

25. Find $\int_2^6 f(x)\, dx$ where $f(x) = \begin{cases} 5 & \text{if } 2 \le x \le 3 \\ 0 & \text{if } 3 < x \le 4 \\ x^3 & \text{if } x > 4 \end{cases}$.

26. Find $\int_3^{10} |4 - x|\, dx$.

27. Find the areas indicated in (a) Fig. 6 (b) Fig. 7.

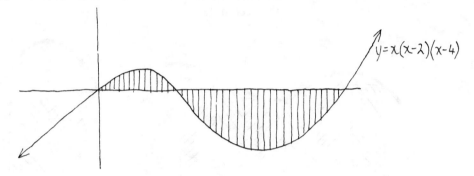

FIG. 6

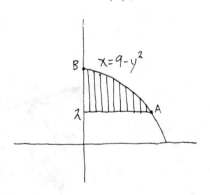

FIG. 7

5.4 Numerical Integration

The evaluation of $\int_a^b f(x)\, dx$ using $F(b) - F(a)$ seems very simple, but it is often *very difficult,* and sometimes *impossible,* to find an (elementary) antiderivative F. In such a case, it may be possible to approximate the integral, a procedure called *numerical integration.* A variety of numerical integration routines exist, each involving much arithmetic, preferably to be done on a calculator or a computer. In fact, some calculators have a button labeled "numerical integration." In order to program the calculator in the first place, a background in numerical analysis is required. This section is a brief introduction.

One way to estimate $\int_a^b f(x)\,dx$ is to use a specific Riemann sum $\Sigma f(x)\,dx$, instead of the *limit* of the Riemann sums. In other words, we can estimate the area under a curve using a sum of areas of rectangles such as those in Fig. 1. (This was actually done by the computer program in (2) of Section 5.2.) The error in the approximation arises from the underlap and overlap created when a horizontal line is used as a substitute "top" instead of the graph of f itself. Frequently, a large number of very thin rectangles is required to force the error down to a reasonable size. There are other numerical methods which require fewer subintervals and are said to *converge more rapidly*. Figure 2 shows chords serving as tops, creating trapezoids. The sum of the areas of the trapezoids is an approximation to $\int_a^b f(x)\,dx$; it is expected to converge faster than a sum of rectangles because the trapezoids seem to fit with less underlap and overlap than the rectangles.

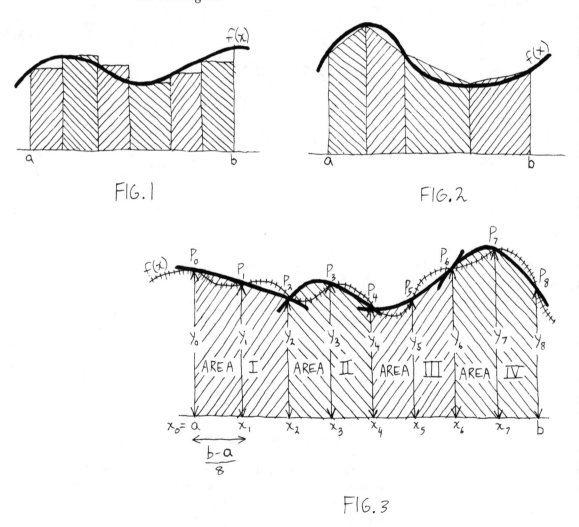

FIG. 1

FIG. 2

FIG. 3

There is yet another top that usually fits even better than a chord. Figure 3 shows 8 subdivisions of $[a, b]$, of the same width. The parabola determined by the points P_0, P_1, P_2 on the graph of f can serve as a top for the first two subintervals, creating area I. Similarly, we use the parabola determined by P_2, P_3, P_4 on the graph of f as a top for the next two sub-intervals, forming area II, and so on. The sum of the areas I, II, III and IV approximates the area under the graph of $f(x)$, and thus is an approximation to $\int_a^b f(x)\,dx$. The approximation using parabolas is viewed by many as the best numerical method within the context of elementary calculus, so we will continue with Fig. 3 and develop the formula for the sum of the areas I, II, III and IV.

As a first step we will derive the formula

$$(1) \qquad\qquad \text{area} = \frac{1}{3}h(Y_0 + Y_2 + 4Y_1)$$

for the area of the parabola-topped region with the three "heights" Y_0, Y_1, Y_2, and two "bases" of length h shown in Fig. 4. The second step will apply the formula to the regions I, II, III and IV in Fig. 3. To derive (1), insert axes in Fig. 4 in a convenient manner; one possibility is shown in Fig. 5. The parabola has an equation of the form $y = Ax^2 + Bx + C$, so the area in Fig. 5 is

$$\int_{-h}^{h} (Ax^2 + Bx + C)\,dx = \frac{1}{3}Ax^3 + \frac{1}{2}Bx^2 + Cx \Big|_{-h}^{h}$$

$$= \frac{2}{3}Ah^3 + 2Ch$$

$$(2) \qquad\qquad\qquad\qquad = \frac{1}{3}h(2Ah^2 + 6C).$$

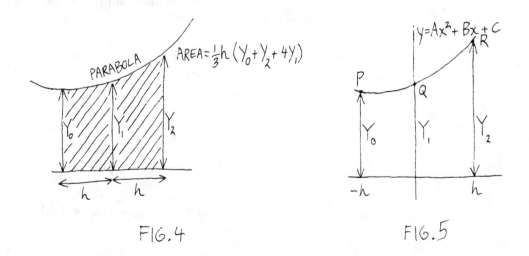

FIG. 4   FIG. 5

The points $P = (-h, Y_0)$, $Q = (0, Y_1)$, $R = (h, Y_2)$ lie on the parabola, and substituting these coordinates into the equation of the parabola gives

$$(3) \qquad Ah^2 - Bh + C = Y_0, \qquad C = Y_1, \qquad Ah^2 + Bh + C = Y_2.$$

From (3), $Y_0 + Y_2 = 2Ah^2 + 2C$ and $Y_1 = C$, so the factor $2Ah^2 + 6C$ in (2) is $Y_0 + Y_2 + 4Y_1$, and (1) follows.

Now apply (1) to I, II, III and IV in Fig. 3. Since the interval $[a, b]$ is divided into 8 equal subdivisions, $h = (b - a)/8$ and

$$I + II + III + IV = \frac{1}{3}h(y_0 + y_2 + 4y_1) + \frac{1}{3}h(y_2 + y_4 + 4y_3)$$

$$+ \frac{1}{3}h(y_4 + y_6 + 4y_5) + \frac{1}{3}h(y_6 + y_8 + 4y_7)$$

$$= \frac{1}{3}h(y_0 + 4y_1 + 2y_2 + 4y_3 + 2y_4 + 4y_5$$

$$+ 2y_6 + 4y_7 + y_8).$$

More generally, using n subintervals *where n is even,*

(4)
$$\int_a^b f(x)\,dx \approx \frac{1}{3}h(y_0 + 4y_1 + 2y_2 + 4y_3 + 2y_4$$

$$+ \cdots + 2y_{n-2} + 4y_{n-1} + y_n)$$

where

$$h = \frac{b - a}{n},$$

(5)
$$y_0 = f(x_0) = f(a)$$
$$y_1 = f(x_1) = f(a + h)$$
$$y_2 = f(x_2) = f(a + 2h)$$
$$y_3 = f(x_3) = f(a + 3h)$$

and so on. The approximation in (4) is known as *Simpson's rule.*

As an example, we will use Simpson's rule with 6 subintervals to approximate $\int_0^1 e^{x^2}\,dx$. We have

$$f(x) = e^{x^2}, \quad a = 0, \quad b = 1, \quad h = \frac{b - a}{n} = \frac{1}{6}.$$

Then,

$$x_0 = 0 \qquad y_0 = f(x_0) = 1$$

$$x_1 = \frac{1}{6} \qquad y_1 = f(x_1) = 1.0281672$$

$$x_2 = \frac{2}{6} \qquad y_2 = f(x_2) = 1.1175191$$

$$x_3 = \frac{3}{6} \qquad y_3 = f(x_3) = 1.2840254$$

$$x_4 = \frac{4}{6} \qquad y_4 = f(x_4) = 1.5596235$$

$$x_5 = \frac{5}{6} \qquad y_5 = f(x_5) = 2.0025962$$

$$x_6 = 1 \qquad y_6 = f(x_6) = 2.7182818$$

and

$$\frac{1}{3}h(y_0 + 4y_1 + 2y_2 + 4y_3 + 2y_4 + 4y_5 + y_6) = 1.4628735.$$

Therefore, $\int_0^1 e^{x^2}\,dx$ is approximately 1.4628735.

It is not easy to find an error estimate for Simpson's rule, that is, to decide how many accurate decimal places the approximation contains. The following procedure is often used instead. To find an approximation to four decimal places, use Simpson's method repeatedly, doubling the value of n each time, obtaining successive approximations $S_2, S_4, S_8, S_{16}, S_{32}, \cdots$. When two successive approximations agree to five decimal places, choose the first four rounded places as the approximation. The accuracy of the four decimal places is *not* guaranteed, but experience shows that if approximations converge rapidly, then when two successive approximations are near each other, they are also near the limit. Therefore, computer users who adopt this rule of thumb have reason to hope for four place accuracy.

Problems for Section 5.4

1. Approximate the integral using Simpson's rule with the given number of subintervals.

(a) $\int_0^1 \sqrt{1 + x^4}\, dx$, $n = 4$ (c) $\int_1^2 \dfrac{1}{1 + x^3}\, dx$, $n = 8$

(b) $\int_0^1 \ln(1 + x^2)\, dx$, $n = 6$ (d) $\int_0^1 e^{-x^4}\, dx$, $n = 6$

2. Approximate $\int_1^2 \dfrac{1}{x^2}\, dx$ using Simpson's rule with $n = 4$, and compare with the exact answer.

5.5   Nonintegrable Functions

So far we have ignored the possibility that a function might not have an integral, and concentrated on the methods that will compute the integral *if* it exists. This section will display two nonintegrable functions to give more insight into the definition of the integral.

Example 1   To understand our first nonintegrable function you must know the difference between rational and irrational numbers, and how they are distributed on a line. The rational numbers are the decimals that either stop or eventually repeat, such as 2.5, 0.33333..., 3.14, 4.78626767676767.... All other decimals are called irrational. For example, 2.1234567891011121314151617181920212223 24... (which has a pattern but doesn't repeat) is irrational; so are π, $\sqrt{2}$ and e. On the number line, the rationals and irrationals are so thoroughly interspersed that there are no solid intervals of rationals and no solid intervals of irrationals; in any interval there are both rationals and irrationals. We can demonstrate this with the interval $(4.2, 4.3)$. The rational number 4.25 is in the interval, and so is the irrational number 4.256789101112131415161718 19.... Thus the interval is neither entirely rational nor entirely irrational.

Now we are ready to define a nonintegrable function. Let

(1)
$$f(x) = \begin{cases} 0 & \text{if } x \text{ is rational} \\ 1 & \text{if } x \text{ is irrational}. \end{cases}$$

Consider two people trying to compute $\int_2^5 f(x)\, dx$. Each divides $[2, 5]$ into many small subintervals (as in Fig. 1 of Section 5.2). Each picks values of x in the subintervals, but she chooses rationals and he chooses irrationals. Her $f(x_1), f(x_2), \cdots$ are all 0, so her Riemann sum $\Sigma f(x) dx$ is 0. His $f(x_1)$, $f(x_2), \cdots$ are all 1, so his Riemann sum is Σdx, which is 3, the length of the interval. They repeat the process with smaller subintervals, but if she keeps picking rationals and he keeps picking irrationals, they again get $\Sigma f(x) dx = 0$ and $\Sigma f(x) dx = 3$, respectively. Since their Riemann sums continue to disagree drastically, $\lim_{dx \to 0} \Sigma f(x) dx$ does not exist, and the function is not integrable on $[2, 5]$, or on any other interval for that matter.

It is the extreme discontinuity of the function in (1) that causes it to be nonintegrable. In fact, the function is discontinuous everywhere. If we try to draw the graph of f, you will see this. We can plot many points on the graph, for instance, $(2, 0)$, $(2.6, 0)$, $(4.1, 0)$, $(e, 1)$, $(\pi, 1)$, and so on. All points of the graph are either at height 0 or height 1. But no part of the graph is a solid line at height 1 or at height 0 because no interval on the x-axis is solidly rational or solidly irrational. So no portion of the graph can be drawn without lifting the pencil from the paper (and the *complete* graph is humanly impossible to draw).

Example 2   Let $f(x) = 1/\sqrt{x}$. Consider two people trying to find $\int_0^1 f(x) dx$ by computing $\Sigma f(x) dx$. Suppose they begin by dividing $[0, 1]$ into 100 subintervals of equal length, so that each dx is $1/100$ (Fig. 1). Then they must choose values of x in the subintervals. If their Riemann sums disagree, and continue to disagree as more and more subintervals of smaller size are used, then f is not integrable on $[0, 1]$. The greatest opportunity for disagreement comes from the first subinterval, where f varies enormously. The product $f(x) dx$ corresponding to the first subinterval is of the form "large $\times$ small" and its value depends on "how large" and "how small." Suppose he picks $x = 1/100$ at the right end of the first subinterval and she picks $x = 1/100^4$ near the left end. Then

$$\text{his } f(x_1) \, dx_1 = f\!\left(\frac{1}{100}\right)\frac{1}{100} = 10 \cdot \frac{1}{100} = \frac{1}{10}$$

while

$$\text{her } f(x_1) \, dx_1 = f\!\left(\frac{1}{100^4}\right)\frac{1}{100} = 100^2 \cdot \frac{1}{100} = 100 \, .$$

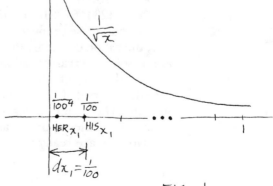

FIG. 1

If they use 10,000 subintervals and he picks $x = 1/10,000$ at the right end of the first subinterval while she picks $x = 1/10,000^4$ near the left end, then

$$\text{his } f(x_1)\,dx_1 = f\left(\frac{1}{10,000}\right)\frac{1}{10,000} = 100 \cdot \frac{1}{10,000} = \frac{1}{100}$$

while

$$\text{her } f(x_1)\,dx_1 = f\left(\frac{1}{10,000^4}\right)\frac{1}{10,000} = 10,000^2 \cdot \frac{1}{10,000} = 10,000\,.$$

Their values of $f(x_1)\,dx_1$ grow more unlike (hers becomes large, his becomes small) as $dx \to 0$. This predicts that their *entire* Riemann sums will also grow more unlike (in fact it can be shown that hers will approach ∞ and his will approach 2), indicating that f is not integrable on $[0, 1]$.

It is the infinite discontinuity of the function $1/\sqrt{x}$ at $x = 0$ that causes it to be nonintegrable. The next section will define a new integral to handle unbounded functions.

5.6 Improper Integrals

The definition of $\int_a^b f(x)\,dx$ involves dividing $[a, b]$ into many small subintervals, and finding Riemann sums $\Sigma f(x)\,dx$. The definition does not apply to intervals of the form $[a, \infty)$, $(-\infty, b]$ and $(-\infty, \infty)$ because it isn't possible to divide infinite intervals into a *finite* number of *small* subintervals. Furthermore, with this definition of $\int_a^b f(x)\,dx$, it can be shown that functions with infinite discontinuities are not integrable; one of the difficulties that can arise is illustrated in Example 2 of the preceding section. New integrals, called *improper integrals,* will be defined to cover the cases of infinite intervals and infinite functions.

Integrating on intervals of the form $[a, \infty)$ and $(-\infty, b]$ As an illustration, we define

$$\int_1^\infty \frac{1}{x}\,dx = \lim_{b \to \infty} \int_1^b \frac{1}{x}\,dx\,.$$

In other words, to integrate on $[1, \infty)$, integrate from $x = 1$ to $x = b$ and then let b approach ∞. Therefore

$$\int_1^\infty \frac{1}{x}\,dx = \lim_{b \to \infty}\left(\ln x \,\Big|_1^b\right) = \lim_{b \to \infty}(\ln b - \ln 1) = \infty - 0 = \infty\,.$$

We interpret this geometrically to mean that the area of the unbounded region in Fig. 1 is infinite. As a convenient shorthand, we write

(1) $$\int_1^\infty \frac{1}{x}\,dx = \ln x \,\Big|_1^\infty = \ln \infty - \ln 1 = \infty - 0 = \infty\,.$$

In general,

$$\int_a^\infty f(x)\,dx = \lim_{b \to \infty} \int_a^b f(x)\,dx$$

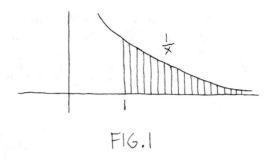

FIG. 1

and

$$\int_{-\infty}^{b} f(x)\, dx = \lim_{a \to -\infty} \int_{a}^{b} f(x)\, dx.$$

In abbreviated notation, if F is an antiderivative for f then

(2)
$$\int_{a}^{\infty} f(x)\, dx = F(x)\Big|_{a}^{\infty} \quad \text{and} \quad \int_{-\infty}^{b} f(x)\, dx = F(x)\Big|_{-\infty}^{b}.$$

Convergence versus divergence   Evaluating an improper integral will always involve computing an ordinary integral and a limit. If the limit is finite, then the improper integral is said to be *convergent*. If the limit is ∞ or $-\infty$, or if no value at all, either finite or infinite, can be assigned to the limit, the integral *diverges*. For example, the integral in (1) is divergent; in particular, it diverges to ∞.

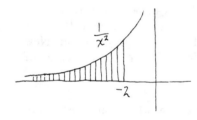

FIG. 2

Example 1
$$\int_{-\infty}^{-2} \frac{1}{x^2}\, dx = -\frac{1}{x}\Big|_{-\infty}^{-2} = \frac{1}{2} + \frac{1}{-\infty} = \frac{1}{2} + 0 = \frac{1}{2}.$$
The integral converges to $\frac{1}{2}$ and the unbounded region in Fig. 2 is considered to have area $\frac{1}{2}$.

  The unbounded regions in Figs. 1 and 2 look similar, but the former has finite area and the latter has infinite area. The function x^2 has a higher order of magnitude than x, the graph of $1/x^2$ approaches the x-axis faster than the graph of $1/x$, and the region in Fig. 2 narrows down fast enough to have a finite area.

Integrating on the interval $(-\infty, \infty)$ The usual definition is

(3)
$$\int_{-\infty}^{\infty} f(x)\, dx = \lim_{\substack{a \to -\infty \\ b \to \infty}} \int_{a}^{b} f(x)\, dx.$$

This is the first appearance of a limit involving *two* independent variables, a and b in this case. When we say that the limit in (3) is L we mean that we can force $\int_{a}^{b} f(x)\, dx$ to be as close as we like to L for all b sufficiently high and all a sufficiently low.

  In abbreviated notation, if F is an antiderivative for f, then

(4)
$$\int_{-\infty}^{\infty} f(x)\, dx = F(x)\Big|_{-\infty}^{\infty}$$

provided that the right-hand side is not of the form $\infty - \infty$. If it is of the form $\infty - \infty$ we assign no value at all (an instance of divergence).

For example,

$$\int_{-\infty}^{\infty} \frac{1}{1 + x^2}\, dx = \tan^{-1}x \,\Big|_{-\infty}^{\infty} = \frac{\pi}{2} - \left(-\frac{\pi}{2}\right) = \pi.$$

As an example of (4) which results in the form $\infty - \infty$, consider

$$\int_{-\infty}^{\infty} x^3\, dx = \frac{x^4}{4} \,\Big|_{-\infty}^{\infty} = \infty - \infty.$$

No specific value can be assigned since as $a \to -\infty$ and $b \to \infty$, the value of $\frac{1}{4}x^4 \big|_a^b = \frac{1}{4}b^4 - \frac{1}{4}a^4$ depends on *how fast* a and b move. Therefore the integral is simply called divergent.

Integrating functions which blow up at the end of the interval of integration The function $1/x^2$ blows up at $x = 0$. To integrate on an interval such as $[0, 1]$ we define

$$\int_0^1 \frac{1}{x^2}\, dx = \lim_{a \to 0+} \int_a^1 \frac{1}{x^2}\, dx.$$

Then

$$\int_0^1 \frac{1}{x^2}\, dx = -\frac{1}{x} \,\Big|_{0+}^1 = -1 + \frac{1}{0+} = -1 + \infty = \infty.$$

In general, let F be an antiderivative of f.

(5)

$$\boxed{\begin{array}{l} \text{If } f \text{ blows up at } x = a \text{ then } \int_a^b f(x)\, dx = F(x) \,\Big|_{a+}^{b}. \\[2em] \text{If } f \text{ blows up at } x = b \text{ then } \int_a^b f(x)\, dx = F(x) \,\Big|_{a}^{b-}. \end{array}}$$

Example 2 The function $1/\sqrt{x}$ has an infinite discontinuity at $x = 0$; Example 2 in the preceding section showed that it is not integrable on $[0, 1]$ using the definition of the integral from Section 5.2. But reconsidered as an *improper* integral,

$$\int_0^1 \frac{1}{\sqrt{x}}\, dx = 2\sqrt{x} \,\Big|_{0+}^1 = 2.$$

Example 3 Find $\int_2^3 \frac{1}{(3 - x)^2}\, dx$.

Solution: The integral is improper because the integrand blows up at $x = 3$. Then

$$\int_2^3 \frac{1}{(3 - x)^2}\, dx = \frac{1}{3 - x} \,\Big|_2^{3-} = \frac{1}{0+} - 1 = \infty - 1 = \infty.$$

Note that when the blowup is located at 3, and 3 is the *upper* limit of integration, it is treated as $3-$ in the calculation. If 3 were the *lower* limit of integration, it would be treated as $3+$ in the calculation.

Warning   Whenever a limit of the form $1/0$ arises in the computation, look closely to see if it is $1/0+$ or $1/0-$.

Integrating functions which blow up within the interval of integration
Suppose f blows up at c between a and b. If F is an antiderivative of f, we define $\int_a^b f(x)\,dx$ by

(6)
$$\int_a^b f(x)\,dx = \int_a^{c-} f(x)\,dx + \int_{c+}^b f(x)\,dx = F(x)\Big|_a^{c-} + F(x)\Big|_{c+}^b .$$

As before, if (6) results in the form $\infty - \infty$, no finite or infinite value is assigned (an instance of divergence).

For example, the function $1/x^4$ blows up at $x = 0$, inside the interval $[-1, 3]$, so

$$\int_{-1}^3 \frac{1}{x^4}\,dx = \frac{-1}{3x^3}\Big|_{-1}^{0-} + \frac{-1}{3x^3}\Big|_{0+}^3$$

$$= \frac{-1}{0-} - \frac{1}{3} - \frac{1}{81} - \frac{-1}{0+} = \infty - \frac{1}{3} - \frac{1}{81} + \infty = \infty .$$

The improper integral diverges to ∞.

Warning   It is *not* correct to write $\displaystyle\int_{-1}^3 \frac{1}{x^4}\,dx = -\frac{1}{3x^3}\Big|_{-1}^3$, and, in general, if f blows up inside $[a, b]$, it is *not* correct to write $\int_a^b f(x)\,dx = F(x)\,|_a^b$. You must use (6) instead.

Example 4   Find $\displaystyle\int_4^7 \frac{1}{(x-5)^3}\,dx$.

Solution: The integrand blows up at $x = 5$, inside the interval $[4, 7]$. So

$$\int_4^7 \frac{1}{(x-5)^3}\,dx = \frac{-1}{2(x-5)^2}\Big|_4^{5-} + \frac{-1}{2(x-5)^2}\Big|_{5+}^7 = \frac{-1}{0+} + \frac{1}{2} + \frac{-1}{8} - \frac{-1}{0+}.$$

This results in the form $-\infty + \infty$ so the integral diverges.

Problems for Section 5.6

1. $\displaystyle\int_3^\infty \frac{1}{x^5}\,dx$

2. $\displaystyle\int_2^\infty \frac{1}{\sqrt[5]{x}}\,dx$

3. $\displaystyle\int_{-\infty}^{-2} \frac{1}{x^3}\,dx$

4. $\displaystyle\int_{-1}^0 \frac{1}{x^2}\,dx$

5. $\displaystyle\int_0^2 \frac{1}{x}\,dx$

6. $\displaystyle\int_{-2}^3 \frac{1}{x^3}\,dx$

7. $\displaystyle\int_{-\infty}^0 \frac{1}{1+x^2}\,dx$

8. $\displaystyle\int_{-\infty}^0 2e^{4x}\,dx$

9. $\displaystyle\int_2^5 \frac{1}{\sqrt[3]{4-x}}\,dx$

10. $\displaystyle\int_{-2}^3 \frac{1}{x^2}\,dx$

11. $\displaystyle\int_0^\infty \frac{1}{x}\,dx$

12. $\displaystyle\int_0^\infty \sin x\,dx$

13. $\displaystyle\int_{-\infty}^\infty e^{-|x|}\,dx$

14. $\displaystyle\int_0^{\pi/2} \tan x\,dx$ given $F(x) = -\ln\cos x$

15. $\displaystyle\int_{-\infty}^\infty \frac{1}{(x^2+1)^2}\,dx$ given $F(x) = \frac{1}{2}\left(\frac{x}{x^2+1} + \tan^{-1}x\right)$

16. $\displaystyle\int_0^1 \ln x\,dx$ given $F(x) = x\ln x - x$

REVIEW PROBLEMS FOR CHAPTER 5

1.

(a) $\displaystyle\int_{-1}^{1} x^6\,dx$ (g) $\displaystyle\int_{0}^{\pi} \cos\frac{1}{2}x\,dx$

(b) $\displaystyle\int_{-1}^{1} \frac{1}{x^6}\,dx$ (h) $\displaystyle\int_{4}^{7} 3\,dx$

(c) $\displaystyle\int_{1}^{\infty} \frac{1}{x^6}\,dx$ (i) $\displaystyle\int_{-1}^{3} e^{-|x|}\,dx$

(d) $\displaystyle\int_{1}^{2} (x^2 + 3)\,dx$ (j) $\displaystyle\int_{-1}^{0} \frac{(2x+5)^5}{4}\,dx$

(e) $\displaystyle\int_{1}^{2} \sqrt{3x+4}\,dx$ (k) $\displaystyle\int_{0}^{1} \frac{4}{2-x}\,dx$

(f) $\displaystyle\int_{2}^{\infty} e^{-3x}\,dx$ (l) $\displaystyle\int_{15}^{17} dx$

2. Let $f(x)$ be the function in Fig. 1. Find $\int_0^6 f(x)\,dx$ (a) using areas and (b) using the Fundamental Theorem.

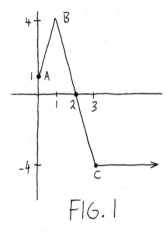

FIG. 1

3. Use Simpson's rule to approximate $\int_0^1 \sqrt[4]{1 + x^2}\,dx$ using 6 subintervals.
4. Let $I = \int_a^b |f(x)|\,dx$ and $II = |\int_a^b f(x)\,dx|$. Which is larger, I or II?
5. Find the average value of $1/x$ on the interval $[1, e]$.
6. Find the area in Fig. 2.

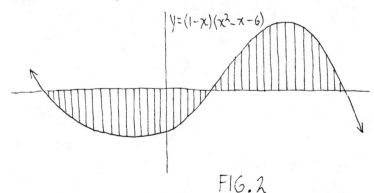

FIG. 2

7. Odd and even functions were defined in Problem 8 of Section 1.2. (a) If f is odd, find $\int_{-3}^{3} f(x)\,dx$. (b) If f is even, compare $\int_{-3}^{3} f(x)\,dx$ and $\int_0^3 f(x)\,dx$.

6/THE INTEGRAL PART II

6.1 Further Applications of the Integral

Section 5.2 included applications to area and average values. This section continues with integral models for many more physical concepts, and the problems will ask you to construct your own models in new situations. It is time-consuming material because the examples and problems are quite varied. On the other hand, it is precisely the wide scope of the applications that makes the material so important. After a while, you will get a feeling for the type of problem that leads to an integral, namely, one that is solved with a sum of the form $\sum f(x)\, dx$.

Example 1 The volume formula "base × height" applies to a cylinder and a box, but not to a cone, pyramid or sphere. To understand why not, consider the full implications of the "base" in the formula. It does not mean the bottom of the solid; instead it refers to the constant cross-sectional area (Fig. 1). The formula really says

(1) $$\text{volume} = \text{cross-sectional area} \times \text{height},$$

provided that the solid has *constant* cross-sectional area.

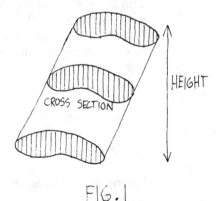

FIG. 1

Consider a cone with radius R and height h. Geometry books declare its volume to be $\frac{1}{3}\pi R^2 h$, and the problem is to derive this volume formula using calculus. Formula (1) does not apply directly because the cone does not have constant cross sections. To get around this difficulty, divide the cone into thin slabs. With the number line in Fig. 2, a typical slab is located around position x and has thickness dx. The significance of the slab is that its cross-sectional area is almost constant. The lower part has smaller radius than the upper part, but the slab is so thin that we take its radius throughout

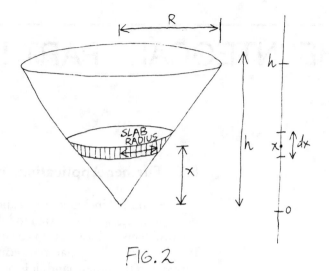

FIG. 2

to be the radius at position x. By similar triangles,

$$\frac{\text{slab radius}}{x} = \frac{R}{h}$$

$$\text{slab radius} = \frac{Rx}{h}.$$

Thus the slab has cross-sectional area $\pi\left(\dfrac{Rx}{h}\right)^2$ and height dx, so, by (1),

$$\text{volume } dV \text{ of the slab} = \pi\left(\frac{Rx}{h}\right)^2 dx.$$

This is only the approximate volume of the slab, but the approximation improves as $dx \to 0$. We want to add the volumes dV to find the total volume of the cone, and use thinner slabs (i.e., let $dx \to 0$) to remove the error in the approximation. The integral will do both of these things. We integrate from 0 to h because the slabs begin at $x = 0$ and end at $x = h$. Thus

$$\text{cone volume} = \int_0^h \pi\left(\frac{Rx}{h}\right)^2 dx = \frac{\pi R^2}{h^2}\int_0^h x^2\, dx = \frac{\pi R^2}{h^2}\frac{x^3}{3}\bigg|_0^h = \frac{1}{3}\pi R^2 h,$$

the desired formula.

Example 2 A flag pole painting company charges customers by the formula

$$\text{cost in dollars} = h^2 l \qquad \text{(Fig. 3)}$$

where h is the height (in meters) of the flagpole above the street and l is the length of the pole. If the pole in Fig. 3 is 4 meters above the ground and 2 meters long, then the paint job costs \$32.†

---

†The units on $h^2 l$ are (meters)^3, so to make the units on each side of the formula agree, it is understood that the right-hand side contains the factor 1 dollar/(meter)^3. It is common in physics for formulas to contain constants in this manner for the purpose of making the units match.

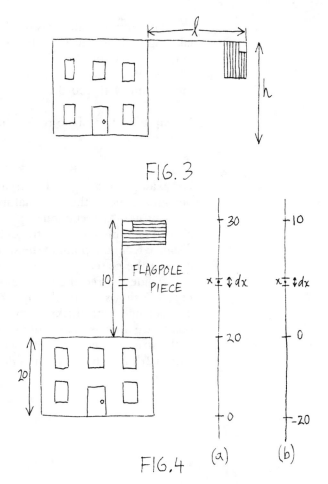

FIG. 3

FIG. 4   (a)   (b)

Now consider the cost of painting the pole in Fig. 4. Its length is 10 meters, but the formula h^2l can't be used directly because the pole is not at one fixed height above the ground. To get around this, divide the pole into pieces. With the number line in Fig. 4(a), a typical piece has length dx and is small enough to be considered (almost) all at height x. Use the formula h^2l to find that the cost of painting the small piece, called $d$cost to emphasize its smallness, is $x^2\,dx$. Then use the integral to add the $d$costs and obtain

(2) $$\text{total cost} = \int_{20}^{30} d\,\text{cost} = \int_{20}^{30} x^2\,dx.$$

(The integration process includes not only a summation but also a limit as dx approaches 0, which removes the error caused by the "almost.") The interval of integration is $[20, 30]$ because that's where the flagpole is located. If you incorrectly integrate from 0 to 30, then you are paying to have a white stripe painted down the front of the house.

If we compute the integral we get the final answer $\left.\dfrac{x^3}{3}\right|_{20}^{30} = \dfrac{19{,}000}{3}$.

However, (2) is considered to be final enough in this section since the emphasis here is on *setting up* the integral that solves the problem, that is, on finding the model.

The number line does not have to be labeled as in Fig. 4(a). Another labeling is shown in Fig. 4(b). In this case, the small piece of flagpole has height $x + 20$ and length dx, so $d\,\text{cost} = (x + 20)^2\,dx$ and the total cost is $\int_0^{10}(x + 20)^2\,dx$. The integral looks different from (2), but its value is the same, namely 19,000/3.

Example 3   If a plane region has constant density, then its total mass is given by

(3) $$\text{mass} = \text{density} \times \text{area}.$$

For example, if a region has area 6 square meters and density 7 kilograms per square meter then its total mass is 42 kilograms.

Consider a rectangular plate with dimensions 2 by 3. Suppose that instead of being constant, the density at a point in the plate is equal to the distance from the point to the shorter side. The problem is to find the total mass of the plate.

Divide the rectangular region into strips parallel to the shorter side. Figure 5 shows a typical strip located around position x on the indicated number line, with thickness dx. The significance of the strip is that all its points are approximately distance x from the shorter side, so the density in the strip may be considered constant, at the value x. The area dA of the strip is $2\,dx$ and, by (3), its mass dm is $2x\,dx$. Therefore, total mass $= \int_0^3 dm = \int_0^3 2x\,dx$.

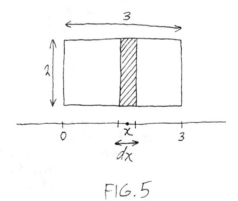

FIG. 5

The general pattern for applying integrals   After three applications in this section, perhaps you already sense the pattern. There will be a formula (base × height from geometry, $h^2 l$ from our imagination, density × area from physics) that applies in a *simple* situation (*constant* cross sections, heights, densities) to compute a total "thing" (volume, cost, mass). In a more complicated situation (*non*constant cross sections, heights, densities) the formula cannot be used directly. However, if a physical entity (the cone, the flagpole, the rectangular plate) is divided into pieces, it may be possible to apply the formula to the pieces and compute "$d$thing" (dV, $d$cost, $d$mass). The integral is then used to add the $d$things and find a total.

The comment on mathematical models in Section 5.2 still applies. We are not *proving* that the integral actually computes the total; the integral is just the best mathematical model presently available.

Warning By the physical nature of the particular problems in this section, the simple factor dx should be contained in the expression for d thing; it should not be missing, nor should it appear in a form such as $(dx)^2$ or $1/dx$. For example, d thing may be $x^3\,dx$, but should not be x^3, or $x^3(dx)^2$, or x^3/dx. The integral is defined to add only terms of the form $f(x)\,dx$. A sum of terms of the form x^3 or $x^3(dx)^2$ or x^3/dx is not an integral, and in particular cannot be computed with $F(b) - F(a)$.

Example 4 The charges of a moving company depend on the weight of your household goods and on the distance they must be shipped. Suppose

$$(4) \qquad\qquad \text{cost} = \text{weight} \times \text{distance},$$

where cost is measured in dollars, weight in pounds and distance in feet. If an object weighing 6 pounds is moved 5 feet, the company charges \$30 (and physicists say that 30 foot pounds of work has been done).

Suppose a cylindrical tank with radius 5 and height 20 is half filled with a liquid weighing 2 pounds per cubic foot. Find the cost of pumping the liquid out, that is, of hiring movers to lift the liquid up to the top of the tank, at which point it spills out.

Solution: Formula (4) doesn't apply directly because different layers of liquid must move different distances; the top layer moves 10 feet but the bottom layer must move 20 feet. Divide the liquid into slabs; a typical slab is shown in Fig. 6, with thickness dx and located around position x on the number line. The significance of the slab is that all of it must be moved up $20 - x$ feet. The slab has radius 5 and height dx, so its volume dV is $25\pi\,dx$. Then

$$d\text{weight} = 2 \text{ pounds/cubic foot} \times 25\pi\,dx \text{ cubic feet} = 50\pi\,dx \text{ pounds},$$

and, by (4),

$$d\text{cost} = 50\pi\,dx \times (20 - x) = 50\pi(20 - x)\,dx.$$

Integrate on the interval $[0, 10]$, since that is the extent of the liquid,

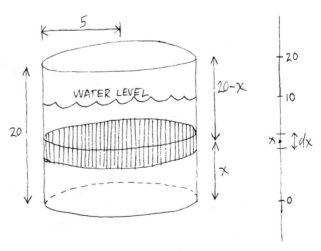

FIG. 6

to obtain

$$\text{total cost} = \int_0^{10} d\text{cost} = \int_0^{10} 50\pi(20 - x)\,dx = 50\pi\left(20x - \frac{1}{2}x^2\right)\Bigg|_0^{10}$$
$$= 7500\pi.$$

If a different number line is used, say with 0 at the top of the cylinder and 20 at the bottom, the integral may look different, but the final answer must be 7500π.

Example 5   Merry-go-round riders all pay the same price and can sit anywhere they like. This is a comparatively unusual policy because most events have different prices for different seats; seats on the 50-yard line at a football game cost more than seats on the 10-yard line. Obviously, some merry-go-round seats are better than others. Seats right next to the center pole give a terrible ride; the best horses, the most sweeping rides, and the gold ring are all on the outside. The price of a ticket should reflect this and depend on the distance to the pole. Furthermore, the price of a ticket should depend on the mass of the rider (airlines don't measure passengers but they do take the amount of luggage into consideration). Suppose the price charged for a seat on the merry-go-round is given by

(5)   price $= md^2$

where m is the mass of the customer and d is the distance from the seat to the center pole. (In physics, md^2 is the *moment of inertia* of a rotating object.)

Consider a solid cylinder with radius R, height h and density δ mass units per unit volume, revolving around its axis as a center pole (Fig. 7). Find the price of the ride.

Solution: Formula (5) doesn't apply directly because different parts of the cylinder are at different distances from the center pole. Dividing the

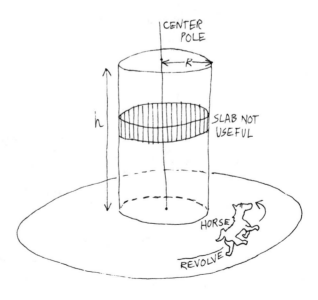

FIG.7

cylinder into slabs, one of which is shown in Fig. 7, doesn't help because the same difficulty persists—different parts of the slab are at different distances from the center pole. Instead, divide the solid cylinder into cylindrical shells. Each shell is like a tin can, and the solid cylinder is composed of nested tin cans; Fig. 8 shows one of the shells with thickness dx, located around position x on the number line. The advantage of the shell is that all its points may be considered at distance x from the pole. The formula $dV = 2\pi rh\, dr$ for the volume of a cylindrical shell with radius r, height h and thickness dr was derived in (9) of Section 4.8. The shell in Fig. 8 has radius x, height h and thickness dx, so $dV = 2\pi xh\, dx$ and

$$d\,\text{mass} = \text{density} \times \text{volume} = 2\pi xh\, \delta\, dx.$$

By (5), when the shell is revolved,

$$d\,\text{price} = 2\pi xh\, \delta\, dx \cdot x^2 = 2\pi x^3 h\, \delta\, dx.$$

Therefore

$$\text{total price} = \int_0^R d\,\text{price} = 2\pi h\, \delta \int_0^R x^3\, dx = 2\pi h\, \delta \left.\frac{x^4}{4}\right|_0^R = \frac{1}{2}\pi h\, \delta R^4.$$

Note that the shell area and volume formulas from Section 4.8 are only approximations. But we anticipated that they would be used in integral problems, such as this one, where the thickness dr (or in this case, dx) approaches 0 as the integral adds. Section 4.8 claimed that under those circumstances, the error in the approximation is squeezed out.

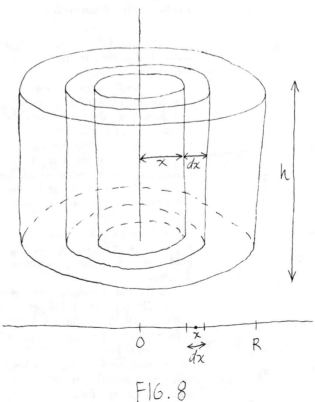

FIG. 8

Example 6   Let's try a reverse example for practice. Usually we conclude that $\int_a^b f(x)\,dx$ is a total. Suppose we begin with the "answer": let $\int_3^7 f(x)\,dx$ be the total number of gallons of oil that has flowed out of the spigot at the end of the Alaska pipeline between hour 3 and 7. Go backwards and decide what was divided into pieces, what dx stands for, and what a term of the form $f(x)\,dx$ represents physically. In general, what does the function $f(x)$ represent?

Solution: The *time* interval $[3, 7]$ was partitioned. A typical dx stands for a small amount of time, such as $1/10$ of an hour. Since the integral adds terms of the form $f(x)\,dx$ to produce total gallons, one such term represents gallons; in particular, one term of the form $f(x)\,dx$ is the (small) number of gallons, more appropriately called d gallons, that has flowed out during the dx hours around time x. Since the units of $f(x)\,dx$ are gallons, and those of dx are hours, $f(x)$ itself must stand for gallons/hour, the rate of flow. If $f(4.5) = 6$, then at time 4.5, the oil is flowing instantaneously at the rate of 6 gallons per hour.

Note that in general, *the integral of a "rate"* (e.g., gallons per hour) *produces a "total."*

Warning   In the preceding example, a term of the form $f(x)\,dx$ represents the d gallons of oil flowing out *during* a time interval of duration dx hours around time x, *not* oil flowing out *at* time x. It is impossible for a positive amount of oil to pour out *at an* instant. Furthermore, if $f(4.5) = 6$ then it is *not* the case that 6 gallons flow out at time 4.5; rather, at this instant, the flow is 6 gallons *per hour*.

Problems for Section 6.1

(The aim of the section was to demonstrate how to produce integral models for physical situations. In the solutions we usually set up the integrals and then stop without computing their values.)

1. If an 8-centimeter wire has a constant density of 9 grams per centimeter then its total mass is 72 grams. Suppose that instead of being constant, the density at a point along the wire is the cube of its distance to the left end. For example, at the middle of the wire the density is 64 grams/cm, and at the right end the density is 512 grams/cm. Find the total mass of the wire.

2. If travelers go at R miles per hour for T hours, then the total distance traveled is RT miles. Suppose the speed on a trip is not constant, but is t^2 miles per hour at time t. For example, the speed at time 3 is 9 miles per hour, the speed at time 3.1 is 9.61 miles per hour, and so on. Find the total distance traveled between times 3 and 5.

3. Suppose that the cost of painting a ceiling of height h and area A is $.01h^2A$. For example, the cost of painting the ceiling in Fig. 9 is $.01(36)(35)$ or 12.60. Find the cost of painting the wall in Fig. 9 (which is not at a constant height h above the floor).

4. Use slabs to derive the formula $\frac{4}{3}\pi R^3$ for the volume of a sphere of radius R.

5. The price of land depends on its area (the more area, the more expensive) and on its distance from the railroad tracks (the closer to the tracks, the less expensive). Suppose the cost of a plot of land is area × distance to tracks. Find the cost of the plot of land in Fig. 10.

6. Suppose a conical tank with radius 5 and height 20 is filled with a liquid weighing 2 pounds per cubic foot. Continue from Example 4 to find the cost of pumping the liquid out.

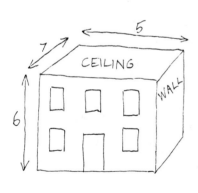

FIG. 9

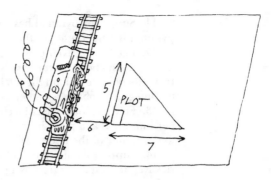

FIG. 10

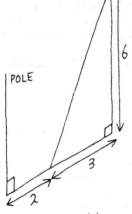

FIG. 11

7. Suppose the right triangular region in Fig. 11 with density δ mass units per unit area revolves around the indicated pole. Continue from Example 5 to find its moment of inertia.

8. If the specific heat of an object of unit mass is constant, then the heat needed to raise its temperature is given by

$$\text{heat} = (\text{specific heat}) \times (\text{desired increase in temperature}).$$

For example, if the object has specific heat 2 and its temperature is to be raised from 72° to 78° then 12 calories of heat are needed. Suppose that the specific heat of the object is not constant, but is the cube of the object's temperature. Thus, the object becomes harder and harder to heat as its temperature increases. Find the heat needed to raise its temperature from 54° to 61°.

9. Suppose $\int_2^{14} f(x)\,dx$ is the total number of words typed by a secretary between minute 2 and minute 14.

(a) What does dx stand for in the physical situation?
(b) What does a term of the form $f(x)\,dx$ represent?
(c) What does the function f represent? If $f(3.2) = 25$, what is the secretarial interpretation?

10. Find the volume of the solid of revolution formed as follows. (First find the volume of the slab obtained by revolving a strip, and then add the slab volumes.)

(a) Revolve the region bounded by $y = x^2$ and the x-axis, $0 \le x \le 2$, around the x-axis (Fig. 12).
(b) Revolve the region bounded by $y = x^2$ and the y-axis, $0 \le y \le 4$, around the y-axis.

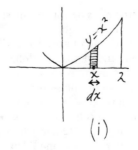

(i)

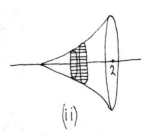

(ii)

FIG. 12

11. Suppose a pyramid has a square base with side a, and the top vertex of the pyramid is height h above the center of the square. Find its volume.

12. Let P be a fixed point on an infinitely long wire. Suppose that the charge density at any point on the wire is e^{-d} charge units per foot, where d is the distance from the point to P. Find the total charge on the wire with an integral, and compute the integral to obtain a numerical answer.

13. Find the total mass of a circular region of radius 6 if the density (mass units per unit area) at a point in the region is the square of the distance from the point to the center of the circle. (Divide the region into circular shells, i.e., washers.)

14. Suppose a solid sphere of radius R and density δ mass units per unit volume revolves around a diameter as a pole. Continue from Example 5 to find its moment of inertia.

15. Suppose $\int_3^7 g(x)\,dx$ is the cost in dollars of building the Alaska pipeline between milemarker 3 and milemarker 7.

(a) What does dx represent in the physical situation?
(b) What does a term of the form $g(x)\,dx$ stand for?
(c) What does the function g represent? If $g(4) = 17{,}000$, what is the physical interpretation?

16. The kinetic energy of an object with mass m grams and speed v centimeters per second is $\frac{1}{2}mv^2$. Suppose a rod with length 10 centimeters and density 3 grams per centimeter rotates around one fixed end (like the hand of a clock) at one revolution per second. The formula $\frac{1}{2}mv^2$ does not apply directly because different portions of the rod are moving at different speeds (the fixed end isn't moving at all and the outer tip is moving fastest). Find the kinetic energy of the rod by using an integral.

17. The area of a circle with radius R is πR^2. If a sector has angle θ (measured in radians) then its area is a fraction of the circle's area, namely the fraction $\theta/2\pi$, so

$$\text{area of sector} = \frac{\theta}{2\pi} \cdot \pi R^2 = \frac{1}{2}\theta R^2.$$

Suppose that we start at point C to draw a sector with angle $\pi/4$ and center at Q (Fig. 13) but the "radius" R varies with the angle θ so that $R = \cos\theta$. Find the area of the "sector" CQB.

18. Find the total mass of a solid cylinder with radius R and height h if its density (mass per unit volume) at a point is equal to (a) the distance from the point to the axis of the cylinder (b) the distance from the point to the base of the cylinder.

19. A machine earns $225 - t^2$ dollars per year when it is t years old. (a) Find the useful lifetime of the machine. (b) Find the total amount of money it earns during its lifetime.

20. The weight w of an object depends on its mass m and on its height h above the (flat) earth. Suppose $w = \dfrac{m}{2 + h^2}$. (The further away from the earth, the lighter the object.) If the mass density of the solid box in Fig. 14 is δ mass units per unit of volume, find its total weight.

21. If a plot of land of area A is at distance d from an irrigation pump, then the cost of irrigating the plot is Ad^3 dollars. Find the cost of irrigating a circular field of radius R if the pump is located at the center of the field.

22. The flat roof of a one-story house acts as a solar collector which radiates heat down to the rooms below. Suppose that the heat collected in a region of volume V at distance d below a collector is $V/(d + 1)$. Find the total heat collected in a room whose ceiling has height 12 and whose floor has dimensions 9 by 10.

23. When water with volume V lands after falling distance d, then a splash of size Vd occurs. For example, if water of volume 6 is poured onto the floor from a height of 7 then the total splash is 42.

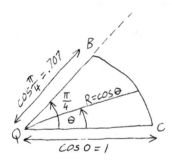

FIG. 13

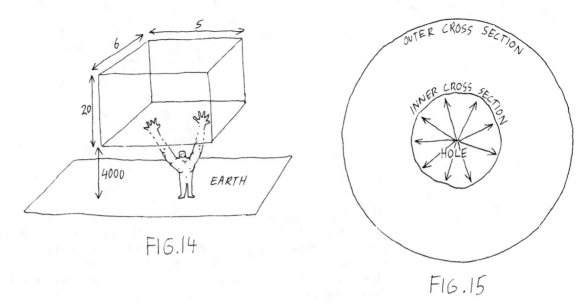

FIG.14

FIG.15

Suppose a cylindrical glass with radius 3 and height 5 is set under a faucet so that the distance from the top of the glass to the faucet is 4. Water drips into the glass until it is full. The falling water creates a splash, but the formula Vd can't be used directly since different slabs of water in the full glass fell through different heights (the lowest slab fell through distance 9 while the top slab fell through distance 4). Express the total splash with an integral.

24. Consider a unit positive charge fixed at point A. Like charges repel so if a second unit positive charge moves toward A, effort is required, and the effort increases as it nears A. Suppose that when the moving charge is d feet from A, the effort required to advance a foot toward A is $1/d^2$; i.e., it takes $1/d^2$ effort units *per foot*. Find the total effort required for the charge to advance (a) from distance 5 to distance 2 from A (b) from distance 5 to point A itself.

25. Snow starts falling at time $t = 0$, and then falls at the rate of $R(t)$ flakes/hour at time t. (a) How much snow will accumulate by time 10?   (b) Some of the flakes melt after they land, and don't live to see time 10. Suppose that only 1/4 of newly landed flakes still exist 3 hours later, only 1/5 still exist 4 hours later and, in general, of F newly fallen flakes, only $F/(x + 1)$ flakes will last x more hours. How much snow accumulates by time $t = 10$?

26. If current flows for distance L through a wire with cross-sectional area A, then the resistance R that it encounters is L/A. Suppose a sphere with radius 10 has a hole of radius 1 at its center, and current flows radially out of the hole through the solid sphere. The formula L/A doesn't apply directly because the current encounters spherical "cross sections" (Fig. 15) with increasing area rather than constant area A; e.g., visualize the current flowing away from the center of an onion through layers of onion shells. Use spherical shells to find an integral formula for R.

6.2   The Centroid of a Solid Hemisphere

This section consists of just one substantial application of integration, primarily of interest to those who will take physics courses.

If an object has constant density, then its balance point is called its *centroid*. For example, to picture the centroid of a wire (Fig. 1) imagine the

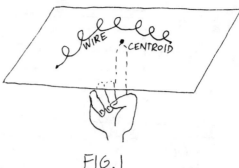

FIG. 1

wire lying in a plane which is weightless except for the wire. The point at which the plane balances is the centroid of the wire. Note that the centroid does not necessarily lie *on* the wire itself. One application of centroids is in the analysis of the behavior of an object in a gravitational force field, where the solid may be replaced by a point mass at its centroid. For some objects, the centroid is obvious. The centroid of a solid sphere is its center; the centroid of a rectangular region is the point of intersection of its diagonals. In this section we will find the centroid of a solid hemisphere of radius R, illustrating a method that may be used for other (symmetric) objects as well.

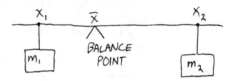

FIG. 2

We need some balancing principles first. Experiments have shown that if masses m_1 and m_2 dangle from a rod at positions x_1 and x_2 (Fig. 2) then the rod will balance at the point $\bar{x}$ where $m_1(\bar{x} - x_1) = m_2(x_2 - \bar{x})$. This is the well-known seesaw principle, which says that the heavier child should move forward on the seesaw to balance with a lighter partner. Solve the equation to obtain

$$m_1\bar{x} - m_1x_1 = m_2x_2 - m_2\bar{x}$$

$$\bar{x}(m_1 + m_2) = m_1x_1 + m_2x_2$$

$$\bar{x} = \frac{m_1x_1 + m_2x_2}{m_1 + m_2}.$$

The terms m_1x_1 and m_2x_2 are called the moments (with respect to the origin) of the masses m_1 and m_2 respectively. In other words, moment = mass × coordinate. More generally, if n masses $m_1, \cdots, m_n$ hang from positions $x_1, \cdots, x_n$ then

(1) $$\bar{x} = \frac{m_1x_1 + \cdots + m_nx_n}{m_1 + \cdots + m_n} = \frac{\text{total moment}}{\text{total mass}}.$$

Now consider a solid hemisphere with radius R and constant density δ mass units per unit volume. By geometric considerations, the centroid must

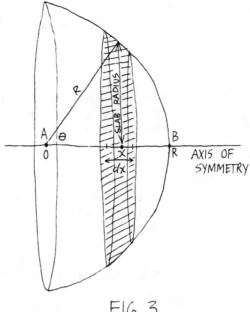

$$FIG. 3$$

lie on the axis of symmetry (Fig. 3). To decide *where* on the axis, divide the hemisphere into slabs. Figure 3 shows a typical slab with thickness dx located around position x on the number line AB. By the Pythagorean theorem, the slab radius is $\sqrt{R^2 - x^2}$. The (cylindrical) slab has height dx, so

$$\text{volume } dV = \text{base} \times \text{height} = \pi(\sqrt{R^2 - x^2})^2 \, dx = \pi(R^2 - x^2) \, dx$$

and

$$d\,\text{mass} = \delta \, dV = \delta\,\pi(R^2 - x^2) \, dx.$$

To simulate the situation in Fig. 2, picture each slab as a mass hanging from the axis of symmetry. Figure 4 shows the mass corresponding to the slab in Fig. 3. For this slab,

$$d\,\text{moment} = x\, d\,\text{mass} = \delta\,\pi(R^2 x - x^3) \, dx.$$

To find the *total* moment of all the slabs for the numerator of the formula in (1), add $d$moments and let dx approach 0 to improve the simulation. Thus

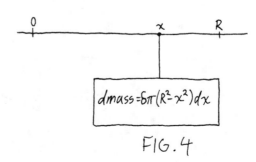

$$FIG. 4$$

$$\text{total moment} = \int_0^R d\,\text{moment} = \delta\pi\int_0^R (R^2x - x^3)\,dx$$

$$= \delta\pi\left(\frac{R^2x^2}{2} - \frac{x^4}{4}\right)\Big|_0^R = \frac{1}{4}\delta\pi R^4.$$

One way to find the *total* mass is to compute $\int_0^R d\,\text{mass} = \int_0^R \delta\pi(R^2 - x^2)\,dx$. Better still, since a sphere with radius R has volume $\frac{4}{3}\pi R^3$, the hemisphere has volume $\frac{2}{3}\pi R^3$ and its total mass is $\frac{2}{3}\delta\pi R^3$. Therefore

$$\bar{x} = \frac{\text{total moment}}{\text{total mass}} = \frac{3}{8}R.$$

The centroid lies on axis AB, three-eighths of the way from A to B. Note that the density δ does not appear in the answer. As long as the density is constant, its actual value is irrelevant for the location of the centroid.

6.3   Area and Arc Length

Section 6.1 constructed integral models for a variety of (sometimes fictional) physical concepts. This section is concerned with the standard models for the area between two curves, and arc length on a curve.

We will continue the policy of not evaluating integrals if antiderivatives are not readily available for the integrands. In such cases, numerical integration can be used, if desired, or you can return to the integrals later, after learning more antidifferentiation techniques in Chapter 7.

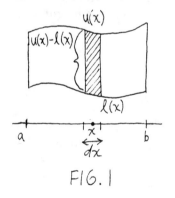

FIG. 1

Area between two curves   So far, integrals have been used to find the area of a region bounded by the x-axis, vertical lines and the graph of a function $f(x)$ (see Figs. 4, 5 and 6 in Section 5.2). Integration can also be used to find the area bounded by vertical lines and *two* curves, an upper function $u(x)$ and a lower function $l(x)$ (Fig. 1). To find the area, divide the region into vertical strips. Figure 1 shows a typical strip located around position x on the x-axis, with thickness dx. The strip has a curved top and bottom, but it is almost a rectangle with base dx and height $u(x) - l(x)$. In Figs. 2 and 3, one or both of $u(x)$ and $l(x)$ is negative, but $u(x) - l(x)$ is positive and *in each case* is the height of the strip. Therefore the area dA of the strip is $(u(x) - l(x))\,dx$. Thus, *for the region between $x = a$ and $x = b$, bounded by an*

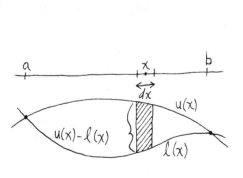

FIG. 2

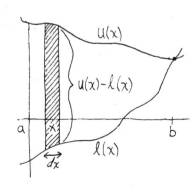

FIG. 3

upper curve u(x) and a lower curve l(x),

$$(1) \qquad \boxed{\ \text{area} = \int_a^b (u(x) - l(x))\, dx\,.\ }$$

The formula holds whether the region is above (Fig. 1), below (Fig. 2) or straddling (Fig. 3) the *x*-axis.

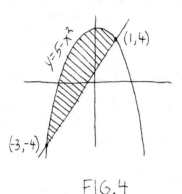

FIG.4

Example 1   Find the area of the region bounded by the parabola $y = 5 - x^2$ and the line through the points $(1, 4)$ and $(-3, -4)$ on the parabola.

   Solution: The line has slope 2, so by the point-slope formula its equation is $y - 4 = 2(x - 1)$, or $y = 2x + 2$. Figure 4 shows that the region has the parabola as its upper boundary, the line as its lower boundary, and lies between $x = -3$ and $x = 1$. Therefore $u(x) = 5 - x^2$, $l(x) = 2x + 2$, and

$$\text{area} = \int_{-3}^{1} [5 - x^2 - (2x + 2)]\, dx = \int_{-3}^{1} (-x^2 - 2x + 3)\, dx$$

$$= \left(-\frac{x^3}{3} - x^2 + 3x \right)\Bigg|_{-3}^{1} = \frac{32}{3}\,.$$

Arc length   To find the arc length s on a curve between points P and Q (Fig. 5), divide the curve into pieces. A typical piece with length ds is approximately the hypotenuse of a right triangle whose legs we label dx and dy. Then $ds^2 = dx^2 + dy^2$ and

$$(2) \qquad\qquad ds = \sqrt{dx^2 + dy^2}\,.$$

The total length of the curve is the sum of the small lengths ds, so, symbolically,

$$(3) \qquad\qquad s = \int_{\text{point } P}^{\text{point } Q} ds\,.$$

The details will depend on the algebraic description of the curve, as the next two examples will show.

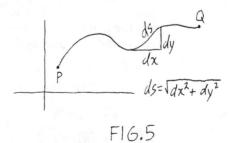

FIG.5

Example 2   Consider the arc length on the curve $y = x^3$ between the points $(-1, -1)$ and $(2, 8)$. Before using the integral in (3) we will express ds in terms of *one* variable. If $y = x^3$ and dy is a change in y then $dy = 3x^2\, dx$ (Section 4.8, (1′)). Therefore

$$ds = \sqrt{dx^2 + dy^2} = \sqrt{dx^2 + (3x^2\, dx)^2} = \sqrt{1 + 9x^4}\, dx,$$

so

$$s = \int_{x=-1}^{x=2} \sqrt{1 + 9x^4}\, dx.$$

Example 3 Suppose a circle of radius a, with a spot of paint on it, rolls along a line. The spot traces out a periodic curve called a *cycloid* (Fig. 6), and the problem is to find the arc length of one arch.

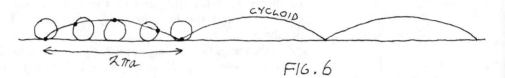

CYCLOID

$2\pi a$

FIG. 6

We'll begin by finding an algebraic description of the cycloid. Insert axes so that the circle rolls down the x-axis and the spot of paint begins at the origin. The x and y coordinates of a point on the cycloid are more easily described in terms of the angle of revolution θ (Fig. 7) than in terms of each other, so we will derive parametric equations for the cycloid instead of a single equation in x and y.

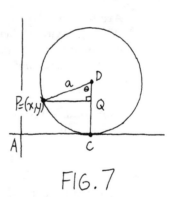

FIG. 7

Figure 7 shows a typical point $P = (x, y)$ on the cycloid with corresponding angle θ. Then $x = \overline{AC} - \overline{PQ}$. Furthermore, the length of *segment AC* is equal to the length of *arc PC* (visualize the arc PC matching segment AC point for point as the circle rolls). So

$$x = \overset{\frown}{PC} - \overline{PQ}$$

$$= a\theta - \overline{PQ} \quad \text{(by the arc length formula } s = r\theta \text{ in (5) of Section 1.3)}$$

$$= a\theta - a \sin \theta \quad \text{(by trigonometry in right triangle } PDQ).$$

Also, $y = \overline{DC} - \overline{DQ} = a - a \cos \theta$. Therefore the cycloid has parametric equations

$$(4) \qquad x = a\theta - a \sin \theta, \qquad y = a - a \cos \theta,$$

where a is the radius of the rolling circle and θ is the parameter. The cycloid is periodic, and the first period begins with $\theta = 0, x = 0$ and concludes with $\theta = 2\pi, x = 2\pi a$ (the circumference of the circle).

To find the length of the first arch using the integral in (3), first express ds in terms of *one* variable, θ in this case. We have

$$dx = x'(\theta)\,d\theta = (a - a \cos \theta)\,d\theta \quad \text{and} \quad dy = y'(\theta)\,d\theta = a \sin \theta \, d\theta.$$

Then (2) becomes

$$ds = \sqrt{(a - a \cos \theta)^2 \, d\theta^2 + a^2 \sin^2\theta \, d\theta^2}$$

$$= \sqrt{a^2 - 2a^2 \cos \theta + a^2 \cos^2\theta + a^2 \sin^2\theta} \; d\theta$$

$$= \sqrt{2a^2 - 2a^2 \cos \theta} \; d\theta \quad (\text{since } \cos^2\theta + \sin^2\theta = 1)$$

and

$$s = \int_{\theta=0}^{2\pi} \sqrt{2a^2 - 2a^2 \cos \theta} \; d\theta$$

$$= 2a \int_0^{2\pi} \sqrt{\frac{1 - \cos \theta}{2}} \; d\theta \quad (\text{by algebra})$$

$$= 2a \int_0^{2\pi} \sin \frac{1}{2}\theta \, d\theta \quad \left(\text{by the identity } \sin^2 \frac{1}{2}\theta = \frac{1 - \cos \theta}{2} \right)^\dagger$$

$$= 2a \left(-2 \cos \frac{1}{2}\theta \right) \Big|_0^{2\pi}$$

$$= 8a.$$

The cycloid has some surprising physical properties (too hard to prove in this course). If a frictionless slide is to be built so that children can slide down under the force of gravity from an arbitrary point A to an arbitrary point B, then one built in the shape of a half an arch of a reflected cycloid will produce the least time for the trip (Fig. 8). Furthermore, if several children slide down the reflected arch from different points, they all arrive at the lowest point at the same time.

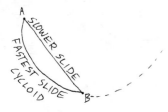

FIG. 8

Credibility of the integral models   As this chapter has shown, to compute a total size (volume, area, arc length) we divide the object into pieces and find d size (dV, dA, ds) of a piece. The formulas we use for dV, dA and ds are not exact. In Figs. 1–3, dA is only approximately $[u(x) - l(x)]\,dx$ since each strip is only approximately rectangular. In Fig. 5, ds is only approximately $\sqrt{dx^2 + dy^2}$ and furthermore in Example 2, the length dy is only approximately $3x^2\,dx$. However, when the integral adds dV's, dA's or ds's, we believe (not prove, but merely believe) that the value of the integral deserves to be called the *exact* value of the total volume V, total area A and total arc length s. The integral not only adds, but also takes a limit as dx approaches 0, and we count on the limit process to wipe out the approximation error.

†It is not true in general that taking square roots on both sides of the identity produces

(*) $$\sin \frac{1}{2}\theta = \sqrt{\frac{1 - \cos \theta}{2}},$$

because the right-hand side of (*) is positive while the left-hand side may be negative. But it *is* true when θ is in the interval $[0, 2\pi]$, the interval of integration, since in that case, $\sin \frac{1}{2}\theta$ *is* positive.

(As further reassurance, whenever a previous formula for size exists, it agrees with the integral. Problem 4 will show that the integral formula for arc length does produce the standard formula for the distance between two points.)

Not every approximation for $d\,$size can be integrated to achieve a reasonable total. In the next section we will have to be careful to avoid a bad model for surface area.

Problems for Section 6.3

1. Find the area of the region with the indicated boundaries.

(a) $y = x^2$, $y = 3x$
(b) $y = x^2$, $x = y^2$
(c) $xy = 8$, line AB where $A = (-2, -4)$ and $B = (-1, -8)$
(d) $y = x^2 - 4x + 3$, the x-axis

2. Find the area of the region in (a) Fig. 9   (b) Fig. 10.

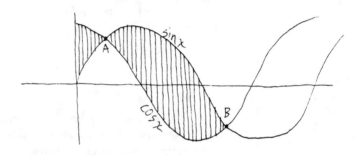

FIG. 9

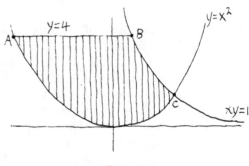

FIG. 10

3. Express with an integral the arc length along the indicated curve.

(a) $y = e^x$ between $(0, 1)$ and $(1, e)$ (c) $xy = 1$ between $(1, 1)$ and $(2, \frac{1}{2})$
(b) $x = y^3$ between $(0, 0)$ and $(64, 4)$ (d) $x = 2t + 1$, $y = t^2$ between the points $(3, 1)$ and $(9, 16)$

4. Use an integral to find the distance between the points $A = (x_1, y_1)$ and $B = (x_2, y_2)$.

6.4 The Surface Area of a Cone and a Sphere

This section will continue the geometric applications of the integral by deriving the surface area formulas for a cone and a sphere. (Its omission will not affect your understanding of any other section of the book.)

A cylinder with height h and radius r may be cut open and unrolled to form a rectangle with one dimension h and the other dimension equal to the perimeter $2\pi r$ of the circular end of the cylinder. Therefore the (lateral) surface area (not including top and bottom) of the cylinder is $2\pi rh$. To find the surface area of *noncylinders*, we need a formula dS for the (lateral) surface area of an *almost*-cylindrical slab. Figure 1(a) shows a typical slab with height dx and "radius" r. It is not precisely cylindrical since the radius varies; in fact Fig. 1(a) deliberately exaggerates the variation in radius to show an accordion-like ridge of length ds. In Example 1 of Section 6.1 we ignored the varying radius and selected the volume formula $dV = 2\pi r^2\,dx$ (Fig. 1(b)). If we were to continue to ignore the varying radius, we would choose dS to be $2\pi r\,dx$. But with this dS, $\int_a^b dS$ produces values which do not match results from geometry. (If a surface is cut open and unit squares drawn on it, the number of squares does not agree with the integral.) The variation of the radius which we successfully ignored in finding dV cannot be ignored in finding dS. *(A wrinkled elephant has about the same volume as, but much more surface area than, an unwrinkled elephant.)* To find an appropriate formula for dS, imagine the accordion (Fig. 1(a)) pulled open to form a genuine cylinder with height ds, not dx (Fig. 1(c)). Then, by the standard formula for the surface area of a cylinder, the newly created cylinder, hence the original almost-cylinder, has surface area

$$(1) \qquad\qquad dS = 2\pi r\,ds.$$

We are now ready to use (1) on cones and spheres.

Surface area of a cone   Consider a cone with radius R, height h and slant height s. To find its (lateral) surface area, begin by dividing the cone into slabs. Figure 2 shows a typical slab with thickness dx around position x on the indicated number line. To use (1), we need the slab radius and ds. By

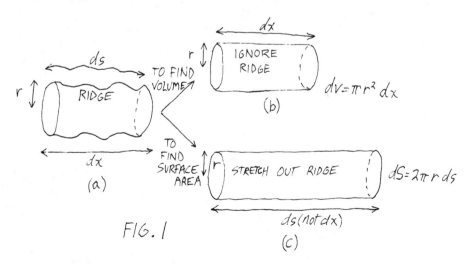

FIG. 1

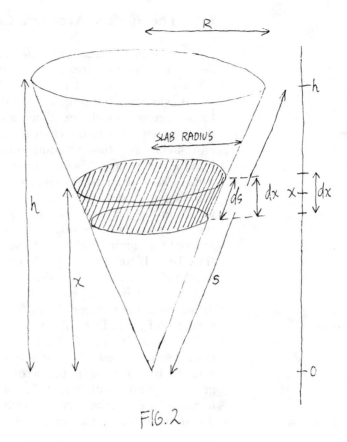

FIG. 2

similar triangles,

$$\frac{\text{slab radius}}{x} = \frac{R}{h},$$

so

$$\text{slab radius} = \frac{Rx}{h}.$$

Again by similar triangles,

$$\frac{ds}{dx} = \frac{s}{h}$$

so

$$ds = \frac{s}{h}\, dx.$$

Then, by (1),

$$dS = 2\pi\left(\frac{Rx}{h}\right)\frac{s}{h}\, dx = \frac{2\pi Rs}{h^2}\, x\, dx,$$

and

$$S = \frac{2\pi Rs}{h^2}\int_{x=0}^{x=h} x\, dx = \frac{2\pi Rs}{h}\frac{x^2}{2}\bigg|_0^h = \pi Rs.$$

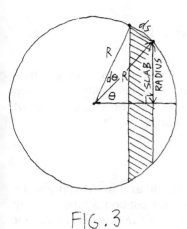

FIG. 3

Surface area of a sphere Consider a sphere with radius R. To find its surface area, divide the sphere into slabs. It will be convenient to locate slabs (shown in cross section in Figs. 3 and 4) using a central angle θ rather than position along a horizontal line. For the typical slab in Fig. 3, $ds = R\,d\theta$ by (5) of Section 1.3, and the slab radius is $R \sin\theta$ by trigonometry. Therefore, by (1),

$$dS = 2\pi R \sin\theta \cdot R\,d\theta = 2\pi R^2 \sin\theta\,d\theta.$$

The sphere is packed with slabs whose corresponding values of θ range from 0 to π (Fig. 4) so

$$S = \int_0^\pi dS = 2\pi R^2 \int_0^\pi \sin\theta\,d\theta = 2\pi R^2(-\cos\theta)\Big|_0^\pi = 4\pi R^2.$$

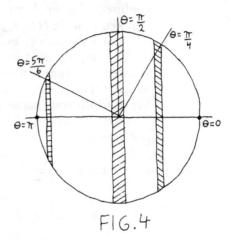

FIG. 4

6.5 Integrals with a Variable Upper Limit

This section describes a new way of creating functions, and discusses applications, computation and derivatives of the new functions.

Introductory example Suppose a particle starts at time 4 and travels with speed $2x$ feet per second at time x. The problem is to find the distance traveled by time 7, and then more generally, the *cumulative distance traveled by time x*, denoted by $s(x)$.

Divide the time interval $[4, 7]$ into subintervals, with a typical subinterval containing time x and of duration dx seconds. The distance ds traveled during the dx seconds is $2x\,dx$ (since distance = speed × time), and the total distance traveled by time 7 is $\int_4^7 2x\,dx = x^2\big|_4^7 = 33$.

More generally,

(1) cumulative distance $s(x)$ traveled up to time x

$$= \int_4^x 2x\,dx = x^2\bigg|_4^x = x^2 - 16.$$

In order to distinguish the independent variable x of the function $s(x)$ from the dummy variable of integration, we usually choose a letter other than x for the dummy variable and rewrite (1) as

$$(1') \qquad s(x) = \int_4^x 2t\,dt = t^2 \Big|_4^x = x^2 - 16\,.$$

Integrals with a variable upper limit   The function $s(x)$ in $(1')$ is given by an integral with an upper limit of integration x. More generally, for a given function f and fixed number a, $\int_a^x f(t)\,dt$ is a function of the upper limit of integration, and we may define a new function $I(x)$ by

$$(2) \qquad I(x) = \int_a^x f(t)\,dt\,.$$

For example, $I(4)$ is the number $\int_a^4 f(t)\,dt$. The integral in (2) can also be written as $\int_a^x f(u)\,du$, $\int_a^x f(r)\,dr$ and so on. However, most books avoid writing $I(x) = \int_a^x f(x)\,dx$ so that the independent variable of the function $I(x)$ is not confused with the dummy variable in the integral, and the student is not tempted to write $I(4) = \int_a^4 f(4)\,d4$, which is meaningless.

The introductory example illustrates one application of the functions in (2). They are used to represent a cumulative total such as the distance traveled until time x, the mass of a rod up to position x, or your income up to age x. The particular lower limit used depends on the time, position or age at which you choose to begin the accumulation.

Some functions of the form (2) are especially useful in mathematics and science:

$$\text{Erf } x = \frac{2}{\sqrt{\pi}} \int_0^x e^{-t^2}\,dt \qquad \text{(the error function)}$$

$$(3) \qquad \text{Ei } x = \int_1^x \frac{e^{-t}}{t}\,dt \qquad \text{(the exponential-integral function)}$$

$$\text{Si } x = \int_0^x \frac{\sin t}{t}\,dt \qquad \text{(the sine-integral function)}\,.$$

The integral in $(1')$ is defined only for $x > 4$ since an integral is defined only on an interval of the form $[a, b]$ where $b > a$. On the other hand, the function $s(x)$ is 0 when $x = 4$ since no distance has yet accumulated. This suggests the definition

$$(4) \qquad \int_a^a f(t)\,dt = 0\,.$$

With this definition, the function in (2) is defined for $x \geq a$, and $I(a) = 0$.

Computing $I(x)$ If $f(t)$ has a readily available antiderivative, then an explicit formula for $I(x)$ may be found using the Fundamental Theorem. For example,

$$(5) \qquad \text{if } I(x) = \int_1^x 3t^2\,dt \quad \text{then} \quad I(x) = t^3 \Big|_1^x = x^3 - 1\,;$$

$$(6) \qquad \text{if } J(x) = \int_2^x 3t^2\,dt \quad \text{then} \quad J(x) = t^3 \Big|_2^x = x^3 - 8\,.$$

Note that $I(x)$ and $J(x)$ differ by only a constant since they begin the same accumulation process but from different starting places, that is, with different lower limits. In particular they differ by the constant $\int_1^2 3t^2\,dt = t^3 \Big|_1^2 = 7$.

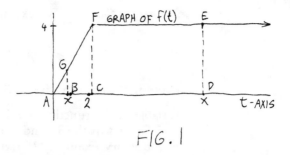

FIG. 1

If the graph of f is simple, it may be possible to find a formula for $I(x)$ using cumulative area. Suppose $f(t)$ is the function shown in Fig. 1, and $I(x) = \int_0^x f(t)\,dt$. Consider a value of x between 0 and 2 (see point B). Since $\overline{AB} = x$, we have $\overline{GB} = 2x$ by similar triangles. So

$$I(x) = \text{area of triangle } ABG = \frac{1}{2}x \cdot 2x = x^2.$$

For a value of x larger than 2 (see point D),

$$I(x) = \text{area of triangle } ACF + \text{area of rectangle } CDEF$$

$$= \frac{1}{2} \cdot 2 \cdot 4 + 4(x - 2)$$

$$= 4x - 4.$$

Therefore

$$I(x) = \begin{cases} x^2 & \text{if } 0 \le x \le 2 \\ 4x - 4 & \text{if } x > 2. \end{cases}$$

On the other hand, it is more difficult to evaluate the functions in (3). It can be shown in advanced courses that it is not possible to find anti-derivatives for e^{-t^2}, $\dfrac{e^{-t}}{t}$ and $\dfrac{\sin t}{t}$ using the basic functions listed in Section 1.1; so Erf, Ei and Si cannot be simplified as in (5) and (6). However, tables of values for Erf, Ei and Si can be produced by numerical integration. For example, Si $\pi = \int_0^\pi \dfrac{\sin t}{t}\,dt$, and its value may be approximated with a numerical integration routine such as Simpson's rule.

As still another method of evaluating an integral with a variable upper limit, given a fixed number a, an electric network can be designed so that if voltage $f(t)$ is fed in at time t, the network will produce, on an oscilloscope, the graph of the function $I(x) = \int_a^x f(t)\,dt$.

The derivative of $I(x)$ When functions of the form $I(x)$ arise, we want to be able to find their derivatives.

Consider the functions $I(x)$ and $J(x)$ defined in (5) and (6). From their explicit formulas we can see that $I'(x)$ and $J'(x)$ are both $3x^2$, the *integrand used in the original formulation of $I(x)$ and $J(x)$.* This is not a coincidence. It can be shown in general that *if $I(x) = \int_a^x f(t)\,dt$ then $I'(x) = f(x)$ at all points where f is continuous.* In other words, *if a continuous function f is integrated with a variable upper limit x, and then the integral is differentiated with respect to x, the original function f is obtained.* This result is called the *Second Fundamental*

Theorem of Calculus. For example,

$$(7) \qquad\qquad D_x \text{ Si } x = \frac{\sin x}{x}.\dagger$$

(Note that the derivative of Si x is $\dfrac{\sin x}{x}$, *not* $\dfrac{\sin t}{t}$ since the independent variable of the function Si x is named x, not t.)

To see why the Second Fundamental Theorem holds, first consider the introductory example. If $f(x)$ is the speed of a particle at time x and $I(x) = \int_a^x f(t)\,dt$, then $I(x)$ is the cumulative mileage traveled by time x (the odometer reading). Therefore $I'(x)$ is the rate of change of mileage with respect to time, which is the speed of the particle.

To understand the Second Fundamental Theorem from a geometric point of view, let x increase by dx and consider the corresponding change dI in $I(x)$. Since $I(x)$ is the cumulative area under the graph of f, Fig. 2 shows that I increases by approximately a rectangular area with base dx and height $f(x)$, so $dI = f(x)\,dx$ (approximately). Equivalently

$$(8) \qquad\qquad \frac{\text{change } dI}{\text{change } dx} = f(x),$$

or, $I'(x) = f(x)$.

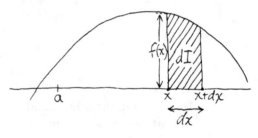

FIG. 2

Backward limits of integration So far it makes no sense to write "backward" limits such as $\int_7^2 f(x)\,dx$, where the upper limit of integration is smaller than the lower limit. The solution of a physical problem (averages, area, arc length and so on) never involves backward limits. However, there is a situation in which backward limits do arise in a natural way. The function $I(x) = \int_a^x f(t)\,dt$ is defined only for $x \geq a$. If $I(x)$ is the cumulative distance traveled by an object starting at time a, then the integral continues to have physical meaning only for $x \geq a$. But in more theoretical circumstances, it may be useful to define $I(x)$ for $x < a$, for example to have Erf x and Si x defined for $x < 0$ and Ei x defined for $x < 1$.

In one sense, the definition of $\int_b^a f(x)\,dx$, where $a < b$, can be anything we like. But it is desirable that the integral with backward limits retain the same properties as the original integral. It can be shown that, for $a < b$, if we define

---

†As already mentioned, it can be shown that $(\sin x)/x$ does not have an elementary antiderivative, that is, an antiderivative expressed in terms of the basic functions. But (7) shows that Si x is an antiderivative for $(\sin x)/x$. Therefore Si x is a *non*elementary function. Similarly, Ei x and Erf x are nonelementary.

(9)
$$\int_b^a f(x)\, dx = -\int_a^b f(x)\, dx,$$

then properties (9)–(11) of Section 5.2 still hold, and so do both fundamental theorems. For example, with the definition in (9),

$$\int_2^1 3x^2\, dx = -\int_1^2 3x^2\, dx = -x^3 \Big|_1^2 = -8 + 1 = -7.$$

But more directly, we can use the Fundamental Theorem with the backward limits and get the same answer:

$$\int_2^1 3x^2\, dx = x^3 \Big|_2^1 = 1 - 8 = -7.$$

Unfortunately, the relationship between integrals and area is different with backward limits of integration. If $a < b$ then

$$\int_a^b f(x)\, dx = \text{area above the } x\text{-axis} - \text{area below the } x\text{-axis},$$

so

$$\int_b^a f(x)\, dx = -\int_a^b f(x)\, dx = \text{area below} - \text{area above}.$$

If $I(x) = \int_2^x f(t)\, dt$ where the graph of f is given in Fig. 1, then $I(0) = \int_2^0 f(t)\, dt$ which is $-(\text{area of triangle } ACF)$, or -4.

Problems for Section 6.5

1. Find an explicit formula for $I(x)$ if $I(x) = \int_2^x (t + 5)\, dt$.

2. A wire beginning at A and extending infinitely in one direction has charge density e^{-x} charge units per foot at a point x feet from A. (a) Find the total charge in the wire. (b) Find a formula for the cumulative charge in the first x feet of the wire.

3. Suppose it begins raining at 3 P.M., and x hours later it is raining at the rate of x^3 inches per hour. For example, at 3:30 P.M. it is raining at the rate of $1/8$ inch per hour. (a) Find the total rainfall by 5 P.M. (b) Find the cumulative rainfall after x hours.

4. Figure 3 gives the graph of $f(x)$. If $I(x) = \int_0^x f(t)\, dt$, find an explicit formula for $I(x)$ for $x \geq 0$.

5. Let $I(x) = \int_1^x f(t)\, dt$ where the graph of f is shown in Fig. 4. Sketch a rough graph of $I(x)$.

6. Let $f(x) = \begin{cases} 1 & \text{if } 0 \leq x \leq 1 \\ \dfrac{1}{x} & \text{if } x > 1 \end{cases}$ and let $I(x) = \int_0^x f(t)\, dt$.

(a) Find $I(\tfrac{1}{2})$.
(b) Find $I(2)$.
(c) Find $I(x)$, in general, for $x \geq 0$.

7. Let $I(x) = \int_1^x \ln t\, dt$ and $J(x) = \int_{1/2}^x \ln t\, dt$. (a) Which is the larger of $I(7)$ and $J(7)$? (b) How do the graphs of $I(x)$ and $J(x)$ compare with one another?

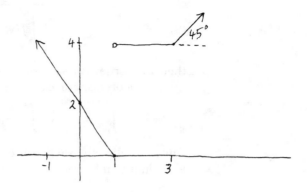

FIG. 3

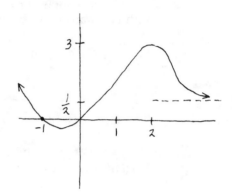

FIG. 4

8. Find (a) $\dfrac{d(\text{Erf } x)}{dx}$ (b) $\dfrac{d(\text{Ei } x)}{dx}$ (c) $\dfrac{d^2(\text{Ei } x)}{dx^2}$.

9. If $I(x) = \int_2^x \sin t^2\, dt$, find $I'(x)$ and $I''(x)$.

10. Where does Si x have relative maxima and minima?

11. (harder) Let $f(t) = \displaystyle\int_2^{x^3} \dfrac{\sin t}{t}\, dt$. (Note that the upper limit is x^3, not x.) Find $f'(x)$.

12. Find $\lim_{x \to 0} \dfrac{\text{Si } x}{x}$.

13. Evaluate the integral (which has backward limits).

(a) $\displaystyle\int_4^2 (x - 5)\, dx$

(b) $\displaystyle\int_2^0 \dfrac{1}{2x + 5}\, dx$

REVIEW PROBLEMS FOR CHAPTER 6

1. A colony of bacteria grows at the rate of $f(t)$ cubic centimeters per day at day t. By how much will it grow between days 3 and 7?

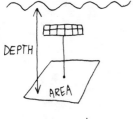

DEPTH

AREA

FIG. 1

2. Refer to Example 5 in Section 6.1 and find the moment of inertia of a solid cone with radius R, height h and density δ mass units per unit volume, which revolves around its axis of symmetry.

3. An empty scale submerged in water will register a weight due to the water pressing on it. The larger the scale and the greater the depth, the higher the scale reading. Suppose that the empty scale reading is depth × scale area (Fig. 1), so that a scale of area 6 submerged at depth 4 reads 24 pounds. If a scale lies on its side (Fig. 2) there is still a reading since water presses as hard sideways as downward, but the simple formula no longer applies since the depth is not constant. Find the scale reading in Fig. 2.

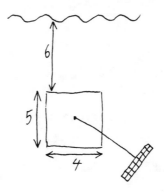

FIG. 2

4. Find the area of the region bounded by the graph of $y = \sin \pi x$ and the segment AB where $A = (\frac{3}{2}, -1)$ and $B = (2, 0)$.

5. Consider the region bounded by the lines $x + y = 12$, $y = 2x$ and the x-axis. Find its area using (a) plane geometry   (b) calculus.

6. A farmer purchases a 2-year-old sheep which produces $100 - t$ pounds of wool per year at age t. (a) Find the total amount of wool it produces for the farmer by age 4.   (b) Find the cumulative amount of wool produced for the farmer by age t.

7. Let $I(x) = \int_2^x f(t)\,dt$. Find an explicit formula for $I(x)$ if (a) $f(x) = 2x + 3$
(b) $f(x) = \begin{cases} 3x^2 & \text{for } x \le 7 \\ 5 & \text{for } x > 7 \end{cases}$ (c) $f(x)$ has the graph in Fig. 3.

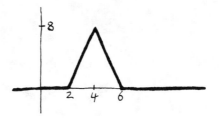

FIG. 3

8. If $I(x) = \int_5^x e^{t^2}\,dt$, find $I'(x)$ and $I''(x)$.

7/ANTIDIFFERENTIATION

7.1 Introduction

Antidifferentiation has many applications, such as finding the path of a bullet (Section 3.8), evaluating integrals (Section 5.3) and solving differential equations (Section 4.9). We began finding antiderivatives in Section 3.8 but were limited to a few standard types of problems. This chapter covers some techniques of antidifferentiation, also called indefinite integration, or simply integration, so that additional functions can be handled.

Let's compare antidifferentiation with differentiation to see what we are up against. Each operation begins with a function, probably arising from a physical problem. If the function is elementary, then differentiation is easy and mechanical. Using the derivatives of the basic functions and the rules for combinations (sums, products, quotients, compositions), we can differentiate *any* elementary function, no matter how complicated. Furthermore, the derivative is another elementary function. The situation for antidifferentiation is very different. First of all, an elementary function might not have an elementary antiderivative. Even if there is an elementary antiderivative, there is no mechanical rule for finding it. There are no product, quotient and chain rules for antiderivatives. The best we can offer so far are the sum and constant-multiple rules (Section 3.8):

$$(1) \qquad \int [f(x) + g(x)]\,dx = \int f(x)\,dx + \int g(x)\,dx$$

$$(2) \qquad \int cf(x)\,dx = c \int f(x)\,dx \qquad \text{where } c \text{ is a fixed constant}.$$

In the absence of sufficient combination rules, it is common practice to consult tables of antiderivatives. However, tables can't contain *every* function because there are infinitely many functions. If a function is not in the tables we try to "reduce" it to one that is in the tables. (This is not a first encounter with incomplete tables. Trigonometry tables only go up to 90°. To find sin 91°, the reduction rule sin 91° = sin 89° is used.) If we learn from the tables that our function has no elementary antiderivative, we quit, with the justification that this course concentrates on elementary functions. If we cannot find our function (reduced or unreduced) in the tables, we are forced to quit again, although it is possible that a larger set of tables or extended reduction techniques would help. (An entire book of tables is usually available in the library.)

Our tables do not contain the following very simple antiderivative formulas which should be in your *mental* tables:

(3)
$$\int x^r \, dx = \frac{x^{r+1}}{r+1} + C \qquad \text{for } r \neq -1$$

(4)
$$\int \frac{1}{x} \, dx = \ln|x| + C$$

(5)
$$\int e^x \, dx = e^x + C$$

(6)
$$\int \sin x \, dx = -\cos x + C$$

(7)
$$\int \cos x \, dx = \sin x + C.$$

Much of this chapter is concerned with procedures for reducing functions not listed in the tables to listed functions. (One of the difficulties here is that there is no precise rule for deciding how to reduce or even if a reduction is possible.) We will also show how some of the formulas in the tables were derived. (In retrospect, each antidifferentiation formula in the table can be checked by differentiating the answer.)

7.2 Substitution

Substitution is a very effective method for reducing a function not listed in the tables to one that is listed. The method involves reversing the chain rule. As with all antidifferentiation methods, you will have to practice to become accustomed to it.

By the chain rule, $D_x \sin x^2 = 2x \cos x^2$, so $\int 2x \cos x^2 \, dx = \sin x^2 + C$. But how can we obtain the antiderivative formula *without* seeing the derivative problem first? To go backwards and find $\int 2x \cos x^2 \, dx$, use the device of letting $u = x^2$, $du = 2x \, dx$. Substitute this into the integral to get

$$\int 2x \cos x^2 \, dx = \int \cos u \, du = \sin u + C$$

$$= \sin x^2 + C \qquad (\text{replace } u \text{ by } x^2).$$

We'll continue to illustrate the technique with some more examples.

Example 1 To find $\int \dfrac{x^3}{(2x^4 + 7)^2} \, dx$ (which is not in the tables), let $u = 2x^4 + 7$, $du = 8x^3 \, dx$. Replace $(2x^4 + 7)^2$ by u^2 and replace $x^3 \, dx$ by $\frac{1}{8} du$ to obtain

$$\int \frac{x^3}{(2x^4 + 7)^2} \, dx = \frac{1}{8} \int \frac{du}{u^2} = \frac{1}{8} \frac{u^{-1}}{-1} + C = -\frac{1}{8(2x^4 + 7)} + C.$$

Example 2 To find $\int \cos^2 x \sin x \, dx$, let $u = \cos x$, $du = -\sin x \, dx$. Then replace $\cos^2 x$ by u^2 and replace $\sin x \, dx$ by $-du$ to get

$$\int \cos^2 x \, \sin x \, dx = -\int u^2 \, du = -\frac{1}{3} u^3 + C = -\frac{1}{3} \cos^3 x + C.$$

Choosing a good substitution   Unfortunately there is no set rule for deciding when or what to substitute. One useful tactic is to search the integrand for an expression whose derivative is a factor in the integrand, and let u be that expression. In Example 1, the expression is $2x^4 + 7$; its derivative x^3 (give or take an 8) is a factor. In Example 2, the expression is $\cos x$; its derivative $\sin x$ (give or take a negative sign) is a factor. It is also possible for more than one substitution to work or for no substitution to help.

Example 3   From Section 3.8, we have $\int e^{3x} \, dx = \frac{1}{3} e^{3x} + C$, by inspection. The extra factor $\frac{1}{3}$ is inserted to counteract the factor 3 produced by the chain rule when we differentiate back. The problem can also be done by substitution. Let $u = 3x$, $du = 3 \, dx$. Then $\int e^{3x} \, dx = \frac{1}{3} \int e^u \, du = \frac{1}{3} e^u + C = \frac{1}{3} e^{3x} + C$. The extra factor $\frac{1}{3}$ is automatically inserted by the substitution process.

Warning   Don't forget to substitute for dx. In the preceding example, dx must be replaced by $\frac{1}{3} du$. The substitution process will give wrong answers if dx is ignored, lost or incorrectly replaced by just du.

Example 4   Find $\int x^5 \cos x^3 \, dx$.
   Solution: Try the tables first, but without success. Then try substituting $u = x^3$, $du = 3x^2 \, dx$ to get

$$\int x^5 \cos x^3 \, dx = \int x^5 \cos u \frac{du}{3x^2} \qquad \text{(replace } x^3 \text{ by } u, \, dx \text{ by } du/3x^2\text{)}$$

$$= \frac{1}{3} \int x^3 \cos u \, du \qquad \text{(cancel } x^2\text{)}$$

$$= \frac{1}{3} \int u \, \cos u \, du \qquad \text{(replace } x^3 \text{ by } u\text{)}$$

$$= \frac{1}{3} (\cos u + u \, \sin u) + C \qquad \text{(formula 49)}$$

$$= \frac{1}{3} (\cos x^3 + x^3 \sin x^3) + C.$$

Remember that every antidifferentiation problem can be checked by differentiating the answer. In this case, you can check to see that the derivative of $\frac{1}{3}(\cos x^3 + x^3 \sin x^3)$ is $x^5 \cos x^3$.

Warning   Don't forget to substitute at the end of a problem to get a final answer in terms of the original variable.

Example 5   Formula 13 in the tables is

$$\int \frac{x \, dx}{\sqrt{a + bx}} \, dx = \frac{2(bx - 2a)}{3b^2} \sqrt{a + bx} + C.$$

The formula can be derived in the first place with the substitution $u = a + bx$ and also with $u = \sqrt{a + bx}$. In the latter case, it is algebraically easier to write x in terms of u and find dx in terms of du rather than du in terms of dx. We have $u^2 = a + bx$, so $x = \dfrac{u^2 - a}{b}$, $dx = \dfrac{2}{b} u\, du$. Therefore

$$\int \frac{x\, dx}{\sqrt{a + bx}} = \int \frac{\dfrac{u^2 - a}{b}}{u} \frac{2}{b} u\, du$$

$$= \frac{2}{b^2} \int (u^2 - a)\, du \qquad \text{(algebra)}$$

$$= \frac{2}{b^2} \left(\frac{u^3}{3} - au \right) + C$$

$$= \frac{2}{3b^2} u(u^2 - 3a) + C$$

$$= \frac{2}{3b^2} \sqrt{a + bx}\, (a + bx - 3a) + C$$

$$= \frac{2}{3b^2} (bx - 2a) \sqrt{a + bx} + C.$$

Warning The tables list the formula

$$(1) \qquad \int u \sin u\, du = \sin u - u \cos u + C.$$

Therefore it is also true that $\int x \sin x\, dx = \sin x - x \cos x + C$ since all we did was change every occurrence of the dummy variable u to x. Similarly, it is also true that $\int t \sin t\, dt = \sin t - t \cos t + C$ and so on. However

$$(2) \qquad \int \frac{x}{2} \sin \frac{x}{2}\, dx \text{ is NOT } \sin \frac{x}{2} - \frac{x}{2} \cos \frac{x}{2} + C$$

because not *all* occurrences of u in (1) have been changed to $x/2$; in particular the occurrence of u in the symbol du did not become $x/2$. Instead, to do the integral in (2), let $u = x/2$. Then $du = \frac{1}{2} dx$ and

$$\int \frac{x}{2} \sin \frac{x}{2}\, dx = 2 \int u \sin u\, du = 2(\sin u - u \cos u) + C$$

$$= 2 \left(\sin \frac{x}{2} - \frac{x}{2} \cos \frac{x}{2} \right) + C.$$

Furthermore, despite (1),

$$(3) \qquad \int x^2 \sin x^2\, dx \text{ is NOT } \sin x^2 - x^2 \cos x^2 + C,$$

because not *every* occurrence of u in (1) has been replaced by x^2. In an attempt to apply (1) to the integral in (3), let $u = x^2$. But then $du = 2x\, dx$,

$$\int x^2 \sin x^2\, dx = \int u \sin u \frac{du}{2x} = \int u \sin u \frac{du}{2\sqrt{u}} = \frac{1}{2} \int \sqrt{u}\, \sin u\, du$$

and it turns out that (1) doesn't apply at all.

Problems for Section 7.2

1. $\int x e^{x^2}\, dx$

2. $\int x\sqrt{3x^2 + 7}\, dx$

3. $\int \sqrt{3 + 5x}\, dx$

4. $\int \dfrac{1}{\sqrt{3 + 7x}}\, dx$

5. $\int \tan^{14}x\ \sec^2 x\, dx$

6. $\int \dfrac{x - 1}{(x + 1)^5}\, dx$

7. $\int \dfrac{\sec\theta\,\tan\theta}{\sqrt{1 + 2\sec\theta}}\, d\theta$

8. $\int \dfrac{1}{x\,\ln x}\, dx$

9. $\int x^3 \sin x^2\, dx$

10. $\int (1 + 3x)^7\, dx$

11. $\int \dfrac{1}{2 - 3x}\, dx$

12. $\int \dfrac{1}{(2 - x)^3}\, dx$

13. $\int \cos\!\left(\dfrac{1}{2}\theta - 1\right) d\theta$

14. $\int x e^{-x}\, dx$

15. $\int \cos^3 x\ \sin x\, dx$

16. $\int e^{-x}\, dx$

17. $\int x \sin 3x\, dx$

18. $\int \sin^2 \pi x\, dx$

19. $\int 3x \sin x\, dx$

20. $\int x^2 \cos 3x\, dx$

21. $\int \ln(2x + 3)\, dx$

22. $\int \dfrac{dx}{\cos x}$

23. We know that $\int \dfrac{1}{1 + x^2}\, dx = \arctan x$. Is the following antidifferentiation correct:

$$\int \frac{1}{1 + 3x^2}\, dx = \int \frac{1}{1 + (\sqrt{3}\,x)^2}\, dx = \arctan \sqrt{3}\,x + C\,?$$

24. Find if possible at this stage (a) $\int \tan^{-1} 3x\, dx$ (b) $\int \tan^{-1} x^2\, dx$.

25. Derive formula 31 for $\int \tan x\, dx$ using substitution on $\int \dfrac{\sin x}{\cos x}\, dx$.

26. Derive formula 33 for $\int \sec x\, dx$ by multiplying numerator and denominator by $\sec x + \tan x$ and using substitution.

27. Derive formula 39 for $\int \sin^2 x\, dx$ using the trigonometric identity $\sin^2 x = \frac{1}{2}(1 - \cos 2x)$.

7.3 Pre-Table Algebra I

If the function to be antidifferentiated is not listed in the tables, sometimes it may be reduced to a listed function by algebra. This section and the next offer algebraic suggestions.

Example 1 Consider $\int \dfrac{1}{\sqrt{6x^2 + 3}}\, dx$. Formula 23 in the tables lists $\int \dfrac{1}{\sqrt{a^2 + u^2}}\, du$ which matches the given problem, except for the 6. Thus we try to eliminate the 6. One possibility is to factor it out to obtain

$$\int \frac{1}{\sqrt{6x^2 + 3}}\, dx = \int \frac{1}{\sqrt{6(x^2 + \frac{1}{2})}}\, dx = \frac{1}{\sqrt{6}} \int \frac{1}{\sqrt{x^2 + \frac{1}{2}}}\, dx.$$

Then use formula 23 with $a^2 = \frac{1}{2}$ to get

(1)
$$\int \frac{1}{\sqrt{6x^2 + 3}}\, dx = \frac{1}{\sqrt{6}} \ln\!\left(x + \sqrt{x^2 + \frac{1}{2}}\right) + C.$$

Another possibility is to write $6x^2$ as $(\sqrt{6}x)^2$ and then let $u = \sqrt{6}x$, $du = \sqrt{6}\, dx$. With this substitution,

$$\int \frac{1}{\sqrt{6x^2 + 3}}\, dx = \int \frac{1}{\sqrt{(\sqrt{6}x)^2 + 3}}\, dx = \int \frac{1}{\sqrt{u^2 + 3}} \frac{du}{\sqrt{6}}$$

$$= \frac{1}{\sqrt{6}} \ln(u + \sqrt{u^2 + 3}) + C \qquad \text{(formula 23)}$$

(2)
$$= \frac{1}{\sqrt{6}} \ln(\sqrt{6}x + \sqrt{6x^2 + 3}) + C.\dagger$$

Warning Don't forget to substitute for dx in carrying out the substitution.

Example 2 $\int \sqrt{3x^2 + 4x - 8}\, dx$ isn't in a small set of tables which concentrates on forms involving $u^2 - a^2$ and $a^2 \pm u^2$ rather than on forms involving $Ax^2 + Bx + C$. In this case, use the algebraic process called *completing the square*. First factor out the leading coefficient to get

$$3x^2 + 4x - 8 = 3\!\left(x^2 + \frac{4}{3}x - \frac{8}{3}\right).$$

Then take half the coefficient of x, square it to obtain $\frac{4}{9}$, and add and subtract that value within the parentheses:

$$3x^2 + 4x - 8 = 3\!\left(x^2 + \frac{4}{3}x + \frac{4}{9} - \frac{4}{9} - \frac{8}{3}\right) = 3\!\left[\left(x + \frac{2}{3}\right)^2 - \frac{28}{9}\right].$$

Thus

$$\int \sqrt{3x^2 + 4x - 8}\, dx = \sqrt{3} \int \sqrt{\left(x + \frac{2}{3}\right)^2 - \frac{28}{9}}\, dx.$$

Now let $u = x + \frac{2}{3}$, $du = dx$ to get

$\dagger$Note that at first glance the two methods do not seem to produce the same answers in (1) and (2). But (2) may be rewritten as

$$\frac{1}{\sqrt{6}} \ln\!\left[\sqrt{6}\!\left(x + \sqrt{x^2 + \frac{1}{2}}\right)\right] + C \qquad \text{(by factoring)}$$

$$= \frac{1}{\sqrt{6}} \ln \sqrt{6} + \frac{1}{\sqrt{6}} \ln\!\left(x + \sqrt{x^2 + \frac{1}{2}}\right) + C \qquad \text{(since } \ln ab = \ln a + \ln b)$$

$$= \frac{1}{\sqrt{6}} \ln\!\left(x + \sqrt{x^2 + \frac{1}{2}}\right) + K \qquad \left(\text{call } \frac{1}{\sqrt{6}} \ln \sqrt{6} + C \text{ a new constant } K\right)$$

which matches (1).

$$\int \sqrt{3x^2 + 4x - 8} = \sqrt{3} \int \sqrt{u^2 - \frac{28}{9}}\, du$$

$$= \frac{\sqrt{3}\,u}{2} \sqrt{u^2 - \frac{28}{9}} - \frac{14}{9}\sqrt{3}\,\ln\left|u + \sqrt{u^2 - \frac{28}{9}}\right| + C$$

(formula 28)

$$= \sqrt{3}\,\frac{\left(x + \frac{2}{3}\right)}{2}\sqrt{\left(x + \frac{2}{3}\right)^2 - \frac{28}{9}}$$

$$- \frac{14\sqrt{3}}{9}\ln\left|x + \frac{2}{3} + \sqrt{\left(x + \frac{2}{3}\right)^2 - \frac{28}{9}}\right| + C.$$

Example 3 *Improper* fractions, such as $\dfrac{x^5}{x^2 + x + 2}$ and $\dfrac{3x^5}{x^5 - 7}$, are those where the degree of the numerator is greater than or equal to the degree of the denominator. *Proper* fractions, such as $\dfrac{3x}{x^2 + 1}$, are those where the degree of the numerator is less than the degree of the denominator. The improper kind are rarely listed in antiderivative tables. To find an antiderivative for an improper fraction that is not listed, begin with *long division*. Consider $\int \dfrac{x^5\, dx}{x^2 + x + 2}$. We have

$$
\begin{array}{r}
x^3 - x^2 - x + 3 \\
x^2 + x + 2\,\overline{)\,x^5 } \\
\underline{x^5 + x^4 + 2x^3} \\
-x^4 - 2x^3 \\
\underline{-x^4 - x^3 - 2x^2} \\
-x^3 + 2x^2 \\
\underline{-x^3 - x^2 - 2x} \\
3x^2 + 2x \\
\underline{3x^2 + 3x + 6} \\
-x - 6.
\end{array}
$$

So

(3) $$\underbrace{\frac{x^5}{x^2 + x + 2}}_{\text{improper fraction}} = \underbrace{x^3 - x^2 - x + 3}_{\text{polynomial}} + \underbrace{\frac{-x - 6}{x^2 + x + 2}}_{\text{proper fraction}}.$$

This illustrates that an improper fraction can be written as the sum of a polynomial and a proper fraction, each of which is easier to antidifferentiate than the original improper fraction. For the polynomial in (3) we have

(4) $$\int (x^3 - x^2 - x + 3)\, dx = \frac{1}{4}x^4 - \frac{1}{3}x^3 - \frac{1}{2}x^2 + 3x + C.$$

To antidifferentiate the proper fraction in (3), first separate it into the sum

(5) $$-\frac{x}{x^2 + x + 2} - \frac{6}{x^2 + x + 2}.$$

Then, for the first term in (5), we have

$$-\int \frac{x}{x^2 + x + 2} \, dx = -\frac{1}{2} \ln|x^2 + x + 2| + \frac{1}{2} \int \frac{dx}{x^2 + x + 2}$$

(formula 2)

$$= -\frac{1}{2} \ln|x^2 + x + 2| + \frac{1}{\sqrt{7}} \tan^{-1} \frac{2x + 1}{\sqrt{7}} + C$$

(formula 1b).

For the second term in (5), use formula 1b to get

(7) $$-6\int \frac{dx}{x^2 + x + 2} = \frac{-12}{\sqrt{7}} \tan^{-1} \frac{2x + 1}{\sqrt{7}} + C$$

Finally, combine (4), (6) and (7) for the final answer

$$\int \frac{x^5}{x^2 + x + 2} \, dx = \frac{x^4}{4} - \frac{x^3}{3} - \frac{x^2}{2} + 3x - \frac{11}{\sqrt{7}} \tan^{-1} \frac{2x + 1}{\sqrt{7}}$$

$$-\frac{1}{2} \ln|x^2 + x + 2| + C.$$

Problems for Section 7.3

1. $\int \dfrac{dx}{\sqrt{2 + 6x - x^2}}$ 5. $\int x\sqrt{2x + x^2} \, dx$

2. $\int \dfrac{1}{\sqrt{x + 2x^2}} \, dx$ 6. $\int \dfrac{x}{2x + 6} \, dx$ three ways (long division, tables, substitution)

3. $\int \dfrac{1}{\sqrt{3x^2 - 5}} \, dx$ 7. $\int \dfrac{x^2}{x^2 + 1} \, dx$

4. $\int \dfrac{x^4 + 2x}{x^2 + 4} \, dx$

7.4   Pre-Table Algebra II: Partial Fraction Decomposition

The preceding section advised dividing out *improper* fractions because they are rarely listed in tables. But tables often omit *proper* fractions as well, when the degree of the denominator is greater than 2. *Partial fraction decomposition* is an algebraic technique that helps in this case.

The *addition* of fractions is a familiar idea from algebra. By finding a least common denominator we have

(1)
$$\frac{2x}{x^2 + 6} + \frac{7}{2x - 9} = \frac{2x(2x - 9) + 7(x^2 + 6)}{(x^2 + 6)(2x - 9)}$$

$$= \frac{11x^2 - 18x + 42}{2x^3 - 9x^2 + 12x - 54}.$$

However, if the aim is to antidifferentiate the expression on the left in (1), it is silly to change to the rightmost fraction. The pieces on the left are easier to handle than the single fraction on the right. In fact, the point is to learn how to *decompose* $\dfrac{11x^2 - 18x + 42}{2x^3 - 9x^2 + 12x - 54}$ back to $\dfrac{2x}{x^2 + 6} + \dfrac{7}{2x - 9}$. In general, we want to *decompose a proper fraction which is not in the tables into a sum of "partial fractions" which are either in the tables* (formulas 1–4) *or which may be antidifferentiated by substitution or inspection.* The decomposition is accomplished in several steps, and it works only for *proper* fractions. We will describe the general steps, and cover the details in the examples. (The proof of the method is beyond the scope of the course.)

Step 1 Factor the denominator as far as possible, which means into linear factors and nonfactorable (also called irreducible) quadratics. A quadratic is taken to be nonfactorable only if its two linear factors involve nonreal numbers. For example $x^2 - 3$ *does* factor, namely into $(x - \sqrt{3})(x + \sqrt{3})$, but $x^2 + 4$, which equals $(x - 2i)(x + 2i)$, is considered nonfactorable. Quadratics can sometimes be factored by trial and error, but the following general rule is available:

If $b^2 - 4ac < 0$ then $ax^2 + bx + c$ doesn't factor.

(2) If $b^2 - 4ac \geq 0$ then

$$ax^2 + bx + c = a\left(x - \frac{-b + \sqrt{b^2 - 4ac}}{2a}\right)\left(x - \frac{-b - \sqrt{b^2 - 4ac}}{2a}\right).$$

There is no easy rule for factoring polynomials of higher degree but they can all be factored into linear and nonfactorable quadratics.

Step 2 The nature of the decomposition depends on the factors in the denominator.

If a *linear* factor such as $2x + 3$ appears in the denominator then a fraction of the form $A/(2x + 3)$ appears as one of the partial fractions in the decomposition.

If a *repeated linear* factor such as $(2x + 3)^3$ appears in the denominator then

$$\frac{A}{2x + 3} + \frac{B}{(2x + 3)^2} + \frac{C}{(2x + 3)^3}$$

appears in the decomposition.

If a *nonfactorable quadratic* such as $x^2 + x + 10$ appears in the denominator then $\dfrac{Ax + B}{x^2 + x + 10}$ appears in the decomposition.

If a *repeated nonfactorable quadratic* such as $(x^2 + x + 10)^4$ appears in the denominator then

$$\frac{Ax + B}{x^2 + x + 10} + \frac{Cx + D}{(x^2 + x + 10)^2} + \frac{Ex + F}{(x^2 + x + 10)^3} + \frac{Gx + H}{(x^2 + x + 10)^4}$$

appears in the decomposition.

Step 3 Determine $A, B, C, \cdots$ in the decomposition by the methods to be shown in the examples.

Decomposition is a useful algebraic tool which has applications in addition to antidifferentiation. It will be used in Section 8.7 to find a power series for a quotient of polynomials, and it occurs in the theory of Laplace Transforms, encountered in advanced engineering mathematics. In each instance it is easier to work separately with the partial fractions than with their sum.

Example 1 Decompose $\dfrac{2x^2 + 3x - 1}{(x + 3)(x + 2)(x - 1)}$ and then antidifferentiate.

Solution: The decomposition has the form

$$\frac{2x^2 + 3x - 1}{(x + 3)(x + 2)(x - 1)} = \frac{A}{x + 3} + \frac{B}{x + 2} + \frac{C}{x - 1}.$$

Before trying to determine A, B and C, simplify by multiplying both sides by $(x + 3)(x + 2)(x - 1)$ to obtain

$$(3) \qquad 2x^2 + 3x - 1 = A(x + 2)(x - 1) + B(x + 3)(x - 1)$$
$$+ C(x + 3)(x + 2).$$

Equation (3) is supposed to be true for all x, so we are allowed to substitute an arbitrary value of x. Use the "good" values $-3, -2, 1$ to facilitate the algebra.

$$\text{If } x = -3 \text{ then } 8 = 4A, A = 2.$$

$$\text{If } x = -2 \text{ then } 1 = -3B, B = -\frac{1}{3}.$$

$$\text{If } x = 1 \text{ then } 4 = 12C, C = \frac{1}{3}.$$

Using good values of x in this manner produces A, B, C immediately. (They are good because they make two of the factors on the right-hand side of (3) become 0.) Using other values of x will produce three equations in the three unknowns A, B, C. The equations can be solved for A, B, C, but this procedure is unnecessarily complicated. Stay with the good values of x as long as they last.

The result is

$$\frac{2x^2 + 3x - 1}{(x + 3)(x + 2)(x - 1)} = \frac{2}{x + 3} - \frac{1/3}{x + 2} + \frac{1/3}{x - 1}.$$

Finally, each term in the decomposition may be antidifferentiated by inspection to obtain

$$\int \frac{2x^2 + 3x - 1}{(x + 3)(x + 2)(x - 1)}\, dx = 2 \ln|x + 3| - \frac{1}{3} \ln|x + 2|$$
$$+ \frac{1}{3} \ln|x - 1| + K.$$

Example 2 Find $\displaystyle\int \frac{x^2 + 2x + 6}{(2x + 3)(x - 2)^2}\, dx.$

Solution: The fraction is proper, but not in the tables. The decomposition has the form

$$\frac{x^2 + 2x + 6}{(2x + 3)(x - 2)^2} = \frac{A}{2x + 3} + \frac{B}{x - 2} + \frac{C}{(x - 2)^2}.$$

Multiply both sides by $(2x + 3)(x - 2)^2$ to simplify:

(4) $x^2 + 2x + 6 = A(x - 2)^2 + B(x - 2)(2x + 3) + C(2x + 3).$

If $x = 2$ then $14 = 7C,$ $C = 2.$

If $x = -\dfrac{3}{2}$ then $\dfrac{21}{4} = \dfrac{49}{4}A,$ $A = 3/7.$

Although the good values of x are exhausted, there are still several ways to find B easily. One possibility is to *use any other value of x.* For example, if $x = 0$ then $6 = 4A - 6B + 3C$. Since we already have A and C, $B = \frac{1}{6}(4A + 3C - 6) = \frac{2}{7}$. Another possibility is to *equate coefficients.* Each side of (4) is a polynomial, and since they agree for all values of x, it can be shown that they must be the *same* polynomial. The polynomial on the left leads with an x^2 term whose coefficient is 1. When the right-hand side is multiplied out and rearranged, its x^2 term is $(A + 2B)x^2$. Equate the two coefficients of x^2 to obtain $1 = A + 2B$, $B = \frac{1}{2}(1 - A) = \frac{2}{7}$. Instead of using the coefficients of x^2 we can also use the coefficients of x. On the left side the coefficient is 2 and on the right-hand side, after simplification, the coefficient is $-4A - B + 2C$. Thus $2 = -4A - B + 2C$, $B = -4A + 2C - 2 = \frac{2}{7}$.

Therefore

$$\frac{x^2 + 2x + 6}{(2x + 3)(x - 2)^2} = \frac{3/7}{2x + 3} + \frac{2/7}{x - 2} + \frac{2}{(x - 2)^2}.$$

Each term on the right can be antidifferentiated by inspection or with a simple substitution to give

$$\int \frac{x^2 + 2x + 6}{(2x + 3)(x - 2)^2}\,dx = \frac{3}{14}\ln|2x + 3| + \frac{2}{7}\ln|x - 2| - \frac{2}{x - 2} + K.$$

Example 3   Find $\displaystyle\int \frac{3x^2 + 2x - 2}{(x - 1)(x^2 + x + 1)}\,dx.$

Solution: First see if the denominator factors further. Since $x^2 + x + 1$ is nonfactorable ($b^2 - 4ac < 0$), we can proceed to the decomposition which is of the form

$$\frac{3x^2 + 2x - 2}{(x - 1)(x^2 + x + 1)} = \frac{A}{x - 1} + \frac{Bx + C}{x^2 + x + 1}.$$

Then

(5) $3x^2 + 2x - 2 = A(x^2 + x + 1) + (Bx + C)(x - 1).$

If $x = 1$ (the only good x) then $3 = 3A, A = 1$. The preceding example illustrated two ways to find the remaining letters if there are not enough good values of x. We prefer not to solve a system of equations to find B and C, and from this point of view, equating coefficients is usually better than using other values of x. The constant term on the left side of (5) is -2. When the right side is multiplied out and simplified, its constant term is $A - C$. Therefore $-2 = A - C, C = A + 2 = 3$. The coefficient of x^2 on the left side is 3. The coefficient of x^2 on the right side is $A + B$. Therefore $3 = A + B, B = 3 - A = 2$. Thus the decomposition is

$$\frac{3x^2 + 2x - 2}{(x - 1)(x^2 + x + 1)} = \frac{1}{x - 1} + \frac{2x + 3}{x^2 + x + 1}$$

and

$$\int \frac{3x^2 + 2x - 2}{(x - 1)(x^2 + x + 1)} \, dx = \int \frac{dx}{x - 1} + 2\int \frac{x}{x^2 + x + 1} \, dx$$
$$+ 3\int \frac{dx}{x^2 + x + 1}.$$

The first integral on the right may be done by inspection or with the substitution $u = x - 1$. Use formula 2 and then 1b on the second integral, and use 1b for the third integral. Thus

$$\int \frac{3x^2 + 2x - 2}{(x - 1)(x^2 + x + 1)} \, dx$$

$$= \ln|x - 1| + \ln|x^2 + x + 1| - \frac{2}{\sqrt{3}} \tan^{-1} \frac{2x + 1}{\sqrt{3}}$$

$$+ \frac{6}{\sqrt{3}} \tan^{-1} \frac{2x + 1}{\sqrt{3}} + K$$

$$= \ln|x - 1| + \ln|x^2 + x + 1| + \frac{4}{\sqrt{3}} \tan^{-1} \frac{2x + 1}{\sqrt{3}} + K.$$

Warning   1. The factor $x^2 - 5$ in a denominator is *factorable* and the decomposition does *not* contain $\dfrac{Ax + B}{x^2 - 5}$. Instead, factor into $(x - \sqrt{5})(x + \sqrt{5})$ and put $\dfrac{A}{x - \sqrt{5}} + \dfrac{B}{x + \sqrt{5}}$ in the decomposition.

2. A numerator of the form $Bx + C$ goes on top of a *non*factorable quadratic only. A factor such as $(x - 3)^2$ in the denominator is a repeated linear factor, not a nonfactorable quadratic, and the decomposition contains $\dfrac{A}{x - 3} + \dfrac{B}{(x - 3)^2}$, NOT $\dfrac{A}{x - 3} + \dfrac{Bx + C}{(x - 3)^2}$. Similarly, the factor x^2 in a denominator is a repeated linear factor, and the decomposition contains $\dfrac{A}{x} + \dfrac{B}{x^2}$.

3. The decomposition technique in this section does not work for *improper* fractions. Use long division on improper fractions first, and then decompose further, if necessary.

Problems for Section 7.4

1. Describe the form of the decomposition without actually computing the values of $A, B, C, \cdots$

(a) $\dfrac{2x^3 + 3}{x^3(x + 1)(2x + 3)}$ (b) $\dfrac{4x^3}{(x^2 + 2x - 2)(x^2 - 2x + 2)}$

2. Decompose into partial fractions

(a) $\dfrac{12}{x^2 - 3}$ (b) $\dfrac{1}{2x^2 - 5x - 12}$ (c) $\dfrac{5x}{(x^2 + 1)(x - 2)}$ (d) $\dfrac{2x + 3}{(x - 2)^2}$

3. Find $\displaystyle\int \frac{3}{(2 - x)(x + 1)} \, dx$ (a) by decomposing and (b) directly from the tables. Confirm that the two answers agree.

4. Find (a) $\int \dfrac{2x + 3}{x^2 - 4x + 4}\, dx$ (b) $\int \dfrac{8x}{x^4 - 1}\, dx$ (c) $\int \dfrac{dx}{x^2(2x - 3)}$.

5. Derive formula 11.

6. Find $\int \dfrac{x^2}{x^2 + 5x + 4}\, dx$ and aim for the answer $x + \tfrac{1}{3}\ln|x + 1| - \tfrac{16}{3}\ln|x + 4|$.

7.5 Integration by Parts

The substitution method in Section 7.2 is a reversal of the chain rule for derivatives. The idea behind integration by parts is to reverse the derivative product rule. Since $D_x uv = uv' + vu'$ we have the integration formula $\int (uv' + vu')\, dx = uv$. But problems don't usually originate in the form $\int (uv' + vu')\, dx$, so we continue on to a more useful version of the integration formula. Write it as $\int uv'\, dx = uv - \int vu'\, dx$, and then to make it easier to apply, use the notation $dv = v'\, dx$, $du = u'\, dx$ to get

(1)
$$\int u\, dv = uv - \int v\, du.$$

This formula can be used to trade one problem (namely, $\int u\, dv$) for another (namely, $\int v\, du$), which may or may not help depending on how good a trader you are. To apply (1), a factor in the integrand must be called u. The rest of the integrand including the "factor" dx is labeled dv. Success of the method, called *integration by parts,* then depends on being able to find v from dv (this in itself is antidifferentiation) and on being able to find $\int v\, du$.

Example 1 We'll show how the tables arrived at the formula for $\int x \sin x\, dx$. We must think of $x \sin x\, dx$ as $u\, dv$. One possibility is to let $u = x$, $dv = \sin x\, dx$. Then $du = dx$ and $v = -\cos x$. (Finding v after choosing dv is a small antidifferentiation problem buried in the overall antidifferentiation problem.) Then, by (1),

$$\int x \sin x\, dx = -x \cos x + \int \cos x\, dx = -x \cos x + \sin x + K.$$

The trade was a good one since the new integral, $\int \cos x\, dx$, was easy to do.

Another possibility (which proves to be a false start) is to let $u = \sin x$, $dv = x\, dx$. Then $du = \cos x\, dx$, $v = \tfrac{1}{2}x^2$ and, by (1),

$$\int x \sin x\, dx = \frac{1}{2}x^2 \sin x - \frac{1}{2}\int x^2 \cos x\, dx.$$

This is *correct* but *not useful* since the new integral looks harder than the original.

Example 2 Derive the formula in the tables for $\int e^x \cos x\, dx$.

Solution: Let $u = e^x$, $dv = \cos x\, dx$ (it would do just as well to begin with $u = \cos x$ and $dv = e^x\, dx$). Then $du = e^x\, dx$, $v = \sin x$ and

$$\int e^x \cos x\, dx = e^x \sin x - \int e^x \sin x\, dx.$$

The new integral is just as bad as the original, but surprisingly if we work on the new one we'll succeed. Let $u = e^x, dv = \sin x\, dx$. (Using $u = \sin x$, $dv = e^x\, dx$ at this stage leads nowhere.) Then $du = e^x\, dx$, $v = -\cos x$ and

$$\int e^x \cos x \, dx = e^x \sin x - \left(-e^x \cos x + \int e^x \cos x \, dx\right).$$

On the right-hand side is the *original* integral which seems circular. But collect the terms involving $\int e^x \cos x \, dx$ to get $2\int e^x \cos x \, dx = e^x \sin x + e^x \cos x$. Thus the final answer is

$$\int e^x \cos x \, dx = \frac{1}{2}(e^x \sin x + e^x \cos x) + C.$$

Problems for Section 7.5

    1. Derive the formulas given in the tables for

    (a) $\displaystyle\int xe^x \, dx$ (b) $\displaystyle\int \tan^{-1}x \, dx$ (c) $\displaystyle\int \sin^{-1}x \, dx$ (d) $\displaystyle\int \ln x \, dx$

    2. Find (a) $\displaystyle\int \cos(\ln x) \, dx$ (b) $\displaystyle\int x^2 e^x \, dx$ (c) $\displaystyle\int x \tan^{-1}x \, dx$.

    3. Problem 26 in Section 7.2 derived the formula for $\int \sec x \, dx$. Use it to find the formula for $\int \sec^3 x \, dx$.

    4. Suppose $Q(x)$ is an antiderivative for e^{-x^2}. Find $\int x^2 e^{-x^2} \, dx$ in terms of $Q(x)$.

7.6 Recursion Formulas

    Some antiderivative formulas, said to be *recursive,* can be applied repeatedly within a problem to help get a final answer. We will illustrate how they are used and how they are derived.

Example of a recursion formula Suppose we want to find $\int x^7 \sin x \, dx$. The tables in this book do not help, and even larger tables will probably not contain this specific integral. However many tables will list the following pertinent formula:

(1) $\displaystyle\int x^n \sin x \, dx = -x^n \cos x + nx^{n-1} \sin x - n(n-1)\int x^{n-2} \sin x \, dx.$

Use (1) with $n = 7$ to obtain

$$\int x^7 \sin x \, dx = -x^7 \cos x + 7x^6 \sin x - 42\int x^5 \sin x \, dx.$$

Then use (1) again with $n = 5$ to get

$$\int x^7 \sin x \, dx = -x^7 \cos x + 7x^6 \sin x$$
$$- 42\left(-x^5 \cos x + 5x^4 \sin x - 20\int x^3 \sin x \, dx\right).$$

And again with $n = 3$ to get

$$\int x^7 \sin x \, dx = -x^7 \cos x + 7x^6 \sin x + 42x^5 \cos x - 210x^4 \sin x$$
$$+ 840\left(-x^3 \cos x + 3x^2 \sin x - 6\int x \sin x \, dx\right).$$

Finally use formula 48 in the tables to finish the job and compute $\int x \sin x\, dx$. The final answer is

$$\int x^7 \sin x\, dx = (-x^7 + 42x^5 - 840x^3 + 5040x) \cos x$$
$$+ (7x^6 - 210x^4 + 2520x^2 - 5040) \sin x + C.$$

Many of the formulas collected in tables are recursion formulas like (1), and are usually found by integration by parts. To derive (1), let $u = x^n$, $dv = \sin x\, dx$. Then $du = nx^{n-1}\, dx$, $v = -\cos x$ and

(2) $$\int x^n \sin x\, dx = -x^n \cos x + n \int x^{n-1} \cos x\, dx.$$

We don't stop here because (2) is not recursive; that is, it can't be used over and over again. If it is used on $\int x^7 \sin x\, dx$, we obtain the new integral $\int x^6 \cos x\, dx$ to which (2) no longer applies. So we integrate by parts again. Let $u = x^{n-1}$, $dv = \cos x\, dx$. Then $du = (n-1)x^{n-2}\, dx$, $v = \sin x$ and

$$\int x^n \sin x\, dx = -x^n \cos x + n\left[x^{n-1} \sin x - (n-1) \int x^{n-2} \sin x\, dx \right],$$

which simplifies to the recursion formula in (1). Typically, a recursion formula lowers an exponent in the integrand. The formula in (1) happens to bring an exponent down by 2. Look at formula 3 in the tables to see an instance where an exponent (called r) is lowered by 1.

The recursion formulas for $\int \sin^m x\ \cos^n x\, dx$ Products of powers of sines and cosines occur frequently, and the tables contain four recursion formulas for them. Formula 52a brings the sine exponent down by 2 and leaves the cosine exponent alone. Formula 52b brings the cosine exponent down by 2 and leaves the sine exponent alone. Similarly, formulas 52c and 52d leave one exponent unchanged and *raise* the other exponent by 2; they are used if an exponent is negative to begin with. For example,

$$\int \sin^4 x\ \cos^4 x\, dx = -\frac{\sin^3 x\ \cos^5 x}{8} + \frac{3}{8} \int \sin^2 x\ \cos^4 x\, dx$$

(by formula 52a with $m = 4$, $n = 4$)

$$= -\frac{\sin^3 x\ \cos^5 x}{8} + \frac{3}{8}\left[-\frac{\sin x\ \cos^5 x}{6} + \frac{1}{6} \int \cos^4 x\, dx \right]$$

(by formula 52a with $m = 2$, $n = 4$)

$$= -\frac{\sin^3 x\ \cos^5 x}{8} - \frac{1}{16} \sin x\ \cos^5 x$$
$$+ \frac{1}{16}\left[\frac{\sin x\ \cos^3 x}{4} + \frac{3}{4} \int \cos^2 x\, dx \right]$$

(by formula 52b with $m = 0$, $n = 4$)

$$= -\frac{\sin^3 x\ \cos^5 x}{8} - \frac{1}{16} \sin x\ \cos^5 x + \frac{1}{64} \sin x\ \cos^3 x$$
$$+ \frac{3}{128}[x + \sin x\ \cos x] + C$$

(by formula 52b or 40).

The special case of $\int \sin^m x \cos^n x \, dx$ where m and/or n is a positive odd integer   One way to find $\int \sin^{100} x \cos x \, dx$ is to use formula 52a fifty times to bring the sine exponent down to 0, and finish by doing $\int \cos x \, dx$. But it is much easier to substitute $u = \sin x$, $du = \cos x \, dx$ to obtain

$$\int \sin^{100} x \, \cos x \, dx = \int u^{100} \, du = \frac{u^{101}}{101} + C = \frac{\sin^{101} x}{101} + C.$$

As another example, consider

$$(3) \qquad\qquad \int \cos^{98} x \, \sin^3 x \, dx.$$

One possibility is to use formula 52b forty-nine times to bring the cosine exponent down to 0, use formula 52a once to bring the sine exponent down to 1, and finish by finding $\int \sin x \, dx$. But it is easier to use formula 52a once to obtain

$$\int \cos^{98} x \, \sin^3 x \, dx = -\frac{\sin^2 x \, \cos^{99} x}{101} + \frac{2}{101} \int \cos^{98} x \, \sin x \, dx,$$

and then substitute $u = \cos x$, $du = -\sin x \, dx$ to get

$$\int \cos^{98} x \, \sin^3 x \, dx = -\frac{\sin^2 x \, \cos^{99} x}{101} - \frac{2}{101} \int u^{98} \, du$$

$$= -\frac{\sin^2 x \, \cos^{99} x}{101} - \frac{2}{101} \frac{u^{99}}{99} + C$$

$$(4) \qquad\qquad = -\frac{\sin^2 x \, \cos^{99} x}{101} - \frac{2}{(101)(99)} \cos^{99} x + C.$$

Another approach to (3) is to use the identity $\cos^2 x + \sin^2 x = 1$ and write

$$\sin^3 x = \sin^2 x \, \sin x = (1 - \cos^2 x) \sin x.$$

Then

$$\int \cos^{98} x \, \sin^3 x \, dx = \int \cos^{98} x (1 - \cos^2 x) \sin x \, dx$$

$$= \int (\cos^{98} x - \cos^{100} x) \sin x \, dx$$

and the problem may be completed with the substitution $u = \cos x$, $du = -\sin x \, dx$ to obtain

$$(5) \qquad\qquad \int \cos^{98} x \, \sin^3 x \, dx = -\frac{\cos^{99} x}{99} + \frac{\cos^{101} x}{101} + K.$$

In general, suppose at least one of the exponents, say n, is a *positive odd integer*. Instead of using the recursion formulas to lower both m and n, it is faster to use 52b or the identity $\sin^2 x + \cos^2 x = 1$ to reduce the problem to $\int \sin^m x \cos x \, dx$, and then finish with the substitution $u = \sin x$, $du = \cos x \, dx$.

Problems for Section 7.6

1. Derive a recursion formula for $\int x^n e^x \, dx$.
2. Derive recursion formula 53 by writing $\tan^n x$ as $\tan^2 x \, \tan^{n-2} x$ and using the identity $\tan^2 x = \sec^2 x - 1$.

3. Derive a recursion formula for $\int (\ln x)^n \, dx$ and then use it to find $\int (\ln x)^3 \, dx$.

4. Use formula 52 to derive a recursion formula for $\int \sin^m x \cos^n x \, dx$ which brings *m and n each* down by 2.

5. Explain why formula 4 is *not* recursive.

6. Find

(a) $\displaystyle\int \sin x \, \cos x \, dx$
(b) $\displaystyle\int \sin x \, \cos^{12} x \, dx$
(c) $\displaystyle\int \sec^5 x \, dx$

(d) $\displaystyle\int \tan^4 x \, dx$
(e) $\displaystyle\int \frac{\cos x}{\sin^2 x} \, dx$
(f) $\displaystyle\int \frac{\sin^2 x}{\cos^3 x} \, dx$

(g) $\displaystyle\int \sin^4 x \, \cos^3 x \, dx$ (try it without tables for practice)
(h) $\displaystyle\int \sin^4 3x \, dx$

7. Show that the answers in (4) and (5) agree.

7.7 Trigonometric Substitution

A collection of integrals in the tables (and similar integrals not listed) can be found using a substitution of a special type called *trigonometric substitution*. We will illustrate the method by deriving formula 26 for $\int \dfrac{1}{x\sqrt{a^2 + x^2}} \, dx$. The expression $\sqrt{a^2 + x^2}$ can be labeled as the hypotenuse of the right triangle in Fig. 1. The triangle will be the basis for the substitution. Let u be one of the acute angles in the triangle (it doesn't matter which angle you choose.) All the relations between x and u that are needed for the substitution will be read directly from the triangle. There are many relations available:

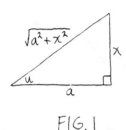

FIG. 1

$$\tan u = \frac{x}{a}, \qquad \cos u = \frac{a}{\sqrt{a^2 + x^2}}, \qquad \sin u = \frac{x}{\sqrt{a^2 + x^2}}.$$

The second relation can be used to replace $\sqrt{a^2 + x^2}$ by an expression involving u alone, namely by $a/\cos u$. But we also have to replace dx and x and for this purpose, the first relation, which is simplest, is most useful. It yields $x = a \tan u$, $dx = a \sec^2 u \, du$. (So far, our substitutions have usually expressed u in terms of x, and du in terms of dx. In trigonometric substitutions it is more convenient to express x in terms of u, and dx in terms of du.) Then

$$\int \frac{1}{x\sqrt{a^2 + x^2}} \, dx = \int \frac{1}{a \tan u \cdot \dfrac{a}{\cos u}} a \sec^2 u \, du = -\frac{1}{a} \int \csc u \, du$$

$$(1) \qquad\qquad = -\frac{1}{a} \ln|\csc u + \cot u| + C \qquad \text{(formula 34)}.$$

To express the integral in terms of x, read directly from the triangle that

$$\csc u = \frac{\text{hypotenuse}}{\text{opposite}} = \frac{\sqrt{a^2 + x^2}}{x} \quad \text{and} \quad \cot u = \frac{\text{adjacent}}{\text{opposite}} = \frac{a}{x}.$$

Substitute these expressions into (1) to obtain the final answer

$$\int \frac{1}{x\sqrt{a^2 + x^2}} \, dx = -\frac{1}{a} \ln\left| \frac{\sqrt{a^2 + x^2}}{x} + \frac{a}{x} \right| + C.$$

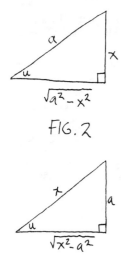

FIG. 2

FIG. 3

In general, *trigonometric substitution applies to integrands containing the expressions* $a^2 + x^2$ *(use Fig. 1),* $a^2 - x^2$ *(use Fig. 2) and* $x^2 - a^2$ *(use Fig. 3). In each case, u can be either of the acute angles in the triangle.* If the antidifferentiation is part of an overall physical problem, it is very likely that the triangle will already be part of the setup, as the following example illustrates.

Example 1 A destroyer detects an enemy battleship 8 km due west (Fig. 4). The destroyer's orders are to follow the battleship, always move toward it, but maintain the 8 km distance between them. The problem is to find the path of the destroyer if the battleship moves north.

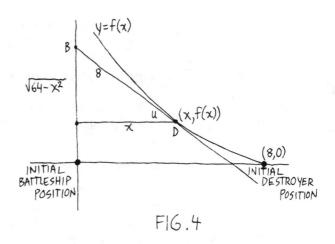

FIG. 4

For convenience draw axes so that initially the battleship is at the origin and the destroyer is at the point $(8, 0)$. Let the unknown path be named $y = f(x)$. Since the destroyer always moves towards the battleship, it is characteristic of the destroyer's path that at any point, the line from the destroyer D to the battleship B is tangent to the destroyer's path. Figure 4 shows a typical point $(x, f(x))$ on the unknown path. To find the unknown function $f(x)$, read from the picture that $f'(x)$, the slope of line BD, is negative, and in particular is $-\sqrt{64 - x^2}/x$. Therefore, to find $f(x)$ we need

$$-\int \frac{\sqrt{64 - x^2}}{x} \, dx.$$

The integral can be found with formula 21, but we'll practice with trigonometric substitution (which was used to derive formula 21 in the first place). The problem already contains a suggestive right triangle; let u be one of its acute angles. From the triangle, $\tan u = \sqrt{64 - x^2}/x$ so the entire integrand becomes $\tan u$. We also have $\cos u = x/8$, so $x = 8 \cos u$, $dx = -8 \sin u \, du$. Therefore,

$$-\int \frac{\sqrt{64 - x^2}}{x} \, dx = -\int \tan u \cdot -8 \sin u \, du = 8 \int \frac{\sin^2 u}{\cos u} \, du.$$

We can continue with formula 52c in the tables, or with the identity $\sin^2 u + \cos^2 u = 1$ as follows:

$$-\int \frac{\sqrt{64 - x^2}}{x}\, dx = 8\int \frac{1 - \cos^2 u}{\cos u}\, du$$

$$= 8\int \left(\frac{1}{\cos u} - \frac{\cos^2 u}{\cos u}\right) du \qquad \text{(by algebra)}$$

$$= 8\int (\sec u - \cos u)\, du$$

$$= 8\ln(\sec u + \tan u) - 8\sin u + C$$

$$\text{(by formula 33)}.$$

(The absolute values in formula 33 may be omitted because sec u and tan u are positive in this problem.) To finish the substitution and express the answer in terms of x, read sec u, tan u and sin u from the triangle to get

(2) $$8\ln\left(\frac{8}{x} + \frac{\sqrt{64 - x^2}}{x}\right) - \sqrt{64 - x^2} + C.$$

The function $f(x)$ must have the form of (2). To determine C, note that the point $(8, 0)$ is on the graph, that is, $f(8) = 0$. Thus if x is set equal to 8 in (2), the result must be 0. So $0 = 8\ln 1 - 0 + C$. Therefore $C = 0$ and the path is $y = 8\ln\left(\dfrac{8 + \sqrt{64 - x^2}}{x}\right) - \sqrt{64 - x^2}$ (an example of a curve called a tractrix).

Problems for Section 7.7

1. Derive the formulas in the tables for

(a) $\displaystyle\int \frac{dx}{\sqrt{x^2 - a^2}}$ (b) $\displaystyle\int \sqrt{a^2 - x^2}\, dx$ (c) $\displaystyle\int \frac{\sqrt{a^2 + x^2}\, dx}{x}$

2. $\displaystyle\int \frac{\sqrt{3 - x^2}}{x^2}\, dx$

3. $\displaystyle\int \frac{dx}{x^2\sqrt{x^2 - 5}}$

4. $\displaystyle\int \frac{dx}{(7 + x^2)^2}$

5. $\displaystyle\int \frac{dx}{(a^2 + x^2)^{3/2}}$

7.8 Choosing a Method

So far, the chapter has dealt with one method at a time. A list of miscellaneous problems is more forbidding, especially since there is no definite set of rules for deciding which method to use. If you have access to a large set of tables, they will be a great comfort. If a function is not listed in the tables, we have a few suggestions.

METHOD

Incomplete list of imperfect strategies (a) Complete the square if the problem involves $ax^2 + bx + c$ but the only similar formula in the tables does not contain the term x.

(b) Substitute if there is an expression in the integrand whose derivative is also a factor in the integrand. Substitutions might (unpredictably) work in other situations too.

(c) Use long division on improper fractions.

(d) Decompose proper fractions if they aren't in the tables.

(e) Use integration by parts to get recursion formulas. Integration by parts may also work when other methods don't seem to apply.

(f) If a problem involving $a^2 \pm x^2$ or $x^2 - a^2$ is not in the tables, try trigonometric substitution.

The perfect strategy The reason that we, the authors, can find antiderivatives is that we have already done so many. Almost any reasonable problem, suitable for a calculus course, is either one we have seen before or similar to one we have seen before. We don't have a secret weapon or inborn ability or a strict set of rules. Our *real* strategy is second sight, and it comes from practice.

Problems for Section 7.8

Outline a method for finding each antiderivative.

1. $\int \dfrac{\sin\sqrt{x}}{\sqrt{x}}\, dx$

2. $\int \dfrac{x}{(1 - x^2)^3}\, dx$

3. $\int \dfrac{1}{3 + x^2}\, dx$

4. $\int \dfrac{1}{\sqrt{2x + 3}}\, dx$

5. $\int x(x - 1)^{20}\, dx$

6. $\int \dfrac{1}{e^x}\, dx$

7. $\int \dfrac{x^2 + 2x + 3}{x + 1}\, dx$

8. $\int \dfrac{x}{\sqrt{4 - x^2}}\, dx$

9. $\int (2x + 9)\, dx$

10. $\int \dfrac{1}{\sqrt{4 - x^2}}\, dx$

11. $\int \dfrac{1}{x(x + 1)}\, dx$

12. $\int 2 \tan 3x\, dx$

13. $\int \dfrac{1}{x(x + 1)^2}\, dx$

14. $\int \dfrac{\sqrt{3 - 2x^2}}{x}\, dx$

15. $\int \dfrac{1}{(x + 3)(x + 1)^2}\, dx$

16. $\int \dfrac{1 - x}{1 + x}\, dx$

17. $\int \dfrac{\cos^4 x}{\sin^2 x}\, dx$

18. $\int \sin \pi x\, dx$

19. $\int \dfrac{1}{(3x + 1)^9}\, dx$

20. $\int \dfrac{x^2}{9 + 4x^3}\, dx$

21. $\int \tan x\, \sin^2 x\, dx$

22. $\int \dfrac{dx}{x\sqrt{3 - x^2}}$

23. $\int (9 + 4x)^3\, dx$

24. $\int \dfrac{1}{\cos^2 x}\, dx$

25. $\int \dfrac{\sqrt{2x^2 - 4}}{x^2}\, dx$

26. $\int \dfrac{\sin^5 x}{\cos^4 x}\, dx$

27. $\int \dfrac{x^2 - 4}{x^2}\, dx$

28. $\int \sec^3 x\, dx$

29. $\int \dfrac{1}{2x + 1}\, dx$

30. $\int x \sin x^2\, dx$

31. $\int \sec^4 x\, dx$

32. $\int x^2 \sin x\, dx$

33. $\int \dfrac{1}{\sqrt{2 + 3x^2}}\, dx$

34. $\int e^{3x}\, dx$

35. $\int \dfrac{x}{2x + 3}\, dx$

36. $\int \sin 5x\, dx$

37. $\int r\sqrt{2 - r^2}\, dr$

38. $\int \dfrac{\cos^2 x}{\sin^2 x}\, dx$

39. $\int \sin^3 x\, dx$

40. $\int 2\, dx$

41. $\int 5 \sec 2x\, dx$

42. $\int \dfrac{2x + 3}{x}\, dx$

43. $\int \sin 3x \cos 2x\, dx$

44. $\int \sin \dfrac{2x}{\pi}\, dx$

45. $\int \cos 2x \sin 2x\, dx$

46. $\int \dfrac{1}{5x - 2}\, dx$

47. $\int \dfrac{(x + 3)(x - 2)}{x - 1}\, dx$

48. $\int x^3 \sqrt{x^2 + 7}\, dx$

49. $\int \dfrac{\sin x}{\cos^2 x}\, dx$

50. $\int x \sin^{-1} x\, dx$

51. $\int x e^{x^2}\, dx$

52. $\int \dfrac{1}{x^3 - 2}\, dx$

53. $\int x e^x\, dx$

54. $\int \dfrac{\sin^2 x}{\cos x}\, dx$

55. $\int \dfrac{1}{\sqrt{2x^2 - 5}}\, dx$

56. $\int 8 \tan(3 - 2x)\, dx$

57. $\int \sqrt{3 - x}\, dx$

58. $\int \left(2 - \dfrac{1}{3}x\right)^4 dx$

59. $\int \dfrac{1}{e^x + e^{-x}}\, dx$

60. $\int \tan x \cos^4 x\, dx$

61. $\int \sqrt{4 - 2x^2}\, dx$

62. $\int \sqrt{x^2 - x + 3}\, dx$

63. $\int \dfrac{3}{x^2 + x + 1}\, dx$

64. $\int 7 \ln(4x + 5)\, dx$

65. $\int e^{\theta/2}\, d\theta$

66. $\int \dfrac{1 + 2x}{1 + x^2}\, dx$

67. $\int x^2 \sqrt{3x^2 - 1}\, dx$

68. $\int \sin^4 x\, dx$

69. $\int \cos^3 x \sin^2 x\, dx$

70. $\int \dfrac{\sin 2x}{\cos 2x}\, dx$

71. $\int (1 + e^x)^2\, dx$

72. $\int \sin^5 x\, dx$

73. $\int \dfrac{1}{\sqrt{6x - x^2}}\, dx$

74. $\int \dfrac{\sin 2x}{9 - \cos^2 2x}\, dx$

75. $\displaystyle\int \frac{x}{(x^2 - 5)(1 - x)}\, dx$

76. $\displaystyle\int x(2 + 3x)^4\, dx$

77. $\displaystyle\int e^{2x} \sin 3x\, dx$

78. $\displaystyle\int \frac{x + 4}{2x^2 + x - 1}\, dx$

79. $\displaystyle\int (\ln x)^3\, dx$

80. $\displaystyle\int \tan^2 3x\, dx$

81. $\displaystyle\int x^3 \sin x\, dx$

82. $\displaystyle\int \frac{\sin x}{(2 + \cos x)^2}\, dx$

83. $\displaystyle\int \cos^2 x\, dx$

84. $\displaystyle\int \cos^3 x\, dx$

85. $\displaystyle\int \cos^2 x \, \sin x\, dx$

7.9 Combining Techniques of Antidifferentiation with the Fundamental Theorem

By the Fundamental Theorem of Section 5.3, to find the (definite) integral $\int_a^b f(x)\, dx$, we first try to find the antiderivative (indefinite integral) $F(x) = \int f(x)\, dx$, and then compute $F(b) - F(a)$. This can be done in two separate steps, or to save time and paper, the two steps can be combined as shown in this section.

Combining substitution and the Fundamental Theorem Consider $\int_2^3 x^2 \cos x^3\, dx$. We'll begin by finding an antiderivative for the integrand as a first step, apply the Fundamental Theorem in a second step, and then see how to merge the two. To antidifferentiate, substitute

(1) $$u = x^3, \qquad du = 3x^2\, dx.$$

Then

$$\int x^2 \cos x^3\, dx = \frac{1}{3} \int \cos u\, du = \frac{1}{3} \sin u + C = \frac{1}{3} \sin x^3 + C.$$

Any antiderivative may be used in applying the Fundamental Theorem; with the antiderivative $\frac{1}{3} \sin x^3$, we have

$$\int_2^3 x^2 \cos x^3\, dx = \frac{1}{3} \sin x^3 \Big|_2^3 = \frac{1}{3}(\sin 27 - \sin 8).$$

To accomplish this in one step, use the substitution in (1) to express the integrand in terms of u, and *write the limits of integration in terms of u*. If $x = 2$ then $u = 8$; if $x = 3$ then $u = 27$. Thus

(2) $$\int_2^3 x^2 \cos x^3\, dx = \frac{1}{3} \int_8^{27} \cos u\, du = \frac{1}{3} \sin u \Big|_8^{27} = \frac{1}{3}(\sin 27 - \sin 8).$$

Switching to u limits produces the same answer as before, but in less space.

Note the difference between a substitution in $\int f(x)\, dx$ versus $\int_a^b f(x)\, dx$. For the former, we must eventually change *back* from u to x so that the final antiderivative is expressed as a function of x. But in (2), the new integral $\int_8^{27} \cos u\, du$ computes to be a *number*, and there is no "changing back" to be done.

Example 1 Find $\displaystyle\int_{7}^{\infty} \frac{1}{5-2x}\,dx$.

Solution: Let $u = 5 - 2x$, $du = -2\,dx$. If $x = 7$ then $u = -9$; if $x \to \infty$ then $u \to -\infty$. Therefore

(3)
$$\int_{7}^{\infty} \frac{1}{5-2x}\,dx = -\frac{1}{2} \int_{-9}^{-\infty} \frac{1}{u}\,du = -\frac{1}{2}\,\ln|u|\,\Big|_{-9}^{-\infty}$$

$$= -\frac{1}{2}\,\ln\infty + \frac{1}{2}\,\ln 9 = -\infty.$$

Note that after the substitution, the lower limit $u = -9$ is larger than the upper limit $u = -\infty$, that is, the limits are backwards. This causes no difficulty. Simply continue with $F(b) - F(a)$, which still holds even for backward limits.

Warning When substituting in a (definite) integral, the limits of integration must be changed to new u limits. In (3), it is *not* correct to write $\int_{7}^{\infty} \frac{1}{5-2x}\,dx = -\frac{1}{2}\int_{7}^{\infty} \frac{1}{u}\,du$ or $\int_{7}^{\infty} \frac{1}{5-2x}\,dx = -\frac{1}{2}\int \frac{1}{u}\,du$. The original x limits cannot be retained, nor can they be dropped in the middle of a problem (even if you intend to restore them later).

Example 2 Without evaluating either integral, show that

$$\int_{0}^{3} e^{x}\sin(3-x)\,dx = \int_{0}^{3} e^{3-x}\sin x\,dx.$$

Solution: Let $u = 3 - x$, $du = -dx$. If $x = 0$ then $u = 3$; if $x = 3$ then $u = 0$. Since $u = 3 - x$, we have $x = 3 - u$. Therefore

$$\int_{0}^{3} e^{x}\sin(3-x)\,dx = -\int_{3}^{0} e^{3-u}\sin u\,du \qquad \text{(substitution)}$$

$$= \int_{0}^{3} e^{3-u}\sin u\,du \qquad \left(\text{use } \int_{b}^{a} f(x)\,dx = -\int_{a}^{b} f(x)\,dx\right)$$

$$= \int_{0}^{3} e^{3-x}\sin x\,dx$$

$$\text{(change the dummy variable from } u \text{ to } x).$$

The last step often bothers students. Remember that $\int_{0}^{3} e^{3-u}\sin u\,du$ is a *number;* the letter u is a dummy variable. We can write the integral as $\int_{0}^{3} e^{3-t}\sin t\,dt$ or $\int_{0}^{3} e^{3-a}\sin a\,da$ or (as we did) $\int_{0}^{3} e^{3-x} \sin x\,dx$. All of these stand for the same number.

Combining integration by parts with the Fundamental Theorem To find $\int_{0}^{\pi/4} x\,\sec^{2}x\,dx$, let $u = x$, $dv = \sec^{2}x\,dx$. Then $du = dx$, $v = \tan x$ and

$$\int_{0}^{\pi/4} x\,\sec^{2}x\,dx = x\,\tan x\,\Big|_{0}^{\pi/4} - \int_{0}^{\pi/4}\tan x\,dx = \frac{\pi}{4} + \ln|\cos x|\,\Big|_{0}^{\pi/4}$$

$$= \frac{\pi}{4} + \ln\frac{1}{2}\sqrt{2}.$$

The limits of integration do *not* change in the process. More generally, the integration by parts rule for (definite) integrals, as opposed to antiderivatives (indefinite integrals) is

$$\int_a^b u\, dv = uv \Big|_a^b - \int_a^b v\, du.$$

Problems for Section 7.9

1. For each integral, perform the indicated substitution, and then stop after reaching an integral involving only u.

(a) $\displaystyle\int_2^5 \sin^5 x\, dx, \qquad u = 3x$

(b) $\displaystyle\int_1^{e^3} \sin(\ln x)\, dx, \qquad u = \ln x$

(c) $\displaystyle\int_2^4 \frac{\sqrt{x^2 - 4}}{x^2}\, dx, \qquad u$ is the angle indicated in Fig. 1

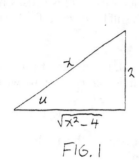

2. Evaluate the integral.

(a) $\displaystyle\int_2^4 x(3x^2 - 1)^{10}\, dx$ (b) $\displaystyle\int_0^\pi e^{-x} \cos x\, dx$ (c) $\displaystyle\int_1^e \frac{(\ln x)^5}{x}\, dx$

(d) $\displaystyle\int_{\pi/2}^\pi \sin^2 x\, \cos^2 x\, dx$ (e) $\displaystyle\int_{-\infty}^2 x^2 e^{x^3}\, dx$ (f) $\displaystyle\int_0^2 x\sqrt{x^2 + 4}\, dx$

3. Show that the integrals are equal without evaluating them.

(a) $\displaystyle\int_0^1 x^m(1 - x)^n\, dx = \int_0^1 x^n(1 - x)^m\, dx$

(b) $\displaystyle\int_0^{10} (x + 20)^2\, dx = \int_{20}^{30} x^2\, dx$

(c) $\displaystyle\int_{2a}^{2b} \sqrt{\sin \frac{1}{2} x}\ dx = 2\int_a^b \sqrt{\sin x}\ dx$

4. Given that $\displaystyle\int_2^3 \frac{x}{\ln x}\, dx = k$, find $\displaystyle\int_2^3 x \ln \ln x\, dx$ in terms of k.

REVIEW PROBLEMS FOR CHAPTER 7

1. Find $\displaystyle\int \frac{x}{x^2 + 1}\, dx$

(a) directly from the tables   (b) by ordinary substitution
(c) with a trigonometric substitution   (d) by integration by parts

2. Find $\displaystyle\int \frac{1}{(x + 2)(x - 4)}\, dx$

(a) using substitution and formula 9
(b) by completing the square and using formula 18
(c) directly from the tables
(d) by partial fractions

3. Indicate a method.

(a) $\displaystyle\int e^{2x} \sin x \, dx$ (h) $\displaystyle\int \frac{x}{x + 3} \, dx$

(b) $\displaystyle\int \frac{1}{3x + 4} \, dx$ (i) $\displaystyle\int \frac{x + 3}{x} \, dx$

(c) $\displaystyle\int \frac{1}{\sqrt{2 - x^2}} \, dx$ (j) $\displaystyle\int \frac{2x + 1}{x + 6} \, dx$

(d) $\displaystyle\int \frac{x^2}{(1 + 2x^3)^4} \, dx$ (k) $\displaystyle\int \frac{x}{\sqrt{3x + 4}} \, dx$

(e) $\displaystyle\int \tan^2 3x \, dx$ (l) $\displaystyle\int \frac{1}{\sqrt{2x^2 + x}} \, dx$

(f) $\displaystyle\int e^{-6x} \, dx$ (m) $\displaystyle\int \frac{1}{x^2(1 + x)} \, dx$

(g) $\displaystyle\int \frac{1}{5x} \, dx$ (n) $\displaystyle\int \frac{1}{x^2 + x + 2} \, dx$

4. Find $\displaystyle\int \sin 3x \, \sin 5x \, dx$

(a) directly from the tables
(b) with the identity $\sin x \, \sin y = \frac{1}{2}[\cos(x - y) - \cos(x + y)]$
(c) with integration by parts

5. If $\displaystyle\int_0^{\pi/3} e^x \sec^2 x \, dx = Q$, find $\displaystyle\int_0^{\pi/3} e^x \tan x \, dx$ in terms of Q.

6. Find $\displaystyle\int_0^1 x(2 + x^2)^5 \, dx$.

7. Find

(a) $\displaystyle\int \sin x \, dx$ (f) $\displaystyle\int \sin^2 x \, \cos^2 x \, dx$

(b) $\displaystyle\int \sin^2 x \, dx$ (g) $\displaystyle\int \frac{1}{x^2} \, dx$

(c) $\displaystyle\int \sin^3 x \, dx$ (h) $\displaystyle\int \frac{1}{x} \, dx$

(d) $\displaystyle\int \sin x \, \cos x \, dx$ (i) $\displaystyle\int \frac{dx}{\sqrt{x}}$

(e) $\displaystyle\int \sin^2 x \, \cos x \, dx$

8/SERIES

8.1 Introduction

In precalculus mathematics, addition can only be done with *finitely* many numbers. Addition of this type is very concrete: $3 + 4 = 7$ because a pile of 3 apples merged with a pile of 4 apples becomes a pile of 7 apples. Addition of *infinitely* many numbers is physically impossible in the apple sense, but this chapter presents a sensible mathematical definition and its consequences. The first application is in the next section, and the main applications are in Sections 8.6 and 8.7.

Series and their sums   The symbol $a_1 + a_2 + a_3 + \cdots$ is called a *series* with *terms* $a_1, a_2, a_3, \cdots$. The series is also written as $\sum_{n=1}^{\infty} a_n$. Frequently we will use $\sum a_n$ as an abbreviation. The *partial sums* of the series are

$$
\begin{aligned}
S_1 &= a_1 \\
S_2 &= a_1 + a_2 \\
S_3 &= a_1 + a_2 + a_3 \\
&\;\;\vdots
\end{aligned}
$$

(1)

If the partial sums approach a number S, that is, if

(2)
$$\lim_{n \to \infty} S_n = S,$$

we call S the *sum* of the series, and write $\sum a_n = S$. In this case the series is called *convergent;* in particular, it converges to S. *The definition of the sum of a series says to start adding and see where the subtotals are heading.*

If the partial sums do not approach a number, the series is *divergent*. There are three types of divergence. If the partial sums approach ∞, we say that the series *diverges to* ∞, and write $\sum a_n = \infty$. Similarly, if the partial sums approach $-\infty$, the series *diverges to* $-\infty$, and $\sum a_n = -\infty$. If the partial sums oscillate so vigorously that they approach neither a limit, nor ∞, nor $-\infty$, we simply say that the series *diverges*.

Example 1   The series $1 - 2 + 1 - 3 + 1 - 4 + 1 - 5 + \cdots$ diverges to $-\infty$, since the partial sums are $1, -1, 0, -3, -2, -6, -5, -10, \cdots$ which approach $-\infty$. In other words, $1 - 2 + 1 - 3 + 1 - 4 + 1 - 5 + \cdots = -\infty$.

Example 2   Consider $\sum_{n=1}^{\infty} \left(\frac{1}{2}\right)^n = \frac{1}{2} + \frac{1}{4} + \frac{1}{8} + \frac{1}{16} + \cdots$. The partial sums are

$$S_1 = \frac{1}{2}$$

$$S_2 = \frac{1}{2} + \frac{1}{4} = \frac{3}{4}$$

$$S_3 = \frac{1}{2} + \frac{1}{4} + \frac{1}{8} = \frac{7}{8}$$

$$S_4 = \frac{1}{2} + \frac{1}{4} + \frac{1}{8} + \frac{1}{16} = \frac{15}{16}$$

$$\vdots$$

Since $\lim_{n \to \infty} S_n = 1$, the series has sum 1, that is, the series converges to 1, and we write $\sum_{n=1}^{\infty} (\frac{1}{2})^n = 1$. (If you eat half a pie, then half of the remaining half-portion, then half of the still remaining quarter-portion, and so on, you are on your way to eating the entire pie.)

Warning If the sum of a series is S, it is not necessarily true that S is ever *reached* as term after term is added in. In the preceding example, if we start adding $\frac{1}{2}, \frac{1}{4}, \frac{1}{8}, \cdots$ we will never *reach* 1. But the subtotals are getting closer and closer to 1, so the definition calls 1 the sum.

Example 3 Consider the series

(3) $$2 - 2 + 2 - 2 + 2 - 2 + \cdots.$$

The partial sums are $S_1 = 2, S_2 = 0, S_3 = 2, S_4 = 0, \cdots$. They do not have a limit as $n \to \infty$, so the series does not have a sum; it diverges.

This example often disturbs students. Some would like the answer to be either 2 or -2 depending on whether the "last" term is odd or even numbered. But there is no last term; they just keep coming. Some would like the answer to be 0 because they visualize the series grouped into pairs and turned into

(4) $$(2 - 2) + (2 - 2) + (2 - 2) + \cdots = 0 + 0 + 0 + 0 + \cdots.$$

Some would like the answer to be 2 because they group the terms into

(5) $$2 + (-2 + 2) + (-2 + 2) + \cdots = 2 + 0 + 0 + 0 + \cdots.$$

It is true that the series in (4) converges to 0 because the partial sums are all 0, and the series in (5) converges to 2 because the partial sums are all 2. But they are not the same as the original divergent series in (3), whose partial sums oscillate between 0 and 2.

Grouping a string of 10 numbers has no effect on their sum. But this example illustrates that grouping the terms of a series may produce a new series with a different sum.

Factoring a series For a sum of two numbers we have the factoring principle $cx + cy = c(x + y)$. Similarly, it can easily be shown that

(6) $$ca_1 + ca_2 + ca_3 + ca_4 + \cdots = c(a_1 + a_2 + a_3 + a_4 + \cdots),$$

or equivalently, $\sum ca_n = c \sum a_n$ (we assume $c \neq 0$). The equation in (6) is intended to mean that either the series $ca_1 + ca_2 + ca_3 + \cdots$ and $a_1 + a_2 + a_3 + \cdots$ *both* converge, in which case the first sum is c times the second, or *both* diverge.

For example,

$$\frac{1}{2}T + \frac{1}{4}T + \frac{1}{8}T + \frac{1}{16}T + \cdots = T\left(\frac{1}{2} + \frac{1}{4} + \frac{1}{8} + \frac{1}{16} + \cdots\right)$$
$$= T \cdot 1 = T.$$

Term by term addition of two convergent series It is not hard to show that if $\sum a_n$ converges to A and $\sum b_n$ converges to B, then $\sum (a_n + b_n)$ converges to $A + B$. In abbreviated form, $\sum (a_n + b_n) = \sum a_n + \sum b_n$.

We offer a numerical illustration although the principle is more useful for theory than for computation. Since Example 2 showed that

$$\frac{1}{2} + \frac{1}{4} + \frac{1}{8} + \frac{1}{16} + \cdots = 1,$$

and the next section (Problem 2) will show that

$$\frac{1}{4} - \frac{1}{16} + \frac{1}{64} - \frac{1}{256} + \cdots = \frac{1}{5},$$

we may add termwise to obtain

$$\frac{3}{4} + \frac{3}{16} + \frac{9}{64} + \frac{15}{256} + \cdots = \frac{6}{5}.$$

Dropping initial terms It can easily be shown that if the first three terms of $\sum_{n=1}^{\infty} a_n$ are dropped, then the new series $\sum_{n=4}^{\infty} a_n$ and the original series will *both* converge or *both* diverge. In other words, chopping off the beginning of a series doesn't change convergence or divergence. Of course, dropping terms *will* change the *sum* of a convergent series.

Dropping terms is useful if a series doesn't begin to exhibit a pattern until say the 100th term. In that case, it is convenient to drop the first 99 terms when the series is tested for divergence versus convergence.

For example, the series $6 + 100 + 2 + 3 + \frac{1}{2} + \frac{1}{4} + \frac{1}{8} + \frac{1}{16} + \cdots$ converges because if the first four terms are dropped, the remainder is the convergent series in Example 2. In particular, the sum of the remaining terms is 1, so the sum of the original series is $6 + 100 + 2 + 3 + 1$, or 112.

Problems for Section 8.1

1. Write the first three terms of the series.

(a) $\sum_{n=3}^{\infty} (-1)^n \frac{1}{2n + 1}$ (b) $\sum_{n=1}^{\infty} n^2 a_n$

2. Decide if the series converges or diverges.

(a) $1 - 2 + 3 - 4 + 5 - 6 + \cdots$ (b) $\frac{1}{2} + \frac{1}{2} + \frac{1}{2} + \frac{1}{2} + \frac{1}{2} + \cdots$

3. Find the terms and the sum of the series given the following partial sums.

(a) $S_n = n$ (b) $S_n = 1$ for all n

4. Find the sum of the series $\sum_{n=1}^{\infty} \left(\frac{1}{n} - \frac{1}{n+1}\right)$ by slowly writing out some partial sums until you see the pattern.

5. There is no term-by-term addition principle for two divergent series; that is, their term-by-term sum is unpredictable. Prove this by finding two divergent series whose term-by-term sum also diverges, and two other divergent series whose term-by-term sum converges.

6. If $\sum a_n$ has partial sums S_n then $S_{100} - S_{99} = a_{\text{what}}$?

8.2 Geometric Series

One particular type of series, called geometric, occurs often in applications, and is easy to sum.

Definition of a geometric series A series of the form

$$a + ar + ar^2 + ar^3 + ar^4 + \cdots, \qquad a \neq 0,$$

is called a *geometric series* with *ratio r*. The series is also denoted by $\sum_{n=0}^{\infty} ar^n$. Each term of a geometric series is obtained from the preceding term by multiplying by r.

For example, $5 + 15 + 45 + 135 + \cdots$ is geometric with $a = 5, r = 3$.

Geometric series test Not only is there a simple criterion for convergence, but if the series converges, the sum can easily be found. We will show:

(A) If $r \geq 1$ or $r \leq -1$ then $\sum_{n=0}^{\infty} ar^n$ diverges.

(B) If $-1 < r < 1$ then $\sum_{n=0}^{\infty} ar^n$ converges to $\dfrac{a}{1-r}$.

To illustrate why (A) holds, we'll look at some series with $r \geq 1$ or $r \leq -1$. For example, the series $2 + 2 + 2 + 2 + \cdots$ has $r = 1$ and diverges to ∞; the series $1 + 2 + 4 + 8 + \cdots$ has $r = 2$ and diverges to ∞. The series $1 - 2 + 4 - 8 + \cdots$ has $r = -2$ and diverges because the partial sums oscillate wildly.

To prove (B) we will find a formula for the partial sums S_n and examine the limit as $n \to \infty$. We have

(1) $\qquad S_n = a + ar + ar^2 + ar^3 + \cdots + ar^{n-1}.$

Multiply by r to obtain

(2) $\qquad rS_n = ar + ar^2 + ar^3 + \cdots + ar^{n-1} + ar^n.$

Subtract (2) from (1) to get

$$(1 - r)S_n = a - ar^n.$$

Finally, divide by $1 - r$, assuming $r \neq 1$, to get

$$S_n = \frac{a - ar^n}{1 - r}.$$

If $n \to \infty$ and $-1 < r < 1$, then $r^n \to 0$. Therefore

$$\lim_{n \to \infty} S_n = \lim_{n \to \infty} \frac{a - ar^n}{1 - r} = \frac{a}{1 - r} \quad \text{for } -1 < r < 1,$$

and the series converges to $a/(1 - r)$.

For example, the series $3 - \frac{3}{5} + \frac{3}{25} - \frac{3}{125} + \cdots$ converges since $r = -1/5$, which is strictly between -1 and 1. The sum S is given by

$$S = \frac{a}{1 - r} = \frac{3}{1 - \left(-\dfrac{1}{5}\right)} = \frac{15}{6} = \frac{5}{2}.$$

In other words, $\sum_{n=0}^{\infty} 3(-\frac{1}{5})^n = \frac{5}{2}$.

Application Consider a game in which players A and B take turns tossing one die, with A going first. The winner of the game is the first player to throw a 4. We want to find the probability that A wins.

Player A wins if A throws a 4 immediately or the results are

(3) non-4 for A, non-4 for B, 4 for A

or

(4) non-4 for A, non-4 for B, non-4 for A, non-4 for B, 4 for A
and so on.

Note that the probability of a non-4 on any toss is $\frac{5}{6}$ and the probability of a 4 is $\frac{1}{6}$. Therefore the probability that A throws a 4 immediately is $\frac{1}{6}$. To find the probability of (3), consider that in five-sixths of the games, A begins by throwing a non-4; then in five-sixths *of those games*, B continues by tossing a non-4; and in one-sixth *of those games* A follows with a 4. Therefore the probability of (3) is the product $\frac{5}{6} \times \frac{5}{6} \times \frac{1}{6}$, that is $(\frac{5}{6})^2 \frac{1}{6}$. Similarly, the probability of (4) is $(\frac{5}{6})^4 \frac{1}{6}$. Therefore the probability that A wins is $\frac{1}{6} + \frac{1}{6}(\frac{5}{6})^2 + \frac{1}{6}(\frac{5}{6})^4 + \frac{1}{6}(\frac{5}{6})^6 + \cdots$. The series is geometric with $a = \frac{1}{6}$ and $r = (\frac{5}{6})^2$ and its sum is $a/(1 - r)$, or $\frac{6}{11}$. So the probability that A wins is $\frac{6}{11}$.

Problems for Section 8.2

Decide if the series converges or diverges. If a series converges, find its sum.

1. $-1 + \dfrac{1}{6} - \dfrac{1}{36} + \dfrac{1}{216} - \cdots$ **6.** $\dfrac{1}{4} + \dfrac{1}{4}\left(\dfrac{2}{3}\right)^2 + \dfrac{1}{4}\left(\dfrac{2}{3}\right)^4 + \dfrac{1}{4}\left(\dfrac{2}{3}\right)^6 + \cdots$

2. $\dfrac{1}{4} - \dfrac{1}{16} + \dfrac{1}{64} - \dfrac{1}{256} + \cdots$ **7.** $.1 + .01 + .001 + .0001 + \cdots$

3. $\dfrac{3}{4} + \dfrac{9}{8} + \dfrac{27}{16} + \dfrac{81}{32} + \cdots$ **8.** $\displaystyle\sum_{n=1}^{\infty} (\sin \theta)^{2n}$ for a fixed θ

4. $3 + 9 + 27 + 81 + \cdots$ **9.** $\displaystyle\sum_{n=0}^{\infty} \dfrac{1}{\pi^{2n+1}}$

5. $\displaystyle\sum_{n=3}^{\infty} \dfrac{1}{4^n}$

8.3 Convergence Tests for Positive Series I

It is important to be able to decide if a given series converges or diverges, and if it converges, we want the sum. We were extraordinarily successful with geometric series, but we will not be so lucky otherwise. This section begins to collect tests for convergence versus divergence. No test supplies an absolute criterion, a condition that is both necessary and sufficient for convergence, and consequently more than one test may have to be tried. Furthermore, even if a series is identified as convergent, it is usually too difficult to find the sum. We often settle for an approximation to the sum, obtained by adding *some* of the terms of the series.

The series that arise most frequently in applications either have all positive terms or else terms that alternate in sign, so we concentrate on these types in the next three sections.

Positive series A series with all positive terms is called a *positive series*. As a by-product of studying positive series, we will be able to test series with all negative terms as well, since in that case a factor of -1 can be pulled out, leaving a positive series. A series which has *some* negative terms, but becomes positive after say a_{1000}, counts as a positive series, since the first 1000 terms can be dropped in testing for convergence versus divergence.

Since the partial sums of a positive series are increasing, a positive series will either converge or else diverge to ∞. The size of the terms of a positive series $\sum a_n$ determines whether the series converges or diverges. *If the series is to converge, the terms a_n must approach 0 and furthermore, must approach 0 rapidly enough. Otherwise, the subtotals will be dragged to ∞ and the series will diverge to ∞.* For example, if a_n approaches 3, rather than 0, then eventually the series is adding terms near 3, such as

$$(1) \qquad\qquad 2.9 + 2.99 + 3.002 + \cdots$$

and will diverge to ∞. As another example, consider the series

$$(2) \quad \underbrace{\frac{1}{2} + \frac{1}{2}}_{\text{two terms}} + \underbrace{\frac{1}{4} + \frac{1}{4} + \frac{1}{4} + \frac{1}{4}}_{\text{four terms}}$$

$$+ \underbrace{\frac{1}{8} + \frac{1}{8} + \frac{1}{8} + \frac{1}{8} + \frac{1}{8} + \frac{1}{8} + \frac{1}{8} + \frac{1}{8}}_{\text{eight terms}} + \underbrace{\frac{1}{16} + \cdots + \frac{1}{16}}_{\text{sixteen terms}} + \cdots.$$

The series diverges because $S_2 = 1, \cdots, S_6 = 2, \cdots, S_{14} = 3, \cdots$, and $S_n \to \infty$. The terms of the series do approach 0, but not rapidly enough. On the other hand,

$$(3) \qquad\qquad \frac{1}{2} + \frac{1}{4} + \frac{1}{8} + \frac{1}{16} + \cdots$$

is geometric ($r = 1/2$) and converges. Its terms approach 0 rapidly enough.

Our general conclusions may be rephrased in the following four statements.

nth term test   Let $\Sigma\, a_n$ be a positive series.

> (A) If a_n doesn't approach 0 then $\Sigma\, a_n$ diverges to ∞ (e.g., (1)).
>
> (B) If $\Sigma\, a_n$ converges then $a_n \to 0$.

(Part (B) follows from (A): Suppose $\Sigma\, a_n$ converges. If a_n does not approach 0 then, by (A), $\Sigma\, a_n$ diverges, contradicting the hypothesis. Thus a_n must approach 0. In fact, (A) and (B) are logically equivalent, since (A) similarly follows from (B).)

> (C) If $a_n \to 0$ then $\Sigma\, a_n$ may converge (see (3)) or may diverge (see (2)). Convergence of the series depends on whether a_n approaches 0 rapidly enough. More testing will be necessary to decide.
>
> (D) If $\Sigma\, a_n$ diverges then a_n may or may not approach 0. Either a_n does not approach 0 at all (see (1)), or a_n approaches 0 too slowly (see (2)).

Example 1   Consider $\Sigma\, \dfrac{n^2}{3n^2 + 2}$. By the highest power rule (Section 2.3), $\lim\limits_{n \to \infty} \dfrac{n^2}{3n^2 + 2} = 1/3$, which is nonzero. Therefore the series diverges by the nth term test. In particular, it diverges to ∞.

Warning   Don't confuse the limit $1/3$ with the sum of the series. The *terms* approach $1/3$, but the *sum* of the terms is ∞.

Example 2   Test $\Sigma\, \dfrac{n^2 + n}{4n^3 + 6}$ for convergence versus divergence.

Solution: By the highest power rule, $\lim\limits_{n \to \infty} \dfrac{n^2 + n}{4n^3 + 6} = 0$. But until we can decide if the terms approach 0 *rapidly enough,* the series can't be categorized. Additional procedures will be necessary before we can finish this example (Section 8.4).

Warning   The nth term test is only a test for divergence. When a_n does not approach 0, the test concludes that the series diverges, but the test can *never* be used to conclude that a series converges. The nth term test is a crude weapon. It identifies the grossly divergent series, where a_n does not approach 0. But if a series passes the nth term test, that is, $a_n \to 0$, then the only conclusion is that the series has a *chance* to converge ((3) does but (2) doesn't), and more refined tests must be applied.

Comparison test   Suppose a positive series has terms that approach 0. One of the ways to decide if the terms approach 0 rapidly enough is to compare them as follows with the terms of a series already categorized.

> Suppose $\Sigma\, a_n$ and $\Sigma\, b_n$ are positive series, and $a_n \leq b_n$ for all n. If $\Sigma\, b_n$ converges, then $\Sigma\, a_n$ converges. If $\Sigma\, a_n$ diverges to ∞, then $\Sigma\, b_n$ diverges to ∞. Thus, *if the series with larger terms converges, then the series with smaller terms converges also. If the series with smaller terms diverges to ∞, then the series with larger terms also diverges to ∞.*

The comparison test isn't useful unless the terms of a given series can be compared with those of a series already *known* to be convergent or *known* to be divergent. Therefore our next task is to produce a collection of *known* standard series, important in their own right and useful for comparison purposes.

Standard series Section 4.3 listed some functions in increasing order of magnitude. The following expanded version of that list, with x replaced by n (representing a nonnegative integer) will be helpful.

$$(4) \quad \ln n, (\ln n)^2, (\ln n)^3, \cdots, \sqrt{n}, n, n^{3/2}, n^2, \cdots, \left(\frac{3}{2}\right)^n, 2^n, 100^n, \cdots, n!$$

The new entry in (4) is the function $n!$. Remember that $n!$ is defined as the product $n(n-1)(n-2)\cdots 1$, so that, for example, $5! = 5 \times 4 \times 3 \times 2 \times 1 = 120$. As a special case, $1!$ and $0!$ are both defined to be 1. To see that $n!$ is indeed of a higher order of magnitude than 100^n, consider the quotient $100^n/n!$ say for $n = 200$:

$$\frac{100^{200}}{200!} = \left(\frac{100 \cdot 100 \cdots \cdot 100}{1 \cdot 2 \cdots \cdot 100}\right)\left(\frac{100 \cdot 100 \cdot 100 \cdots \cdot 100}{101 \cdot 102 \cdot 103 \cdots \cdot 200}\right).$$

We have written the result as the product of two factors; note that the second factor is very small. As $n \to \infty$, we may continue to write $100^n/n!$ as the product of two factors, one remaining fixed and the other approaching 0. Therefore $100^n/n!$ approaches 0, showing that $n!$ grows faster than 100^n. Similarly, it may be shown that $n!$ has a higher order of magnitude than any exponential function b^n.

Next, consider the reciprocals of the functions in (4):

$$(5) \quad \frac{1}{\ln n}, \frac{1}{(\ln n)^2}, \frac{1}{(\ln n)^3}, \cdots, \frac{1}{\sqrt{n}}, \frac{1}{n}, \frac{1}{n^{3/2}}, \frac{1}{n^2}, \frac{1}{n^3}, \cdots, \frac{1}{(1.5)^n}, \frac{1}{2^n}, \frac{1}{100^n}, \cdots, \frac{1}{n!}.$$

The entries in (5) approach 0 as $n \to \infty$, as opposed to (4) where the entries approach ∞. Section 4.3 discussed orders of magnitude for functions which approach ∞. Similar ideas hold for functions approaching 0. If a_n and b_n both approach 0 as $n \to \infty$, their quotient takes on the indeterminate form $0/0$, and its value depends on the particular a_n and b_n. If $a_n/b_n \to \infty$, or equivalently $b_n/a_n \to 0$, we say that a_n approaches 0 more slowly than b_n and has a *higher order of magnitude* than b_n. If $a_n/b_n \to L$, where L is a positive number, (not 0 or ∞) then a_n and b_n are said to have the *same order of magnitude*. The orders of magnitude in (5) decrease reading from left to right. Equivalently, *the entries in (5) approach 0 more rapidly reading from left to right*.

Finally, consider the series in Table 1, corresponding to the terms in (5). Some, such as $\Sigma 1/2^n$, are geometric series. The series of the form $\Sigma 1/n^p$ are called *p-series*. For example, $\Sigma 1/n^2 = 1 + \frac{1}{4} + \frac{1}{9} + \frac{1}{16} + \cdots$ is a p-series with $p = 2$. The p-series with $p = 1$,

$$\sum \frac{1}{n} = 1 + \frac{1}{2} + \frac{1}{3} + \frac{1}{4} + \cdots,$$

is called the *harmonic series*. All the series in the table are given a chance to converge by the nth term test, since their terms do approach 0 as $n \to \infty$. When the terms approach 0 slowly, the series will diverge; when the terms

Table 1   Standard Series

Diverge		Converge	
$\sum \dfrac{1}{\ln n}, \sum \dfrac{1}{(\ln n)^2}, \cdots, \sum \dfrac{1}{\sqrt{n}}, \sum \dfrac{1}{n}$		$\cdots, \sum \dfrac{1}{n^{3/2}}, \sum \dfrac{1}{n^2}, \cdots, \sum \dfrac{1}{(1.5)^n}, \sum \dfrac{1}{2^n}, \cdots, \sum \dfrac{1}{n!}$	
$\sum \dfrac{1}{(\ln n)^p},$	$\sum \dfrac{1}{n^p}, 0 < p \leq 1$	$\sum \dfrac{1}{n^p}, p > 1$	$\sum r^n, 0 < r < 1$
	p-series	*p*-series	geometric series

approach 0 rapidly, the series will converge. We will show at the end of the section that *a p-series converges if p > 1 and diverges if p ≤ 1*; in particular, *the harmonic series diverges*. Thus the dividing line in Table 1 comes after $\sum 1/n$. The series in the table to the left of the series $\sum 1/n$ have terms which are respectively larger than $1/n$ so they too diverge, by comparison. Similarly, the series to the right of the convergent *p*-series where $p > 1$ converge by comparison with their neighbors on the left, since they have correspondingly smaller terms. (Table 1 does not contain *all* series. In particular, there are divergent series between $\sum 1/n$ and the dividing line, albeit not *p*-series, and there are convergent series between the dividing line and the *p*-series with $p > 1$. There is no "last" series before the line and no first series after the line.)

For example, $\sum \dfrac{1}{\sqrt[4]{n}} = 1 + \dfrac{1}{\sqrt[4]{2}} + \dfrac{1}{\sqrt[4]{3}} + \dfrac{1}{\sqrt[4]{4}} + \cdots$ is a *p*-series with $p = \frac{1}{4}$, and diverges.

Warning   Don't confuse a *p*-series such as

$$\sum \frac{1}{n^3} = 1 + \frac{1}{8} + \frac{1}{27} + \frac{1}{64} + \frac{1}{125} + \cdots \qquad (p = 3, \text{ series converges})$$

with a geometric series such as

$$\sum \frac{1}{3^n} = 1 + \frac{1}{3} + \frac{1}{9} + \frac{1}{27} + \frac{1}{81} + \cdots \qquad \left(r = \frac{1}{3}, \text{ series converges} \right).$$

Example 3   Test $\sum \dfrac{1}{n^n} = 1 + \dfrac{1}{4} + \dfrac{1}{27} + \dfrac{1}{256} + \cdots$ for convergence versus divergence.

Solution: The series is not a *p*-series because the exponent n is not fixed, and is not a geometric series because the base n is not fixed. However, it can be successfully compared to either type. If $n > 2$, the terms of $\sum 1/n^n$ are respectively less than those of the convergent *p*-series $\sum 1/n^2 = 1 + \frac{1}{4} + \frac{1}{9} + \frac{1}{16} + \cdots$, that is $1/n^n < 1/n^2$ for $n > 2$. Therefore $\sum 1/n^n$ converges by the comparison test.

Subseries of a positive convergent series   If $\sum a_n$ is a positive convergent series, then every subseries also converges. In other words, *if the original terms produce a finite sum then any subcollection will also produce a finite sum.* For example, $1 + \frac{1}{4} + \frac{1}{16} + \frac{1}{36} + \frac{1}{64} + \cdots$ converges since it consists of every other term of the convergent *p*-series $\sum 1/n^2$.

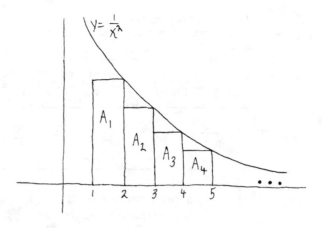

FIG. 1

Proof of the p-series principle   We conclude the section with a proof that a p-series converges for $p > 1$ and diverges for $p \leq 1$.

We'll begin with the case of $p = 2$, that is, with $\sum 1/n^2$. The trick is to assign geometric significance to the terms of the series using the graph of $1/x^2$ and the rectangles in Fig. 1. The first rectangle has base 1 and height $\frac{1}{4}$, so area A_1 is $\frac{1}{4}$. Similarly, $A_2 = \frac{1}{9}$, $A_3 = \frac{1}{16}$, and so on. Therefore,

(6) $$\sum \frac{1}{n^2} = 1 + \frac{1}{4} + \frac{1}{9} + \frac{1}{16} + \cdots = 1 + A_1 + A_2 + A_3 + \cdots.$$

But the sum of the rectangular areas in Fig. 1 is less than the area under the graph of $1/x^2$ for $x \geq 1$, so

(7) $$A_1 + A_2 + A_3 + \cdots < \int_1^\infty \frac{1}{x^2}\,dx = -\frac{1}{x}\Big|_1^\infty = 1.$$

Therefore, by (6) and (7), $\sum 1/n^2$ converges (to a sum which is less than 2).

The general proof for $\sum 1/n^p$, $p > 1$, is similar, but with the exponent 2 replaced by p.

Next, consider the case where $p = 1$. As a first attempt, see the graph of $1/x$ in Fig. 2 which shows that

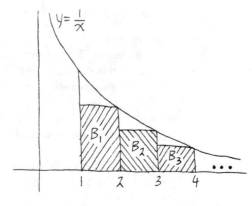

FIG. 2

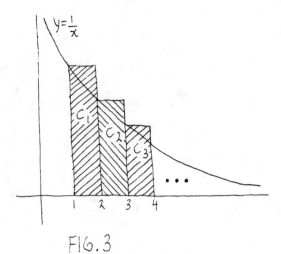

FIG.3

$$\sum \frac{1}{n} = 1 + \frac{1}{2} + \frac{1}{3} + \frac{1}{4} + \cdots = 1 + B_1 + B_2 + B_3 + \cdots .$$

The area $B_1 + B_2 + B_3 + \cdots$ is less than the total area under the graph of $1/x$, $x \geq 1$. The latter area is $\int_1^\infty (1/x)\,dx = \ln x \big|_1^\infty = \infty$. But this is useless since it does not reveal if the *smaller* area $B_1 + B_2 + B_3 + \cdots$ is finite or infinite. As a second attempt, consult Fig. 3 to see that

$$\sum \frac{1}{n} = 1 + \frac{1}{2} + \frac{1}{3} + \frac{1}{4} + \cdots$$

$$= C_1 + C_2 + C_3 + \cdots \geq \int_1^\infty \frac{1}{x}\,dx = \ln x \bigg|_1^\infty = \infty .$$

Therefore, the second attempt shows that $\sum 1/n$ diverges to ∞.

Finally, the p-series with $p < 1$ (which are to the left of $\sum 1/n$ in Table 1) diverge by comparison with $\sum 1/n$, since their terms are respectively larger.

Problems for Section 8.3

1. Suppose $\sum a_n$ is a positive series. Decide if the statement is true or false.

(a) If $a_n \to 0$ then $\sum a_n$ converges.   (c) If $\sum a_n$ diverges then a_n does not
(b) If a_n does not approach 0 then   approach 0.
     $\sum a_n$ diverges.   (d) If $\sum a_n$ converges then $a_n \to 0$.

2. What conclusion can you draw from the nth term test about the convergence or divergence of the series?

(a) $\sum \dfrac{n!}{4^n}$ (b) $\sum \dfrac{n^2}{4^n}$

3. In Problems (a)–(q), decide if the series converges or diverges.

(a) $1 + \dfrac{1}{3} + \dfrac{1}{9} + \dfrac{1}{27} + \dfrac{1}{81} + \cdots$ (c) $-\dfrac{1}{\sqrt{3}} - \dfrac{1}{\sqrt{4}} - \dfrac{1}{\sqrt{5}} - \cdots$

(b) $1 + \dfrac{1}{4} + \dfrac{1}{9} + \dfrac{1}{16} + \dfrac{1}{25} + \cdots$ (d) $\sqrt{3} + \sqrt{4} + \sqrt{5} + \sqrt{6} + \cdots$

(e) $\dfrac{3}{5!} + \dfrac{3}{6!} + \dfrac{3}{7!} + \cdots$ (l) $\dfrac{1}{7^5} + \dfrac{1}{8^5} + \dfrac{1}{9^5} + \cdots$

(f) $\dfrac{1}{8} + \dfrac{1}{27} + \dfrac{1}{64} + \dfrac{1}{125} + \cdots$ (m) $\dfrac{1}{5^7} + \dfrac{1}{5^8} + \dfrac{1}{5^9} + \cdots$

(g) $\dfrac{1}{3\sqrt{3}} + \dfrac{1}{5\sqrt{5}} + \dfrac{1}{7\sqrt{7}} + \cdots$ (n) $\dfrac{1}{2e^3} + \dfrac{1}{3e^4} + \dfrac{1}{4e^5} + \cdots$

(h) $5 + 6 + \dfrac{1}{6} + \dfrac{1}{7} + \dfrac{1}{8} + \dfrac{1}{9} + \cdots$ (o) $\left(\dfrac{1}{8}\right)^2 + \left(\dfrac{1}{9}\right)^2 + \left(\dfrac{1}{10}\right)^2$

(i) $\dfrac{3}{4} + \dfrac{4}{5} + \dfrac{5}{6} + \dfrac{6}{7} + \dfrac{7}{8} + \cdots$ $+ \left(\dfrac{1}{11}\right)^2 + \cdots$

(j) $\sum \dfrac{1}{2^n n!}$ (p) $\dfrac{1}{8} + \dfrac{1}{88} + \dfrac{1}{888} + \cdots$

(k) $\sum \dfrac{1}{n 2^n}$ (q) $\dfrac{1}{4!} + \dfrac{1}{6!} + \dfrac{1}{8!} + \cdots$

4. Suppose $\sum a_n$ is a positive convergent series. Decide, if possible, if each of the following series converges or diverges.

(a) $\sum \dfrac{1}{a_n}$ (b) $\sum \dfrac{a_n}{n!}$ (c) $\sum n! a_n$ (d) $\sum \cos a_n$

8.4 Convergence Tests for Positive Series II

This section continues with two more tests for positive series.

Limit comparison test We'll begin with a preliminary example to introduce the idea behind the test. You may prefer to skip directly to the test itself (next page) which most students find plausible without proof. Consider the series $\sum 1/(2n + 3)$. Since $\sum 1/n$ diverges, it might appear that we can test the given series by comparison. But $2n + 3 > n$, so

(1) $$\dfrac{1}{2n + 3} < \dfrac{1}{n}$$

which is not a useful inequality; if the terms of a series are respectively *smaller* than the terms of a *divergent* series, no conclusion can be drawn. However, we can find another comparison by first finding a limit. We have

(2) $\displaystyle\lim_{n\to\infty} \dfrac{1/(2n + 3)}{1/n} = \lim_{n\to\infty} \dfrac{n}{2n + 3}$ (by algebra) $= \dfrac{1}{2}$ (highest power rule).

Numbers which approach 1/2 must eventually go above and remain above .4, so eventually $\dfrac{1/(2n + 3)}{1/n} > .4$.

Thus, eventually,

(3) $$\dfrac{1}{2n + 3} > \dfrac{.4}{n}.$$

But the series $\sum .4/n$ is $.4 \sum 1/n$, which diverges (harmonic series). Therefore, $\sum 1/(2n + 3)$ diverges by comparison with $\sum .4/n$.

Let's summarize the results. Although the original comparison in (1) did not help, the impulse to compare the given series with $\sum 1/n$ was sound, and in (3), we found a useful comparison with a multiple of $\Sigma\, 1/n$. The procedure worked because the limit in (2) was a positive number rather than 0 or ∞. In essence, $\Sigma\, 1/(2n + 3)$ diverges because $1/(2n + 3)$ and $1/n$ have the same order of magnitude and $\Sigma\, 1/n$ diverges. In general, we have the following *limit comparison test.*

> *Suppose that a_n and b_n, both positive, have the same order of magnitude. Then $\Sigma\, a_n$ and $\Sigma\, b_n$ act alike in the sense that either both converge or both diverge.*

Intuitively, the test claims that for *positive* series, if a_n and b_n have the same order of magnitude, they are similar enough in size so that $\Sigma\, a_n$ and $\Sigma\, b_n$ behave alike. The preliminary example showed why this is the case for $\Sigma\, 1/2n + 3$ and $\Sigma\, 1/n$. We omit the more general proof.

To apply the limit comparison test to a positive series $\Sigma\, a_n$, try to find a standard series $\Sigma\, b_n$ such that b_n has the same order of magnitude as a_n. One way to do this is to use the fact (whose uninteresting proof we omit) *that if a_n is a fraction then a_n has the same order of magnitude as the new fraction*

$$\frac{\text{term of highest order of magnitude in the numerator}}{\text{term of highest order of magnitude in the denominator}}.$$

For example, $(n^2 + n)/(4n^3 + 6)$ has the same order of magnitude as $n^2/4n^3$, or $1/4n$. Therefore

$$\Sigma\, \frac{n^2 + n}{4n^3 + 6} \quad \text{acts like} \quad \Sigma\, \frac{1}{4n} = \frac{1}{4} \Sigma\, \frac{1}{n}.$$

Since the latter is the divergent harmonic series, the first series diverges also.

Ratio test   Series such as

$$(4) \qquad\qquad \Sigma\, \frac{n^3}{2^n}, \qquad \Sigma\, \frac{n^3}{n!}, \qquad \Sigma\, \frac{3^n}{n!}$$

are not standard series, nor can they be compared to standard series via the limit comparison test. The ratio test is a general method for testing positive series and is particularly useful for the series in (4). We'll state the test first, give examples, and then prove it.

> *Let $\Sigma\, a_n$ be a positive series. Consider $\lim\limits_{n \to \infty} \dfrac{a_{n+1}}{a_n}$.*
>
> (A) *If the limit is less than 1 then $\Sigma\, a_n$ converges.*
> (B) *If the limit is either greater than 1 or is ∞ then $\Sigma\, a_n$ diverges.*
> (C) *If the limit is 1 then no conclusion can be drawn. Try another test.*

For example, consider $\Sigma\, 2^n/n!$. Then $a_n = 2^n/n!$, $a_{n+1} = 2^{n+1}/(n + 1)!$ and

$$\frac{a_{n+1}}{a_n} = \frac{2^{n+1}/(n + 1)!}{2^n/n!} = \frac{2^{n+1}}{(n + 1)!} \cdot \frac{n!}{2^n} \text{ (by algebra) } = \frac{2}{n + 1} \text{ (cancel)}.$$

Therefore,

$$\lim_{n \to \infty} \frac{a_{n+1}}{a_n} = \lim_{n \to \infty} \frac{2}{n + 1} = 0.$$

Since the limit is less than 1, the series $\Sigma\, 2^n/n!$ converges by the ratio test.

As another example, we'll test $\sum n^3/2^n$. We have

$$\lim_{n \to \infty} \frac{a_{n+1}}{a_n} = \lim_{n \to \infty} \frac{(n+1)^3}{2^{n+1}} \cdot \frac{2^n}{n^3} = \lim_{n \to \infty} \frac{1}{2} \left(\frac{n+1}{n}\right)^3 = \frac{1}{2}.$$

Since the limit is less than 1, the given series converges by the ratio test.

Proof of the ratio test

(A) We assume that $\lim_{n \to \infty} a_{n+1}/a_n$ is less than 1. Suppose the limit is .97. If the ratios approach .97, eventually they must go below and remain below .98. We'll discard initial terms until we reach this eventuality, so that we may consider that *all* the ratios a_{n+1}/a_n under consideration are less than .98. Then $a_{n+1} < .98a_n$, and if we imagine multiplying our way from one term of the series to the next, we have to multiply by something *less* than .98 each time:

(5) $a_1 \qquad + \qquad a_2 \qquad + \qquad a_3 \qquad + \qquad a_4 + \cdots.$

   multiply by less   multiply by less   multiply by less
   than .98   than .98   than .98

The multiples in (5) may all be different, but each is less than .98. If we multiply by *precisely* .98 each time we have

(6) $a_1 \qquad + \qquad .98a_1 \qquad + \qquad (.98)^2a_1 \qquad + \qquad (.98)^3a_1$

   multiply by .98   multiply by .98   multiply by .98

$$+ \cdots.$$

The series in (6) is a convergent geometric series ($r = .98$), and the terms in (5) are respectively smaller than the terms in (6). Therefore, (5) converges by comparison.

The proof, in general, is handled in the same way with .97 replaced by an arbitrary positive number r, $r < 1$, and .98 by a number between r and 1.

(B) If a_{n+1}/a_n approaches ∞, or any number greater than 1, then eventually a_{n+1} must be larger than a_n. Therefore the terms of $\sum a_n$ *increase* and cannot approach 0, and the series diverges by the nth term test. In fact, any series in case (B) can more easily be identified as divergent by the nth term test.

(C) We will produce both convergent and divergent series with $\lim_{n \to \infty} a_{n+1}/a_n = 1$. Consider the harmonic series $\sum 1/n$, which we know diverges. We have

$$\lim_{n \to \infty} \frac{a_{n+1}}{a_n} = \lim_{n \to \infty} \frac{\dfrac{1}{n+1}}{\dfrac{1}{n}} = \lim_{n \to \infty} \frac{n}{n+1} = 1.$$

On the other hand, consider $\sum 1/n^2$, which we know converges. In this case,

$$\lim_{n \to \infty} \frac{a_{n+1}}{a_n} = \lim_{n \to \infty} \frac{n^2}{(n+1)^2} = 1.$$

Since both convergent and divergent series can have ratio limits of 1, such a limit does not help categorize a series.

Choosing a test   There is no decisive rule for selecting a convergence test. The more problems you do, the more expert you will become, because being an "expert" usually means that you have seen the problem, or a similar problem, before. We have the following recommendations.

1) See if the series is standard or acts like a standard series.
2) Apply the nth term test. Examine a_n to see if it approaches 0 (inconclusive) or does not approach 0 (series diverges).

   These methods are accomplished by a quick inspection of the series. If the inspection produces no immediate results, keep going.
3) Try the ratio test, especially if a_{n+1}/a_n looks like it will cancel nicely so that its limit is easy to find. The ratio test is usually more successful with ingredients such as $n!$ or 5^n than with $\sin n$ or $\ln n$. In particular, it can be used to show that series such as those in (4) converge.
4) Perhaps the comparison test can be used with your series and a standard series.
5) As a last resort, you might try using integrals as in the proofs in Section 8.3 that $\sum 1/n$ diverges and $\sum 1/n^2$ converges. Or you may be able to find a formula for the partial sums as we did for a geometric series.

There are other tests for convergence that are not included in the book, but more tests still give no guarantee of success. On the other hand, you now have enough methods to test many, although not all, series. In fact, it is quite possible for more than one method to work in a particular problem.

So far, this chapter has been mainly concerned with distinguishing convergent from divergent series. The results will be used in the important applications beginning in Section 8.6.

Problems for Section 8.4

In Problems 1–35, decide if the series converges or diverges.

1. $\sum \dfrac{1}{2n^2 + n}$

2. $\sum \dfrac{(2n)!}{(3n)!}$

3. $\sum \dfrac{1}{\sqrt[3]{n}}$

4. $\sum \dfrac{1}{3^{n-1}}$

5. $\sum \dfrac{n!}{10^n}$

6. $\dfrac{1}{\sqrt{2}} + \dfrac{1}{\sqrt[3]{2}} + \dfrac{1}{\sqrt[4]{2}} + \cdots$

7. $\sum \dfrac{1}{(n-2)^3}$

8. $-\dfrac{1}{4} - \dfrac{2}{9} - \dfrac{3}{16} - \dfrac{4}{25} - \cdots$

9. $\sum n^2 \left(\dfrac{3}{4}\right)^n$

10. $\sum \dfrac{10^n}{n!}$

11. $\sum \dfrac{\ln n}{\sqrt{n}}$

12. $\sum \dfrac{n-1}{n}$

13. $\sum \dfrac{2}{n+7}$

14. $\sum \dfrac{n^2}{5^n}$

15. $.3 + .03 + .003 + .0003 + \cdots$

16. $\dfrac{1}{3} + \dfrac{1 \cdot 3}{3 \cdot 6} + \dfrac{1 \cdot 3 \cdot 5}{3 \cdot 6 \cdot 9} + \cdots$

17. $\sum \dfrac{1}{5^n}$

18. $\dfrac{2!}{1\cdot 3} + \dfrac{3!}{1\cdot 3\cdot 5} + \dfrac{4!}{1\cdot 3\cdot 5\cdot 7} + \cdots$

19. $\sum \left(\dfrac{e}{3}\right)^n$

20. $\sum \left(\dfrac{e}{2}\right)^n$

21. $\dfrac{1}{9} + \dfrac{1}{25} + \dfrac{1}{49} + \cdots$

22. $\dfrac{1}{1\cdot 2} + \dfrac{1}{2\cdot 3} + \dfrac{1}{3\cdot 4} + \cdots$

23. $\sum \dfrac{n+1}{n\sqrt{n}}$

24. $\sum \dfrac{n}{n-1}$

25. $\sum \dfrac{(n!)^2}{(2n)!}$

26. $\sum \dfrac{\sqrt{n}}{3^n}$

27. $\dfrac{2\cdot 4}{5!} + \dfrac{2\cdot 4\cdot 6}{7!} + \dfrac{2\cdot 4\cdot 6\cdot 8}{9!} + \cdots$

28. $\sum \dfrac{\sqrt{n}}{n + \ln n}$

29. $\sum \dfrac{3^n}{4^n}$

30. $\dfrac{3}{4} + \dfrac{4}{9} + \dfrac{5}{16} + \dfrac{6}{25} + \cdots$

31. $\dfrac{1}{2} + \dfrac{1}{6} + \dfrac{1}{10} + \dfrac{1}{14} + \cdots$

32. $\sum \dfrac{1}{n4^n}$

33. $\sum \dfrac{\ln n}{n^2}$

34. $\sum \dfrac{1}{n^2 \ln n}$

35. $\sum_{n=2}^{\infty} \dfrac{1}{n \ln n}$ (use integrals)

36. The harmonic series $1 + \frac{1}{2} + \frac{1}{3} + \frac{1}{4} + \cdots$ diverges to ∞. (a) Show that the two subseries created by using every other term of the harmonic series also diverge. (b) Find a subseries that converges.

37. (a) Show that if $\sum a_n$ converges then $\sum na_n$ may converge or may diverge. (b) Show that if $\sum a_n$ converges *by the ratio test* then $\sum na_n$ also converges.

8.5 Alternating Series

Let a_n be positive. A series of the form

(1) $$\sum_{n=1}^{\infty} (-1)^{n+1}a_n = a_1 - a_2 + a_3 - a_4 + \cdots$$

is called an *alternating series*. The partial sums of a *positive* series are increasing, so a positive series either converges or else diverges to ∞. But the partial sums of an *alternating* series rise and fall since terms are alternately added and subtracted; therefore an alternating series either converges, diverges to ∞, diverges to $-\infty$, or diverges but not to ∞ or $-\infty$. For example, the series

(2) $$3 - 3 + 3 - 3 + 3 - 3 + \cdots$$

diverges (but not to ∞ or $-\infty$) since the partial sums oscillate from 3 to 0; the series

(3) $$3 - 4 + 3 - 5 + 3 - 6 + 3 - 7 + 3 - 8 + \cdots$$

diverges to $-\infty$ since the partial sums are $3, -1, 2, -3, 0, -6, -3, -10, \cdots$ which approach $-\infty$.

There are two major tests for alternating series. We have an *nth term test* for divergence which is very similar to the nth term test for positive series.

(In fact, an nth term test holds for arbitrary series, not necessarily positive or alternating.) Also there is an *alternating series test* for convergence.

nth term test   Consider the alternating series in (1).

> (A) If a_n doesn't approach 0 then the series diverges. (The partial sums oscillate but are not damped, and hence do not approach a limit—see (2) and (3).)
> (B) If the series converges then $a_n \to 0$.
> (C) If a_n does approach 0 then the alternating series may converge or may diverge. More testing will be necessary to make a decision.
> (D) If the series diverges then a_n may or may not approach 0.

As before, the nth term test is only a test for divergence. When a_n does not approach 0, the test concludes that the series diverges, but the test can *never* be used to conclude that a series converges. Again, it identifies the grossly divergent series.

Alternating series test   The alternating harmonic series

$$1 - \frac{1}{2} + \frac{1}{3} - \frac{1}{4} + \cdots$$

passes the nth term test, and as an introduction to the next test we will show that the series converges. Furthermore, although we can't find the sum, we can do the next best thing by producing a bound on the error when a partial sum is used to approximate the sum of the series. Then we will state the alternating series test in general.

Consider the partial sums, plotted on a number line in Fig. 1. Begin with $S_1 = 1$. Move down $\frac{1}{2}$ to plot S_2; move up $\frac{1}{3}$ to get S_3; move down $\frac{1}{4}$ to locate S_4; and so on. As successive terms are added and subtracted, the swing of oscillation of the partial sums is (consistently) decreasing because *each new term added or subtracted is less than the one before.* Figure 1 suggests that the partial sums oscillate their way to a limit S between 0 and 1. (Surprisingly, the formal proof requires quite sophisticated mathematics.) In other words, the series converges to a sum S between 0 and the first term a_1. Furthermore, note that S_5 is *above* the sum S, but the gap between S_5 and S is less than $\frac{1}{6}$ because subtracting $\frac{1}{6}$ sends us below S. In other words, if S_5 is used to approximate S then the approximation is an overestimate and the error is less than $\frac{1}{6}$. Similarly S_6 is an underestimate and the approximation error is less than $\frac{1}{7}$.

The key to the argument above is that the terms $1, \frac{1}{2}, \frac{1}{3}, \frac{1}{4}, \cdots$ being alternately added and subtracted do not merely approach 0 casually but *decrease (steadily) toward* 0. If this is *not* the case, then the alternating series

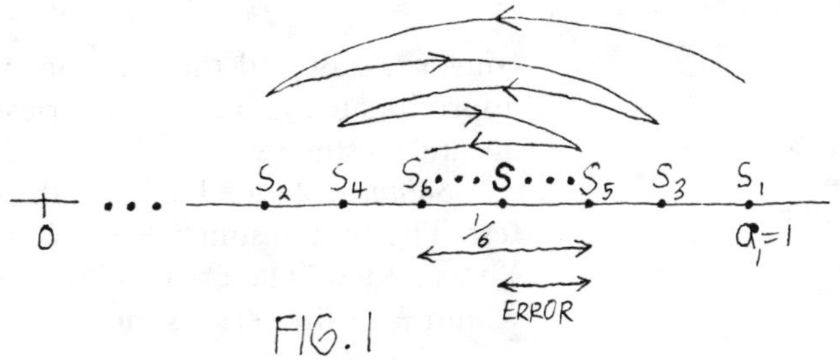

FIG. 1

may be (but is not necessarily) divergent. As an example of the latter possibility, consider

$$(4) \qquad 1 - \frac{1}{10} + \frac{1}{2} - \frac{1}{100} + \frac{1}{3} - \frac{1}{1,000} + \frac{1}{4} - \frac{1}{10,000} + \cdots.$$

The numbers $1, \frac{1}{10}, \frac{1}{2}, \frac{1}{100}, \cdots$ do approach 0 but *do not decrease* (steadily). They go up and down and up and down as they wend their way toward 0. The partial sums of the series in (4) do not oscillate with decreasing swing as in Fig. 1, and the argument used to show that the alternating harmonic series converges simply does not apply to (4). As a matter of fact, the positive terms alone in (4) amount to a harmonic series which diverges to ∞; the negative terms alone are a geometric series which converges to $-1/9$; and it can be shown that the partial sums are dragged to ∞ by the positive terms. Hence the series in (4) diverges to ∞.

If a_n not only approaches 0 but decreases (that is, decreases "steadily"), meaning that each term is smaller than the preceding one, then we write $a_n \downarrow 0$. As an example, for the alternating harmonic series we do have $a_n \downarrow 0$ but for the series in (4) we have $a_n \to 0$ but *not* $a_n \downarrow 0$. With this terminology we are ready for the following general conclusions, called the *alternating series test.*

> *Consider the alternating series* $\Sigma (-1)^{n+1} a_n$. *Suppose* $a_n \downarrow 0$. *Then the series converges to a sum S between 0 and* a_1.
>
> *Furthermore, if the last term of a subtotal involves addition, then the subtotal is greater than S; if the last term of a subtotal involves subtraction then the subtotal is less than S. In either case if only the first n terms are used, then the error, the difference between the subtotal* S_n *and the series sum S, is less than the first term not considered. In other words,* $|S - S_n| < a_{n+1}$.

The nth term test and the alternating series test are adequate to test most alternating series as follows.

If a_n does not approach 0 then the alternating series diverges by the nth term test.

If $a_n \downarrow 0$ then the alternating series converges by the alternating series test. For most alternating series, one of these two cases occurs.

It is unusual to have $a_n \to 0$ and not also have $a_n \downarrow 0$ so that neither test applies. For all practical purposes, if $a_n \to 0$ and there aren't separate formulas for $a_{\text{odd } n}$ *and* $a_{\text{even } n}$ *as in (4), then it will also be true that* $a_n \downarrow 0$. For example, if $a_n = n^3/2^n$ then not only does $a_n \to 0$ but also $a_n \downarrow 0$ eventually and $\Sigma (-1)^{n+1} n^3/2^n$ converges by the alternating series test.

Example 1 Show that the series

$$\Sigma (-1)^{n+1}/n^2 = 1 - \tfrac{1}{4} + \tfrac{1}{9} - \tfrac{1}{16} + \cdots$$

converges. Bound the error in using the sum of the first three terms to approximate the sum of the series. Is the approximation an overestimate or an underestimate?

Solution: Since $1/n^2 \downarrow 0$, the series converges by the alternating series test. The partial sum $1 - \tfrac{1}{4} + \tfrac{1}{9} = \tfrac{31}{36}$ is above the sum S since the last term, $\tfrac{1}{9}$, was added. The error is less than the next term, $\tfrac{1}{16}$. In other words, $\tfrac{31}{36}$ is within $\tfrac{1}{16}$ of the series sum.

Warning 1. The alternating series test is just a test for convergence. When $a_n \downarrow 0$, the test concludes that $\sum (-1)^{n+1} a_n$ converges. But if we do *not* have $a_n \downarrow 0$, the test does not conclude that the series diverges.

2. If a_n and b_n, both positive, have the same order of magnitude then the limit comparison test states that the two *positive* series $\sum a_n$ and $\sum b_n$ act alike. But the two *alternating* series $\sum (-1)^{n+1} a_n$ and $\sum (-1)^{n+1} b_n$ do *not* necessarily act alike. It is possible for an alternating series to converge so gingerly, because of a delicate balance of positive and negative terms, that another alternating series with terms of the same order of magnitude may behave differently. In other words, the limit comparison test does not apply to alternating series.

Absolute convergence   Another way to test the alternating series

(5) $a_1 - a_2 + a_3 - a_4 + a_5 - a_6 + \cdots,$ where $a_n > 0$,

is to remove the alternating signs and test the positive series

(6) $a_1 + a_2 + a_3 + a_4 + a_5 + a_6 + \cdots.$

We will prove that *if* (6) *converges then* (5) *also converges*. For the proof, consider the two new series

(7) $a_1 + 0 + a_3 + 0 + a_5 + 0 + a_7 + 0 + \cdots$

and

(8) $0 + a_2 + 0 + a_4 + 0 + a_6 + 0 + a_8 + \cdots.$

The terms in (7) and (8) are positive (and zero), and in each case are respectively less than or equal to the terms of (6). Since (6) converges by hypothesis, the series in (7) and (8) converge by the comparison test. If (8) is multiplied by -1, it still converges, by the factoring rule in Section 8.1, and the sum of (7) and $-(8)$ converges by the term by term addition rule in that section. But (7) $-$ (8) *is* (5), so (5) converges.

More generally, a similar proof can show that for *any* series (with any pattern of signs),

(9)   *if $\sum |a_n|$ converges then $\sum a_n$ converges.*

If $\sum |a_n|$ converges then the original series $\sum a_n$ is called *absolutely convergent*, so (9) shows that *absolute convergence implies convergence*.

For example, $1 - \frac{1}{2} - \frac{1}{4} + \frac{1}{8} - \frac{1}{16} - \frac{1}{32} + \frac{1}{64} - \cdots$ is neither alternating nor positive. It converges by (9) since the series of its absolute values is a convergent geometric series. As another example, consider $\sum (-1)^{n+1}/n^2$. It converges by the alternating series test since $1/n^2 \downarrow 0$. Alternatively, its series of absolute values is a convergent p-series, $p = 2$, so the original series converges by (9).

Conditional convergence   If $\sum |a_n|$ diverges it is still possible for $\sum a_n$ to converge. In this case, $\sum a_n$ is called *conditionally convergent*. The alternating harmonic series is conditionally convergent, since it converges but the series of its absolute values, i.e., the harmonic series, diverges.

So far we have been concerned with distinguishing convergent from divergent series. Now we have *three* categories since every convergent series

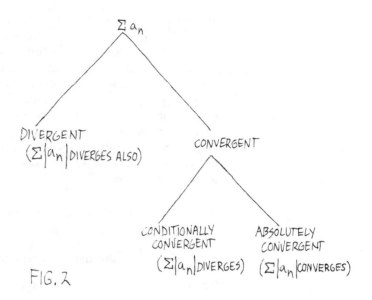

FIG. 2

$\sum a_n$ can be further categorized as either absolutely convergent ($\sum |a_n|$ converges) or conditionally convergent ($\sum |a_n|$ diverges). Divergent series cannot be subcategorized in this manner; if $\sum a_n$ diverges then, by (9), $\sum |a_n|$ cannot converge. Figure 2 shows the three possibilities for a series: divergent, conditionally convergent, absolutely convergent.

Conditionally convergent and absolutely convergent series both do converge, but absolute convergence is more desirable for several reasons, one of which we will mention here. It can be shown that if the terms of an absolutely convergent series are rearranged, that is, added in a different order, then the new series still converges to the same sum as before. On the other hand, if $\sum a_n$ is conditionally convergent then, given any number, the series can be rearranged to converge to that number. Furthermore, the series can be rearranged to diverge to ∞, and rearranged to diverge to $-\infty$.†

†We will illustrate with the conditionally convergent alternating harmonic series $1 - \frac{1}{2} + \frac{1}{3} - \frac{1}{4} + \cdots$, which converges to a sum between 0 and 1. We will rearrange the series to converge to 37. First note that the subseries of positive terms $1 + \frac{1}{3} + \frac{1}{5} + \cdots$ diverges to ∞ and the subseries of negative terms $-\frac{1}{2} - \frac{1}{4} - \frac{1}{6} - \cdots$ diverges to $-\infty$ (Problem 36a, Section 8.4). Then begin the rearrangement of the alternating harmonic series by adding positive terms until the subtotal goes over 37. (How do we know that the subtotal will *ever* get that large? The positive subseries diverges to ∞, so surely if enough positive terms are added, the subtotal passes 37.) Then add negative terms until the subtotal goes below 37. (How do we know that the subtotal can be brought down below 37? Because the negative terms add to $-\infty$.) Then add positive terms to bring the subtotal back over 37, add negative terms to bring the subtotal back below 37, and so on. The partial sums oscillate around 37 and the overall swing of oscillation is approaching 0 because $a_n \to 0$. It can be shown in fact that the rearrangement converges to 37. The alternating harmonic series can also be rearranged to diverge to ∞. First add positive terms until the subtotal is larger than 1, possible because the positive terms themselves add to ∞. Then feed in one negative term to avoid being accused of leaving out the negatives. Then add positive terms until the subtotal is larger than 2, followed by one more negative term, and so on. This produces a rearrangement, since all terms are eventually used, although each partial sum contains many more positive than negative terms. Furthermore, the partial sums approach ∞, so the rearrangement diverges to ∞. Similarly, the series can be rearranged to diverge to $-\infty$. On the other hand, the absolutely convergent geometric series $\sum_{n=0}^{\infty} (-\frac{1}{2})^n$ has sum $\frac{2}{3}$ and every rearrangement converges to $\frac{2}{3}$; if a rearrangement has 1,000 positive terms followed by one negative term, followed by 1,000,000 positive terms followed by one negative term, and so on, the rearrangement still converges to $\frac{2}{3}$.

Problems for Section 8.5

1. Show that $1 - \dfrac{1}{\sqrt{2}} + \dfrac{1}{\sqrt{3}} - \dfrac{1}{\sqrt{4}} + \cdots$ converges, and estimate the error if the sum is approximated by S_{24}. Is the approximation an overestimate or an underestimate?

2. Show that the series converges and approximate the sum so that the error is at most .001. Is your estimate over or under?

(a) $\displaystyle\sum_{n=1}^{\infty} (-1)^{n+1} \dfrac{1}{n!}$ (b) $\dfrac{1}{4^4} - \dfrac{1}{5^5} + \dfrac{1}{6^6} - \cdots$

3. True or false?

(a) If we do not have $a_n \downarrow 0$ then $\sum (-1)^{n+1} a_n$ diverges.
(b) If we do not have $a_n \to 0$ then $\sum (-1)^{n+1} a_n$ diverges.

4. Test the series for divergence versus convergence.

(a) $\displaystyle\sum (-1)^{n+1} \dfrac{n^2}{n!}$ (b) $\displaystyle\sum (-1)^{n+1} \dfrac{n!}{n^2}$ (c) $\displaystyle\sum (-1)^{n+1} \dfrac{1}{n \ln n}$

(d) $\displaystyle\sum (-1)^{n-1} \dfrac{2n}{n^2 + 4}$ (e) $.1 - .01 + .001 - \cdots$

(f) $\dfrac{3}{2} - \dfrac{4}{3} + \dfrac{5}{4} - \dfrac{6}{5} + \cdots$ (g) $\dfrac{\sqrt{2}}{3} - \dfrac{\sqrt{3}}{4} + \dfrac{\sqrt{4}}{5} - \cdots$

5. True or False?

(a) If $\sum b_n$ is a convergent positive series then $\sum b_n^2$ converges also.
(b) If $\sum (-1)^{n+1} b_n$ is a convergent alternating series then $\sum b_n^2$ converges also.

6. Table 1 in Section 8.3 lists some standard positive series, some convergent and some divergent. Consider all the corresponding *alternating* series, namely,

$$\sum (-1)^{n+1} \dfrac{1}{\ln n}, \cdots, \sum (-1)^{n+1} \dfrac{1}{n!}.$$

(a) Test them for convergence versus divergence.   (b) Of the convergent series in part (a), test for conditional versus absolute convergence.

7. Test for conditional convergence versus absolute convergence versus divergence.

(a) $\displaystyle\sum (-1)^n \dfrac{n}{1 + n^2}$ (b) $\displaystyle\sum (-1)^{n+1} \dfrac{n + 2}{n^3 + 3}$

8. What conclusions can be drawn about $\sum a_n$ if

(a) $\sum |a_n|$ diverges   (b) $\sum |a_n|$ converges

9. What conclusions can be drawn about $\sum |a_n|$ if

(a) $\sum a_n$ diverges   (b) $\sum a_n$ converges?

10. Test the series for convergence versus divergence using the alternating series test, and then again using the series of absolute values.

(a) $\displaystyle\sum (-1)^{n+1} \dfrac{1}{n!}$ (b) $\displaystyle\sum (-1)^{n+1} \dfrac{1}{\sqrt{n}}$

11. Decide, if possible, whether the series converges absolutely or conditionally.
(a) a convergent geometric series   (b) a convergent p-series

8.6 Power Series Functions

Polynomials such as $ax^2 + bx + c$ are familiar elementary functions. The generalization of a polynomial is a series of the form

(1) $$a_0 + a_1x + a_2x^2 + a_3x^3 + a_4x^4 + \cdots,$$

called a *power series*. For example, $5 + 6x + 7x^2 + 8x^3 + 9x^4 + \cdots$ is a power series. A power series is a function of x, often *non*elementary. The rest of the chapter discusses power series and their applications.

Application Power series may be used to create new functions when the elementary functions are inadequate. It can be shown that the differential equation

(2) $$xy'' + y = 0$$

cannot be satisfied by an elementary function. Thus it is necessary to invent a new function to solve the equation. Consider the power series

$$y = a_0 + a_1x + a_2x^2 + a_3x^3 + \cdots.$$

We will determine the coefficients so that y satisfies (2). We have

(3) $$y' = a_1 + 2a_2x + 3a_3x^2 + 4a_4x^3 + 5a_5x^4 + \cdots$$

$$y'' = 2a_2 + 3 \cdot 2a_3x + 4 \cdot 3a_4x^2 + 5 \cdot 4a_5x^3 + \cdots.$$

Substitute y and y'' into (2) to obtain

$$x(2a_2 + 3 \cdot 2a_3x + 4 \cdot 3a_4x^2 + 5 \cdot 4a_5x^3 + \cdots) + a_0 + a_1x + a_2x^2 + a_3x^3 + \cdots = 0.$$

Collect terms to get

(4) $$a_0 + (2a_2 + a_1)x + (3 \cdot 2a_3 + a_2)x^2 + (4 \cdot 3a_4 + a_3)x^3 + \cdots = 0.$$

(We write $4 \cdot 3$ instead of 12, and $3 \cdot 2$ instead of 6, because we want to discover patterns, and the combined form conceals patterns.) Now choose $a_0, a_1, a_2, \cdots$ so that (2) holds. We can do this by forcing all coefficients on the left side of (4) to be 0. Therefore, let $a_0 = 0$. Then let $2a_2 + a_1 = 0$, which doesn't determine either a_1 or a_2 but can be written as $a_2 = -\frac{1}{2}a_1$. Then choose $3 \cdot 2a_3 + a_2 = 0$ so that

$$a_3 = -\frac{a_2}{3 \cdot 2} = -\frac{-\frac{1}{2}a_1}{3 \cdot 2} = \frac{a_1}{3 \cdot 2 \cdot 2}.$$

Continue with $4 \cdot 3a_4 + a_3 = 0$ so that

$$a_4 = -\frac{a_3}{4 \cdot 3} = -\frac{\frac{a_1}{3 \cdot 2 \cdot 2}}{4 \cdot 3} = -\frac{a_1}{4 \cdot 3 \cdot 3 \cdot 2 \cdot 2}.$$

The pattern is now established. We have $a_5 = \dfrac{a_1}{5 \cdot 4 \cdot 4 \cdot 3 \cdot 3 \cdot 2 \cdot 2}$ and, in general,

$$a_n = (-1)^{n+1} \frac{a_1}{n!(n-1)!}.$$

Coefficient a_1 isn't determined, so we conclude that for *every* value of a_1,

$$y = \sum_{n=1}^{\infty} (-1)^{n+1} \frac{a_1}{n!(n-1)!} x^n$$

is a solution to (2). The factor a_1 serves as an arbitrary constant. Equivalently, the power series function

$$(5) \qquad y = x - \frac{1}{2} x^2 + \frac{1}{3 \cdot 2 \cdot 2} x^3 - \frac{1}{4 \cdot 3 \cdot 3 \cdot 2 \cdot 2} x^4 + \cdots,$$

and all multiples of it, are solutions to the differential equation in (2).

Interval of convergence   The domain of a power series function is the set of all x for which the series converges. For example, if $g(x) = 7 + x + 2x^2 + 3x^3 + 4x^4 + \cdots$ then $g(0) = 7 + 0 + 0 + 0 + \cdots = 7$ but there is no $g(1)$ because the series $7 + 1 + 2 + 3 + 4 + \cdots$ diverges. If we're going to work with power series functions we must be able to decide when the power series converges. The preceding sections were designed in part to provide that capability.

In general, a power series $\sum a_n x^n$ converges absolutely (hence converges) for x in an interval $(-r, r)$ centered about 0, and diverges for $x > r$ and $x < -r$. (Anything may happen for $x = \pm r$.) The series is said to have *radius of convergence* r and *interval of convergence* $(-r, r)$ (see Fig. 1). This includes the possibility that a power series may converge only for $x = 0$, in which case it has radius of convergence 0, or may converge absolutely for all x, in which case it has radius of convergence ∞ and interval of convergence $(-\infty, \infty)$. The value of r depends on the particular power series.

To illustrate the validity of these claims, and to actually find the interval of convergence of any given power series, we will use a version of the ratio test extended to include series that are not necessarily positive.

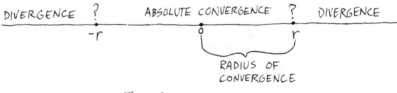

FIG. 1

Ratio test   Given a series $\sum b_n$, not necessarily positive, consider

$$\lim_{n \to \infty} \frac{|b_{n+1}|}{|b_n|}.$$

(a) If the limit is less than 1, then $\sum b_n$ converges absolutely (and therefore converges).

(b) If the limit is greater than 1, or is ∞, then $\sum b_n$ diverges.

(c) If the limit is 1, we have no conclusion.

To prove (a), note that if the limit is less than 1 then $\sum |b_n|$ converges by the ratio test for *positive* series (Section 8.4). Therefore the original series is absolutely convergent.

To prove (b), note that if $|b_{n+1}|/|b_n|$ approaches a number larger than 1 then eventually $|b_{n+1}| > |b_n|$. Therefore the terms $|b_n|$ are increasing and hence do not approach 0. Therefore b_n does not approach 0 either, and so Σb_n diverges by the nth term test.

Finding the interval of convergence   Consider $\Sigma \left(-\dfrac{1}{2}\right)^n \dfrac{1}{n+1} x^n$. To find the interval of convergence, compute

$$\lim_{n \to \infty} \frac{|x^{n+1}\text{term}|}{|x^n\text{term}|} = \lim_{n \to \infty} \left| \frac{\left(-\dfrac{1}{2}\right)^{n+1} x^{n+1}}{n+2} \right| \left| \frac{n+1}{\left(-\dfrac{1}{2}\right)^n x^n} \right| = \lim_{n \to \infty} \frac{1}{2} \cdot \frac{n+1}{n+2} |x|.$$

Since $n \to \infty$ while x is fixed, the limit is $\frac{1}{2}|x|$. By the ratio test, the series converges absolutely if $\frac{1}{2}|x| < 1$, $|x| < 2$, $-2 < x < 2$; and diverges if $\frac{1}{2}|x| > 1$, that is, $x > 2$ or $x < -2$. Therefore there *is* an interval of convergence, namely $(-2, 2)$. (If $x = 2$ then the series is $\Sigma (-1)^n \dfrac{1}{n+1}$, which converges by the alternating series test. If $x = -2$ then the series is the divergent harmonic series. Thus the series converges at the right end of the interval of convergence and diverges at the left end.)

As another example, consider

$$\Sigma (-1)^{n+1} \frac{1}{n!(n-1)!} x^n,$$

the power series in (5) that solved the differential equation $xy'' + y = 0$. We have

(6) $$\frac{|x^{n+1}\text{term}|}{|x^n\text{term}|} = \frac{|x|^{n+1}}{(n+1)!n!} \frac{n!(n-1)!}{|x|^n}$$

Note that

$$\frac{(n-1)!}{(n+1)!} = \frac{(n-1)(n-2)(n-3)\cdots 1}{(n+1)n(n-1)(n-2)\cdots 1} = \frac{1}{(n+1)n}$$

so (6) cancels to

$$\frac{|x|}{(n+1)n}.$$

For any fixed x, the limit is 0 as $n \to \infty$. Therefore the limit is less than 1 for any x, and the series converges for all x. The interval of convergence is $(-\infty, \infty)$ and the radius of convergence is ∞.

In practice, *the interval of convergence of a power series is the set of x for which* $\lim_{n \to \infty} \dfrac{|x^{n+1}\text{term}|}{|x^n\text{term}|}$ *is less than 1.*

Problems for Section 8.6

For each power series, find the interval of convergence.

1. $\Sigma (-1)^n (n+1) x^n$ **2.** $\Sigma \dfrac{x^n}{3^n n^2}$ **3.** $\Sigma n! x^n$ **4.** $\Sigma \dfrac{x^n}{n!}$

5. $x - x^3 + x^5 - x^7 + \cdots$ **6.** $2^2 x^3 + 2^4 x^5 + 2^6 x^7 + \cdots$

7. $3x + \dfrac{9x^2}{2} + \dfrac{27x^3}{3} + \dfrac{81x^4}{4} + \cdots$

8.7 Power Series Representations for Elementary Functions I

The solution to the differential equation in (2) of Section 8.6 illustrated why it is useful to invent *new* functions using power series. But it is useful to have power series expansions for *old* functions as well. Polynomials are pleasant functions, and representing an old function as an "infinite polynomial" can make that function easier to handle. In this section and the next we will find power series expansions for some elementary functions.

A power series for $1/(1 - x)$ The power series $1 + x + x^2 + x^3 + x^4 + \cdots$ is a geometric series with $a = 1$ and $r = x$. Therefore it converges for $-1 < x < 1$, that is, its interval of convergence is $(-1, 1)$, and the sum is $1/(1 - x)$. Thus

(1)
$$\frac{1}{1 - x} = 1 + x + x^2 + x^3 + x^4 + \cdots \qquad \text{for } -1 < x < 1,$$

and we have a power series expansion for $1/(1 - x)$. The function $1/(1 - x)$ exists for all $x \neq 1$ but its expansion is valid only for $-1 < x < 1$. The series has a smaller domain than the function $1/(1 - x)$, but when the series and the function are both defined, they agree.

Binomial series There is an entire class of familiar elementary functions whose power series expansions we can *guess*. Recall (Appendix A4) that

$$(1 + x)^5 = (1 + x)(1 + x)(1 + x)(1 + x)(1 + x)$$

$$= 1 + 5x + \frac{5 \cdot 4}{2!}x^2 + \frac{5 \cdot 4 \cdot 3}{3!}x^3 + \frac{5 \cdot 4 \cdot 3 \cdot 2}{4!}x^4$$

$$+ \frac{5 \cdot 4 \cdot 3 \cdot 2 \cdot 1}{5!}x^5$$

(2)
$$= 1 + 5x + 10x^2 + 10x^3 + 5x^4 + x^5.$$

Functions such as $(1 + x)^{-5}$ and $(1 + x)^{1/2}$ cannot be similarly written as polynomials because the exponents -5 and $1/2$ are not positive integers. However, we might suspect that these functions can be written as *infinite* polynomials, in the same pattern exhibited by the polynomial expansion for $(1 + x)^5$. In other words, we guess that the function $(1 + x)^q$ has the power series expansion

(3)
$$1 + qx + \frac{q(q - 1)}{2!}x^2 + \frac{q(q - 1)(q - 2)}{3!}x^3 + \cdots$$

for x in the interval of convergence of the series. We omit the proof that confirms the guess.

For example,

$$\sqrt{1 + x} = (1 + x)^{1/2}$$

$$= 1 + \frac{1}{2}x + \frac{\left(\frac{1}{2}\right)\left(-\frac{1}{2}\right)}{2!}x^2 + \frac{\left(\frac{1}{2}\right)\left(-\frac{1}{2}\right)\left(-\frac{3}{2}\right)}{3!}x^3 + \cdots$$

$$= 1 + \frac{1}{2}x - \frac{1}{2^2 2!}x^2 + \frac{1 \cdot 3}{2^3 3!}x^3 - \frac{1 \cdot 3 \cdot 5}{2^4 4!}x^4 + \cdots.$$

We still must find the interval of convergence in (3). If q is a positive integer then (3) collapses to a polynomial (as in (2) where $q = 5$) and "converges" for all x. If q is not a positive integer, the interval of convergence can be found with the ratio test. We have

$$n\text{th term} = \frac{q(q - 1) \cdots (q - [n - 1])\}}{n!}x^n$$

and

$$(n + 1)\text{st term} = \frac{q(q - 1) \cdots (q - [n - 1])(q - n)}{(n + 1)!}x^{n+1}.$$

So

$$\frac{|(n + 1)\text{st term}|}{|n\text{th term}|} = \frac{|q - n|}{n + 1}|x|.$$

The limit as $n \to \infty$ is $|x|$; solve $|x| < 1$ to get the interval of convergence $(-1, 1)$. Thus

(4)
$$\boxed{\begin{aligned} (1 + x)^q = 1 + qx + \frac{q(q - 1)}{2!}x^2 + \frac{q(q - 1)(q - 2)}{3!}x^3 + \cdots \\ \text{for } -1 < x < 1. \end{aligned}}$$

The series in (4) is called the *binomial series*.

Application We will show why it may be useful to approximate a function by the first few terms of its series expansion.

An inverse square law states that if two unit positive charges are distance r apart, then each is repelled by a force $F = 1/r^2$; if a unit positive charge and a unit negative charge are distance r apart, then they are attracted by a force $F = 1/r^2$. Now suppose that one negative charge and two positive charges are situated as shown in Fig. 1, where d is much smaller than r. The problem is to find the total force on charge C.

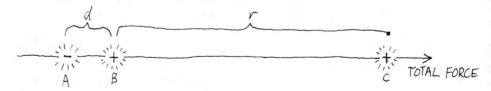

FIG. 1

Since C is repelled by B and attracted by A, we have

(5) $$\text{total force on } C = \frac{1}{r^2} - \frac{1}{(r + d)^2}.$$

This is an accurate description of the force on C, but it is difficult to tell from (5) just how the force varies with d and r. So we continue by rewriting the second fraction in (5). Factor to get

$$\frac{1}{(r + d)^2} = \frac{1}{\left(r\left[1 + \dfrac{d}{r}\right]\right)^2} = \frac{1}{r^2\left(1 + \dfrac{d}{r}\right)^2} = \frac{1}{r^2}\left(1 + \frac{d}{r}\right)^{-2}.$$

Since d is less than r, d/r is in the interval $(-1, 1)$, so we may expand $[1 + (d/r)]^{-2}$ in a binomial series by setting $q = -2$, $x = d/r$ to obtain

$$\frac{1}{(r + d)^2} = \frac{1}{r^2}\left[1 + (-2)\left(\frac{d}{r}\right) + \frac{(-2)(-3)}{2!}\left(\frac{d}{r}\right)^2 \right.$$
$$\left. + \frac{(-2)(-3)(-4)}{3!}\left(\frac{d}{r}\right)^3 + \cdots \right].$$

If d is *much* less than r, as intended in Fig. 1, then $(d/r)^2$, $(d/r)^3, \cdots$ are so small that

$$\frac{1}{(r + d)^2} = \frac{1}{r^2}\left(1 - 2\frac{d}{r} + \text{negligible terms}\right)$$
$$= \frac{1}{r^2} - 2\frac{d}{r^3} \quad \text{(approximately)}.$$

Thus, back in (5), we have (approximately)

$$\text{total force on } C = \frac{1}{r^2} - \left(\frac{1}{r^2} - 2\frac{d}{r^3}\right) = 2\frac{d}{r^3}.$$

Therefore, the force on C may be succinctly (albeit approximately) described as directly proportional to d and inversely proportional to r^3.

Making replacements in an old series to find a new series So far we have expansions for $1/(1 - x)$ and $(1 + x)^q$. We continue the problem of finding expansions for functions by showing how new series may be obtained from existing series.

Suppose we want an expansion for the function $1/(1 + 2x)$. Rewrite the function as $\dfrac{1}{1 - (-2x)}$ so that it resembles the left-hand side of (1). Then replace x by $-2x$ in (1) to obtain

$$\frac{1}{1 - (-2x)} = 1 + (-2x) + (-2x)^2 + (-2x)^3 + (-2x)^4 + \cdots$$

$$\text{for } -1 < -2x < 1.$$

To solve the inequality, divide each member by -2 to get $\frac{1}{2} > x > -\frac{1}{2}$ (multiplying or dividing by a negative number reverses an inequality), which may be written as $-\frac{1}{2} < x < \frac{1}{2}$. Thus we have an expansion for the function $1/(1 + 2x)$ and its interval of convergence, namely,

(6) $\dfrac{1}{1 + 2x} = 1 - 2x + 4x^2 - 8x^3 + 16x^4 - \cdots$ for $-\dfrac{1}{2} < x < \dfrac{1}{2}$.

As you can see, the replacement method involves solving an inequality to obtain the new interval of convergence. Table 1 lists some inequalities and their solutions, typical of those that occur most frequently.

Table 1

Inequality		Solution
$-r < \dfrac{2}{3}x < r,$	$-r < -\dfrac{2}{3}x < r$	$-\dfrac{3}{2}r < x < \dfrac{3}{2}r$
$-r < \dfrac{3}{4}x^n < r,$	$-r < -\dfrac{3}{4}x^n < r$	$-\sqrt[n]{\dfrac{4}{3}}r < x < \sqrt[n]{\dfrac{4}{3}}r$

As another example of replacement, we will find an expansion for $1/(3 - x^2)$. First do some factoring:

$$\frac{1}{3 - x^2} = \frac{1}{3\left(1 - \dfrac{1}{3}x^2\right)} = \frac{1}{3}\frac{1}{1 - \dfrac{1}{3}x^2}.$$

Then replace x by $\frac{1}{3}x^2$ in (1) to obtain

$$\frac{1}{3 - x^2} = \frac{1}{3}\left[1 + \left(\frac{1}{3}x^2\right) + \left(\frac{1}{3}x^2\right)^2 + \left(\frac{1}{3}x^2\right)^3 + \cdots\right]$$

(7) $\qquad\qquad\qquad\qquad\qquad\qquad\qquad$ for $-1 < \dfrac{1}{3}x^2 < 1$.

Some students are bothered by the inequality in (7) because the left-hand part, $-1 < \frac{1}{3}x^2$, is *vacuous* (it is *always* true that $\frac{1}{3}x^2$ is greater than -1). Nevertheless it is *not wrong*. The inequality may be rewritten simply as $\frac{1}{3}x^2 < 1$, and its solution, as indicated by the second line in the table, is $-\sqrt{3} < x < \sqrt{3}$. Therefore the final answer is

$$\frac{1}{3 - x^2} = \frac{1}{3} + \frac{1}{3^2}x^2 + \frac{1}{3^3}x^4 + \frac{1}{3^4}x^6 + \frac{1}{3^5}x^8 + \cdots$$

$$\text{for } -\sqrt{3} < x < \sqrt{3}.$$

Adding and multiplying old series to find new series Suppose $f(x)$ has a power series expansion with interval of convergence $(-r_1, r_1)$, and $g(x)$ has an expansion with interval of convergence $(-r_2, r_2)$. It can be shown that if the two series are multiplied like polynomials, then the product series is an expansion for $f(x)g(x)$. Similarly, if the two series are added like polynomials, the sum series is an expansion for $f(x) + g(x)$. Furthermore, the intervals of convergence of the product and sum series are at least the smaller of the two intervals $(-r_1, r_1)$ and $(-r_2, r_2)$, and, for all practical purposes, *are* the smaller of $(-r_1, r_1)$ and $(-r_2, r_2)$.

As an example, suppose we want an expansion for $\dfrac{1}{(1 - x)(1 + 2x)}$.

From (1) and (6) we have expansions for $\dfrac{1}{1-x}$ and $\dfrac{1}{1+2x}$ on $(-1, 1)$ and $(-\frac{1}{2}, \frac{1}{2})$, respectively. The smaller of the intervals is $(-\frac{1}{2}, \frac{1}{2})$. Therefore,

$$\frac{1}{(1-x)(1+2x)} = \frac{1}{1-x}\frac{1}{1+2x}$$

$$= (1 + x + x^2 + x^3 + x^4 + \cdots)(1 - 2x + 4x^2 - 8x^3 + 16x^4 - \cdots)$$

$$\text{for } x \text{ in } \left(-\frac{1}{2}, \frac{1}{2}\right).$$

As with polynomials, multiply each term in the first parentheses by each term in the second parentheses, and collect terms to get

$$\frac{1}{(1-x)(1+2x)} = 1 - 2x + x + 4x^2 - 2x^2 + x^2 - 8x^3 + 4x^3 - 2x^3$$

$$+ x^3 + 16x^4 - 8x^4 + 4x^4 - 2x^4 + x^4 + \cdots$$

$$= 1 - x + 3x^2 - 5x^3 + 11x^4 - 21x^5 + \cdots$$

$$\text{for } x \text{ in } \left(-\frac{1}{2}, \frac{1}{2}\right).$$

For another approach to the same problem, use partial fraction decomposition (Section 7.4) to get

$$(8) \qquad \frac{1}{(1-x)(1+2x)} = \frac{\frac{1}{3}}{1-x} + \frac{\frac{2}{3}}{1+2x}.$$

To find a series for $\dfrac{\frac{1}{3}}{1-x}$, multiply on both sides of (1) by $1/3$, and *keep* the interval of convergence $(-1, 1)$. Similarly, to find a series for $\dfrac{\frac{2}{3}}{1+2x}$, multiply on both sides of (6) by $2/3$, and *keep* the interval of convergence $(-1/2, 1/2)$. The smaller of the two intervals is $(-1/2, 1/2)$, so (8) becomes

$$\frac{1}{(1-x)(1+2x)} = \frac{1}{3}(1 + x + x^2 + x^3 + \cdots)$$

$$+ \frac{2}{3}(1 - 2x + 4x^2 - 8x^3 + \cdots) \qquad \text{for } -\frac{1}{2} < x < \frac{1}{2}$$

$$= 1 - x + 3x^2 - 5x^3 + 11x^4 - 21x^5 + \cdots$$

$$\text{for } -\frac{1}{2} < x < \frac{1}{2}.$$

In this example, the second method is better. No pattern seems to be revealed by the first method, whereas the second method easily predicts any term in the series; e.g., the coefficient of x^{199} is $\frac{1}{3} + \frac{2}{3}(-2)^{199}$.

Differentiating and antidifferentiating old series to find new series
Suppose $f(x)$ has a power series expansion with interval of convergence
$(-r, r)$. It can be shown that if the series is differentiated like a polynomial,
the new series is an expansion for $f'(x)$ (we already anticipated this in (3) of
Section 8.6); and if the series is antidifferentiated like a polynomial, and the
arbitrary constant of integration appropriately evaluated, the new series
represents any desired antiderivative of $f(x)$. Furthermore, it can be shown
that both the differentiated and antidifferentiated series have the same
interval of convergence as the original.

As an illustration, suppose we want an expansion for $1/(1 - x)^2$. We can
get it by squaring the series for $1/(1 - x)$, and also by using the binomial
series with $q = -2$ and x replaced by $-x$. For a third method, use the fact
that $1/(1 - x)^2$ is the derivative of $1/(1 - x)$. Differentiate on both sides of
(1), and keep the interval of convergence, to get

$$(9) \quad \frac{1}{(1 - x)^2} = 1 + 2x + 3x^2 + 4x^3 + 5x^4 + \cdots \qquad \text{for } x \text{ in } (-1, 1).$$

As another example, suppose we want to expand $\ln(1 + x)$. First find
an expansion for $1/(1 + x)$ by replacing x by $-x$ in (1) to get

$$\frac{1}{1 + x} = 1 + (-x) + (-x)^2 + (-x)^3 + (-x)^4 + \cdots \qquad \text{for } -1 < -x < 1$$

$$= 1 - x + x^2 - x^3 + x^4 - \cdots \qquad \text{for } -1 < x < 1.$$

Then antidifferentiate to get

$$\ln(1 + x) = C + x - \frac{x^2}{2} + \frac{x^3}{3} - \frac{x^4}{4} + \frac{x^5}{5} - \cdots \qquad \text{for } -1 < x < 1.$$

To determine C, substitute a value of x for which both sides can be
computed. The best value to use is $x = 0$, in which case we have
$\ln(1 + 0) = C + 0 + 0 + 0 + \cdots$, $0 = C$. Therefore,

$$(10) \quad \boxed{\ln(1 + x) = x - \frac{x^2}{2} + \frac{x^3}{3} - \frac{x^4}{4} + \frac{x^5}{5} - \cdots \qquad \text{for } -1 < x < 1.}$$

Summary of procedures for finding the new interval of convergence If
a new series is obtained from a known series by differentiation, antidifferentiation, multiplication by a constant, or, more generally, multiplication by a polynomial, keep the original interval of convergence.

If a new series is obtained from *two* known series by addition or multiplication, keep the smaller of the two original intervals.

If a new series is obtained from a known series by replacement, make
the same replacement in the inequality describing the original interval, and
solve for x to find the new interval. (If the known series converges for *all*
x, then after any replacement, the new series also converges for *all* x.)

Application We can use the binomial series to estimate $\displaystyle\int_0^{1/4} \frac{1}{(1 + x^2)^{3/2}} \, dx$
so that the error is less than .0001.

First, use (4) with $q = -3/2$ and x replaced by x^2 to get

$$\frac{1}{(1 + x^2)^{3/2}} = 1 - \frac{3}{2}x^2 + \frac{\left(-\frac{3}{2}\right)\left(-\frac{5}{2}\right)}{2!}(x^2)^2$$

$$+ \frac{\left(-\frac{3}{2}\right)\left(-\frac{5}{2}\right)\left(-\frac{7}{2}\right)}{3!}(x^2)^3 + \cdots \quad \text{for } -1 < x^2 < 1$$

$$= 1 - \frac{3}{2}x^2 + \frac{3 \cdot 5}{2^2 \cdot 2!}x^4 - \frac{3 \cdot 5 \cdot 7}{2^3 \cdot 3!}x^6 + \frac{3 \cdot 5 \cdot 7 \cdot 9}{2^4 \cdot 4!}x^8 - \cdots$$

$$\text{for } -1 < x < 1.$$

Since the interval of integration $[0, 1/4]$ is inside the interval of convergence $(-1, 1)$, it can be shown that we may integrate term by term to obtain

$$\int_0^{1/4} \frac{1}{(1 + x^2)^{3/2}} \, dx = x \Big|_0^{1/4} - \frac{3}{2}\frac{x^3}{3}\Big|_0^{1/4} + \frac{3 \cdot 5}{2^2 \cdot 2!}\frac{x^5}{5}\Big|_0^{1/4}$$

(11)
$$- \frac{3 \cdot 5 \cdot 7}{2^3 \cdot 3!}\frac{x^7}{7}\Big|_0^{1/4} + \cdots.$$

The series in (11) is not a *power* series and does not have an interval of convergence. It is a convergent series of *numbers* whose sum is the integral on the left-hand side. Continuing, we have

$$\int_0^{1/4} \frac{1}{(1 + x^2)^{3/2}} \, dx = .25 - .0078125 + .0003662 - .0000191 + \cdots.$$

By the alternating series test, if we stop adding after two terms, the error is less than .0003662, not enough of a guarantee. But we use the sum of the first three terms, .2425537, as the approximation (an overestimate), then the error is less than .0000191, which *is* less than .0001, as desired.

Problems for Section 8.7

1. Find a power series for each function, and find the interval of convergence of the series.

(a) $\sqrt[3]{1 + x}$ (b) $\frac{x}{1 - x}$ (c) $\frac{1}{(1 + x)^3}$ (d) $\frac{1}{2 - 3x}$ (e) $\frac{1}{(3 + x)^6}$

(f) $\frac{x}{(1 - x)(1 - 3x)}$ (g) $\frac{1}{x - 2}$ (h) $\ln(2 + x)$

2. Find an expansion for $\sqrt{1 - 3x^2}$ and the interval of convergence. Find the term containing x^{34} to illustrate the pattern, and then express the series in summation notation.

3. Find an expansion and its interval of convergence for $1/(1 - x^2)$ by

(a) using the binomial series   (b) using the series for $1/(1 - x)$
(c) multiplying series   (d) adding series   (e) using long division

4. Rederive (9) by (a) using the binomial series   (b) multiplying series.

5. (a) Find a series for $\tan^{-1}x$ and find the interval of convergence. (b) Approximate $\int_0^{1/2} \tan^{-1}x^2 \, dx$ so that the error is less than .0001. Do you have an underestimate or an overestimate?

6. What function has the expansion $x + 2x^2 + 3x^3 + 4x^4 + \cdots$? (Consider how the series is related to the series in (1).)

7. Let $f(x) = x + \dfrac{x^2}{2\cdot 2} + \dfrac{x^3}{3\cdot 2^2} + \dfrac{x^4}{4\cdot 2^3} + \cdots$.

(a) Write the series in summation notation.
(b) Find an expansion for $f'(x)$.
(c) Identify $f'(x)$ and $f(x)$ (they are familiar elementary functions).

8. (a) Write $\sqrt{19}$ as $\sqrt{16+3} = 4\sqrt{1+\frac{3}{16}}$ and use the binomial series with $q = \frac{1}{2}$, $x = \frac{3}{16}$ to approximate $\sqrt{19}$ so that the error is less than .01. (b) What is wrong with writing $\sqrt{19}$ as $\sqrt{1+18}$ and using the binomial series with $q = \frac{1}{2}$, $x = 18$?

8.8 Power Series Representations for Elementary Functions II (Maclaurin Series)

We continue with the task of finding series expansions for functions. The preceding section showed that if a connection can be found between $f(x)$ and a function (or functions) with a known expansion, then the connection can be exploited to find an expansion, along with its interval of convergence, for $f(x)$. But sometimes too much cleverness is required to find such a connection, and sometimes there simply is no connection. It isn't possible to use the preceding section to find an expansion for $\sin x$, since $\sin x$ is not related to $1/(1-x)$ or $(1+x)^q$, our functions with known expansions. This section considers a second method for finding an expansion for a function, based on an explicit formula for the coefficients.

The Maclaurin series for a function Suppose
$$f(x) = a_0 + a_1 x + a_1 x^2 + a_3 x^3 + a_4 x^4 + \cdots.$$
Set $x = 0$ to obtain $a_0 = f(0)$, a formula for the coefficient a_0. Differentiate to get
$$f'(x) = a_1 + 2a_2 x + 3a_3 x^2 + 4a_4 x^3 + \cdots$$
and substitute $x = 0$ to obtain $a_1 = f'(0)$, a formula for a_1. Differentiate again to get
$$f''(x) = 2a_2 + 3\cdot 2a_3 x + 4\cdot 3a_4 x^2 + 5\cdot 4a_5 x^3 + \cdots,$$
and substitute $x = 0$ to obtain $f''(0) = 2a_2$, or $a_2 = \dfrac{f''(0)}{2}$. We'll continue until we are sure of the pattern. Differentiating again, we have
$$f'''(x) = 3\cdot 2a_3 + 4\cdot 3\cdot 2a_4 x + 5\cdot 4\cdot 3a_5 x^2 + \cdots.$$
Let $x = 0$ to get $f'''(0) = 3\cdot 2a_3$, or $a_3 = \dfrac{f'''(0)}{3\cdot 2}$. Similarly, $a_4 = \dfrac{f^{(4)}(0)}{4\cdot 3\cdot 2}$, and, in general,

(1) $$a_n = \frac{f^{(n)}(0)}{n!}.$$

(Remember that $0! = 1$, $1! = 1$ and $f^{(n)}$ means the nth derivative of f.)

We have shown that given a function $f(x)$, there are two possibilities. Either f has *no* power series expansion of the form $\sum a_n x^n$, or

(2)

$$f(x) = \frac{f(0)}{0!} + \frac{f'(0)}{1!}x + \frac{f''(0)}{2!}x^2 + \frac{f'''(0)}{3!}x^3 + \cdots .$$

Certain functions fall into the "no series" category because they and/or their derivatives blow up at $x = 0$. In that case, the coefficients in (2) can't even be computed, so the series doesn't exist. Some functions of this type are $\ln x$, $\sqrt{x}$ and $1/x$. Otherwise, *every function occurring in practice (provided it does not blow up or have derivatives which blow up at $x = 0$) has the expansion in (2), called the Maclaurin series for f.* In this case, the expansion holds on the interval of convergence of the series, which can be found by the ratio test. (There are functions $f(x)$, rarely encountered, whose Maclaurin coefficients exist but whose Maclaurin series regrettably converge to something other than $f(x)$. However, such functions will play no role in this book.)

If a series is found for f using a method from the preceding section, or using several methods from that section, the answer(s) will inevitably be the Maclaurin series for f; no other series is possible. Regardless of how it is obtained, the coefficient a_n is given by (1). All series found in the preceding section are Maclaurin series although they were not computed directly from (2).

We'll use (2) to find a power series for $\sin x$. We have

$$
\begin{array}{ll}
f(x) = \sin x & f(0) = 0 \\
f'(x) = \cos x & f'(0) = 1 \\
f''(x) = -\sin x & f''(0) = 0 \\
f'''(x) = -\cos x & f'''(0) = -1 \\
f^{(4)}(x) = \sin x & f^{(4)}(0) = 0 \\
f^{(5)}(x) = \cos x & f^{(5)}(0) = 1 \\
\vdots &
\end{array}
$$

Thus the Maclaurin series in (2) is

$$x - \frac{x^3}{3!} + \frac{x^5}{5!} - \frac{x^7}{7!} + \cdots .$$

To find the interval of convergence, consider

$$\frac{|x^{2n+1}\text{term}|}{|x^{2n-1}\text{term}|} = \frac{|x^{2n+1}|}{(2n + 1)!} \cdot \frac{(2n - 1)!}{|x^{2n-1}|} = \frac{|x|^2}{(2n + 1)2n} .$$

For any fixed x, the limit as $n \to \infty$ is 0. So the series converges for all x, and

(3)

$$\sin x = x - \frac{x^3}{3!} + \frac{x^5}{5!} - \frac{x^7}{7!} + \cdots \qquad \text{for all } x .$$

As a corollary, we can differentiate (3) to find a series for $\cos x$. Note that the derivative of a term such as $x^5/5!$ is $5x^4/5!$, or $x^4/4!$. Thus

(4)

$$\cos x = 1 - \frac{x^2}{2!} + \frac{x^4}{4!} - \frac{x^6}{6!} + \cdots \qquad \text{for all } x .$$

We can also use (2) to find a series for e^x. If $f(x) = e^x$ then any derivative $f^{(n)}(x)$ is e^x again, and $f^{(n)}(0) = 1$. Therefore the Maclaurin series for e^x is $1 + x + x^2/2! + x^3/3! + x^4/4! + \cdots$. The ratio test will show that the series has interval of convergence $(-\infty, \infty)$, so

(5)
$$e^x = 1 + x + \frac{x^2}{2!} + \frac{x^3}{3!} + \frac{x^4}{4!} + \cdots \qquad \text{for all } x.$$

Using (2) to find a series for $f(x)$ works well if the nth derivatives of f are easy to compute, as with $\sin x$ and e^x. It would not be easy with functions such as $\sqrt{1 + x^2}$ and $x/(1 - x)(1 - 3x)$, whose derivatives become increasingly messy; the methods of the preceding section are preferable in such cases. Note that when (2) is used (as for $\sin x$), the interval of convergence must be found with the ratio test. When a series is found using a known series for a related function (as for $\cos x$, related to $\sin x$), the interval of convergence is found easily from the interval for the known series.

Maclaurin polynomials The discussion in Example 2 of Section 4.3 showed that for x near 0, $\sin x$ is approximately the same size as x. The power series for $\sin x$ goes many steps further and shows that we can get a better approximation using the polynomial $x - x^3/3!$, a still better approximation using $x - x^3/3! + x^5/5!$, and so on. In general, the partial sums of the Maclaurin series in (2) are called *Maclaurin polynomials*. We will show graphically how f is approximated by its Maclaurin polynomials. Consider the graph of f versus the graph of its Maclaurin polynomial of degree 1, that is, f versus

(6)
$$y = \frac{f(0)}{0!} + \frac{f'(0)}{1!}x.$$

Equation (6) is a line, and a line does not usually approximate a curve very well. But (6) is special; it is the line tangent to the graph of f at the point $(0, f(0))$ (Fig. 1). To confirm this, note that the tangent line has slope $f'(0)$, and so, using the point-slope form $y = mx + b$, the tangent line has equation $y = f'(0)x + f(0)$, which is (6).

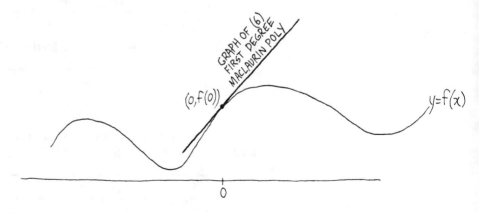

FIG. 1

Consider

(7)
$$y = \frac{f(0)}{0!} + \frac{f'(0)}{1!}x + \frac{f''(0)}{2!}x^2,$$

the Maclaurin polynomial of degree 2. Its graph is a parabola (Fig. 2) which passes through the point $(0, f(0))$, and hugs the graph of f more closely than

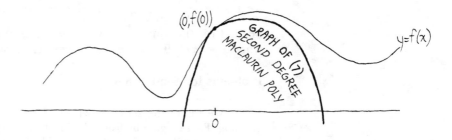

FIG. 2

the tangent line in Fig. 1. Similarly, the graph of

(8)
$$y = \frac{f(0)}{0!} + \frac{f'(0)}{1!}x + \frac{f''(0)}{2!}x^2 + \frac{f'''(0)}{3!}x^3$$

passes through the point $(0, f(0))$ and does still a better job of staying close to the graph of f (Fig. 3).

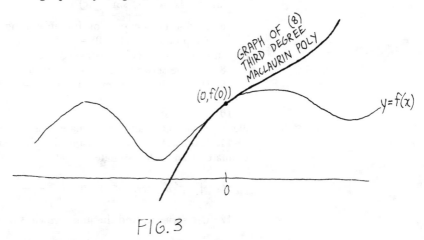

FIG. 3

In general, graphs of successive Maclaurin polynomials provide better and better approximations to the graph of f. At first (that is, after adding only a few terms of the Maclaurin series for f), the polynomials approximate the graph of f nicely only if x is near 0. After a while (that is, after adding many terms), the polynomials approximate f nicely even if x is far from 0, near the end of the interval of convergence.

If the sum of just a few terms of a series produces a good approximation to the sum of the series, the convergence is said to be *fast;* if many terms must be added before the approximation error becomes small, the convergence is *slow.* The graphs of the Maclaurin polynomials in Figs. 1–3 illustrate that the power series expansion for $f(x)$ converges more rapidly if x is near 0 and more slowly if x is far from 0.

Application Suppose we want to approximate $\sin 1°$ so that the error is less than 10^{-7}. Switching to radian measure so that we may use (3), we have

$$\sin 1° = \sin \frac{\pi}{180} = \frac{\pi}{180} - \frac{1}{3!}\left(\frac{\pi}{180}\right)^3 + \frac{1}{5!}\left(\frac{\pi}{180}\right)^5 - \cdots.$$

Since the series alternates, and the third term is the first one less than 10^{-7} we take $\dfrac{\pi}{180} - \dfrac{1}{3!}\left(\dfrac{\pi}{180}\right)^3$ as the approximation. Only two terms were needed for the approximation; the series in (3) converges rapidly to $\sin x$ when $x = \pi/180$ since $\pi/180$ is very close to 0.

Problems for Section 8.8

1. We found the series expansion for $(1 + x)^q$ in the preceding section by guessing. Find it again by using the Maclaurin series.

2. We found series for $1/(1 - x)$ and $\ln(1 + x)$ in the preceding section ((1) and (10)). Find them again using the Maclaurin series formula.

3. Find a series expansion for the function, and the interval of convergence of the series, by using the Maclaurin series and then again by using established series.

(a) $\dfrac{1}{2}(e^x - e^{-x})$ (b) $\dfrac{1}{3 - 2x}$

4. Write the series for $\sin x$ and $\cos x$ using the notation $\sum_{n=0}^{\infty} a_n x^n$.

5. Find a series expansion and the interval of convergence.

(a) $\cos 3x$ (b) $x^3 \sin x$ (c) e^{4x}

6. Find a series expansion for $\sin^2 x$ using $\sin^2 x = \frac{1}{2}(1 - \cos 2x)$.

7. Suppose $f(0) = 1$, $g(0) = 0$, $f'(x) = g(x)$ and $g'(x) = f(x)$. Find a series for $f(x)$ and find its interval of convergence.

8. Use the series for $\sin x$ to confirm that $\sin(-x) = -\sin x$.

9. Differentiate the series for e^x to see what happens. (In a sense, nothing should happen since the derivative of e^x is e^x again.)

10. Use the series for $\sin x$ to estimate $\sin 1$ (radian) so that the error is less than .0001. Do you have an overestimate or an underestimate?

11. Estimate the integral using the given error bound. Do you have an over- estimate or an underestimate?

(a) $\displaystyle\int_0^1 e^{-x^2}\,dx$, error $< .1$ (b) $\displaystyle\int_0^{1/3} \frac{1}{(1 + x^2)^4}\,dx$, error $< .01$

12. Use series to find the limit, which is of the indeterminate form 0/0.

(a) $\displaystyle\lim_{x\to 0} \frac{\ln(1 + x^2)}{1 - \cos x}$ (b) $\displaystyle\lim_{x\to 0} \frac{\sin x}{x}$

13. Use the power series for e^x to find the sum of the standard convergent series $\sum_{n=0}^{\infty} 1/n!$.

8.9 The Taylor Remainder Formula and an Estimate for the Number e

If we set $x = 1$ in the power series for e^x (see (5) of the preceding section), we have

(1)
$$e = 1 + 1 + \frac{1}{2!} + \frac{1}{3!} + \frac{1}{4!} + \cdots .$$

We can approximate e by partial sums of the series, but since the series does not alternate we do not have an error bound. The aim of this section is to introduce an error bound for the Maclaurin series for $f(x)$ in general, and then use it in the special case of e^x.

Suppose x is fixed and $f(x)$ is approximated by the beginning of its Maclaurin series, that is, by a Maclaurin polynomial, say of degree 8:

$$\frac{f(0)}{0!} + \frac{f'(0)}{1!}x + \frac{f''(0)}{2!}x^2 + \cdots + \frac{f^{(8)}(0)}{8!}x^8 .$$

If the series alternates, then the first term omitted supplies an error bound. But whether or not the series alternates, the error in the approximation may be bounded as follows. Consider all possible values of $\left| \frac{f^{(9)}(m)}{9!}x^9 \right|$ for m between 0 and x, and find the maximum of the values. *Taylor's remainder formula* states that the error, in absolute value, is less than or equal to that maximum.

In general, *the error (in absolute value) in approximating $f(x)$ by its Maclaurin polynomial of degree n is less than or equal to the maximum value of*

(2)
$$\left| \frac{f^{(n+1)}(m)}{(n+1)!}x^{n+1} \right|$$

for m between 0 and x. We omit the proof.

Returning to the problem of approximating e, we will obtain a first estimate using areas, and then use it, along with power series, to find a sharper estimate.

In Fig. 1, the shaded region has area $\int_1^2 (1/x)\, dx = \ln 2 - \ln 1 = \ln 2$. The rectangular region $ABCD$ within the shaded region has area $1/2$, so $\ln 2 > 1/2$. Therefore $\ln 4 = \ln 2^2 = 2 \ln 2 > 1$. But $\ln e = 1$, so $\ln 4 > \ln e$. Since $\ln x$ is an increasing function, we have $4 > e$. Similarly, since $\ln e = 1$ and $\ln 1 = 0$, we have $e > 1$. Thus, a first estimate of e is $1 < e < 4$.

Now let's return to (1). Suppose the first five terms are added to obtain the approximation

(3)
$$e = 1 + 1 + \frac{1}{2!} + \frac{1}{3!} + \frac{1}{4!} = 2.708 .$$

To estimate the error, consider (2) with $f(x) = e^x$, $x = 1$, $n = 4$ (since we added through the x^4 term in the series for e^x), and $0 \le m \le 1$. Then $f^{(5)}(x) = e^x$ and

$$\left| \frac{f^{(5)}(m)}{5!} 1^5 \right| = \frac{e^m}{5!} .$$

Since $1 < e < 4$, the maximum occurs when $m = 1$, and that maximum is less than $4/5!$ or $1/30$. Therefore the error in the approximation in (3) is less than $1/30$. Furthermore, when the expansion in (1) stops somewhere, all the terms omitted are positive, so the approximation in (3) is an underestimate. Thus,

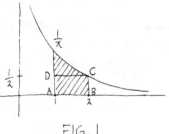

FIG. 1

$$2.708 < e < 2.708 + \frac{1}{30} < 2.742.$$

In a similar fashion, by adding more terms, it can be shown that $2.718281 < e < 2.718282$.

8.10   Power Series in Powers of $x - b$ (Taylor Series)

Certain basic functions such as $\ln x$, $\sqrt{x}$ and $1/x$ cannot be expressed in the form $\Sigma a_n x^n$ because they and/or their derivatives blow up at $x = 0$ (Section 8.8). Also, other functions have power series which converge too slowly if x is far from 0. We attempt to overcome these difficulties by considering power series of the form

$$(1) \qquad \sum_{n=0}^{\infty} a_n(x - b)^n = a_0 + a_1(x - b) + a_2(x - b)^2 + a_3(x - b)^3 + \cdots.$$

We call (1) a *power series about b*. The power series we have considered so far are the special case where $b = 0$. In this section we will show how a function $f(x)$ can be expanded about b with a generalization of the Maclaurin series formula, or, better still, using known series about 0.

In Section 8.8 we showed that if f has an expansion of the form $\Sigma a_n x^n$, then $a_n = f^{(n)}(0)/n!$. A similar argument shows that if f has an expansion of the form $\Sigma a_n(x - b)^n$, then $a_n = f^{(n)}(b)/n!$. This leads to the following generalization of Maclaurin series.

Every function $f(x)$ encountered in practice, which does not blow up or have derivatives which blow up at $x = b$, has the expansion

$$(2) \qquad \boxed{f(x) = \frac{f(b)}{0!} + \frac{f'(b)}{1!}(x - b) + \frac{f''(b)}{2!}(x - b)^2 + \frac{f'''(b)}{3!}(x - b)^3 + \cdots,}$$

called the *Taylor series for f about b*. The expansion holds on an interval of convergence *centered about b* and found with the ratio test. The partial sums of the Taylor series are called *Taylor polynomials*. Graphs of successive Taylor polynomials are a line, a parabola, a cubic, and so on, tangent to the graph of $f(x)$ at the point $(b, f(b))$; they supply better and better approximations to the graph. The Taylor series converges more rapidly if x is near b, and more slowly if x is far from b.

The Maclaurin series, with interval of convergence centered about 0, and the Maclaurin polynomials, tangent to the graph of $f(x)$ at the point $(0, f(0))$, are the special case of Taylor polynomials when $b = 0$.

One method for expanding a given $f(x)$ in powers of $x - b$ is to use (2) directly, along with the ratio test to determine the interval of convergence. Another method is to write $f(x)$ as $f([x - b] + b)$ and maneuver algebraically, as illustrated in examples, until it is ultimately possible to make use of a known series in powers of x, but with x replaced by $x - b$. With this approach, the interval of convergence can be obtained from the interval for the known series. No matter which method is used, the answer will agree with (2); no other series in powers of $x - b$ is possible.

Example 1   Find an expansion for $\cos x$ in powers of $x - \frac{1}{2}\pi$, and find the interval of convergence.

Solution: For a first approach, use (2) with $f(x) = \cos x$, $b = \frac{1}{2}\pi$. Then $f'(x) = -\sin x$, $f''(x) = -\cos x$, $f'''(x) = \sin x$, $f^{(4)}(x) = \cos x, \cdots$; and $f(\frac{1}{2}\pi) = 0$, $f'(\frac{1}{2}\pi) = -1$, $f''(\frac{1}{2}\pi) = 0$, $f'''(\frac{1}{2}\pi) = 1$, $f^{(4)}(\frac{1}{2}\pi) = 0$, and so on. Therefore

$$\cos x = \frac{f(\frac{1}{2}\pi)}{0!} + \frac{f'(\frac{1}{2}\pi)}{1!}(x - \tfrac{1}{2}\pi) + \frac{f''(\frac{1}{2}\pi)}{2!}(x - \tfrac{1}{2}\pi)^2 + \cdots$$

$$= -(x - \tfrac{1}{2}\pi) + \frac{1}{3!}(x - \tfrac{1}{2}\pi)^3 - \frac{1}{5!}(x - \tfrac{1}{2}\pi)^5 + \cdots.$$

To find the interval of convergence, use the ratio test. We have

$$\frac{|(x - \tfrac{1}{2}\pi)^{2n+1}\text{term}|}{|(x - \tfrac{1}{2}\pi)^{2n-1}\text{term}|} = \left|\frac{(x - \tfrac{1}{2}\pi)^{2n+1}}{(2n + 1)!}\right| \left|\frac{(2n - 1)!}{(x - \tfrac{1}{2}\pi)^{2n-1}}\right|$$

$$= \frac{|x - \tfrac{1}{2}\pi|^2}{(2n + 1)2n}.$$

The limit as $n \to \infty$ is 0 so the series converges for all x.

As a second approach, write $\cos x = \cos([x - \tfrac{1}{2}\pi] + \tfrac{1}{2}\pi)$, and, for convenience, let $u = x - \tfrac{1}{2}\pi$. Then

$$\cos x = \cos[u + \tfrac{1}{2}\pi]$$

$$= \cos u \cos \tfrac{1}{2}\pi - \sin u \sin \tfrac{1}{2}\pi$$

(by a trig identity, Section 1.3)

$$= -\sin u \quad \text{(since } \cos \tfrac{1}{2}\pi = 0,\ \sin \tfrac{1}{2}\pi = 1\text{)}$$

$$= -\left(u - \frac{u^3}{3!} + \frac{u^5}{5!} - \cdots\right)$$

for all u (using the series for $\sin u$, Section 8.8).

Now replace u by $x - \tfrac{1}{2}\pi$ to obtain the final answer

$$(3) \quad \cos x = -[x - \tfrac{1}{2}\pi] + \frac{[x - \tfrac{1}{2}\pi]^3}{3!} - \frac{[x - \tfrac{1}{2}\pi]^5}{5!} + \cdots$$

for all $x - \tfrac{1}{2}\pi$, that is, for all x.

Warning Consider an incorrect approach to the preceding example. Begin with

$$(4) \qquad \cos x = 1 - \frac{x^2}{2!} + \frac{x^4}{4!} - \cdots$$

and replace x by $x - \tfrac{1}{2}\pi$ to obtain

$$\cos(x - \tfrac{1}{2}\pi) = 1 - \frac{(x - \tfrac{1}{2}\pi)^2}{2!} + \frac{(x - \tfrac{1}{2}\pi)^4}{4!} - \cdots.$$

This is a series expansion in powers of $x - \tfrac{1}{2}\pi$ for the function $\cos(x - \tfrac{1}{2}\pi)$, but it is *not* an expansion for $\cos x$, as requested.

Application Suppose we want to estimate cos 80°, that is, $\cos(80\pi/180)$, so that the error is less than .0001. We can set $x = 80\pi/180$ in any series for cos x, say the series about 0 in (4) or the series about $\frac{1}{2}\pi$ in (3). Since 80° is nearer to 90° than to 0°, the convergence will be faster if we use the series about $\frac{1}{2}\pi$. So using (3), we have

$$\cos 80° = \cos \frac{80\pi}{180} = -\left(\frac{80\pi}{180} - \frac{\pi}{2}\right) + \frac{1}{3!}\left(\frac{80\pi}{180} - \frac{\pi}{2}\right)^3 - \frac{1}{5!}\left(\frac{80\pi}{180} - \frac{\pi}{2}\right)^5$$

$$+ \frac{1}{7!}\left(\frac{80\pi}{180} - \frac{\pi}{2}\right)^7 - \cdots$$

$$= \frac{\pi}{18} - \frac{1}{3!}\left(\frac{\pi}{18}\right)^3 + \frac{1}{5!}\left(\frac{\pi}{18}\right)^5 - \frac{1}{7!}\left(\frac{\pi}{18}\right)^7 + \cdots$$

$$= .1745329 - .0008861 + .0000013 - \cdots.$$

The series alternates, and the first term less than .0001 is the third term of the series. Therefore we use two terms as the approximation and have cos 80° = .1736468 (an underestimate) with error less than .0000013.

Example 2 Expand $1/(2 - x)$ in powers of $x + 4$; that is, expand about -4.

Solution: Write the function as $\dfrac{1}{2 - ([x + 4] - 4)}$ and simplify by letting $u = x + 4$. Then

$$\frac{1}{2 - x} = \frac{1}{2 - (u - 4)} = \frac{1}{6 - u} = \frac{1}{6} \frac{1}{1 - \dfrac{u}{6}}.$$

Now use the expansion for $1/(1 - x)$ (Section 8.7, (1)) with x replaced by $u/6$ to get

$$\frac{1}{2 - x} = \frac{1}{6}\left[1 + \frac{u}{6} + \left(\frac{u}{6}\right)^2 + \left(\frac{u}{6}\right)^3 + \cdots\right] \qquad \text{for } -1 < \frac{u}{6} < 1$$

$$= \frac{1}{6}\left[1 + \frac{x + 4}{6} + \left(\frac{x + 4}{6}\right)^2 + \left(\frac{x + 4}{6}\right)^3 + \cdots\right]$$

$$\text{for } -1 < \frac{x + 4}{6} < 1$$

$$= \frac{1}{6} + \frac{1}{6^2}(x + 4) + \frac{1}{6^3}(x + 4)^2 + \frac{1}{6^4}(x + 4)^3 + \cdots$$

$$\text{for } -10 < x < 2.$$

Note that the interval of convergence is centered about -4.

Example 3 Expand ln x in powers of $x - 2$, that is, about 2.

Solution: Write ln x as $\ln([x - 2] + 2)$, and for convenience, let $u = x - 2$. Then

$$\ln x = \ln(u + 2) = \ln 2(1 + \tfrac{1}{2}u) \qquad \text{(factor)}$$

$$= \ln 2 + \ln(1 + \tfrac{1}{2}u) \qquad \text{(using ln } ab = \ln a + \ln b\text{)}.$$

Now use the established series for $\ln(1 + x)$ (Section 8.7, Eq. (10)) with x replaced by $\frac{1}{2}u$ to get

$$\ln x = \ln 2 + \tfrac{1}{2}u - \frac{(\frac{1}{2}u)^2}{2} + \frac{(\frac{1}{2}u)^3}{3} - \cdots$$

$$\text{for } -1 < \tfrac{1}{2}u < 1$$

$$= \ln 2 + \frac{x-2}{2} - \frac{1}{2}\left(\frac{x-2}{2}\right)^2 + \frac{1}{3}\left(\frac{x-2}{2}\right)^3 - \cdots$$

$$\text{for } -1 < \frac{x-2}{2} < 1$$

$$= \ln 2 + \frac{1}{2}(x-2) - \frac{1}{2^2 \cdot 2}(x-2)^2 + \frac{1}{2^3 \cdot 3}(x-2)^3 - \cdots$$

$$\text{for } 0 < x < 4.$$

The term $\ln 2$ is the constant term in the series. Note that the interval of convergence is centered about 2.

Warning Don't combine numbers if by doing so you conceal the pattern. In Example 3, the coefficients should be left as $\frac{1}{2^2 \cdot 2}, \frac{1}{2^3 \cdot 3}, \frac{1}{2^4 \cdot 4}, \cdots$ to indicate the pattern, rather than written as $\frac{1}{8}, \frac{1}{24}, \frac{1}{64}, \cdots$ which obscures the pattern.

Problems for Section 8.10

1. Find the interval of convergence of $\sum \frac{(x-4)^n}{n\,3^n}$.

2. Consider expanding each function in powers of $x - b$. For which value(s) of b is it impossible?

(a) $\dfrac{1}{(x+8)^5}$ (b) $\ln x$

3. Find the series expansion and its interval of convergence. For parts (a) and (g), try both methods. Otherwise, use known series.

(a) $\ln x$ in powers of $x - 1$

(b) $\sin x$ in powers of $x - \pi$

(c) e^x in powers of $x - 1$

(d) $\dfrac{1}{-6-x}$ in powers of $x + 1$

(e) $\dfrac{1}{x}$ in powers of $x + 2$

(f) $\sqrt{x}$ in powers of $x - 9$, and find the coefficient of $(x-9)^{50}$

(g) $\dfrac{1}{(x+8)^5}$ in powers of $x - 1$, and find the coefficient of $(x-1)^{19}$

(h) $\cos 2x$ in powers of $x + \frac{1}{2}\pi$

(i) $\ln 3x$ in powers of $x - 2$

(j) $\dfrac{1}{1+2x}$ in powers of $x + 4$

REVIEW PROBLEMS FOR CHAPTER 8

1. Test the series for convergence versus divergence.

(a) $\sum \dfrac{1}{7^n}$

(h) $\sum \dfrac{6^n}{(n-1)!}$

(b) $\sum \dfrac{1}{n^7}$

(i) $\dfrac{3}{4} - \dfrac{4}{5} + \dfrac{5}{6} - \dfrac{6}{7} + \cdots$

(c) $\sum \dfrac{1}{(\frac{1}{2})^n}$

(j) $\dfrac{1}{7} + \dfrac{1}{78} + \dfrac{1}{789} + \dfrac{1}{7890} + \dfrac{1}{78901}$

(d) $\sum \dfrac{1}{n^{1/2}}$

$\quad + \dfrac{1}{789012} + \cdots$

(e) $-\dfrac{1}{7} - \dfrac{1}{8} - \dfrac{1}{9} - \cdots$

(k) $\dfrac{1 \cdot 3 \cdot 5}{2 \cdot 4 \cdot 8} + \dfrac{1 \cdot 3 \cdot 5 \cdot 7}{2 \cdot 4 \cdot 8 \cdot 16} + \cdots$

(f) $\sum (-1)^n \dfrac{3n}{n^2 + n}$

(l) $\sum \dfrac{n^2}{5^n}$

(g) $\sum \dfrac{3n}{n^2 + n}$

(m) $1 - 3 + 1 - 3 + 1 - 3 + \cdots$

2. Find the sum of the series $(\frac{1}{9})^5 + (\frac{1}{9})^7 + (\frac{1}{9})^9 + \cdots$.

3. Estimate the sum of $\sum_{n=0}^{\infty} (-1)^n (2^n/n!)$ so that the error is less than .01. Do you have an overestimate or an underestimate?

4. Decide if the series converges absolutely, converges conditionally, or diverges.

(a) $\sum (-1)^n \dfrac{1}{(\ln n)^2}$ (b) $\sum \dfrac{1}{3^n}$

5. Suppose $\sum a_n$ is a positive convergent series. Decide, if possible, if the given series converges or diverges.

(a) $\sum (-1)^{n+1} a_n$ (b) $\sum n^2 a_n$

6. Suppose $e^{a_1} + e^{a_2} + e^{a_3} + \cdots$ converges. Decide, if possible, whether $a_1 + a_2 + a_3 + \cdots$ also converges.

7. (a) Show that if $\sum a_n$ and $\sum b_n$ converge, then $\sum a_n b_n$ does not necessarily converge.
(b) Show that if $\sum a_n$ and $\sum b_n$ are *positive* convergent series, then $\sum a_n b_n$ also converges.

8. Find the interval of convergence of $x^5/4^4 + x^6/4^5 + x^7/4^6 + \cdots$.

9. Expand the function in powers of x, and find the interval of convergence.

(a) $\dfrac{1}{3 - x}$

(c) $\dfrac{1}{(1 + x)^6}$

(b) $\dfrac{1}{(x - 1)(1 - 2x)}$

(d) $\dfrac{1}{1 + x^6}$

10. Find the first three terms of the power series for $x^2 e^x$, first using the Maclaurin series formula and then again using an established series.

11. Use power series to find $\lim_{x \to 0} (1 - \cos x)/x^2$, which is of the indeterminate form $0/0$.

12. Find an expansion and its interval of convergence for (a) $\cos x$ in powers of $x - \frac{1}{4}\pi$ (b) $\sqrt[3]{x}$ in powers of $x - 8$.

13. Approximate $\int_0^1 x^3 e^{-x^3}$ so that the error is less than .001. Is your estimate over or under?

14. Find a series in powers of x for $\sin^{-1} x$ by antidifferentiating $1/\sqrt{1 - x^2}$.

9/VECTORS

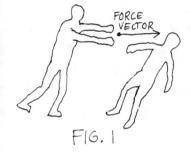

FIG. 1

9.1   Introduction

Certain quantities in physical applications of mathematics are represented by arrows; we refer to the arrows as *vectors*. For example, a force is represented by a vector (Fig. 1); the direction of the vector describes the direction in which the force is applied, and the length (magnitude) of the vector indicates its strength (in units such as pounds). The velocity of a car is represented by a vector which points in the direction of motion, and whose length indicates the speed of the car (Fig. 2). If an object moves from point A to point B (Fig. 3), its displacement is depicted by a vector drawn from A to B. In the context of vector mathematics, numbers are usually referred to as *scalars*. We say that velocity, force, displacement and so on, which are represented by arrows, are *vector quantities*, while speed, weight, time, temperature, distance and so on, which are described by numbers, are *scalar quantities*. We will use letters with overhead arrows, such as $\vec{u}$ and $\vec{v}$, to denote vectors. For a vector whose tail is point A and head is point B, as in Fig. 3, we often use the notation $\overrightarrow{AB}$.

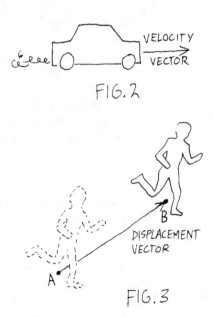

FIG. 2

FIG. 3

Rectangular coordinate systems in 3-space   We will draw vectors in space, as well as in a plane, so we begin by establishing a 3-dimensional coordinate system for reference. You are familiar with the use of a rectangular coordinate system to assign coordinates to a point in a plane. A similar coordinate

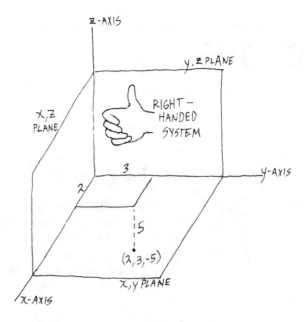

FIG. 4

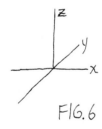

FIG. 5

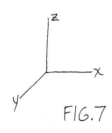

FIG. 6

FIG. 7

system may be used in space; see Fig. 4 where the point $(2, 3, -5)$ is plotted as an illustration. The plane determined by the x-axis and y-axis is called the x, y plane; Fig. 4 also shows the y, z plane and the x, z plane.

For a 2-dimensional coordinate system it is traditional to draw a horizontal x-axis and a vertical y-axis, but several different sets of axes are commonly used in 3-space. Figures 5–7 show three more coordinate systems. Each coordinate system in 3-space is called either *right-handed* or *left-handed* according to the following criterion. Hold your right hand so that your fingers curl from the positive x-axis toward the positive y-axis. If your thumb points in the direction of the positive z-axis then the system is right-handed (Figs. 4–6). Otherwise, the system is left-handed (Fig. 7). For certain purposes (Section 9.4) right-handed systems are necessary, so we use right-handed systems throughout the book.

In 2-space, the distance between the points (x_1, y_1) and (x_2, y_2) is

(*)
$$\sqrt{(x_2 - x_1)^2 + (y_2 - y_1)^2}.$$

It may similarly be shown that the distance in 3-space between the points (x_1, y_1, z_1) and (x_2, y_2, z_2) is

$$\sqrt{(x_2 - x_1)^2 + (y_2 - y_1)^2 + (z_2 - z_1)^2}.$$

For example, if $D = (3, 4, 7)$ and $E = (-5, -2, 5)$ then $\overline{DE} = \sqrt{64 + 36 + 4} = \sqrt{104}$.

Components of a vector A vector in 2-space has two *components*, indicating the changes in x and y from tail to head. The vector $\vec{u}$ in Fig. 8a has x-component -2 and y-component 3, and we write $\vec{u} = (-2, 3)$. In 2-space, the coordinates of a point and the components of a vector both measure

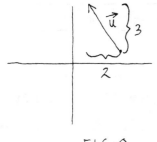

FIG. 8a

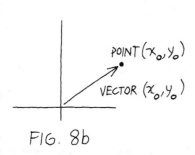

FIG. 8b

"over" and "up." However, the coordinates of a point measure over and up from the origin to the point, while the components of a vector measure over and up from the tail to the head. Note that if the vector (x_0, y_0) is drawn with its tail at the origin then the coordinates of the head are the same as the components of the vector (Fig. 8(b)).

A vector in 3-space has three components, indicating the changes in x, y and z from tail to head. For vector $\overrightarrow{AD}$ in Fig. 9, to move from tail A to head D we must go 4 in the negative x direction, 5 in the positive y direction and 3 in the positive z direction. Thus $\overrightarrow{AD} = (-4, 5, 3)$.

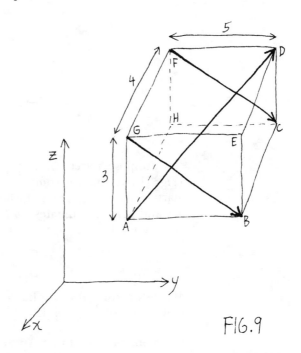

FIG. 9

Any vectors $\vec{u}$ and $\vec{v}$ with the same length and direction will have the same components, and in that case we write $\vec{u} = \vec{v}$. In Fig. 9, $\overrightarrow{GB} = \overrightarrow{FC} = (0, 5, -3)$, $\overrightarrow{AH} = \overrightarrow{BC} = \overrightarrow{GF} = \overrightarrow{ED} = (-4, 0, 0)$.

The vectors $(0, 0)$ and $(0, 0, 0)$ are thought of as arrows with zero length and arbitrary direction, and called zero vectors. Both are denoted by $\vec{0}$.

Suppose the tail of a vector is the point $(6, -1)$ and its head is the point $(2, 4)$ (Fig. 10). Examine the changes in x and y from tail to head to see that the vector has components $(-4, 5)$. In general

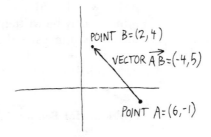

FIG. 10

(1)

$$\boxed{\text{vector components} = \text{head coordinates} - \text{tail coordinates},}$$

which we abbreviate by writing

(2)
$$\text{vector } \overrightarrow{AB} = \text{point } B - \text{point } A.$$

For example, the vector with tail at $(3, 5, 1)$ and head at $(2, 1, 5)$ has components $(2 - 3, 1 - 5, 5 - 1)$, or $(-1, -4, 4)$.

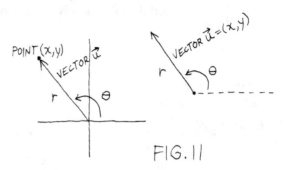

FIG. 11

Suppose a vector $\vec{u}$ in 2-space has length r and angle of inclination θ (Fig. 11). To find the components (x, y) of $\vec{u}$, note that if the vector is drawn starting at the origin then the head of the arrow has rectangular coordinates x, y and polar coordinates r, θ (Appendix A6). Since the two sets of coordinates are related by $x = r \cos \theta$, $y = r \sin \theta$, we have

(3)
$$\boxed{\vec{u} = (r \cos \theta, r \sin \theta).}$$

If a 2-dimensional vector has length 6 and angle of inclination 127°, then its components are $(6 \cos 127°, 6 \sin 127°)$.

n-dimensional vectors   An arrow in space with a triple of components (u_1, u_2, u_3) is called a 3-dimensional vector. More generally, *a 3-dimensional vector is any phenomenon described with an ordered triple of numbers,* such as position in space, or a weather report which lists, in order, temperature, humidity and windspeed. Similarly, an ordered string of seven numbers, such as $(4, 8, 6, 2, 0, -1, 6)$ is said to be a 7-dimensional vector (or point). For example, $(0, 0, 0, 0, 0, 0, 0)$ is the 7-dimensional zero vector, or, alternatively, the origin in 7-space. If an experiment involves reading five strategically placed thermometers each day then a result can be recorded as a 5-dimensional vector $(T_1, T_2, T_3, T_4, T_5)$. If a system of equations with four unknowns has the solution $x_1 = 2$, $x_2 = -4$, $x_3 = 0$, $x_4 = 2$ then the solution may be written as the 4-dimensional vector $(2, -4, 0, 2)$. If $n > 3$ then the n-dimensional vector $(u_1, \cdots, u_n)$ cannot be pictured *geometrically* as an arrow or a point, but (with the exception of the cross product in Section 9.4) vector *algebra* will be the same whether the vector has 2, 3 or 100 components.

Problems for Section 9.1

1. Let $P = (2, 3, -7)$. Find the following distances.

(a) P to point $Q = (1, 5, 2)$ (c) P to the x, y plane
(b) P to the origin   (d) P to the y, z plane

(e) P to the z-axis   (g) point (x, y, z) to the x-axis

(f) P to the y-axis   (h) point (x, y, z) to the z-axis

2. In Fig. 9, find the components of $\overrightarrow{AF}, \overrightarrow{HB}, \overrightarrow{HE}$.

3. Find the components of $\vec{u}$ if $\vec{u}$ points like the positive y-axis and has length 2 in 2-space.

4. Find several vectors parallel to the line $2x + 3y + 4 = 0$.

5. Find the components of $\overrightarrow{AB}$ if $A = (2, 7)$ and $B = (-1, 4)$.

6. If the vector $(3, 1, 6)$ has tail $(1, 0, 4)$, find the coordinates of its head.

7. Find the components of the 2-dimensional vector $\vec{u}$ with length 3 and angle of inclination 120°.

9.2   Vector Addition, Subtraction, Scalar Multiplication and Norms

In this section we will develop some vector algebra along with the corresponding vector geometry.

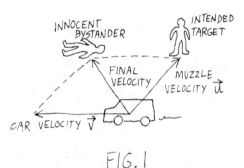

FIG. 1

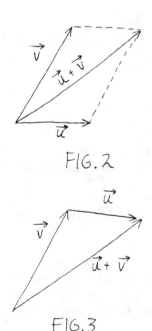

FIG. 2

FIG. 3

Vector addition Let the vector $\vec{u}$ in Fig. 1 be the muzzle velocity of a bullet fired toward a target. Suppose further that the gun is fired from a car moving with velocity $\vec{v}$. Experiments show that the bullet does not head toward the intended target; instead, the car velocity and muzzle velocity combine (physicists call it "addition") to produce the final bullet velocity shown in Fig. 1. In general, the sum of two vectors is defined by the parallelogram law of Fig. 2, or equivalently, the triangle law in Fig. 3 (the triangle is half the parallelogram). Figure 4 shows addition of parallel vectors, and Fig. 5 shows a sum of three vectors.

To find the algebraic counterpart of the parallelogram law, we want the components of $\vec{u} + \vec{v}$ given the components of $\vec{u}$ and $\vec{v}$. Suppose $\vec{u} = (2, 3)$ and $\vec{v} = (5, 1)$. Figure 6 shows $\vec{u}, \vec{v}$ and $\vec{u} + \vec{v}$; we can read the changes in x and y from tail to head of $\vec{u} + \vec{v}$ to see that $\vec{u} + \vec{v} = (7, 4)$. Each component

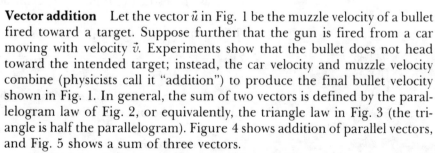

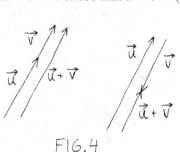

FIG. 4

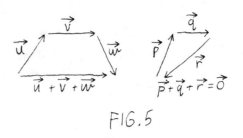

FIG. 5

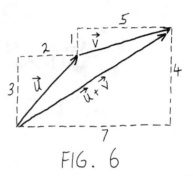

FIG. 6

of $\vec{u} + \vec{v}$ is the sum of the corresponding components of $\vec{u}$ and v. In general, if $\vec{u} = (u_1, \cdots, u_n)$ and $\vec{v} = (v_1, \cdots, v_n)$ then

(1) $$\vec{u} + \vec{v} = (u_1 + v_1, \cdots, u_n + v_n).$$

If $\vec{u}$ and $\vec{v}$ are vectors in 2-space or 3-space, then (1) accompanies the geometric parallelogram rule. If $\vec{u}$ and $\vec{v}$ are higher dimensional, then (1) serves as an abstract definition of vector addition.

The vector $-\vec{u}$ If $\vec{u} = (u_1, \cdots, u_n)$ we define

(2) $$-\vec{u} = (-u_1, \cdots, -u_n).$$

For example, if $\vec{u} = (4, 2, -1, 3)$ then $-\vec{u} = (-4, -2, 1, -3)$.

If $\vec{u}$ is a vector in 2-space or 3-space then $-\vec{u}$ has the same length as $\vec{u}$ but points in the opposite direction (Fig. 7).

Vector subtraction   If $\vec{u} = (u_1, \cdots, u_n)$ and $\vec{v} = (v_1, \cdots, v_n)$, we define

(3) $$\vec{u} - \vec{v} = (u_1 - v_1, \cdots, u_n - v_n).$$

For example, if $\vec{u} = (2, -1)$ and $\vec{v} = (1, 7)$ then $\vec{u} - \vec{v} = (1, -8)$.

If $\vec{u}$ and $\vec{v}$ are drawn as vectors with a common tail (Fig. 8a) then the vector $\vec{u} - \vec{v}$ can be drawn by reversing $\vec{v}$ and adding, that is, by finding $\vec{u} + -\vec{v}$ (Fig. 8b). The final result, the triangle law for vector subtraction, is shown in Fig. 9: the head of $\vec{u} - \vec{v}$ is the head of $\vec{u}$, and the tail of $\vec{u} - \vec{v}$ is the head of $\vec{v}$.

Note that to *add* two vectors geometrically, they can either be placed with a common tail and added with the parallelogram law in Fig. 2, or can be drawn head to tail and added with the triangle law in Fig. 3. But to *subtract* two vectors geometrically, they should be placed with a common tail so that the triangle rule of Fig. 9 can be applied. The parallelo-

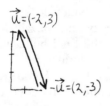

FIG. 7

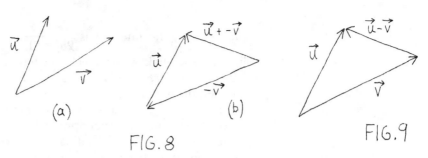

FIG. 8

FIG. 9

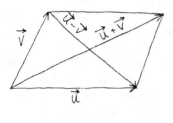

FIG. 10

gram in Fig. 10 neatly displays the vectors $\vec{u}, \vec{v}, \vec{u} + \vec{v}$ and $\vec{u} - \vec{v}$ all in the same diagram.

Properties of vector addition and subtraction   As expected, the vector operations behave like addition and subtraction of numbers.

(4) $\vec{u} + \vec{v} = \vec{v} + \vec{u}$

(5) $(\vec{u} + \vec{v}) + \vec{w} = \vec{u} + (\vec{v} + \vec{w})$

(6) $\vec{u} + \vec{0} = \vec{0} + \vec{u} = \vec{u}$

(7) $\vec{u} + -\vec{u} = \vec{0}$

Scalar multiplication   If $\vec{u} = (u_1, \cdots, u_n)$ and c is a scalar, we define

(8) $$c\vec{u} = (cu_1, \cdots, cu_n)$$

and call the operation *scalar multiplication*. For example, if $\vec{u} = (2, -3)$ then $5\vec{u} = (10, -15)$. If $\vec{u}$ is a vector in 2-space or 3-space then $2\vec{u}$ and $\vec{u}$ have the same direction, but $2\vec{u}$ is twice as long. A car with velocity $2\vec{u}$ is traveling in the same direction as a car with velocity $\vec{u}$, but with twice the speed. The vectors $\vec{u}$ and $-\frac{1}{2}\vec{u}$ have opposite directions, and $-\frac{1}{2}\vec{u}$ is half as long as $\vec{u}$. In general, *two vectors are parallel if one is a multiple of the other;* they are parallel with the *same* direction if the multiple is *positive,* and parallel with *opposite* directions if the multiple is *negative* (Fig. 11).

PARALLEL
VECTORS

FIG. 11

Parallel lines   In 3-space, two lines are either parallel, intersecting or skew. The pyramid in Fig. 12 illustrates parallel lines BE and CD, intersecting lines AB and AD, and skew lines AE and CD. We will consider coincident lines as a special case of parallel lines; the lines BF and BA are parallel, and furthermore are coincident.

Vectors may be used to detect parallel lines: *the lines PQ and RS are parallel if and only if the vectors $\vec{PQ}$ and $\vec{RS}$ are multiples of one another.* For example, let $A = (1, 2, 3)$, $B = (4, 8, -1)$, $P = (6, 1, 3)$ and $Q = (-4, 0, 2)$. Then $\vec{AB} = B - A = (3, 6, -4)$ and $\vec{PQ} = Q - P = (-10, -1, -1)$. The vectors are not multiples of one another, so the lines are not parallel. (Section 10.3 will give a method for distinguishing between the two remaining possibilities, skew versus intersecting, and show how to find the point of intersection if it exists.)

In 2-space, both slopes *and* vectors may be used to detect parallel lines. In fact we will show that the two techniques have much in common. Let's decide if the lines AB and CD are parallel, where $A = (1, 2)$, $B = (3, 5)$, $C = (21, -3)$ and $D = (25, 3)$. The slope of the line AB is $\dfrac{5 - 2}{3 - 1}$ or $\frac{3}{2}$, while

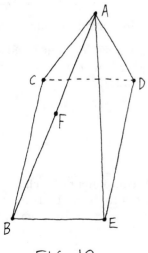

FIG. 12

the slope of the line CD is $\dfrac{3 - -3}{25 - 21}$ or $\frac{6}{4}$. Since $\frac{3}{2}$ and $\frac{6}{4}$ are equal, the lines are parallel. Alternatively, we have $\overrightarrow{AB} = (3 - 1, 5 - 2) = (2, 3)$ and $\overrightarrow{CD} = (25 - 21, 3 - -3) = (4, 6)$. Since $(2, 3)$ and $(4, 6)$ are multiples of one another, the lines are parallel. Both methods involve subtraction to find the key numbers $3, 2$ and $6, 4$. But one method uses them to form quotients, called slopes, and the other approach uses them to form ordered pairs, the components of vectors. Deciding if the two quotients are equal is equivalent to deciding if the two vectors are multiples of one another; the two methods accomplish the same purpose, but in different notation. The slope of a line AB is a convenient way of combining the *two* components of the vector $\overrightarrow{AB}$ into *one* number, without losing information about the direction of the line. Since there is no useful way of combining the *three* components of a 3-dimensional vector into *one* number, slopes are not defined in space. Questions about parallelism, perpendicularity, angles and direction will be answered in 3-space using vectors. In 2-space we may choose between vectors and slopes.

Properties of scalar multiplication

(9) $c(\vec{u} + \vec{v}) = c\vec{u} + c\vec{v}$ (For example, $2(\vec{u} + \vec{v}) = 2\vec{u} + 2\vec{v}$.)

(10) $a\vec{u} + b\vec{u} = (a + b)\vec{u}$ (For example, $2\vec{u} + 3\vec{u} = 5\vec{u}$.)

(11) $a(b\vec{u}) = (ab)\vec{u}$ (For example, $2(3\vec{u}) = 6\vec{u}$.)

Properties (9)–(11) are similar to familiar algebraic identities for scalars. We omit the straightforward proofs.

Example 1 Precalculus algebra courses show that if $A = (x_1, y_1)$ and $B = (x_2, y_2)$ then the midpoint of the segment AB is $\left(\dfrac{x_1 + x_2}{2}, \dfrac{y_1 + y_2}{2}\right)$. In vector notation, the midpoint is $\dfrac{A + B}{2}$. We can use vectors to find the two trisection points, C and D, of segment AB (Fig. 13). We have $\overrightarrow{AC} = \frac{1}{3}\overrightarrow{AB}$, so $C - A = \frac{1}{3}(B - A)$, and $C = \dfrac{2A + B}{3}$. The midpoint formula computes an average of the endpoints. The formula for the trisection point C takes a *weighted* average of the endpoints, with A weighted twice as much as B, since C is the trisection point nearer to A. Similarly, $\overrightarrow{AD} = \frac{2}{3}\overrightarrow{AB}$ and $D = \dfrac{A + 2B}{3}$. If $A = (2, 3)$ and $B = (-1, 6)$ then the trisection point nearest A is $\dfrac{2A + B}{3} = (1, 4)$.

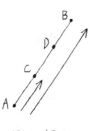

FIG. 13

The norm of a vector If $\vec{u} = (u_1, u_2)$ then the x component of $\vec{u}$ changes by u_1 and the y component changes by u_2 from tail to head. Thus by the Pythagorean theorem, the length of the vector is $\sqrt{u_1^2 + u_2^2}$. In general, if $\vec{u} = (u_1, \cdots, u_n)$ we define the *norm* or *magnitude* of $\vec{u}$ by

(12) $$\boxed{\|\vec{u}\| = \sqrt{u_1^2 + \cdots + u_n^2}.}$$

If the vector $\vec{u}$ is 2-dimensional or 3-dimensional then $\|\vec{u}\|$ is the length of $\vec{u}$. If a point has coordinates $(u_1, \cdots, u_n)$ then the square root in (12) is the distance from the point to the origin.

For example, if $\vec{u} = (2, 3, -5)$ then $\|\vec{u}\| = \sqrt{4 + 9 + 25} = \sqrt{38}$. The length of the vector $\vec{u}$ is $\sqrt{38}$ and the distance from the point $(2, 3, -5)$ to the origin is $\sqrt{38}$.

Properties of the norm   It follows from the interpretation of $\|\vec{u}\|$ as the length of a vector that

(13) $$\|\vec{u}\| \geq 0$$

and

(14) $$\|\vec{u}\| = 0 \text{ if and only if } \vec{u} = \vec{0}.$$

We have already observed that the vectors $3\vec{u}$ and $-3\vec{u}$ are each 3 times as long as $\vec{u}$. In the language of norms, $\|3\vec{u}\| = \|-3\vec{u}\| = 3\|\vec{u}\|$, and in general,

(15) $$\|c\vec{u}\| = |c|\,\|\vec{u}\|.$$

Geometrically, (15) says that the length of the vector $c\vec{u}$ is the absolute value of c times the length of $\vec{u}$. Algebraically, (15) claims that the scalar c can be extracted from inside the norm signs in the expression $\|c\vec{u}\|$, provided that its absolute value is taken.

Example 2   Suppose a force $\vec{f}$ acts at point A due to a nearby disturbance. Let $\vec{r}$ be the vector from the disturbance to A (Fig. 14). Describe the direction and magnitude of $\vec{f}$ if $\vec{f} = \dfrac{\vec{r}}{\|\vec{r}\|^3}$.

FIG. 14

Solution: The denominator $\|\vec{r}\|^3$ is a positive scalar, so $\vec{f}$ has the same direction as $\vec{r}$. Thus the disturbance creates a *repelling* force at A. To find the magnitude of $\vec{f}$, use (15): since $\dfrac{1}{\|\vec{r}\|^3}$ is a positive scalar, the length of $\dfrac{\vec{r}}{\|\vec{r}\|^3}$ is $\dfrac{1}{\|\vec{r}\|^3}$ times the length of $\vec{r}$. Therefore,

$$\|\vec{f}\| = \frac{1}{\|\vec{r}\|^3}\|\vec{r}\| = \frac{1}{\|\vec{r}\|^2} = \frac{1}{(\text{distance from } A \text{ to the disturbance})^2}.$$

Thus the magnitude of the repelling force is inversely proportional to the square of the distance to the disturbance. (The electrical force felt by a positive charge at point A due to a nearby positive charge is an example of a repelling, inverse square force.)

Normalized vectors   By (15), the norm of $\dfrac{\vec{u}}{3}$ is $\frac{1}{3}$ times the norm of $\vec{u}$. Similarly, the norm of $\dfrac{\vec{u}}{\|\vec{u}\|}$ is $\dfrac{1}{\|\vec{u}\|}$ times the norm of $\vec{u}$. Thus $\dfrac{\vec{u}}{\|\vec{u}\|}$ is a *unit vector,* that is, has norm 1. Furthermore it has the *same direction as $\vec{u}$* since the scalar multiple $\dfrac{1}{\|\vec{u}\|}$ is positive. The process of dividing $\vec{u}$ by $\|\vec{u}\|$ is called *normalizing* the vector $\vec{u}$. We will use the notation $\vec{u}_{\text{normalized}}$ so that

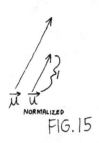

FIG. 15

(16)
$$\boxed{\vec{u}_{\text{normalized}} = \frac{\vec{u}}{\|\vec{u}\|} = \left(\frac{u_1}{\|\vec{u}\|}, \cdots, \frac{u_n}{\|\vec{u}\|}\right).}$$

For example, if $\vec{u} = (4, 5)$ then $\|\vec{u}\| = \sqrt{41}$ and $\vec{u}_{\text{normalized}} = \left(\dfrac{4}{\sqrt{41}}, \dfrac{5}{\sqrt{41}}\right)$ (Fig. 15).

The normalized $\vec{u}$ will be a useful geometric tool because of its unit length.

Warning A *norm* is a *scalar*, but as the name implies, a *normalized vector* is a *vector*. In other words, $\|\vec{u}\|$ is a scalar but $\vec{u}_{\text{normalized}}$ is a vector.

Finding a vector with a given direction and norm Suppose $\vec{u}$ has length 3 and the same direction as a given vector $\vec{v}$. Then $\vec{u} = 3\vec{v}_{\text{normalized}}$ since tripling the *unit* vector $\vec{v}_{\text{normalized}}$ produces a vector with length 3, still pointing like $\vec{v}$. In general, *if $\|\vec{u}\| = l$ and $\vec{u}$ has the same direction as a given vector $\vec{v}$, then*

(17)
$$\boxed{\vec{u} = l\vec{v}_{\text{normalized}} = l\frac{\vec{v}}{\|\vec{v}\|}.}$$

For example, if $\vec{u}$ has length 4 and the same direction as $\vec{w} = (1, 3, 2)$ then

$$\vec{u} = 4\vec{w}_{\text{normalized}} = 4\left(\frac{1}{\sqrt{14}}, \frac{3}{\sqrt{14}}, \frac{2}{\sqrt{14}}\right) = \left(\frac{4}{\sqrt{14}}, \frac{12}{\sqrt{14}}, \frac{8}{\sqrt{14}}\right).$$

Example 3 If you start at point $A = (1, 6)$ and walk 2 units toward point $B = (4, 10)$, at what point do you stop?

Solution: Let the final destination be named C (Fig. 16). Then $\overrightarrow{AC}$ has length 2 and the same direction as $\overrightarrow{AB} = (3, 4)$, so $\overrightarrow{AC} = 2\overrightarrow{AB}_{\text{normalized}} = (\frac{6}{5}, \frac{8}{5})$. Therefore $C - A = (\frac{6}{5}, \frac{8}{5})$ and $C = A + (\frac{6}{5}, \frac{8}{5}) = (\frac{11}{5}, \frac{38}{5})$.

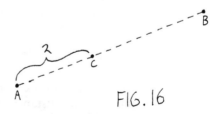

FIG. 16

The vectors $\vec{i}, \vec{j}, \vec{k}$ In 2-space, the special vectors $\vec{i}$ and $\vec{j}$ are defined by $\vec{i} = (1, 0)$ and $\vec{j} = (0, 1)$. Both are unit vectors, and if attached to the origin they point along the coordinate axes (Fig. 17). Every 2-dimensional vector can be easily written in terms of $\vec{i}$ and $\vec{j}$. For example, $(2, 3) = 2(1, 0) + 3(0, 1) = 2\vec{i} + 3\vec{j}$ (Fig. 17). The notation $\vec{u} = u_1\vec{i} + u_2\vec{j}$ is often used in place of $\vec{u} = (u_1, u_2)$. From now on, we will use both representations.

Similarly, in 3-space, $\vec{i} = (1, 0, 0)$, $\vec{j} = (0, 1, 0)$ and $\vec{k} = (0, 0, 1)$ (Fig. 18). The vector (u_1, u_2, u_3) can be written as $u_1\vec{i} + u_2\vec{j} + u_3\vec{k}$. For example, if $\vec{u} = 2\vec{i} - 7\vec{j} + 3\vec{k}$ and $\vec{v} = \vec{i} + \vec{j} + 2\vec{k}$ then $\vec{u} + \vec{v} = 3\vec{i} - 6\vec{j} + 5\vec{k}$, $3\vec{u} = 6\vec{i} - 21\vec{j} + 9\vec{k}$, $\|\vec{u}\| = \sqrt{4 + 49 + 9} = \sqrt{62}$.

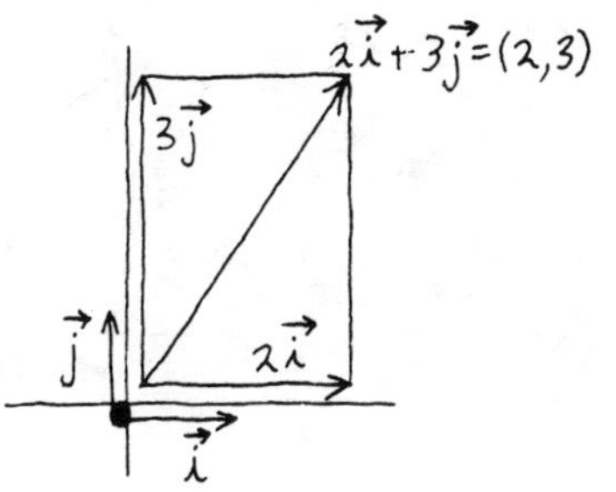

$$2\vec{i} + 3\vec{j} = (2, 3)$$

FIG. 17

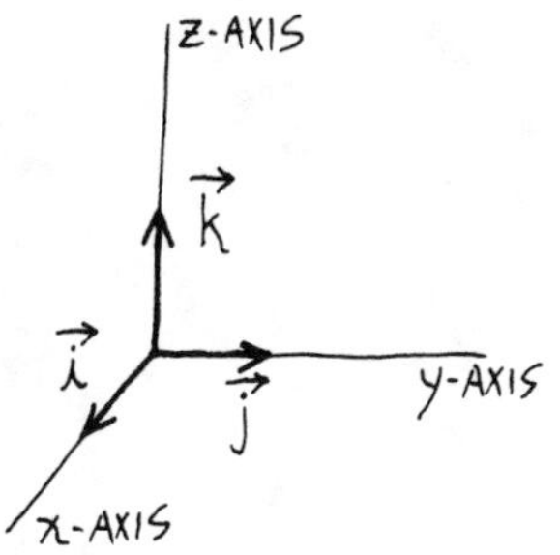

FIG. 18

Warning If $\vec{u}$ has components 2 and 8, you may write $\vec{u} = (2, 8)$ or $\vec{u} = 2\vec{i} + 8\vec{j}$, but $\vec{u}$ is *not* $(2\vec{i}, 8\vec{j})$.

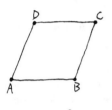

FIG. 19

Problems for Section 9.2

1. Use the parallelogram in Fig. 19 to find

(a) $\overrightarrow{DC} + \overrightarrow{DA}$ (d) $\overrightarrow{AB} - \overrightarrow{CB}$
(b) $\overrightarrow{AB} - \overrightarrow{AD}$ (e) $\overrightarrow{AB} + \overrightarrow{CD}$
(c) $\overrightarrow{AB} + \overrightarrow{CB}$

2. Let $A = (2, 4, 6), B = (1, 2, 3), C = (5, 5, 5)$. Find point D so that $ABCD$ is a parallelogram.

3. Let $A = (1, 4, 5), B = (2, 8, 1), C = (8, 8, 8), D = (6, 0, 16)$. Are the lines AB and CD parallel?

4. Let $A = (1, 2, 3), B = (4, 8, -1), P = (6, y, z), Q = (-4, 0, 2)$. Find y and z so that the lines PQ and AB are parallel.

5. Are the points $A = (3, 6, -1), B = (2, 0, 3), C = (-1, 3, -4)$ collinear?

6. Of the nine points that divide the segment PQ into ten equal parts, find the three nearest to P.

7. Figure 20 shows vectors $\vec{u}, \vec{v}, \vec{w}$ lying in the plane of the page. Find scalars a and b so that $\vec{w} = a\vec{u} + b\vec{v}$.

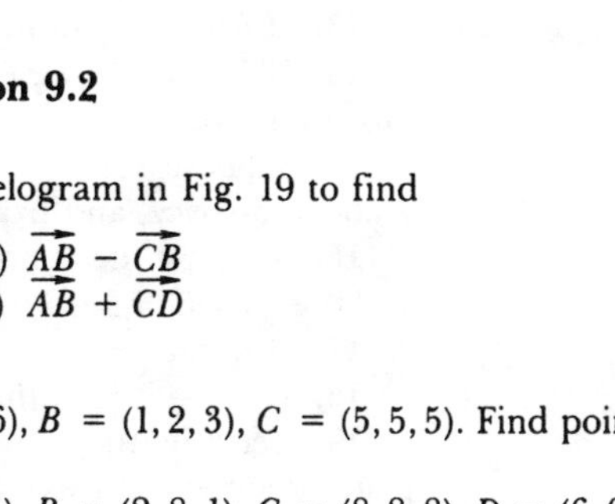

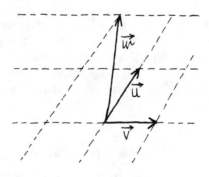

FIG. 20

8. A median vector of a triangle is a vector from a vertex to the midpoint of the opposite side. Show that the sum of the three median vectors is $\vec{0}$ (Fig. 21).

Suggestions: For one method note that $E = \dfrac{B + C}{2}$ since E is the midpoint of segment BC. For another method note that $\overrightarrow{AE} = \overrightarrow{AB} + \overrightarrow{BE}$.

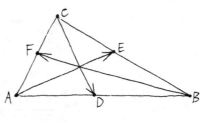

FIG. 21

9. Find $\|\vec{u}\|$ if (a) $\vec{u} = (3, -1, 5)$ (b) $\vec{u} = (\pi, \pi, \pi, \pi, \pi)$.

10. Find the unit vector in the direction of $(2, -6, 8)$.

11. If $\vec{v}$ and $\vec{u}$ have opposite directions and $\|\vec{v}\| = 5$, express $\vec{v}$ as a multiple of $\vec{u}$.

12. Suppose that you walk on a line for 12 meters from point $B = (1, 2, 6)$ to point C, passing through the point $A = (1, 1, 2)$ along the way. Find the coordinates of C.

13. If $\vec{u} = (2, 3, 5)$, find the norm of $217\vec{u}$.

14. If $\vec{u}$ makes angle θ with the positive x-axis in 2-space, find a unit vector in the direction of $\vec{u}$.

15. Suppose u has tail at point $(4, 5, 6)$, is directed perpendicularly toward the y-axis in 3-space, and has norm 3. Find its components.

16. Suppose the tail of $\vec{u}$ is at the point $A = (5, 6, 7)$, $\vec{u}$ points toward the origin, and the length of $\vec{u}$ is $1/(\text{distance from } A \text{ to the origin})^2$. Find the components of $\vec{u}$.

17. If $\vec{u} = 2\vec{i} + 3\vec{j} - \vec{k}$ and $\vec{v} = \vec{i} - \vec{j} + \vec{k}$, find $\vec{u} - 2\vec{v}$, $\|\vec{u}\|$ and $\vec{u}_{\text{normalized}}$.

18. If $\|\vec{r}\| = r$, find the norm of $r^3\vec{r}$.

19. Let θ be the angle determined by $\vec{u}$ and $\vec{v}$ drawn with a common tail. Use plane geometry to explain why $\vec{u} + \vec{v}$ does not necessarily bisect angle θ, but $\dfrac{\vec{u}}{\|\vec{u}\|} + \dfrac{\vec{v}}{\|\vec{v}\|}$ does bisect the angle.

9.3 The Dot Product

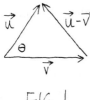

FIG. 1

We'll begin by finding a formula for the angle between two vectors. This leads to a new vector product and further applications.

If two vectors $\vec{u}$ and $\vec{v}$ are drawn with the same tail, they determine an angle θ (Fig. 1). If the vectors are parallel with the same direction, the angle is 0°; if the vectors are parallel with opposite directions, the angle is 180°. Otherwise, the angle is taken to be between 0° and 180°. We want to find the angle θ in terms of the components of $\vec{u}$ and $\vec{v}$. In Fig. 1, the vector $\vec{u} - \vec{v}$ completes a triangle with sides $\|\vec{u}\|$, $\|\vec{v}\|$ and $\|\vec{u} - \vec{v}\|$. By the law of cosines (Section 1.3),

$$(1) \qquad \|\vec{u} - \vec{v}\|^2 = \|\vec{u}\|^2 + \|\vec{v}\|^2 - 2\|\vec{u}\|\,\|\vec{v}\|\cos\theta.$$

If $\vec{u} = (u_1, u_2, u_3)$ and $\vec{v} = (v_1, v_2, v_3)$ then (1) becomes

$$(u_1 - v_1)^2 + (u_2 - v_2)^2 + (u_3 - v_3)^2 = u_1^2 + u_2^2 + u_3^2 + v_1^2 + v_2^2 + v_3^2 - 2\|\vec{u}\|\,\|\vec{v}\|\cos\theta.$$

This simplifies to $u_1v_1 + u_2v_2 + u_3v_3 = \|\vec{u}\|\,\|\vec{v}\|\cos\theta$, so

$$\cos\theta = \frac{u_1v_1 + u_2v_2 + u_3v_3}{\|\vec{u}\|\,\|\vec{v}\|}.$$

We single out the numerator of the cosine formula for special attention.

The dot product   If $\vec{u} = (u_1, \cdots, u_n)$ and $\vec{v} = (v_1, \cdots, v_n)$ then the *dot product* or *inner product* of $\vec{u}$ and $\vec{v}$ is defined by

(2)
$$\vec{u} \cdot \vec{v} = u_1 v_1 + \cdots + u_n v_n .$$

For example, if $\vec{u} = 2\vec{i} + 3\vec{j} - 4\vec{k}$ and $\vec{v} = 5\vec{i} - 3\vec{j} + 2\vec{k}$ then $\vec{u} \cdot \vec{v} =$ $(2)(5) + (3)(-3) + (-4)(2) = 10 - 9 - 8 = -7$.

With this definition, if θ is the angle determined by the nonzero vectors $\vec{u}$ and $\vec{v}$ drawn with the same tail, then

(3)
$$\cos \theta = \frac{\vec{u} \cdot \vec{v}}{\|\vec{u}\| \|\vec{v}\|}$$

or, equivalently,

(4)
$$\vec{u} \cdot \vec{v} = \|\vec{u}\| \|\vec{v}\| \cos \theta .$$

If $\vec{u} = 2\vec{i} + 5\vec{k}$ and $\vec{v} = -2\vec{i} + 2\vec{j} - 7\vec{k}$ then $\cos \theta = \dfrac{-39}{\sqrt{29}\sqrt{57}}$, which is approximately $-.959$. Since the angle is always taken to be between $0°$ and $180°$, an approximation for θ is $\cos^{-1}(-.959)$, or about $164°$.

The sign of $\cos \theta$ determines whether θ is acute or obtuse. This sign in turn is determined by the sign of $\vec{u} \cdot \vec{v}$ since the denominator in (3) is always positive. In particular,

(5)
$$\text{if } \vec{u} \cdot \vec{v} \text{ is positive} \quad \text{then } 0° \le \theta < 90°$$
$$\text{if } \vec{u} \cdot \vec{v} = 0 \quad \text{then } \theta = 90°$$
$$\text{if } \vec{u} \cdot \vec{v} \text{ is negative} \quad \text{then } 90° < \theta \le 180° .$$

As a corollary of (5), for nonzero vectors $\vec{u}$ and $\vec{v}$,

$$\vec{u} \cdot \vec{v} = 0 \textit{ if and only if } \vec{u} \textit{ and } \vec{v} \textit{ are perpendicular} .$$

More generally, $\vec{u} \cdot \vec{v} = 0$ if and only if $\vec{u} = \vec{0}$ or $\vec{v} = \vec{0}$ or $\vec{u}$ and $\vec{v}$ are nonzero perpendicular vectors.

Example 1   Let $A = (1, 2, 3), B = (3, 5, -1), C = (5, -1, 0), D = (11, -1, 3)$. Are the lines AB and CD perpendicular?

Solution: We have $\overrightarrow{AB} = (2, 3, -4)$ and $\overrightarrow{CD} = (6, 0, 3)$. Then $\overrightarrow{AB} \cdot \overrightarrow{CD} =$ $12 + 0 - 12 = 0$, so the vectors are perpendicular. Therefore the lines are considered perpendicular although we cannot tell from the dot product alone whether they are perpendicular and *intersecting* (such as a telephone pole and the taut telephone wire) or perpendicular and *skew* (such as a telephone pole and a railroad track).

Warning   Note that for $\overrightarrow{AB} \cdot \overrightarrow{CD}$ is *not* $(12, 0, -12)$; it is 12 *plus* 0 *plus* -12. The dot product is a *scalar*.

Free vectors versus fixed points and lines   Suppose $A = (1, 2)$ and $B = (5, 0)$. Then the points A and B are fixed in the plane, line AB is fixed in the plane, but the vector $\overrightarrow{AB} = (4, -2)$ is said to be *free* in the sense that an arrow with components 4 and -2 can be drawn starting at any point in the plane.

Similarly, two vectors $\vec{u}$ and $\vec{v}$ can be drawn with a common tail to display the angle they determine (Fig. 1), but the *same* vectors can also be drawn apart. It makes sense to ask if two vectors are parallel or nonparallel, perpendicular or nonperpendicular, but it makes no sense to refer to vectors as skew or as intersecting.

Properties of the dot product Several dot product rules are similar to familiar algebraic identities for the multiplication of numbers:

(6) $\quad \vec{u} \cdot \vec{v} = \vec{v} \cdot \vec{u}$

(7) $\quad \vec{u} \cdot (\vec{v} + \vec{w}) = \vec{u} \cdot \vec{v} + \vec{u} \cdot \vec{w},$

$\quad (\vec{u} + \vec{v}) \cdot (\vec{p} + \vec{q}) = \vec{u} \cdot \vec{p} + \vec{v} \cdot \vec{p} + \vec{u} \cdot \vec{q} + \vec{v} \cdot \vec{q}$

(8) $\quad \vec{u} \cdot \vec{0} = \vec{0} \cdot \vec{u} = 0 .$

We omit the proofs, which are straightforward.

If $\vec{u} = (u_1, \cdots, u_n)$ then $\vec{u} \cdot \vec{u} = u_1^2 + \cdots + u_n^2$. But this sum of squares is also $\|\vec{u}\|^2$, so

(9)
$$\boxed{\vec{u} \cdot \vec{u} = \|\vec{u}\|^2 .}$$

Still another property is

(10)
$$(c\vec{u}) \cdot \vec{v} = \vec{u} \cdot (c\vec{v}) = c(\vec{u} \cdot \vec{v})$$

which states that a scalar multiplying one factor in a dot product may be switched to the other factor or taken to multiply the dot product itself. For the proof of (10), let $\vec{u} = (u_1, \cdots, u_n)$ and $\vec{v} = (v_1, \cdots, v_n)$. Then

$$c(\vec{u} \cdot \vec{v}) = c(u_1 v_1 + \cdots + u_n v_n) = c u_1 v_1 + \cdots + c u_n v_n$$
$$(c\vec{u}) \cdot \vec{v} = (c u_1) v_1 + \cdots + (c u_n) v_n = c u_1 v_1 + \cdots + c u_n v_n$$
$$\vec{u} \cdot (c\vec{v}) = u_1(c v_1) + \cdots + u_n(c v_n) = c u_1 v_1 + \cdots + c u_n v_n .$$

Therefore, (10) holds. Note that three kinds of multiplication appear in (10), dot multiplication, scalar multiplication (in the products $c\vec{u}$ and $c\vec{v}$) and multiplication of two numbers (in the product $c(\vec{u} \cdot \vec{v})$ since both c and $\vec{u} \cdot \vec{v}$ are scalars).

Example 2 By (9), (7) and (6),
$$\|\vec{u} + \vec{v}\|^2 = (\vec{u} + \vec{v}) \cdot (\vec{u} + \vec{v}) = \vec{u} \cdot \vec{u} + 2\vec{u} \cdot \vec{v} + \vec{v} \cdot \vec{v}$$
$$= \|\vec{u}\|^2 + 2\vec{u} \cdot \vec{v} + \|\vec{v}\|^2 .$$

Example 3 Show that $\vec{u}$ is perpendicular to $\vec{v} - \dfrac{\vec{v} \cdot \vec{u}}{\|\vec{u}\|^2} \vec{u}$.

(Note that $\dfrac{\vec{v} \cdot \vec{u}}{\|\vec{u}\|^2}$ is the quotient of two scalars, so it too is a scalar, multiplying the vector $\vec{u}$.)

Solution: For the vectors to be perpendicular, their dot product must be 0. We have

$$\vec{u} \cdot \left(\vec{v} - \frac{\vec{v} \cdot \vec{u}}{\|\vec{u}\|^2}\vec{u}\right) = \vec{u} \cdot \vec{v} - \vec{u} \cdot \left(\frac{\vec{v} \cdot \vec{u}}{\|\vec{u}\|^2}\vec{u}\right) \quad \text{(by (7))}$$

$$= \vec{u} \cdot \vec{v} - \frac{\vec{v} \cdot \vec{u}}{\|\vec{u}\|^2}(\vec{u} \cdot \vec{u}) \quad \left(\text{by (10) with } c \text{ taken to be } \frac{\vec{v} \cdot \vec{u}}{\|\vec{u}\|^2}\right)$$

$$= \vec{u} \cdot \vec{v} - \vec{v} \cdot \vec{u} \quad \begin{array}{l}\text{(cancel the scalars } \|u\|^2 \text{ and } \vec{u} \cdot \vec{u},\\ \text{by (9))}\end{array}$$

$$= 0 \quad \text{(by (6))}.$$

Warning Don't write meaningless combinations. For example, $(\vec{u} \cdot \vec{v}) + \vec{w}$ is the sum of a scalar and a vector, which is impossible. Similarly, expressions such as $\vec{u}^2$, $\vec{u}\vec{v}$ and $\vec{u}/\vec{v}$ make no sense.

The (scalar) component of $\vec{u}$ in a direction We'll begin with an example to introduce a new and important application of the dot product. Suppose a boxer is vulnerable to the knockout force $\overrightarrow{KO} = (1, 2, 3)$. If a fist has the direction of the vector $\overrightarrow{KO}$ as it lands on his chin, and has $\sqrt{14}$ units of force behind it, he will be knocked out. More units of force will also knock him out, but not less. Suppose he is hit by the blow $\vec{u} = (1, 4, 2)$. There is sufficient strength, namely $\|\vec{u}\| = \sqrt{21}$, in the blow but it isn't in the $\overrightarrow{KO}$ direction. The problem is to decide whether he is knocked out. Think of $\vec{u}$ as the sum of two vectors, $\vec{a}$, parallel to $\overrightarrow{KO}$, and $\vec{b}$, perpendicular to $\overrightarrow{KO}$ (Fig. 2). Physical experiments show that applying the force $\vec{u}$ is equivalent to simultaneously applying $\vec{a}$ and $\vec{b}$. Furthermore, the vector $\vec{b}$ is harmlessly tangent to his chin and can be ignored. In other words, the blow that has *effectively* been struck is $\vec{a}$, and the possibility of a knockout depends on whether the magnitude of $\vec{a}$ is at least $\sqrt{14}$. This is a geometry problem. We want to find the length of the projection of $\vec{u}$ onto the $\overrightarrow{KO}$ direction. (Figure 2 is drawn with $\|\vec{a}\| > \sqrt{14}$, that is, with $\vec{a}$ longer than $\overrightarrow{KO}$. The problem is to decide if this is indeed the case.) In the right triangle in Fig. 3, the length of the projection is labeled p. Then $\cos \theta = \frac{p}{\|\vec{u}\|}$ so

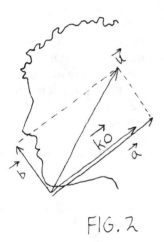

FIG. 2

$$p = \|\vec{u}\| \cos \theta = \|\vec{u}\| \frac{\vec{u} \cdot \overrightarrow{KO}}{\|\vec{u}\|\|\overrightarrow{KO}\|} \quad \text{(by (3))}$$

$$(11) \qquad = \frac{\vec{u} \cdot \overrightarrow{KO}}{\|\overrightarrow{KO}\|} \quad \text{(cancel)} = \frac{15}{\sqrt{14}} = \frac{15}{14}\sqrt{14}.$$

Since $p > \sqrt{14}$ (barely), the force $\vec{u}$ does knock him out.

Let's extract some general results from the example. By (11), if the angle θ between $\vec{u}$ and $\vec{v}$ is acute (as in Fig. 3) then the length p of the projection of $\vec{u}$ onto the $\vec{v}$ direction is given by $p = \frac{\vec{u} \cdot \vec{v}}{\|\vec{v}\|}$. In the case where θ is obtuse (Fig. 4) then, instead of (11),

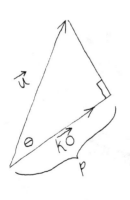

FIG. 3

$$p = \|\vec{u}\| \cos(\pi - \theta) = -\|\vec{u}\| \cos \theta \ [\text{since } \cos(\pi - \theta) = -\cos \theta] = -\frac{\vec{u} \cdot \vec{v}}{\|\vec{v}\|}$$

(This is *positive*, as expected, since $\vec{u} \cdot \vec{v}$ is negative in this case.) We summarize as follows.

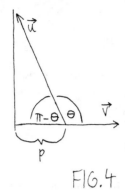

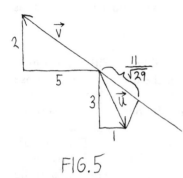

FIG.4

FIG.5

(12) The scalar $\dfrac{\vec{u} \cdot \vec{v}}{\|\vec{v}\|}$ is called the *component of $\vec{u}$ in the direction of $\vec{v}$*. If $\vec{u}$ and $\vec{v}$ are drawn with a common tail then this component may be thought of as the "signed projection" of $\vec{u}$ onto a line through $\vec{v}$. It is positive if the angle between $\vec{u}$ and $\vec{v}$ is acute, negative if the angle is obtuse, and in either case, its absolute value is the length of the projection.

Example 4 Let $\vec{u} = \vec{i} - 3\vec{j}$ and $\vec{v} = -5\vec{i} + 2\vec{j}$. Find the component of $\vec{u}$ in the direction of $\vec{v}$ and show its geometric significance in a sketch.

Solution: We have $\dfrac{\vec{u} \cdot \vec{v}}{\|\vec{v}\|} = \dfrac{-11}{\sqrt{29}}$. The negative sign indicates that $\vec{u}$ makes an obtuse angle with $\vec{v}$, and the absolute value, $\dfrac{11}{\sqrt{29}}$, is the length of the projection in Fig. 5.

Example 5 Figure 6 shows a rectangular box with edges 10, 7 and 2. Find the length of the projection of segment GF on the line CA. (One way to visualize the projection is to imagine the foot of the perpendicular from F to line AC, and the foot of the perpendicular from G to AC, which happens to be C; the projection is the distance between the two feet. In Fig. 6 the projection of $\vec{GF}$ may also be visualized as the projection of $\vec{CB}$.)

Solution: If ray DA is taken as the positive x-axis, ray DC as the positive y-axis and ray DH as the positive z-axis then $\vec{GF} = (2, 0, 0)$, $\vec{CA} = (2, -10, 0)$ and

$$\frac{\vec{GF} \cdot \vec{CA}}{\|\vec{CA}\|} = \frac{4}{\sqrt{104}}.$$

Therefore the length of the projection is $4/\sqrt{104}$.

The vector component of $\vec{u}$ in a direction We have already identified $\dfrac{\vec{u} \cdot \vec{v}}{\|\vec{v}\|}$ as the (scalar) component of $\vec{u}$ in the direction of $\vec{v}$. We now examine the

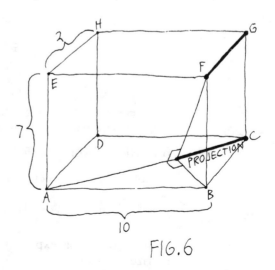

FIG.6

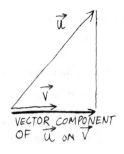

VECTOR COMPONENT
OF $\vec{u}$ ON $\vec{v}$

FIG.7

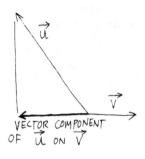

VECTOR COMPONENT
OF $\vec{u}$ ON $\vec{v}$

FIG. 8

vector obtained by projecting $\vec{u}$ onto $\vec{v}$ (Figs. 7 and 8); it is called the *vector component or projection of $\vec{u}$ in the direction of $\vec{v}$.*

The vector component in Fig. 7 has the same direction as $\vec{v}$ and its length is the scalar component $\dfrac{\vec{u} \cdot \vec{v}}{\|\vec{v}\|}$. Therefore, by (17) in Section 9.2, the vector component is

(13) $$\frac{\vec{u} \cdot \vec{v}}{\|\vec{v}\|} \frac{\vec{v}}{\|\vec{v}\|}.$$

Let's see if (13) applies to Fig. 8 as well, where the angle between $\vec{u}$ and $\vec{v}$ is obtuse. In this case, the scalar component $\dfrac{\vec{u} \cdot \vec{v}}{\|\vec{v}\|}$ is negative. When it multiplies $\dfrac{\vec{v}}{\|\vec{v}\|}$ in (13), it has the effect of reversing direction *as desired* for Fig. 8 where the vector component has a direction opposite to $\vec{v}$. Thus (13) is the vector component in both Figs. 7 and 8. Simplifying (13) produces the following conclusion.

(14)
> The vector component of $\vec{u}$ in the direction of $\vec{v}$ may be written as $\dfrac{\vec{u} \cdot \vec{v}}{\|\vec{v}\|^2}\vec{v}$ or, equivalently, as $\dfrac{\vec{u} \cdot \vec{v}}{\vec{v} \cdot \vec{v}}\vec{v}$. In other words, the vector component is a multiple of $\vec{v}$, and the multiple is the scalar $\dfrac{\vec{u} \cdot \vec{v}}{\vec{v} \cdot \vec{v}}$.

For example, if $\vec{u} = \vec{i} - 3\vec{j}$ and $\vec{v} = -5\vec{i} + 2\vec{j}$ then the vector component of $\vec{u}$ in the direction of $\vec{v}$ is

$$\frac{\vec{u} \cdot \vec{v}}{\vec{v} \cdot \vec{v}}\vec{v} = \frac{-11}{29}(-5\vec{i} + 2\vec{j}) = \frac{55}{29}\vec{i} - \frac{22}{29}\vec{j} \qquad \text{(Fig. 9)}.$$

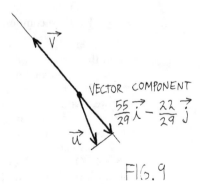

VECTOR COMPONENT
$\frac{55}{29}\vec{i} - \frac{22}{29}\vec{j}$

FIG.9

Problems for Section 9.3

1. Decide if the angle between $\vec{u} = \vec{i} + 2\vec{j} - 3\vec{k}$ and $\vec{v} = 5\vec{i} + 6\vec{j} + 5\vec{k}$ is acute, right or obtuse.

2. Find $\vec{u} \cdot \vec{v}$ if $\|\vec{u}\| = 5$, $\|\vec{v}\| = 6$ and $\vec{u}$ and $\vec{v}$ have opposite directions.

3. Find angle A in the triangle with vertices $A = (1, 4, -3)$, $B = (2, 1, 6)$ and $C = (4, 3, 2)$.

4. Let $\vec{u} = (u_1, u_2, u_3)$ and let $\theta_1, \theta_2, \theta_3$ be the angles between $\vec{u}$ and the positive x-axis, y-axis and z-axis, respectively.

(a) Find $\cos \theta_1$, $\cos \theta_2$, $\cos \theta_3$ (called the direction cosines of $\vec{u}$)
(b) Show that $(\cos \theta_1, \cos \theta_2, \cos \theta_3)$ is the unit vector in the direction of $\vec{u}$.

5. Let $A = (2, 3)$, $B = (5, 8)$, $C = (-1, 4)$, $D = (4, 1)$. Show that lines AB and CD are perpendicular using (a) slopes (b) dot products.

6. Suppose that you walk from point $A = (2, 4)$ to point $B = (8, 9)$ and then make a left turn and walk 7 feet to point C. Use vectors to find the coordinates of C.

7. Find the acute angle determined by two lines with slopes $-7/2$ and 4.

8. Show that $(\vec{u} \cdot \vec{u})\vec{v} - (\vec{v} \cdot \vec{u})\vec{u}$ is perpendicular to $\vec{u}$.

9. If $\|\vec{u}\| = 3$, $\|\vec{v}\| = 2$ and $\vec{u} \cdot \vec{v} = 5$, find $\|-6\vec{u}\|$, $\vec{u} \cdot 3\vec{u}$ and $\|\vec{u} - \vec{v}\|$.

10. Let $\vec{u} = (5, 2, 3, -4)$ and $\vec{v} = (-4, 3, -1, 4)$. Compute whichever of the following are meaningful

(a) $|\vec{u} \cdot \vec{v}|$ (e) $\dfrac{2}{\|\vec{u}\|}$

(b) $\|\vec{u} \cdot \vec{v}\|$ (f) $(\vec{u} \cdot \vec{v})\vec{v}$

(c) $\|\vec{v}\|\vec{u}$ (g) $(\vec{u} \cdot \vec{v}) \cdot \vec{v}$

(d) $\dfrac{2}{\vec{u}}$

11. Give (i) a geometric argument and then (ii) an algebraic argument for the following.

(a) If $\vec{u} \cdot \vec{v} = 0$ then $\|\vec{u} + \vec{v}\| = \|\vec{u} - \vec{v}\|$.
(b) If $\|\vec{u}\| = \|\vec{v}\|$ then $\vec{u} + \vec{v}$ is perpendicular to $\vec{u} - \vec{v}$.

12. If $\vec{u} = 4\vec{i} + 2\vec{j} + 3\vec{k}$ and $\vec{v} = -\vec{i} - 3\vec{j} + \vec{k}$ find (a) the component of $\vec{u}$ in the direction of $\vec{v}$ and (b) the component of $\vec{v}$ in the direction of $\vec{u}$.

13. In Fig. 6, find the length of the projection of segment FH on the line AG.

14. Suppose the component of $\vec{u}$ in the direction of $\vec{v}$ is 6.

(a) Find the component of $\vec{u}$ in the direction of $4\vec{v}$.
(b) Find the component of $\vec{u}$ in the direction of $-\vec{v}$.
(c) Find the component of $4\vec{u}$ in the direction of $\vec{v}$.

15. If $\|\vec{u}\| = 6$, $\|\vec{v}\| = 4$ and the angle between $\vec{u}$ and $\vec{v}$ is 120°, find the component of $\vec{u}$ in the direction of $\vec{v}$.

16. The 100 meter dash is run on a track in the direction of the vector $t = \vec{i} + 2\vec{j}$. The wind velocity $\vec{w}$ is $2\vec{i} + 2\vec{j}$; that is, the wind is blowing from the southwest with windspeed $\sqrt{8}$. The rules say that a legal wind speed, measured in the direction of the dash, must not exceed 2. If the dash results in a world record, will it be disqualified because of an illegal wind?

17. A spike being hammered into a mountain is represented by the vector $(2, 3, -4)$. One more blow with magnitude at least 10 (in the direction of the spike) will finish the job. Is the force $(9, 8, -1)$ enough?

18. Find the direction in which the component of $\vec{u}$ is maximum, and find that maximum value.

19. If $\vec{v} = 2\vec{i} + 3\vec{j}$ and $q = 5\vec{i} - 2\vec{j}$, find the vector component of $\vec{v}$ in the direction of q.

20. In Fig. 10, which of $\vec{p}$ and $\vec{q}$ has the larger component in the direction of $\vec{u}$?

FIG. 10

9.4 The Cross Product

We will begin with a result from physics that introduces the cross product, a new vector multiplication. Consider a unit positive electric charge in a magnetic field, which we simplistically view as a charged marble near a bar magnet lying on a table. If the charge is stationary then it is not affected by the magnet, but suppose the charge rolls along the table. Figure 1 shows the velocity $\vec{v}$ of the charge; the magnet is represented by a vector $\vec{m}$ directed from the south pole to the north pole; the length of $\vec{m}$ indicates the strength of the magnet. Experiments show that the moving charge feels a force $\vec{f}$ which points like the thumb of your right hand when your fingers curl from $\vec{v}$ toward $\vec{m}$. Furthermore, the strength of the force depends on the speed of the charge, the strength of the magnet, and the angle θ between $\vec{v}$ and $\vec{m}$. In particular, $\|\vec{f}\| = \|\vec{v}\|\|\vec{m}\| \sin \theta$. The force $\vec{f}$ is denoted by $\vec{v} \times \vec{m}$ and suggests the following definition.

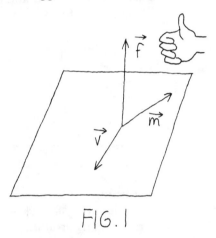

FIG. 1

The cross product   Given 3-dimensional vectors $\vec{u}$ and $\vec{v}$, the cross product $\vec{u} \times \vec{v}$ is a vector characterized geometrically by two properties.

> (1)   (direction of $\vec{u} \times \vec{v}$) The cross product $\vec{u} \times \vec{v}$ is perpendicular to both $\vec{u}$ and $\vec{v}$. In particular (Fig. 2) it points in the direction of your thumb if the fingers of your right-hand curl from $\vec{u}$ to $\vec{v}$ (right-hand rule). Equivalently, the cross products points in the direction in which a screw advances if it is turned from $\vec{u}$ to $\vec{v}$.
>
> (2)   (length of $\vec{u} \times \vec{v}$) If θ is the angle between $\vec{u}$ and $\vec{v}$ then
> $$\|\vec{u} \times \vec{v}\| = \|\vec{u}\|\|\vec{v}\| \sin \theta.$$

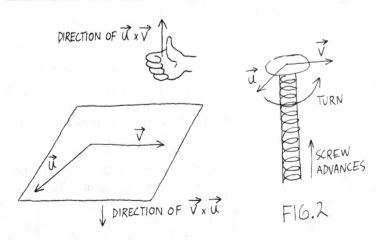

FIG. 2

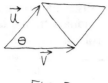

FIG. 3

The formula in (2) has a nice corollary. From trigonometry, the area of a triangle is half the product of any two sides with the sine of the included angle (Section 1.3, Eq. (19)), so the area of the triangle determined by $\vec{u}$ and $\vec{v}$ in Fig. 3 is $\frac{1}{2}\|\vec{u}\|\|\vec{v}\| \sin \theta$. Therefore $\|\vec{u}\|\|\vec{v}\| \sin \theta$ is twice the area of the triangle. Thus

(3) $\boxed{\|\vec{u} \times \vec{v}\| \text{ is the area of the parallelogram determined by } \vec{u} \text{ and } \vec{v} \quad \text{(Fig. 3).}}$

The result in (3) shows that for nonzero $\vec{u}$ and $\vec{v}$, $\|\vec{u} \times \vec{v}\| = 0$ (equivalently $\vec{u} \times \vec{v} = \vec{0}$) if and only if the parallelogram degenerates to zero area. Therefore, for nonzero $\vec{u}$ and $\vec{v}$,

$\boxed{\vec{u} \times \vec{v} = \vec{0} \text{ if and only if } \vec{u} \text{ and } \vec{v} \text{ are parallel.}}$

As a special case,

$\boxed{\text{the cross product of a vector with itself is } \vec{0}.}$

More generally, $\vec{u} \times \vec{v} = \vec{0}$ if and only if $\vec{u} = \vec{0}$ or $\vec{v} = \vec{0}$ or $\vec{u}$ and $\vec{v}$ are nonzero parallel vectors.

Warning If you intend to write $\vec{u} \times \vec{v} = \vec{0}$, make sure you write $\vec{0}$, not 0. If you write $\|\vec{u} \times \vec{v}\| = \|\vec{u}\|\|\vec{v}\| \sin \theta =$ parallelogram area, don't omit the norm signs around the vectors. Otherwise you will be writing meaningless equations.

Properties of the cross product By (3), $\vec{u} \times \vec{v}$ and $\vec{v} \times \vec{u}$ must have the same length, namely the area of the parallelogram determined by $\vec{u}$ and $\vec{v}$. But by the right-hand rule they have opposite directions. Therefore

(4) $\boxed{\vec{u} \times \vec{v} = -(\vec{v} \times \vec{u}).}$

By (2), or (3), $\|\vec{u} \times \vec{0}\|$ and $\|\vec{0} \times \vec{u}\|$ are both 0. Therefore,

(5) $$\vec{u} \times \vec{0} = \vec{0} \times \vec{u} = \vec{0}.$$

We state another property without proof:

$$\vec{u} \times (\vec{v} + \vec{w}) = \vec{u} \times \vec{v} + \vec{u} \times \vec{w}$$

(6) $$(\vec{u} + \vec{v}) \times \vec{w} = \vec{u} \times \vec{w} + \vec{v} \times \vec{w}$$

$$(\vec{u} + \vec{v}) \times (\vec{p} + \vec{q}) = \vec{u} \times \vec{p} + \vec{u} \times \vec{q} + \vec{v} \times \vec{p} + \vec{v} \times \vec{q}.$$

This property is the familiar distributive law, but note that on the right side of the vector identities in (6), the vectors in the cross product must appear in the same order as they did on the left side. It is *not* correct to expand $\vec{u} \times (\vec{v} + \vec{w})$ to $\vec{v} \times \vec{u} + \vec{w} \times \vec{u}$.

To discover another law, note that $\vec{u} \times \vec{v}$ and $\vec{u} \times 2\vec{v}$ have the same direction by the right-hand rule; but $\vec{u} \times 2\vec{v}$ is twice as long because the parallelogram determined by $\vec{u}$ and $2\vec{v}$ has twice the area of the parallelogram determined by $\vec{u}$ and $\vec{v}$. Therefore $\vec{u} \times 2\vec{v} = 2(\vec{u} \times \vec{v})$. In general,

(7) $$\vec{u} \times (c\vec{v}) = (c\vec{u}) \times \vec{v} = c(\vec{u} \times \vec{v}).$$

As an example, consider $(\vec{u} + \vec{v}) \times (\vec{u} - \vec{v})$. By (6) and (7), its expansion is $\vec{u} \times \vec{u} - \vec{u} \times \vec{v} + \vec{v} \times \vec{u} - \vec{v} \times \vec{v}$. The cross product of a vector with itself is $\vec{0}$, so $\vec{u} \times \vec{u}$ and $\vec{v} \times \vec{v}$ are $\vec{0}$. Then, by (4), the remaining terms combine rather than cancel, so $(\vec{u} + \vec{v}) \times (\vec{u} - \vec{v})$ is $2(\vec{v} \times \vec{u})$ or, equivalently, $-2(\vec{u} \times \vec{v})$.

The components of the cross product We would like to derive a formula for the components of $\vec{u} \times \vec{v}$ in terms of the components of $\vec{u}$ and $\vec{v}$. But first we must deal with an unusual situation involving the type of rectangular coordinate system used. Consider $\vec{i} \times \vec{j}$ in the right-handed system in Fig. 4 and in the left-handed system in Fig. 5. In each case, $\|\vec{i} \times \vec{j}\| = \|\vec{i}\|\|\vec{j}\| \sin 90° = 1$, but by the right-hand rule, $\vec{i} \times \vec{j}$ points up in Fig. 4 and down in Fig. 5. So $(1, 0, 0) \times (0, 1, 0)$ is either $(0, 0, 1)$ or $(0, 0, -1)$ depending on whether the vectors are plotted in a right-handed or left-handed system. This illustrates that the components of $\vec{u} \times \vec{v}$ depend on the type of coordinate system. By convention, *only right-handed systems are used,* and in this case we will derive a unique formula for the components of $\vec{u} \times \vec{v}$.

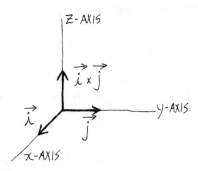

FIG. 4

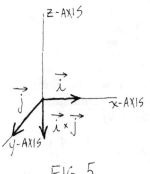

FIG. 5

Let $\vec{u} = u_1\vec{i} + u_2\vec{j} + u_3\vec{k}$ and $\vec{v} = v_1\vec{i} + v_2\vec{j} + v_3\vec{k}$. By (6) and (7),

$$\vec{u} \times \vec{v} = (u_1\vec{i} + u_2\vec{j} + u_3\vec{k}) \times (v_1\vec{i} + v_2\vec{j} + v_3\vec{k})$$

(8)
$$= u_1v_1(\vec{i} \times \vec{i}) + u_2v_2(\vec{j} \times \vec{j}) + u_3v_3(\vec{k} \times \vec{k})$$
$$+ u_1v_2(\vec{i} \times \vec{j}) + u_2v_1(\vec{j} \times \vec{i})$$
$$+ u_1v_3(\vec{i} \times \vec{k}) + u_3v_1(\vec{k} \times \vec{i})$$
$$+ u_2v_3(\vec{j} \times \vec{k}) + u_3v_2(\vec{k} \times \vec{j}).$$

The cross product of a vector with itself is $\vec{0}$, so $\vec{i} \times \vec{i} = \vec{j} \times \vec{j} = \vec{k} \times \vec{k} = \vec{0}$. We have already seen that in a right-handed system, $\vec{i} \times \vec{j} = \vec{k}$. Similarly, $\vec{j} \times \vec{i} = -\vec{k}, \vec{k} \times \vec{i} = \vec{j}, \vec{i} \times \vec{k} = -\vec{j}, \vec{j} \times \vec{k} = \vec{i}, \vec{k} \times \vec{j} = -\vec{i}$. Therefore (8) simplifies to

(9) $$\vec{u} \times \vec{v} = (u_2v_3 - u_3v_2)\vec{i} + (u_3v_1 - u_1v_3)\vec{j} + (u_1v_2 - u_2v_1)\vec{k}.$$

The formula in (9) looks formidable to memorize, but we will give some simple routines for finding cross products easily. It is convenient to use the determinant notation

$$\begin{vmatrix} a & b \\ c & d \end{vmatrix} = ad - bc$$

to write (9) as

(10)
$$\vec{u} \times \vec{v} = \begin{vmatrix} u_2 & u_3 \\ v_2 & v_3 \end{vmatrix} \vec{i} + \begin{vmatrix} u_3 & u_1 \\ v_3 & v_1 \end{vmatrix} \vec{j} + \begin{vmatrix} u_1 & u_2 \\ v_1 & v_2 \end{vmatrix} \vec{k}.$$

To apply (10), line up the components of $\vec{u}$ and $\vec{v}$ so that the components of the first factor $\vec{u}$ appear in the first row as follows:

(11)
$$\begin{matrix} u_1 & u_2 & u_3 \\ v_1 & v_2 & v_3 \end{matrix}.$$

Ignoring the *first* column in (11) leaves the configuration

$$\begin{matrix} \cdot & u_2 & u_3 \\ \cdot & v_2 & v_3 \end{matrix}$$

whose determinant is the *first* component of $\vec{u} \times \vec{v}$. Ignoring the *second* column of (11) leaves

$$\begin{matrix} u_1 & \cdot & u_3 \\ v_1 & \cdot & v_3 \end{matrix}$$

whose *negated* determinant produces the *second* component of $\vec{u} \times \vec{v}$. Finally, disregarding the *third* column in (11) leaves

$$\begin{matrix} u_1 & u_2 & \cdot \\ v_1 & v_2 & \cdot \end{matrix}$$

whose determinant is the *third* component of $\vec{u} \times \vec{v}$. The procedure just described can also be carried out by writing

(12)
$$\vec{u} \times \vec{v} = \begin{vmatrix} \vec{i} & \vec{j} & \vec{k} \\ u_1 & u_2 & u_3 \\ v_1 & v_2 & v_3 \end{vmatrix}$$

and expanding the determinant across the first row. This immediately produces (10). (Appendix A5 contains a review of determinants.)

For example, if $\vec{u} = 2\vec{i} + \vec{j} - 4\vec{k}$ and $\vec{v} = 3\vec{i} - 2\vec{j} + 5\vec{k}$ then

$$\vec{u} \times \vec{v} = \begin{vmatrix} 2 & 1 & -4 \\ 3 & -2 & 5 \end{vmatrix} \vec{i} - \begin{vmatrix} 2 & 1 & -4 \\ 3 & -2 & 5 \end{vmatrix} \vec{j} + \begin{vmatrix} 2 & 1 & -4 \\ 3 & -2 & 5 \end{vmatrix} \vec{k}$$

$$= -3\vec{i} - 22\vec{j} - 7\vec{k}.$$

Alternatively,

$$\vec{u} \times \vec{v} = \begin{vmatrix} \vec{i} & \vec{j} & \vec{k} \\ 2 & 1 & -4 \\ 3 & -2 & 5 \end{vmatrix} = \vec{i} \begin{vmatrix} 1 & -4 \\ -2 & 5 \end{vmatrix} - \vec{j} \begin{vmatrix} 2 & -4 \\ 3 & 5 \end{vmatrix} + \vec{k} \begin{vmatrix} 2 & 1 \\ 3 & -2 \end{vmatrix}$$

$$= -3\vec{i} - 22\vec{j} - 7\vec{k}.$$

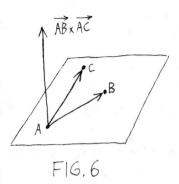

FIG. 6

Warning When the second column in (11) is ignored to compute the second component of the cross product, a minus sign must also be inserted to obtain $-\begin{vmatrix} u_1 & u_3 \\ v_1 & v_3 \end{vmatrix}$.

Example 1 Find a vector perpendicular to the plane determined by the points $A = (1, 2, 3)$, $B = (4, 5, 6)$, $C = (-2, 0, 3)$.

Solution: By (1), the vector $\overrightarrow{AB} \times \overrightarrow{AC}$ is perpendicular to both $\overrightarrow{AB}$ and $\overrightarrow{AC}$, and hence is perpendicular to the plane (Fig. 6). Therefore an answer is

$$\overrightarrow{AB} \times \overrightarrow{AC} = (3, 3, 3) \times (-3, -2, 0) = (6, -9, 3) = 6\vec{i} - 9\vec{j} + 3\vec{k}.$$

Another answer, with simpler components, is $\frac{1}{3}(6\vec{i} - 9\vec{j} + 3\vec{k})$ or $2\vec{i} - 3\vec{j} + \vec{k}$. (In fact we could have used $\frac{1}{3}\overrightarrow{AB}$ in the original cross product instead of $\overrightarrow{AB}$.) Still another answer is $-2\vec{i} + 3\vec{j} - \vec{k}$. There are many vectors perpendicular to the plane but, by geometry, all are multiples of one another.

The cross product of 2-dimensional vectors The vector operations in earlier sections originated from geometric considerations, and were extended algebraically to n-dimensional vectors in general. For example, $\|\vec{u}\|$ was inspired by the length of an arrow, and the 2-dimensional formula $\sqrt{u_1^2 + u_2^2}$ generalized easily to $\sqrt{u_1^2 + \cdots + u_n^2}$, independent of geometry. Similarly, $\vec{u} + \vec{v}$, $c\vec{u}$ and $\vec{u} \cdot \vec{v}$ are defined for n-dimensional vectors and used extensively in mathematics and applications (such as the theory of systems of equations with n variables). The cross product was defined geometrically in (1) and (2) for *three*-dimensional vectors, and this is the first operation we do not find profitable to extend to n-space. It remains a tool in 3-space only. However, for the purpose of taking a cross product, a *two*-dimensional vector such as $(2, 3)$, lying in the x, y plane, can be regarded as the *three*-dimensional vector $(2, 3, 0)$ lying in (or parallel to) the x, y plane *in* 3-space.

Example 2 Find the area of the triangle determined by the points $A = (2, 3)$, $B = (4, 6)$, $C = (-1, 2)$.

Solution: The triangle is determined by the vectors $\overrightarrow{AB} = (2, 3)$ and $\overrightarrow{AC} = (-3, -1)$. Then

$$\overrightarrow{AB} \times \overrightarrow{AC} = (2, 3, 0) \times (-3, -1, 0) = (0, 0, 7),$$

and $\|\overrightarrow{AB} \times \overrightarrow{AC}\| = 7$. Thus, the area of the triangle, half a parallelogram, is $\frac{7}{2}$.

Problems for Section 9.4

1. The vectors in Fig. 7 lie in the plane of the page. The vector $\vec{p}$, not shown, points perpendicularly into the page. Find the directions of $\vec{u} \times \vec{v}$, $\vec{p} \times \vec{q}$ and $\vec{s} \times \vec{t}$.

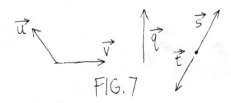

FIG. 7

2. What can you conclude about $\vec{u}$ and $\vec{v}$ if $\vec{u} \times \vec{v} = \vec{0}$ *and* $\vec{u} \cdot \vec{v} = 0$?

3. If $\vec{u} = 3\vec{i} + 987\vec{j} + 38\vec{k}$, find $\vec{u} \times \vec{u}_{\text{normalized}}$.

4. If $\vec{a} \cdot \vec{b} \neq 0$, show that the equation $\vec{a} \times \vec{x} = \vec{b}$ has no solution for $\vec{x}$.

5. An expression of the form $\vec{u} \cdot \vec{v} \times \vec{w}$ must mean $\vec{u} \cdot (\vec{v} \times \vec{w})$ rather than $(\vec{u} \cdot \vec{v}) \times \vec{w}$.

(a) Explain why. (b) Find $\vec{u} \cdot \vec{v} \times \vec{u}$. (c) Find $\vec{u} \cdot \vec{v} \times \vec{v}$.

6. Find $(\vec{u} + \vec{v}) \times (\vec{u} + \vec{v})$.

7. If the vectors $\vec{u}$ and $\vec{v}$ lie on the floor in Room 321 and the vectors $\vec{p}$ and $\vec{q}$ lie on the floor in Room 432, find $(\vec{u} \times \vec{v}) \times (\vec{p} \times \vec{q})$.

8. Simplify $3\vec{u} \times (4\vec{u} + 5\vec{v})$.

9. If all vectors are drawn with a common tail, show that $\vec{u} \times (\vec{v} \times \vec{w})$ lies in the plane determined by $\vec{v}$ and $\vec{w}$.

10. Find $\vec{u} \times \vec{v}$ if

(a) $\vec{u} = (6, -1, 2), \vec{v} = (3, 4, 3)$ (c) $\vec{u} = (6, 1), \vec{v} = (3, 4)$

(b) $\vec{u} = -2\vec{i} - 3\vec{j} + 5\vec{k}, \vec{v} = \vec{i} + \vec{j} + 4\vec{k}$ (d) $\vec{u} = 5\vec{i} - \vec{j} - 2\vec{k}, \vec{v} = \vec{i} + 2\vec{j}$

11. Find $\vec{w} \times \vec{v}$ and $\vec{v} \times \vec{w}$ if $\vec{v} = (-1, -2, -3)$ and $\vec{w} = (3, 3, -2)$.

12. Let $\vec{u} = 3\vec{i} + 2\vec{j} - \vec{k}$ and $\vec{v} = -\vec{i} + 5\vec{j} + 2\vec{k}$. If θ is the angle determined by $\vec{u}$ and $\vec{v}$, find $\cos\theta$ and $\sin\theta$ independently and then check to see that $\cos^2\theta + \sin^2\theta = 1$.

13. Let $\vec{u} = 2\vec{i} - \vec{j} + 3\vec{k}$ and $\vec{v} = 5\vec{i} + 3\vec{j} - 6\vec{k}$.

(a) Find four nonparallel vectors perpendicular to $\vec{u}$.

(b) Find a vector perpendicular to both $\vec{u}$ and $\vec{v}$.

14. Find the area of the triangle determined by the points $A = (0, 2, -1)$, $B = (4, -4, 2)$, $C = (-1, -4, 6)$.

9.5 The Scalar Triple Product

We have already seen that the area of the parallelogram determined by $\vec{u}$ and $\vec{v}$ is $\|\vec{u} \times \vec{v}\|$. Let's go one dimension further and find the volume of the parallelepiped determined by $\vec{u}, \vec{v}$ and $\vec{w}$ (Fig. 1). The base indicated in Fig. 1 is a parallelogram whose area is $\|\vec{v} \times \vec{w}\|$. The height is the length of the projection of $\vec{u}$ onto a line perpendicular to the base. The vector $\vec{v} \times \vec{w}$ has this perpendicular direction, so by (12) of Section 9.3, the height is the absolute value of the component of $\vec{u}$ in the direction of $\vec{v} \times \vec{w}$, that is, the height is $\left| \dfrac{\vec{u} \cdot (\vec{v} \times \vec{w})}{\|\vec{v} \times \vec{w}\|} \right|$. Note that both the numerator and denominator are

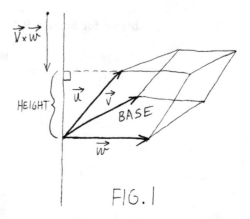

FIG. 1

numbers, and since the denominator is positive we may write the height as $\dfrac{|\vec{u} \cdot (\vec{v} \times \vec{w})|}{\|\vec{v} \times \vec{w}\|}$. Then

$$(1) \quad \text{volume} = (\text{base})(\text{height}) = \|\vec{v} \times \vec{w}\| \frac{|\vec{u} \cdot (\vec{v} \times \vec{w})|}{\|\vec{v} \times \vec{w}\|} = |\vec{u} \cdot (\vec{v} \times \vec{w})|.$$

If $\vec{u}, \vec{v}, \vec{w}$ are 3-dimensional vectors then

$$\vec{u} \cdot (\vec{v} \times \vec{w})$$

is called a *scalar triple product.* Without ambiguity we may omit the parentheses and write the scalar triple product as $\vec{u} \cdot \vec{v} \times \vec{w}$. (It cannot be misinterpreted as $(\vec{u} \cdot \vec{v}) \times \vec{w}$ since the latter expression is the proposed cross product of a scalar and a vector, which is meaningless.) As its name implies, $\vec{u} \cdot \vec{v} \times \vec{w}$ is a scalar. For example, if $\vec{u} = (1, 2, 1)$, $\vec{v} = (2, 4, 6)$ and $\vec{w} = (1, 3, -1)$ then $\vec{u} \cdot \vec{v} \times \vec{w} = (1, 2, 1) \cdot (-22, 8, 2) = -4$.

If $\vec{u} = (u_1, u_2, u_3)$, $\vec{v} = (v_1, v_2, v_3)$ and $\vec{w} = (w_1, w_2, w_3)$, the configuration

$$(2) \quad \begin{matrix} u_1 & u_2 & u_3 \\ v_1 & v_2 & v_3 \\ w_1 & w_2 & w_3 \end{matrix}$$

will help keep track of the arithmetic involved in computing $\vec{u} \cdot \vec{v} \times \vec{w}$. We use the last two rows of (2) to find

$$\vec{v} \times \vec{w} = \left(\begin{vmatrix} v_2 & v_3 \\ w_2 & w_3 \end{vmatrix}, -\begin{vmatrix} v_1 & v_3 \\ w_1 & w_3 \end{vmatrix}, \begin{vmatrix} v_1 & v_2 \\ w_1 & w_2 \end{vmatrix} \right),$$

and then dot with the first row to get

$$(3) \quad \vec{u} \cdot \vec{v} \times \vec{w} = u_1 \begin{vmatrix} v_2 & v_3 \\ w_2 & w_3 \end{vmatrix} - u_2 \begin{vmatrix} v_1 & v_3 \\ w_3 & w_3 \end{vmatrix} + u_3 \begin{vmatrix} v_1 & v_2 \\ w_1 & w_2 \end{vmatrix}.$$

But (3) may also be viewed as the expansion (along the first row) of the determinant of (2). Therefore,

$$(3) \quad \boxed{ \vec{u} \cdot \vec{v} \times \vec{w} = \begin{vmatrix} u_1 & u_2 & u_3 \\ v_1 & v_2 & v_3 \\ w_1 & w_2 & w_3 \end{vmatrix}. }$$

The determinant formula is a compact expression for the scalar triple product.

By (1), the absolute value of the scalar triple product is a volume; in particular,

$$(4) \quad \boxed{ |\vec{u} \cdot \vec{v} \times \vec{w}| \text{ is the volume of the parallelepiped determined by } \vec{u}, \vec{v} \text{ and } \vec{w}. }$$

For example, if $\vec{p} = 2\vec{i} - 3\vec{j} + 5\vec{k}$, $\vec{q} = -6\vec{i} + \vec{j} - \vec{k}$ and $\vec{r} = 2\vec{i} + \vec{k}$ then

$$\vec{p} \cdot \vec{q} \times \vec{r} = \begin{vmatrix} 2 & -3 & 5 \\ -6 & 1 & -1 \\ 2 & 0 & 1 \end{vmatrix} = -20,$$

so the volume of the parallelepiped determined by $\vec{p}, \vec{q}$ and $\vec{r}$ is 20.

The result in (4) shows that for nonzero $\vec{u}, \vec{v}, \vec{w}, |\vec{u} \cdot \vec{v} \times \vec{w}| = 0$ (equivalently $\vec{u} \cdot \vec{v} \times \vec{w} = 0$) if and only if the parallelepiped degenerates to zero volume. Therefore, for nonzero $\vec{u}, \vec{v}, \vec{w}$,

> $\vec{u} \cdot \vec{v} \times \vec{w} = 0$ *if and only if* $\vec{u}, \vec{v}, \vec{w}$ *are coplanar when drawn with a common tail*.

More generally, $\vec{u} \cdot \vec{v} \times \vec{w} = 0$ if and only if $\vec{u} = \vec{0}$ or $\vec{v} = \vec{0}$ or $\vec{w} = \vec{0}$ or $\vec{u}, \vec{v}, \vec{w}$ are nonzero coplanar vectors.

To conclude this section we investigate the effect of a switch in the order of the factors in a scalar triple product. There are six possible arrangements:

$$\vec{u} \cdot \vec{v} \times \vec{w}, \qquad \vec{w} \cdot \vec{u} \times \vec{v}, \qquad \vec{v} \cdot \vec{w} \times \vec{u}$$

(5)
$$\vec{v} \cdot \vec{u} \times \vec{w}, \qquad \vec{w} \cdot \vec{v} \times \vec{u}, \qquad \vec{u} \cdot \vec{w} \times \vec{v}.$$

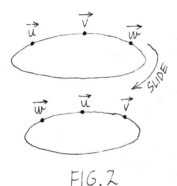

FIG. 2

All six have the same absolute value, namely, the volume of the parallelepiped determined by $\vec{u}, \vec{v}$ and $\vec{w}$. We will prove that *the three in the first row are equal, the three in the second row are equal, and the value from the first row is the negative of the value from the second row.* Before offering the proof we will give a device for *remembering* which rearrangements have the same value and which have opposite values. Picture the letters $\vec{u}, \vec{v}, \vec{w}$ as beads on a bracelet. If a new order is produced by *sliding* the beads on the bracelet, the new arrangement is called a *cyclic permutation* of the original. Figure 2 shows that $\vec{w} \cdot \vec{u} \times \vec{v}$ is a cyclic permutation of $\vec{u} \cdot \vec{v} \times \vec{w}$ since it can be obtained by sliding $\vec{w}$ around to the front. *If a new arrangement is obtained by a cyclic permutation of the letters, the value of the scalar triple product is unchanged. Otherwise the value is negated.* For example, if $\vec{u} \cdot \vec{p} \times \vec{v} = -7$ then $\vec{p} \cdot \vec{v} \times \vec{u}$ is also -7 since it can be obtained by cyclic permutation, while $\vec{v} \cdot \vec{p} \times \vec{u}$ is 7 since it cannot be obtained from the original by cyclic permutation.

One proof of the rearrangement principle uses the fact that if two rows of a determinant are interchanged, then the sign of the determinant changes (Appendix A5). Compare the determinants for $\vec{u} \cdot \vec{v} \times \vec{w}$ and its cyclic permutation $\vec{w} \cdot \vec{u} \times \vec{v}$:

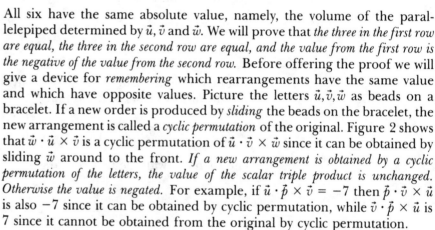

If rows 1 and 3 are interchanged in the first determinant, and then rows 2 and 3 interchanged, the result is the second determinant. Each interchange of rows changes the sign, so two interchanges restore the original value. Therefore the cyclic permutation has the same value as the original. On the other hand, only one interchange of rows is required to go from the determinant for $\vec{u} \cdot \vec{v} \times \vec{w}$ to the determinant of any *non*cyclic permutation. Thus permuting *non*cyclically negates the scalar triple product.

Problems for Section 9.5

1. Find $\vec{u} \cdot \vec{v} \times \vec{w}$ if $\vec{u} = (1, 2, 3)$, $\vec{v} = (-1, 1, 1)$, $\vec{w} = (0, 3, 4)$.

2. Find the volume of the parallelepiped determined by $\vec{u} = \vec{i} + \vec{k}$, $\vec{v} = 2\vec{j} + 3\vec{k}$, $\vec{w} = 3\vec{i} - 5\vec{k}$.

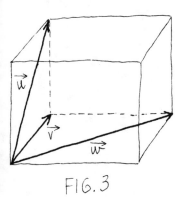

FIG. 3

3. Are the points $A = (1, 1, 2)$, $B = (2, 3, 5)$, $C = (2, 0, 4)$, $D = (2, -3, -1)$ coplanar?

4. Figure 3 shows $\vec{u}, \vec{v}, \vec{w}$. Is $\vec{u} \cdot \vec{v} \times \vec{w}$ positive, negative or zero?

5. Suppose $\vec{u}$ lies on the floor in Room 223, $\vec{v}$ lies on the floor in Room 224 and $\vec{w}$ lies on a desk top in Room 347. Find $\vec{u} \cdot \vec{v} \times \vec{w}$.

6. If $\vec{q} \cdot \vec{p} \times \vec{r} = -5$ find

(a) $\vec{p} \cdot \vec{r} \times \vec{q}$ (d) $3\vec{q} \cdot 4\vec{p} \times 5\vec{r}$
(b) $\vec{r} \cdot \vec{p} \times \vec{q}$ (e) $\vec{q} \cdot \vec{p} \times \vec{q}$
(c) $\vec{q} \cdot \vec{r} \times \vec{p}$ (f) $\vec{q} \cdot \vec{r} \times \vec{r}$

9.6 The Velocity Vector

(Appendix A6 is a prerequisite for this section.)

In Section 3.5 we found the velocity and acceleration of a particle moving on a number line. In this section and the next we extend the topic to motion in a plane and space. For convenience we measure distance in meters and time in seconds throughout.

Equations of motion; the position vector   An equation such as $x = t^2 + 2t$ describes the position x, at time t, of a particular particle on a number line. Similarly, a pair of equations such as $x = t^2 + 2t, y = 3t - t^3$ describes the position (x, y), at time t, of a particular particle moving in a plane. More generally, position in 2-space at time t is described by a pair of parametric equations of the form $x = x(t), y = y(t)$, and position in 3-space at time t is given by $x = x(t), y = y(t), z = z(t)$.

For example, consider

(1) $$x = t, \qquad y = t^2 - 3 .$$

The table in (2) lists some values of t with corresponding points. (Remember that a negative time such as $t = -3$ simply means 3 seconds before the fixed time designated as $t = 0$.)

(2)

time t	position (x, y)
-3	$(-3, 6)$
-2	$(-2, 1)$
-1	$(-1, -2)$
0	$(0, -3)$
1	$(1, -2)$
2	$(2, 1)$
3	$(3, 6)$

If the points are plotted, and connected in a reasonable fashion, we have the path in Fig. 1. Each point is labeled with its associated value of t; the timing indicates that the particle travels from left to right along the path. Soon we will use calculus to identify its speed and acceleration at any instant.

In addition to plotting points to produce the anonymous path in Fig. 1, we can find a direct connection between x and y, a process called *eliminating the parameter*. Since $x = t$, we can substitute x for t in the second

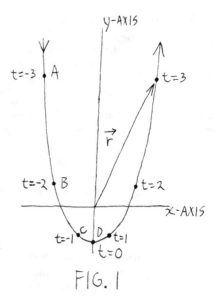

FIG. 1

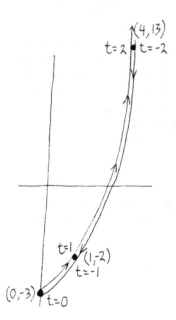

FIG. 2

equation in (1) to obtain $y = x^2 - 3$. Therefore the curve in Fig. 1 is the parabola $y = x^2 - 3$.

The vector drawn from the origin to the curve is called the *position vector $\vec{r}(t)$*. For the path $x = x(t)$, $y = y(t)$, we have $\vec{r}(t) = x(t)\vec{i} + y(t)\vec{j}$. The position vector for the path in (1) is $\vec{r}(t) = t\vec{i} + (t^2 - 3)\vec{j}$; if $t = 3$ then $\vec{r} = 3\vec{i} + 6\vec{j}$ and the particle is at the point $(3, 6)$ (Fig. 1).

As another example, let $\vec{r}(t) = t^2\vec{i} + (t^4 - 3)\vec{j}$, that is, let $x = t^2$, $y = t^4 - 3$. We can eliminate the parameter to obtain $y = x^2 - 3$, so again the particle travels on the parabola $y = x^2 - 3$. However, if we plot a few points we see that it does not travel along the *entire* parabola (Fig. 2). It moves from right to left during negative time until it reaches the point $(0, -3)$ and then turns around and goes back the way it came. (Even before we plot individual points we can tell that the particle can't travel on the entire parabola since the first coordinate, t^2, is never negative.)

This latter example illustrates that *if the parameter is eliminated from the equations $x = x(t)$, $y = y(t)$ to obtain a single equation in x and y, then the particle must travel along the graph of the single equation, but does not necessarily traverse the entire graph.* It is necessary to plot a few points to capture the timing, direction and extent of the motion. *Similarly, suppose the parameter is eliminated from the equations $x = x(t)$, $y = y(t)$, $z = z(t)$ to obtain a single equation in x, y and z. The graph of the single equation is a surface in 3-space (Chapter 10 will discuss this further), and the path of the particle is a curve lying on the surface.* There is no single method for eliminating the parameter. One possibility is to try to solve one equation for t and substitute in the other. On the other hand, in some instances it may not be desirable or practical to eliminate the parameter.

Circular motion at constant speed Let

(3) $$x = 6 \cos t, \qquad y = 6 \sin t,$$

or, equivalently, $\vec{r}(t) = (6 \cos t, 6 \sin t)$. A method for eliminating the parameter is not obvious here. But we can take advantage of the identity

$\cos^2 t + \sin^2 t = 1$ to get

(4) $\qquad x^2 + y^2 = 36 \cos^2 t + 36 \sin^2 t = 36(\cos^2 t + \sin^2 t) = 36$.

Therefore the path lies along the circle $x^2 + y^2 = 36$, with center at the origin and radius 6. Another (better) way to identify the path is to compare (3) with the equations $x = r \cos \theta$, $y = r \sin \theta$ which relate polar coordinates r, θ with rectangular coordinates x, y (Appendix A6). The comparison shows that any point (x, y) satisfying (3) has polar coordinate $r = 6$ and consequently lies on a circle centered at the origin with radius 6. Furthermore, the parameter t representing time is the polar coordinate angle θ. At time $t = 0$ the particle is on the circle with $\theta = 0$; at time $t = \pi/2$ the particle has moved to the point on the circle with $\theta = \pi/2$ (Fig. 3). In general, *the equations $x = r_0 \cos t$, $y = r_0 \sin t$ describe counterclockwise motion around the origin with radius r_0, making one revolution every 2π seconds.*

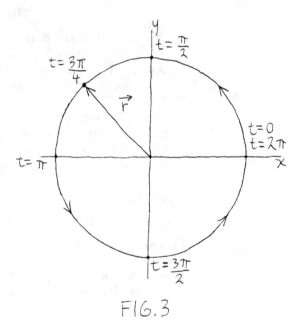

FIG. 3

Similarly, the path $x = 2 \cos 3t$, $y = 2 \sin 3t$ is circular motion with radius 2 but this time the angle θ is $3t$, not t. The particle still moves counterclockwise but makes one revolution in $2\pi/3$ seconds or, equivalently, makes three revolutions in 2π seconds.

Velocity and speed Suppose the equations of motion are $x = x(t)$, $y = y(t)$ or, equivalently, the position vector is $\vec{r}(t) = x(t)\vec{i} + y(t)\vec{j}$. Then $x(t)$ may be regarded as the horizontal position of the particle at time t, so $x'(t)$ is its horizontal velocity; similarly $y'(t)$ is the vertical velocity. If a particle travels horizontally at $x'(t)$ meters per second and simultaneously travels vertically at $y'(t)$ meters per second (Fig. 4) then it is really traveling in the direction of the vector $x'(t)\vec{i} + y'(t)\vec{j}$ at the rate of

(5) $\qquad \sqrt{[x'(t)]^2 + [y'(t)]^2}$ meters per second .

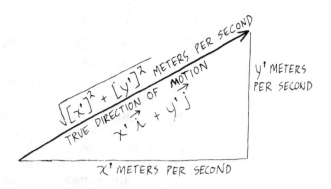

FIG. 4

> If $\vec{r}(t) = x(t)\vec{i} + y(t)\vec{j}$ we define the *velocity vector* $\vec{v}$ by
>
> $$\vec{v}(t) = x'(t)\vec{i} + y'(t)\vec{j}$$
>
> and refer to $\vec{v}$ as the derivative, $\vec{r}'$, of $\vec{r}$. *If $\vec{v}$ is drawn with its tail on the curve, it points in the instantaneous direction of motion* (hence is tangent to the path). Furthermore, by (5), *the instantaneous speed of the particle at time t is $\|\vec{v}(t)\|$ meters per second.*

We will illustrate the velocity vector and its norm, the speed, by returning to the equations of motion in (1) and the path in Fig. 1. Since $x = t$, $y = t^2 - 3$, we have

$$\vec{v}(t) = x'(t)\vec{i} + y'(t)\vec{j} = \vec{i} + 2t\vec{j}.$$

Equivalently,

$$\vec{r}(t) = t\vec{i} + (t^2 - 3)\vec{j}, \quad \vec{v}(t) = \vec{r}'(t) = \vec{i} + 2t\vec{j}.$$

Let's examine a few specific instances. At time $t = 2$ we have $\vec{r} = 2\vec{i} + \vec{j}$ so the particle is at the point (2, 1). The velocity vector is $\vec{v} = \vec{i} + 4\vec{j}$, which we draw with its tail at the point (2, 1) (Fig. 5). As predicted, $\vec{v}$ is tangent to the path, and of the two tangent directions, $\vec{v}$ points in the instantaneous direc-

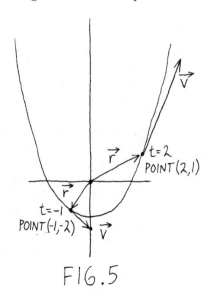

FIG. 5

tion of motion. Furthermore, at time 2, the particle's speed is $\|\vec{v}\| = \sqrt{17}$ meters per second. Similarly, if $t = -1$ then $\vec{r} = -\vec{i} - 2\vec{j}$ and $\vec{v} = \vec{i} - 2\vec{j}$. The particle is at the point $(-1, -2)$. The vector $\vec{v}$ attached to this point on the curve is tangent to the curve and points in the instantaneous direction of motion. At this instant, the speed is $\|\vec{v}\| = \sqrt{5}$. Note that the position vector $\vec{r}$ is drawn with its tail at the origin, while the velocity vector $\vec{v}$ is pictured with its tail on the curve, at the head of $\vec{r}$.

Problems for Section 9.6

1. Sketch the path and indicate the direction of motion. Eliminate the parameter if feasible to identify the path more thoroughly.

(a) $x = t^2 + 5, y = t$ (d) $\vec{r} = (2 + t^2)\vec{i} + (4 - 2t^2)\vec{j}$
(b) $x = t^2 + 5, y = -t$ (e) $\vec{r} = e^t\vec{i} + 2e^t\vec{j}$
(c) $x = 2 + t, y = 4 - 2t$

2. Sketch the path and indicate the direction of motion (without trying to eliminate the parameter).

(a) $x = 4 \cos \frac{1}{3}t, y = 4 \sin \frac{1}{3}t$

(b) $\vec{r} = \dfrac{\cos t}{t}\vec{i} + \dfrac{\sin t}{t}\vec{j}, t \geq 0$

(c) $\vec{r} = \cos t\vec{i} + \sin t\vec{j} + t\vec{k}$

3. Find equations of motion for the circular path.

(a) radius 3, around the origin, *clockwise*, one revolution per 2π seconds
(b) radius 3, *around point* $(2, 7)$, counterclockwise, one revolution per 2π seconds
(c) radius 3, around the origin, counterclockwise, *one revolution per second*

4. If $\vec{r}(t) = (2 - t)\vec{i} + (3 + t^2)\vec{j} + 6t\vec{j}$, does the particle pass through the following points, and if so, at what times?

(a) $(-3, 28, 4)$ (b) $(3, 4, -6)$

5. Find the connection between the paths with respective position vectors $\vec{r}_1(t) = t^3\vec{i} + t^2\vec{j}$ and $\vec{r}_2(t) = (t - 5)^3\vec{i} + (t - 5)^2\vec{j}$.
6. Suppose the position vector is $\vec{r}(t)$ where $\|\vec{r}(t)\| = 7$ for all t. Describe the path if (a) the motion is in 2-space   (b) the motion is in 3-space.
7. Find $\vec{v}(t)$ if $\vec{r} = t^3\vec{i} + 2t\vec{j} + \cos t\vec{k}$.
8. If $\vec{r} = t \cos t\vec{i} + t \sin t\vec{j}$, sketch the path, find $\vec{v}$ at time $t = \pi$ and attach $\vec{v}$ to the appropriate point on the path.
9. Consider the circular motion with position vector $\vec{r} = 6 \cos t\vec{i} + 6 \sin t\vec{j}$.

(a) Find the speed $\|\vec{v}\|$ to see that it agrees with one revolution per 2π seconds.
(b) Find $\vec{v}$ at time $t = \pi/2$ to see that it agrees with counterclockwise motion.

10. Let $x = t, y = t^2 - 3$ (Fig. 1). Examine $\|\vec{v}\|$ to see that the particle decelerates until time $t = 0$ and then accelerates.
11. Show that $x = \cos t^2, y = \sin t^2$, describes circular motion with *decreasing* speed until time $t = 0$ and *increasing* speed after $t = 0$.
12. Let $\vec{r}(t) = (-1 + 3t)\vec{i} + (1 - 2t)\vec{j} + 4t\vec{k}$.

(a) Use the velocity vector to show that the path is a line.
(b) Find the speed.
(c) Change $\vec{r}$ so that the particle moves on the same line but with speed 2.

13. Describe the path with position vector $\vec{r}$ if (a) $\vec{r}(t) = \vec{0}$ for all t (b) $\vec{r}'(t) = \vec{0}$ for all t.

14. Suppose $\vec{v}(t) = 2t\vec{i} + 5t^2\vec{j} + 6\vec{k}$ and the particle passes through the point $(1, 4, 6)$ at time $t = 3$.

(a) Find $\vec{r}(t)$.
(b) Find a unit tangent to the path at the point where $t = 2$.

15. Let $x = 3 \cos t$, $y = 2 \sin t$.

(a) Use a variation of the technique in (4) to show that the path is an ellipse.
(b) Show that the speed is not constant, and find the maximum and minimum speeds.

9.7   The Acceleration Vector

So far we have ignored one important aspect of motion. What is it that *makes* particles move around on curves? Newton postulated in his first law of motion that particles do not voluntarily move on circles or parabolas: *A particle initially at rest remains at rest, and a particle initially in motion will continue to move in a line with its direction and speed unchanged, unless acted on by an external force.* Therefore an external force is required except for straight line motion at constant speed, and another of Newton's laws singles out the precise force for any prescribed path, as follows.

The acceleration vector   Let $\vec{r}(t) = x(t)\vec{i} + y(t)\vec{j}$.

> The second derivative $\vec{r}''(t) = x''(t)\vec{i} + y''(t)\vec{j}$ is called the *acceleration vector* and is denoted by $\vec{a}(t)$.
>
> Newton's second law postulates that *if a particle with mass m has position vector $\vec{r}(t)$ at time t, then the propelling force $\vec{f}$ is given by $\vec{f} = m\vec{a}$. The acceleration vector $\vec{a}(t)$ itself is therefore the force per unit of mass.* It is pictured as a vector attached to the path; the particle must be pushed in the direction of $\vec{a}(t)$ with $m\|\vec{a}(t)\|$ units of force if it is to traverse the path.

If the force is suddenly removed at time t_0 then, by Newton's first law, the particle will fly off along a line in the direction of the vector $\vec{v}(t_0)$ with constant speed $\|\vec{v}(t_0)\|$ instead of continuing on the original path.

For example, consider the position vector $\vec{r}(t) = t\vec{i} + (t^2 - 3)\vec{j}$ from (1) of the preceding section. Then $\vec{v}(t) = \vec{i} + 2t\vec{j}$ and $\vec{a}(t) = 2\vec{j}$. At time $t = -1$,

$$\vec{r} = -\vec{i} - 2\vec{j}, \quad \vec{v} = \vec{i} - 2\vec{j}, \quad \vec{a} = 2\vec{j} \quad \text{(Fig. 1)}.$$

The particle is at the point $(-1, -2)$ and moving instantaneously in the direction of the vector $\vec{i} - 2\vec{j}$. It is acted on by the force $m\vec{a} = 2m\vec{j}$ where m is the mass of the particle; in other words, it is pushed north by $2m$ units of force. The direction of the force is such that the particle is pulled back toward the path and away from its natural inclination to leave the path in the direction of $\vec{v}$. Furthermore, the force is at an obtuse angle with $\vec{v}$ so it acts as a drag and decelerates the particle (decreases its speed). At time $t = 2$,

$$\vec{r} = 2\vec{i} + \vec{j}, \quad \vec{v} = \vec{i} + 4\vec{j}, \quad \vec{a} = 2\vec{j} \quad \text{(Fig. 1)}.$$

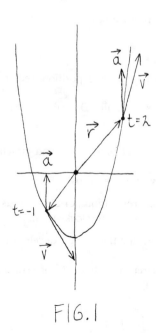

FIG. 1

The particle is at the point (2, 1) and moving instantaneously in the direction of the vector $\vec{i} + 4\vec{j}$. The force $m\vec{a} = 2m\vec{j}$ is acting on the particle, pulling it toward the parabola and preventing it from flying off on a tangent. The force accelerates the particle since it pulls at an acute angle with $\vec{v}$. (In this particular example, the acceleration vector is the same for all times t, but usually $\vec{a}$ varies with t.)

The tangential component of the acceleration vector The word acceleration has more than one meaning. The acceleration *vector* is the force per unit mass. However, a driver considers acceleration to be the *rate of change of speed*. With this second meaning, the acceleration of a particle is a *scalar*. It is positive if the particle is speeding up and negative if it is slowing down. In the preceding example we observed that the size of the angle between $\vec{a}$ and $\vec{v}$ determines whether the particle/car accelerates or decelerates. Now we wish to go further and compute the car's precise acceleration.

The acceleration of the particle/car is the rate of change of its speed, that is, the derivative of $\|\vec{v}(t)\|$. If the position vector is $\vec{r} = x(t)\vec{i} + y(t)\vec{j}$ then $\vec{v} = x'(t)\vec{i} + y'(t)\vec{j}$ and

$$\text{car's acceleration} = D_t\|\vec{v}(t)\| = D_t\sqrt{[x'(t)]^2 + [y'(t)]^2}$$

$$= \frac{1}{2}([x'(t)]^2 + [y'(t)]^2)^{-1/2} D_t([x'(t)]^2 + [y'(t)]^2)$$

(chain rule)

$$= \frac{2x'(t)x''(t) + 2y'(t)y''(t)}{2\sqrt{[x'(t)]^2 + [y'(t)]^2}}.$$

After the 2's are cancelled, the denominator is $\|\vec{v}(t)\|$ and the numerator is the dot product $\vec{a} \cdot \vec{v}$. Therefore the particle/car's acceleration is $\dfrac{\vec{a} \cdot \vec{v}}{\|\vec{v}\|}$, a formula given geometric significance in (12) of Section 9.3:

> *The car's acceleration is the component of $\vec{a}$ in the direction of $\vec{v}$.* It is called the *tangential component of acceleration,* or the *tangential acceleration,* and often denoted by $a_{\tan}$. In other words
>
> (1) $\qquad a_{\tan} = \dfrac{\vec{a} \cdot \vec{v}}{\|\vec{v}\|} = \text{rate of change of speed}.$

As predicted, *if the angle between $\vec{a}$ and $\vec{v}$ is acute* (so that the force is an impetus) *then $a_{\tan}$ is positive and the car is accelerating; if the angle is obtuse* (so that the force is a drag) *then $a_{\tan}$ is negative and the car is decelerating.* Figure 2 shows $a_{\tan} = 4$ if $t = 1$, and $a_{\tan} = -3$ if $t = 7$. The particle is accelerating at time $t = 1$ by 4 meters per second per second, and decelerating at time $t = 7$ by 3 meters per second per second.

Warning If $a_{\tan} = -3$ then either write that the particle is decelerating by 3 meters/second per second (this is the clearest report) or write that it is accelerating by -3 meters/second per second, but don't use a double negative and say that it is decelerating by -3 meters/second per second.

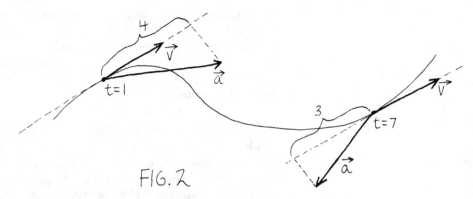

FIG. 2

Example 1 Suppose the position vector is $\vec{r}(t) = t^2\vec{i} + (t^4 - 12t)\vec{j}$. Then $\vec{v}(t) = 2t\vec{i} + (4t^3 - 12)\vec{j}$ and $\vec{a}(t) = 2\vec{i} + 12t^2\vec{j}$. At time $t = 1$ we have

$$\vec{r} = \vec{i} - 11\vec{j}, \quad \vec{v} = 2\vec{i} - 8\vec{j}, \quad \vec{a} = 2\vec{i} + 12\vec{j},$$

$$\|\vec{v}\| = \sqrt{68}, \quad \|\vec{a}\| = \sqrt{148}, \quad a_{\text{tan}} = \frac{\vec{a} \cdot \vec{v}}{\|\vec{v}\|} = -\frac{92}{\sqrt{68}}.$$

The particle is at the point $(1, -11)$, moving instantaneously in the direction of the vector $2\vec{i} - 8\vec{j}$ with speed $\sqrt{68}$ meters per second. A force acts on the particle in the direction of the vector $2\vec{i} + 12\vec{j}$; the magnitude of the force is $m\sqrt{148}$, where m is the mass of the particle. The particle is decelerating at the moment by $92/\sqrt{68}$ meters/second per second.

Warning It is a_{tan} and *not* $\|\vec{a}\|$ which indicates whether the particle is speeding up or slowing down.

The normal (radial) component of acceleration In 2-space there are two directions perpendicular to a velocity vector $\vec{v}$; the *inward* perpendicular is called the *normal* or *radial* direction (Fig. 3). It is more difficult to describe the radial direction in 3-space where there are infinitely many directions perpendicular to a velocity vector $\vec{v}$. In fact, the radial direction in 3-space

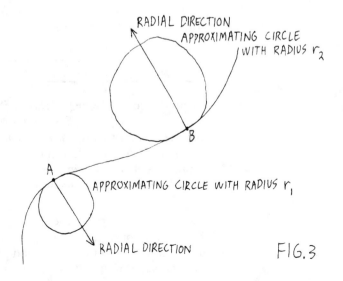

FIG. 3

is usually not defined geometrically, but in an algebraic manner which we will not pursue.

In addition to selecting a radial direction we also assign (but omit the details) an instantaneous radius of curvature r at each point by approximating the curve with a circle through the point. Figure 3 shows $r = r_1$ at A, $r = r_2$ at B. For the extreme case of a line, $r = \infty$.†

The component of $\vec{a}$ in the radial direction is called the *radial* or *normal component of acceleration*, and is often denoted by a_{rad}. Figure 4 shows $a_{rad} = 2$ and $a_{tan} = -3$. The tangential component may be positive or negative, but the radial component is never negative; the angle between $\vec{a}$ and the inward normal is never obtuse.

The radial component is taken as that aspect of $\vec{a}$ which changes the direction of the particle; $a_{rad} = 0$ if and only if the particle moves on a line. (We have already shown that *the tangential component is that aspect which changes speed; $a_{tan} = 0$ if and only if the particle moves at constant speed.*) It can be shown that, at any point,

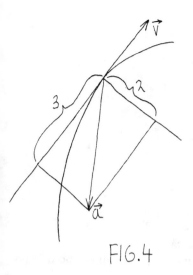

FIG.4

$$(2) \qquad a_{rad} = \frac{(\text{speed})^2}{r} = \frac{\|\vec{v}\|^2}{r}$$

where r is the instantaneous radius of curvature. If the radial force (per unit mass) supplied by friction between tires and road, and by the bank of the road, is not enough to satisfy (2), then the car plunges off the road. At a sharp curve, r is small, and the small denominator tends to increase the required a_{rad}. Thus drivers are warned to compensate by slowing down to decrease the numerator $\|\vec{v}\|^2$, so that the available radial force (per unit mass) will be sufficient to match (2).

Problems for Section 9.7

1. Let $\vec{r} = t\vec{i} + (1/t)\vec{j}$. Sketch the path. Then draw $\vec{v}$ and $\vec{a}$ at time $t = -1$. Is the particle accelerating or decelerating at this moment, and by how much? How many pounds of force act on the particle at time $t = -1$?

2. Let $x = e^t$, $y = e^{-t}$. Sketch the path. For $t = -1$, draw $\vec{v}$ and $\vec{a}$. Is the particle speeding up or slowing down at this moment?

3. Suppose the position vector is $\vec{r}(t)$ and the force on the particle is directed toward the origin for all t. Show that $\vec{r} = \vec{r}'' = \vec{0}$ for all t.

4. A particle with mass m is launched in 2-space at time $t = 0$ from the point $(1, 2)$ with initial velocity $4\vec{i} + 2\vec{j}$ (Fig. 5). Newton's law of gravity states that a force acts down at every instant of time, with magnitude mg where g is a constant (whose value depends on the system of units used to measure distance and mass). Find the position vector $\vec{r}(t)$.

5. Suppose $s(t)$ is the distance traveled by a particle from time 0 to time t.

(a) Find the physical significance of $\dfrac{ds}{dt}$ and $\dfrac{d^2s}{dt^2}$ and express them in terms of $\vec{v}$ and $\vec{a}$.

(b) We know that $\dfrac{d\vec{r}}{dt}$ is the velocity vector. Use the chain rule to show that $\dfrac{d\vec{r}}{ds}$ is a unit tangent vector.

†If the instantaneous radius of curvature is r then $1/r$ is called the instantaneous curvature κ. For example, a line has $r = \infty$ and $\kappa = 0$. At the other extreme, at a tight turn, the approximating circle is small, so r is small and κ is large.

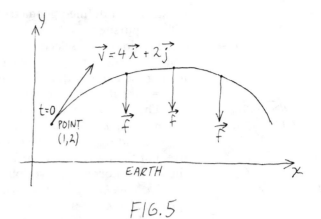

FIG.5

6. Consider the circular motion $x = 5 \cos t$, $y = 5 \sin t$.

(a) Find $\|\vec{v}(t)\|$ to verify that the speed is constant.
(b) Confirm that $a_{\text{tan}} = 0$, as appropriate for a particle whose speed is not changing.
(c) Find a_{rad} and confirm that (2) holds.

7. Newton's first law states that a particle moves on a (frictionless) line with constant speed if and only if no force acts, that is, $\vec{a} = \vec{0}$. Describe $\vec{a}$ if the particle moves (a) on a line but at nonconstant speed   (b) at constant speed but not on a line.

8. Let $x = 4 - t^3$, $y = \frac{1}{4}t^4 + t$ be the position of a particle at time t. Consider time $t = 1$. Where is the particle? What is its instantaneous direction and speed? Is it speeding up or slowing down and by how much? How many pounds of force are acting on it and in what direction?

REVIEW PROBLEMS FOR CHAPTER 9

1. If $\vec{u} = 2\vec{i} + 3\vec{j} + \vec{k}$ and $\vec{v} = -4\vec{i} + 5\vec{j} - 2\vec{k}$ find

(a) $\vec{u} \cdot \vec{v}$
(b) $\|\vec{u}\|$
(c) $\vec{u} \times \vec{v}$
(d) the cosine of the angle determined by $\vec{u}$ and $\vec{v}$
(e) the component of $\vec{u}$ in the direction of $\vec{v}$
(f) the vector component of $\vec{u}$ in the direction of $\vec{v}$
(g) the unit vector in the direction of $\vec{v}$
(h) a vector with length 6 in the direction of $\vec{u}$

2. Let $A = (2, 1, 6)$, $B = (4, -2, 7)$. Is point $C = (6, -5, 9)$ on line AB?

3. (a) If $\vec{u} \times \vec{v} = -\vec{j} + 5\vec{k}$, find $\vec{v} \times \vec{u}$.   (b) If $\|\vec{u} \times \vec{v}\| = 6$ find $\|\vec{v} \times \vec{u}\|$.

4. The associative law for multiplication of numbers states that $(xy)z = x(yz)$ for all x, y, z. Decide if the following are true or false and explain.

(a) $(\vec{u} \cdot \vec{v}) \cdot \vec{w} = \vec{u} \cdot (\vec{v} \cdot \vec{w})$ for all 3-dimensional vectors $\vec{u}, \vec{v}, \vec{w}$.
(b) $(\vec{u} \times \vec{v}) \times \vec{w} = \vec{u} \times (\vec{v} \times \vec{w})$ for all 3-dimensional vectors $\vec{u}, \vec{v}, \vec{w}$.

5. Let $P = (1, 3, 0)$, $Q = (3, 7, z)$, $A = (10, 1, 3)$, $B = (16, -1, 5)$. Find z so that the lines AB and PQ are perpendicular.

6. Show that the area of the parallelogram determined by the vectors $\vec{u} = u_1\vec{i} + u_2\vec{j}$ and $\vec{v} = v_1\vec{i} + v_2\vec{j}$ is the absolute value of the determinant $\begin{vmatrix} u_1 & u_2 \\ v_1 & v_2 \end{vmatrix}$.

7. Show (a) geometrically and (b) algebraically that if $\vec{u} \cdot \vec{v} = 0$ then $\|\vec{u}\|^2 + \|\vec{v}\|^2 = \|\vec{u} + \vec{v}\|^2$.

8. Show that if $\vec{u}$ and $\vec{v}$ are perpendicular unit vectors then $\vec{u} \times \vec{v}$ is a unit vector perpendicular to both $\vec{u}$ and $\vec{v}$.

9. Draw pictures to show why $\|\vec{u} + \vec{v}\| \leq \|\vec{u}\| + \|\vec{v}\|$. Under what conditions does equality hold?

10. (a) Show that $\|u + v\|^2 + \|u - v\|^2 = 2\|u\|^2 + 2\|v\|^2$. (b) Find the geometric significance of part (a) for a parallelogram.

11. Suppose the wind velocity is $\vec{u} = 2\vec{i} + 3\vec{j} - \vec{k}$ and a plane flies in the direction of $\vec{v} = 4\vec{i} - 5\vec{j} + \vec{k}$. Does the plane experience a head wind or a tail wind? By how much is the plane's speed increased or decreased because of the wind?

12. When is it true that $\|\vec{u} \times \vec{v}\| = \|\vec{u}\|\|\vec{v}\|$?

13. Let $\vec{r} = t \cos t\vec{i} + \sin t\vec{j}$. It is not feasible to eliminate the parameter; nevertheless sketch the path for $t \geq 0$ and include $\vec{r}(\pi)$, $\vec{v}(\pi)$ and $\vec{a}(\pi)$ in the picture. Find the speed at time π. Is the particle speeding up or slowing down at time $t = \pi$, and by how much?

14. Describe the motion if, for all t,

(a) $\vec{r} \cdot \vec{r}' = 0$ (c) $\vec{r}' \times \vec{r}'' = \vec{0}$

(b) $\vec{r} \times \vec{r}' = \vec{0}$ (d) $\vec{r}' \cdot \vec{r}'' = 0$

10/TOPICS IN THREE-DIMENSIONAL ANALYTIC GEOMETRY

SURFACE
CURVE

FIG. 1

In 2-space, a curve can be described with an equation in x and y or with a pair of parametric equations giving x and y in terms of the parameter t (Sections 3.8 and 9.6). *For all practical purposes, in 3-space, the graph of an equation in x, y and z is a surface, while a curve is described with parametric equations giving x, y and z in terms of a parameter t* (Fig. 1). This chapter discusses some special surfaces along with the most important type of curve, the line. The preceding chapter on vectors provides the basis for much of the geometry of this chapter, and both will be needed when we continue with calculus in the next chapter.

10.1 Spheres

The circle in 2-space with center (x_0, y_0) and radius r has equation $(x - x_0)^2 + (y - y_0)^2 = r^2$. Similarly, an equation of the sphere with center (x_0, y_0, z_0) and radius r is

(1)
$$(x - x_0)^2 + (y - y_0)^2 + (z - z_0)^2 = r^2.$$

In particular, an equation of the sphere with center at the origin and radius r is

(2)
$$x^2 + y^2 + z^2 = r^2.$$

Example 1 An equation of the sphere with center $(3, -4, 6)$ and radius 7 is

$$(x - 3)^2 + (y + 4)^2 + (z - 6)^2 = 49.$$

Example 2 The graph of $x^2 + y^2 + z^2 = 4$ is a sphere with center at the origin and radius 2. The upper hemisphere in Fig. 1, where $z \geq 0$, has equation $z = \sqrt{4 - x^2 - y^2}$. Similarly, the lower hemisphere where $z \leq 0$ has equation $z = -\sqrt{4 - x^2 - y^2}$. The interior of the sphere is described by $x^2 + y^2 + z^2 < 4$, and the exterior by $x^2 + y^2 + z^2 > 4$.

Problems for Section 10.1

1. Find an equation of the sphere with center $(4, -3, 5)$ and passing through the origin.

FIG. 1

297

2. Complete the square to identify the graph of $x^2 + y^2 + z^2 + z = 0$.

3. Is the origin on, inside, or outside the sphere $(x + 2)^2 + (y - 3)^2 + (z - 2)^2 = 2$?

4. Find an equation of the sphere with center $(3, 5, 6)$ and tangent to the x, y plane.

10.2 Planes

Just as lines are fundamental curves in 2-space, planes are fundamental surfaces in 3-space.

An equation of a plane parallel to a coordinate plane The graph of an equation of the form $x = c$ (where c is a fixed constant) is a plane parallel to the y, z plane (Fig. 1). Similarly, $y = c$ is a plane parallel to the x, z plane, and $z = c$ is a plane parallel to the x, y plane.

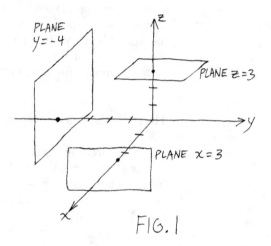

FIG. 1

The point-normal equation of a plane A vector perpendicular to a plane is said to be *normal to the plane* and is called a *normal vector,* or simply a *normal.* (*Warning:* Don't confuse a normal vector (a perpendicular) with the norm of a vector (its length) or with a normalized vector (a unit vector).)

Figure 2 shows the plane through the point (x_0, y_0, z_0) and with normal vector (a, b, c). To find its equation, note that the point (x, y, z) is on the plane

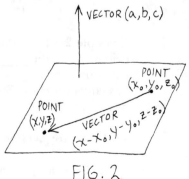

FIG. 2

if and only if the vectors (a, b, c) and $(x - x_0, y - y_0, z - z_0)$ are perpendicular or, equivalently, if and only if $(a, b, c) \cdot (x - x_0, y - y_0, z - z_0) = 0$. Therefore, we have the following conclusion.

(1)

> An equation of the plane that contains the point (x_0, y_0, z_0) and has normal vector (a, b, c) is
> $$a(x - x_0) + b(y - y_0) + c(z - z_0) = 0.$$

Examples will illustrate that no matter how a plane is determined, it is almost always feasible to extract a point and a normal vector from the given data and use (1), the *point-normal form*, to find its equation.

Example 1 Find an equation of the plane determined by the points $A = (1, 2, 3)$, $B = (4, 5, 6)$, and $C = (-2, 0, 3)$.

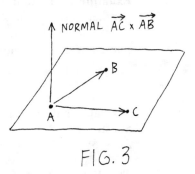

NORMAL $\vec{AC} \times \vec{AB}$

FIG. 3

Solution: The vector $\vec{AC} \times \vec{AB} = (-3, -2, 0) \times (3, 3, 3) = (-6, 9, -3)$ is a normal to the plane (Fig. 3). To simplify the components, divide by 3 to get the normal $(-2, 3, -1)$. Any of the points A, B, or C can be used in the point-normal equation. With point A, (1) becomes

(2) $$-2(x - 1) + 3(y - 2) - (z - 3) = 0.$$

With point B, we have $-2(x - 4) + 3(y - 5) - (z - 6) = 0$; with C the equation is $-2(x + 2) + 3y - (z - 3) = 0$. Each equation simplifies to

(3) $$-2x + 3y - z - 1 = 0.$$

If we use the original normal $(-6, 9, -3)$ along with point A we have $-6(x - 1) + 9(y - 2) - 3(z - 3) = 0$, which is a multiple of (2) and also simplifies to (3).

The general equation of a plane When the point-normal equation in (2) is rewritten as (3), the point $(1, 2, 3)$ is no longer prominently displayed, but the components of the normal $-2\vec{i} + 3\vec{j} - \vec{k}$ still appear in the equation as the coefficients of x, y and z. In general, the graph of the equation $az + by + cz + d = 0$ is a plane with normal vector $a\vec{i} + b\vec{j} + c\vec{k}$. It is called the *general equation* of the plane.

For example, the graph of $2x + 3y + 4z - 24 = 0$ is a plane with normal $2\vec{i} + 3\vec{j} + 4\vec{k}$. Figure 4 shows (a portion of) the plane, obtained by plotting the three convenient points $(0, 0, 6)$, $(0, 8, 0)$ and $(12, 0, 0)$.

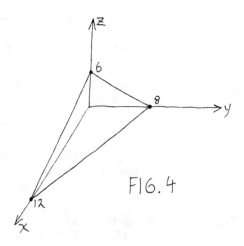

FIG. 4

Example 2   Find an equation of the plane through the point $(1, 4, 5)$ and parallel to the plane $2x + 3y - 4z = 7$.

Solution: The given plane has normal $2\vec{i} + 3\vec{j} - 4\vec{k}$, and the parallel plane has the same normal. Therefore an equation for the parallel plane is

$$2(x - 1) + 3(y - 4) - 4(z - 5) = 0, \quad \text{or} \quad 2x + 3y - 4z = -6.$$

As a second method, note that since parallel planes have a common normal, the desired plane has an equation of the form $2x + 3y - 4z = d$. Substitute $x = 1$, $y = 4$, $z = 5$ to obtain $d = -6$. Therefore the plane is $2x + 3y - 4z = -6$, as before.

The distance from a point to a plane   Consider the plane $ax + by + cz + d = 0$ and the point $P = (x_0, y_0, z_0)$. To find the distance between the point and the plane, pick any point in the plane, say $Q = \left(0, 0, -\dfrac{d}{c}\right)$. Then (Fig. 5), the desired distance is the length of the projection of the vector $\overrightarrow{QP} = \left(x_0, y_0, z_0 + \dfrac{d}{c}\right)$ in the direction of the normal vector $\vec{n} = (a, b, c)$. By (12) of Section 9.3, this is the absolute value of $\dfrac{\overrightarrow{QP} \cdot \vec{n}}{\|\vec{n}\|}$. We have the following result.

> The distance from point (x_0, y_0, z_0) to the plane $ax + by + cz + d = 0$ is
> $$\frac{|ax_0 + by_0 + cz_0 + d|}{\sqrt{a^2 + b^2 + c^2}}.$$

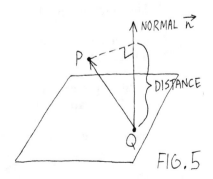

FIG. 5

For example, to find the distance from the point $(5, 2, 9)$ to the plane $z = x + 3y - 4$, rewrite the equation of the plane in the general form $x + 3y - z - 4 = 0$ (equivalently, $-x - 3y + z + 4 = 0$). Then the distance is $\dfrac{|5 + 6 - 9 - 4|}{\sqrt{11}}$, or $\dfrac{2}{\sqrt{11}}$.

Lines in 2-space versus planes in 3-space   The equations and formulas involving *planes* in *three*-space have abridged versions applying to *lines* in *two*-space. Similarly, we may generalize from lines in 2-space to planes in 3-space except when slope is involved, since slope remains a 2-dimensional concept. The following summary lists results from this chapter and Appendix A2, and in (4), (6) and (7′) produces new results by analogy.

Lines in 2-space	Planes in 3-space
(4) The line through the point (x_0, y_0) with normal vector $a\vec{i} + b\vec{j}$ is $$a(x - x_0) + b(y - y_0) = 0.$$	(4′) The plane through the point (x_0, y_0, z_0) with normal vector $a\vec{i} + b\vec{j} + c\vec{k}$ is $$a(x - x_0) + b(y - y_0) + c(z - z_0) = 0.$$
(5) The general equation of a line is $ax + by + c = 0$.	(5′) The general equation of a plane is $ax + by + cz + d = 0$.
(6) The distance between the point (x_0, y_0) and the line $ax + by + c = 0$ is $$\frac{|ax_0 + by_0 + c|}{\sqrt{a^2 + b^2}}.$$	(6′) The distance between the point (x_0, y_0, z_0) and the plane $ax + by + cz + d = 0$ is $$\frac{|ax_0 + by_0 + cz_0 + d|}{\sqrt{a^2 + b^2 + c^2}}.$$
(7) The line with x-intercept a and y-intercept b is $$\frac{x}{a} + \frac{y}{b} = 1.$$	(7′) The plane with x-intercept a, y-intercept b and z-intercept c is $$\frac{x}{a} + \frac{y}{b} + \frac{z}{c} = 1.$$

Problems for Section 10.2

1. Find an equation of the plane

(a) containing point $(5, 5, 4)$ and perpendicular to the line AB, where $A = (4, 5, 6)$ and $B = (9, 8, 7)$
(b) determined by points $A = (2, 0, 0)$, $B = (0, 5, 0)$, $C = (0, 0, 7)$
(c) containing point $(3, 4, 5)$ and perpendicular to the x-axis
(d) containing point $(3, 4, 5)$ and parallel to the x, y plane
(e) containing point $(3, \pi, 7)$ and parallel to the plane $2x + 9y - 6z + 4 = 0$

2. Let $A = (1, 3, -2)$, $B = (2, -1, 0)$, $C = (4, 4, 3)$ and $D = (1, 2, 3)$. Use an equation of plane ABC to decide if the four points are coplanar.

3. Find the distance from the point $(2, 3, -4)$ to the plane $3x - 4y + 2z = 6$.

4. Find the distance from the origin to the line $y = 3x + 4$ in 2-space.

5. Find an equation of the sphere with center $(1, 3, -1)$ and tangent to the plane $2x + y - z = 4$.

6. Show that the planes $2x - y + 3z = 6$ and $2x - y + 3z = 8$ are parallel, and then find the distance between them.

7. If a plane has intercepts a, b, c, and its distance to the origin is D, show that
$$\frac{1}{D^2} = \frac{1}{a^2} + \frac{1}{b^2} + \frac{1}{c^2}.$$

8. Let $A = (-1, 2, 4)$, $B = (0, 3, 3)$, $C = (1, -8, 2)$ and $D = (4, 5, 5)$. Find an equation of the plane containing line AB and parallel to line CD (Fig. 6).

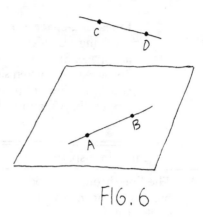

FIG. 6

10.3 Lines

A line in *two*-space is determined by a point and a parallel direction; the direction is usually conveyed by a slope m. Similarly, a line in *three*-space is determined by a point (x_0, y_0, z_0) and a parallel direction, but the direction is conveyed by a parallel vector (a, b, c) rather than a slope, since a line in 3-space does not have a slope. To find an equation for the line (Fig. 1) note that the point (x, y, z) is on the line if and only if the vector $(x - x_0, y - y_0, z - z_0)$ is parallel to the vector (a, b, c), that is, if and only if there is some scalar t such that $(x - x_0, y - y_0, z - z_0) = t(a, b, c)$. Equate respective components to obtain $x - x_0 = ta$, $y - y_0 = tb$, $z - z_0 = tc$. Therefore, we have the following result.

(1)

> The line determined by the point (x_0, y_0, z_0) and parallel to the vector $a\vec{i} + b\vec{j} + c\vec{k}$ has parametric equations
>
> $$x = x_0 + at$$
> $$y = y_0 + bt$$
> $$z = z_0 + ct.$$
>
> Similarly the line in 2-space through point (x_0, y_0) and parallel to $a\vec{i} + b\vec{j}$ has equations
>
> $$x = x_0 + at$$
> $$y = y_0 + bt.$$

For example, the equations $x = 2 + 4t$, $y = 5 - 3t$, $z = 6 + 7t$ describe a line containing the point $(2, 5, 6)$ and parallel to the vector $4\vec{i} - 3\vec{j} - 7\vec{k}$. In particular, the point $(2, 5, 6)$ corresponds to the parameter value $t = 0$. Every value of t produces a point on the line; if $t = 2$ then the corresponding point is $(10, -1, 20)$. To decide if a point is on the line, search for a value

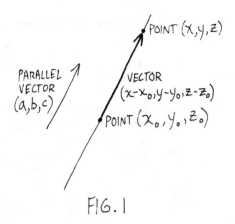

FIG. 1

of t that produces it. The point $(3, 4, -6)$ is *not* on the line since $x = 3$ requires that t be $\frac{1}{4}$ (from the first of the parametric equations), but $t = \frac{1}{4}$ does not produce $y = 4$ and $z = -6$.

Example 1 Consider the line determined by the points $A = (3, 5, 6)$ and $B = (1, 6, 8)$. The line has parallel vector $\vec{AB} = -2\vec{i} + \vec{j} + 2\vec{k}$ and passes through point A, so it has parametric equations

(2) $\qquad x = 3 - 2t, \qquad y = 5 + t, \qquad z = 6 + 2t.$

With the same parallel vector, but using point B this time, we have

(3) $\qquad x = 1 - 2t, \qquad y = 6 + t, \qquad z = 8 + 2t,$

another parametrization of the same line. The *same* points are produced by the two sets of parametric equations, but *for different values of t*. Point A corresponds to $t = 0$ in (2), and to $t = -1$ in (3). We can also use the new point $C = (-1, 7, 10)$ obtained by setting $t = 1$ in (3) (or $t = 2$ in (2)) along with the parallel vector $4\vec{i} - 2\vec{j} - 4\vec{k}$ to get the third parametrization

$$x = -1 + 4t, \qquad y = 7 - 2t, \qquad z = 10 - 4t.$$

There are infinitely many ways to parametrize the line. (From the point of view of Section 9.6 where (x, y, z) is the position of a particle at time t, the different parametrizations represent motions with different timing and velocities.)

Example 2 Find equations for the line containing the point $A = (4, -3, 6)$ and perpendicular to the plane $3x - 7y + 2z + 9 = 0$.

  Solution: The plane's normal, $3\vec{i} - 7\vec{j} + 2\vec{k}$, is parallel to the line (Fig. 2). Therefore a parametrization of the line is $x = 4 + 3t$, $y = -3 - 7t$, $z = 6 + 2t$.

The intersection of two lines in 3-space In 2-space, two lines are either parallel and different, parallel and actually the same line (that is, coincident) or intersecting. In 3-space, the lines may also be skew. A good way to decide among the four possibilities is to test for parallelism first. If they are parallel, decide if the lines are the same or different. If they are not parallel, decide if they are intersecting or skew. We will illustrate the procedure and show how to find the point of intersection, if it exists.

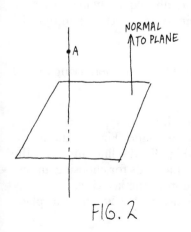

FIG. 2

Consider the lines

(4)
$$x = 2 + 3t \qquad x = 9 - 2t$$
$$y = 4 - t \quad \text{and} \quad y = -1 + 2t$$
$$z = 8 + 3t \qquad z = 5 + 3t.$$

The lines have respective parallel vectors $\vec{u} = (3, -1, 3)$ and $\vec{v} = (-2, 2, 3)$. Since $\vec{u}$ is not a multiple of $\vec{v}$, the lines are not parallel. Now we decide between skew and intersecting. To find a common point, it is tempting but *incorrect* to equate the expressions for x, y and z in (4) to obtain

$$2 + 3t = 9 - 2t$$
$$4 - t = -1 + 2t$$
$$8 + 3t = 5 + 3t. \text{ INCORRECT}$$

This is *not* correct because it unnecessarily demands that a common point be produced by the *same* value of t, whereas it is possible for a common point to appear with the label $t = 6$ on one line, and $t = 117$ on the other line. To allow for this possibility we switch to the letter s as the parameter for one of the lines, and rewrite (4) as

(5)
$$x = 2 + 3t \qquad x = 9 - 2s$$
$$y = 4 - t \quad \text{and} \quad y = -1 + 2s$$
$$z = 8 + 3t \qquad z = 5 + 3s.$$

Then equate expressions for x, y and z to get

(6)
$$2 + 3t = 9 - 2s$$
$$4 - t = -1 + 2s$$
$$8 + 3t = 5 + 3s.$$

To solve this system of three equations in two variables, write the first two equations in the standard form

$$2s + 3t = 7$$
$$-2s - t = -5$$

and solve to get $s = 2$, $t = 1$. These values satisfy the third equation in (6) so there *is* a point of intersection. Substitute $t = 1$ in (5), and as a check set $s = 2$, to obtain $x = 5$, $y = 3$, $z = 11$, the point of intersection. If it had turned out that $s = 2$, $t = 1$ did *not* satisfy the third equation in (6) then we would have concluded that there was no point of intersection and that the lines were skew.

Warning Two parameter letters, s and t, must be used when trying to find a point of intersection of two lines.

The intersection of a line and a plane For a line and a plane in 3-space, there are three possibilities: the line lies in the plane (Fig. 3a) the line is parallel to but not contained in the plane (Fig. 3b) the line intersects the plane at one point (Fig. 3c). We will illustrate methods for choosing among them, and for finding the point of intersection in the last case.

Consider the line $x = 1 + 2t$, $y = 3 - t$, $z = 2 + 2t$ and the plane $2x + 6y + 3z = 6$. Substitute to obtain

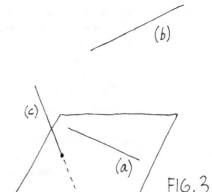

FIG. 3

$$2(1 + 2t) + 6(3 - t) + 3(2 + 2t) = 6 \tag{7}$$

and solve to get $t = -5$. Since there is a unique solution for t, the line and plane intersect in one point. Substitute $t = -5$ in the equation of the line to find the point of intersection $(-9, 8, -8)$. If it had turned out that the t's canceled in (7) leaving $0 = 0$, then every value of t would be a solution; in that case the line would be contained in the plane. If the t's had canceled out leaving say $0 = 27$, then there would be no solution for t; in that case the line would be parallel to but not contained in the plane.

The intersection of two planes Two planes are either the same, parallel but not coincident, or intersect in a line (Fig. 4). We will illustrate how to decide which situation holds and give a method for finding the line of intersection in the third case.

The planes $2x + 3y + 4z + 5 = 0$ and $4x + 6y + 8z + 10 = 0$ are the same because the second equation is simply a multiple of the first.

The planes $2x + 3y + 4z + 5 = 0$ and $4x + 6y + 8z + 11 = 0$ are parallel (since their normals, $2\vec{i} + 3\vec{j} + 4\vec{k}$ and $4\vec{i} + 6\vec{j} + 8\vec{k}$, are multiples of one another) but not coincident (since the equations are not multiples of one another).

The planes

$$2x + y + z = 2$$
$$x + y + 3z = -6 \tag{8}$$

are neither identical nor parallel since their normals, $\vec{u} = 2\vec{i} + \vec{j} + \vec{k}$ and $\vec{v} = \vec{i} + \vec{j} + 3\vec{k}$, are not parallel. Therefore the planes intersect in a line. In fact, the pair of equations in (8) can be regarded as a set of equations for the line. We want to find a standard set of parametric equations for the line, using (1). For a vector parallel to the line (Fig. 4), use $\vec{u} \times \vec{v} = 2\vec{i} - 5\vec{j} + \vec{k}$. To find a point on the line, set one of the variables in (8) equal to any specific value and then solve to find the corresponding values of the other two variables. With the convenient choice of $z = 0$, we have

$$2x + y = 2$$
$$x + y = -6$$

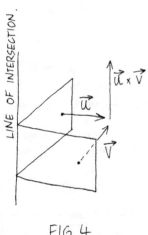

LINE OF INTERSECTION.

FIG. 4

whose solution is $x = 8$, $y = -14$. Therefore $(8, -14, 0)$ lies on *both* planes and hence is on the line of intersection. Therefore the line has parametric equations $x = 8 + 2t$, $y = -14 - 5t$, $z = t$.

When we chose to set $z = 0$ in (8), we found the particular point where the line of intersection crosses the x, y plane. If the line happens to be parallel to the x, y plane, then setting $z = 0$ in (8) will result in a system of two equations in x and y with no solution. In this case set $x = 0$ or $y = 0$.

Problems for Section 10.3

1. Find equations for the line

(a) containing the point $(1, 2, 3)$ and parallel to the line $x = 3 + t$, $y = 4 - 2t$, $z = 1 + 5t$
(b) containing the point $(1, 4, 5)$ and perpendicular to the plane $3x - 4y + 6z = 1$
(c) through the point $(2, 3, 4)$ and parallel to the x-axis
(d) through the point $(2, 3, 4)$ and perpendicular to the x, z plane
(e) through the origin and the point $(7, -1, 16)$
(f) through the point $(1, 5, 7)$ and parallel to the planes $2x - y + z = 0$ and $3x + y + 4z = 2$

2. Let $A = (-1, 3)$ and $B = (13, -2)$. Find parametric equations for line AB.
3. Find equations for (a) the x, z plane (b) the z-axis.
4. Find an equation of the plane containing the point $A = (3, -1, 2)$ and the line $x = 6 + 4t$, $y = 2 - t$, $z = 7 + 8t$.
5. Consider the line $x = 2 + t$, $y = 3 - 4t$, $z = 6 + 5t$ and the line $x = 6 + 2t$, $y = -6 - t$, $z = 10 - 6t$.

(a) Verify that the lines intersect at the point $(4, -5, 16)$ but are not the same line.
(b) Find the equation of the plane determined by the intersecting lines.
(c) Find the equations of the line perpendicular to both given lines through their point of intersection.

6. Where does the line $x = 2 - t$, $y = 3 + 4t$, $z = -5 + 2t$ intersect the y, z plane?
7. Confirm that the lines $x = 2 - 3t$, $y = 5 + t$, $z = 4 + 2t$ and $x = -7 - 3t$, $y = 6 + t$, $z = 2t$ are parallel but different, and find the equation of the plane they determine.
8. Let $A = (1, 3, -2)$, $B = (4, 5, 0)$, $C = (3, 3, 5)$. Find equations of line AB and use them to decide if the points A, B and C are collinear.
9. Consider the line $x = x_0 + at$, $y = y_0 + bt$, $z = z_0 + ct$ and the plane $Ax + By + Cz + D = 0$ where $a, b, c, A, B, C, D, x_0, y_0, z_0$ are fixed constants. Identify the geometric connection between the line and the plane if both (∗) $aA + bB + cC = 0$ and (∗∗) $Ax_0 + By_0 + Cz_0 + D = 0$ hold.
10. Are the lines coincident, parallel, skew or intersecting?

(a) $x = 1 - 6t$, $y = 2 + t$, $z = 4 + 3t$;
   $x = -5 + 12t$, $y = 3 - 2t$, $z = 13 - 6t$
(b) $x = 2 - t$, $y = 3 + 2t$, $z = 5 - 3t$;
   $x = t$, $y = 5 - 4t$, $z = -1 + 6t$
(c) $x = 2 - t$, $y = 3 + t$, $z = 5 + 2t$;
   $x = 3 - t$, $y = 4 + 2t$, $z = 1 + t$
(d) $x = 2 + 3t$, $y = 5 - t$, $z = 3 + t$;
   $x = -4 - 3t$, $y = 7 + t$, $z = 1 - t$

11. Find the intersection of the line $x = 1 + 2t, y = 3 - t, z = 2 + 2t$ and the plane (a) $2x + 6y + z = 8$ (b) $2x + 6y + z = 22$.

12. Find the line of intersection of the planes $2x + y + 3z = 5$ and $x - y + z = 4$.

13. Show that the line of intersection of the planes $x + 2y + 8z = 20$ and $x - y + 2z = 8$ is parallel to but not contained in the plane $3x - 2y + 8z = 5$.

14. Find the projection P of the point $(-1, 1, 1)$ on the plane $2x - 3y + z + 1 = 0$, that is, the foot of the perpendicular from the point to the plane. (First find the line through the point and perpendicular to the plane.)

15. Consider the line L with equations $x = 1 + t, y = 2 - t, z = 3 + 4t$ and the point $Q = (4, -1, 4)$. Find the projection P of Q on L (the foot of the perpendicular from Q to L) and the distance from Q to L. (First find the equation of the plane through Q and perpendicular to L.)

16. Show that the lines $x = 2 - t$, $y = 3 + t$, $z = 5 + 2t$ and $x = 3 - t$, $y = 4 + 2t$, $z = 1 + 7t$ are skew and find the distance between them. (First find the plane containing the second line and parallel to the first.)

17. Let $A = (3, 0, 2)$, $B = (2, 4, 0)$, $C = (4, 5, 6)$, $D = (7, -7, 12)$.

(a) Find the equations of lines AB and CD and show that the lines are parallel but not the same line.

(b) Find the distance between the lines. (First find the plane through A and perpendicular to both lines.)

10.4 Cylindrical and Quadric Surfaces

The graph in 3-space of an equation in x, y and z is (with rare, contrived exceptions) a surface. Often the graph is too difficult to draw, but in this section we single out a few special types of equations which occur frequently and whose graphs are easy to sketch.

Cylindrical surfaces   The graph in 3-space of an equation containing only one or two of the three variables x, y, z is called a *cylindrical surface*. As an illustration consider the equation $y = x^2$ whose graph in an x, y plane, that is, in *two*-space, is a parabola. We wish to draw its graph in *three*-space. If we extend 2-space to 3-space by adding a z-axis (Fig. 1), a point such as $(2, 4)$ on the parabola is now $(2, 4, 0)$ and still satisfies the equation $y = x^2$. In fact, the z-coordinate of a point is irrelevant in deciding if the point satisfies the equation. Thus if we move any point on the parabola up or down so that its x and y coordinates remain unchanged, we obtain more points which satisfy $y = x^2$. Figure 1 shows a portion of the final graph, a parabolic cylinder, obtained by sliding the original parabola up and down.

To draw the surface $z = y^2$ in 3-space first sketch the curve $z = y^2$ in the y, z plane. Then slide the curve in the x direction (draw several replicas of it) to create the cylindrical surface in Fig. 2.

The graph of $x^2 + z^2 = 5$ in the x, z plane is a circle. Slide the circle in the y direction to obtain the graph of the equation in *three*-space, a circular cylinder (Fig. 3).

Quadric surfaces   The graph in *two*-space of a second-degree equation in x and y is an ellipse, parabola or hyperbola (or a degenerate form thereof). In *three*-space, the graph of a second degree equation in x, y and z is one of

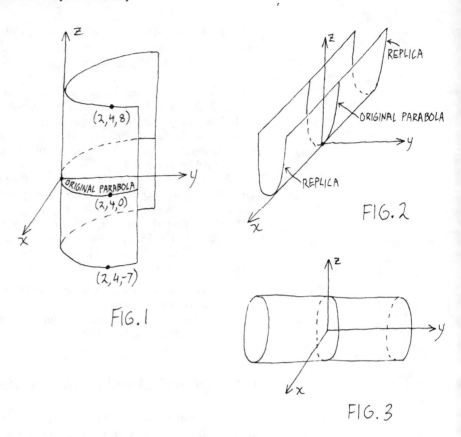

FIG. 1

FIG. 2

FIG. 3

six *quadric surfaces* (barring degeneracies). We will illustrate the six with examples and give a method for matching each type of quadratic equation with the appropriate surface. Table 1 summarizes the results.

Consider the surface

(1) $$6x^2 + y^2 - 2z^2 = 4.$$

To sketch a graph in two-space, we usually plot some points and connect them in a reasonable fashion to form a curve. But the graph of (1) is a surface, and plotting a few points is not enough to suggest the shape of the surface. Instead we will sketch cross sections. The cross section in the x, y plane where $z = 0$ is the ellipse $6x^2 + y^2 = 4$ (Fig. 4). To find the cross section at height 2, set $z = 2$ in (1) to obtain the larger, but similar, ellipse $6x^2 + y^2 = 12$. The cross section at $z = -2$ is the same as at $z = 2$, while those at $z = \pm 3$ are the still larger ellipses $6x^2 + y^2 = 22$. Figure 4 shows the cross sections, piled up like poker chips. The overall shape of the surface is emerging from the stack of elliptical slices, but we need "sides" to complete the picture. For this purpose, set $x = 0$ in (1) to obtain the cross section in the y, z plane, the hyperbola $y^2 - 2z^2 = 4$. Similarly, the cross section in the x, z plane is the hyperbola $6x^2 - 2z^2 = 4$. Figure 5 shows a final graph of (1), called an *elliptic hyperboloid of one sheet*.

Consider the surface

(2) $$6x^2 - y^2 - 2z^2 = 4.$$

To find its cross section in the y, z plane, set $x = 0$ to obtain $y^2 + 2z^2 = -4$, a null ellipse. In other words, the surface does not intersect the y, z plane.

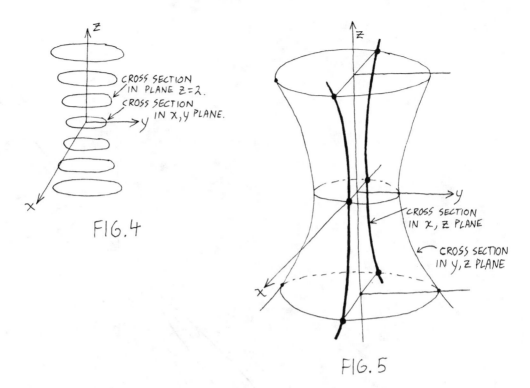

FIG.4

FIG. 5

The cross sections at $x = \pm\sqrt{\frac{2}{3}}$ are the point ellipse $y^2 + 2z^2 = 0$. At $x = \pm 1$ the cross sections are the ellipse $y^2 + 2z^2 = 2$, and similarly at $x = \pm 2$ the cross sections are the larger ellipse $y^2 + 2z^2 = 20$. Figure 6 shows the elliptical cross sections. To determine the sides, set $z = 0$ in (2) to get the cross section in the x, y plane, the hyperbola $6x^2 - y^2 = 4$; and set $y = 0$ for the cross section in the x, z plane, the hyperbola $6x^2 - 2z^2 = 4$. The surface (Fig. 6) is an *elliptic hyperboloid of two sheets.*

Note that to graph (1) we began by setting z constant and found cross sections parallel to the x, y plane. However for (2) we first set x constant and found cross sections parallel to the y, z plane. Why the switch? The purpose of the initial cross sectioning is to produce a stack of curves which suggests the shape of the surface. For both (1) and (2), two sets of cross sections are hyperbolas and one set are ellipses. A stack of ellipses is easier to draw and more indicative of an overall shape than a stack of hyperbolas (or parabolas). We choose to begin each example by looking for *elliptical* cross sections if they exist. The variable that is set constant to accomplish this depends on the particular equation. Look for two square terms with the same sign on one side of the equation and substitute constants for the remaining variable.

Consider the equation $6x^2 + y^2 - 2z^2 = 0$, or

$$(3) \qquad z^2 = 3x^2 + \tfrac{1}{2}y^2.$$

Setting $z = z_0$ leaves x^2 and y^2 terms with the same sign, hence produces elliptical slices. If $z = 0$, we have the point ellipse $3x^2 + \tfrac{1}{2}y^2 = 0$. If $z = \pm 1$, the cross sections are the ellipse $3x^2 + \tfrac{1}{2}y^2 = 1$. If $z = \pm 3$, the cross sections are the larger ellipse $3x^2 + \tfrac{1}{2}y^2 = 9$. Figure 7 shows the elliptical cross sections. To complete the picture we need another cross section. Setting $y = 0$

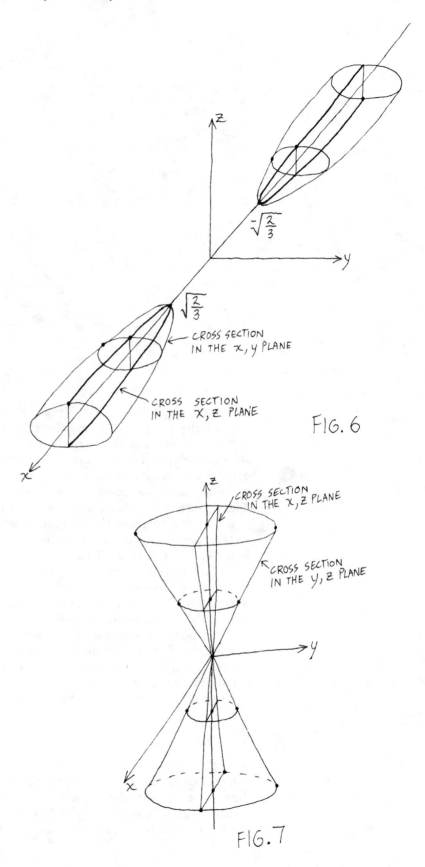

$-\sqrt{\dfrac{2}{3}}$

$\sqrt{\dfrac{2}{3}}$

CROSS SECTION
IN THE x, y PLANE

CROSS SECTION
IN THE x, z PLANE

FIG. 6

CROSS SECTION
IN THE x, z PLANE

CROSS SECTION
IN THE y, z PLANE

FIG. 7

ELLIPTIC HYPERBOLOID OF ONE SHEET $\dfrac{x^2}{a^2} + \dfrac{y^2}{b^2} - \dfrac{z^2}{c^2} = 1$

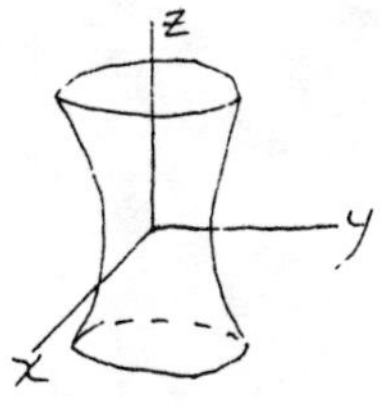

FIG. 8

ELLIPTIC HYPERBOLOID OF TWO SHEETS $-\dfrac{x^2}{a^2} - \dfrac{y^2}{b^2} + \dfrac{z^2}{c^2} = 1$

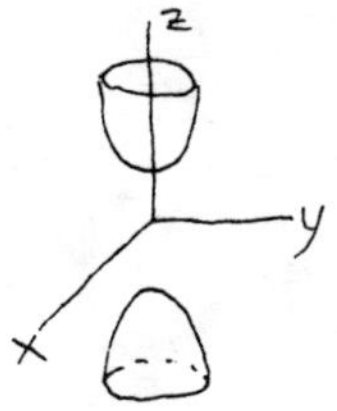

FIG. 9

ELLIPTIC CONE $z^2 = \dfrac{x^2}{a^2} + \dfrac{y^2}{b^2}$

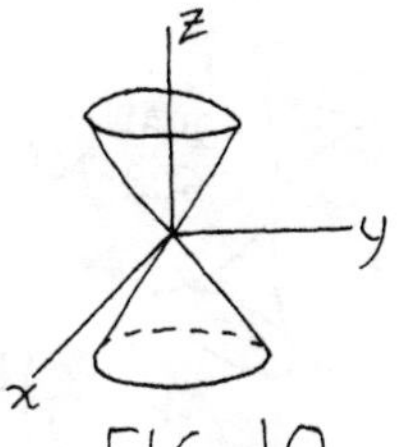

FIG. 10

ELLIPSOID $\dfrac{x^2}{a^2} + \dfrac{y^2}{b^2} + \dfrac{z^2}{c^2} = 1$

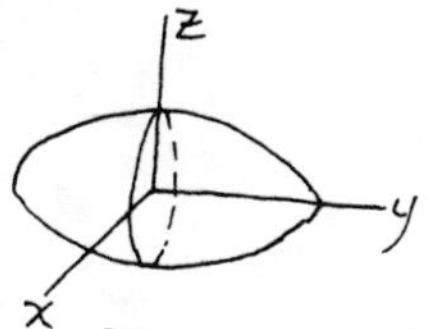

FIG. 11

ELLIPTIC PARABOLOID $z = \dfrac{x^2}{a^2} + \dfrac{y^2}{b^2}$

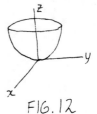

FIG. 12

HYPERBOLIC PARABOLOID $z = \dfrac{y^2}{b^2} - \dfrac{x^2}{a^2}$

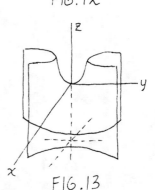

FIG. 13

in (3) gives $z^2 = 3x^2$, or $z = \pm\sqrt{3}\,x$, a pair of intersecting lines in the x, z plane. Similarly, if $x = 0$, the cross section is the pair of lines $z = \pm\sqrt{\tfrac{1}{2}}\,y$ in the y, z plane. The surface in Fig. 7 is an *elliptic cone*.

Table 1 displays the six quadric surfaces and corresponding equations. The table is incomplete in the sense that the surfaces may appear in different alignments. The graph of $-\dfrac{x^2}{a^2} + \dfrac{y^2}{b^2} + \dfrac{z^2}{c^2} = 1$ is the elliptic hyperboloid of one sheet as in Fig. 8 but aligned along the x-axis rather than the z-axis. To identify the graph of a quadratic equation keep the six possibilities in mind and sketch a few cross sections (preferably ellipses) to single out the particular surface and its alignment) rather than memorize the table.

Warning   1. The cylindrical and quadric surfaces are *surfaces,* not solids. For example, the graph of $6x^2 + y^2 + 2z^2 = 4$ is an ellipsoidal *surface.* The points *inside* the ellipsoid do *not* satisfy the equation (they satisfy the inequality $6x^2 + y^2 + 2z^2 < 4$). Furthermore, the graphs do not include circular "lids." For example, the cylinder in Fig. 3 is an unbounded tube with no "ends"; only a portion of it is shown in the diagram.

2. The graph of an equation such as $5x^2 - y^2 = 0$ in the x, y plane is *not* a hyperbola. It is the pair of intersecting lines $y = \pm\sqrt{5}\,x$.

3. The graph of a second degree equation never looks like Fig. 14; i.e., it never is a surface with curved "sides" *and* coming to a point in the "middle."

NOT A QUADRIC SURFACE

FIG. 14

Problems for Section 10.4

In Problems 1–16, sketch the graph in 3-space.

1. $y^2 + z^2 = 4$
2. $z = \sin x$
3. $3x + 7y = 21$
4. $x^2 + 2 = z$
5. $y^2 + z^2 \le 6$
6. $4x^2 - 25y^2 + z^2 = 100$
7. $4x^2 - 25y^2 + z^2 = 0$
8. $y^2 + 4z^2 = 2x$

9. $9x^2 - y^2 - z^2 = -4$
10. $x^2 + y^2 + 2z^2 = 1$
11. $x^2 = y^2 + 4z^2$
12. $-9x^2 + 4y^2 - z^2 = 5$
13. $z = 2x^2 + y^2 + 7$
14. $z = -\sqrt{x^2 + y^2}$
15. $xy = 1$
16. $z = 2x^2 - y^2$

17. Choose a and b so that $x^2/a^2 + y^2/b^2 = z^2$ is a circular cone with radius 3 at height 6.

10.5   Cylindrical and Spherical Coordinates

The cylindrical coordinate system   Polar coordinates in the x, y plane (Appendix A6) together with the rectangular coordinate z, are called *cylindrical coordinates* (Fig. 1). The angle θ, considered as a *polar coordinate* in the x, y plane, is a counterclockwise rotation from the positive x-axis. The *cylindrical* coordinate θ measures rotation from the x, z plane. In each case, θ can be positive, negative or zero. The *polar* coordinate r in the x, y plane is the distance to the origin. The *cylindrical coordinate r is the distance to the z-axis.* In each case, we take $r \ge 0$. Thus, in polar coordinates, the graph of $r = 0$ is the origin, while in cylindrical coordinates, $r = 0$ describes the z-axis.

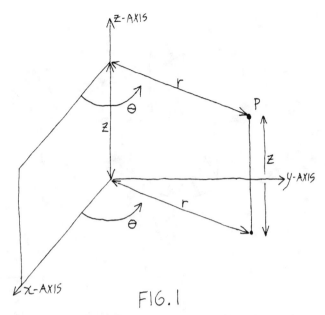

FIG. 1

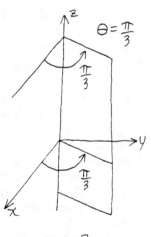

FIG. 2

Cylindrical coordinates are often used when the distance to a fixed line in 3-space is a determining factor. If an electric field is created by a line of charge, then the strength of the field at a point depends on the distance from the point to the line. In this situation, the natural system in which to solve problems is a cylindrical coordinate system with the line of charge on the z-axis, so that the all-important distance to the line is the coordinate r.

If θ_0 is a constant then the graph of $\theta = \theta_0$ in *polar* coordinates is a ray; its graph in *cylindrical* coordinates is a half-plane (Fig. 2). The graph of $z = z_0$ is a plane in both rectangular and cylindrical coordinates (Fig. 3). The graph of $r = r_0$ in *polar* coordinates is a circle; its graph in cylindrical coordinates is a cylinder (Fig. 4). The equation of a cylinder is much simpler in cylindrical than in rectangular coordinates ($r = r_0$ versus $x^2 + y^2 = r_0^2$); in Chapter 12 we take advantage of this by switching to cylindrical coordinates when it becomes necessary to compute triple integrals over cylindrical regions.

The cylindrical coordinates r, θ and the rectangular coordinates x, y continue to be related as they were in the polar coordinate system. Thus

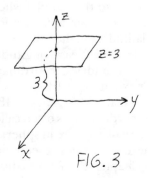

FIG. 3

$$(1) \qquad \boxed{x = r \cos \theta, \qquad y = r \sin \theta, \qquad z = z,}$$

where the last equation simply expresses the fact that the cylindrical coordinate z is identical to the rectangular coordinate z. Furthermore, as with polar coordinates,

$$(2) \qquad r = \sqrt{x^2 + y^2}, \qquad r^2 = x^2 + y^2, \qquad \tan \theta = \frac{y}{x}.$$

The spherical coordinate system   A point P in 3-space has *spherical coordinates* ρ, ϕ, θ determined as follows. (Figure 5 includes rectangular, cylindrical and spherical for comparison.) The coordinate ρ is the distance from P to the origin. Thus $\rho \geq 0$, and $\rho = 0$ only at the origin. If the origin is

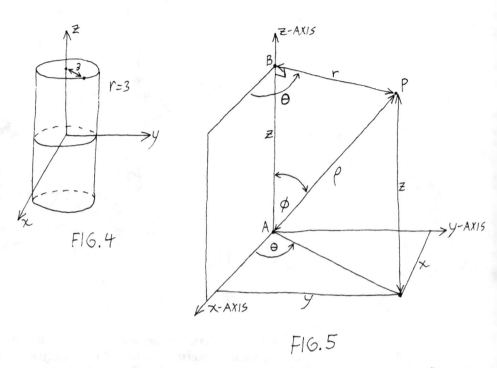

FIG. 4

FIG. 5

denoted by A then the coordinate ϕ is the angle determined by $\overrightarrow{AP}$ and the positive z-axis. Thus $0 \leq \phi \leq 180°$. The spherical coordinate θ is the same as the cylindrical coordinate θ; it measures rotation from the x, z plane.

Spherical coordinates are often used when the distance to a fixed point in 3-space is of particular importance. If an electric field is created by a point charge, then the strength of the field depends on the distance to the point charge. In this situation, the natural system in which to solve problems is a spherical coordinate system with the point charge at the origin, so that the distance to the charge is the spherical coordinate ρ.

Figure 6 may be used to picture spherical coordinates geographically. If $\rho = \rho_0$ then the point lies on a sphere centered at the origin, an "earth." On the earth, ϕ measures "down" from the north pole N. The great circle $NASBN$ illustrates various ϕ coordinates. At N, $\phi = 0°$; at A, $\phi = 90°$; at S, $\phi = 180°$; at B, $\phi = 90°$. A parallel of latitude (including the equator) is a circle on which ϕ is fixed. On the earth, θ measures "around" from the prime meridian in the x, z plane. Each meridian of longitude is a great semicircle on which θ is fixed. (Cartographers refer to the parallel where $\phi = 30°$ as 60° north latitude, to the equator where $\phi = 90°$ as 0° latitude, and to the parallel where $\phi = 160°$ as 70° south latitude. Further, they refer to $\theta = -20°$ as 20° west longitude and to $\theta = 25°$ as 25° east longitude.)

The graph of $\theta = \theta_0$ is a half-plane, as in cylindrical coordinates (Fig. 2). The graph of $\phi = \phi_0$ is the positive z-axis if ϕ_0 is 0°; a cone if $0 < \phi_0 < 90°$ (Fig. 7); the x, y plane if ϕ_0 is 90°; a cone if $90° < \phi_0 < 180°$ (Fig. 8); and the negative z-axis if ϕ_0 is 180°. The coordinate ϕ is often called the *cone angle*. As we have already seen, the equation of a sphere in spherical coordinates is $\rho = \rho_0$, much simpler than the corresponding equation $x^2 + y^2 + z^2 = \rho_0^2$ in rectangular coordinates; in Chapter 12 we take advantage of this by switching to spherical coordinates to compute triple integrals over spherical regions.

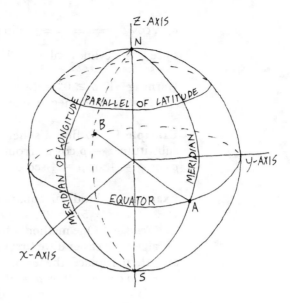

FIG. 6 $(\rho = \rho_0)$

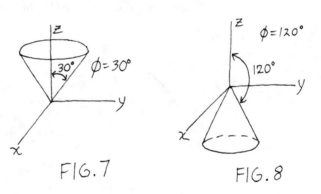

FIG. 7   FIG. 8

We will demonstrate the following connections between rectangular and spherical coordinates

(3)
$$x = \rho \sin \phi \cos \theta, \qquad y = \rho \sin \phi \sin \theta, \qquad z = \rho \cos \phi,$$
$$\rho = \sqrt{x^2 + y^2 + z^2}$$

and

$$\phi = \cos^{-1} \frac{z}{\sqrt{x^2 + y^2 + z^2}}, \qquad \tan \theta = \frac{y}{x}.$$

From the right triangle ABP in Fig. 5, $z = \rho \cos \phi$, verifying the third equation in (3). The same triangle shows that $r = \rho \sin \phi$; substitute for r in (1) to get the first and second equations in (3). The fourth formula expresses the fact that ρ is the distance from (x, y, z) to the origin. Substituting for ρ in $z = \rho \cos \phi$ gives $\cos \phi = \dfrac{z}{\sqrt{x^2 + y^2 + z^2}}$; since $0 \le \phi \le 180°$,

$\phi = \cos^{-1}\dfrac{z}{\sqrt{x^2 + y^2 + z^2}}$. Finally, $\tan\theta = \dfrac{y}{x}$ as in cylindrical coordinates since θ is a cylindrical as well as a spherical coordinate.

Warning 1. $\rho \geq 0$, and we always take $r \geq 0$
  2. ϕ is never negative and is never larger than $180°$.

Example 1 To find a spherical coordinate equation of the plane $z = 4$, substitute $z = \rho \cos\phi$ from (3) in the equation of the plane to get $\rho \cos\phi = 4$, or $\rho = 4\sec\phi$.

Example 2 Find an equation in spherical coordinates of the sphere $x^2 + y^2 + (z - 5)^2 = 25$.
  Solution: One method is to substitute for x, y and z using (3). However, the algebra is easier if we first write the equation as $x^2 + y^2 + z^2 - 10z = 0$, and then replace the entire expression $x^2 + y^2 + z^2$ by ρ^2 to obtain $\rho^2 - 10\rho\cos\phi = 0$ or $\rho = 10\cos\phi$.
  To see this geometrically note that the sphere has radius 5 and center $(0, 0, 5)$ (Fig. 9). A point P is on the sphere if and only if angle APB is inscribed in a hemisphere and thus is $90°$. Therefore P is on the sphere if and only if $\cos\phi = \rho/10$, or $\rho = 10\cos\phi$.

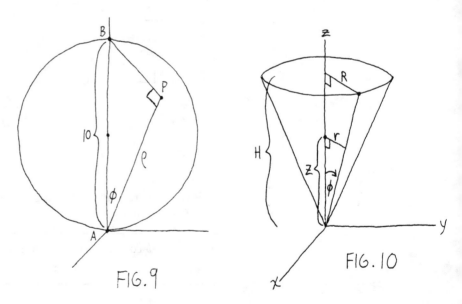

FIG. 9

FIG. 10

Equation of a cone Consider the (half) cone in Fig. 10 with radius R and height H. Let P be a typical point on the cone, with rectangular coordinates x, y, z; cylindrical coordinates r, θ, z; and spherical coordinates ρ, ϕ, θ. By similar triangles, $r/R = z/H$ so the cone has equation

$$(4) \qquad z = \frac{H}{R}r$$

in *cylindrical coordinates*. Since $r = \sqrt{x^2 + y^2}$, the cone's equation in *rectangular coordinates* is

$$(5) \qquad\qquad z = \frac{H}{R}\sqrt{x^2 + y^2}$$

(the double cone is $z^2 = \dfrac{H^2}{R^2}(x^2 + y^2)$). In *spherical coordinates,* the cone is

$$(6) \qquad\qquad \phi = \tan^{-1}\frac{R}{H}.$$

Problems for Section 10.5

1. Find by inspection (a) cylindrical coordinates if $x = 0$, $y = 2$, $z = 3$ (b) spherical coordinates if $x = 0$, $y = 5$, $z = 0$.

2. Find cylindrical coordinates given rectangular coordinates (a) $(3, 2, 5)$ (b) $(-3, -2, 5)$.

3. Find rectangular coordinates given (a) cylindrical coordinates $r = 2$, $\theta = 150°$, $z = 7$ (b) spherical coordinates $\rho = 2$, $\phi = 30°$, $\theta = 120°$.

4. Sketch in cylindrical coordinates.

(a) $1 \le r \le 2$ (b) $r = 3$, $z = 2$ (c) $\theta = \pi/3$, $z = 7$ (d) $\theta = \pi/3$, $r = 3$

5. Find the equation in cylindrical and in spherical coordinates (and identify the graph).

(a) $x^2 + y^2 = 4z^2$ (b) $x^2 + y^2 + z^2 = 10$

6. Find the equation of the x, y plane and the z-axis in (a) cylindrical and (b) spherical coordinates.

7. (a) Find the distance to the origin in rectangular, spherical and cylindrical coordinates.   (b) Find the distance to the z-axis in cylindrical, rectangular and spherical coordinates.

REVIEW PROBLEMS FOR CHAPTER 10

1. Let $A = (2, 3, -3)$, let line L_1 have equations $x = 2 - t$, $y = 2 + t$, $z = -4 + 3t$, and let line L_2 be $x = 7 + 2t$, $y = -t$, $z = 4 + t$. Find equations for

(a) the line through A parallel to L_1
(b) the line through A perpendicular to both L_1 and L_2
(c) the plane through A perpendicular to L_1
(d) the plane through A parallel to L_1 and L_2

2. Find the intersection, if it exists.

(a) line $x = t$, $y = 3 + 2t$, $z = 2 - t$ and plane $x + 5y - z = 1$
(b) lines $x = 2 + t$, $y = 1 - 2t$, $z = 3 + 3t$ and $x = 2 - t$, $y = 3 + t$, $z = 1 - 2t$
(c) planes $3x - y + z = 5$ and $x + y - 6z = 3$

3. Find equations of line AB if $A = (9, 8, 7)$ and $B = (6, 4, 3)$.

4. Find the distance from the origin to the plane $5x + 2y - 6z = 8$.

5. Sketch the graph in 3-space.

(a) $y = \ln x$ (c) $x = y + 2z$
(b) $x = y^2 + 2z^2$ (d) $x^2 = y^2 + 2z^2$

6. Show that the point $(-1, 1, \sqrt{3})$ lies on the sphere $x^2 + y^2 + z^2 = 5$ and find an equation of the plane tangent to the sphere at the point.

7. Suppose a point has cylindrical coordinates $r = 2$, $\theta = 90°$, $z = 5$. Draw a diagram and find the rectangular and spherical coordinates by inspection.

8. Consider a cylinder with radius 2 and the z-axis as its axis of symmetry. Find its equation in (a) rectangular, (b) cylindrical and (c) spherical coordinates.

9. Let line L have equations $x = 2 + 4t$, $y = 3 - 2t$, $z = 1 + t$, and let plane p_1 have the equation $x + y - 2z = 3$. Show that L lies in p_1 and find the equation of the plane p_2 containing L and perpendicular to p_1.

11/PARTIAL DERIVATIVES

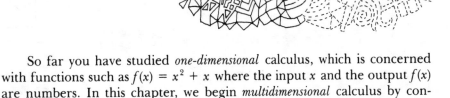

So far you have studied *one-dimensional* calculus, which is concerned with functions such as $f(x) = x^2 + x$ where the input x and the output $f(x)$ are numbers. In this chapter, we begin *multidimensional* calculus by considering functions such as

$$f(x, y) = 2xy + 7, \qquad g(x, y, z) = e^x + 2yz, \qquad h(x_1, x_2, x_3, x_4) = \frac{3x_1 x_2 x_3}{x_4}.$$

For example, $f(2, 4) = 23$, $g(2, 4, 5) = e^2 + 40$, $h(1, 2, 3, 7) = 18/7$. The input may be viewed as an n-dimensional point or as a collection of n independent variables; the output is a number, a single dependent variable. The function f might represent the temperature at the point (x, y). The function g could be the air pressure at the point (x, y, z). The function h might be the blood pressure of a patient taking x_1, x_2, x_3, x_4 units respectively of four particular medicines. We call f a *function of two variables*, g a *function of three variables* and h a *function of four variables*. This chapter discusses the differential calculus of such functions, and the next chapter considers integrals. The details in this chapter will be new but the overall outline should be familiar. Many of the topics from one-dimensional calculus (graphs, rates of change, maximization) are repeated, but with added dimensions.

11.1   Graphs and Level Sets

The purpose of a picture is to convey information easily. This section shows two ways to visualize a function of n variables.

Graphs   As you know, the graph of $f(x) = x^2$ is defined as the graph of the equation $y = x^2$, a parabola in 2-space. Similarly, if $g(x, y) = 3 - x^2 - 2y^2$, the graph of g is defined as the graph of the equation $z = 3 - x^2 - 2y^2$ in 3-space. The methods of Section 10.4 may be used to produce a sketch. The cross section in the x, y plane (where $z = 0$) is the ellipse $x^2 + 2y^2 = 3$. The cross section in the plane $z = 3$ is the point ellipse $x^2 + 2y^2 = 0$, and the cross section in the plane $z = -3$ is the ellipse $x^2 + 2y^2 = 6$. Some cross sections are shown in Fig. 1 along with the final graph of g, an elliptic paraboloid.

If $h(x, y, z) = x + 2y - 3z$, then the graph of h is (theoretically) the graph of the equation $w = x + 2y - 3z$ in *four*-space. We can list points (x, y, z, w) which belong to the graph, such as $(1, 2, 3, -4)$, but we can't *draw* the graph because the world doesn't have room for a geometric 4-dimensional space.

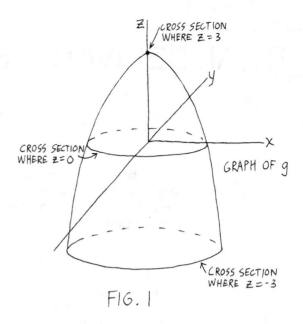

FIG. 1

In general, *the graph of a function $f(x)$ of one variable is the graph of the equation $y = f(x)$, a curve in 2-space. The graph of a function $f(x, y)$ of two variables is the graph of the equation $z = f(x, y)$, a surface in 3-space. The graph of a function $f(x, y, z)$ is a set of points in 4-space, and we make no attempt to construct it.*

Level sets The graph of a function of two variables might be difficult to draw, and the graph of a function of three or more variables can't be drawn at all, so another way of picturing functions will be helpful.

Consider $g(x, y) = 3 - x^2 - 2y^2$. The points where the value of g is (say) -6 satisfy $3 - x^2 - 2y^2 = -6$; that is, the points constitute the ellipse $x^2 + 2y^2 = 9$. If $g(x, y)$ is the temperature at the point (x, y), then the ellipse is called an isotherm. It contains precisely those points at which the temperature is -6. If g represents air pressure at (x, y), then the ellipse is called an isobar. If g is the potential energy at (x, y), then the ellipse is an equipotential curve; if g is utility, then economists call the ellipse an indifference curve. No matter what g represents, the ellipse is a *level set,* in particular, the -6 level. Similarly, the 2 level set is the set of points where $3 - x^2 - 2y^2 = 2$, that is, the ellipse $x^2 + 2y^2 = 1$. The 3 level set is $3 - x^2 - 2y^2 = 3$, the point ellipse $x^2 + 2y^2 = 0$. The 4 level set is $4 = 3 - x^2 - 2y^2$, a null ellipse. Figure 2 shows some of the level sets of g. (If *all* the level sets were drawn, instead of only a few, they would merge into an uninformative solid blob of black print.)

In general, *the level sets of a function $f(x, y)$ are drawn in two-space by considering equations of the form $f(x, y) = C$ for all possible numbers C; they are* usually, but not necessarily, curves. *The level sets of a function $f(x, y, z)$ are drawn in three-space by considering equations of the form $f(x, y, z) = C$ for all C;* they are usually, but not necessarily, surfaces.

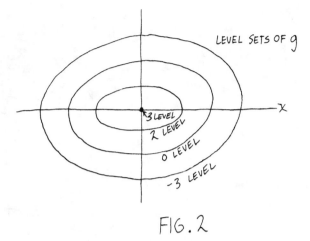

LEVEL SETS OF g

3 LEVEL
2 LEVEL
0 LEVEL
-3 LEVEL

FIG. 2

For example, let $f(x, y, z) = 2x + 3y + 6z - 10$. The 14 level set has equation $2x + 3y + 6z - 10 = 14$, i.e., the plane $2x + 3y + 6z = 24$ (see plane ABC in Fig. 3). The 15 level set is $2x + 3y + 6z - 10 = 15$, the plane $2x + 3y + 6z = 25$, parallel to the 14 level set. In general, the level sets are parallel planes (Fig. 3).

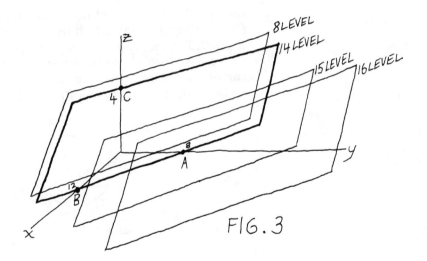

FIG. 3

The connection between the graph and the level sets of a function of two variables   A function $f(x, y)$ has a collection of level sets in *two*-space with equations of the form $f(x, y) = C$. The function also has a unique graph in *three*-space with equation $z = f(x, y)$. The cross section at height $z = 5$ on the graph in 3-space is identical with the 5 level set in 2-space (Fig. 4). In general, *the cross sections on the graph in 3-space in the various planes $z = C$ are the same as the level sets drawn in 2-space.* Figure 1 shows the graph of $g(x, y) = 3 - x^2 - 2y^2$. The level sets in Fig. 2 and the cross sections in Fig. 1 are the same.

One of the best ways to visualize the relation between the graph and the level sets is to think of the x, y plane as the earth's sea level, and let $f(x, y)$ be

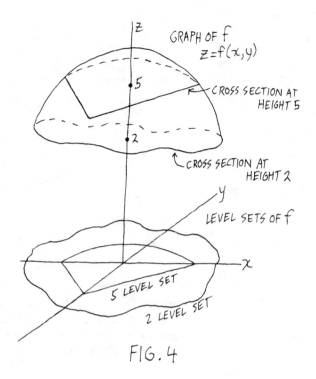

GRAPH OF f
$z = f(x,y)$

CROSS SECTION AT HEIGHT 5

CROSS SECTION AT HEIGHT 2

LEVEL SETS OF f

5 LEVEL SET

2 LEVEL SET

FIG. 4

the altitude (in meters) of the earth above the point (x,y). For example, if $f(2,3) = 4,000$ then the earth rises to an altitude of 4,000 meters above the point $(2,3)$ in the x,y plane. *The graph of f is a plaster model of the earth's surface while the level sets are the contour curves on a topographic map.* The contour curves, that is, the level sets of altitude, in Fig. 5 show a mountain peak at point A and a valley at point B.

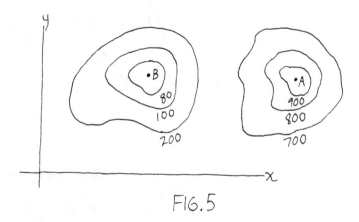

FIG. 5

Graphs of equations versus graphs of functions versus level sets of functions   The plane in Fig. 6 is the graph of the equation $x + 2y + z = 4$. It is also (solve for z) the graph of the function $f(x,y) = 4 - x - 2y$. It is also the 4-level set of the function $g(x,y,z) = x + 2y + z$. It is also the 0-level set of the function $h(x,y,z) = x + 2y + z - 4$, and the 5-level set of $k(x,y,z) = x + 2y + z + 1$. (In fact, it can be described in infinitely many ways as a level set of a function of three variables.)

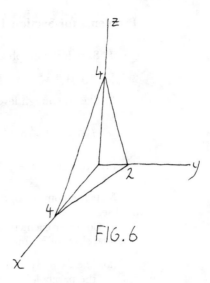

FIG. 6

The sphere with radius 2 in Fig. 7 is the graph of the equation $x^2 + y^2 + z^2 = 4$. It is *not* the graph of a function $f(x, y)$ because the point (x_0, y_0) in the x, y plane is paired with two values of z, and a function cannot produce more than one output for a given input. Equivalently, it is not the graph of a function $f(x, y)$ since its equation cannot be uniquely solved for z. However, the upper hemisphere is the graph of the function $f(x, y) = \sqrt{4 - x^2 - y^2}$ and the lower hemisphere is the graph of $g(x, y) = -\sqrt{4 - x^2 - y^2}$. The entire sphere is the 4-level set of $h(x, y, z) = x^2 + y^2 + z^2$, and also the 0-level set of $k(x, y, z) = x^2 + y^2 + z^2 - 4$, and so on.

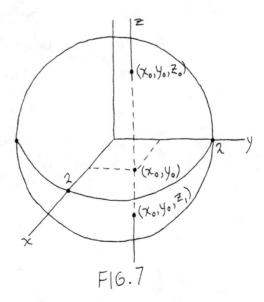

FIG. 7

Warning   The level sets of a function of *two* variables are drawn in *two*-space, while its graph is a surface in *three*-space. The level sets of a function of *three* variables are drawn in *three*-space, and its graph exists only as an abstraction in *four*-space.

Problems for Section 11.1

1. Sketch the graph and enough level sets to suggest the pattern.

(a) $f(x, y) = x^2 + y^2$ (b) $f(x, y) = x^2 - y^2$ (c) $f(x, y) = x + y$

2. Sketch enough level sets to indicate the pattern.

(a) $f(x, y, z) = x^2 + y^2 + z^2$ (c) $f(x, y) = \dfrac{1}{x + y}$

(b) $f(x, y, z) = x^2 + y^2 - z^2$ (d) $f(x, y) = \begin{cases} 3 & \text{if } y \geq x \\ 4 & \text{if } y < x \end{cases}$

3. Is the point $(2, 6)$ on a level set of $f(x, y) = xy$? If so, identify and sketch the level set.

4. Consider each curve or surface. Is it the graph of a function $f(x)$ or $f(x, y)$? Is it a level set of a function? If so, identify the function(s).

(a) line $2x + 3y = 6$ (c) the paraboloid $x = z^2 + 2y^2$
(b) the paraboloid $z = x^2 + 2y^2$ (d) the circular cylinder with radius 2 and the z-axis as its axis.

5. Let $f(x, y)$ be the distance from (x, y) to the x-axis. Sketch some level sets of f without attempting to find a formula for f.

6. Suppose f is a constant function such that $f(x, y) = 6$ for all (x, y). Sketch the graph and level sets of f.

7. Sketch the level sets.

(a) $f(x, y, z) = y^2 - x$ (b) $f(x, y) = y$ (c) $f(x, y, z) = y$

8. If the potential energy at the point (x, y, z) is

$$\frac{2}{\sqrt{(x + 2)^2 + (y - 1)^2 + (z - 3)^2}},$$

identify the equipotential surfaces, that is, the level sets.

9. Suppose the surface in Fig. 8 is the graph of $f(x, y)$. (a) Decide which is the largest of $f(2, 2)$, $f(2, 3)$ and $f(3, 2)$.   (b) Sketch some level sets of f.

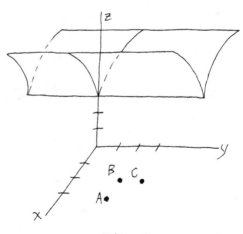

FIG. 8

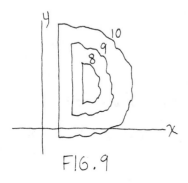

FIG. 9

10. Figure 9 shows some level sets of f. Sketch the graph of f.

11. Can two level sets of f intersect? In other words, can a point Q be on more than one level set of f?

11.2 Partial Derivatives

This section defines and computes the partial derivatives of a function of several variables, and concludes with some applications.

First-order partial derivatives A function $f(x)$ of *one* variable has a first-order derivative denoted by $f'(x)$ or df/dx. A function $f(x,y)$ of *two* variables has two first-order partial derivatives. *The partial derivative $\partial f/\partial x$ is defined as the derivative of f with respect to x, with y treated as a constant. The partial derivative $\partial f/\partial y$ is the derivative of f with respect to y, with x treated as a constant.* Similarly a function $f(x,y,z)$ has three first-order partial derivatives, $\partial f/\partial x$, $\partial f/\partial y$ and $\partial f/\partial z$, and so on.

For example, to find the partial of x^2y^3 with respect to x, think of y^3 as a constant coefficient of x^2 so that the problem resembles cx^2. Since $D_x cx^2 = 2cx^2$, we have $\dfrac{\partial(x^2y^3)}{\partial x} = 2xy^3$. Similarly, $\dfrac{\partial(x^2y^3)}{\partial y} = 3x^2y^2$.

As another example, if $f(x,y,z) = ze^{2x+3y+4z}$ then $\dfrac{\partial f}{\partial x} = 2ze^{2x+3y+4z}$, $\dfrac{\partial f}{\partial y} = 3ze^{2x+3y+4z}$ and, by the product rule, $\dfrac{\partial f}{\partial z} = 4ze^{2x+3y+4z} + e^{2x+3y+4z}$.

Higher order partial derivatives If f is a function of several variables, among them x and y, then $\dfrac{\partial^2 f}{\partial x^2}$ is the derivative of f twice with respect to x, $\dfrac{\partial^2 f}{\partial x \, \partial y}$ is the derivative of f first with respect to y and then with respect to x, $\dfrac{\partial^3 f}{\partial x \, \partial y^2}$ is the derivative of f twice with respect to y and then with respect to x, and so on.

For example, if $f(x,y) = x^3y^5 + x^3 + y^4 + 7$ then

$$\frac{\partial f}{\partial x} = 3x^2y^5 + 3x^2, \qquad \frac{\partial f}{\partial y} = 5x^3y^4 + 4y^3,$$

$$\frac{\partial^2 f}{\partial x^2} = 6xy^5 + 6x, \qquad \frac{\partial^2 f}{\partial y^2} = 20x^3y^3 + 12y^2;$$

also

(1)
$$\frac{\partial^2 f}{\partial y \, \partial x} = \frac{\partial}{\partial y}(3x^2y^5 + 3x^2) = 15x^2y^4,$$

$$\frac{\partial^2 f}{\partial x \, \partial y} = \frac{\partial}{\partial x}(5x^3y^4 + 4y^3) = 15x^2y^4.$$

The two partials in (1) are called *mixed partials* and it is not a coincidence that they came out equal. *For all functions $f(x,y)$ encountered in practice, the mixed partials are equal.* (We omit a precise statement and proof of this result.) Similarly, if f is a function of x_1, x_2, x_3, x_4 then

$$\frac{\partial^5 f}{\partial x_1^2 \, \partial x_3^2 \, \partial x_4} = \frac{\partial^5 f}{\partial x_4 \, \partial x_1^2 \, \partial x_3^2} = \frac{\partial^5 f}{\partial x_1 \, \partial x_4 \, \partial x_1 \, \partial x_3^2}$$

and so on. All that counts is how *often* the differentiation is done with respect to each variable; the *order* in which it is done is immaterial.

Notation Some other commonly used partial derivative symbols are f_x for $\frac{\partial f}{\partial x}$, f_y for $\frac{\partial f}{\partial y}$, f_{xx} for $\frac{\partial^2 f}{\partial x^2}$, f_{yy} for $\frac{\partial^2 f}{\partial y^2}$, f_{xy} for $\frac{\partial^2 f}{\partial y\,\partial x}$ and f_{yx} for $\frac{\partial^2 f}{\partial x\,\partial y}$. The notation f_{xy} means $(f_x)_y$, so the partial is found by differentiating first with respect to x and then with respect to y; the order of differentiation is found by reading the subscripts from left to right. The notation $\frac{\partial^2 f}{\partial x\,\partial y}$ means $\frac{\partial}{\partial x}\left(\frac{\partial f}{\partial y}\right)$, so the partial is found by differentiating first with respect to y and then with respect to x; the order of differentiation is found by reading the "denominator" from right to left. It is true that the same final answer results regardless of the order of differentiation since mixed partials are equal; nevertheless, every notation does indicate a particular order.

The partial derivative $\frac{\partial f}{\partial x}$ evaluated at the point $x = 2,\ y = 1$ is often denoted by $\frac{\partial f}{\partial x}\Big|_{x=2, y=1}$. For example, if $f(x,y) = \frac{x^2}{y}$ then $\frac{\partial f}{\partial y} = -\frac{x^2}{y^2}$ and $\frac{\partial f}{\partial y}\Big|_{x=1, y=3} = -\frac{1}{9}$.

Warning 1. Don't write $f'(x,y)$ because it is not clear whether the derivative is intended to be with respect to x or y. In engineering and physics, if y is a function of the two variables x (for position) and t (for time) then $\frac{\partial y}{\partial x}$ is often denoted by y', and $\frac{\partial y}{\partial t}$ by $\dot{y}$. However we will not use this notation.

2. The partial $\frac{\partial^2 u}{\partial x\,\partial y}$ should not be written as $\frac{\partial u}{\partial x\,\partial y}$ or $\frac{\partial^2 u}{\partial xy}$ or $\frac{\partial^2 u}{\partial^2 xy}$. The partial $\frac{\partial^2 u}{\partial x^2}$ should not be written as $\frac{\partial^2 u}{\partial^2 x}$ or as $\frac{\partial u^2}{\partial x^2}$.

Application to rates of change Partial derivatives are used in much the same manner as the derivative of a function of one variable. The partial of $f(x,y)$ with respect to x is the rate of change of f with respect to x as y stays fixed. Suppose $\frac{\partial f}{\partial x}\Big|_{x=2, y=3} = -4$; at the moment when $x = 2,\ y = 3$, if y stays fixed at 3 and x increases, f will decrease by 4 units for every unit increase in x. Similarly, $\frac{\partial f}{\partial y}$ is the rate of change of f with respect to y as x stays fixed.

For example, let $f(x,y)$ be the resting pulse of a person smoking x cigarettes and jogging y minutes per day. If $f(30, 40) = 75$, then smoking 30 cigarettes and wheezing through a 40 minute run daily will result in a resting pulse of 75 beats per minute. Suppose $\frac{\partial f}{\partial x}\Big|_{x=30, y=40} = 2$ and $\frac{\partial f}{\partial y}\Big|_{x=30, y=40} = -1/2$. Then with 40 minutes of jogging and 30 cigarettes daily, an increase in cigarette consumption will increase the pulse by 2 beats per minute for each additional cigarette. On the other hand, if smoking is fixed and the run lengthened, each additional minute of running lowers the pulse by $\frac{1}{2}$ beat per minute.

As another example, suppose a rope is wiggled so as to make each point move up and down in the following way: at time t, the shape of the rope is $y = \sin(x - 2t)$, time in seconds and distance in meters. Thus, at time $t = \pi/2$, its shape is $y = \sin(x - \pi)$ (Fig. 1). We will find the partials of y at $x = \pi$, $t = \frac{1}{2}\pi$ and interpret the results physically. We have $\frac{\partial y}{\partial x} = \cos(x - 2t)$ and $\frac{\partial y}{\partial t} = -2\cos(x - 2t)$, so $\left.\frac{\partial y}{\partial x}\right|_{x=\pi,\, t=(1/2)\pi} = 1$ and $\left.\frac{\partial y}{\partial t}\right|_{x=\pi,\, t=(1/2)\pi} = -2$. Since x and y are the usual rectangular coordinates, $\frac{\partial y}{\partial x}$ is a slope; in particular, the slope at point A in Fig. 1 is 1. Since y is vertical position at time t, the partial $\frac{\partial y}{\partial t}$ is a vertical velocity; point A in Fig. 1 is in the process of moving down instantaneously at the rate of 2 meters per second.

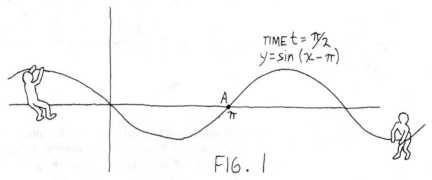

TIME $t = \pi/2$
$y = \sin(x - \pi)$

FIG. 1

As a third example, let $f(x, y)$ be the temperature at the point (x, y) (distance in meters, temperature in degrees). Suppose

$$\left.\frac{\partial f}{\partial x}\right|_{x=7,\, y=2} = -3 \quad \text{and} \quad \left.\frac{\partial f}{\partial y}\right|_{x=7,\, y=2} = 4.$$

If a particle at point $(7, 2)$ moves east (so that y is fixed and x increases) the temperature drops momentarily by 3 degrees per meter. Similarly, if a particle moves north through $(7, 2)$, the temperature rises momentarily by 4 degrees per meter. From this point of view, the partial derivatives are called *directional derivatives*. In particular, $\partial f/\partial x$ is the rate of change of f per meter in direction $\vec{i}$ and $\partial f/\partial y$ is the rate of change of f per meter in direction $\vec{j}$. In Section 11.6, we will develop directional derivatives more generally, so that we can find the rate of change of temperature in directions such as southwest, northeast, and so on.

Connection between the partials and the graph of $f(x, y)$ Consider the graph of $f(x, y)$, the surface $z = f(x, y)$ in 3-space. To interpret the partial derivative

(2)
$$\left.\frac{\partial f}{\partial x}\right|_{x=x_0,\, y=y_0},$$

we must first fix y at y_0, which geometrically restricts us to the curve of

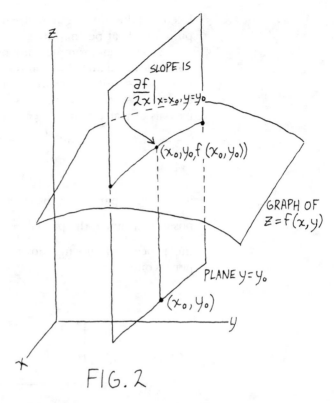

FIG. 2

intersection of the surface with the plane $y = y_0$ (Fig. 2). Then (2) is the rate of change of height z with respect to x, so (2) is the *slope on the curve of intersection of the graph of f and the plane $y = y_0$ at the point $(x_0, y_0, f(x_0, y_0))$.*

Similarly, $\dfrac{\partial f}{\partial y}\Big|_{x=x_0, y=y_0}$ is the *slope of the curve of intersection of the graph of f and the plane $x = x_0$ at the point $(x_0, y_0, f(x_0, y_0))$.* We still do not admit slopes for arbitrary curves in 3-space. The partial in (2) gives the slope of a curve in an x, z coordinate system in the plane $y = y_0$, that is, in a *two*-dimensional setting within 3-space.

Connection between the partials and the level sets of $f(x, y)$ or $f(x, y, z)$ We will illustrate how to determine the sign of the partial derivatives from a

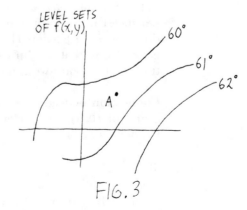

FIG. 3

picture of the level sets. Figure 3 shows level sets of a function $f(x, y)$, which we will think of as the temperature at the point (x, y). A particle moving east through point A will cross from lower to higher levels, and thus experiences rising temperatures. In other words, as x increases and y stays fixed, $f(x, y)$ increases. Therefore $\partial f / \partial x$ is positive at point A. On the other hand, a particle moving north through point A will move from higher to lower levels and feel the temperature decreasing. That is, as x stays fixed and y increases, $f(x, y)$ decreases. Therefore $\partial f / \partial y$ is negative at point A.

Problems for Section 11.2

1. Find $\partial z / \partial x$ and $\partial z / \partial y$ if

(a) $z = x^2 + 2x^3 y^2$ (d) $z = x(2x + 5y)^4$

(b) $z = xe^{-y}$ (e) $z = \dfrac{3y}{x}$

(c) $z = \dfrac{x}{x + y}$

2. Find $\partial^2 f / \partial y^2$ and $\partial^2 f / \partial x\, \partial y$ if

(a) $f(x, y) = \ln(2x + 3y)$ (b) $f(x, y) = \tan^{-1} \dfrac{y}{x}$ (c) $f(x, y) = \dfrac{x + y}{x - y}$

3. The partial $\partial f / \partial x$ can also be written as f_x. Rewrite the following partials in a similar fashion.

(a) $\dfrac{\partial^3 g}{\partial a\, \partial b\, \partial c}$ (b) $\dfrac{\partial^3 u}{\partial x^2\, \partial t}$

4. If $f(x, y, z) = z \sin x / y$ find

(a) $\dfrac{\partial f}{\partial x}$ (c) $\dfrac{\partial f}{\partial z}$

(b) $\dfrac{\partial f}{\partial y}$ (d) $\dfrac{\partial^2 f}{\partial x\, \partial z}$

5. Find

(a) $\dfrac{\partial^3 (e^x \cos y)}{\partial y^3}$ (d) $\dfrac{\partial^3 (x \sin y)}{\partial y^3}$

(b) $\dfrac{\partial^3 (e^x \cos y)}{\partial x^3}$ (e) $\dfrac{\partial^3 (x \sin y)}{\partial x\, \partial y^2}$

(c) $\dfrac{\partial^3 (x \sin y)}{\partial x^3}$

6. If $x = \rho \sin \phi \cos \theta$ find (a) $\dfrac{\partial^2 x}{\partial \rho^2}$ (b) $\dfrac{\partial^2 x}{\partial \phi\, \partial \theta}$

7. Let x be the price of a camera, y the price of a roll of film, and $f(x, y)$ the number of cameras sold daily. For example, if $f(30, 1) = 100$ then 100 cameras will be sold if a camera cost \$30 and film is \$1 a roll. Find the sign of $\partial f / \partial x$ and of $\partial f / \partial y$.

8. Let a and b be the prices of a round-trip ticket between New York and Los Angeles charged by airlines A and B, respectively. If $f(a, b)$ is the number of passengers who fly airline A each week, find the signs of $\partial f / \partial a$ and $\partial f / \partial b$.

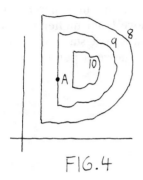

FIG.4

9. Let $f(x, y)$ be the day's profit in a company whose employees jog x miles and bicycle y miles during their lunch hour. For example, if $f(1, 2) = 1,000$ then the company makes a profit of $1,000 when its employees healthfully run a mile and cycle two miles that day. Interpret the following experimental observation:

$$\left. \frac{\partial f}{\partial x} \right|_{x=1/2, y=1} > \left. \frac{\partial f}{\partial y} \right|_{x=1/2, y=1} > 0.$$

10. Suppose the temperature at the point (x, y) is $(2x - 3y)^4$. What is the rate of change of temperature if a particle moves north through the point $(4, 3)$? east? south?

11. Use the level sets of f in Fig. 4 to find the signs of the partial derivatives of f at A.

12. Figure 5 shows the graph of $f(x, y)$. Find the signs of (a) $\partial f / \partial x$ and (b) $\partial f / \partial y$ at $(2, 5)$.

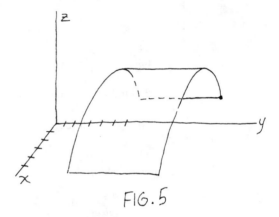

FIG.5

13. Figure 6 shows level sets of $f(x, y, z)$. (Each is an unbounded circular cylinder, with no lids.) Find the signs of $\partial f / \partial x$, $\partial f / \partial y$ and $\partial f / \partial z$ at the point $P = (-2, 3, 4)$.

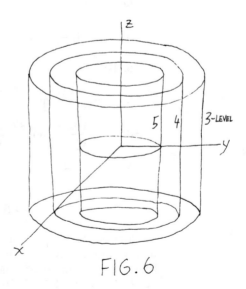

FIG.6

14. Figure 3 shows level sets of f. Suppose the level sets of g are the same curves but for different levels as indicated in Fig. 7. Which is larger at point A, $\partial f / \partial x$ or $\partial g / \partial x$?

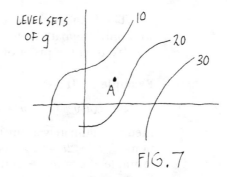

FIG. 7

11.3 Chain Rules for First-Order Partial Derivatives

The *one*-dimensional chain rule is familiar: if y is a function of u, denoted $y = y(u)$, and u is a function of x, denoted $u = u(x)$, then (Section 3.6)

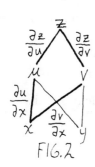

FIG. 1

(1)
$$\frac{dy}{dx} = \frac{dy}{du}\frac{du}{dx}.$$

For a *two*-dimensional version, suppose z is a function of u and v, denoted $z = z(u,v)$, and u and v are functions of x and y, denoted $u = u(x,y)$, $v = v(x,y)$. For example, we might have $z = uv^3$ with $u = xy$, $v = 2x + 3y$, so that $z = xy(2x + 3y)^3$. Figure 1 shows a *dependence diagram* picturing the hierarchy of variables. A chain typically arises with a switch to a new coordinate system; if z is a function of x and y and we change to polar coordinates r and θ, then $z = z(x,y)$ where $x = r\cos\theta$, $y = r\sin\theta$. This section will give chain rules which express the ultimate derivatives, such as $\partial z/\partial x$ and $\partial z/\partial y$ in Fig. 1, in terms of intermediate derivatives, such as $\partial z/\partial u$, $\partial z/\partial v$, $\partial u/\partial x$, $\partial u/\partial y$, $\partial v/\partial x$ and $\partial v/\partial y$. An application will be discussed after the technique has been described.

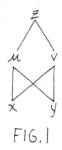

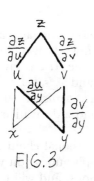

FIG. 2

The pattern of a chain rule Each dependence diagram in multi-dimensional calculus has its own chain rule. We will illustrate the pattern so that you can write them in any situation.

As a device for writing the chain rule for $\partial z/\partial x$ in Fig. 1, consider all (two) paths in the diagram from z to x, and label each branch with a corresponding derivative (Fig. 2). Then multiply down each path and add the results to obtain

(2)
$$\frac{\partial z}{\partial x} = \frac{\partial z}{\partial u}\frac{\partial u}{\partial x} + \frac{\partial z}{\partial v}\frac{\partial v}{\partial x}.$$

The chain rule contains a term for each path from z to x, and the factors in each term are the derivatives written along the path. Figure 3 shows the two paths from z to y that are used to write the chain rule

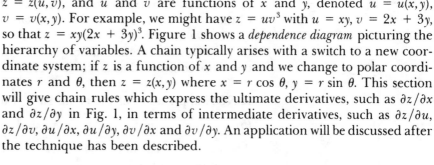

FIG. 3

(3)
$$\frac{\partial z}{\partial y} = \frac{\partial z}{\partial u}\frac{\partial u}{\partial y} + \frac{\partial z}{\partial v}\frac{\partial v}{\partial y}.$$

As a check, note that each term on the right side of (2) and of (3) "cancels" to the left side.

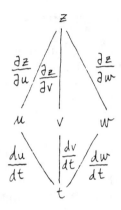

FIG.4

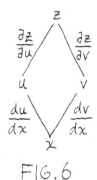

FIG.5

The familiar chain rule in (1) is a special case of the general pattern, with only one path from y to x (Fig. 4).

We omit the proof of the chain rules.

Example 1 If

$$z = z(u, v, w), \qquad u = t^3, \qquad v = t^2 - 2t, \qquad w = e^{2t} \qquad \text{(Fig. 5)},$$

then z is ultimately a function of t, but in a special way. Only the combinations t^3, $t^2 - 2t$ and e^{2t} occur in the formula for z, not "loose" t's. For example, if $z = uv^4 + w$ then $z = t^3(t^2 - 2t)^4 + e^{2t}$. All paths in Fig. 5 lead from z to t, and

$$(4) \quad \frac{dz}{dt} = \frac{\partial z}{\partial u}\frac{du}{dt} + \frac{\partial z}{\partial v}\frac{dv}{dt} + \frac{\partial z}{\partial w}\frac{dw}{dt} = 3t^2\frac{\partial z}{\partial u} + (2t - 2)\frac{\partial z}{\partial v} + 2e^{2t}\frac{\partial z}{\partial w}.$$

Warning 1. Use the partial derivative symbol ∂ and the ordinary derivative symbol d appropriately by thinking about the existence or nonexistence of *other* variables. In Fig. 5 the top branches are labeled with *partial* derivatives because at that level, z is a function of the *three* variables u, v, w. The lower branches are labeled with *ordinary* derivatives because at that level, u, v and w are each functions of the *single* variable t. The derivative dz/dt in (4) is written with d rather than ∂ because when z is considered as a function of the variable t, it is a function of t *alone*.

2. We never mix tiers in a dependence diagram. In Fig. 5, z is a function of u, v, w and from another point of view, z is a function of t. However z is *never* considered as a function of u, v, w and t *simultaneously*.

A NON-application of the chain rule Suppose

$$(5) \qquad z = uv^2, \qquad u = \sin x \quad \text{and} \quad v = x^2 \qquad \text{(Fig. 6)}.$$

Then $z = x^4 \sin x$ and *without* any multidimensional chain rule we have (by the product rule)

$$(6) \qquad \frac{dz}{dx} = x^4 \cos x + 4x^3 \sin x.$$

With the chain rule, using the formulas in (5) and the two paths from z to x in Fig. 6,

$$(7) \quad \frac{dz}{dx} = \frac{\partial z}{\partial u}\frac{du}{dx} + \frac{\partial z}{\partial v}\frac{dv}{dx} = v^2 \cdot \cos x + 2uv \cdot 2x$$

$$= (x^2)^2 \cos x + 2(\sin x)(x^2)2x = x^4 \cos x + 4x^3 \sin x.$$

The direct method in (6) and the chain rule in (7) produce the same answer, which verifies the chain rule for this particular example. It also illustrates an important aspect of the multidimensional chain rule. If z is given as a *particular*, rather than an *arbitrary*, function of u and v, say $z = uv^2$ as opposed to $z = z(u, v)$, and, in turn, u and v are given as *particular* functions of x, then z may be written directly in terms of x, and a derivative found *without* the chain rule.

It is not the primary purpose of the multidimensional chain rule to produce derivatives of a *particular* function. The applications you will encounter in later courses will involve arbitrary functions, and will typically produce a *general* result. Compare this with a similar situation in algebra. You learned that

(8)
$$(x + y)^2 = x^2 + 2xy + y^2$$

but the main purpose of (8) is not to *compute* $(3 + 4)^2$ by writing it as $3^2 + 2 \cdot 3 \cdot 4 + 4^2$. For computational purposes, $(3 + 4)^2$ is called 7^2 and found directly. However (8) is often used in postalgebra courses in applications involving an *arbitrary x* and *y*.

An application to differential equations Section 4.9 introduced (ordinary) differential equations, in which the unknown is a function $y(x)$, a function of one variable. The equation

(9)
$$\frac{\partial u}{\partial z} = 3\frac{\partial u}{\partial x} - 3\frac{\partial u}{\partial y}$$

is a (partial) differential equation in which the unknown is a function $u(x, y, z)$, a function of three variables. We will show how the chain rule is used to confirm that a certain type of function is a solution to the particular equation in (9).

We claim that the functions

(10)
$$u = (x + 3z)^2 + 5(y - 3z), \qquad u = \frac{\sin(x + 3z)}{5(y - 3z)},$$
$$u = (y - 3z)e^{-2(x+3z)}$$

are solutions and, *in general,* any function of x, y and z in which the variables occur only in the combinations $x + 3z$ and $y - 3z$ is a solution. (We will not conjecture how this was discovered originally.) Let's check the first function in (10). Its three partials are

$$\frac{\partial u}{\partial x} = 2(x + 3z), \qquad \frac{\partial u}{\partial y} = 5,$$
$$\frac{\partial u}{\partial z} = 2(x + 3z) \cdot 3 - 15 = 6(x + 3z) - 15,$$

and it can be seen that they do satisfy (9). But to confirm our *general* claim, we cannot test functions one at a time, but must perform a *general* test. An arbitrary function of x, y and z containing only the combinations $x + 3z$ and $y - 3z$ is referred to as a function of $x + 3z$ and $y - 3z$ and denoted by $u(x + 3z, y - 3z)$. Equivalently, and particularly useful for the chain rule,

(11) $u = u(p, q), \qquad p = x + 3z, \qquad q = y - 3z$ (Fig. 7).

For example, the first function in (10) is of the form $p^2 + 5q$ where $p = x + 3z$, and $q = y - 3z$. We can test all the functions in (11) at once with the chain rule. Figure 7 shows one path from u to x (not two, because q is not a function of x), one path from u to y (not two, because p is not a function of y) and two paths from u to z. Therefore

(12)
$$\frac{\partial u}{\partial x} = \frac{\partial u}{\partial p}\frac{\partial p}{\partial x} = \frac{\partial u}{\partial p}, \qquad \frac{\partial u}{\partial y} = \frac{\partial u}{\partial q}\frac{\partial q}{\partial y} = \frac{\partial u}{\partial q},$$
$$\frac{\partial u}{\partial z} = \frac{\partial u}{\partial p}\frac{\partial p}{\partial z} + \frac{\partial u}{\partial q}\frac{\partial q}{\partial z} = 3\frac{\partial u}{\partial p} - 3\frac{\partial u}{\partial q}.$$

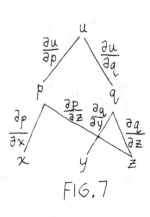

FIG. 7

The partials in (12) do satisfy (9). This verifies that all the functions in (11), including the specific ones in (10), are solutions to the differential equation.

Problems for Section 11.3

1. Write a chain rule for $\partial w/\partial s$ if $w = w(x, y, z)$, $x = x(r, s, t)$, $y = y(r, s, t)$, $z = z(r, s, t)$.

2. Write a chain rule for du/dt if $u = u(x, y, z)$, $x = x(a, b)$, $y = y(a, b)$, $z = z(a, b)$, $a = a(t)$, $b = b(t)$.

3. Let $p = p(t)$, $t = t(x, y, z)$. Write a chain rule for $\partial p/\partial y$.

4. Find the following derivatives directly and then, for practice, find them again using the multidimensional chain rule.

(a) dw/dt if $w = \sin xy$, $x = \ln t$, $y = t^3$ (b) $\partial w/\partial x$ if $w = 1/u$, $u = x \sin y$

5. Let $w = w(x, y)$ be the temperature at the point (x, y) and let $x = x(t)$, $y = y(t)$ be the position of a particle at time t. (a) Find dw/dt.   (b) Suppose dw/dt is -2 at time 3. What is the physical significance for the particle?

6. Find $\partial z/\partial u$ if $z = z(x, y)$, $x = \sin t$, $y = 2t^3$, $t = t(u, v)$.

7. If u is a function of $\sqrt{x^2 + y^2 + z^2}$, that is, $u = u(\rho)$ where $\rho = \sqrt{x^2 + y^2 + z^2}$, show that $\left(\dfrac{du}{d\rho}\right)^2 = \left(\dfrac{\partial u}{\partial x}\right)^2 + \left(\dfrac{\partial u}{\partial y}\right)^2 + \left(\dfrac{\partial u}{\partial z}\right)^2$.

8. If $u = u\left(\dfrac{y - x}{xy}, \dfrac{z - x}{xz}\right)$ show that $x^2 u_x + y^2 u_y + z^2 u_z = 0$.

9. Show *all at once* that the three functions $z = (x^2 + y^2)^3$, $z = \sqrt{x^2 + y^2}$, $z = e^{x^2+y^2} \sin(3 + x^2 + y^2)$ satisfy $y\dfrac{\partial z}{\partial x} = x\dfrac{\partial z}{\partial y}$.

10. Let $u = x^2 w\left(\dfrac{y}{x}, \dfrac{z}{x}\right)$, that is, u is the product of x^2 with the arbitrary function w of x, y, z which contains only the combinations y/x and z/x. Show that $xu_x + yu_y + zu_z = 2u$. (Treat u as a product, and use the chain rule when it becomes necessary to find derivatives of w.)

11.4   Chain Rules for Second-Order Partial Derivatives

Most of the applications of the multidimensional chain rules involve *second-order* partials. For example, if v is a function of x and y then the expression $\dfrac{\partial^2 v}{\partial x^2} + \dfrac{\partial^2 v}{\partial y^2}$ is called the Laplacian of v and is used extensively in mathematics and mathematical physics. (If v is the temperature at the point (x, y) then the Laplacian of $-v$ is the heat flux density, which measures the flow of calories in or out of a region.) If v is defined in a circular region, it is preferable to switch to polar coordinates and express the Laplacian in terms of r and θ using a chain rule for the second-order partials $\dfrac{\partial^2 v}{\partial x^2}$ and $\dfrac{\partial^2 v}{\partial y^2}$. This section will show that the same techniques that worked for first-order partials can still be used, but more *care* and *patience* (and often product rules) are needed as well.

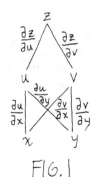

FIG. 1

Example 1   Let

$$z = z(u, v), \qquad u = x^2 y, \qquad v = 3x + 2y \qquad \text{(Fig. 1)}.$$

We will find $\dfrac{\partial^2 z}{\partial y^2}$. From Fig. 1,

(1)
$$\frac{\partial z}{\partial y} = \frac{\partial z}{\partial u}\frac{\partial u}{\partial y} + \frac{\partial z}{\partial v}\frac{\partial v}{\partial y} = x^2\frac{\partial z}{\partial u} + 2\frac{\partial z}{\partial v}.$$

Now differentiate again with respect to y to obtain

$$\frac{\partial^2 z}{\partial y^2} = \frac{\partial}{\partial y}\left(x^2\frac{\partial z}{\partial u}\right) + \frac{\partial}{\partial y}\left(2\frac{\partial z}{\partial v}\right) \qquad \text{(sum rule)}$$

(2)
$$= x^2\frac{\partial}{\partial y}\left(\frac{\partial z}{\partial u}\right) + 2\frac{\partial}{\partial y}\left(\frac{\partial z}{\partial v}\right) \qquad \text{(constant-multiple rule)}.$$

Before continuing, note that z is initially a function of u and v, and ultimately a function of x and y (Fig. 1). *The partial derivatives $\dfrac{\partial z}{\partial u}$ and $\dfrac{\partial z}{\partial v}$ are also initially functions of u and v and eventually functions of x and y.* In other words, $\dfrac{\partial z}{\partial u}$ and $\dfrac{\partial z}{\partial v}$ *are variables with the same dependence diagrams as the original variable z* (Figs. 2 and 3). In those dependence diagrams, the branches are labeled with partial derivatives in the usual way. For example, the branches in Fig. 2 from $\dfrac{\partial z}{\partial u}$ to u and v correspond to $\dfrac{\partial}{\partial u}\left(\dfrac{\partial z}{\partial u}\right)$ and $\dfrac{\partial}{\partial v}\left(\dfrac{\partial z}{\partial u}\right)$ or, equivalently, $\dfrac{\partial^2 z}{\partial u^2}$ and $\dfrac{\partial^2 z}{\partial v\,\partial u}$ as indicated.

Back in (2), to find the derivative of $\partial z/\partial u$ with respect to y, consider the two paths in Fig. 2 from $\partial z/\partial u$ to y and use the chain rule. Similarly, the derivative of $\partial z/\partial v$ with respect to y requires a chain rule using the two paths in Fig. 3 from $\partial z/\partial v$ to y. Therefore

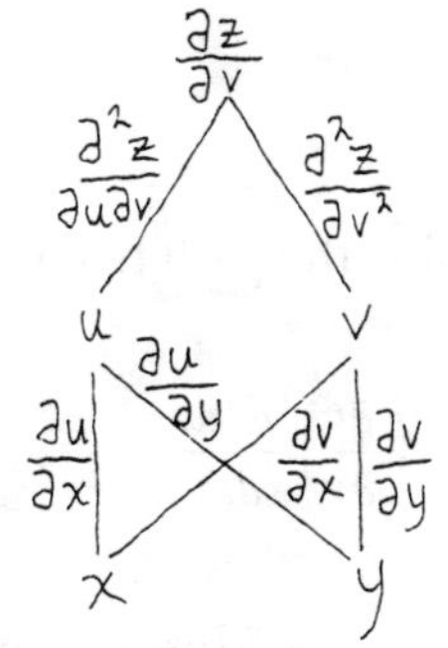

$$\frac{\partial^2 z}{\partial y^2} = x^2\left(\frac{\partial^2 z}{\partial u^2}\frac{\partial u}{\partial y} + \frac{\partial^2 z}{\partial v\,\partial u}\frac{\partial v}{\partial y}\right) + 2\left(\frac{\partial^2 z}{\partial u\,\partial v}\frac{\partial u}{\partial y} + \frac{\partial^2 z}{\partial v^2}\frac{\partial v}{\partial y}\right)$$

$$= x^2\left(x^2\frac{\partial^2 z}{\partial u^2} + 2\frac{\partial^2 z}{\partial v\,\partial u}\right) + 2\left(x^2\frac{\partial^2 z}{\partial u\,\partial v} + 2\frac{\partial^2 z}{\partial v^2}\right).$$

The mixed partials are equal, so the answer simplifies to

$$\frac{\partial^2 z}{\partial y^2} = x^4\frac{\partial^2 z}{\partial u^2} + 4x^2\frac{\partial^2 z}{\partial u\,\partial v} + 4\frac{\partial^2 z}{\partial v^2}.$$

Warning The indicated partial $\dfrac{\partial}{\partial y}\left(\dfrac{\partial z}{\partial u}\right)$ in (2) is never left in the form, or even temporarily denoted by $\dfrac{\partial^2 z}{\partial y\,\partial u}$, because z is never considered as a function of y and u simultaneously.

Example 2 Let's continue from (1) to find $\dfrac{\partial^2 z}{\partial x\,\partial y}$ by differentiating with respect to x. By the sum rule,

(3)
$$\frac{\partial^2 z}{\partial x\,\partial y} = \frac{\partial}{\partial x}\left(x^2\frac{\partial z}{\partial u}\right) + \frac{\partial}{\partial x}\left(2\frac{\partial z}{\partial v}\right).$$

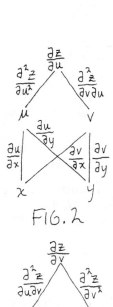

FIG. 2

FIG. 3

The second term on the right side of (3) is the derivative of a constant multiple of $\partial z/\partial v$, but the first term requires the product rule since both x^2 and $\partial z/\partial u$ are functions of x. Therefore

$$\frac{\partial^2 z}{\partial x\,\partial y} = x^2\frac{\partial}{\partial x}\left(\frac{\partial z}{\partial u}\right) + 2x\frac{\partial z}{\partial u} + 2\frac{\partial}{\partial x}\left(\frac{\partial z}{\partial v}\right).$$

To find the partials of $\partial z/\partial u$ and $\partial z/\partial v$ with respect to x, consider paths to x in Figs. 2 and 3 and use the chain rule to obtain

$$\frac{\partial^2 z}{\partial x\,\partial y} = x^2\left(\frac{\partial^2 z}{\partial u^2}\frac{\partial u}{\partial x} + \frac{\partial^2 z}{\partial v\,\partial u}\frac{\partial v}{\partial x}\right) + 2x\frac{\partial z}{\partial u} + 2\left(\frac{\partial^2 z}{\partial u\,\partial v}\frac{\partial u}{\partial x} + \frac{\partial^2 z}{\partial v^2}\frac{\partial v}{\partial x}\right)$$

$$= x^2\left(2xy\frac{\partial^2 z}{\partial u^2} + 3\frac{\partial^2 z}{\partial v\,\partial u}\right) + 2x\frac{\partial z}{\partial u} + 2\left(2xy\frac{\partial^2 z}{\partial u\,\partial v} + 3\frac{\partial^2 z}{\partial v^2}\right)$$

$$= 2x^3 y\frac{\partial^2 z}{\partial u^2} + (3x^2 + 4xy)\frac{\partial^2 z}{\partial u\,\partial v} + 2x\frac{\partial z}{\partial u} + 6\frac{\partial^2 z}{\partial v^2}.$$

Example 3 Let $w = w(t)$, $t = t(x,y)$. Find $\dfrac{\partial^2 w}{\partial x^2}$.

Solution: From Fig. 4,

$$\frac{\partial w}{\partial x} = \frac{dw}{dt}\frac{\partial t}{\partial x}.$$

Now differentiate again with respect to x. Since t is a function of x and y, $\partial t/\partial x$ is also a function of x and y. The function dw/dt has the same dependence diagram as w (Fig. 5) so it, too, is a function (eventually) of x and y. Therefore, use the product rule to obtain

$$\frac{\partial^2 w}{\partial x^2} = \frac{dw}{dt}\frac{\partial^2 t}{\partial x^2} + \frac{\partial t}{\partial x}\frac{\partial}{\partial x}\left(\frac{dw}{dt}\right).$$

The derivative of dw/dt with respect to x is found using the path from dw/dt to x in Fig. 5. Thus

(4) $$\frac{\partial^2 w}{\partial x^2} = \frac{dw}{dt}\frac{\partial^2 t}{\partial x^2} + \frac{\partial t}{\partial x}\frac{d^2 w}{dt^2}\frac{\partial t}{\partial x} = \frac{dw}{dt}\frac{\partial^2 t}{\partial x^2} + \left(\frac{\partial t}{\partial x}\right)^2\frac{d^2 w}{dt^2}.$$

Warning 1. The expression $\left(\dfrac{\partial t}{\partial x}\right)^2$ in (4) is the square of the first-order partial $\dfrac{\partial t}{\partial x}$ and is not the same as the second-order partial $\dfrac{\partial^2 t}{\partial x^2}$.

2. Don't forget to use the product rule when necessary.

Example 4 Let $w = t^3$, $t = x^3 y^2$. Find $\dfrac{\partial^2 w}{\partial x^2}$ using the chain rule in (4) and then verify by finding the partial directly.

Solution: We have

$$\frac{\partial t}{\partial x} = 3x^2 y^2, \qquad \frac{\partial^2 t}{\partial x^2} = 6xy^2, \qquad \frac{dw}{dt} = 3t^2, \qquad \frac{d^2 w}{dt^2} = 6t.$$

Substituting in (4) gives

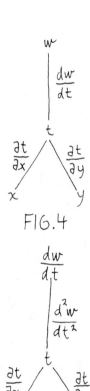

FIG.4

FIG.5

$$\frac{\partial^2 w}{\partial x^2} = 3t^2 \cdot 6xy^2 + (3x^2y^2)^2 \cdot 6t$$

(5)
$$= 3(x^3y^2)^2 \cdot 6xy^2 + 9x^4y^4 \cdot 6x^3y^2 = 72x^7y^6.$$

Directly, $w = (x^3y^2)^3 = x^9y^6$, so $\dfrac{\partial w}{\partial x} = 9x^8y^6$ and $\dfrac{\partial^2 w}{\partial x^2} = 72x^7y^6$, which agrees with (5). As noted in the preceding section, it is not the purpose of the chain rule to produce derivatives of *particular* functions, as in this example, since the direct approach is usually faster. It is meant for more general problems such as computing the Laplacian of an arbitrary function v in polar coordinates (see Problem 5 and its solution).

Problems for Section 11.4

1. Let $p = p(a, b)$, $a = 3u + 4v$, $b = 5u + 6v$. Find $\dfrac{\partial^2 p}{\partial u^2}$.

2. If $z = z(x, y)$ where $x = 3t$, $y = 4t$, find $\dfrac{d^2 z}{dt^2}$.

3. Let $u = u(x, y)$, $x = 2a + 3b$, $y = a^2 b$. Find $\dfrac{\partial^2 u}{\partial b\, \partial a}$.

4. Let $w = w(x, y)$, $x = t^3$, $y = t^2$. Find $\dfrac{d^2 w}{dt^2}$.

5. Let $v = v(r, \theta)$ where $r = \sqrt{x^2 + y^2}$ and $\theta = \tan^{-1} y/x$.

    (a) Show that $\dfrac{\partial r}{\partial x} = \dfrac{x}{r}$ and $\dfrac{\partial \theta}{\partial x} = -\dfrac{y}{r^2}$.

    (b) Show that $\dfrac{\partial^2 r}{\partial x^2} = \dfrac{y^2}{r^3}$ and $\dfrac{\partial^2 \theta}{\partial x^2} = \dfrac{2xy}{r^4}$.

    (c) Find $\dfrac{\partial^2 v}{\partial x^2}$ and simplify using parts (a) and (b).

6. Let $v = v(x, y)$, $x = x(t)$, $y = y(t)$. Find v_{tt}.

7. Show that if w is a function of x and t but contains them only in the forms $p = x - ct$, $q = x + ct$ where c is a constant, then $c^2 w_{xx} - w_{tt} = 4c^2 w_{pq}$.

11.5 Maxima and Minima

The process of maximizing and minimizing functions of two variables is similar to the one-dimensional case. We will begin with relative maxima and minima, and then concentrate on absolute extrema.

Relative extrema   The function $f(x, y)$ has a *relative maximum* at (x_0, y_0) if $f(x_0, y_0) \geq f(x, y)$ for all points (x, y) near (x_0, y_0). Similarly, f has a *relative minimum* at (x_0, y_0) if $f(x_0, y_0) \leq f(x, y)$ for all points (x, y) near (x_0, y_0).

Suppose the function $f(x, y)$ has a relative maximum value at the point (x_0, y_0). Then the graph of f has a peak there (Fig. 1). If $f(x, y)$ is the temperature at the point (x, y), then the point (x_0, y_0) is a hot spot surrounded by lower levels of temperature (Fig. 2).

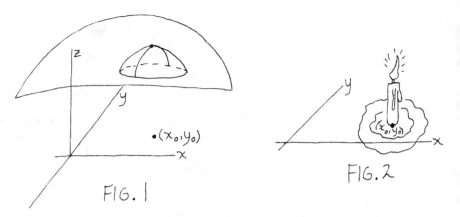

FIG. 1

FIG. 2

As in the one-dimensional case, there is a connection between relative extrema and zero derivative.† Consider Fig. 2 which shows a relative maximum of f at (x_0, y_0). If a particle moves east through (x_0, y_0) so that the temperature is a function of x alone, it experiences a relative maximum temperature at $x = x_0$. Therefore, by the theory of relative extrema for functions of one variable,

$$(1) \qquad \frac{\partial f}{\partial x}\bigg|_{x=x_0, y=y_0} = 0.$$

Similarly, if the particle moves north through the point (x_0, y_0), so that the temperature is a function of y, it feels a relative maximum when $y = y_0$. Thus,

$$(2) \qquad \frac{\partial f}{\partial y}\bigg|_{x=x_0, y=y_0} = 0.$$

Alternatively, (1) and (2) hold because the partials of f are the slopes on the curves of intersection of the graph of f (Fig. 1) with the planes $x = x_0$ and $y = y_0$, respectively, and each curve of intersection has a peak with a zero slope when $x = x_0$, $y = y_0$.

In general, *if $f(x, y)$ has a relative extreme value at (x_0, y_0) then both $\partial f/\partial x$ and $\partial f/\partial y$ are zero at (x_0, y_0); that is, both (1) and (2) hold. Equivalently, if either of the partials is nonzero at (x_0, y_0), then f cannot have a relative extreme value at (x_0, y_0). On the other hand, if both partials are zero at (x_0, y_0), then a relative extreme value may* (Figs. 1 and 2), *but need not, occur.* As an example of the latter, let f be the function with the level sets in Fig. 3 and the graph in Fig. 4. If a particle moves east through (x_0, y_0) in Fig. 3, it experiences a relative *minimum* value of f at the point, so (1) holds. Alternatively (1) holds because the curve of intersection of the graph of f with the plane $y = y_0$, namely, the curve *BAC,* has a *valley* at the point $A = (x_0, y_0, 6)$, with slope 0. On the other hand, (2) holds for the *opposite reason.* A particle moving north through the point (x_0, y_0) in Fig. 3 feels a relative *maximum* value of f at the point, and the curve of intersection of the graph of f with the plane $x = x_0$, namely, the curve *DAE,* has a *peak* at the point A. Therefore (2) holds in addition to (1), but f does not have a relative extreme value at (x_0, y_0).

If both $\partial f/\partial x$ and $\partial f/\partial y$ are zero at (x_0, y_0) then (x_0, y_0) is called a *critical point,* and $f(x_0, y_0)$ a *critical value,* of f. The preceding discussion shows that

---

†In one sense, we take a narrower approach in the two-dimensional case. We assume that $f(x, y)$ is finite at every point in the plane and that the partial derivatives always exist, assumptions we did not make for $f(x)$.

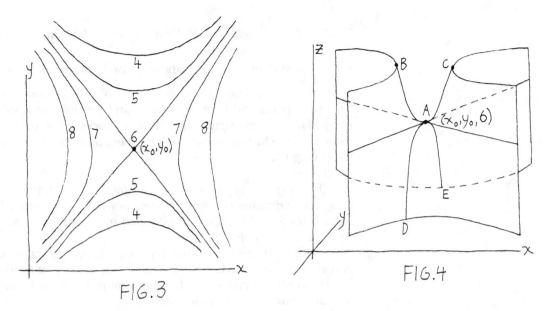

FIG.3

FIG.4

the list of critical points includes all the relative maxima, all the relative minima, and possibly nonextrema as well. In other words, *the critical points are the only possible candidates for relative extrema.* There is a second derivative test, involving f_{xx}, f_{yy} and f_{xy} for determining whether a critical point is a relative maximum, a relative minimum, or neither. We omit the test since our intention is to use relative extrema to help locate absolute extrema, and for that purpose, as in the one-dimensional case, it is not necessary to classify the critical points.

Absolute extrema   The standard *one*-dimensional extremum problem is to find the largest and smallest values of $f(x)$ in an interval $[a, b]$. The standard *two*-dimensional problem is to find the largest and smallest values, that is, the *absolute extrema,* of $f(x, y)$ in a region in the plane. We will refer to absolute extrema simply as extrema.

In physical problems, a region arises from restrictions on the independent variables. If $f(x, y)$ is the profit when a factory hires x women and y men, then $x \geq 0$, $y \geq 0$, and perhaps $x + y \leq 500$ by Fire Department safety regulations. In that case, Fig. 5 shows the region of interest.

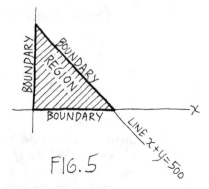

FIG.5

The procedure for finding extrema of $f(x, y)$ in a region in the plane is analogous to the method used for $f(x)$ in $[a, b]$. An extremum occurs either on the boundary of the region (Fig. 5) or at one of the relative extrema. Thus, to locate the extrema, choose from the following candidates.

(A) *Critical values of f:* Find the critical points by solving $\frac{\partial f}{\partial x} = 0$, $\frac{\partial f}{\partial y} = 0$, a system of two equations in two unknowns. *Ignore critical points not in the region.* The corresponding list of critical values contains all the relative maxima and relative minima, and possibly some values of f of no particular max/min significance. It is not necessary to decide which critical value serves which purpose. Include them all in the candidate list without classifying them.

(B) *Boundary values of f:* For a function $f(x)$ defined on an interval $[a, b]$, the end values $f(a)$ and $f(b)$ are among the candidates. The analogous candidates for a function $f(x, y)$ defined in a 2-dimensional region are the boundary values of f.

A difficulty arises here that did not occur for a function of one variable. A region in the plane has *infinitely many* boundary points, while an interval $[a, b]$ has only *two* endpoints. Instead of putting *all* the boundary points on the candidate list, we will use a select few, as demonstrated in the examples.

Example 1 Let

(3) $$f(x, y) = x^2 y - 80x.$$

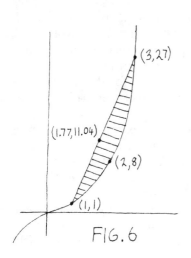

FIG. 6

We will find the maximum and minimum values of f in the region bounded by the curve $y = x^3$ and the line through the points $(1, 1)$ and $(3, 27)$ (Fig. 6). If $f(x, y)$ is the temperature at the point (x, y), we are finding the highest and lowest temperatures in the region. We are also finding the highest and lowest points on the graph of f over the region. In this example, both the function f and the region in Fig. 6 were chosen arbitrarily. In Example 3, we will begin with a physical situation which gives rise to a function and a region. In any event, the region is *not* the graph of the function (the graph is a surface in 3-space over and/or under the region), nor in any way related to it. You may think of the region as a city whose residents wish to find the hottest and coldest points within the city limits.

There are two types of candidates, critical points and boundary points.

Critical points The partial derivatives are $\frac{\partial f}{\partial x} = 2xy - 80$ and $\frac{\partial f}{\partial y} = x^2$. The critical points are the solutions of the system of equations $2xy - 80 = 0, x^2 = 0$. The second equation is satisfied only when $x = 0$, but substituting $x = 0$ in the first equation produces $-80 = 0$. Therefore the system has *no* solution and there are no candidates from this source.

The boundary $y = x^3$ To avoid having infinitely many boundary points on the candidate list, we will choose only the best, and use as candidates the hottest and coldest *of the boundary points.* To find them, proceed as follows. *Solve the equation of the boundary for x or y and substitute into (3), so that the temperature on the boundary is expressed in terms of only one variable. Choose whichever variable makes the algebra easier.* In this case, if y is replaced by x^3 (rather than x replaced by $y^{1/3}$) we have

(4) $$f = x^5 - 80x.$$

The function in (4) gives the temperature at the point (x, y) on the curve $y = x^3$. To find the extreme temperature on the curve between the points $(1, 1)$ and $(3, 27)$, maximize and minimize the function of *one* variable in (4), for $1 \le x \le 3$. *In general, each boundary curve gives rise to a standard one-dimensional subproblem for a function of one variable on an interval. The candidates for the extrema are the critical numbers and the ends of the interval.* To find the critical numbers here, differentiate (4) to get $f'(x) = 5x^4 - 80$ and solve $5x^4 - 80 = 0$. The solutions are $x = \pm 2$, but -2 is outside the interval $[1, 3]$ so it is discarded. Therefore the candidates in the subproblem are the critical number $x = 2$, and the endpoints $x = 1, 3$. *Each candidate determines a point on the boundary; the other coordinate of the point may be found by substituting in the equation of the boundary,* $y = x^3$ in this instance. If $x = 1$ then $y = 1$; if $x = 2$ then $y = 8$; if $x = 3$ then $y = 27$. Therefore the candidates for the hottest and coldest points on the boundary $y = x^3$ are $(1, 1)$, $(2, 8)$ and $(3, 27)$. Put them on the list of candidates in the original problem.

The straight line boundary The line has slope 13 and equation $y - 1 = 13(x - 1)$, or

$$(5) \qquad\qquad y = 13x - 12.$$

Substituting $13x - 12$ for y in (3) produces

$$(6) \qquad f = x^2(13x - 12) - 80x = 13x^3 - 12x^2 - 80x.$$

Since the boundary is only that portion of the line between the points $(1, 1)$ and $(3, 27)$, the subproblem involves (6) restricted to the interval $1 \le x \le 3$. (If we had solved (5) for x and switched to f as a function of y alone, the interval would be $1 \le y \le 27$.) Then $f'(x) = 39x^2 - 24x - 80$, and the critical numbers are the solutions of $39x^2 - 24x - 80 = 0$. By the quadratic formula, the solutions are $x = \dfrac{24 \pm \sqrt{13056}}{78}$. The value of x corresponding to the minus sign is negative, hence outside the interval $[1, 3]$ and irrelevant. The other solution is approximately 1.77. Therefore the candidates in the subproblem are the critical number 1.77 and the endpoints $x = 1, 3$. The corresponding values of y are found by substituting in (5), the equation of the boundary. If $x = 1$ then $y = 1$; if $x = 1.77$ then $y = 11.04$ approximately; if $x = 3$ then $y = 27$. Therefore the candidates for the hottest and coldest points on the straight line boundary are $(1, 1)$, $(1.77, 11.04)$ and $(3, 27)$.

The list of candidates There are no critical points so the list contains only the boundary candidates (Fig. 6). For each candidate, the corresponding value of f is found from (3).

point	value of f
$(1, 1)$	-79
$(2, 8)$	-128
$(3, 27)$	3
$(1.77, 11.04)$	-107.11 approximately

The value of f at a boundary point may also be found from the one-dimensional version of f on the boundary. For example, instead of substituting $x = 2$, $y = 8$ into (3) to find $f = -128$, it is also possible to merely substitute $x = 2$ into (4).

The table shows that the maximum value of f in the region is 3 and the minimum value is -128.

Warning 1. A standard two-dimensional extremum problem involves a function $f(x, y)$ and a region with one or more boundaries, each with an equation in x and y. *The boundary equations are never differentiated.* A boundary equation is used to express f as a function of one variable, and *it is the function f that is differentiated.*

2. Each boundary subproblem involves a function of one variable on an interval, and the ends of the interval are automatically candidates in that one-dimensional problem. Don't forget them. As a consequence, the vertices of a region (points at which the boundary changes equation) are always candidates.

3. Check to see that each candidate is *in* the region. It is silly to locate the region's hottest spot at a point outside the region.

Example 2 Suppose we want to find the maximum and minimum values of

(7)
$$f(x, y) = x^2 + 3y^2 - 2x$$

in the x, y plane, an *unbounded* region. In one-dimensional calculus, to find the extreme values of $f(x)$ on the unbounded interval $[a, \infty)$, we consider the ordinary end value $f(a)$, critical values, and the "end" value $f(\infty)$. Similarly, to find extreme values of $f(x, y)$ on an unbounded region, we consider ordinary boundaries (if there are any), critical points, and a hypothetical *boundary at infinity,* represented in Fig. 7 by the jagged curve "very far" from the origin. It is frequently the case that a function blows up somewhere on the boundary at infinity, and this possibility should be examined first.

FIG. 7

Consider a north path to the boundary at infinity along the y-axis. Set $x = 0$ in (7) and let $y \to \infty$ to see that the limiting value of f is $\lim_{y \to \infty} 3y^2 = \infty$.

In fact, $f \to \infty$ on any path toward the boundary at infinity because x^2 and y^2 are always positive, and x^2 has a higher order of magnitude than $2x$. Therefore, we have identified the maximum value of f, namely ∞, but we still need the minimum. The only place it can occur is at a critical point. The partial derivatives are $\dfrac{\partial f}{\partial x} = 2x - 2$, $\dfrac{\partial f}{\partial y} = 6y$. The system of equations $2x - 2 = 0$, $6y = 0$ has the solution $x = 1$, $y = 0$. Therefore f has its minimum at the point $(1, 0)$. Since $f(1, 0) = -1$, the minimum value is -1.

The maximum and minimum can also be found without calculus. Complete the square to obtain $f(x, y) = (x - 1)^2 + 3y^2 - 1$. By inspection, the maximum is ∞ when x or y approach $\pm\infty$; and the minimum occurs when the square terms are 0, that is, the minimum is -1 when $x = 1$ and $y = 0$. The graph of $z = (x - 1)^2 + 3y^2 - 1$ is a paraboloid (Fig. 8) with vertex $(1, 0, -1)$.

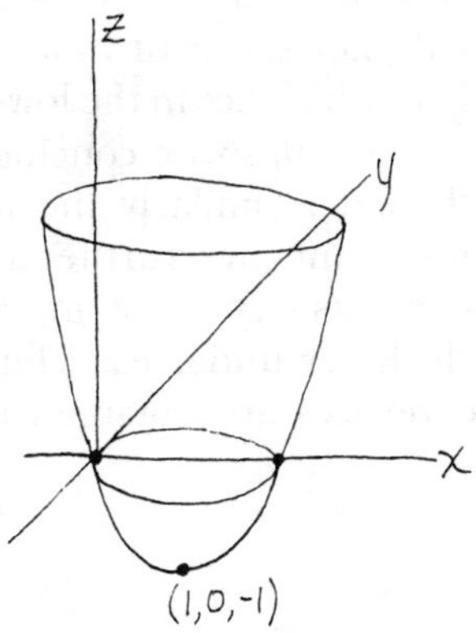

$(1, 0, -1)$

FIG. 8

Example 3   A company making clocks and radios earns a profit of $40 per clock and $60 per radio. Three machines, A, B and C, are involved in the manufacturing process. In its construction, a clock requires 2 hours of time from machine A, 1 hour from machine B and 1 hour from machine C. A radio requires 1 hour from A, 1 hour from B and 3 hours from C. Machine A is available for at most 70 hours a week, B for at most 40 hours a week and C for at most 90 hours. Table 1 summarizes the data.

Table 1

	clock $40 profit	radio $60 profit
machine A available 70 hours	2 hours	1 hour
machine B available 40 hours	1 hour	1 hour
machine C available 90 hours	1 hour	3 hours

The problem is to decide how many clocks and radios to manufacture each week so as to maximize the company profits.

To begin, let x be the number of clocks to be made and y the number of radios. Then the profit p is given by the function

(8) $p = 40x + 60y$.

With no restrictions on x and y, p can be made unboundedly large. However x and y are restricted by the limited availability of the machines. If each of the x clocks requires 2 hours from A, and each of the y radios requires 1 hour from A, but A cannot be used for more than 70 hours, then

(9) $2x + y \leq 70$.

Similarly,

(10) $x + y \leq 40$ and $x + 3y \leq 90$.

Furthermore,

(11) $$x \geq 0 \quad \text{and} \quad y \geq 0.$$

The inequalities in (9)–(11) determine a region in the plane. To sketch the region, first draw the lines $2x + y = 70$, $x + y = 40$ and $x + 3y = 90$ (Fig. 9). The line $2x + y = 70$ determines two half planes, one of which is $2x + y < 70$ and the other $2x + y > 70$. The origin, which lies in the lower half plane, happens to satisfy the inequality $2x + y < 70$, so we conclude that (9) is the line together with the *lower* half plane. Similarly the inequalities in (10) are lines plus lower half planes. Points are further restricted to quadrant I by (11). Figure 9 shows the points (x, y) satisfying all the inequalities, namely, the polygon *ABCDE* which lies under each line, and within quadrant I. The coordinates of the vertices are obtained by finding the intersections of pairs of lines.

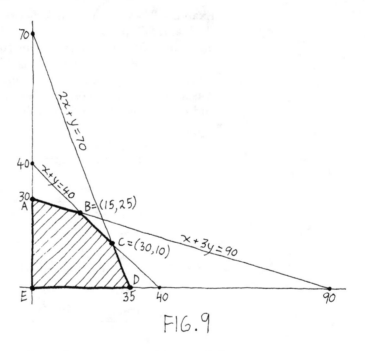

FIG. 9

We have finally reached the stage where Examples 1 and 2 began; we want to maximize the function $p(x, y)$ where (x, y) is restricted to the region *ABCDE*.

The partial derivative $\partial p / \partial x$ is 40, never 0, so even without considering $\partial p / \partial y$, there are no critical points. It remains to examine the boundaries.

Consider the line *AB* with equation $x + 3y = 90$. Solve the equation for x and substitute in (8) to obtain

$$p = 40(90 - 3y) + 60y = -60y + 3600.$$

The boundary *AB* is that part of the line between the points $(0, 30)$ and $(15, 25)$, so y is restricted to the interval $[25, 30]$. Since $p'(y) = -60$, the derivative is never 0, and there are no critical numbers in this subproblem. The only candidates are the endpoints $y = 25$ and $y = 30$, which produce the candidates $(15, 25)$ and $(0, 30)$ in the original problem. Similarly, the other boundary subproblems produce no critical numbers and the only candidates that emerge are the vertices A, B, C, D and E.

For the final decision, compute $p(0, 30) = 1800$, $p(15, 25) = 2100$, $p(30, 10) = 1800$, $p(35, 0) = 1400$, $p(0, 0) = 0$. The list shows that the maximum profit ($2100) is obtained when the company manufactures 15 clocks and 25 radios each week.

The maximum can also be located using the level sets of p, a collection of parallel lines. For example, the 0 level set is the line $40x + 60y = 0$; the 1000 level set is the line $40x + 60y = 1000$ (Fig. 10). To superimpose the level lines on top of the polygon, note that the level lines have slope $-2/3$, line AB has slope $-1/3$, and line BC has slope -1. Therefore the level lines are not as steep as BC, but steeper than AB. The higher levels do not intersect the polygonal region at all, indicating that the factory cannot make a million dollar profit. From the inclinations of the various lines, we see that the first level line to hit the region from above does so at point B. Therefore, of all points in the region, B has the largest value of p, and the company should manufacture 15 clocks and 25 radios.

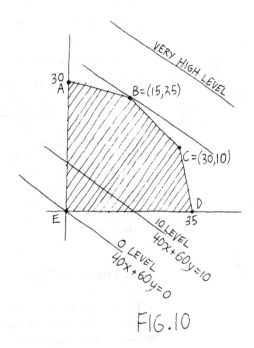

FIG. 10

Maximizing and minimizing a linear function with linear inequalities restricting the variables is called *linear programming*. Such problems occur frequently in economics and operations research. In the two-dimensional case, the inequalities determine a polygonal region, and the example shows that the maximum value occurs at a vertex of the region. In most applications, the number of variables is much larger than 2, and entire courses are devoted to techniques of solution.

Warning Example 3 asks for the *location* of the maximum, so the answer is $x = 15$, $y = 25$. If the example had asked for the maximum *value* then the answer would be 2100. Make your answer fit the question.

FIG. 11

Problems for Section 11.5

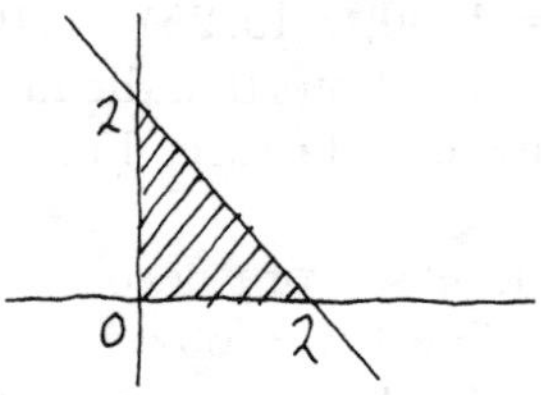

1. Find the maximum and minimum values of $f(x,y)$ in the indicated region.

(a) $3xy - 2x^2 + 2y + 8$, region in quadrant I bounded by the axes and $y = 3 - x^2$

(b) $3xy - 2x^2 + 2y + 8$, region in Fig. 11

(c) $2x - 3y$, region bounded by $y = x^2$ and $y = 4$

(d) $x^2 + x^2y - y$, region where $x^2 + 2y^2 \le 4$

2. At what points does the function have maximum and minimum values in the region.

(a) $x^2 + y^2 + 3xy + 10x$ in the triangular region with vertices $(0,3)$, $(5,3)$, $(-1,-3)$

(b) $x^2 + 2y^2 + x$ in the region bounded by circle $x^2 + y^2 = 1$

3. Find the maximum and minimum values of the function in the x, y plane.

(a) $6y - 2x - x^2 - y^2$ (c) $x^2 - xy + y^2 + 2x + 2y - 4$

(b) $x^2 + xy + 3x + 2y + 5$ (d) $x^2 - 2xy - y^2 + y$

4. Find the point in the plane $3x + 2y + z = 14$ which is nearest the origin.

5. Find the distance from the point $(1, 3, 0)$ to the plane $2x - 2y + z + 10 = 0$

(a) using formula (6') in Section 10.2   (b) by solving a minimization problem.

6. Consider the skew lines $x = 2 + 2t, y = 3 - 2t, z = 4 + 4t$ and $x = 1 + t$, $y = 2 + t, z = 7 + 3t$. Find the points of closest contact on the lines. In other words, of all points A on the first line and all points B on the second line, find the two such that distance AB is minimum.

7. A rectangular tank with an open top is to be built to hold 256 cubic feet. The builder is anxious to conserve materials, so the worst option maximizes surface area and the best option minimizes surface area. (a) Find the worst option. (b) Find the dimensions of the tank with minimum surface area. (c) Suppose the tank must stand on a 4-by-20 plot of land so that the dimensions x and y of the base are now restricted to $0 \le x \le 4$ and $0 \le y \le 20$. Find the dimensions that minimize surface area under these conditions.

11.6 The Gradient

In this section we will frequently think of $f(x,y)$ as the temperature at the point (x,y) so that we can express results more concretely. Assume that distance is measured in meters.

In Section 11.2 we saw that $\partial f/\partial x$ at a point is the rate of change of f (degrees per meter) along an *east* path through the point; similarly, $\partial f/\partial y$ is degrees/meter along a *north* path. In this section we will find rates of change in an *arbitrary* direction.

Directional derivatives Let f be a function of two or three (or more) variables. The instantaneous rate of change of f (degrees per meter) in the direction of a vector $\vec{u}$ is called the *directional derivative* of f in the direction of $\vec{u}$ and is denoted by $D_{\vec{u}}f$ (or $\frac{\partial f}{\partial \vec{u}}$, or $\frac{\partial f}{\partial s}$ where s represents distance). If f is a function of two variables then the partial derivatives $\partial f/\partial x$ and $\partial f/\partial y$ are the special directional derivatives $D_{\vec{i}}f$ and $D_{\vec{j}}f$.

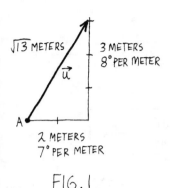

√13 METERS

3 METERS
8° PER METER

$\vec{u}$

A

2 METERS
7° PER METER

FIG. 1

We will develop a formula for $D_{\vec{u}}f$ by considering a specific example first. Let f be a function of two variables, with $\partial f/\partial x = 7$ and $\partial f/\partial y = 8$ at point A in the plane. Then, on an east path through A, the temperature is instantaneously increasing by 7° per meter, and on a north path, the temperature is instantaneously increasing by 8° per meter. Suppose we want the rate of change in the direction of $\vec{u} = 2\vec{i} + 3\vec{j}$. Figure 1 shows a step of $\sqrt{13}$ meters in the direction of $\vec{u}$, visualized as the superposition of a 2 meter east step and a 3 meter north step. If the temperature rises by 7° per meter on the east leg and by 8° per meter on the north leg, then, on the original step of $\sqrt{13}$ meters, the temperature rises by $7 \times 2 + 8 \times 3$ degrees, at the *rate* of

(1) $$\frac{7 \times 2 + 8 \times 3}{\sqrt{13}}$$ degrees per meter.

The analysis is incomplete since we want the *instantaneous* rate of change at A in the direction of $\vec{u}$, and a step of $\sqrt{13}$ meters is too large. In fact, the 7°/meter and 8°/meter rates are themselves *instantaneous* rates *at* A and do not necessarily persist for the entire 2 meter east leg and 3 meter north leg. So consider the situation in Fig. 2 with steps a tenth as large as before. Then the rate of change of temperature along the hypotenuse, in the direction of $\vec{u}$, is

$$\frac{7 \times \dfrac{2}{10} + 8 \times \dfrac{3}{10}}{\dfrac{\sqrt{13}}{10}}$$ degrees per meter

which simplifies to (1) again. As the computation is repeated for smaller and smaller steps, (1) is obtained each time, so we may take it to be the *instantaneous* rate. As a generalization of (1), if $\vec{u} = u_1\vec{i} + u_2\vec{j}$ then

(2) $$D_{\vec{u}}f = \frac{\dfrac{\partial f}{\partial x}u_1 + \dfrac{\partial f}{\partial y}u_2}{\|\vec{u}\|}.$$

We will digress briefly so that we may rewrite (2) in a more useful form and give it geometric significance.

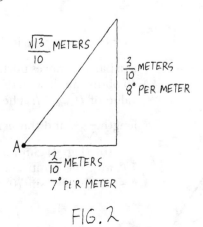

$\dfrac{\sqrt{13}}{10}$ METERS

$\dfrac{3}{10}$ METERS
8° PER METER

A

$\dfrac{2}{10}$ METERS
7° PER METER

FIG. 2

If f is a function of x and y, the vector ∇f, called the *gradient* of f, is defined by

(3)
$$\nabla f = \left(\frac{\partial f}{\partial x}, \frac{\partial f}{\partial y}\right) = \frac{\partial f}{\partial x}\vec{i} + \frac{\partial f}{\partial y}\vec{j}.$$

Similarly, if f is a function of x, y and z, then

(4)
$$\nabla f = \left(\frac{\partial f}{\partial x}, \frac{\partial f}{\partial y}, \frac{\partial f}{\partial z}\right) = \frac{\partial f}{\partial x}\vec{i} + \frac{\partial f}{\partial y}\vec{j} + \frac{\partial f}{\partial z}\vec{k}.$$

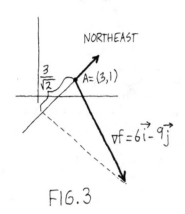

NORTHEAST

$\frac{3}{\sqrt{2}}$

$A = (3,1)$

$\nabla f = 6\vec{i} - 9\vec{j}$

FIG. 3

To illustrate the idea, suppose $f(x,y) = \frac{x^2}{y}$. Then $\nabla f = \frac{2x}{y}\vec{i} - \frac{x^2}{y^2}\vec{j}$. *There is a gradient vector at every point, which we picture as an arrow attached to the point.* At the point $A = (3,1)$ for example the gradient vector ∇f is $6\vec{i} - 9\vec{j}$; we use the notation $\nabla f|_{x=3,y=1} = 6\vec{i} - 9\vec{j}$ and draw the gradient as an arrow with its tail at the point $(3,1)$ (Fig. 3).

We may now write the numerator in (2) as the dot product $\nabla f \cdot \vec{u}$, and use (12) of Section 9.3:

(5)
$$D_{\vec{u}}f = \frac{\nabla f \cdot \vec{u}}{\|\vec{u}\|} = \text{component of } \nabla f \text{ in the direction of } \vec{u}.$$

$D_{\vec{u}}f$ may be visualized as the signed projection of ∇f onto $\vec{u}$. It is positive if ∇f makes an acute angle with $\vec{u}$, and negative if ∇f makes an obtuse angle with $\vec{u}$; its absolute value is the length of the projection of ∇f on a line in the direction of $\vec{u}$.

Example 1 Let's continue with $f(x,y) = x^2/y$, $A = (3,1)$ and $\nabla f|_A = 6\vec{i} - 9\vec{j}$. We will find several directional derivatives of f at the point A.

(a) (Fig. 3) Consider a northeast path through the point, that is, a path in the direction of the vector $\vec{u} = \vec{i} + \vec{j}$. The directional derivative is

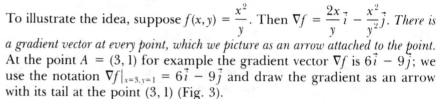

A 70°

$\vec{v}$

∇f

FIG. 4

$$D_{\text{northeast}}f = \frac{\nabla f \cdot \vec{u}}{\|\vec{u}\|} = -\frac{3}{\sqrt{2}},$$

so the temperature is dropping instantaneously by $\frac{3}{\sqrt{2}}$ degrees per meter as a particle moves northeast through point A. Figure 3 shows that ∇f forms an obtuse angle with the northeast direction, corresponding to the negative value of $D_{\text{northeast}}f$. The projection of ∇f onto the northeast direction has length $\frac{3}{\sqrt{2}}$, and the *signed* projection is $\frac{-3}{\sqrt{2}}$.

(b) (Fig. 4) Suppose a particle moves through point A at an angle of $70°$ with the positive x-axis. A unit vector in the direction of the path is $\vec{v} = \cos 70°\,\vec{i} + \sin 70°\,\vec{j}$ and, since $\|\vec{v}\| = 1$, we have

$$D_{\vec{v}}f = \nabla f \cdot \vec{v} = 6\cos 70° - 9\sin 70°.$$

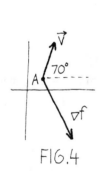

$A = (3,1)$

$\frac{60}{\sqrt{68}}$

$\vec{u}$ ∇f

POINT$(1,-7)$

FIG. 5

(c) (Fig. 5) Consider the path from A toward the point $(1, -7)$, that is, a path in the direction of the vector $\vec{u} = -2\vec{i} - 8\vec{j}$. Then

$$(6) \qquad D_{\vec{u}}f = \frac{\nabla f \cdot \vec{u}}{\|\vec{u}\|} = \frac{60}{\sqrt{68}}$$

so a particle moving through point A and heading toward the point $(1, -7)$ feels the temperature rising instantaneously by $60/\sqrt{68}$ degrees per meter.

Warning 1. The rate of change in (6) is an instantaneous rate *at A*. Once a particle takes a small step past A, it is at a new point with a new gradient, and a new rate of change prevails. The temperature does *not* continue to rise by $60/\sqrt{68}$ degrees per meter as the particle moves along.
 2. Note that in (6), $\vec{u}$ is $(-2, -8)$, *not* $(1, -7)$. In the formula for $D_{\vec{u}}f$, $\vec{u}$ must be a *direction,* that is, a *vector, not a point* toward which the particle moves.

Zero directional derivatives and maximum directional derivatives The projection of a vector in a direction perpendicular to that vector is 0. Since $D_{\vec{u}}f$ is a signed projection of ∇f, $D_{\vec{u}}f$ is 0 *in a direction perpendicular to* ∇f. In other words:

> *If a particle moves through a point in a direction perpendicular to ∇f at that point, it instantaneously feels no change in the temperature.*

In 2-space there are only two directions perpendicular to ∇f; in 3-space, there are infinitely many such directions (Figs. 6 and 7).

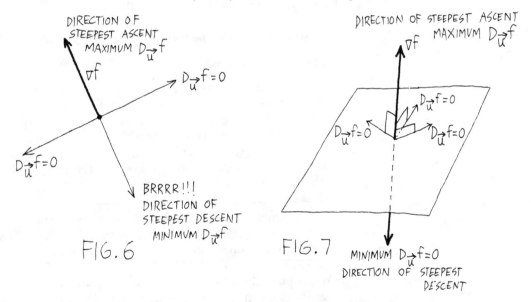

The projection of a vector is maximum in the direction of the vector itself; the maximum value is the vector's own length. Since the directional derivative is a signed projection of ∇f, we have the following results.

> $D_{\vec{u}}f$ at a point is maximum in the direction of ∇f itself. The value of the maximum directional derivative at a point is $\|\nabla f\|$. In other words, *if a particle moves in the direction of ∇f, the temperature rises by $\|\nabla f\|$ degrees per meter, the maximum possible rate. The gradient is said to point in the direction of steepest ascent of f. In the direction of $\vec{u} = -\nabla f$, $D_{\vec{u}}f$ is minimum;* the temperature drops by $\|\nabla f\|$ degrees per meter, the steepest possible drop (Figs. 6 and 7).

Consider Example 1 again where $\nabla f = 6\vec{i} - 9\vec{j}$ at the point $A = (3, 1)$ (Fig. 8). On a path through A in the direction of $6\vec{i} - 9\vec{j}$, the temperature rises by $\|\nabla f\| = \sqrt{117}$ degrees per meter, and this is the maximum rate available at A. In the direction of $-6\vec{i} + 9\vec{j}$, the temperature falls by $\sqrt{117}$ degrees per meter, the maximum drop. In the two directions which are perpendicular to ∇f, namely, $3\vec{i} + 2\vec{j}$ and $-3\vec{i} - 2\vec{j}$, the directional derivative is 0.

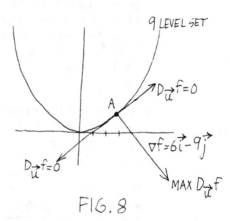

FIG. 8

Warning In Example 1, to attain the maximum rate of change of temperature at A, a particle should move from A in the direction of the *vector* $6\vec{i} - 9\vec{j}$, *not* toward the *point* $(6, -9)$.

The gradient of f and the level sets of f In Example 1, $f(3, 1) = 9$ so point A lies on the 9-level set, $\dfrac{x^2}{y} = 9$, or $y = \frac{1}{9}x^2$, a parabola (Fig. 8). If a particle moves *along* the level set, the temperature f does not change. On the other hand, if it moves from A in the direction of ∇f, the temperature changes maximally. This suggests that ∇f points "directly away" from the level set, that is, in a perpendicular direction. To see this from another point of view, note that in directions perpendicular to ∇f, $D_{\vec{u}}f$ is 0, so these directions coincide with tangents to the level set; therefore ∇f itself is perpendicular to the level set.

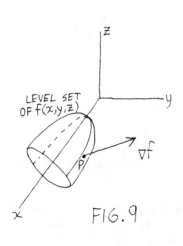

FIG. 9

> In general, if f is a function of two (Fig. 8) or three (Fig. 9) variables, ∇f *at a point is perpendicular to the level set of f through the point.* Of the two perpendicular directions in each case, f points toward higher levels.

(At any point, heat flows in the direction opposite to ∇f at that point, since heat flows down temperature hills from hot to cold; in particular, it flows in the direction in which temperature drops most rapidly. Therefore heat flows on paths perpendicular to level sets of temperature. In 2-space, the heat flow lines are the orthogonal trajectories (Section 4.9) of the temperature level curves.)

Normal vectors to a surface As a by-product of the directional derivative, given a surface in space, we can find a normal vector (that is, a perpendicular) at any point. To illustrate, we will find a normal to the paraboloid

$x = y^2 + z^2 + 4$ at the point $P = (6, 1, 1)$, and then find the tangent plane at the point. Rewrite the equation as

(7) $$y^2 + z^2 + 4 - x = 0$$

so that we can think of the paraboloid as the 0 level set of the function $h(x, y, z) = y^2 + z^2 + 4 - x$ (Fig. 9). Then $\nabla h = -\vec{i} + 2y\vec{j} + 2z\vec{k}$ and $\nabla h|_P = -\vec{i} + 2\vec{j} + 2\vec{k}$. Since the gradient of a function at a point is perpendicular to the level set through the point, the vector $-\vec{i} + 2\vec{j} + 2\vec{k}$ is perpendicular to the paraboloid at P. (There are two directions normal to the paraboloid, "in" versus "out." From Fig. 9, in which the vector is plotted fairly accurately, we see that $-\vec{i} + 2\vec{j} + 2\vec{k}$ happens to be an outer normal, a distinction that is irrelevant in the problem). Besides (7), we can write the equation of the paraboloid in other ways, say

(8) $$\underbrace{y^2 + z^2 - x}_{h_1(x, y, z)} = -4, \quad \underbrace{x - y^2 - z^2}_{h_2(x, y, z)} = 4, \quad \underbrace{x - y^2 - z^2 + 17}_{h_3(x, y, z)} = 21.$$

Therefore the paraboloid can also be viewed as the -4 level set of h_1, the 4 level set of h_2, the 21 level set of h_3, and so on. Since ∇h_1 is the same as ∇h while ∇h_2 and ∇h_3 equal $-\nabla h$ these methods produce either the same outward normal as before, or its negative, an equally acceptable inward normal. The tangent plane at P has $-\vec{i} + 2\vec{j} + 2\vec{k}$ as a normal vector, so an equation for the plane is $-(x - 6) + 2(y - 1) + 2(z - 1) = 0$ (Section 10.2, (1)), or $-x + 2y + 2z + 2 = 0$.

In general, given a surface in 3-space containing point P, to find a normal vector to the surface at P, write the equation of the surface so that all the variables appear on one side (so that the surface is a level set). The normal is the gradient of that side, evaluated at the given point.

$D_{\vec{u}}z$ as a slope Let $z = f(x, y) = 9 - x^2 - 2y^2$, $P = (-2, -1)$. Then $\nabla f = \nabla z = -2x\vec{i} - 4y\vec{j}$, and at P we have $z = 3$, $\nabla z = 4\vec{i} + 4\vec{j}$. The graph of f is a mountain surface containing the point $Q = (-2, -1, 3)$; the level sets in Fig. 10 are the contour curves of the mountain. Climbers at Q can move in many directions on the mountain, such as north, southeast, WSW and so on. Note that these are *two*-dimensional directions (vectors) although the climbers are in *three*-space. The altitude z rises maximally on the path in the direction of ∇z, that is, on a northeast path through Q, at the rate of $\|\nabla z\| = 4\sqrt{2}$ meters up per northeast meter. In other words, of all paths through Q on the mountain, the northeast path ascends most steeply, and its slope is $4\sqrt{2}$. (Note that we still do not admit slopes in 3-space in general. The slope of the northeast path is taken with respect to the indicated *plane* in Fig. 10.)

Figure 11 shows the path through Q in the direction of $\vec{u} = -5\vec{i} + \vec{j}$, mostly west and slightly north. The altitude z on the path is changing at the rate of $D_{\vec{u}}z = \dfrac{\nabla z \cdot \vec{u}}{\|\vec{u}\|} = -\dfrac{16}{\sqrt{26}}$ meters up per meter in the $\vec{u}$ direction.

The climber in Fig. 11 is descending; the path has slope $-\dfrac{16}{\sqrt{26}}$ at Q.

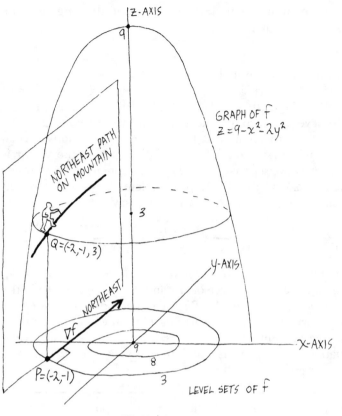

FIG. 10

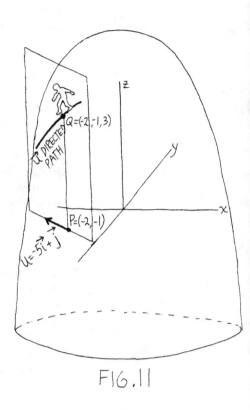

FIG. 11

In general, suppose a surface (i.e., mountain) in 3-space contains the point $Q = (x_0, y_0, z_0)$. Let $\vec{u}$ be a 2-dimensional vector. Consider the path on the mountain through Q in direction $\vec{u}$ (Fig. 11). To find the slope on the path at Q, solve the equation of the surface for z; the slope is $D_{\vec{u}} z = \dfrac{\nabla z \cdot \vec{u}}{\|\vec{u}\|}$ evaluated at $x = x_0$, $y = y_0$. Furthermore, the 2-dimensional vector $\nabla z|_{x=x_0, y=y_0}$ is the direction of steepest ascent up the mountain at Q and $\|\nabla z\|$ itself is the slope of that steepest path.

Warning ∇z is not a 3-dimensional vector and does not point "up the mountain" or perpendicular to the mountain. It is a 2-dimensional vector *lying in the x, y plane* determining (*above it*) the path on the mountain of steepest ascent.

Summary of normal vectors, slopes, direction of steepest ascent Consider a surface in 3-space; as a special case we have the graph of $f(x, y)$ with equation $z = f(x, y)$.

To find a normal to the surface, write the equation of the surface with all variables on one side and take the (3-dimensional) gradient of that side.

Given a 2-dimensional vector $\vec{u}$, to find the slope at a point on the $\vec{u}$-directed path on the surface (Fig. 11) solve the equation for z and find

$D_{\vec{u}} z = \dfrac{\nabla z \cdot \vec{u}}{\|\vec{u}\|}$. The (2-dimensional) vector ∇z itself points in the direction of steepest ascent on the surface.

Example 2 Let $f(x,y) = x^2 + y^2$ and let $P = (1,2,5)$. Then P lies on the graph of f since $f(1,2) = 5$.

(a) Find a normal vector to the graph of f at P.
(b) Find the direction of steepest ascent at P on the graph of f.
(c) Find the slope at P on the northwest path on the graph of f.

Solution: (a) The equation of the graph of f is

(9) $$z = x^2 + y^2.$$

Rewrite the equation as $z - x^2 - y^2 = 0$. Then $\nabla(z - x^2 - y^2) = -2x\vec{i} - 2y\vec{j} + \vec{k}$, and $\nabla(z - x^2 - y^2)\big|_P = -2\vec{i} - 4\vec{j} + \vec{k}$, a normal to the graph of f at P.

(b) From (9), $\nabla z = 2x\vec{i} + 2y\vec{j}$, and $\nabla z\big|_{x=1, y=2} = 2\vec{i} + 4\vec{j}$, the direction of steepest ascent on the graph of f at P.

(c) Let $\vec{u} = -\vec{i} + \vec{j}$, a vector pointing northwest. With the vector ∇z from part (b), the desired slope is $D_{\vec{u}} z = \dfrac{\nabla z \cdot \vec{u}}{\|\vec{u}\|} = \dfrac{2}{\sqrt{2}}$.

Warning Consider a function $f(x,y)$ and its graph $z = f(x,y)$, a mountain surface.

The vector ∇f is *two*-dimensional. It is not perpendicular to the mountain. Rather, perpendiculars are *three*-dimensional and are found, as in (a) above, by rewriting $z = f(x,y)$ as $z - f(x,y) = 0$ or as $f(x,y) - z = 0$ and taking the gradient of the lefthand side.

The vector ∇f (i.e., ∇z) lies in the x,y plane, perpendicular to a level set of f. It is called the direction of steepest ascent up the mountain but does not point up the mountain. It is a direction on the 2-dimensional map on the floor of the expedition tent (as illustrated in Fig. 10).

Problems for Section 11.6

1. Suppose the temperature at the point (x,y) is $xy^2 + 6x + 3$. Find the rate of change of temperature per meter experienced by a particle at point $P = (1,2)$ if it moves through P (a) southwest (b) toward the point $Q = (3, -4)$ (c) toward the x-axis (d) in the direction of ∇ temp (e) WNW, i.e., halfway between W and NW.

2. Let the temperature at a point (x,y) be $x^2 y$. A relay runner going northeast passes the baton at the point $(2,3)$ to a teammate who continues northeast down the track. What rate of change of temperature does each runner experience at the handoff?

3. Let $f(x,y,z) = xy - y^2 + z$ and let $A = (5,2,1)$.

(a) If a particle moves through A away from the z-axis, what rate of change of f is experienced?
(b) In what direction(s) from A is f instantaneously not changing?
(c) Suppose a particle at A and a second particle at $B = (6,4,2)$ start moving toward one another. What rate of change of f does each feel initially? If they move until they meet midway, what rate of change of f will each experience as they pass one another?

(d) If $f(x, y, z)$ is the air pressure at (x, y, z) then a cloud at point A will move in the direction in which air pressure is decreasing most rapidly. Which way does it move initially, and what rate of change of air pressure does it experience?

4. If a particle moves on the path in Fig. 12, does it feel the temperature increasing or decreasing as it passes through P?

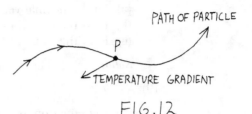

FIG. 12

5. Suppose that at the point P, $D_{\bar{u}} f$ is maximum in the direction of $3\vec{i} + 2\vec{j}$, and the maximum value is 2. Find $D_{\bar{u}} f$ in the north direction at P.

6. Let $A = (1, 2)$. If a particle moves through A toward the point $(1, 1)$, the temperature is rising initially by 2° per meter; if it moves through A toward the point $(7, 10)$, the temperature is initially dropping by 4° per meter. In what direction from A is the temperature rising most rapidly? Find the maximum rate of change of temperature at A.

7. Find ∇f at the point P and sketch the gradient vector and the level set of f through P if (a) $f(x, y) = x^2 y$, $P = (-1, 2)$ (b) $f(x, y, z) = x^2 + 2y^2 - z^2 + 4$, $P = (1, 2, 1)$.

8. Figure 13 shows some level sets of f and g. Find the direction of ∇f at A and of ∇g at B, and decide which gradient is longer.

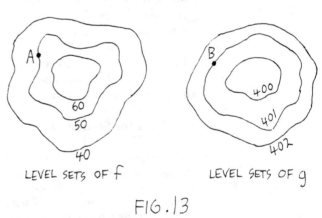

LEVEL SETS OF f LEVEL SETS OF g

FIG. 13

9. Suppose $f(x, y)$ has a relative maximum value at P. Find ∇f at P.

10. Let $f(x, y, z)$ be the distance from (x, y, z) to a fixed line L. Without finding a formula for f, sketch enough level sets and gradient vectors of f to indicate the pattern.

11. Consider the surface $xyz = 12$ and the point $P = (1, 2, 6)$.

(a) Verify that P lies on the surface.
(b) If mountain climbers at P want to climb on the steepest route possible, in what direction should they begin, and what slope do they encounter?
(c) If the climbers move southeast from P, are they ascending or descending the surface? What is the slope at P on their path?
(d) Find a vector normal to the surface at P, an equation for the tangent plane at P, and equations for the normal line through P.

12. Repeat problem 11 with the graph of $f(x,y) = 3x^2 - 2y^2$ as the surface, and $P = (1, -1, 1)$.

13. Let $f(x,y) = 2x^2 + y^2$.

(a) Find ∇f when $x = 1$, $y = 3$.
(b) Find $\nabla(z - 2x^2 - y^2)$ when $x = 1$, $y = 3$.
(c) The gradients in (a) and (b) are perpendicular to what curves or surfaces?

14. Suppose the earth's surface is the ellipsoid $x^2 + 2y^2 + 3z^2 = 15$, and the temperature in space at the point (x, y, z) is $2xz + y^2 + 6$. You are in a space ship at the point $P = (1, 1, 2)$ on the earth's surface, about to be launched perpendicularly into space away from the earth (Fig. 14).

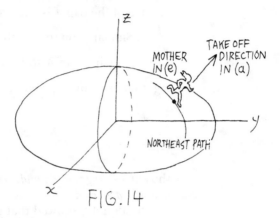

FIG. 14

(a) Find this outward perpendicular direction.
(b) What rate of change of temperature do you feel just as the ship begins to take off?
(c) Your mother worries about you, and she wants the ship to take off from P not perpendicularly away from the earth, but in a direction in which temperature is not initially changing (she thinks that changes in temperature cause colds). Find three specific directions (of the infinitely many available) which NASA can use to get no temperature change initially, if they care as much about your health as your mother does.
(d) You may not have thought of this in part (c), but you want the three specific directions to take you out into space and not burrowing into the earth (your mother would rather see you catch cold than crash). Check that your answers to (c) do not burrow, and change them if necessary.
(e) Your mother has just proudly watched you take off and now walks northeast off the field from point P. Find the slope at P of her path on the earth's surface.

11.7 Differentials and Exact Differential Equations

Section 4.9 discussed separable differential equations. This section will solve another type of differential equation, called exact.

Approximating a change in $f(x,y)$ If $y = f(x)$ and x changes by dx then Section 4.8 defined the differential of f by $dy = f'(x)\,dx$, and showed that dy approximates the corresponding change in y. Now suppose that $z = f(x,y)$ and x and y change slightly by dx and dy. We want to approximate the corresponding change in z. If $f(x,y)$ is the temperature at the point (x,y)

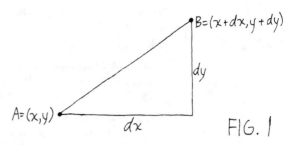

FIG. 1

then the change in z is the change in temperature as a particle moves from the point $A = (x, y)$ to the point $B = (x + dx, y + dy)$ (Fig. 1). On the east leg of the trip, f is a function of x alone, so z changes by approximately $\frac{\partial f}{\partial x} dx$. Similarly, on the north leg, z changes by approximately $\frac{\partial f}{\partial y} dy$. Therefore on a trip from A to B, the superposition of the east leg and the north leg, z changes by approximately $\frac{\partial f}{\partial x} dx + \frac{\partial f}{\partial y} dy$. We define the *differential* of $z = f(x, y)$ by

$$(1) \qquad dz = \frac{\partial f}{\partial x} dx + \frac{\partial f}{\partial y} dy;$$

thus, *if x changes by dx and y changes by dy then the corresponding change in z is approximated by dz.*

Section 4.8 showed that if $y = f(x)$, then approximating the change in y by $f'(x) dx$ amounts to approximating a change in the height of the graph of f by the change in the tangent line. Similarly, it can be shown that if $z = f(x, y)$ then approximating the change in z by (1) corresponds to approximating a change in the height of the graph of f by the change in the tangent plane.

Mathematicians use the notation ∇z for the change in z and use dz for the differential in (1) which approximates the change in z. In applied fields, and in this text, the distinction between $\frac{\partial f}{\partial x} dx + \frac{\partial f}{\partial y} dy$ and the change in z is often blurred, and both are referred to as dz; i.e., we often take the liberty of claiming that

$$(1') \qquad \boxed{\begin{aligned} dz &= \frac{\partial f}{\partial x} dx + \frac{\partial f}{\partial y} dy \\ &= \text{change in } z \text{ when } x \text{ changes by } dx \text{ and } y \text{ changes by } dy. \end{aligned}}$$

Example 1 Let $z = x^2 y^3$. Then we write

$$dz = \frac{\partial z}{\partial x} dx + \frac{\partial z}{\partial y} dy = 2xy^3 dx + 3x^2 y^2 dy,$$

meaning that if x and y change by dx and dy, respectively, there is a corresponding change in z given approximately by $2xy^3 dx + 3x^2 y^2 dy$.

Example 2 To find $d(3q^2)$, use the one-dimensional differential formula $dy = f'(x) dx$, since only one varible is involved, to get $d(3q^2) = 6q \, dq$.

Sum, product, quotient and chain rules for differentials Let u and v be functions of one or more variables. Then

(2) $$d(u + v) = d(u) + d(v)$$

(3) $$d(uv) = u\,d(v) + v\,d(u)$$

(4) $$d\left(\frac{u}{v}\right) = \frac{v\,d(u) - u\,d(v)}{v^2}$$

(5) $$d[f(u)] = f'(u)\,d(u).$$

For example, $d(\ln u) = \dfrac{1}{u}\,d(u)$, $d(\sin u) = \cos u\,d(u)$.

A differential can always be found directly, using (1), but sometimes (2)–(5) are more convenient. For example to find $d\,\ln(2x + 3y)$ by (1), we have

$$d\,\ln(2x + 3y) = \frac{\partial\,\ln(2x + 3y)}{\partial x}\,dx + \frac{\partial\,\ln(2x + 3y)}{\partial y}\,dy$$

$$= \frac{2}{2x + 3y}\,dx + \frac{3}{2x + 3y}\,dy.$$

But also

$$d\,\ln(2x + 3y) = \frac{1}{2x + 3y}\,d(2x + 3y) \quad \text{(by (5))}$$

$$= \frac{1}{2x + 3y}(2\,dx + 3\,dy) \quad \text{(by (2))}$$

$$= \frac{2\,dx + 3\,dy}{2x + 3y}.$$

Exact differentials Example 1 began with a function $f(x, y) = x^2 y^3$ and found its differential, $df = 2xy^3\,dx + 3x^2 y^2\,dy$. To identify and solve exact differential equations we will be concerned with the opposite problem: given the differential expression $2xy^3\,dx + 3x^2 y^2\,dy$, find a function $f(x, y)$ with that differential. In general, an expression of the form

(6) $$p(x, y)\,dx + q(x, y)\,dy$$

is called a *differential form*. It is possible (indeed likely) that (6) simply is not df for any f. If there does exist a function $f(x, y)$ such that

(7) $$df = p(x, y)\,dx + q(x, y)\,dy$$

then the differential is called *exact*. In other words, (6) *is exact if there is an* $f(x, y)$ *such that*

(8) $$\frac{\partial f}{\partial x} = p(x, y) \quad \text{and} \quad \frac{\partial f}{\partial y} = q(x, y).$$

For example, consider the differential

(9) $$\underbrace{(3x^2 y^2 + 2y^3 + x)}_{p}\,dx + \underbrace{(2x^3 y + 6xy^2 + \cos y + 7)}_{q}\,dy$$

The problem is to find $f(x,y)$, if possible, so that (7) and (8) hold. Begin by antidifferentiating p with respect to x to obtain the terms

$$(10) \qquad x^3y^2 + 2xy^3 + \tfrac{1}{2}x^2.$$

The derivative with respect to y of this tentative answer is

$$(11) \qquad 2x^3y + 6xy^2.$$

Compare this result with q in (9). Since (11) lacks $\cos y + 7$, expand (10) by adding $\sin y + 7y$ to obtain $x^3y^2 + 2xy^3 + \tfrac{1}{2}x^2 + \sin y + 7y$. Note that expanding the answer does not change its partial derivative with respect to x, since *the additional terms do not contain the variable x*. Thus, the final answer, including the standard arbitrary constant, is $f(x,y) = x^3y^2 + 2xy^3 + \tfrac{1}{2}x^2 + \sin y + 7y + C$. Check the answer by finding its partials to see that p and q are obtained.

Example 3 Let

$$(12) \qquad p = 3x^2y^2 + 2y^3 \quad \text{and} \quad q = 2x^3y + 6xy^2 + 8xy^3.$$

Try, but find it impossible, to obtain an f such that $df = p\,dx + q\,dy$. In other words, show that $p\,dx + q\,dy$ is *not* exact.

  Solution: Antidifferentiate p to obtain the terms

$$(13) \qquad x^3y^2 + 2xy^3.$$

Differentiate this tentative answer with respect to y to obtain

$$(14) \qquad 2x^3y + 6xy^2$$

and compare with q. The term $8xy^3$ is missing from (14) and can be produced only if (13) is expanded to $x^3y^2 + 2xy^3 + 2xy^4$. However, $2xy^4$ *contains the variable x*, so the expanded "answer" no longer has the desired partial p with respect to x. We conclude that it is not possible to find f with partials p and q; the differential $p\,dx + q\,dy$ is not exact.

A criteria for exactness Given $p\,dx + q\,dy$, one way to decide if there exists an f such that (7) holds, is to simply try to find it as in the preceding examples. It is also possible to develop a test for determining *in advance* if f exists. Then the antidifferentiation process for *finding f* need be used only when the criterion guarantees the *existence* of f. We will find the criterion and then use it in examples.

  If (7) holds then $\dfrac{\partial f}{\partial x} = p$ and $\dfrac{\partial f}{\partial y} = q$, so $\dfrac{\partial q}{\partial x} = \dfrac{\partial^2 f}{\partial x\,\partial y}$ and $\dfrac{\partial p}{\partial y} = \dfrac{\partial^2 f}{\partial y\,\partial x}$;

hence $\dfrac{\partial q}{\partial x} = \dfrac{\partial p}{\partial y}$. In more advanced courses the converse can be proved: if $\dfrac{\partial q}{\partial x} = \dfrac{\partial p}{\partial y}$ then (7) holds. We restate these results as our criterion:

$$(15) \qquad \boxed{If\ \dfrac{\partial q}{\partial x} \neq \dfrac{\partial p}{\partial y}\ then\ p\,dx + q\,dy\ is\ not\ exact.}$$

$$(16) \qquad \boxed{If\ \dfrac{\partial q}{\partial x} = \dfrac{\partial p}{\partial y}\ then\ p\,dx + q\,dy\ is\ exact.}$$

For example, let $p(x, y) = xy$ and $q(x, y) = x^5 + 2xy^3$. Then $\dfrac{\partial q}{\partial x} = 5x^4 + 2y^3$ and $\dfrac{\partial p}{\partial y} = x$. The two are not identical, so $p\,dx + q\,dy$ is *not* exact. If you choose to not use the criterion and simply try to find f so that $p\,dx + q\,dy = df$, you will find it impossible (as in Example 3); thus you will, in a more roundabout fashion, conclude that $p\,dx + q\,dy$ is not exact.

A table of exact differentials You have already seen that integral tables are available to help in antidifferentiation problems. Similarly, there are tables of exact differentials to facilitate finding f such that $df = p\,dx + q\,dy$. In (17)–(22), we select some items from the tables for reference:

$$(17) \qquad \frac{y\,dx - x\,dy}{y^2} = d\!\left(\frac{x}{y}\right)$$

$$(18) \qquad \frac{x\,dy - y\,dx}{x^2} = d\!\left(\frac{y}{x}\right)$$

$$(19) \qquad \frac{-2x\,dx - 2y\,dy}{(x^2 + y^2)^2} = d\!\left(\frac{1}{x^2 + y^2}\right)$$

$$(20) \qquad \frac{x\,dx + y\,dy}{\pm\sqrt{x^2 + y^2}} = d(\pm\sqrt{x^2 + y^2})$$

$$(21) \qquad \frac{2x\,dx + 2y\,dy}{x^2 + y^2} = d\,\ln(x^2 + y^2)$$

$$(22) \qquad \frac{-y\,dx + x\,dy}{x^2 + y^2} = d\!\left(\tan^{-1}\frac{y}{x}\right).$$

Exact differential equations Consider the equation $y' = \dfrac{2x - y^3}{3xy^2}$. Then $\dfrac{dy}{dx} = \dfrac{2x - y^3}{3xy^2}$ so $3xy^2\,dy = (2x - y^3)\,dx$; the equation is not separable (Section 4.9). Now we try a second approach. Write the equation as

$$(23) \qquad \underbrace{(y^3 - 2x)}_{p}\,dx + \underbrace{3xy^2}_{q}\,dy = 0.$$

Since $\dfrac{\partial q}{\partial x} = \dfrac{\partial p}{\partial y}$ $(= 3y^2)$, the left side of (23) is an exact differential df. To find f, antidifferentiate p with respect to x to obtain the terms $xy^3 - x^2$. The derivative of this tentative f with respect to y is $3xy^2$, precisely q, so the tentative f is final. Therefore the differential equation may be written as $d(xy^3 - x^2) = 0$. Since the differential is 0, if x changes by dx and y changes by dy, the function $xy^3 - x^2$ itself does not change. Therefore it is a constant function. (In general $f(x, y)$ *is constant if and only if* $df = 0$, analogous to the one-dimensional result that $f'(x)$ is constant if and only if $f' = 0$.) Thus $xy^3 - x^2 = K$ where K is an arbitrary constant, and this describes an *implicit* solution y to the original differential equation. The *explicit* solution is found by solving for y to obtain $y = \sqrt[3]{\dfrac{x^2 + K}{x}}$.

> In general, *consider the differential equation* $p(x, y)\, dx\, + \, q(x,y)\, dy \,=\, 0$ *where* $\dfrac{\partial q}{\partial x} = \dfrac{\partial p}{\partial y}$. *The left side of the equation is an exact differential, the equation is called exact, and there is a function* $f(x, y)$ *such that the equation can be written as* $df = 0$. *The solution* $y(x)$ *to the differential equation is given implicitly by* $f(x, y) = K$. *An explicit solution is found by solving the implicit solution for* y *if possible.*
>
> If a differential equation can be written as $df = dg$ (rather than simply as $df = 0$) then its solution is given implicitly by $f(x, y) = g(x, y) + K$.

Example 4 (a) Solve $y' = \dfrac{x^2 - y}{x}$. (b) Find the particular solution satisfying the condition $y(3) = 1$.

Solution: (a) Write the equation as $\dfrac{dy}{dx} = \dfrac{x^2 - y}{x}$, or

$$\underbrace{(x^2 - y)\, dx}_{p} \,\, \underbrace{- \, x\, dy}_{q} \, = \, 0 \, .$$

Since $\dfrac{\partial q}{\partial x}$ and $\dfrac{\partial p}{\partial y}$ both equal -1, the equation is exact. In particular, it may be written as $d(\tfrac{1}{3}x^3 - xy) = 0$, so the solution is given implicitly by

$$(24) \qquad\qquad \frac{1}{3}x^3 - xy = K \, ,$$

and explicitly by

$$(25) \qquad\qquad y = \frac{1}{3}x^2 - \frac{K}{x} \, .$$

(b) To determine K, substitute $x = 3$, $y = 1$ in either (24) or (25). Using (24) which is more convenient, we have $9 - 3 = K$, $K = 6$, so the solution is $y = \tfrac{1}{3}x^2 - 6/x$.

Warning The solution to (a) is *not* $f(x, y) = \tfrac{1}{3}x^3 - xy$ and is *not* $\tfrac{1}{3}x^3 - xy$. The solution is the function $y(x)$ defined implicitly by the equation $\tfrac{1}{3}x^3 - xy = K$.

Integrating factors Consider the equation $y\, dx - x\, dy = y^3\, dy$. The right side is an exact differential, namely $d(\tfrac{1}{4}y^4)$, but the left side is not exact since $p(x, y) = y$, $q(x, y) = -x$ and $\dfrac{\partial q}{\partial x} \neq \dfrac{\partial p}{\partial y}$. However, compare the left side with (17) to see that it can be made exact by multiplying by $1/y^2$. Therefore, we multiply on both sides to obtain $\dfrac{y\, dx - x\, dy}{y^2} = y\, dy$. The left side is now an exact differential, and fortunately, the right side remains exact. The equation may be written as $d\left(\dfrac{x}{y}\right) = d(\tfrac{1}{2}y^2)$; the implicit solution is $\dfrac{x}{y} = \tfrac{1}{2}y^2 + K$. It is not convenient to solve for y and obtain the explicit solution, so we settle for the implicit version.

A factor, $1/y^2$ in this case, which changes a differential equation from nonexact to exact is called an *integrating factor*. There is no precise rule for finding integrating factors, or for determining if one exists at all, but the exact differentials in (17)–(22) often serve as goals.

Problems for Section 11.7

1. Check formulas (18)–(22) by finding the differential indicated on the right-hand side.

2. Suppose a point has polar coordinates r, θ and rectangular coordinates x, y. If r changes by dr and θ changes by $d\theta$, find dx and dy.

3. Decide if the expression is an exact differential df, and if so, find f.

(a) $2xy\,dx + y\,dy$ (b) $(x^3 + 3x^2y)\,dx + (x^3 + y^3)\,dy$ (c) $\dfrac{y}{x^2}dx + \left(5 - \dfrac{1}{x}\right)dy$

4. Find q so that $xy^3\,dx + q\,dy$ is exact.

5. Solve the differential equation if it is exact. Find the explicit solution whenever possible.

(a) $(6x^2 + y^2)\,dx + (2xy + 3y^2)\,dy = 0$ (e) $(2r\cos\theta - 1)\,dr = r^2\sin\theta\,d\theta$

(b) $(3x^2 + y)\,dx + x\,dy = 0$ (f) $(x + y)\,dx + (x^2 + y^2)\,dy = 0$

(c) $y' = \dfrac{x - y\cos x}{y + \sin x}$ (g) $\cos x\cos y\,dx - \sin x\sin y\,dy$
$\qquad\qquad\qquad\qquad = x^3\,dx$

(d) $y' = e^{xy}$ (h) $(ye^{-x} - \sin x)\,dx = (e^{-x} + 2y)\,dy$

6. Find the particular solution satisfying the given condition.

(a) $2xy\,dx + (x^2 + y)\,dy = 0$, $y(1) = 4$

(b) $2\sin(2x + 3y)\,dx + 3\sin(2x + 3y)\,dy = 0$, $y(0) = \pi/2$

(c) $\dfrac{1}{x + y}dx + \dfrac{1}{x + y}dy = dx$, $y = 1$ when $x = 0$

7. The equation $(x^2 + 2)\,dx + 3y\,dy = 0$ is both exact and separable. Solve it twice.

8. Find an integrating factor which makes the equation exact, and then solve.

(a) $(x^2 + y^2)\,dx = x\,dy - y\,dx$ (c) $\sqrt{x^2 + y^2}\,dy = x\,dx + y\,dy$

(b) $y\,dx - x\,dy = y^2\,dx$ (d) $x\,dx + y\,dy = (x^2 + y^2)\,dy$

REVIEW PROBLEMS FOR CHAPTER 11

1. Sketch the graph and some level sets if

(a) $$f(x,y) = 2x + 3y$$

(b) $$f(x,y) = \sqrt{x^2 + y^2}.$$

2. Sketch enough level sets of f to indicate the pattern if

(a) $$f(x,y,z) = y^2 + z^2$$

(b) $$f(x,y,z) = 5 - x^2 - 2y^2 - 3z^2.$$

3. If $f(x,y,z) = xyz$ find (a) $\dfrac{\partial^2 f}{\partial x\,\partial z}$ (b) $\dfrac{\partial^3 f}{\partial x^2\,\partial z}$.

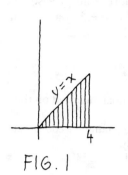

FIG. 1

4. Show that if $z = z(x, y)$ where $x = r \cos \theta$ and $y = r \sin \theta$ then

$$\left(\frac{\partial z}{\partial r}\right)^2 + \frac{1}{r^2}\left(\frac{\partial z}{\partial \theta}\right)^2 = \left(\frac{\partial z}{\partial x}\right)^2 + \left(\frac{\partial z}{\partial y}\right)^2.$$

5. If $u = u(x, y, z)$, $x = 2a + 3b$, $y = 3b + 4c$, $z = ac$, find $\dfrac{\partial^2 u}{\partial a \, \partial c}$.

6. Find the maximum and minimum values of $xy + 2y^2 - 12y$ in the indicated triangular region (Fig. 1).

7. Let $f(x, y) = 3x^2 + 4y^2$. Of all the directional derivatives of f at all points on the unit circle $x^2 + y^2 = 1$, find the maximum. In what direction and at what point does the maximum occur?

8. Let $f(x, y, z) = x^2yz$, $P = (1, 2, 3)$, $Q = (0, -1, 1)$, $R = (2, -3, 5)$. If a particle arrives at P from Q and then leaves for R, find the directional derivative of f upon arrival and upon departure at P.

9. Suppose the temperature T at the point (x, y) is $x^2 - y$.

(a) Find the level set of T which passes through the point $(-2, 2)$ and find a vector perpendicular to the level set at that point.

(b) Let $Q = (-2, 2, 2)$.

    (i) Verify that Q lies on the graph of T.
    (ii) Find a vector perpendicular to the graph at Q.
    (iii) Find the slope at Q on the northwest path on the graph.
    (iv) Find the path on the graph through Q which rises most steeply.

10. The surface area S of a cylinder with radius r and height h is $2\pi rh + 2\pi r^2$. Find dS if r changes by dr and h changes by dh.

11. Find the explicit solution to $y' = \dfrac{2x - y}{x}$ satisfying the condition $y(1) = 2$.

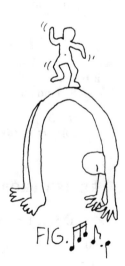

FIG.♪♪.

12/MULTIPLE INTEGRALS

12.1 Definition and Some Applications of the Double Integral

In this section we will review the definition of the integral of a function of *one* variable, $\int_a^b f(x)\,dx$, and then define a new integral involving a function of *two* variables. A comparison of the two integrals will reveal similar constructions and common applications. In the next section we will begin computing the new integral.

Definition of the integral of $f(x)$ on the interval $[a, b]$ Given a function $f(x)$ and an interval $[a, b]$ on a line, divide the interval into many subintervals, not necessarily of the same length. Let dx be the length of a typical subinterval, and let x be a number in the subinterval (Fig. 1). For each subinterval, compute the value of f at x and multiply by dx. Add the results from all the subintervals to obtain $\sum f(x)\,dx$. Repeat the process with smaller and smaller values of dx, which requires more and more subintervals. The integral of $f(x)$ on $[a, b]$ is defined by

(1)
$$\int_a^b f(x)\,dx = \lim_{dx \to 0} \sum f(x)\,dx.$$

We think of the integral as adding many representative values of f from the interval, each weighted by the length of the subinterval it represents.

With a new integral about to be defined we will refer to (1), which involves a function of one variable on a one-dimensional interval, as the *single* integral.

Definition of the integral of $f(x, y)$ on a region in the plane (the double integral) Given a function $f(x, y)$ and a region in the plane, divide the region into many small subregions, not necessarily of the same area. Let a typical subregion contain the point (x, y) and have area dA (Fig. 2). For each subregion, find the value of f at (x, y), and multiply by dA. Add the results from all the subregions to obtain $\sum f(x, y)\,dA$. Repeat the process with smaller and smaller values of dA, which requires more and more subregions. It is likely that the resulting sums will be close to one particular number eventually, that is, the sums will approach a limit. The limit is called the *double integral* of $f(x, y)$ on the region and is denoted by $\int_{\text{region}} f(x, y)\,dA$.

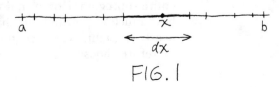

FIG. 1

363

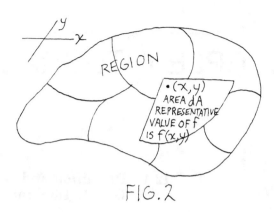

FIG. 2

---

(2)

> In summary,
>
> $$\int_{\text{region}} f(x, y) \, dA = \lim_{dA \to 0} \sum f(x, y) \, dA .$$
>
> We think of the double integral as adding many representative values of f from the region, each weighted by the area of the subregion it represents.

The function $f(x, y)$ is called the *integrand* and the region is called the *region of integration*.

Common application of the two integrals to average value   Suppose a city is divided into three boroughs, I, II, and III, with temperatures 68°, 72°, and 69°, respectively. Then the average temperature in the city is the weighted average of the three borough temperatures. Weigh each temperature by the area of the borough, add, and divide by the sum of the weights, that is, by the total area of the city, to obtain

$$\text{average temperature} = \frac{68° \times \text{area I} + 72° \times \text{area II} + 69° \times \text{area III}}{\text{area of the city}} .$$

Now suppose that the temperature in the city (a region in the plane) is $f(x, y)$ at the point (x, y), not necessarily constant over each borough. The problem is to find the average temperature in the city in this more general situation. Divide the city into many subregions (Fig. 2). Let a typical subregion contain the point (x, y) and have area dA. Let the temperature f at the point (x, y) represent the entire subregion. To find the average temperature, weight each subregion's representative temperature by the area of the region, add, and divide by the sum of the weights, that is, the total city area. Therefore the average temperature in the city is approximately $\frac{\sum f(x, y) \, dA}{\text{total area}}$. This is only an approximation because the representative temperature chosen from each subregion does not necessarily prevail over the entire subregion. However, if the subregions are very small, then f doesn't have much opportunity to change within a subregion and remains more nearly constant. We expect the approximation to improve as $dA \to 0$, and therefore choose

$$\text{average city temperature} = \frac{\lim\limits_{dA \to 0} \sum f(x,y)\, dA}{\text{total area}} = \frac{\displaystyle\int_{\text{city}} f(x,y)\, dA}{\text{area of city}}.$$

In general, we have the following two-dimensional model for average value:

(3) $\displaystyle \text{average value of } f(x,y) \text{ in a region in 2-space} = \frac{\displaystyle\int_{\text{region}} f(x,y)\, dA}{\text{area of region}}.$

This extends the result in (3) of Section 5.2:

(4) $\displaystyle \text{average value of } f(x) \text{ on the interval } [a,b] = \frac{\displaystyle\int_a^b f(x)\, dx}{b - a}.$

Common application to computing a total amount given a variable density   We will illustrate the idea with total charge and charge density. Similar results hold for mass and mass density, population and population density, and so on.

If the area of a region in 2-space is 10 square centimeters and it has a constant charge density of 4 coulombs per square centimeter, then its total charge Q is 40 coulombs (total charge is density $\times$ area). Now suppose that a region has *variable* charge density $f(x,y)$. To find the total charge, divide the region into many subregions and let a typical subregion contain the point (x,y) and have area dA (Fig. 2). Let the density of the subregion be represented by the value of f at the point (x,y). Then, by the formula density $\times$ area, the charge dQ of the subregion is given approximately by $dQ = f(x,y)\, dA$. This is only an approximation to dQ because the representative density $f(x,y)$ chosen from the subregion does not necessarily prevail over the entire subregion. To find the precise total charge Q, add dQ's, and let $dA \to 0$ to squeeze out the approximation error. Integration accomplishes both tasks, so

(5) $\displaystyle \text{total charge of a region} = \int_{\text{region}} \text{charge density } f(x,y)\, dA.$

Similarly, if the charge density of a rod along the interval $[a,b]$ is $f(x)$ coulombs per centimeter at the point x, then

(6) $\displaystyle \text{total charge of the rod} = \int_a^b \text{charge density } f(x)\, dx.$

Common application to the size of the region of integration   If we let $f(x) = 1$ in (2) then the double integral adds dA's and produces the total area of the region of integration. In other words,

(7) $\displaystyle \int_{\text{region}} dA = \text{area of the region}.$

Analogous to (7),

(8) $$\int_a^b dx = b - a = \text{length of the interval } [a,b].$$

Common application involving the graphs of $f(x)$ and $f(x,y)$ Assume that $f(x) \geq 0$ for x in $[a,b]$ and extend Fig. 1 to include the graph of $f(x)$ above the x-axis (Fig. 3). The rectangle in Fig. 3 has base dx, height $f(x)$ and area $f(x)\,dx$. The sum of the areas of such rectangles approximates the area under the curve. The approximation improves as $dx \to 0$, so

if $f(x) \geq 0$ then

(9) $$\int_a^b f(x)\,dx = \begin{array}{l}\text{area under the graph of } f(x) \text{ and over the}\\ \text{interval } [a,b] \text{ on the } x\text{-axis}.\end{array}$$

Similarly, assume that $f(x,y) \geq 0$ for (x,y) in a Region R in 2-space, and extend Fig. 2 to include the graph of $f(x,y)$ above the x,y plane (Fig. 4). The tube in Fig. 4 has base dA, height $f(x,y)$ and volume $f(x,y)\,dA$. The sum of the volumes of such tubes approximates the volume under the graph of f. The approximation improves as $dA \to 0$ so we have the following result.

(10)
> If $f(x,y) \geq 0$ then
>
> $$\int_R f(x,y)\,dA = \begin{array}{l}\text{volume under the graph of } f(x,y) \text{ and over}\\ \text{the region } R \text{ in the } x,y \text{ plane}.\end{array}$$

Once you learn to compute double integrals, you can use (7) to find areas and (10) to find volumes.

More generally, if the restriction $f(x) \geq 0$ is removed, then

(11)
$$\int_a^b f(x)\,dx = \begin{bmatrix}\text{area } above \text{ the } x\text{-axis}\\ \text{between the graph of } f\\ \text{and the interval } [a,b]\end{bmatrix}$$
$$- \begin{bmatrix}\text{area } below \text{ the } x\text{-axis}\\ \text{between the graph of } f\\ \text{and the interval } [a,b]\end{bmatrix}$$

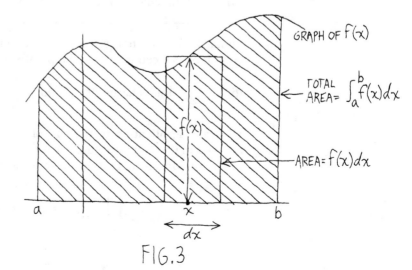

FIG.3

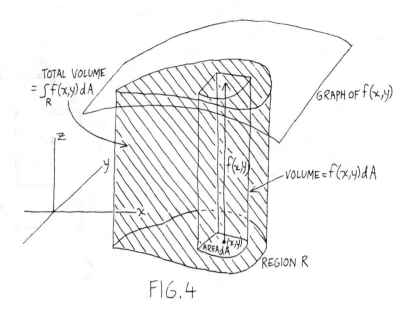

FIG. 4

Similarly, if the restriction $f(x,y) \geq 0$ is removed, then

(12)

$$\int_R f(x,y)\,dA \;=\; \begin{bmatrix} \text{volume } above \text{ the } x,y \text{ plane} \\ \text{between the graph of } f \\ \text{and the region } R \end{bmatrix}$$
$$-\; \begin{bmatrix} \text{volume } below \text{ the } x,y \text{ plane} \\ \text{between the graph of } f \\ \text{and the region } R \end{bmatrix}$$

Example 1   Express with a double integral the volume of the solid in Fig. 5, a box topped by a portion of the sphere centered at the origin with radius 6.

Solution: The sphere has equation $x^2 + y^2 + z^2 = 36$, and the top of the sphere, where z is positive, has equation $z = \sqrt{36 - x^2 - y^2}$. Therefore, the box lies under the graph of $f(x,y) = \sqrt{36 - x^2 - y^2}$, over the square region R in the x,y plane (Fig. 5), and, by (10), its volume is $\int_R \sqrt{36 - x^2 - y^2}\,dA$.

The nonconnection between $f(x,y)$ and the region of integration   When we compute double integrals in the next section, the function and the region will be chosen arbitrarily, to provide practice. In applications, the function and the region are chosen to achieve a desired result. To find total charge, the integrand $f(x,y)$ is the charge density in the x,y plane and the region of integration is the region whose total charge is desired. To find the volume of a solid, the integrand is the function $f(x,y)$ whose graph is the roof of the solid, and the region of integration is the floor of the solid in the x,y plane. To find an area in the x,y plane, the integrand is 1 and the region of integration is the region whose area is desired. In any case, the region of integration is *not* the graph of the integrand $f(x,y)$; rather, the graph of $f(x,y)$ is a surface in 3-space lying above and/or below the region in the x,y plane.

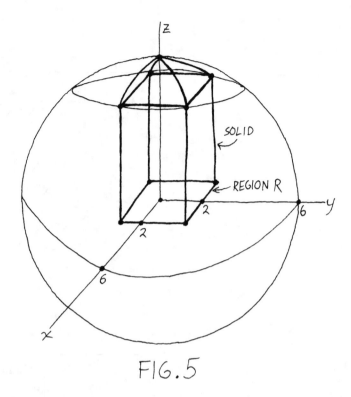

FIG. 5

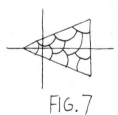

FIG. 6

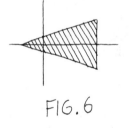

FIG. 7

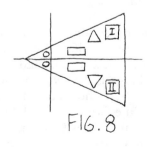

FIG. 8

Example 2 To reinforce the definition of the double integral as a weighted sum, we will show how it may be possible to estimate the sign of a double integral without actually computing its value. Let R be the region in Fig. 6, and consider $\int_R xy^2\,dA$. Suppose the region is divided into small subregions (Fig. 7). Each subregion has a positive area dA and a representative value of xy^2, which is negative if the region is to the left of the y-axis and positive if the subregion is to the right of the y-axis. From the location of the region R we see that $\sum xy^2\,dA$ is positive, since the positive terms outweigh the negative terms. Therefore $\int_R xy^2\,dA$ is positive.

Alternatively, consider the graph of $z = xy^2$ in 3-space. Since it is neither a cylindrical nor a quadratic surface, the graph is not easy to draw, but by considering the sign of z we can see that the graph lies below that portion of R to the left of the y-axis and lies above that portion of R to the right of the y-axis. There is more volume above than below so, by (12), $\int_R xy^2\,dA$ is positive.

Example 3 Continue with the Region R in Fig. 6 and find the sign of $\int_R x^2 y\,dA$.

Solution: Figure 8 shows a revealing subdivision of R. For each subregion above the x-axis, such as I where the value of $x^2 y\,dA$ is positive, there is a corresponding subregion II below the x-axis where $x^2 y\,dA$ takes on the negative of that value. Therefore $\sum x^2 y\,dA = 0$, and hence $\int_R x^2 y\,dA = 0$. Alternatively, the integral is 0 because the surface $z = x^2 y$ determines as much volume above as below the region R.

Some properties of the double integral As with the single integral, a constant factor may be pulled out of an integral sign. In other words,

(13) $$\int_R cf(x,y)\,dA \;=\; c\int_R f(x,y)\,dA\,.$$

A single integral on the interval $[2,7]$ may be computed, if desired, by integrating separately on say $[2,4]$ and $[4,7]$ and then adding the results. Similarly if a region is divided into two parts, R_1 and R_2, like a city divided into two precincts, then

(14) $$\int_{\text{region}} f(x,y)\,dA \;=\; \int_{R_1} f(x,y)\,dA \;+\; \int_{R_2} f(x,y)\,dA\,.$$

Problems for Section 12.1

1. Let R be a circular region with radius 2. Find $\int_R 5\,dA$.

2. Express with a double integral the volume of (a) a hemisphere with radius 3   (b) a cylinder with radius 2 and height 5.

3. Let R_1 and R_2 be the square and circular regions in Fig. 9. Describe the solid whose volume is (a) $\int_{R_1} 5\,dA$ (b) $\int_{R_2}(x^2 + y^2)\,dA$.

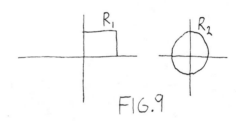

FIG. 9

4. Refer to Fig. 10 and decide if the integral is positive, negative or zero.

(a) $\displaystyle\int_{\text{III}} xy\,dA$ (d) $\displaystyle\int_{\text{III}} xy^2\,dA$

(b) $\displaystyle\int_{\text{I and II}} xy\,dA$ (e) $\displaystyle\int_{\text{IV}} dA$

(c) $\displaystyle\int_{\text{I and II}} xy^2\,dA$

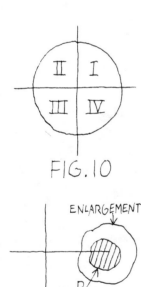

FIG. 10

ENLARGEMENT

FIG. 11

5. Suppose $\int_R f(x,y)\,dA = 7$. If the region R is enlarged (Fig. 11), will the value of the integral increase?

6. Find $\int_R 0\,dA$.

7. True or False?

(a) If $f(x,y) > g(x,y)$ for all (x,y) in R, then $\int_R f(x,y)\,dA > \int_R g(x,y)\,dA$.

(b) If the region R_1 is larger in area then the region R_2, then $\int_{R_1} f(x,y)\,dA > \int_{R_2} f(x,y)\,dA$.

(c) If the region R lies in quadrant I, then $\int_R f(x,y)\,dA > 0$.

(d) If the region R_1 is larger in area than the region R_2 then $\int_{R_1} dA > \int_{R_2} dA$.

(e) If $f(x,y) > 0$ for all (x,y) in R then $\int_R f(x,y)\,dA > 0$.

8. The average value of x in a circular region is clearly the x-coordinate of its center. Use this, together with (3), to find $\int_R x\,dA$ over the circular region R with center $(4,6)$ and radius 3.

9. Refer to Fig. 10 and decide if $\int x\,dA$ on the entire circular region can be computed by finding $\int x\,dA$ on region I (one-quarter of the circle) and then multiplying by 4.

12.2 Computing Double Integrals

We'll begin by evaluating a specific integral. The methods we develop will then be summarized and used in general.

Example 1 Consider $\int_R xy^2\, dA$ where R is the semicircular region of radius 2 in Fig. 1. Divide the region into rectangular subregions using horizontal and vertical lines. A typical subregion contains the point (x, y) and has dimensions dx by dy, so that $xy^2\, dA = xy^2\, dx\, dy$. To find the integral, we will first add $xy^2\, dA$'s across a typical horizontal strip and then add the strip sums from bottom to top. On the typical horizontal strip at height y in Fig. 2, the left end is $x = -\sqrt{4 - y^2}$ (choose the *negative* square root since x is negative on the left half of the circle) and the right end is $x = 0$. Adding $xy^2\, dx\, dy$'s across this strip is a one-dimensional integral problem involving the interval $[-\sqrt{4 - y^2}, 0]$ on an x-line, so

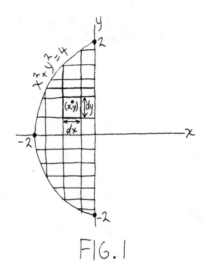

FIG. 1

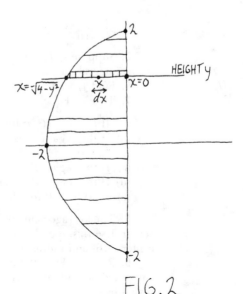

FIG. 2

$$(1) \qquad \text{typical horizontal strip sum} = \int_{x=-\sqrt{4-y^2}}^{x=0} xy^2 \, dx \, dy.$$

The horizontal strips stretch from $y = -2$ to $y = 2$, so adding the strip sums to get a final total is a one-dimensional integral problem involving the interval $[-2, 2]$ on a y-line. Therefore,

$$(2) \qquad \int_R xy^2 \, dA = \int_{y=-2}^{y=2} \left(\int_{x=-\sqrt{4-y^2}}^{x=0} xy^2 \, dx \right) dy.$$

The right side of (2) is called an *iterated integral* and consists of *two consecutive single integrals;* the parentheses are usually omitted. Note that in the inner integral, the upper limit of integration is $x = 0$ because the right end of every horizontal strip is $x = 0$. However, the lower limit of integration is $x = -\sqrt{4 - y^2}$, not $x = -2$, because a typical horizontal strip does not begin at $x = -2$; the value at the left end of the strip depends on the height y. On the other hand, in the outer integral, the limits are the extreme values $y = -2$ and $y = 2$ because the outer integral must add all the horizontal strip sums, from bottom to top.

To compute (2), first find the inner integral in which the integration is performed with respect to x, with y held constant. We have

$$\text{inner integral} = \int_{x=-\sqrt{4-y^2}}^{x=0} xy^2 \, dx$$

$$= \frac{1}{2} x^2 y^2 \Big|_{x=-\sqrt{4-y^2}}^{x=0} \qquad \text{(antidifferentiate with respect to } x)$$

$$= -\frac{1}{2}(4 - y^2)y^2 = \frac{1}{2}y^4 - 2y^2.$$

This result is the integrand of the outer integral, so

$$\text{outer integral} = \int_{y=-2}^{y=2} \left(\frac{1}{2}y^4 - 2y^2 \right) dy = \left(\frac{1}{10}y^5 - \frac{2}{3}y^3 \right) \Big|_{-2}^{2} = -\frac{64}{15}.$$

Therefore $\int_R xy^2 \, dA = -64/15$.

We can also evaluate the double integral by first adding $xy^2 \, dA$'s across a typical vertical strip (Fig. 3) and then adding the strip sums from left to right. On a vertical strip located at horizontal position x, the lower end is $y = -\sqrt{4 - x^2}$ and the upper end is $y = \sqrt{4 - x^2}$. Adding $xy^2 \, dx \, dy$'s along the strip is a one-dimensional integral problem involving the interval $[-\sqrt{4 - x^2}, \sqrt{4 - x^2}]$ on a y-line. Adding the resulting strip sums from left to right is a one-dimensional integral problem involving the interval $[-2, 0]$ on an x-line. Therefore, as a second method,

$$(3) \qquad \int_R xy^2 \, dA = \int_{x=-2}^{x=0} \left(\int_{y=-\sqrt{4-x^2}}^{y=\sqrt{4-x^2}} xy^2 \, dy \right) dx.$$

Again, the right-hand side contains consecutive single integrals. We write $dy \, dx$ in that order to indicate that the inner integration is with respect to y and the outer integration is with respect to x. Note that the inner limits of integration are *not* the extreme values $y = -2$ and $y = 2$ because a typical vertical strip does not begin at $y = -2$ and end at $y = 2$; the lower and upper ends depend on its horizontal position x. However the outer limits *are*

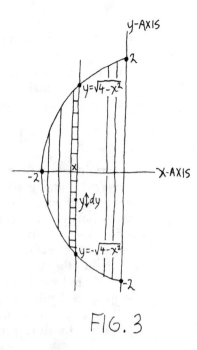

FIG. 3

the extreme values $x = -2$ and $x = 0$ because the vertical strip sums must be added from extreme left to extreme right to catch *all* of them.

One computation per double integral is sufficient, but we will evaluate (3) for practice. First,

$$\text{inner integral} = \int_{y=-\sqrt{4-x^2}}^{y=\sqrt{4-x^2}} xy^2\,dy$$

$$= \frac{1}{3}xy^3 \Big|_{y=-\sqrt{4-x^2}}^{y=\sqrt{4-x^2}} \qquad \text{(antidifferentiate with respect to } y\text{)}$$

$$= \frac{2}{3}x(4 - x^2)^{3/2}.$$

Then

$$\text{outer integral} = \frac{2}{3}\int_{x=-2}^{x=0} x(4 - x^2)^{3/2}\,dx.$$

Substitute $u = 4 - x^2$, $du = -2x\,dx$ to obtain

$$\text{outer integral} = -\frac{1}{3}\int_0^4 u^{3/2}\,du = -\frac{1}{3}\frac{2}{5}u^{5/2}\Big|_0^4 = -\frac{64}{15},$$

as before.

Of what use is the result? If xy^2 is the charge density in the plane, then the total charge in the semicircular region R is $-\dfrac{64}{15}$. If the graph of xy^2 is sketched in 3-space, it lies entirely below the region R (since xy^2 is negative at all points in R) and the volume between the graph and the region is $64/15$. If xy^2 is the temperature in the plane then

$$\text{average temperature in } R = \frac{-64/15}{\text{area of } R} = \frac{-64/15}{2\pi} = -\frac{32}{15\pi}.$$

This section is concerned primarily with computing integrals, but we will return to these and other applications in the next sections.

Warning   The double integral in Example 1 does *not* compute the area of the semicircular region. The integrand must be 1, that is, the double integral must be $\int_R dA$, to compute area.

Computing a double integral   The methods established in Example 1 may be applied in general. A double integral over a region R can be evaluated with two single integrals in two ways.

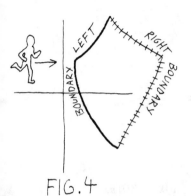

FIG. 4

> For one method, identify the left and right boundaries of R by imagining a horizontal walk from left to right through the region (Fig. 4). Solve the boundary equations for x to find x on the left and on the right boundary. Further, find the lowest y and the highest y in R. Then
>
> (4) $$\int_R f(x, y)\, dA = \int_{\text{lowest } y \text{ in } R}^{\text{highest } y \text{ in } R} \int_{x \text{ on left boundary of } R}^{x \text{ on right boundary of } R} f(x, y)\, dx\, dy.$$

We write dA as $dx\, dy$ in that order to indicate that the inner integration is with respect to x, and the outer integration is with respect to y. (The integral in (2) is of this form.)

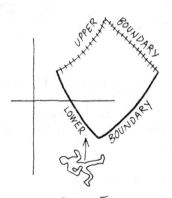

FIG. 5

> For a second method, identify the lower and upper boundaries of R by imagining a vertical walk through the region (Fig. 5). Solve the boundary equations for y to obtain y on the lower and on the upper boundary. Further, find the leftmost x and the rightmost x in R. Then
>
> (5) $$\int_R f(x, y)\, dA = \int_{\text{leftmost } x \text{ in } R}^{\text{rightmost } x \text{ in } R} \int_{y \text{ on lower boundary}}^{y \text{ on upper boundary}} f(x, y)\, dy\, dx.$$

We write dA as $dy\, dx$ in that order to indicate that the inner integration is with respect to y and the outer integration is with respect to x. (The integral in (3) is of this form.)

A double integral can be evaluated with either order of integration, but sometimes one is easier to compute than the other. In Example 1, the antidifferentiation in the outer integral was easier when the integration was done first with respect to x. We will frequently set up double integrals twice, using both orders of integration for practice.

Note that in both (4) and (5), the *inner* and *outer* limits of integration follow *different* patterns. *The limits on the inner integral are found by solving boundary equations for one of the variables;* the inner limits will contain the other variable, unless the boundary in question is a vertical or horizontal line with an equation as simple as $x = 3$ or $y = 2$. On the other hand, *the limits on the outer integral are always constants,* the extreme values of the other variable.

Example 2   Find $\int_R (4x - y)\, dA$ where R is the triangular region bounded by the y-axis, the line $y = 2$, and the line $2x + y = 8$.

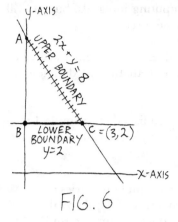

FIG. 6

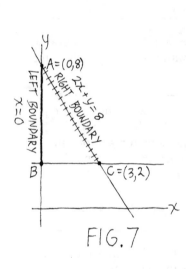

FIG. 7

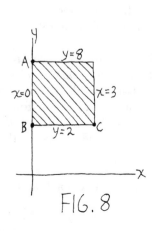

FIG. 8

First solution: Set up the double integral so that y goes first, that is, so that the inner integration is with respect to y. The lower boundary of the region is the line $y = 2$ (Fig. 6). The upper boundary is the line $2x + y = 8$; solve for y to obtain $y = 8 - 2x$. The leftmost x in the region is $x = 0$ and the rightmost x is the x-coordinate of point C, $x = 3$ (obtained by substituting $y = 2$ in $2x + y = 8$). Therefore

$$\int_R (4x - y)\, dA = \int_{x=0}^{x=3} \int_{y=2}^{y=8-2x} (4x - y)\, dy\, dx.$$

Then

$$\text{inner integral} = \int_{y=2}^{y=8-2x} (4x - y)\, dy = \left(4xy - \frac{1}{2}y^2\right)\Big|_{y=2}^{y=8-2x}$$

$$= 4x(8 - 2x) - \frac{1}{2}(8 - 2x)^2 - 8x + 2$$

$$= -10x^2 + 40x - 30,$$

and

$$\text{outer integral} = \int_{x=0}^{x=3} (-10x^2 + 40x - 30)\, dx$$

$$= \left(-\frac{10}{3}x^3 + 20x^2 - 30x\right)\Big|_0^3 = 0.$$

Therefore $\int_R (4x - y)\, dA = 0$.

Second solution: Let the inner integration be with respect to x. The left boundary of the region is the y-axis where $x = 0$ (Fig. 7). The right boundary is the line $2x + y = 8$; solve for x to obtain $x = 4 - \frac{1}{2}y$. The lowest y in the region is $y = 2$ and the highest y is the y-coordinate of point A, $y = 8$ (obtained by setting $x = 0$ in $2x + y = 8$). Thus

$$\int_R (4x - y)\, dA = \int_{y=2}^{y=8} \int_{x=0}^{x=4-y/2} (4x - y)\, dx\, dy.$$

Then

$$\text{inner integral} = \int_{x=0}^{x=4-y/2} (4x - y)\, dx = (2x^2 - xy)\Big|_{x=0}^{x=4-y/2}$$

$$= 2\left(4 - \frac{1}{2}y\right)^2 - \left(4 - \frac{1}{2}y\right)y = y^2 - 12y + 32,$$

and

$$\text{outer integral} = \int_{y=2}^{y=8} (y^2 - 12y + 32)\, dy = \left(\frac{1}{3}y^3 - 6y^2 + 32y\right)\Big|_2^8 = 0.$$

Warning The inner limits are *boundary* values while the outer limits are *extreme* values.

1. If you mistakenly use extremes for both sets of limits in Example 2, and write $\int_{y=2}^{y=8} \int_{x=0}^{x=3}$ or $\int_{x=0}^{x=3} \int_{y=2}^{y=8}$, then, instead of integrating over the triangular region R in Figs. 6 and 7, you are integrating over the rectangular region in Fig. 8.

2. Suppose you mistakenly use boundary values for both sets of limits in Example 2, and write $\int_{x=0}^{x=4-y/2} \int_{y=2}^{y=8-2x}$ or $\int_{y=2}^{y=8-2x} \int_{x=0}^{x=4-y/2}$. Since the outer limits in the first instance contain y, the first setup produces a result containing y. Similarly, the second setup leads to a result containing the variable x. But a double integral is a *number,* so neither setup calculates a double integral over the region in Figs. 6 and 7. In fact they do not correspond to a double integral over *any* region whatsoever. *The outer limits of the iterated integral must be constants.*

Warning   The region R in Example 2 (Figs. 6 and 7) is not the graph of the integrand $4x - y$. It is the region over which the function is being integrated, just as $\int_a^b f(x)\,dx$ integrates $f(x)$ over the interval $[a, b]$. The graph of $4x - y$ is the plane $z = 4x - y$ in 3-space, lying partly above and partly below R. To evaluate $\int_R f(x, y)\,dA$ it is not necessary to sketch the graph of $f(x, y)$, but it is important to sketch R so that the boundary values and extreme values may be identified for the limits of integration.

Integrating on a region with a two-curve boundary   Consider $\int_R f(x, y)\,dA$ where R is the region bounded by the line $x + y = 6$ and the parabola $x = y^2$. We will express the double integral in terms of single integrals in two ways, to practice setting up the limits of integration.

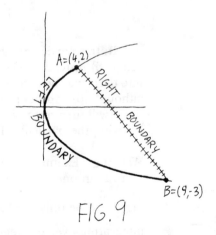

FIG. 9

Figure 9 shows that the left boundary of the region is the parabola $x = y^2$. The right boundary is the line $x + y = 6$; solve the equation for x to get $x = 6 - y$. To find the extreme values of y we need points A and B. Substitute $x = y^2$ into $x + y = 6$, obtaining $y^2 + y = 6$, $(y + 3)(y - 2) = 0$, $y = -3, 2$. Therefore $A = (4, 2)$, $B = (9, -3)$, the lowest y is -3 at B, and the highest y is 2, at A. Thus

(6) $$\int_R f(x, y)\,dA = \int_{y=-3}^{y=2} \int_{x=y^2}^{x=6-y} f(x, y)\,dx\,dy.$$

Consider the other order of integration. The lower boundary is the parabola $x = y^2$ (Fig. 10); solve the equation for y to obtain $y = -\sqrt{x}$ (choose the negative square root because y is negative on the lower portion of the parabola). However, *the upper boundary consists of two curves,* the parabola and the line. It is not possible to find *one* expression for y on the upper bound-

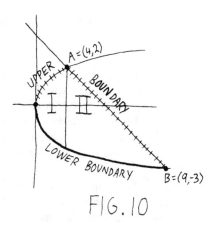

FIG. 10

ary. To continue with this order of integration, divide the region into the two indicated parts, I and II. For region I, the lower boundary is that portion of the parabola where $y=-\sqrt{x}$, and the upper boundary is $y=\sqrt{x}$; the extreme values of x are 0 and 4. For region II, the lower boundary is $y=-\sqrt{x}$ again and the upper boundary is the line $y=6-x$; the extreme values of x are 4 and 9. Therefore

$$(7) \quad \int_R f(x,y)\,dA = \int_I f(x,y)\,dA + \int_{II} f(x,y)\,dA$$

$$= \int_{x=0}^{x=4}\int_{y=-\sqrt{x}}^{y=\sqrt{x}} f(x,y)\,dy\,dx + \int_{x=4}^{x=9}\int_{y=-\sqrt{x}}^{y=6-x} f(x,y)\,dy\,dx.$$

In this particular example, the two-curve boundary can be avoided by using (6) instead of (7). However, in some problems, dividing the region into parts may be unavoidable (see Problem 2c, for example). Furthermore, although (6) appears preferable to (7), it may be that the antidifferentiation in (7) will turn out to be easier. We can't state a preference for (6) without knowing the particular function f.

Warning Examine boundaries carefully to catch the ones consisting of more than one curve. If you mistakenly consider the upper boundary in Fig. 10 to be only the line, and write $\int_{x=0}^{x=9}\int_{y=-\sqrt{x}}^{y=6-x} f(x,y)\,dy\,dx$, then you are integrating over the region in Fig. 11, not Fig. 10.

Problems for Section 12.2

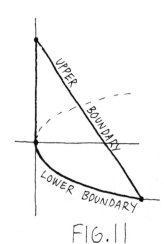

FIG. 11

1. Evaluate

(a) $\int_R x^3\,dA$ where R is the region bounded by the lines $y = 2x$, $y = 3$ and the y-axis
(b) $\int_R 3\,dA$ where R is the region in quadrant I enclosed by $y = x^3$ and $y = x$
(c) $\int_R 2xy\,dA$ where R is the region between the parabola $y = x^2$ and the line $y = 10$

2. Express $\int_R f(x,y)\,dA$ in terms of single integrals for the given region R. Set up each integral twice, using the two orders of integration.

(a) R is bounded by the parabolas $y = \frac{1}{8}x^2$ and $x = y^2$
(b) R is bounded by $y = e^x$, the line $y = 2$ and the y-axis

(c) R is the triangular region with vertices $A = (0,0), B = (2,4)$ and $C = (3,1)$
(d) R is the set of points (x,y) such that $2 \le x \le 7$ and $0 \le y \le 5$
(e) R is bounded by the parabola $4y = x^2$ and the line $x - 2y + 4 = 0$
(f) R is bounded by the x-axis, lines $x = 2$, $x = 3$, and the hyperbola $xy = 1$
(g) R is the (unbounded) region inside the parabola $y = x^2$
(h) R is the (unbounded) region between the two branches of the hyperbola $xy = 1$
(i) R is the region in Fig. 12
(j) R is the interior of the ellipse $2x^2 + y^2 = 4$

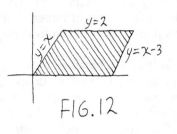

FIG. 12

3. Let R be the region in Fig. 13. (a) Set up $\int_R f(x,y)\, dA$ using the order of integration $dy\, dx$. (b) Can $\int_R f(x,y)\, dA$ be computed by integrating on only the quarter of the region in quadrant I and then multiplying that answer by 4?

4. Each of the following represents a double integrals already expressed in terms of single integrals. Sketch the region of integration R and then set up the double integral using the reverse order of integration

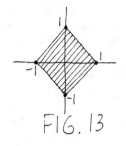

FIG. 13

(a) $\displaystyle\int_{y=0}^{y=1} \int_{x=(1/2)y}^{x=1} f(x,y)\, dx\, dy$ (d) $\displaystyle\int_{x=0}^{\infty} \int_{y=2x}^{\infty} f(x,y)\, dy\, dx$

(b) $\displaystyle\int_{0}^{1} \int_{-x}^{x^2} f(x,y)\, dy\, dx$ (e) $\displaystyle\int_{0}^{1} \int_{\sin^{-1}x}^{2} f(x,y)\, dy\, dx$

(c) $\displaystyle\int_{0}^{1} \int_{-\sqrt{1-x^2}}^{1-x^2} f(x,y)\, dy\, dx$ (f) $\displaystyle\int_{y=0}^{\infty} \int_{x=0}^{x=e^y} f(x,y)\, dx\, dy$

5. Let R be the triangular region with vertices $A = (0,0), B = (0,1), C = (2,1)$. Set up $\int_R e^{y^2}\, dA$ in both orders of integration. Then choose the order in which the calculation will be easier and evaluate the integral.

12.3   Double Integration in Polar Coordinates

We'll begin by evaluating a specific integral using polar coordinates to see how it is done and why it may have an advantage over rectangular coordinates. The method we develop will then be summarized and used in general.

Example 1   Consider $\int_R x^2\, dA$, where R is the region in quadrant I between two circles with radii 1 and 5 as indicated in Fig. 1. The region poses

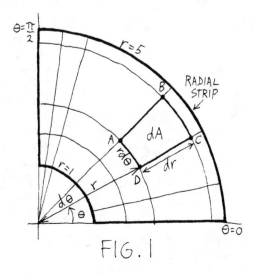

FIG. 1

two difficulties. First, the lower boundary consists of *two* curves, the x-axis and the smaller circle; similarly, the left boundary consists of *two* curves, the y-axis and the smaller circle. Therefore, no matter which order of integration is tried, the region must be divided into two parts. Second, the circles have equations of the form $x^2 + y^2 = r_0^2$. When inner limits of integration are obtained by solving for x or y, unpleasant square roots will result. Polar coordinates will alleviate both difficulties.

The integration method of the preceding section was obtained by using vertical and horizontal lines to divide a region into many small rectangles, with typical area $dA = dx\,dy$. The procedure evaluates a double integral with two single integrals, which add $f(x,y)\,dA$'s on (say) a horizontal strip, and then add strip sums from bottom to top. To integrate in polar coordinates we use *circles and rays* as shown in Fig. 1 to divide the region into many small subregions. The plan is to express $x^2\,dA$ in polar coordinates, add on the typical radial strip in Fig. 1, and then add the strip sums counterclockwise. The typical subregion with area dA, configuration $ABCD$ in Fig. 1, is not a rectangle but it may be considered almost a rectangle since a radial line intersects a circle perpendicularly. The subregion was created by drawing two circles separated by dr, and two rays separated by angle $d\theta$. Therefore side DC has length dr, but $d\theta$ is an angular dimension, not a side of the box, so dA is *not* $dr\,d\theta$. By (5) of Section 1.3, the arc length AD is $r\,d\theta$. Then the (almost) box has dimensions $r\,d\theta$ by dr, so $dA = r\,d\theta\,dr = r\,dr\,d\theta$. Furthermore, $x = r\cos\theta$ (Appendix A6) so

$$x^2\,dA = (r\cos\theta)^2 r\,dr\,d\theta = r^3\cos^2\theta\,dr\,d\theta.$$

The single integral $\int_{r=1}^{r=5} r^3\cos^2\theta\,dr\,d\theta$ adds $x^2\,dA$'s on the typical radial strip in Fig. 1. The many radial strips fan out counterclockwise from $\theta = 0$ to $\theta = \pi/2$, so we add the strip sums with another integral to obtain

(1)
$$\int_R x^2\,dA = \int_{\theta=0}^{\theta=\pi/2}\int_{r=1}^{r=5} r^3\cos^2\theta\,dr\,d\theta.$$

Then

$$\text{inner integral} = \int_{r=1}^{r=5} r^3\cos^2\theta\,dr = \frac{1}{4}r^4\cos^2\theta\Big|_{r=1}^{r=5} = 156\cos^2\theta,$$

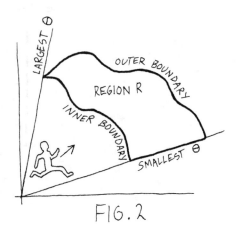

FIG. 2

and (with formula (40) from the integral tables)

$$\text{outer integral} = \int_{\theta=0}^{\theta=\pi/2} 156 \cos^2\theta \, d\theta$$

$$= 156 \cdot \frac{1}{2}(\theta + \sin\theta\cos\theta)\Big|_{\theta=0}^{\theta=\pi/2} = 39\pi.$$

Computing a double integral in polar coordinates   The method established in Example 1 may be applied in general.

> A double integral over a region R can be evaluated with two single integrals. Change x, y, and dA to polar coordinates using
>
> $$x = r\cos\theta, \qquad y = r\sin\theta, \qquad x^2 + y^2 = r^2,$$
> $$dA = r\,dr\,d\theta.$$
>
> (2)
>
> To find the limits of integration, identify the inner and outer boundaries of the region by imagining a radial walk from the origin (Fig. 2). Solve the polar coordinate equation of each boundary for r. Further, swing a ray counterclockwise through the region and find the first θ and last θ encountered, that is, the smallest and largest values of θ. Then
>
> (3)
> $$\int_R f(x,y)\,dA = \int_{\text{smallest }\theta}^{\text{largest }\theta} \int_{r \text{ on inner boundary}}^{r \text{ on outer boundary}} f(r\cos\theta, r\sin\theta)\,r\,dr\,d\theta.$$

Note that, in general, the *inner* and *outer* limits of integration follow *different* patterns. The limits on the inner integral are found by solving boundary equations for r, so the inner limits will contain θ, unless a boundary is a circle centered at the origin with an equation as simple as $r = r_0$ (as in (1)). On the other hand, the outer limits are always constants, the extreme values of θ. It is possible to use the reverse order of integration, but equations in polar coordinates are usually easier to solve for r than for θ and experience shows that the one version is sufficient.

   Integration in polar coordinates is used primarily to handle regions of integration which involve circles, such as the region in Fig. 1. Polar coordinates should also be considered if the integrand involves expressions such as $x^2 + y^2$, $\sqrt{x^2 + y^2}$, $(x^2 + y^2)^3$, since these forms become simpler (namely, r^2, r, r^6) in polar coordinates.

Example 2   Let R_1 be the circular region with radius 4 in Fig. 3 and let R_2 be the lower semicircular portion. Then

$$\int_{R_1} f(x,y)\,dA = \int_{\theta=0}^{\theta=2\pi} \int_{r=0}^{r=4} f(r\cos\theta, r\sin\theta)\,r\,dr\,d\theta$$

and

$$\int_{R_2} f(x,y)\,dA = \int_{\theta=\pi}^{\theta=2\pi} \int_{r=0}^{r=4} f(r\cos\theta, r\sin\theta)\,r\,dr\,d\theta.$$

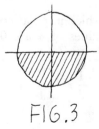

FIG.3

Warning In the latter integral, it is also correct to use $\displaystyle\int_{\theta=-\pi}^{\theta=0}$ and $\displaystyle\int_{\theta=3\pi}^{\theta=4\pi}$ but it is *not* correct to write $\displaystyle\int_{\theta=\pi}^{\theta=0}$. The values of θ must be measured in a continuous increasing fashion.

Example 3 A six mile square housing development, *ABCD*, contains only one bus stop, at point *D*. The management does not want to reveal that some residents may have to walk $6\sqrt{2}$ miles (from *B*) to catch a bus, while others (at *D*) are so close that their homes will reek of bus fumes. To gloss over the unpleasant extremes, the management wants to advertise the *average* distance to the bus stop. The problem is to find this average value.

Insert a coordinate system as in Fig. 4. Then, at the point (x, y), the distance to the bus stop is $\sqrt{x^2 + y^2}$. Therefore we want to find the average value of $\sqrt{x^2 + y^2}$ over the square region. By (3) of Section 12.1,

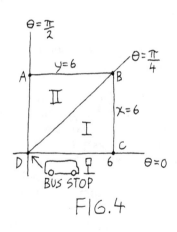

$\theta = \dfrac{\pi}{2}$

$\theta = \dfrac{\pi}{4}$

$y = 6$

A B

II

$x = 6$

I

C

D 6 $\theta = 0$

BUS STOP

FIG. 4

(4) $$\text{average value} = \frac{\displaystyle\int_{\text{development}} \sqrt{x^2 + y^2}\, dA}{\text{area of development}}.$$

If we try to compute the integral in rectangular coordinates, the limits of integration are simple but the antidifferentiation is difficult. If we switch to polar coordinates, the limits of integration require care, but the integrand is simpler since

$$\sqrt{x^2 + y^2}\, dA = r\, r\, dr\, d\theta = r^2\, dr\, d\theta.$$

On balance, polar coordinates will prove somewhat more efficient.

The inner boundary of the development is the point circle where $r = 0$, but the outer boundary consists of the two segments *AB* and *BC*. To set up the integral in (4) we must divide the region into the two parts indicated in Fig. 4. However, in this particular problem we can backtrack and avoid the extra work. By the symmetry of the situation, the average distance from the entire development to the bus stop is the same as the average distance from region I (or II) to the bus stop. So the answer can be found with

(5) $$\frac{\displaystyle\int_{I} \sqrt{x^2 + y^2}\, dA}{\text{area of I}}.$$

For region I, the outer boundary is the line $x = 6$, which in polar coordinates is $r \cos \theta = 6$, or $r = 6 \sec \theta$; the extreme values of θ are 0 and $\pi/4$. So

$$\int_{I} \sqrt{x^2 + y^2}\, dA = \int_{\theta=0}^{\theta=\pi/4} \int_{r=0}^{r=6\sec\theta} r^2\, dr\, d\theta = \int_{\theta=0}^{\theta=\pi/4} \frac{1}{3} r^3 \Big|_{r=0}^{r=6\sec\theta}\, d\theta$$

$$= \int_{\theta=0}^{\theta=\pi/4} 72 \sec^3 \theta\, d\theta$$

$$= 72 \left[\frac{1}{2} \sec \theta \tan \theta + \frac{1}{2} \ln|\sec \theta + \tan \theta| \right]\Big|_{0}^{\pi/4}$$

(Tables, (43))

$$= 36\sqrt{2} + 36 \ln(\sqrt{2} + 1).$$

The area of region I is 18, so the value of (5), the final answer, is $2\sqrt{2} + 2\ln(\sqrt{2} + 1)$.

Warning   Remember to use $r\,dr\,d\theta$ for dA. *Don't forget the extra r.*

Problems for Section 12.3

1. Consider the circular region in Fig. 3 with radius 4. Evaluate

(a) $\int x\,dA$ over the right semicircular portion.

(b) $\int xy\,dA$ over the first quadrant portion.

(c) $\int \dfrac{1}{1 + x^2 + y^2}\,dA$ over the entire region.

2. Find $\int e^{-(x^2+y^2)}\,dA$ over quadrant I.

3. Find $\displaystyle\int_{x=-3}^{x=3}\int_{y=0}^{y=\sqrt{9-x^2}} \ln(1 + x^2 + y^2)\,dy\,dx$ by switching to polar coordinates.

4. Find the limits of integration in polar coordinates for the following regions.

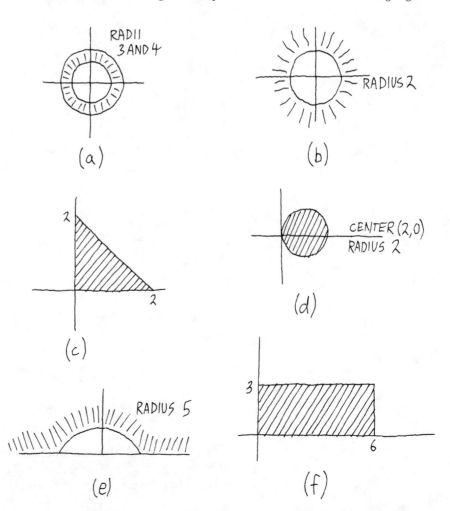

12.4   Area and Volume

We will summarize the methods for using integrals to compute area and volume, and then illustrate with examples.

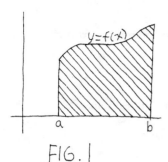

FIG. 1

Area   (A) If a region lies under the graph of $f(x)$ and *above* an interval $[a, b]$ on the x-axis (Fig. 1) then its area is $\int_a^b f(x)\,dx$ (Section 5.2).

(B) Suppose a region lies between an upper curve $u(x)$ and a lower curve $l(x)$ and stretches from $x = a$ to $x = b$; that is, the projection of the region on the x-axis is $[a, b]$ (Fig. 2). Then, whether the region lies above, below, or straddles the x-axis, its area is $\int_a^b (u(x) - l(x))\,dx$ (Section 6.3).

(C) For *any* region R, no matter what its boundaries, and whether it lies above, below or straddles the x-axis, the area of R is $\int_R dA$ (Section 12.1).

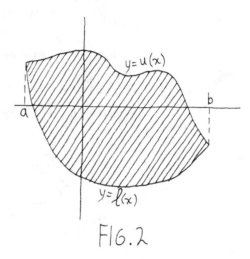

FIG. 2

Example 1   Find the area of the region R bounded by the parabola $y = x^2 - 9$ and the line $y = 2x - 6$ (Fig. 3).

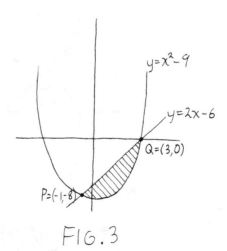

FIG. 3

Solution: First find the coordinates of the points of intersection, P and Q. Solve $2x - 6 = x^2 - 9$ to get $x^2 - 2x - 3 = 0$, $(x - 3)(x + 1) = 0$, $x = -1, 3$. Therefore $P = (-1, -8)$ and $Q = (3, 0)$.

We may use (B) with $u(x) = 2x - 6$ and $l(x) = x^2 - 9$ to get

(1)
$$\text{area} = \int_{-1}^{3} (2x - 6 - [x^2 - 9])\, dx$$

$$= \int_{-1}^{3} (2x + 3 - x^2)\, dx = \left(x^2 + 3x - \frac{1}{3}x^3\right)\Bigg|_{-1}^{3} = \frac{32}{3}.$$

We may also use (C). The lower boundary of the region is $y = x^2 - 9$, the upper boundary is $y = 2x - 6$; the extreme values of x are -1 and 3. Therefore

$$\text{area} = \int_{R} dA = \int_{x=-1}^{x=3} \int_{y=x^2-9}^{y=2x-6} dy\, dx.$$

Since $\int_{a}^{b} dx = b - a$, the inner integral is $2x - 6 - [x^2 - 9]$. Thus the outer integral is $\int_{x=-1}^{x=3} (2x - 6 - [x^2 - 9])\, dx$, the same as (1). Methods (B) and (C) merge, leading to the same final computation and answer.

Volume (A') If a solid lies under the graph of $f(x,y)$ and *above* a region R in the x,y plane (Fig. 4) then its volume is $\int_{R} f(x,y)\, dA$ (Section 12.1).

(B') Suppose a solid lies between an upper surface $u(x,y)$ and a lower surface $l(x,y)$, and its projection in the x,y plane is a region R (Fig. 5). Then, whether the solid lies above, below, or crosses the x,y plane, its volume is $\int_{R} (u(x,y) - l(x,y))\, dA$.

Section 12.6 will give (C') to complete the analogy between area in 2-space and volume in 3-space.

(Note that in addition to the double integrals in (A') and (B'), and the forthcoming triple integral in (C'), certain volumes were computed in Section 6.1 with a *single* integral.)

We may restate (A') and (B') as follows.

(A') If the "floor" of a solid lies in the x,y plane then

$$\text{volume} = \int_{\text{floor}} \text{roof}\, dA$$

(B') If a solid lies between a roof and floor which is not necessarily in the x,y plane (and not necessarily flat) then

$$\text{volume} = \int_{\text{projection in the } x,y \text{ plane}} (\text{roof} - \text{floor})\, dA.$$

The formula in (A') is the special case of (B') when the floor has equation $z = 0$ and roof $-$ floor simplifies to roof.

Example 2 Consider the solid bounded by the cylinder $y = x^2$, the plane $y + z = 5$ and the x,y plane (Fig. 6). The solid lies under the plane $y + z = 5$, that is, under the graph of $f(x,y) = 5 - y$, and above the region R in the x,y plane shown in Fig. 7. (The line $y = 5$ in Fig. 7 is the inter-

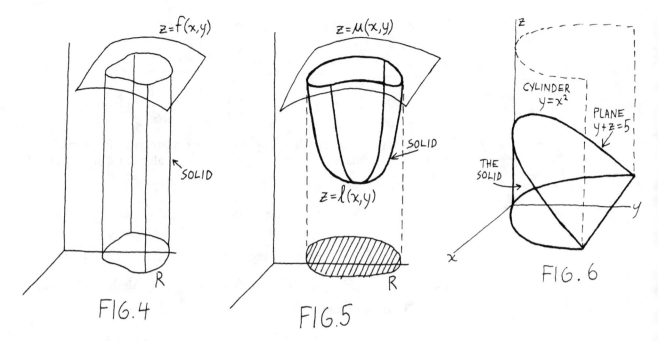

FIG.4

FIG.5

FIG. 6

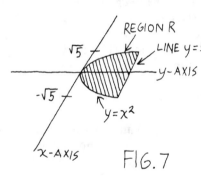

FIG.7

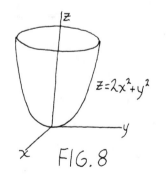

FIG.8

section of the plane $y + z = 5$ with the x, y plane where $z = 0$.) Therefore, by (A′),

$$\text{volume of solid} = \int_R (5 - y)\, dA = \int_{x=-\sqrt{5}}^{x=\sqrt{5}} \int_{y=x^2}^{y=5} (5 - y)\, dy\, dx$$

$$= \int_{x=-\sqrt{5}}^{x=\sqrt{5}} \left(5y - \frac{1}{2}y^2\right) \Big|_{y=x^2}^{y=5} dx$$

$$= \int_{x=-\sqrt{5}}^{x=\sqrt{5}} \left(\frac{25}{2} - 5x^2 + \frac{1}{2}x^4\right) dx$$

$$= \left(\frac{25}{2}x - \frac{5}{3}x^3 + \frac{1}{10}x^5\right) \Big|_{x=-\sqrt{5}}^{x=\sqrt{5}} = \frac{40}{3}\sqrt{5}.$$

Example 3 The elliptic paraboloid

(2) $$z = 2x^2 + y^2 \qquad \text{(Fig. 8)}$$

and the parabolic cylinder

(3) $$z = 4 - y^2 \qquad \text{(Fig. 9)}$$

determine the solid in Fig. 10. By (B′),

(4) $$\text{volume of solid} = \int_R (4 - y^2 - [2x^2 + y^2])\, dA$$

$$= \int_R (4 - 2x^2 - 2y^2)\, dA$$

where R is the projection of the solid in the x, y plane. Often a projection

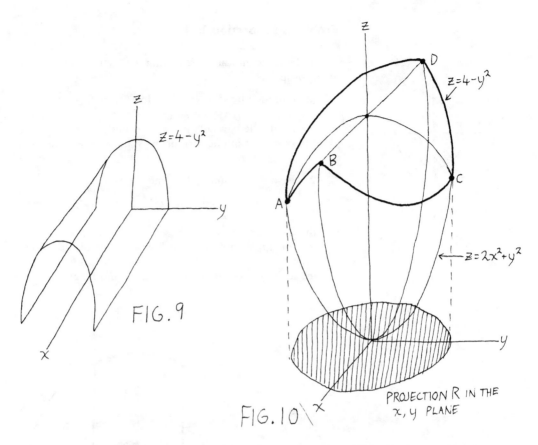

FIG. 9

FIG. 10

PROJECTION R IN THE
x, y PLANE

is easy to visualize, but in this example the projection is not geometrically obvious and we will give an algebraic method for identifying it. Consider the curve $ABCD$ of intersection of the two surfaces. Equate the expressions for z in (2) and (3) to obtain $2x^2 + y^2 = 4 - y^2$ or

$$(5) \qquad\qquad x^2 + y^2 = 2 .$$

The equation in (5) is satisfied by the x and y coordinates of any point on the curve $ABCD$. Since the x and y coordinates do not change as a point is projected onto the x, y plane, the projection of $ABCD$ continues to satisfy (5). Therefore the projection of $ABCD$ in the x, y plane is a circle with radius $\sqrt{2}$, and the projection R of the solid itself is the circle plus its interior.

Now that R has been identified as circular, we choose to evaluate (4) using polar coordinates. Since $x^2 + y^2 = r^2$, the integrand $4 - 2x^2 - 2y^2$ becomes $4 - 2r^2$. Then

$$
\begin{aligned}
\text{volume of solid} &= \int_{\theta=0}^{\theta=2\pi} \int_{r=0}^{r=\sqrt{2}} (4 - 2r^2)\, r\, dr\, d\theta \\
&= \int_{\theta=0}^{\theta=2\pi} \int_{r=0}^{r=\sqrt{2}} (4r - 2r^3)\, dr\, d\theta \\
&= \int_{\theta=0}^{2\pi} \left(2r^2 - \frac{1}{2} r^4 \right) \Bigg|_{r=0}^{r=\sqrt{2}} d\theta \\
&= \int_{\theta=0}^{\theta=2\pi} 2\, d\theta = 2(2\pi - 0) = 4\pi .
\end{aligned}
$$

Problems for Section 12.4

1. Find the indicated area and, for practice, use as many of (A), (B) and (C) as are appropriate.

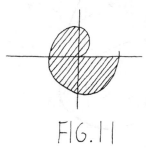

(a) bounded by $y^2 = x + 1$ and $x + y = 1$
(b) bounded by $xy = 4$ and $x + y = 5$
(c) inside the parabola $y = x^2$ and under the line $y = 4$
(d) swept out in one turn of the spiral whose equation in polar coordinates is $r = \theta$ (Fig. 11)

2. Use a double integral to find the volume.

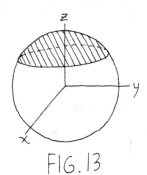

FIG. 11

(a) the solid bounded by the cylinder $x^2 + y^2 = 4$, the x, y plane and the plane $x + 6y + 2z = 12$
(b) a sphere with radius R
(c) the apple core of radius 3 in an apple of radius 6 (Fig. 12)
(d) a circular cone with radius R and height h
(e) a circular cylinder with radius R and height h

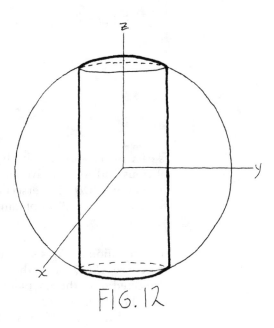

FIG. 12

3. Express the volume with a double integral and then go one step further to write the double integral as two single integrals.

(a) the solid inside $z = 2x^2 + 2y^2$ and under the plane $z = 12$
(b) the solid bounded by the paraboloids $z = x^2 + y^2$ and $z = 8 - 3x^2 - 3y^2$
(c) the polar cap of radius 2 in a sphere of radius 5 (Fig. 13)

4. Two cylindrical pipes of radius 3 intersect so that their axes cross perpendicularly. Find the volume of the solid determined by the intersecting cylinders. (Figure 14 shows half the solid.)

5. Consider $\int_R \sqrt{x^2 + y^2} \, dA$ where R is the circular region in the x, y plane with radius 5 and center at the origin. Is it computing an area? a volume? If so, sketch the figure.

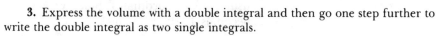

FIG. 13

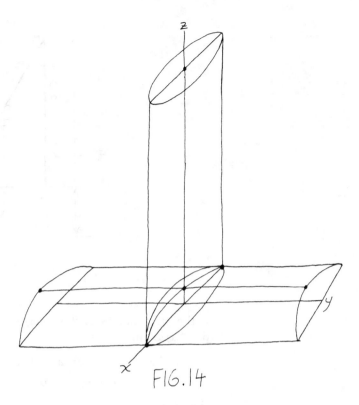

FIG.14

12.5 Further Applications of the Double Integral

The preceding section discussed area and volume; this section will illustrate some nongeometric applications. As with area and volume, we will show that some problems can be done with both single and double integrals.

Example 1 The cost of clearing jungle land is directly proportional to both the area of the land and its distance from the supply road. (As the area increases, naturally the cost goes up. As the distance to the supply road increases, the equipment must be hauled further on jungle trails and the cost increases.) In particular, suppose that

(1)   cost = area × distance to the supply road .

Consider a right triangular region with legs 6 and 3, with the supply road running along its longer leg (Fig. 1). The problem is to find the cost of clearing the land.

We can't use (1) directly because different parts of the region are at different distances from the road. One approach is to establish a number line along the shorter leg as in Fig. 1. Consider a typical strip at position x with thickness dx. The distance from the strip to the longer leg may be considered to be constant, with value x; this is the reason for choosing strips parallel to the road. By similar triangles,

$$\frac{\text{strip height}}{3 - x} = \frac{6}{3},$$

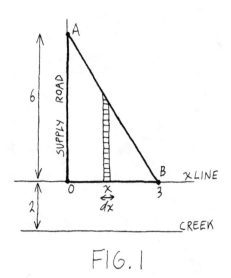

FIG. 1

so the strip height is $2(3 - x)$, or $6 - 2x$. Therefore, the strip area is $(6 - 2x)\,dx$ and

$$d\,\text{cost} = \text{strip area} \times \text{distance to road} = (6 - 2x)\,dx \times x$$
$$= (6x - 2x^2)\,dx.$$

Therefore

(2) $$\text{total cost} = \int_0^3 (6x - 2x^2)\,dx = \left(3x^2 - \frac{2}{3}x^3\right)\Bigg|_0^3 = 9.$$

As a second approach, establish the 2-dimensional coordinate system in Fig. 2 and divide the region into many small subregions. Consider a subregion containing the point (x, y) and with area dA. The distance from (x, y) to the road is x, so the entire subregion is taken to be at distance x from the road. Then

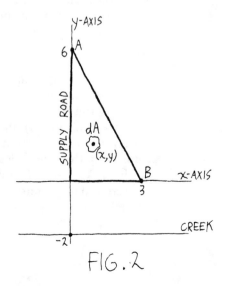

FIG. 2

$$d\,\text{cost} = dA \times \text{distance to road} = x\,dA,$$

and the total cost is $\int_{\text{triangular region}} x\,dA$. The lower boundary of the region is the x-axis where $y = 0$, and the upper boundary is line AB, with slope -2, y-intercept 6 and equation $y = -2x + 6$. The extreme values of x are 0 and 3, so

$$\text{total cost} = \int_{x=0}^{x=3} \int_{y=0}^{y=-2x+6} x\,dy\,dx.$$

The inner integral is $xy \Big|_{y=0}^{y=-2x+6} = 6x - 2x^2$. Thus the outer integral is $\int_{x=0}^{x=3} (6 - 2x^2)\,dx$, the same as (2), so the two approaches merge to produce the same final answer.

Example 2   Continue with the triangular plot of land in Example 1. Suppose that the cost of clearing the land is not only proportional to the area of the land and its distance to the supply road but also inversely proportional to its distance from the creek, running parallel to the shorter leg 2 miles away. (Land far from the creek is less mosquito-ridden and clearing costs are lower.) In particular, suppose that

(3) $$\text{cost} = \frac{\text{area} \times \text{distance to supply road}}{\text{distance to creek}}$$

The problem is to express, with an integral, the cost of clearing the triangular plot.

The strip in Fig. 1 is no longer useful since the points in the strip are at varying distances from the creek. On the other hand, the small subregion in Fig. 2 may be considered to have constant distance x from the supply road and constant distance $y + 2$ from the creek. Therefore, by (3),

$$d\,\text{cost} = \frac{x\,dA}{y + 2}$$

and

$$\text{total cost} = \int_{\text{triangular region}} \frac{x\,dA}{y + 2}$$

$$= \int_{x=0}^{x=3} \int_{y=0}^{y=-2x+6} \frac{x}{y + 2}\,dy\,dx.$$

The aim of the section is to demonstrate how to produce integral models for physical situations. In this example, and in many of the solutions to the problems, we set up the integral and then stop without computing its value.

A general pattern for applying integrals   Suppose a formula (such as (1) or (3)) applies to a plane region in a *simple* situation (*constant* distances to road and creek). In a more complicated situation (*non*constant distances), the formula cannot be used directly. However, if the region is divided into thin rectangular or circular strips or into small subregions, we may be able to apply the formula to the pieces and compute "d thing" (d cost). An integral, single if strips are used, double if small subregions are used, will add the d things and find a total thing (total cost).

Warning   By the physical nature of the particular problems in this section, if a region is divided into many small subregions, the simple factor dA should be contained in the expression for d thing; *it should not be missing* nor should it appear in a form such as $(dA)^2$ or $1/dA$. For example, d thing may be $xy^2\,dA$, but should not be xy^3 or $xy^3(dA)^2$ or xy^3/dA. The double integral is designed to add terms of the form $f(x, y)\,dA$. A sum of terms of the form xy^3 or $xy^3(dA)^2$ or xy^3/dA is not found with an integral.

Problems for Section 12.5

1. Consider a right triangle with legs 5 and 10. (a) Find the average distance from points in the triangular region to the longer leg.   (b) Find the total mass of the region if the mass density (grams per square centimeter) at any point in the region is the product of its distances to the legs.

2. A sandstorm blows in from a desert (Fig. 3) so that the number of particles of sand per square meter deposited at a point in town is $1/d$, where d is the distance from the point to the edge of the desert. Find the total amount of sand in the town.

3. A revolving sprinkler deposits water in a semicircular region of radius 6 centered at the sprinkler, so that by the end of the watering period, the water density (liters per square meter) at a point is the cube of the distance from the point to the sprinkler. Find the total amount of water delivered to the region.

4. The price of land in the city depends on the acreage and on the distance to the town dump (real estate prices are lower near the dump). Suppose that the cost of land is the product of its area and its distance to the dump. Express with an integral the cost of (a) a circular region of radius 2, centered at the dump   (b) the land in Fig. 4.

5. Suppose $\int_R f(x, y)\,dA$ computes the total number of people living in the region R. (a) What does one term of the form $f(x, y)\,dA$ represent physically? (b) What does the function $f(x, y)$ signify physically? For example, what can you conclude if $f(2, 3) = 8$?

6. If a group of persons is d feet from a smokestack then the amount of induced disease among the group is (number of people)$/d$. (The more people exposed, the more disease, and the further they are from the smokestack, the less the amount of disease.) Suppose the region in Fig. 5 has a population density of $f(x, y)$ people per square mile and a smokestack is at point C. Express the total amount of induced disease in the region with an integral.

7. The heat concentration of a region which is distance d from a hot wire is area$/d$. Find the heat concentration in the triangular region of Fig. 6.

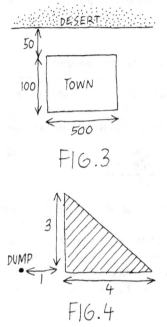

FIG. 3

FIG. 4

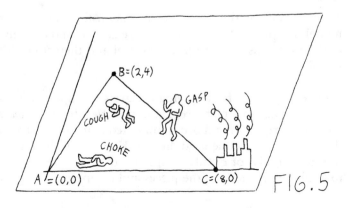

FIG. 5

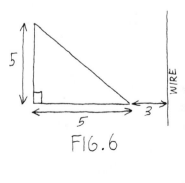

FIG. 6

8. The price of land in a town depends on its area, on its distance to the railroad tracks (land closer to the tracks is cheaper) and on its distance to prestigious Tree Drive (the closer to TD, the more expensive). In particular, the cost of a plot of land is (area × distance to tracks)/distance to TD. Express the cost of the triangular plot in Fig. 7 with an integral.

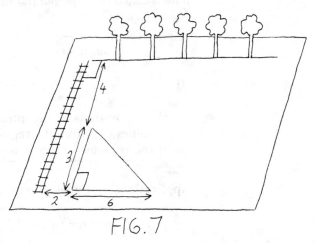

FIG. 7

9. The energy collected in a plane region depends on its area and its distance to the energy source. Suppose that energy is area/d where d is the distance to the source. Find the total energy in a circular region of radius 3 if the energy source is (a) at the center of the region   (b) distance 5 above the center of the region.

10. The cost of painting a section of a billboard depends on the section's height h above the ground, its area A, and its distance d to the ladder. Suppose the cost is Ah^2d. Find the cost of painting the entire billboard in Fig. 8.

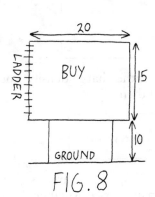

FIG. 8

12.6 Triple Integrals

The definition, applications and computation of triple integrals are similar to those for single and double integrals.

Definition of the integral of $f(x, y, z)$ over a solid region R in 3-space (the triple integral) Given a function $f(x, y, z)$ and a solid region R in 3-space, divide R into many small subregions, not necessarily of the same volume. Let a typical subregion contain the point (x, y, z) and have volume dV (Fig. 1). For each subregion, find the value of f at (x, y, z) and multiply by

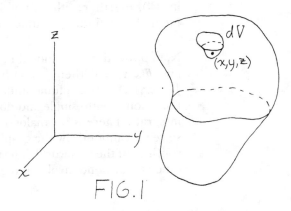

FIG. 1

dV. Add the results from all the subregions to obtain $\sum f(x, y, z)\, dV$. Repeat the process with smaller and smaller values of dV, which requires more and more subregions. It is likely that the resulting sums will be close to one particular number eventually, that is, the sums will approach a limit. The limit is called the *triple integral* of $f(x, y, z)$ over the region and is denoted by $\int_{\text{region}} f(x, y, z)\, dV$.

(1)

> In summary,
>
> $$\int_{\text{region}} f(x, y, z)\, dV = \lim_{dV \to 0} \sum f(x, y, z)\, dV.$$
>
> We think of the integral as adding many representative values of f from the region, each weighted by the volume of the subregion it represents.

Applications By analogy with (3) and (4) of Section 12.1,

(2)

> average value of $f(x, y, z)$ in a region in 3-space
>
> $$= \frac{\displaystyle\int_{\text{region}} f(x, y, z)\, dV}{\text{volume of region}}.$$

As in (5) and (6) of Section 12.1, if $f(x, y, z)$ is the charge density (coulombs per cubic centimeter) at the point (x, y, z) then

(3)

> total charge in a region $= \displaystyle\int_{\text{region}} f(x, y, z)\, dV.$

We already know that if R is a region in 2-space, its area is $\int_R dA$ (Section 12.4, (C)). Similarly, if R is a solid region in 3-space, whether above, below, or crossing the x, y plane, then

(C′)

> $$\int_R dV = \text{volume of the solid region } R;$$

the triple integral in (C′) adds dV's to produce the total volume. The result in (C′), together with (A′) and (B′) in Section 12.4, completes the list of methods for finding volumes using multiple integrals.

Nonapplication We noted in (9) and (10) of Section 12.1 that if $g(x) \geq 0$ and $h(x, y) \geq 0$ then $\int_a^b g(x)\, dx$ is the area under the graph of g and $\int_R h(x, y)\, dA$ is the volume under the graph of h. For the first time, a viewpoint common to single and double integrals has no counterpart for triple integrals. There is no analogous geometric interpretation of $\int_R f(x, y, z)\, dV$ since we cannot draw the graph of a function of three variables within the confines of the real world. The triple integral $\int_R f(x, y, z)\, dV$ has many applications, but none involving the "graph" of $f(x, y, z)$.

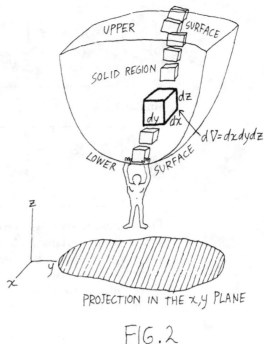

PROJECTION IN THE x,y PLANE

FIG. 2

Computing triple integrals   To devise a method for finding $\int_R f(x, y, z)\, dV$, divide the solid region R into many small boxes. Let a typical box containing the point (x, y, z) have dimensions dx, dy and dz, so that its volume dV is $dx\, dy\, dz$. To evaluate the integral, add $f(x, y, z)\, dV$'s along a typical vertical strip (Fig. 2), then add the strip sums from left to right, and finally add the subtotals from back to front; or add the vertical strip sums from back to front and then add the subtotals from left to right. This manner of addition corresponds to using three single integrals, in the order of integration $dz\, dy\, dx$ or $dz\, dx\, dy$, with limits of integration obtained as follows.

Identify the lower and upper boundary surfaces of R (Fig. 2). Solve the equation of each boundary for z to obtain the inner limits of integration. Use the projection of the solid in the x, y plane to insert x and y limits on the middle and outer integrals as if you were doing a double integral over the projection. In other words,

(4)
$$\int_R f(x, y, z)\, dV = \int \int_{\substack{\text{double integrate}\\ \text{over the projection}\\ \text{of } R \text{ in the } x,y\\ \text{plane}}} \int_{z \text{ on lower boundary}}^{z \text{ on upper boundary}} f(x, y, z) \quad \begin{array}{l} dz\, dy\, dx \\ \text{or} \\ dz\, dx\, dy \end{array}$$

Similarly, the integration may be performed first with respect to x. In this case, identify the rear and forward boundary surfaces of the solid R. Solve the equation of each boundary for x to obtain the inner limits of integration. Use the projection of the solid in the y, z plane to insert y and z limits on the two outer integrals. In other words,

$$\int_R f(x,y,z)\,dV = \int\!\!\!\int_{\substack{\text{double integrate}\\\text{over the projection}\\\text{of } R \text{ in the } y,z\\\text{plane}}} \int_{\substack{x \text{ on rear boundary}}}^{\substack{x \text{ on forward boundary}}} f(x,y,z) \quad \begin{array}{c} dx\,dy\,dz \\ \text{or} \\ dx\,dz\,dy \end{array}$$

(5)

Similarly,

(6)
$$\int_R f(x,y,z)\,dV = \int\!\!\!\int_{\substack{\text{double integrate}\\\text{over the projection}\\\text{of } R \text{ in the } x,z\\\text{plane}}} \int_{\substack{y \text{ on left boundary}}}^{\substack{y \text{ on right boundary}}} f(x,y,z) \quad \begin{array}{c} dy\,dx\,dz \\ \text{or} \\ dy\,dz\,dx \end{array}$$

Thus there are six ways to triple integrate in rectangular coordinates. Note that after one variable is chosen to "go first" and provide the inner limits of integration, the solid is projected into the plane of the other two variables. (Projections are often geometrically clear. If not, the method of Example 3 in Section 12.4 for identifying a projection in the x,y plane may be adapted to find a projection in any coordinate plane.)

Example 1 Consider $\int_R z(x^2 + y^2)\,dV$ where R is the solid polar cap in Fig. 3 bounded by a sphere of radius 2 and the plane $z = 1$. We will set up the triple integral in several ways.

The lower boundary of the region is the plane $z = 1$ and the upper boundary is that portion of the sphere on which $z = \sqrt{4 - x^2 - y^2}$; these may serve as inner limits of integration. For the corresponding middle and outer limits we need the projection of R in the x,y plane. When the plane $z = 1$ intersects the sphere $x^2 + y^2 + z^2 = 4$, we have $x^2 + y^2 + 1 = 4$, $x^2 + y^2 = 3$. Therefore the projection of R is a circular region with radius $\sqrt{3}$. Using the order of integration $dz\,dy\,dx$ we have

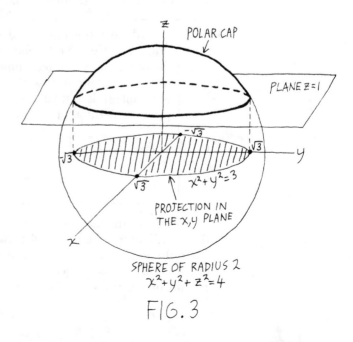

FIG. 3

$$\int_R z(x^2 + y^2)\,dV = \int_{x=-\sqrt{3}}^{x=\sqrt{3}} \int_{y=-\sqrt{3-x^2}}^{y=\sqrt{3-x^2}} \int_{z=1}^{z=\sqrt{4-x^2-y^2}} z(x^2 + y^2)\,dz\,dy\,dx.$$

For the version in which the inner integration is done with respect to x, note that the rear boundary of R is the portion of the sphere where $x = -\sqrt{4 - y^2 - z^2}$ and the front boundary is the portion where $x = \sqrt{4 - y^2 - z^2}$. The projection of R in the y, z plane (Fig. 4) is bounded by the line $z = 1$ and the circle $y^2 + z^2 = 4$. Therefore,

$$\int_R z(x^2 + y^2)\,dV = \int_{y=-\sqrt{3}}^{y=\sqrt{3}} \int_{z=1}^{z=\sqrt{4-y^2}} \int_{x=-\sqrt{4-y^2-z^2}}^{x=\sqrt{4-y^2-z^2}} z(x^2 + y^2)\,dx\,dz\,dy,$$

and also

$$\int_R z(x^2 + y^2)\,dV = \int_{z=1}^{z=2} \int_{y=-\sqrt{4-z^2}}^{\sqrt{4-z^2}} \int_{x=-\sqrt{4-y^2-z^2}}^{x=\sqrt{4-y^2-z^2}} z(x^2 + y^2)\,dx\,dy\,dz.$$

The version with inner y limits of integration is so similar to the setup with inner x limits that we omit it, to avoid repetition.

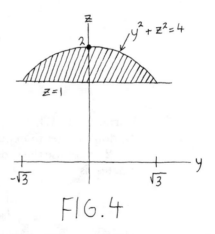

FIG.4

Warning The integral $\int_R z(x^2 + y^2)\,dV$ may be interpreted as the total charge in R if the charge density is $z(x^2 + y^2)$, or the total mass if the mass density is $z(x^2 + y^2)$, but it is *not* the volume of R (the volume is $\int_R dV$).

Triple integration in cylindrical coordinates Consider Example 1 again. Since the projection of R in the x, y plane is a circular region, it is convenient to use polar coordinates for the middle and outer limits, along with inner z limits. Then, instead of $z(x^2 + y^2)\,dz\,dy\,dx$ we write

$$zr^2\,dz\ r\,dr\,d\theta = r^3 z\,dz\,dr\,d\theta,$$

and the upper surface $z = \sqrt{4 - x^2 - y^2}$ becomes $z = \sqrt{4 - r^2}$. Therefore,

$$\int_R z(x^2 + y^2)\,dV = \int_{\theta=0}^{2\pi} \int_{r=0}^{\sqrt{3}} \int_{z=1}^{z=\sqrt{4-r^2}} r^3 z\,dz\,dr\,d\theta.$$

(7)

Evaluating a triple integral using r, θ and z is called *integration in cylindrical coordinates*. In general, the inner limits are z's on the lower and upper boundaries, and the middle and outer limits are found with polar coordinates over the projection in the x,y plane; the integrand, the inner z limits and dV (Fig. 5) are switched to the new variables using

$$x = r \cos \theta, \qquad y = r \sin \theta, \qquad r^2 = x^2 + y^2,$$
$$dV = r \, dz \, dr \, d\theta .$$

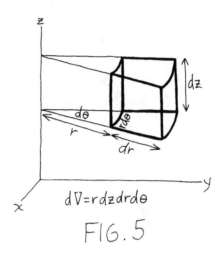

$$dV = r \, dz \, dr \, d\theta$$

FIG. 5

Warning 1. When using polar coordinates for the middle and outer limits, don't forget to express the inner limits in terms of r and θ also.

2. Suppose the order of integration, in rectangular or cylindrical coordinates, is

$$\int_{\text{third variable}} \int_{\text{second variable}} \int_{\text{first variable}}$$

The inner limits of integration may contain the second and third variables. The middle limits of integration may contain the third variable but *not* the first. The outer limits must always be constants.

Example 2   The formula $\frac{1}{3}\pi R^2 h$ for the volume of a cone with radius R and height h was derived in Example 1 of Section 6.1 with a single integral, and in Problem 2d of Section 12.4 with a double integral. Derive it again with a triple integral.

Solution: By (C'), the volume is $\int_{\text{solid cone}} dV$. The lower boundary of the solid cone in Fig. 6 is the conical surface itself, whose equation in cylindrical coordinates, $z = rh/R$, was derived in (4) of Section 10.5; the upper boundary is the plane $z = h$. The projection of the solid in the x,y plane is a circular region of radius R, so

$$\text{volume} = \int_{\theta=0}^{\theta=2\pi} \int_{r=0}^{r=R} \int_{z=rh/R}^{z=h} r \, dz \, dr \, d\theta .$$

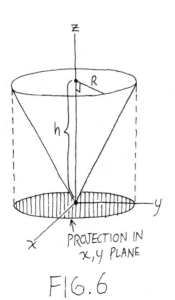

PROJECTION IN
x,y PLANE

FIG. 6

Then

$$\text{inner integral} = \int_{z=rh/R}^{z=h} r\,dz = rz\Big|_{z=rh/R}^{z=h} = r\left(h - \frac{rh}{R}\right),$$

$$\text{middle integral} = \int_{r=0}^{r=R} \left(rh - \frac{r^2h}{R}\right) dr = \left(\frac{1}{2}r^2h - \frac{r^3h}{3R}\right)\Big|_{r=0}^{r=R} = \frac{1}{6}R^2h,$$

and

$$\text{outer integral} = \frac{1}{6}R^2h \int_{\theta=0}^{\theta=2\pi} d\theta = \frac{1}{6}R^2h \cdot 2\pi = \frac{1}{3}\pi R^2h.$$

Warning The lower z limit in Example 2 is *not* the extreme value $z = 0$; rather, it is found by solving the equation of the lower boundary surface, the cone, for z. The limits 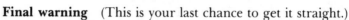, which are *incorrect* for the solid *cone*, correspond to the solid *cylinder* in Fig. 7 whose lower boundary surface is the plane $z = 0$.

Final warning (This is your last chance to get it straight.)
Consider $\int_R f(x, y)\,dA$ over a plane region R.
It is *not* the area of the region R *unless* $f(x, y)$ is identically 1.
If $f(x, y) \geq 0$, it is the volume of the solid with floor R and roof $z = f(x, y)$.
(More generally, if $f(x, y)$ is not ≥ 0, the integral is the volume above minus the volume below; see (12) of Section 12.1.)
Consider $\int_R f(x, y, z)\,dV$ over a solid region R.
If $f(x, y, z)$ is identically 1, the integral is the volume of R. Otherwise, it is not the volume (or area) of anything.
In any case, an integral has many uses other than area and volume. In fact the same integral may be used by one person to compute volume, by another to compute a total mass, by another to find a total cost, by another to find a moment of inertia, etc.

Problems for Section 12.6

1. Set up the triple integral three ways, using the three projections of R. Then evaluate the integral (once).

(a) $\int_R x^2\,dV$ where R is the solid prism bounded by the coordinate planes and the planes $x + y = 1$ and $z = 2$
(b) $\int_R x^2 z\,dV$ where R is the solid cylinder $x^2 + y^2 = 4$ between the planes $z = 0$ and $z = 5$

2. Set up $\int_R f(x, y, z)\,dV$ over the region R using as many projections as feasible and interesting.

(a) the tetrahedron $ABCD$ in Fig. 8
(b) the solid sphere with center at the origin and radius R
(c) the solid cone with radius R and height h in Fig. 6
(d) the quarter-cylinder in Fig. 9 with radius R and height h
(e) the solid in Fig. 10 in the first octant, formed by the intersection of two cylinders of radius 3

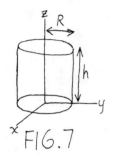

FIG. 7

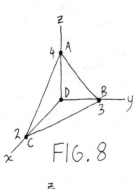

FIG. 8

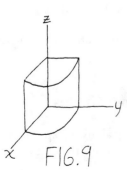

FIG. 9

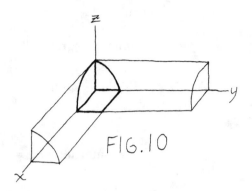

FIG.10

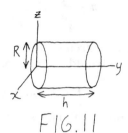

FIG.11

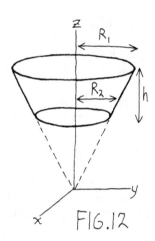

FIG.12

(f) the cylindrical solid in Fig. 11
(g) the region bounded by $z = 2x^2 + y^2$ and $z = 4 - y^2$
(h) the region bounded by $z = x^2$ and planes $z = 5$, $y = -2$, $y = 3$
(i) the region bounded by the ellipsoid $x^2 + 2y^2 + 3z^2 = 12$
(j) the region bounded by the planes $x = 1$, $y = 0$, $y = 2$, $z = 0$, and $x = z$
(k) the region inside the cylinder $x^2 + y^2 = 1$ and between the planes $z = 0$ and $y + z = 5$
(l) the first octant region inside the cylinder $x^2 + y^2 = 1$ and under the plane $z = x$
(m) the solid frustrum in Fig. 12

3. Each problem has already been solved with a single or double integral. Express the solution again with a triple integral.

(a) (Section 6.1, problem 18) Find the total mass of a cylinder of radius R, and height h if its density (mass per unit volume) at a point is

(i) the distance from the point to the axis of the cylinder
(ii) the distance from the point to the base of the cylinder.

(b) (Review problem 2, Chapter 6) If a solid of mass m is revolved around a line, its moment of inertia is md^2 where d is its distance to the line. Find the moment of inertia of a solid cone of radius R and height h which revolves around its axis, if its mass density is constant at δ grams/cm³.

(c) (Section 6.1, Example 4) The work done when an object of weight w moves distance d is wd. Suppose a cylindrical tank of radius 5 and height 20 is half filled with liquid of density 2 pounds per cubic foot. Find the work done in pumping the liquid out, that is, of moving the liquid up to the top of the tank at which point it spills over.

(d) (Section 12.4, problem 3(a)) Find the volume inside the paraboloid $z = 2x^2 + 2y^2$ and under the plane $z = 12$.

4. Express with an integral the volume of the solid bounded by $x^2 + 2y^2 - 3z^2 = 6$ and the planes $z = 0$, $z = 2$.

5. Is it correct to find $\int_{\text{solid sphere}} f(x, y, z)\, dV$ by integrating over a hemisphere and doubling the result?

12.7 Triple Integration in Spherical Coordinates

We will begin by evaluating a specific integral using spherical coordinates. The method we develop will then be summarized and used in general.

Example 1 Consider $\int_R z^2\,dV$ where R is the upper hemispherical region with radius 4, centered at the origin (Fig. 1). The integration can be done with the methods of the preceding section, but just as double integrals over circular regions are usually easier in polar coordinates, triple integrals over spherical and conical regions are often easier in spherical coordinates.

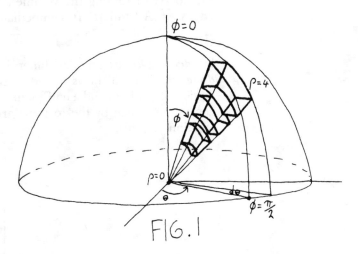

FIG. 1

The integration method of the preceding section was derived by dividing the solid region into many small boxes using planes parallel to the coordinate planes (Section 12.6, Fig. 2). The procedure evaluates a triple integral with three single integrals which add $f(x,y,z)\,dV$'s on (say) a vertical strip, add strip sums from back to front, and finally add subtotals from left to right. To integrate in spherical coordinates we divide the region into many small subregions using spheres, cones and half-planes. Figure 2 shows a typical subregion, called a spherical coordinate box. The face $ABFE$ lies on a cone with cone angle ϕ; the opposite face $DCGH$ lies on a cone with

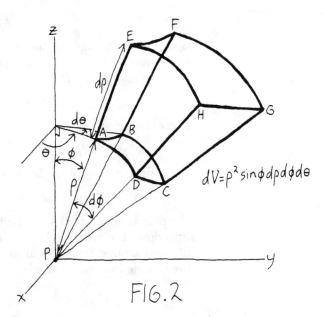

$$dV = \rho^2 \sin\phi\,d\rho\,d\phi\,d\theta$$

FIG. 2

angle $\phi + d\phi$. The face $ADCB$ lies on a sphere of radius ρ; the opposite face $EHGF$ lies on a sphere with radius $\rho + d\rho$. The face $ADHE$ lies on a half-plane, hinged along the z-axis at angle θ; the opposite face $BCGF$ lies on a half-plane with angle $\theta + d\theta$. Not all the walls of the spherical coordinate box are plane, but its faces do intersect perpendicularly so we will take the liberty of finding the volume dV by taking the product of the three edges, $\overline{AE}$, $\widehat{AB}$ and $\widehat{AD}$. It is immediate from the construction of the box that

$$(1) \qquad\qquad \overline{AE} = d\rho,$$

but $d\phi$ and $d\theta$ are angular dimensions, not edges of the box, so dV is *not* $d\rho\,d\phi\,d\theta$. Figure 3 shows the edge AD, an arc on the great circle of intersection of a sphere and a half-plane. The circle has radius ρ and the arc has central angle $d\phi$; by the formula arc length = radius × angle in radians (Section 1.3, (5)),

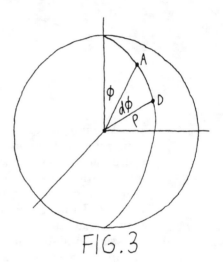

$$\text{FIG. 3}$$

$$(2) \qquad\qquad \widehat{AD} = \rho\,d\phi.$$

Figure 4 shows edge AB, an arc on the circle of intersection of the sphere and a cone. The circle has radius $\overline{QA}$ and the arc has central angle $d\theta$. From the right triangle QAP we have $\overline{QA} = \rho \sin \phi$; by the arc length formula,

$$(3) \qquad\qquad \widehat{AB} = \overline{QA}\,d\theta = \rho \sin \phi\,d\theta.$$

Therefore, from (1)–(3),

$$dV = d\rho \times \rho\,d\phi \times \rho \sin \phi\,d\theta = \rho^2 \sin \phi\,d\rho\,d\phi\,d\theta.$$

Then, since $z = \rho \cos \phi$ (Section 10.5, (3)),

$$z^2\,dV = (\rho \cos \phi)^2 \rho^2 \sin \phi\,d\rho\,d\phi\,d\theta = \rho^4 \cos^2\phi \sin \phi\,d\rho\,d\phi\,d\theta.$$

The plan is to add $z^2\,dV$'s on the typical radial strip in Fig. 1, add the strip sums down the great circle from $\phi = 0$ to $\phi = \pi/2$, and finally add those subtotals around from $\theta = 0$ to $\theta = 2\pi$. The addition is accomplished by three single integrals, namely,

$$\int_R z^2\,dV = \int_{\theta=0}^{\theta=2\pi} \int_{\phi=0}^{\phi=\pi/2} \int_{\rho=0}^{\rho=4} \rho^4 \cos^2\phi \sin \phi\,d\rho\,d\phi\,d\theta.$$

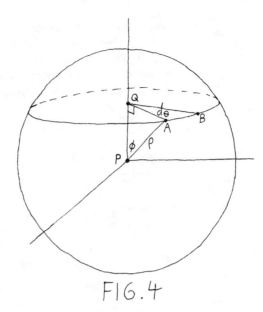

FIG. 4

A routine calculation of the inner, middle and finally outer integral produces the answer $2\pi(2/3)(4^5/5)$, or $4^6\pi/15$.

Computing a triple integral in spherical coordinates The method established in Example 1 may be applied more generally. A triple integral over a solid region R can be evaluated with three single integrals in the following manner. Change x, y, z and dV to spherical coordinates using

(4)
$$x = \rho \sin\phi \cos\theta, \qquad y = \rho \sin\phi \sin\theta, \qquad z = \rho \cos\phi,$$
$$\rho^2 = x^2 + y^2 + z^2, \qquad dV = \rho^2 \sin\phi \, d\rho \, d\phi \, d\theta.$$

To find the limits for the inner integral, identify the inner and outer boundaries of R by imagining a radial walk from the origin; the entrance surface is the inner boundary and the exit surface is the outer boundary. Solve the spherical coordinate equation of each boundary for ρ to obtain the inner limits of integration. For the limits on the middle integral, imagine the positive z-axis, hinged at the origin, falling until it enters the region, and continuing to fall until it exits the region. For all practical purposes, spherical coordinates are used for a limited number of (important) conical and spherical regions, and in these cases, the falling z-axis will enter at a constant angle ϕ_1 whether it falls forward, backwards or sideways, and similarly exit at another constant angle ϕ_2. These extreme values of ϕ are the middle limits. Finally, the outer limits of integration are the extreme values of θ in the region. In other words, for the limited number of regions in which spherical coordinates are advisable, the pattern for the limits is

(5)
$$\int_{\text{smallest } \theta}^{\text{largest } \theta} \int_{\text{smallest } \phi}^{\text{largest } \phi} \int_{\rho \text{ on inner boundary}}^{\rho \text{ on outer boundary}}$$

Other orders of integration are possible, but the version in (5) will be sufficient. The limits on the inner integral will contain θ and/or ϕ unless a boundary is a sphere centered at the origin with an equation as simple as $\rho = \rho_0$. Theoretically, the middle limits may contain θ, but in practice they will be constants, as indicated in (5). The outer limits are always constants. Figures 5–11 show some common regions (with spherical boundaries) and give the corresponding limits.

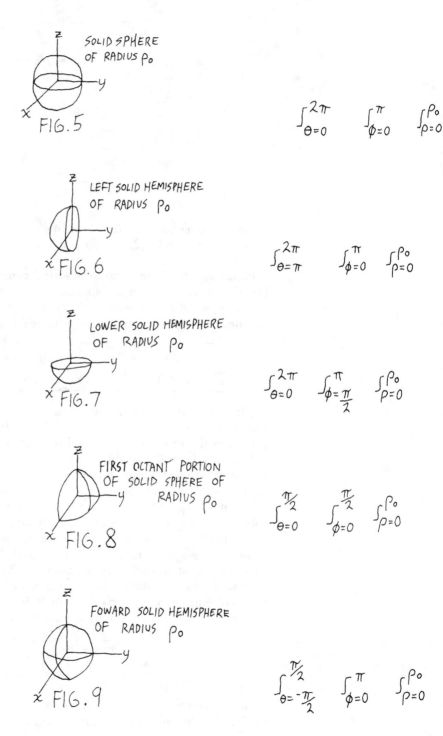

SOLID SPHERE OF RADIUS ρ_0

FIG. 5

$$\int_{\theta=0}^{2\pi} \quad \int_{\phi=0}^{\pi} \quad \int_{\rho=0}^{\rho_0}$$

LEFT SOLID HEMISPHERE OF RADIUS ρ_0

FIG. 6

$$\int_{\theta=\pi}^{2\pi} \quad \int_{\phi=0}^{\pi} \quad \int_{\rho=0}^{\rho_0}$$

LOWER SOLID HEMISPHERE OF RADIUS ρ_0

FIG. 7

$$\int_{\theta=0}^{2\pi} \quad \int_{\phi=\frac{\pi}{2}}^{\pi} \quad \int_{\rho=0}^{\rho_0}$$

FIRST OCTANT PORTION OF SOLID SPHERE OF RADIUS ρ_0

FIG. 8

$$\int_{\theta=0}^{\pi/2} \quad \int_{\phi=0}^{\frac{\pi}{2}} \quad \int_{\rho=0}^{\rho_0}$$

FOWARD SOLID HEMISPHERE OF RADIUS ρ_0

FIG. 9

$$\int_{\theta=-\frac{\pi}{2}}^{\frac{\pi}{2}} \quad \int_{\phi=0}^{\pi} \quad \int_{\rho=0}^{\rho_0}$$

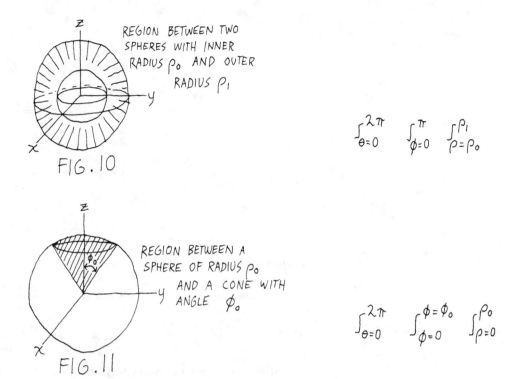

REGION BETWEEN TWO
SPHERES WITH INNER
RADIUS ρ_0 AND OUTER
RADIUS ρ_1

$$\int_{\theta=0}^{2\pi} \quad \int_{\phi=0}^{\pi} \quad \int_{\rho=\rho_0}^{\rho_1}$$

FIG. 10

REGION BETWEEN A
SPHERE OF RADIUS ρ_0
AND A CONE WITH
ANGLE ϕ_0

$$\int_{\theta=0}^{2\pi} \quad \int_{\phi=0}^{\phi=\phi_0} \quad \int_{\rho=0}^{\rho_0}$$

FIG. 11

Warning 1. The θ limits in Fig. 6 may also be written as $\int_{-\pi}^{0}$ or $\int_{3\pi}^{4\pi}$ but *not* in the discontinuous form $\int_{\pi}^{0}$.

2. The spherical coordinate ϕ is always between 0 and π. Therefore, ϕ limits in *any* integral problem are never negative and never larger than π.

Example 2 Consider the solid polar cap bounded by a sphere with radius 4 and center at A and a plane whose distance to A is 3. If the energy density at any point in the cap is $1/d$ where d is the distance from the point to A, find the total energy in the cap.

Solution: Establish the coordinate system in Fig. 12. The distance to the origin in spherical coordinates is ρ, so the total energy, the integral of the energy density, is $\int_{cap} 1/\rho \, dV$. The inner boundary is the plane $z = 3$ which in spherical coordinates is $\rho \cos \phi = 3$, or $\rho = 3 \sec \phi$; the outer boundary is the sphere $\rho = 4$. The smallest value of ϕ is 0 and the largest is ϕ_0, shown in Fig. 12. The extreme values of θ are 0 and 2π. Therefore

$$\text{total energy} = \int_{\theta=0}^{\theta=2\pi} \int_{\phi=0}^{\phi=\phi_0} \int_{3 \sec \phi}^{\rho=4} \frac{1}{\rho} \rho^2 \sin \phi \, d\rho \, d\phi \, d\theta.$$

Then

$$\text{inner integral} = \frac{1}{2} \rho^2 \sin \phi \Big|_{\rho=3 \sec \phi}^{\rho=4} = 8 \sin \phi - \frac{9}{2} \frac{\sin \phi}{\cos^2 \phi}.$$

An antiderivative of $\dfrac{\sin \phi}{\cos^2 \phi}$ is $\dfrac{1}{\cos \phi}$; it may be obtained using the

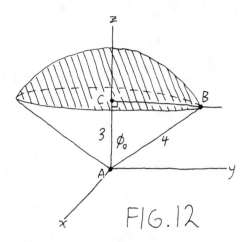

FIG. 12

substitution $u = \cos \phi$. Then

$$\text{middle integral} = \left(-8 \cos \phi - \frac{9}{2} \frac{1}{\cos \phi}\right)\Big|_{\phi=0}^{\phi=\phi_0}$$

$$(6) \qquad\qquad = -8 \cos \phi_0 + 8 - \frac{9}{2}\left(\frac{1}{\cos \phi_0} - 1\right).$$

To continue, it is not necessary to find the angle ϕ_0 itself. All we need is $\cos \phi_0$ which, according to the right triangle ABC, is $3/4$. Therefore

$$\text{middle integral} = -8\left(\frac{3}{4}\right) + 8 - \frac{9}{2}\left(\frac{1}{3/4} - 1\right) = \frac{1}{2}$$

and

$$\text{outer integral} = \frac{1}{2} \int_{\theta=0}^{2\pi} d\theta = \frac{1}{2}(2\pi) = \pi.$$

Problems for Section 12.7

Use spherical coordinates in each problem.

1. Confirm that the volume of a sphere of radius R is $\frac{4}{3}\pi R^3$.
2. Confirm that the volume of a cone with radius R and height h is $\frac{1}{3}\pi R^2 h$.
3. Find the total mass of a spherical region of radius R if the mass density at a point is $1/d^2$ where d is the distance from the point to the center.
4. Find the limits corresponding to the (unbounded) half-space, where (a) $x \geq 0$ (b) $z \geq 0$ (c) $y \geq 0$.
5. Find the volume of the solid in Fig. 11 if the radius of the sphere is 3 and the radius of the circle of intersection of sphere and cone is 2.
6. Set up problem 3(b) in Section 12.6 using spherical coordinates.
7. Find the mass of the frustrum in Fig. 13 if the mass density at a point is $1/d$ where d is the distance from the point to A.

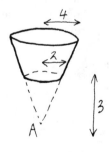

FIG. 13

12.8 Center of Mass

In (1) of Section 6.2 we found that if n masses $m_1, \cdots, m_n$ hang from positions $x_1, \cdots, x_n$ on a line (Fig. 1) then the balance point $\bar{x}$ is given by

(1) $\quad \bar{x} = \dfrac{m_1 x_1 + \cdots + m_n x_n}{m_1 + \cdots + m_n} = \dfrac{\text{total moment with respect to the origin}}{\text{total mass}}$

We used (1) to find the balance point of a solid hemisphere, a symmetric object with constant density. In this section we extend (1) to find balance points of plane and solid objects in general, including nonsymmetric objects with variable density.

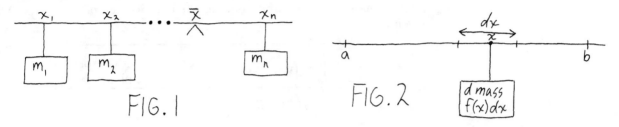

FIG. 1

FIG. 2

Center of mass We'll begin with a one-dimensional situation. Suppose the density in a rod along the interval $[a,b]$ is $f(x)$ mass units per unit of distance. Divide the rod into many small pieces and let a typical piece containing point x have length dx (Fig. 2). The density of the small piece may be considered constant at value $f(x)$, so $d\text{mass} = f(x)\,dx$. To simulate the situation in Fig. 1, picture the entire mass of each small piece concentrated at position x, as indicated in Fig. 2. Then

$$d\text{moment} = \text{coordinate} \times d\text{mass} = xf(x)\,dx,$$

$$\text{total moment} = \int_a^b d\text{moment} = \int_a^b xf(x)\,dx,$$

$$\text{total mass} = \int_a^b d\text{mass} = \int_a^b f(x)\,dx$$

and

(2) $\quad \bar{x} = \dfrac{\displaystyle\int_a^b xf(x)\,dx}{\displaystyle\int_a^b f(x)\,dx} = \dfrac{\text{moment of rod with respect to the origin}}{\text{mass of the rod}}$

The balance point $\bar{x}$ is called the *center of mass* of the rod.

Similarly, if the density at (x,y) in a plane region R is $f(x,y)$ mass units per unit area then the center of mass $(\bar{x}, \bar{y})$ is given by

(3) $\quad \bar{x} = \dfrac{\displaystyle\int_R xf(x,y)\,dA}{\displaystyle\int_R f(x,y)\,dA} = \dfrac{\text{moment of the region with respect to the } y\text{-axis}}{\text{mass of the region}}$

(4) $\quad \bar{y} = \dfrac{\displaystyle\int_R yf(x,y)\,dA}{\displaystyle\int_R f(x,y)\,dA} = \dfrac{\text{moment of the region with respect to the } x\text{-axis}}{\text{mass of the region}}$

If the entire plane is weightless except for the region R then the plane balances at the point $(\bar{x}, \bar{y})$ (which is not necessarily contained in R). The numerator in (3) is called the moment *with respect to the y-axis* because the factor x in the integrand is the signed distance to the y-axis; the numerator in (4) is called the moment *with respect to the x-axis* because the factor y in the integrand is the signed distance to the x-axis.

In the same manner, if the density at (x, y, z) in a solid region R is $f(x, y, z)$ mass units per unit volume, then the center of mass $(\bar{x}, \bar{y}, \bar{z})$ is given by

(5)
$$\bar{x} = \frac{\displaystyle\int_R x f(x, y, z)\, dV}{\displaystyle\int_R f(x, y, z)\, dV} = \frac{\text{moment of } R \text{ with respect}}{\text{mass of } R}$$
$$\text{to the } y, z \text{ plane}$$

(6)
$$\bar{y} = \frac{\displaystyle\int_R y f(x, y, z)\, dV}{\displaystyle\int_R f(x, y, z)\, dV} = \frac{\text{moment of } R \text{ with respect}}{\text{mass of } R}$$
$$\text{to the } x, z \text{ plane}$$

(7)
$$\bar{z} = \frac{\displaystyle\int_R z f(x, y, z)\, dV}{\displaystyle\int_R f(x, y, z)\, dV} = \frac{\text{moment of } R \text{ with respect}}{\text{mass of } R}$$
$$\text{to the } x, y \text{ plane}$$

One application of the center of mass is to the analysis of the behavior of solids in a gravitational force field, where the solid may be replaced by a point mass located at its center of mass.

Example 1   Find the center of mass of a hemispherical solid with radius R if the density is the square of the distance from the point to the center.

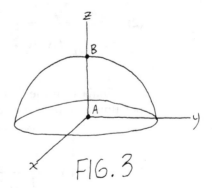

FIG. 3

Solution: Insert a coordinate system as in Fig. 3 so that the density at a point is ρ^2. By symmetry, the center of mass lies on the z-axis so $\bar{x} = \bar{y} = 0$. It remains to find $\bar{z}$. For the numerator of (7) we have

$$\int_{\theta=0}^{\theta=2\pi} \int_{\phi=0}^{\phi=\pi/2} \int_{\rho=0}^{\rho=R} \rho \cos \phi \cdot \rho^2 \cdot \rho^2 \sin \phi \, d\rho \, d\phi \, d\theta.$$

Then

$$\text{inner integral} = \frac{1}{6}\rho^6 \cos\phi \sin\phi \bigg|_{\rho=0}^{\rho=R} = \frac{R^6}{6}\cos\phi\sin\phi,$$

$$\text{middle integral} = \frac{R^6}{12}\sin^2\phi \bigg|_{\phi=0}^{\phi=\pi/2} = \frac{R^6}{12},$$

$$\text{outer integral} = \frac{R^6}{12}\int_{\theta=0}^{\theta=2\pi} d\theta = \frac{R^6}{12}\cdot 2\pi = \frac{\pi R^6}{6}.$$

For the denominator of (7) we have

$$\int_{\theta=0}^{2\pi}\int_{\phi=0}^{\phi=\pi/2}\int_{\rho=0}^{\rho=R}\rho^2\cdot\rho^2\sin\phi\,d\rho\,d\phi\,d\theta.$$

Then

$$\text{inner integral} = \frac{1}{5}\rho^5\sin\phi\bigg|_{\rho=0}^{\rho=R} = \frac{R^5}{5}\sin\phi,$$

$$\text{middle integral} = -\frac{R^5}{5}\cos\phi\bigg|_{\phi=0}^{\phi=\pi/2} = \frac{R^5}{5},$$

$$\text{outer integral} = \frac{R^5}{5}\int_{\theta=0}^{\theta=2\pi} = \frac{R^5}{5}\cdot 2\pi = \frac{2\pi R^5}{5}.$$

Finally,

$$\bar{z} = \frac{\dfrac{1}{6}\pi R^6}{\dfrac{2}{5}\pi R^5} = \frac{5}{12}R.$$

The center of mass lies on the axis AB, five-twelfths of the way from A to B.

Centroids   If the density of a plane region R is *constant*, the center of mass is called a *centroid*. In that case, the constant f may be pulled out of each integral in (3) and (4) and cancels out, producing the centroid formula

(8)
$$\boxed{\bar{x} = \frac{\displaystyle\int_R x\,dA}{\text{area of } R}, \qquad \bar{y} = \frac{\displaystyle\int_R y\,dA}{\text{area of } R}.}$$

Compare (8) with (3) of Section 12.1 to see that the coordinates of the centroid are the average x and y coordinates in R. Similarly, the coordinates of the centroid of a solid region R are given by

(9)
$$\boxed{\bar{x} = \frac{\displaystyle\int_R x\,dV}{\text{volume of } R}, \qquad \bar{y} = \frac{\displaystyle\int_R y\,dV}{\text{volume of } R}, \qquad \bar{z} = \frac{\displaystyle\int_R z\,dV}{\text{volume of } R},}$$

the average x, y and z coordinates in R.

The centroid is a "geometric center" and in certain instances can be identified by inspection. For example, the centroid of a solid sphere is its center; the centroid of a rectangle is the intersection of its diagonals. On the other hand, the centroid of a solid hemisphere or a triangle is not obvious, but can be found with integrals.

Problems for Section 12.8

1. In Section 6.2 we found the centroid of a hemispherical solid of radius R using a single integral. Do it again using a triple integral.
2. Find the centroid of (a) a solid cone of radius R and height h (b) a semicircular region with radius R.
3. Express with integrals the centroid of (a) the solid polar cap of radius 2 in a sphere of radius 6   (b) a right triangle with sides $3, 4, 5$.
4. Express with integrals the center of mass of

(a) a semicircular region of radius 2 if the density at a point is proportional to the square of the distance to the center;
(b) a solid cylinder of radius R and height h if the density at a point is proportional to (i) the distance to the top of the cylinder   (ii) the distance to the axis of the cylinder.

REVIEW PROBLEMS FOR CHAPTER 12

1. Consider $\int x^3 y \, dA$ over the region inside the circle $x^2 + y^2 = 2$ and under the line $y = 1$.

(a) Set it up in rectangular coordinates using both orders of integration.
(b) Set it up in polar coordinates.
(c) Without doing any calculating, decide whether the integral is positive, negative or zero.

2. If the regions R_1 and R_2 are congruent (same size and shape) will $\int_{R_1} f(x,y) \, dA$ and $\int_{R_2} f(x,y) \, dA$ be equal?
3. Express with a double integral the area of the region bounded by $y = x^2$ and $y = x + 2$.
4. Find $\int e^{-x^2-y^2} \, dA$ over the exterior of a circle with center at the origin and radius 3.
5. The cost per square foot of land in a community is proportional to its distances to the (noisy) airport runways and, in particular, is the product of the two distances. Find the cost of the wedge of land in Fig. 1 with radius 2, bordering on the two runways.
6. Let R be the solid bounded by $z = 16 - y^2$ and the planes $z = 0$, $x = 0$, $x = 3$. Set up $\int_R f(x,y,z) \, dV$ three ways, using the three projections of R.
7. Set up $\int_R f(x,y,z) \, dV$ over the solid bounded by $z = 4 - x^2 - y^2$ and $z = 4 - 2y$.
8. Find $\int 3 \, dA$ (easily) over the circular region with center at $(\pi, \sqrt{7})$ and radius 9.
9. Express with (a) a double integral and   (b) a triple integral the volume above the x, y plane inside the cylinder $x^2 + y^2 = 4$ and under the paraboloid $z = 10 - 3x^2 - 3y^2$.
10. If a particle with mass m is at distance d from a line then its moment of inertia as it revolves about the line is md^2. Find the moment of inertia of a solid sphere of radius R and mass density δ if it revolves about a diameter.

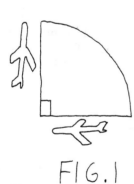

FIG. 1

APPENDIX

A1 Distance and Slope

Distance The distance between the points (x_1, y_1) and (x_2, y_2) is

(1)
$$\sqrt{(x_2 - x_1)^2 + (y_2 - y_1)^2}.$$

If $A = (2, 7)$ and $B = (5, 1)$ then distance AB is $\sqrt{(5 - 2)^2 + (1 - 7)^2}$, or $\sqrt{45}$.

Slope If a nonvertical line contains points (x_1, y_1) and (x_2, y_2), its slope m is given by

(2)
$$m = \frac{\text{change in } y}{\text{change in } x} = \frac{y_2 - y_1}{x_2 - x_1}.$$

If a line passes through $(-1, 1)$ and $(5, 13)$ then $m = \dfrac{13 - 1}{5 - (-1)} = 2$ or, equivalently, $m = \dfrac{1 - 13}{-1 - 5} = 2$.

A line with positive slope rises to the right; a line with negative slope falls to the right (Fig. 1). If a line is horizontal, that is, parallel to the x-axis, then its slope is 0. If a line is vertical, that is, parallel to the y-axis, the formula in (2) cannot be applied because it results in a zero in the denominator, and we say that the line has no slope.

Two nonvertical lines are parallel if and only if they have the same slope. Two nonvertical lines are perpendicular if and only if the product

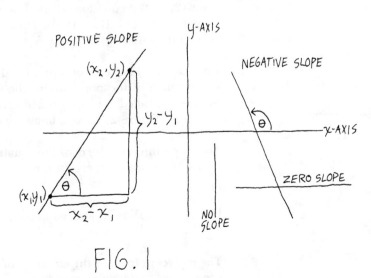

FIG. 1

of their slopes is -1, that is, their slopes are negative reciprocals, such as $\frac{2}{3}$ and $-\frac{3}{2}$.

The angle θ (taken between $0°$ and $180°$) that a line makes with the horizontal is called its *angle of inclination* (Fig. 1). The slope m and angle θ are related by

(3)
$$m = \tan \theta.$$

If a line is inclined at an angle of $100°$ then $m = \tan 100° = -5.67$.

Problems for Section A1

1. Find the slope of line AB and the distance between A and B if

(a) $A = (-2, -3), B = (4, -1)$ (b) $A = (4, 5), B = (6, 5)$
(c) $A = (1, 3), B = (-1, 2)$

2. Let $A = (2, 2), B = (1, 4), C = (5, 1), D = (-3, y)$. Find y so that the lines AB and CD are (a) perpendicular (b) parallel.

A2 Equations of Lines

Horizontal and Vertical lines The graph of an equation of the form $x = c$, where c is a fixed constant, is a line parallel to the y-axis. Similarly, $y = c$ is a line parallel to the x-axis (Fig. 1).

The point-slope form of the equation of a line The line with slope m and passing through the point (x_0, y_0) has equation

(1)
$$y - y_0 = m(x - x_0).$$

For example, the line through $(-2, 5)$ with slope -3 has equation $y - 5 = -3(x + 2)$, or $3x + y = -1$.

As another example, consider the line determined by the two points $(2, 7)$ and $(5, 1)$. Then $m = \dfrac{1 - 7}{5 - 2} = -2$ and, by (1), using the point $(2, 7)$, the equation of the line is $y - 7 = -2(x - 2)$. Alternatively, using the point $(5, 1)$ in (1), the equation of the line is $y - 1 = -2(x - 5)$. Both equations simplify to $2x + y = 11$.

Intercepts The x-intercept of a line, usually denoted by a, is the x-coordinate of the point where the line crosses the x-axis. Similarly, the y-intercept, called b, is the y-coordinate of the point where the line crosses the y-axis. Figure 2 shows a line with $a = 2$ and $b = 1$.

The slope-intercept form of the equation of a line An equation of the line with slope m and y-intercept b is

(2)
$$y = mx + b.$$

If a line has slope 6 and y-intercept -7 then an equation of the line is $y = 6x - 7$.

The intercept form of the equation of a line An equation of the line with x-intercept a and y-intercept b is

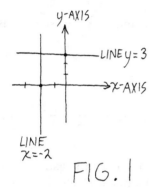

LINE $y = 3$

y-AXIS

x-AXIS

LINE $x = -2$

FIG. 1

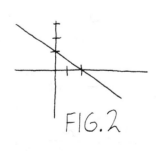

FIG. 2

(3)
$$\frac{x}{a} + \frac{y}{b} = 1.$$

If a line has x-intercept 2 and y-intercept 5 then an equation of the line is $x/2 + y/5 = 1$, or $5x + 2y = 10$.

The general form of the equation of a line   The equations in (1)–(3) can be rewritten in the form

(4)
$$Ax + By + C = 0,$$

called *general* form. All lines, and only lines, have equations of this form.

Example 1   Find the intercepts of the line $2x + 3y - 4 = 0$.
    Solution: If $y = 0$ then $x = 2$, so the x-intercept is 2. If $x = 0$ then $y = \frac{4}{3}$, so the y-intercept is $\frac{4}{3}$.

Example 2   Find the slope of the line $4x + 3y + 2 = 0$.
    Solution: Rewrite the equation as $y = -\frac{4}{3}x - \frac{2}{3}$, its slope-intercept form. By inspection, the slope if $-\frac{4}{3}$ (and the y-intercept is $-\frac{2}{3}$).

Example 3   Sketch the graph of $x + 2y - 2 = 0$.
    Solution: The x-intercept is 2 (set $y = 0$) and the y-intercept is 1 (set $x = 0$). This gives enough information for the sketch in Fig. 2.

Problems for Section A2

   1. Find an equation of the line

   (a) determined by the points $(1, 5)$ and $(7, -2)$
   (b) through the point $(1, 7)$ and parallel to the line $2y - 6x - 7 = 0$
   (c) through the point $(-2, 8)$ and perpendicular to the line $y = 3x - 7$
   (d) perpendicularly bisecting the line segment AB if $A = (1, 3)$ and $B = (-3, 5)$
   (e) with x-intercept 6 and slope 4
   (f) through the point $(1, 7)$ and perpendicular to the line $x/3 + y/6 = 1$
   (g) through the point $(2, 6)$ and parallel to the x-axis

   2. Find the slope of the line with equation

   (a) $2x - 5y + 8 = 0$ (b) $x/3 + y/4 = 1$

   3. Sketch the graph of

   (a) $x - 3y = 4$ (b) $y = 2x + 1$

A3   Circles, Ellipses, Hyperbolas and Parabolas

The circle   An equation of the circle with center at the origin and radius r is

(1)
$$x^2 + y^2 = r^2.$$

More generally, an equation of the circle with center at the point (x_0, y_0) and radius r is

(2)
$$(x - x_0)^2 + (y - y_0)^2 = r^2.$$

For example, the circle with center $(4, -2)$ and radius 3 has equation $(x - 4)^2 + (y + 2)^2 = 9$.

The ellipse Let A, B and C be positive, with $A \neq B$. The graph of an equation of the form

(3) $$Ax^2 + By^2 = C$$

is an ellipse centered at the origin. For example, consider the ellipse $2x^2 + 3y^2 = 12$. To help sketch the graph, plot the intercepts: if $x = 0$ then $y = \pm 2$, and if $y = 0$ then $x = \pm\sqrt{6}$ (Fig. 1).

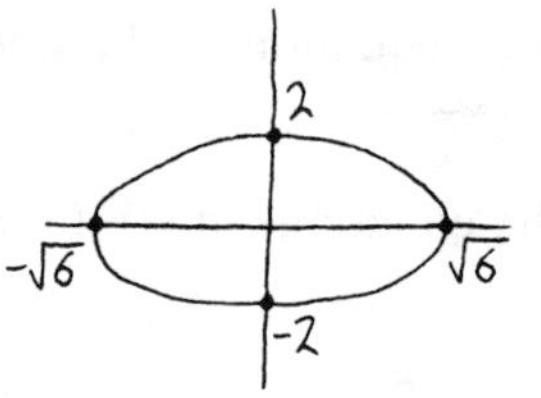

The hyperbola Let A, B and C be positive. The graph of an equation of the form

(4) $$Ax^2 - By^2 = C$$

or

(5) $$Ay^2 - Bx^2 = C$$

is a hyperbola. For example, consider $2x^2 - 3y^2 = 12$. If $x = 0$ then there is no corresponding y since $-3y^2 = 12$ is impossible. If $y = 0$ then $x = \pm\sqrt{6}$. Figure 2 shows the graph. As another example, consider $2y^2 - 3x^2 = 12$. If $x = 0$ then $y = \pm\sqrt{6}$, but if $y = 0$ there is no corresponding value of x. Figure 3 shows the graph.

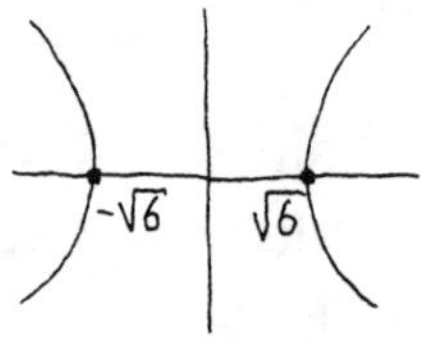

The parabola The graph of an equation of the form $y = ax^2$ or $x = ay^2$ is a parabola. Figure 4 shows the four possibilities. For example, the graph of $x = -3y^2$ is shown in Fig. 5.

Problems for Section A3

Sketch the graph.

1. $3x^2 + y^2 = 6$ **2.** $3x^2 - y^2 = 6$ **3.** $y^2 - 3x^2 = 6$ **4.** $y = -3x^2$
5. $x = -40y^2$ **6.** $y = \frac{1}{10}x^2$ **7.** $x = 3y^2$

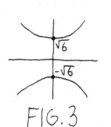

FIG. 1

FIG. 2

FIG. 3

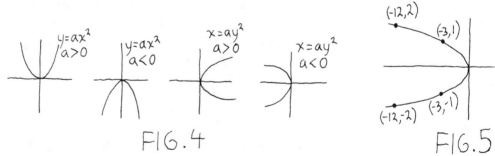

FIG. 4

FIG. 5

A4 The Binomial Theorem

The binomial theorem describes the expansion of $(x + y)^n$ where n is a positive integer. To begin, we have

$$(x + y)^2 = x^2 + 2xy + y^2,$$

$$(x + y)^3 = x^3 + 3x^2y + 3xy^2 + y^3,$$

$$(x + y)^4 = x^4 + 4x^3y + 6x^2y^2 + 4xy^3 + y^4.$$

In each term of the expansion of $(x + y)^n$, the sum of the exponents is n and the coefficient pattern can be obtained from a device called *Pascal's triangle:*

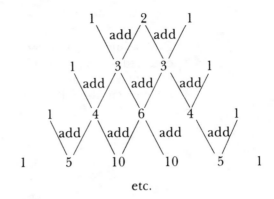

etc.

According to the triangle,

$$(x + y)^5 = x^5 + 5x^4y + 10x^3y^2 + 10x^2y^3 + 5xy^4 + y^5.$$

In general,

$$(x + y)^n = x^n + nx^{n-1}y + \frac{n(n - 1)}{2!}x^{n-2}y^2 + \frac{n(n - 1)(n - 2)}{3!}x^{n-3}y^3$$

$$+ \cdots + \frac{n(n - 1)(n - 2)\cdots 2 \cdot 1}{n!}y^n.$$

where, for example, 5! means $5 \times 4 \times 3 \times 2 \times 1$. For instance, in the expansion of $(x + y)^8$, the coefficient of the term x^5y^3 is $\dfrac{8 \cdot 7 \cdot 6}{3!} = \dfrac{8 \cdot 7 \cdot 6}{3 \cdot 2 \cdot 1} = 56$; the coefficient of the term x^3y^5 is $\dfrac{8 \cdot 7 \cdot 6 \cdot 5 \cdot 4}{5!} = 56$. The two coefficients are equal and, in general, the coefficient of x^ix^j in (1) is the same as the coefficient of x^jy^i. In other words, the coefficients at the end of the expansion match the coefficients at the beginning, as Pascal's triangle illustrates.

Problems for Section A4

1. Expand $(x + y)^7$.
2. Expand $(2p + q)^4$.
3. Find the coefficient of x^3y^{11} in the expansion of $(x + y)^{14}$.
4. Expand $(1 - x)^6$.
5. Find the coefficient of x^9y^2 in the expansion of $(x + y)^{11}$.

A5   Determinants

Evaluating a 2-by-2 determinant   The symbol $\left|\begin{smallmatrix} a & b \\ c & d \end{smallmatrix}\right|$ is called a 2-by-2 determinant, and its value is given by

(1)
$$\begin{vmatrix} a & b \\ c & d \end{vmatrix} = ad - bc.$$

For example,

$$\begin{vmatrix} 2 & 5 \\ 4 & 3 \end{vmatrix} = (2)(3) - (5)(4) = 6 - 20 = -14.$$

Evaluating an $n \times n$ determinant   Every entry in the array has a *cofactor* obtained as follows. Find the minor determinant of the entry by deleting that entry's row and column and attach a sign according to the location of the entry in the following checkerboard pattern:

(2)
$$\begin{matrix} + & - & + & - & + \\ - & + & - & + & - & \cdots \\ + & - & + & - & + \\ & & \vdots & & \end{matrix}$$

For example, consider $\begin{vmatrix} 1 & 2 & 7 & 8 \\ 4 & 3 & 1 & 9 \\ 5 & 2 & 9 & 6 \\ 8 & 1 & 3 & 7 \end{vmatrix}$. The cofactor of the entry 7 in row 1, column 3 is $+\begin{vmatrix} 4 & 3 & 9 \\ 5 & 2 & 6 \\ 8 & 1 & 7 \end{vmatrix}$ (omit row 1 and column 3 from the original and choose the sign in row 1, column 3 of (2)). The cofactor of the entry 1 in row 2, column 3 is $-\begin{vmatrix} 1 & 2 & 8 \\ 5 & 2 & 6 \\ 8 & 1 & 7 \end{vmatrix}$.

To evaluate a 3×3 determinant, pick any row (or column). Then

det = first entry in that row × its cofactor
+ second entry in that row × its cofactor
+ third entry in that row × its cofactor.

For example, using the first row for the expansion, we have

$$\begin{vmatrix} 10 & 2 & 3 \\ 4 & 5 & 6 \\ 7 & 8 & 9 \end{vmatrix} = +10\underbrace{\begin{vmatrix} 5 & 6 \\ 8 & 9 \end{vmatrix}}_{-3} - 2\underbrace{\begin{vmatrix} 4 & 6 \\ 7 & 9 \end{vmatrix}}_{-6} + 3\underbrace{\begin{vmatrix} 4 & 5 \\ 7 & 8 \end{vmatrix}}_{-3} = -27.$$

This method is referred to as *expanding across the first row*. For the same determinant, by expanding down the second column we have

$$\begin{vmatrix} 10 & 2 & 3 \\ 4 & 5 & 6 \\ 7 & 8 & 9 \end{vmatrix} = -2\underbrace{\begin{vmatrix} 4 & 6 \\ 7 & 9 \end{vmatrix}}_{-6} + 5\underbrace{\begin{vmatrix} 10 & 3 \\ 7 & 9 \end{vmatrix}}_{69} - 8\underbrace{\begin{vmatrix} 10 & 3 \\ 4 & 6 \end{vmatrix}}_{48} = -27.$$

A similar method holds for an $n \times n$ determinant. Again, pick any row (or column). The determinant is the sum of entries of that row, each multiplied by its corresponding cofactor.

For example, using the last column, we have

$$\begin{vmatrix} 1 & 2 & 3 & 0 \\ 1 & -1 & 3 & 2 \\ 0 & 1 & 2 & 3 \\ 2 & 1 & 4 & 0 \end{vmatrix} = +2\begin{vmatrix} 1 & 2 & 3 \\ 0 & 1 & 2 \\ 2 & 1 & 4 \end{vmatrix} - 3\begin{vmatrix} 1 & 2 & 3 \\ 1 & -1 & 3 \\ 2 & 1 & 4 \end{vmatrix}$$

Now work on each 3×3 minor determinant. If we expand the first one down column 1 and the second one across row 1 we have

$$\det = 2\left(1\begin{vmatrix} 1 & 2 \\ 1 & 4 \end{vmatrix} + 2\begin{vmatrix} 2 & 3 \\ 1 & 2 \end{vmatrix}\right) - 3\left(1\begin{vmatrix} -1 & 3 \\ 1 & 4 \end{vmatrix} - 2\begin{vmatrix} 1 & 3 \\ 2 & 4 \end{vmatrix} + 3\begin{vmatrix} 1 & -1 \\ 2 & 1 \end{vmatrix}\right)$$

$$= 2(2 + 2) - 3(-7 + 4 + 9) = -10.$$

Some properties of determinants   1. If two rows, or two columns, are interchanged, the sign of the determinant is changed. For example, if $\begin{vmatrix} a & b & c \\ d & e & f \\ g & h & i \end{vmatrix} = -7$, then, after interchanging rows 1 and 3, we have

$$\begin{vmatrix} g & h & i \\ d & e & f \\ a & b & c \end{vmatrix} = 7.$$

2. Multiplying a row or a column by a number will multiply the entire determinant by that number. For example, if the second row is tripled then the new determinant is three times the old determinant. In other words

(3) $$\begin{vmatrix} a & b & c \\ 3d & 3e & 3f \\ g & h & i \end{vmatrix} = 3\begin{vmatrix} a & b & c \\ d & e & f \\ g & h & i \end{vmatrix}.$$

Equation (3) may be thought of as a factoring rule which allows a common factor to be pulled out of a row or a column.

3. If a row or a column consists entirely of zeros, then the determinant is 0.

4. If two rows, or two columns, are identical, then the determinant is 0. For example, $\begin{vmatrix} a & b & c \\ d & e & f \\ a & b & c \end{vmatrix} = 0$.

5. If one row is a multiple of another row, or one column is a multiple of another column, then the determinant is 0. For example, $\begin{vmatrix} 1 & 2 & 5 \\ 3 & 6 & -1 \\ 7 & 14 & \pi \end{vmatrix} = $

0 because the second column is a multiple of the first column.

Problems for Section A5

1. Evaluate $\begin{vmatrix} 2 & -3 \\ 4 & 5 \end{vmatrix}$

2. Find $\begin{vmatrix} 10 & 2 & 3 \\ 4 & 5 & -6 \\ 1 & -3 & 7 \end{vmatrix}$ by expanding (a) across row 2   (b) down column 3.

3. Find $\begin{vmatrix} 1 & -1 & 3 & 4 \\ 0 & 3 & 0 & 1 \\ 2 & 1 & 0 & 3 \\ 1 & 1 & -1 & 2 \end{vmatrix}$ by expanding down column 3.

A6 Polar Coordinates

FIG. 1

FIG. 2

The polar coordinate system locates points using two coordinates named r and θ. The coordinate r is the distance from a point to the origin, and θ is the angle used in trigonometry which measures counterclockwise rotation from the positive x-axis (Fig. 1). In applications, the coordinate r is never negative, and is 0 only at the origin.† As in trig, a positive θ describes counterclockwise rotation while a negative θ describes a clockwise rotation. Figure 2 shows the point with polar coordinates $r = 2$, $\theta = 120°$. The same point also has coordinates $r = 2$, $\theta = -240°$; $r = 2$, $\theta = 480°$; and so on.

The graph of the equation $r = 3$ is a circle with center at the origin and radius 3. The graph of $\theta = 45°$ is a ray from the origin inclined at angle 45°. Polar coordinate graph paper contains a collection of circles (where $r = r_0$) and rays (where $\theta = \theta_0$) as an aid to graphing (Fig. 3).

Connection between rectangular and polar coordinates In Fig. 1, by definition of sine and cosine, $\cos \theta = x/r$ and $\sin \theta = y/r$. So

(1) $$x = r \cos \theta, \qquad y = r \sin \theta.$$

For example, if $r = 2$ and $\theta = 120°$ then $x = 2 \cos 120° = 2 \times -\frac{1}{2} = -1$ and $y = 2 \sin 120° = 2 \times \frac{1}{2}\sqrt{3} = \sqrt{3}$.

---

†Mathematicians are sometimes reluctant to allow a pair such as $r = -2$, $\theta = 30°$, to remain unplotted because of the negative value of r, and adopt the following convention. Find the ray where $\theta = 30°$ but instead of moving distance 2 along that ray, move "backwards," i.e., along the ray $\theta = 210°$, as shown in the figure.

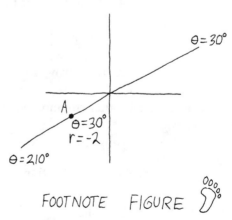

FOOTNOTE FIGURE

In a rectangular coordinate system it is crucial to plot with negative values of x and y as well as positive values since otherwise points outside of quadrant I will have no coordinates. On the other hand, in polar coordinates it may be satisfying to invent a meaning for negative r but it is not crucial in the sense that point A in the figure already has the coordinates $r = 2$, $\theta = 210°$. When polar coordinates are used in mathematics and in applied work, invariably, r is taken to be nonnegative.

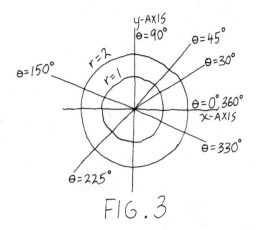

FIG. 3

To find equations which try to express r and θ in terms of x and y, use the definition of $\tan \theta$ as y/x in Fig. 1, and the fact that r is the distance from (x,y) to the origin, to get

(2) $$r = \sqrt{x^2 + y^2}, \qquad r^2 = x^2 + y^2, \qquad \tan \theta = \frac{y}{x}.$$

Sometimes $\tan \theta = y/x$ is written as $\theta = \tan^{-1} y/x$ but this is correct only if $\tan^{-1} y/x$ is loosely interpreted in a nonfunction sense to mean *an* angle whose tangent is y/x. But mathematicians and calculators don't think this way, and you will not necessarily get θ by pushing the $\tan^{-1}$ button on a calculator. Officially, $\tan^{-1} y/x$ stands for *the* angle between $-90°$ and $90°$ whose tangent is y/x, and θ might not be in this range. For example, suppose $x = -4$ and $y = -5$ (Fig. 4). Then $r = \sqrt{16 + 25} = \sqrt{41}$. Furthermore, $\tan^{-1} y/x = 1.25$, which a calculator gives as approximately $51°$. But there are other angles whose tangent is 1.25 since the tangent function repeats every $180°$; the one we want, in quadrant III, is $51° + 180°$ or $231°$. In general, to find θ, first find $\tan^{-1} y/x$; this is θ immediately if the point lies in quadrants I or IV, but you must add $180°$ if the point lies in quadrants II or III. (Many calculators have a button distinct from the $\tan^{-1}$ button which produces θ directly, given x and y.)

We can use (1) to switch equations of curves from rectangular to polar coordinates. For example, the line $x + 3y = 4$ has polar coordinate equation $r \cos \theta + 3r \sin \theta = 4$.

Graphs in polar coordinates Consider the equation $r = \sin \theta$. To draw its graph in polar coordinates, begin with a table of values:

θ	0	30°	45°	60°	90°	120°	150°	180°	←——————→	360°
r	0	$\frac{1}{2}$	$\frac{1}{2}\sqrt{2}$	$\frac{1}{2}\sqrt{3}$	1	$\frac{1}{2}\sqrt{3}$	$\frac{1}{2}$	0	negative	0

Plot points (Fig. 5) until a pattern emerges. If $180° < \theta < 360°$, r is negative and we do not plot any points. The table stops at $\theta = 360°$ because at this stage further entries do not produce new points. For example, if $\theta = 390°$ then $r = \frac{1}{2}$ which describes the same point as $\theta = 30°$, $r = \frac{1}{2}$ already in the table. The graph resembles a circle and we can show that it is indeed a circle by switching to rectangular coordinates. Use (1) and (2) to substitute in

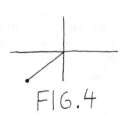

FIG. 4

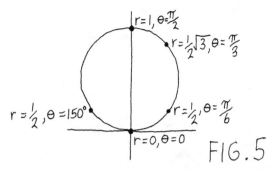

FIG.5

$r = \sin\theta$ to get

$$\sqrt{x^2 + y^2} = \frac{y}{r} = \frac{y}{\sqrt{x^2 + y^2}},$$

or $x^2 + y^2 = y$. Complete the square to get $x^2 + (y - \frac{1}{2})^2 = \frac{1}{4}$, a circle with center at $(0, \frac{1}{2})$ and radius $\frac{1}{2}$.

As another example, consider $r = 2\sin 3\theta$. As θ increases from 0° to 30°, 3θ increases from 0° to 90°, $\sin 3\theta$ increases from 0 to 1 and r increases from 0 to 2. As θ increases from 30° to 60°, 3θ increases from 90° to 180° and r decreases from 2 back to 0, completing a thin loop (see quadrant I in Fig. 6). For θ between 60° and 120°, r is negative and no points are plotted. As θ increases from 120° to 150°, and then to 180°, r increases from 0 to 2, and then decreases to 0 again, completing another loop (see quadrant II in Fig. 6). For θ between 180° and 240°, r is negative, and we do not plot any corresponding points. As θ increases from 240° to 270°, and then to 300°, r increases from 0 to 2, and then decreases to 0 again for still a third loop. Finally, for θ between 300° and 360°, r is negative and we have no points. Figure 6 shows the final graph, a 3-leaved rose.

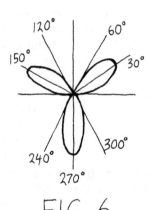

FIG. 6

Problems for Section A6

1. Find x and y if

(a) $r = 3$, $\theta = 60°$ (c) $r = 3$, $\theta = -\pi/4$
(b) $r = 2$, $\theta = 90°$ (d) $r = 2$, $\theta = 0$

2. Find r and θ if

(a) $x = 2$, $y = 4$ (d) $x = 0$, $y = -3$
(b) $x = -2$, $y = -4$ (e) $x = -2$, $y = 4$
(c) $x = -3$, $y = -3$

3. Find the equation of the line $x = 3$ in polar coordinates.
4. Plot the points where

(a) $2 \le r \le 3$ (b) $\pi/6 \le \theta \le \pi/2$ (c) $3 \le r \le 5$ and $\pi/4 \le \theta \le \pi$

5. Show that the distance between the points r_1, θ_1 and r_2, θ_2 is $\sqrt{r_1^2 + r_2^2 - 2r_1 r_2 \cos(\theta_2 - \theta_1)}$ (a) by switching to rectangular coordinates (b) directly with the law of cosines.
6. Sketch the graph

(a) $r = 2 - 2\sin\theta$ (d) $r = 4\sin\frac{1}{2}\theta$
(b) $r = 2\cos 3\theta$ (e) $r = \theta$
(c) $r^2 = 4\sin 2\theta$

Solutions
to the
Problems

1/FUNCTIONS

Section 1.1 (page 4)

1. (a) $f(0) = 2 - 0^2 = 2$
   (b) $f(1) = 2 - 1^2 = 2 - 1 = 1$
   (c) $2 - (b^3)^2 = 2 - b^6$
   (d) 9
   (e) 4
   (f) $(b^3 - 3)^2$
   (g) $g(b) = (b - 3)^2$ so $[g(b)]^3 = [(b - 3)^2]^3 = (b - 3)^6$
   (h) $2 - (2a + b)^2$
   (i) Range of f contains all numbers less than or equal to 2, i.e., range is the interval $(-\infty, 2]$. Range of g is $[0, \infty)$ since g produces all, and only, non-neg numbers.

2. (a) $f(-7) = |-7|/(-7) = 7/(-7) = -1, f(3) = 1$
   (b) $x \neq 0$
   (c) Range contains only 1 and -1.
   (d) $f(2 + 3) = f(5) = 1, f(2) = 1, f(3) = 1, f(2) + f(3) = 2$. No!
   (e) $f(-2 + 6) = f(4) = 1, f(-2) = -1, f(6) = 1, f(-2) + f(6) = 0$. No!
   (f) No! Parts (d) and (e) illustrate what happens when a and b are both positive, and when one is positive and the other negative. If both a and b are negative then $f(a + b) = f(\text{neg}) = -1$, $f(a) + f(b) = -1 + -1 = -2$; still not equal.

3. (a) $1, -1$ (b) All integers (c) $0, 1$
   (d) To get fixed points, need x such that $x^2 + 4 = x$, $x^2 - x + 4 = 0$. But equation has no real solutions so there are no fixed points.

4. $f(a^2) = 2a^2 + 1, (f(a))^2 = (2a + 1)^2$. To see if they are ever equal solve $2a^2 + 1 = (2a + 1)^2$, $2a^2 + 4a = 0, 2a(a + 2) = 0, a = 0, -2$. So $f(a^2) = (f(a))^2$ iff (if and only if) $a = 0, -2$.

5. (a) $f(f(x)) = f(x^3) = (x^3)^3 = x^9$
   (b) Int(Int x) simplifies to plain Int x because after the first Int is taken, the result is an integer and that integer is unchanged by the second Int.
   (c) $f(f(x)) = f(-x + 1) = -(-x + 1) + 1 = x$, $f(f(f(x))) = f(\text{last answer}) = f(x) = -x + 1$, $f(f(f(f(x)))) = f(\text{last answer}) = f(-x + 1) = -(-x + 1) + 1 = x$

In general, if there are an even number of f's used successively then the result is x. If there are an odd number of f's then the result is $-x + 1$.

6. The number of passengers over 200 is $p - 200$, price per ticket is $300 - (p - 200)$, so $A = p[300 - (p - 200)] = 500p - p^2$ for $200 \leq p \leq 350$.

Section 1.2 (page 9)

1. (See figs.) (a) Line through the origin with slope 2; increasing, one-to-one, continuous.
   (b) $x + |x|$ is $2x$ if $x \geq 0$, and is 0 if $x < 0$, continuous.
   (c) $|x|/x$ is 1 if $x > 0$, and is -1 if $x < 0$, disc at $x = 0$, continuous otherwise.
   (d) For example $f(7) = 7, f(2) = 3$. In general, $f(x)$ is x if $x \geq 3$, and is 3 if $x < 3$, continuous.

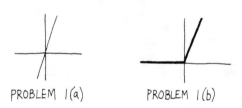

PROBLEM 1(a)   PROBLEM 1(b)

PROBLEM 1(c)   PROBLEM 1(d)

2.

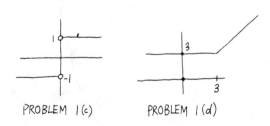

PROBLEM 2

3. (a) $f(-1) = 0$ since $(-1, 0)$ is on the graph, $f(0) = 2$, $f(6) = 2$.
   (b) y is 4 when x is a little larger than 1, and again when x is about 4.
   (c) $x < -1$ (since the graph lies below the x-axis when $x < -1$)

4. Decreases

5. (a) Probably not, since the rate usually jumps at certain weights.
   (b) Yes

6. (a) Graph lies above the x-axis.
   (b) Graph lies above the line $y = x$. (See fig.)

PROBLEM 6(b)

7.

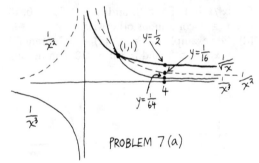

PROBLEM 7(a)

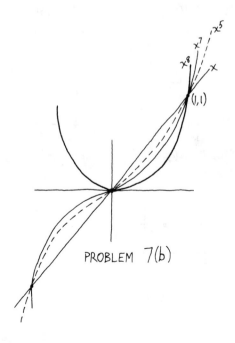

PROBLEM 7(b)

8. (See figs.)
   (a) Symmetric w.r.t. (with respect to) y-axis

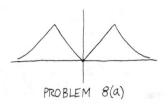

PROBLEM 8(a)

   (b) Symmetric w.r.t. the origin.

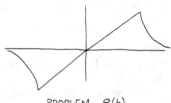

PROBLEM 8(b)

9. Line has slope 3, equ $y - 2 = 3(x - 1), y = 3x - 1$. So $f(x) = 3x - 1$.

10. (a) Doesn't move. Remains at one point forever.
   (b) Moves to the right at 1 mph. (See fig.)

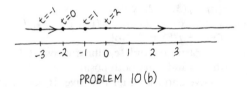

PROBLEM 10(b)

   (c) Moves to the left; as time increases, position decreases.
   (d) Stays to the right of the 0 position on the line. (See fig.)

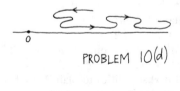

PROBLEM 10(d)

Section 1.3 (page 18)

1. (a) $\pi/5 \times 180/\pi = 36°$ (b) $\frac{5}{6} \times 180 = 150°$
   (c) $-60°$

2. (a) $12 \times \frac{1}{180}\pi = \frac{1}{15}\pi$ (b) $-\pi/2$ (c) $\frac{100}{180}\pi = \frac{5}{9}\pi$

3. (a) $\sin 210° = -\sin 30° = -\frac{1}{2}$
   (b) $\cos 3\pi = \cos \pi = -1$
   (c) $\tan \pi/4 = 1$

4.

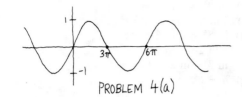

PROBLEM 4(a)

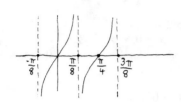

PROBLEM 4 (b)

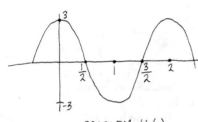

PROBLEM 4 (c)

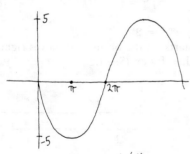

PROBLEM 4 (d)

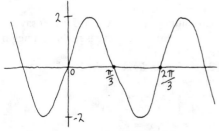

PROBLEM 4 (e)

5. (a) $-\sin x = -a$ (b) $\cos y = b$ (c) $-a$ (d) $-b$ (e) a^2 (f) Can't do.

6.

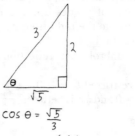

$\cos \theta = \dfrac{\sqrt{5}}{3}$

PROBLEM 6(a)

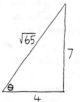

$\sin \theta = \dfrac{7}{\sqrt{65}}$

PROBLEM 6(b)

7.

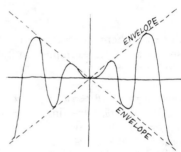

PROBLEM 7(a)

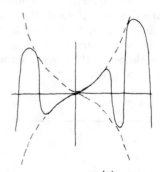

PROBLEM 7 (b)

Section 1.4 (page 23)

1. $f^{-1}(4) = 3$, $f^{-1}(2) = 5$

2. (a) $x + 3$ (b) Not 1-1, no inverse (c) $1/x$, i.e., function is its own inverse since the reverse of taking reciprocal is to take reciprocal again. (d) $-x$ (reverse of negating is to negate again)

3. If $y = 2x - 9$ then $x = \frac{1}{2}(y + 9)$ so $f^{-1}(x) = \frac{1}{2}(x + 9)$.

4. 17

5. An increasing function must be 1-1 so has an inverse. Inverse is increasing: since f increases, as x goes up, y goes up also. So in reverse, if y goes up, x goes up. Alternatively, if a curve rises to the right then its reflection in line $y = x$ also rises to the right.

6. True. If the f graph is unbroken, its reflection is also unbroken.

7. (a) Not unless x^2 is restricted to $x \geq 0$. (b) Yes

8. (a) $\frac{1}{2}\pi$ since $\cos \frac{1}{2}\pi = 0$ and $\frac{1}{2}\pi$ is between $-\frac{1}{2}\pi$ and $\frac{1}{2}\pi$.

  (b) 0

  (c) Not possible since there is no θ such that $\sin \theta = 2$; sines are between -1 and 1.

  (d) 150° since $\cos 150° = -\frac{1}{2}\sqrt{3}$ and 150° is between 0° and 180°.

  (e) $-60°$

  (f) 45°

  (g) $-45°$

9. Just under $\pi/2$, where tangents are very large.

10. (a) False. As a counterexample, $\sin 2\pi = 0$ but $\sin^{-1}0$ is not 2π.

  (b) True.

11. (a) Choose $-\frac{1}{2}\pi \leq \pi\theta \leq \frac{1}{2}\pi$, $-\frac{1}{2} \leq \theta \leq \frac{1}{2}$. Then $2(z - 3) = \sin \pi\theta$, $\pi\theta = \sin^{-1}2(z - 3)$, $\theta = (1/\pi)\sin^{-1}2(z - 3)$

  (b) Choose $0 \leq 2\theta - \frac{1}{3}\pi \leq \pi$, $\frac{1}{6}\pi \leq \theta \leq \frac{2}{3}\pi$. Then $\cos^{-1}\frac{1}{5}x = 2\theta - \frac{1}{3}\pi$, $\theta = \frac{1}{2}(\cos^{-1}\frac{1}{5}x + \frac{1}{3}\pi)$.

12. An even function can't have an inverse since it is not 1-1; $f(3) = f(-3)$ for instance. An odd function may have an inverse (sine does not, x^3 does). If an odd function f does have an inverse then f^{-1} is also odd: the graph of an odd function is characterized by symmetry w.r.t. the origin. After reflection in the line $y = x$ the curve still is symmetric w.r.t. the origin.

Section 1.5 (page 28)

1. (a) e^{10} is very large, $-e^{10}$ is negatively large, $e^{-10} = 1/e^{10}$ is near 0. So in order, $-e^{10}$, e^{-10}, e^{10}

  (b) The larger the exponent on e, the larger the result. So in order, e^{-5}, e^{-3}, $e^{-1/2}$, $e^{1/3}$, e^{6}

  (c) $e^7 > e^6$ so $-e^7 < -e^6$

2. (a) 7

  (b) 4

  (c) $e^{\ln 26} = 64$

  (d) $\ln e^{1/2} = \frac{1}{2}$

  (e) $e^{\ln(1/2)^{-1}} = e^{\ln 2} = 2$

  (f) $e^{1}e^{\ln 4} = 4e$

  (g) $e^{\ln x + \ln y} = e^{\ln x}e^{\ln y} = xy$

3. (a) $\ln 2 + \ln 3 = a + b$

  (b) $\ln 2^3 = 3 \ln 2 = 3a$

  (c) $\frac{1}{2}\ln 3 = \frac{1}{2}b$

  (d) $\ln 3^4 = 4 \ln 3 = 4b$

  (e) $-\ln 2 = -a$

  (f) $\ln 3 - \ln 2 = b - a$

  (g) $a + b$

  (h) ab

  (i) a/b

  (j) a^3

  (k) $3 \ln 2 = 3a$

4. (a) $2x + 3 > 0$, $x > -3/2$

  (b) $\sin \pi x > 0$, $-2 < x < -1$, $0 < x < 1$, $2 < x < 3$, etc.

  (c) All x

  (d) $\ln x > 0$, $x > 1$

  (e) $\ln \ln x > 0$, $\ln x > 1$, $x > e$

  (f) $\ln \ln \ln x > 0$, $\ln \ln x > 1$, $\ln x > e$ (now take exp on both sides again), $x > e^e$

5. $-\ln(\sqrt{2} - 1)$

  $= \ln \dfrac{1}{-1 + \sqrt{2}}$

  $= \ln\left(\dfrac{1}{-1 + \sqrt{2}} \cdot \dfrac{-1 - \sqrt{2}}{-1 - \sqrt{2}}\right)$

  (rationalize denominator)

  $= \ln(1 + \sqrt{2})$

6. (a) True, because ln is 1-1.

  (b) True, because exp is 1-1.

  (c) False; $\sin 0 = \sin 2\pi$ but $0 \neq 2\pi$.

7. $(e^{4 - 2\ln 3 - \ln 2})^{1/3} = (e^{4}e^{-2\ln 3}e^{-\ln 2})^{1/3}$
$= (e^{4}e^{\ln 3^{-2}}e^{\ln 1/2})^{1/3} = (e^4 \cdot 3^{-2} \cdot \frac{1}{2})^{1/3} = \sqrt[3]{e^4/18}$
$= e\sqrt[3]{e/18}$

8. $e^{x \ln 2} = e^{\ln 2^x} = 2^x$

9. Car "starts" at the origin and moves right, slowly at first and then faster. (See fig.)

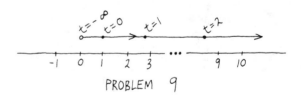

PROBLEM 9

10. (a) $e^{-x} = \frac{3}{2}$, $-x = \ln \frac{3}{2}$, $x = -\ln \frac{3}{2} = \ln \frac{2}{3}$

  (b) $2x + 7 = e^{-1}$, $x = \frac{1}{2}(e^{-1} - 7)$

  (c) No solution. e^x is never neg. Can't say solution is $x = \ln(-5)$ since there is no ln of a neg number.

  (d) $e^{-2} < x < e^{8}$

  (e) $2x + 7 > \ln 5$, $x > \frac{1}{2}(\ln 5 - 7)$

  (f) $\ln x = -4$, $x = e^{-4}$

  (g) $-x = e^4$, $x = -e^4$

  (h) $5x + 3 = 2x$, $x = -1$

  (i) $\ln x = e^{-2}$, $x = e^{e^{-2}}$

  (j) $e^x = \sin \pi/6 = \frac{1}{2}$, $x = \ln \frac{1}{2}$

(k) $\ln x^4 + \ln 2x = 3$, $\ln 2x^5 = 3$, $2x^5 = e^3$,
$x = \sqrt[5]{\tfrac{1}{2}e^3}$

(l) $5x - 3 = 2x$, $x = 1$

(m) $5x + 3 = 2x$, $x = -1$. Impossible since $\ln 2x$ doesn't exist if $x = -1$. No solutions.

(n) $\ln x(x+1) = 2$, $x(x+1) = e^2$, $x^2 + 2x - e^2 = 0$, $x = \tfrac{1}{2}(-2 \pm \sqrt{4 + 4e^2}) = -1 \pm \sqrt{1 + e^2}$. But x and $x + 1$ must be positive. So only solution is $-1 + \sqrt{1 + e^2}$.

(o) $x = -x$, $x = 0$

(p) $x = 0$ (impossible) or $\ln x = 0$, solution is $x = 1$

(q) $e^x(x + 2) = 0$, e^x is never 0, so solution is $x = -2$.

(r) e^x is never 0 so must have $\ln x = 0$, $x = 1$.

(s) $25 = 10 + 5 \ln 3x$, $\ln 3x = 3$, $3x = e^3$, $x = \tfrac{1}{3}e^3$

11. $\ln \tfrac{1}{2}\sqrt{2} = \ln \tfrac{1}{2} + \ln \sqrt{2} = -\ln 2 + \ln 2^{1/2} = -\ln 2 + \tfrac{1}{2}\ln 2 = -\tfrac{1}{2}\ln 2$

12. If $\ln T = -\tfrac{2}{3}\ln V$ then $\ln T = \ln V^{-2/3}$, $T = V^{-2/3}$, $TV^{2/3} = 1$. So $TV^{2/3}$ is constant (namely, always 1).

13. $(\ln x)(4 + 2\ln x) = 0$, $\ln x = 0$ or $4 + 2\ln x = 0$, $x = 1$ or $\ln x = -2$, solutions are 1, e^{-2}.

14. (a) True (b) False (What *is* true is that $e^{a+b} = e^c$.)

15. $\ln \tfrac{1}{2}$ is a neg number since $\tfrac{1}{2} < 1$. Dividing by $\ln \tfrac{1}{2}$ reverses the inequality and produces $2 > 1$.

Section 1.6 (page 31)

1. (a) f is discontinuous at $x = 3$ (where f isn't defined) and f is 0 when $10 - 10x^2 = 0$, $x = +1$. Look at intervals $(-\infty, -1)$, $(-1, 1)$, $(1, 3)$, $(3, \infty)$; $f(-2)$ is neg so f is neg on $(-\infty, -1)$; $f(0)$ is pos so f is pos on $(-1, 1)$; $f(2)$ is neg so f is neg on $(1, 3)$; $f(10)$ is neg so f is neg on $(3, \infty)$.

(b) f is discontinuous if $x = 1$ (where f isn't defined) and f is 0 when $x = -1$. Consider intervals $(-\infty, -1)$, $(-1, 1)$, $(1, \infty)$ where the function is continuous and nonzero; $f(-2)$ is pos so f is positive on $(-\infty, -1)$; $f(0)$ is neg so f is negative on $(-1, 1)$; $f(2)$ is pos so f is positive on $(1, \infty)$.

(c) f is never discontinuous. Equ $x^2 - x + 2 = 0$ has no real solutions, so f is never 0. So on the interval $(-\infty, \infty)$, f has one sign; $f(0)$ is pos so f is pos in $(-\infty, \infty)$.

(d) No fancy theory necessary. e^x is always positive so fraction takes sign of denominator and is pos in $(0, \infty)$, neg in $(-\infty, 0)$.

(e) Never discontinuous, zero if $x = -3, 2$. $f(-100)$ is positive so f is positive in $(-\infty, -3)$. $f(0) = -6$ so f is neg in $(-3, 2)$; $f(100)$ is pos so f is pos in $(2, \infty)$.

2. (a) f is discontinuous at $x = 0$ and is 0 if $16x + 54 = 0$, $x = -\tfrac{27}{8}$. In $(-\infty, -\tfrac{27}{8})$, f is pos (test $f(-10)$ which is $.16 - .054$). In $(-\tfrac{27}{8}, 0)$, f is neg since $f(-1) = 16 - 54$. In $(0, \infty)$, f is positive since $f(1) = 16 + 54$. Solution is $x < -\tfrac{27}{8}$ or $x > 0$.

(b) Problem is $1/2x + 9/(6x + 4) - 3 < 0$. Let $f(x) = 1/2x + 9/(6x + 4) - 3$. First find where f is zero. Solve $1/2x + 9/(6x + 4) = 3$, $6x + 4 + 18x = 6x(6x + 4)$, $36x^2 = 4$, $x = \pm\tfrac{1}{3}$. f is discontinuous at $x = 0$, $-\tfrac{2}{3}$ where f isn't defined. Now look in between. On $(-\infty, -\tfrac{2}{3})$ f is neg (test $x = -100$ for instance). On $(-\tfrac{2}{3}, -\tfrac{1}{3})$ f is pos (test $x = -\tfrac{1}{2}$ for instance). On $(-\tfrac{1}{3}, 0)$ f is neg, on $(0, \tfrac{1}{3})$ f is pos, on $(\tfrac{1}{3}, \infty)$ f is neg. Solution to the given inequality is $x < -\tfrac{2}{3}$ or $-\tfrac{1}{3} < x < 0$ or $x > \tfrac{1}{3}$.

(c) $1/(x^2 - 4)$ is discontinuous when $x = 2, -2$; never is 0. Function is pos in $(-\infty, -2)$, neg in $(-2, 2)$ and pos in $(2, \infty)$. Solution is $x < -2$ or $x > 2$.

Section 1.7 (page 37)

1.

PROBLEM 1(a)

PROBLEM 1(b)

PROBLEM 1(c)

PROBLEM 1(d)

PROBLEM 1(e)

PROBLEM 1(f)

2.

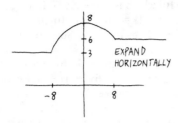

PROBLEM 2 (a)

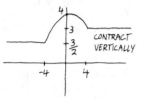

PROBLEM 2 (b)

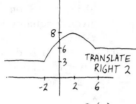

PROBLEM 2 (c)

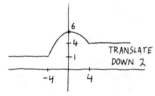

PROBLEM 2 (d)

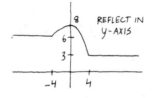

PROBLEM 2 (e)

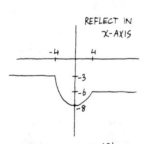

PROBLEM 2 (f)

3. (a) $y = 2(x + 2)^7 + (2[x + 2] + 3)^6$
(b) $y = 2x^7 + (2x + 3)^6 - 5$

4. (See figs.) (a) To make y's pos, keep that part of the sin x graph where $y > 0$ and reflect the other part in the x-axis.

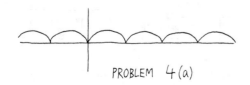

PROBLEM 4 (a)

(b) Reflect that part of ln x which is below the x-axis, and keep the rest.
(c) e^x is always positive so graph is just $y = e^x$.
(d) $e^{|x|}$ is e^x if $x \geq 0$ and is e^{-x} if $x < 0$.
(e) $\ln|x|$ is ln x if $x > 0$ and is $\ln(-x)$ if $x < 0$.

PROBLEM 4 (b)

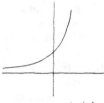

PROBLEM 4 (c)

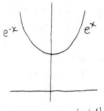

PROBLEM 4 (d)

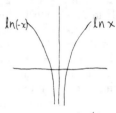

PROBLEM 4 (e)

5.

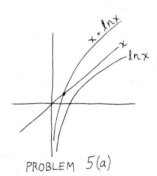

PROBLEM 5 (a)

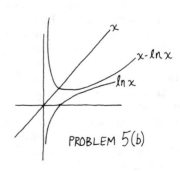

PROBLEM 5 (b)

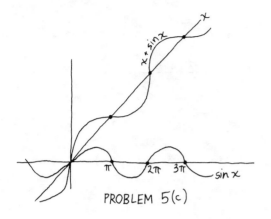

PROBLEM 5(c)

6. The square (similarly cube) of a number between 0 and 1 is smaller than the original number. The cube root of a number between 0 and 1 is larger than the original. (See fig.)

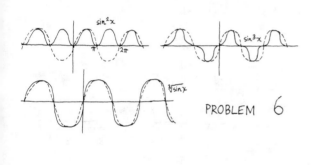

PROBLEM 6

Chapter 1 Review Problems (page 38)

1. (a) $\sqrt{9} = 3$
(b) Defined for $x \leq 5$. Range is $[0, \infty)$.
(c) $f(a^2) = \sqrt{5 - a^2}$, $(f(a))^2 = (\sqrt{5 - a})^2 = 5 - a$
(d) f is half of parabola $x = 5 - y^2$. (See fig.)

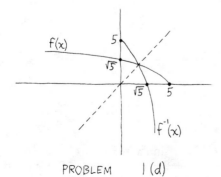

PROBLEM 1(d)

2. (a) Plot points.

x	0	1.3	2.8	3	3.7	5.6	6
y	0	1.3	2.8	0	.7	2.6	0

(See fig.)

(b) $[0, 3)$
(c) Not 1-1, no inverse.
(d) For example, $f(10) = 1$ and $f(f(10)) = f(1) = 1$. In general, $f(f(x)) = f(x)$; i.e., second application of f has no further effect.

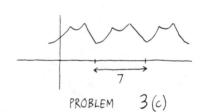

PROBLEM 2(a)

3. (a) $f(x) = x$, graph is line $y = x$.
(b) f is 1-1, graph passes a horizontal line test.
(c) Graph is periodic, repeats every 7 units. (See fig.)

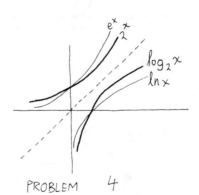

PROBLEM 3(c)

4.

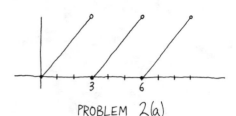

PROBLEM 4

5. $-45°$

6. (a) $e^{y/2} = 3x + 4$, $x = \frac{1}{3}(e^{y/2} - 4)$
(b) $y - 4 = e^{3x}$, $\ln(y - 4) = 3x$, $x = \frac{1}{3}\ln(y - 4)$

7.

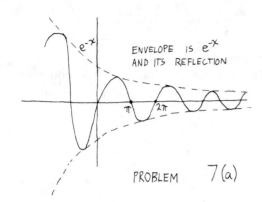

ENVELOPE IS e^{-x}
AND ITS REFLECTION

PROBLEM $7(a)$

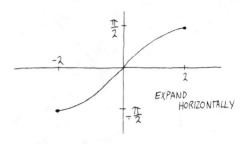

EXPAND
HORIZONTALLY

PROBLEM $7(e)$

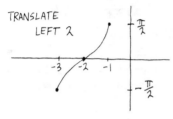

TRANSLATE
LEFT 2

PROBLEM $7(b)$

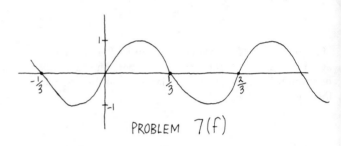

PROBLEM $7(f)$

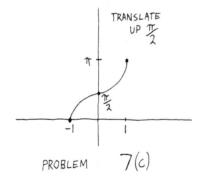

TRANSLATE
UP $\frac{\pi}{2}$

PROBLEM $7(c)$

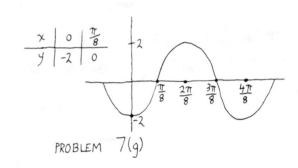

PROBLEM $7(g)$

8. (a)

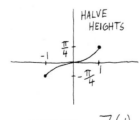

HALVE
HEIGHTS

PROBLEM $7(d)$

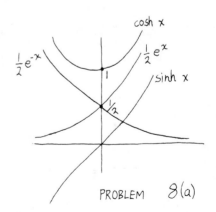

PROBLEM $8(a)$

(b) $\cosh^2 x - \sinh^2 x = \frac{1}{4}(e^x + e^{-x})^2 - \frac{1}{4}(e^x - e^{-x})^2$
$= \frac{1}{4}(e^{2x} + 2 + e^{-2x}) - \frac{1}{4}(e^{2x} - 2 + e^{-2x}) = \frac{1}{2} + \frac{1}{2}$
$= 1$

9. (a) $\ln[x/(2x - 3)] = 4$, $x/(2x - 3) = e^4$, $x = 2e^4x - 3e^4$, $x(2e^4 - 1) = 3e^4$, $x = 3e^4/(2e^4 - 1)$

(b) $x < e^{-8}$

(c) $e^x < -4$. No solutions since e^x is never less than 0. Solution is *not* $x < \ln(-4)$ since there is no ln of -4.

(d) Let $f(x) = 1/(x - 3) - 1/4x$. f is discontinuous

when $x = 0, 3$ and is 0 if $x - 3 = 4x$, $x = -1$; $f(-100)$ is neg so f is neg in $(-\infty, -1)$; $f(-\frac{1}{2})$ is pos so f is pos in $(-1, 0)$; $f(2)$ is neg so f is neg in $(0, 3)$; $f(10)$ is pos so f is pos in $(3, \infty)$. Solution is $-1 < x < 0$ or $x > 3$.

10. $5e^{\ln 3^2} = 5e^{\ln 9} = 5 \times 9 = 45$

11. $\ln(x/5x) = \ln\frac{1}{5} = -\ln 5$

2/LIMITS

Section 2.1 (page 44)

1. (a) 9 (Since x^2 is continuous, just plug in $x = 3$.)

(b) ∞

(c) 1 (Plug in $x = 0$ since $\cos x$ is continuous.)

(d) $-\frac{1}{2}\pi$ (Left half of $\tan^{-1}x$ graph has asymptote $y = -\frac{1}{2}\pi$; see Fig. 7, Section 1.4.)

(e) For example, $(\frac{1}{3})^{10000}$ is very near 0. Limit *is* 0.

(f) See fig. It shows fictitious points $\frac{1}{2}\pi+$ (where tan is $-\infty$) and $\frac{1}{2}\pi-$ (where tan is ∞). Left-hand and right-hand limits do not agree; no overall limit as $x \to \frac{1}{2}\pi$.

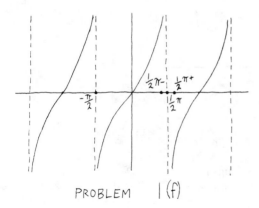

PROBLEM 1 (f)

(g) 9 (Set $x = 2$ since function is continuous.)

2. For example, Int $2.99 = 2$, Int $3.01 = 3$. (a) As $x \to 3-$, Int x remains 2 and limit is 2. (b) As $x \to 3+$, Int x remains 3 and limit is 3.

3. See Problem 1c, Section 1.2. (a) -1 (b) 1

4. (a) From Problem 1(f), $\lim_{x \to (1/2)\pi-} \tan x = \infty$

(b) Left-hand limit (as $x \to (\frac{1}{2}\pi)-$) is ∞, right-hand limit (as $x \to (-\frac{1}{2}\pi)+$) is $-\infty$. No (plain) limit.

5.

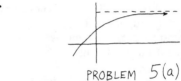

PROBLEM 5(a)

PROBLEM 5(b)

6.

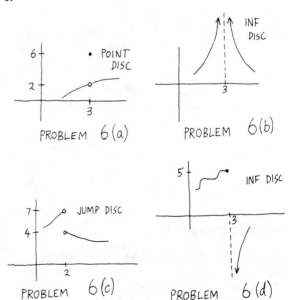

PROBLEM 6(a)

PROBLEM 6(b)

PROBLEM 6(c)

PROBLEM 6(d)

7. Yes *if f* is continuous and we find the limit by setting $a = 0$. But no if f is discontinuous. In the diagram, $f(2) = 5$ but as $a \to 0$, $2 + a \to 2$ and $f(2 + a) \to 4$.

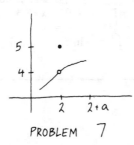

PROBLEM 7

8. $\lim_{x \to 3^-} f(x) = \infty$, $\lim_{x \to -\infty} f(x) = 1$ (or more specifically, $1+$)

9. (a) If x is near 65 then x is not a power of 10 and $f(x) = 1$; so limit is 1.

(b) If x is near, but not at, 100 then x is not a power of 10, $f(x) = 1$ and limit is 1.

(c) No limit. Most f values are 1, but as $x \to \infty$ there are infrequent but persistent 0 values, f never settles down.

10. (See figs.) (a) No limit (violent oscillations).

(b) Limit is 0 (damped oscillations approaching the x-axis asymptotically).

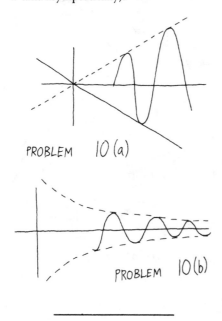

PROBLEM 10 (a)

PROBLEM 10 (b)

___

Section 2.2 (page 47)

1. (a) 0 (b) $-\infty$ (c) ∞ (d) ∞ (e) 0 (f) $e^0 = 1$ (g) $1/\infty = 0$ (h) ∞ (i) 0 (j) $-\infty$

2. (a) $(\ln 0+)^2 = (-\infty)^2 = \infty$

(b) $1/\infty = 0$

(c) $0 - (-\infty) = \infty$

(d) $e^0 = 1$

(e) $\ln 1 = 0$

(f) $1/0$. Examine denominator more carefully. If $x \to (-4)+$, fraction is $1/0+ = \infty$. If $x \to (-4)-$, fraction is $1/0- = -\infty$. No limit as $x \to -4$ since left-hand and right-hand limits disagree.

(g) $2 \times 6 = 12$

(h) $e^{-\infty} = 0$

(i) As $x \to \frac{1}{2}\pi$, $\sin x \to 1-$ and fraction is $3/((1-) - 1) = 3/0- = -\infty$.

(j) $\infty \times \cos 1/\infty = \infty \times \cos 0 = \infty \times 1 = \infty$

(k) $1/-\infty = 0$

3. (See figs.) (a) $1/0+ = \infty$

(b) If $x \to 0+$, limit is $1/0+ = \infty$. If $x \to 0-$, limit is $1/0- = -\infty$.

(c) If $x \to 1+$, x^3 is larger than x, $x - x^3 \to 0-$, limit is $2/0- = -\infty$. If $x \to 1-$ then x^3 is smaller than x (cube of a number between 0 and 1 is less than the number), $x - x^3 \to 0+$, limit is $2/0+ = \infty$.

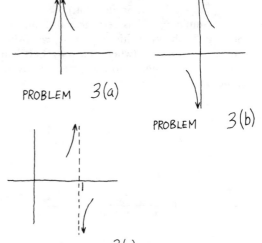

PROBLEM 3(a)

PROBLEM 3(b)

PROBLEM 3(c)

4. (See figs.)

(a) $\lim_{x \to \infty} = e^{-\infty} - 2 = 0 - 2 = -2$, $\lim_{x \to -\infty} = e^{\infty} - 2 = \infty - 2 = \infty$. If $x = 0$, $y = 1 - 2 = -1$.

(b) $\lim_{x \to \infty} = 3 + 2e^{\infty} = 3 + \infty = \infty$, $\lim_{x \to -\infty} = 3 + 0 = 3$. If $x = 0$, $y = 3 + 2 = 5$.

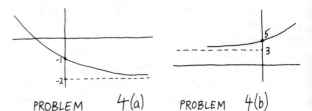

PROBLEM 4(a)   PROBLEM 4(b)

5. $\lim_{x\to 0+} f(x) = e^{\infty} = \infty$, $\lim_{x\to 0-} f(x) = e^{-\infty} = 0$. Infinite discontinuity, not a point disc. Not removable.

6. (a) If $x \to 0+$, then $1/x \to \infty$ and $\sin 1/x$ oscillates between 1 and -1.

(b) $\sin 1/\infty = \sin 0 = 0$

(c) Plot a few points to help. (See fig.)

x	$\frac{1}{x}$	$\sin\frac{1}{x}$
$\frac{1}{\pi}$	π	0
$\frac{1}{\frac{3}{2}\pi}$	$\frac{3\pi}{2}$	-1
$\frac{1}{2\pi}$	2π	0
$\frac{1}{3\pi}$	3π	0

ETC..

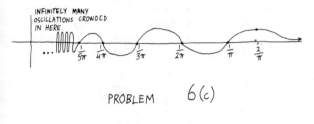

INFINITELY MANY OSCILLATIONS CROWDED IN HERE

PROBLEM $6(c)$

Section 2.3 (page 51)

1. (highest power rule) $\lim_{x\to\infty}(-x^4) = -\infty$

2. (a) (highest power rule)
$\lim_{x\to\infty} 2x^{99}/x^{34} = \lim 2x^{65} = \infty$

(b) (plug in $x = 0$) $-\frac{7}{2}$

(c) (plug in $x = 1$) $-\frac{4}{3}$

3. (a) $\lim_{x\to\infty}(x/-x) = \lim(-1) = -1$

(b) $1/0- = -\infty$

(c) $1/0+ = \infty$

(d) $\lim(x/-x) = \lim(-1) = -1$

4. (a) $\lim(3x^4/x^4) = \lim 3 = 3$

(b) If x is just less than 1 then x^4 is smaller than x, $x^4 - x \to 0-$, lim is $5/0- = -\infty$.

5. $\lim(-x^2/2x^2) = \lim(-\frac{1}{2}) = -\frac{1}{2}$

6. Temporarily get $0/0$, but $(x^2 - 4)/(x - 2) = (x - 2)(x + 2)/(x - 2) = x + 2$. Limit is 4.

7. $\lim(2x/3x^2) = \lim(2/3x) = 0$

Chapter 2 Review Problems (page 51)

1. (a) $0 \cos 0 = 0 \times 1 = 0$

(b) $0 + 1 = 1$

(c) Damped oscillations. Limit is 0. (See fig.)

(d) $\lim(2x/3x) = \lim \frac{2}{3} = \frac{2}{3}$

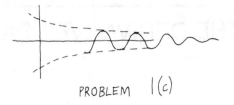

PROBLEM $1(c)$

2. (a) $\lim(2x^4/x^2) = \lim 2x^2 = \infty$

(b) $(32 + 6)/(4 + 5) = 38/9$

3. (a) $\lim(2/x^3) = 2/\infty = 0$

(b) $\frac{2}{0}$. Examine denominator again.
Method 1: If x is pos and near 0 then x^2 is larger than x^3, $x^3 - x^2 \to 0-$ so $\lim_{x\to 0+} = 2/0- = -\infty$. If x is neg and near 0 then x^3 is neg, $x^3 - x^2 \to 0-$ so $\lim_{x\to 0-} = 2/0- = -\infty$. So $\lim_{x\to 0} = -\infty$.
Method 2: $\lim(x^3 - x^2) = \lim x^2(x - 1) = 0 \times -1 = 0-$ so answer is $2/0- = -\infty$.

(c) $\frac{2}{0}$. Look again. If $x \to 1+$ then x^3 is larger than x^2, $x^3 - x^2 \to 0+$, lim $= 2/0+ = \infty$. If $x \to 1-$, x^3 is smaller than x^2, $x^3 - x^2 \to 0-$, lim $= 2/0- = -\infty$. No limit as $x \to 1$.

4. (a) (highest power rule) $\lim(-4x^3) = -\infty$

(b) $4 - 32 = -28$

5. (a) If $x = 0$ or $x \to 0-$ or $x = \pi$ or $x \to \pi+$ then $\sin x$ is 0 or neg, and there is no $\ln \sin x$.

(b) $\ln \sin 0+ = \ln 0+ = -\infty$

(c) $\ln \sin \pi- = \ln 0+ = -\infty$

6. $\lim_{x\to\infty} = 1 - e^{\infty} = 1 - \infty = -\infty$, $\lim_{x\to-\infty} = 1 - e^{-\infty} = 1 - 0 = 1$. If $x = 0$ then $y = 1 - 1 = 0$. (See fig.)

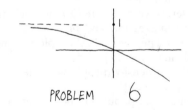

PROBLEM 6

7.

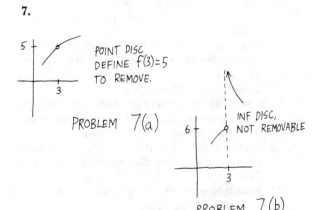

POINT DISC DEFINE $f(3)=5$ TO REMOVE.

PROBLEM $7(a)$

INF DISC, NOT REMOVABLE

PROBLEM $7(b)$

3/THE DERIVATIVE   PART I

Section 3.2 (page 61)

1. Slope at $x = 0$ is 0, so $f'(0) = 0$. At $x = -100$, curve is falling gently so $f'(-100)$ is a small neg number. Similarly $f'(100)$ is small positive number. Slope at $x = 2$ is positive and is the largest slope on the curve so $f'(2)$ is pos and the graph of f' peaks at $x = 2$. (See fig.)

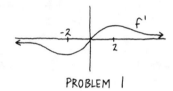

PROBLEM 1

2. If p increases, S decreases. Expect neg dS/dp.

3. Since derivative is negative, as t increases y decreases. Distance to top is decreasing so bucket is moving up, 2 feet per second.

4. If x increases, y increases. So if x decreases, y *decreases.*

5. (a) At age 13.7 you're growing by 2 inches per year, so you'll grow by about 2 inches per year × .3 year = .6 inches by age 14.
  (b) The derivative is 2 *at* age 13.7 but might not stay 2 between ages 13.7 and 14.

6. (a) $[0, x + \Delta x]$ people − $[0, x]$ people = number of people in $[x, x + \Delta x]$.
  (b) Population density (people per mile) in the interval $[x, x + \Delta x]$
  (c) Instantaneous population density at position x
  (d) Population density can't be negative.

7. (a) Brown's salary increases twice as much as Smith's.
  (b) Brown increases only half as much as Smith.
  (c) Brown decreases at the same rate as Smith increases.
  (d) Brown doesn't change.

8. Units on $f'(x)$ are gallons per mile (not miles per gallon), always positive. Van needs more gallons to go a mile so f' is larger for van than for motorcycle.

9. (a) Sentence says that if two curves have same height at $x = 2$ then they have the same slope. False.
  (b) False. Slopes need not increase even if heights do increase. (See fig.)

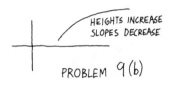

PROBLEM 9(b)

  (c) True. If curve repeats, slopes repeat.
  (d) False. If f is even then its graph is symmetric w.r.t. the y-axis. Slope at $x = -2$ for instance is opposite, not equal to, slope at $x = 2$. In fact f' is odd since $f'(-x) = -f'(x)$. (See fig.)

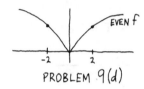

PROBLEM 9(d)

10. Car's speed at time 6 is > the speed limit at position $f(6)$ so $|f'(6)| > L(f(6))$ (*not* $f'(6) > L(6)$).

11. (a) $\lim_{\Delta x \to 0} \dfrac{f(x + \Delta x) - f(x)}{\Delta x} = \lim \dfrac{x + \Delta x - x}{\Delta x}$

  $= \lim 1 = 1$

  (b) The graph of f is the line $y = x$. Slope at any point is 1, so $f'(x) = 1$.
  (c) If a particle on a line has position $f(t)$ at time t then it is moving to the right at 1 mph. So $f'(t) = 1$.

12. From given graph of g' read that $g'(x)$ is near 3 when x is $-10\,000$, decreases to 0 when $x = 0$, then increases again and $\to 3$ as $x \to \infty$. So graph of g has slope near 3 at the left end, slope decreases to 0 and increases toward 3 again. No information about *heights* on the graph of g. (See fig.)

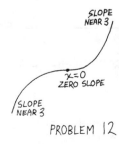

PROBLEM 12

13. (a) Temp is rising by 6° per hour (bad). But the 6° per hour figure is dropping by 4° per hour per hour (good).

(b) Temp is dropping by 6° per hour (good) and the −6° per hour figure is in the process of going down still further, by 4° per hour per hour (better).

(c) Temp isn't changing. It's staying hot. Ugh!

14. (a) Moving left at 4 mph, accelerating by 1 mph per hour.

(b) Moving right at 5 mph, decelerating by 2 mph per hour.

(c) Momentarily halted, but in the process of accelerating by 2 mph per hour.

(d) Moving right at 2 mph, speed not changing at the moment.

15. Graph rises on $[2,8]$, falls on $[8,10]$, concave up on $[2,6]$, concave down on $[6,10]$. (See fig.)

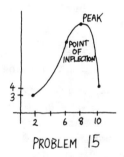

PROBLEM 15

16. The first derivative f' is the "steering ratio," wheel turning per steering wheel turning. It is undesirable for the steering to be more sensitive in one position than another; i.e., want *constant* f', so need $f'' = 0$.

17. f' is positive in $[3,4]$ so f *increases*, more rapidly at first (by 5 f units per x unit) and then more slowly (by only 1 f unit per x unit). Since f' decreases in the interval $[3,4]$, f'' is *negative* in the interval.

18. (a) First 60 barrels cost a total of $400 to produce.

(b) $f'(x)$ is the instantaneous production cost per barrel at the moment that x barrels have just rolled off the assembly line. For the given data, after 60 barrels have been produced, the instantaneous cost of producing more is $21 per barrel, but after 100 barrels have been made the refinery has become more efficient and the instantaneous cost for new barrels is only $10 per barrel. (The 101st barrel is cheaper to produce than the 61st.)

(c) Total cost of the first 10 barrels is $200, an average of $20 per barrel. Instantaneous cost of the barrels after the 10th is only $3 per barrel, much less than the overall average so far. Refinery is more efficient if it produces more than 10 barrels.

Section 3.3 (page 69)

1. (a) $6x^5$

(b) $D_x x^{-6} = -6x^{-7} = -6/x^7$

(c) $\frac{8}{7}x^{1/7}$

(d) $D_u u^{1/3} = \frac{1}{3}u^{-2/3}$

(e) $d(x^{-1/2})/dx = -\frac{1}{2}x^{-3/2}$

(f) $\frac{2}{3}x^{-1/3}$

(g) 0

(h) e^t

(i) 0

2. $1/z$

3. $y' = 1$

4. $f'(x) = 0$

5. $\sec^2 t$

6. (a) $y' = 1/x, y'' = D_x x^{-1} = -x^{-2} = -1/x^2$

(b) $y' = \cos x, y'' = -\sin x$

(c) $y' = e^x, y'' = e^x$

7. $f'(x) = D_x x^{-1/2} = -\frac{1}{2}x^{-3/2} = -1/2x^{3/2}$, $f'(17) = -1/2\sqrt{17^3} = -1/(2 \cdot 17\sqrt{17}) = -1/(34\sqrt{17})$

8. $f(\pi) = \sin \pi = 0, f'(x) = \cos x, f'(\pi) = \cos \pi = -1$

9. (a) $-3x^{-4}$

(b) $14x^{13}$

(c) $D_x x^{1/5} = \frac{1}{5}x^{-4/5}$

(d) $D_x x^{-5} = -5x^{-6} = -5/x^6$

(e) 1

(f) $1/x$

(g) $-\frac{1}{3}x^{-4/3}$

(h) $4x^3$

(i) $D_x x^{-4} = -4x^{-5} = -4/x^5$

(j) $D_x x^{-1} = -x^{-2} = -1/x^2$

(k) $D_x x^{-2} = -2x^{-3} = -2/x^3$

10. Curve rises slowly at B so the slope is a small positive number; the value of $1/x$ at B is also small positive since x at B is large. Curve rises steeply at A the slope is a large positive number; value of $1/x$ at A is also large positive since x at A is slightly larger than 0. (See fig.)

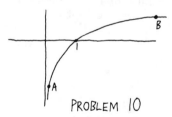

PROBLEM 10

11. Let $y = \tan^{-1}x$. Then $x = \tan y$ where $-\frac{1}{2}\pi < y < \frac{1}{2}\pi$, and $dy/dx = 1/(dx/dy) = 1/\sec^2 y = 1/(1 + \tan^2 y)$ (trig identity) $= 1/(1 + x^2)$.

12. (a) Reflect the piece of $\sin x$ between $\frac{1}{2}\pi$ and $3\pi/2$. (See fig.)

(b) $D_x \text{II}\sin^{-1}x$ can't be $1/\sqrt{1 - x^2}$ since all the slopes on $\text{II}\sin^{-1}x$ are negative while $1/\sqrt{1 - x^2}$ is positive.

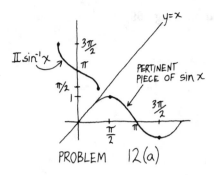

PROBLEM 12(a)

One way to get the correct derivative is to go back to (11). But now y is an angle between $\frac{1}{2}\pi$ and $3\pi/2$, so its cos is *negative*. Therefore $\cos y = -\sqrt{1-x^2}$ and $D_x \text{II}\sin^{-1}x = -1/\sqrt{1-x^2}$.

13. $da/db = -4b^{-5}$. Since $b = a^{-1/4}$, $db/da = -\frac{1}{4}a^{-5/4}$. Then $1/(db/da) = -4a^{5/4} = -4(b^{-4})^{5/4}$ (since $a = b^{-4}$) $= -4b^{-5}$, which does equal da/db. So the inverse derivatives are reciprocals of one another.

14. (a) Diff to get velocity $\cos t$, which is $-\frac{1}{2}$ at $t = 2\pi/3$. Speed is $\frac{1}{2}$ m/sec.

(b) Diff again to get acceleration $-\sin t$, which is $-\frac{1}{2}\sqrt{3}$ if $t = 2\pi/3$. Velocity and acceleration have same sign so particle is speeding up, namely by $\frac{1}{2}\sqrt{3}$ m/sec per sec.

(c) Speed is $|\cos t|$. Max when $t = 0, \pi, 2\pi$, etc. (max value is 1). Min when $t = -\pi/2, \pi/2, 3\pi/2$ etc. (min value is 0).

15. $y' = 4x^3$, slope at $(-2, 16)$ is $4(-2)^3 = -32$. Tan line is $y - 16 = -32(x + 2)$. Perpendicular line has slope $\frac{1}{32}$, equation is $y - 16 = \frac{1}{32}(x + 2)$.

Section 3.5 (page 77)

1. (a) $18x^5 - \sin x$ (b) $10x^4 - 18x^2 - 4$

2. (a) $y = x^{-1}$, $y' = -x^{-2}$, $y'' = 2x^{-3}$, $y''' = -6x^{-4}$, $y^{(4)} = 24x^{-5} = 24/x^5$

(b) $y' = \cos x$, $y'' = -\sin x$, $y''' = -\cos x$, $y^{(4)} = \sin x$

(c) $y' = 1$, $y'' = 0$, $y''' = 0$, $y^{(4)} = 0$

3. (a) $\frac{3}{2}x^2$

(b) $6x^2$

(c) $D2x^{-3} = -6x^{-4} = -6/x^4$

(d) $D\frac{1}{2}x^{-3} = -\frac{3}{2}x^{-4} = -3/2x^4$

(e) $\frac{1}{3}(3x^2 + 2)$

(f) (product rule) $-2x^3 \sin x + 6x^2 \cos x$

(g) $\sqrt{x} \cdot \frac{1}{x} + \frac{1}{2}x^{-1/2} \ln x = \frac{1}{\sqrt{x}} + \frac{\ln x}{2\sqrt{x}}$

(h) $\sec x \cdot \sec^2 x + \tan x \cdot \sec x \tan x$
$= \sec^3 x + \tan^2 x \sec x$

(i) $2e^x(1/x) + 2e^x \ln x + 10x$

(j) $2e^x + 1/x$

(k) $4x^2/(1 + x^2) + 8x \tan^{-1}x$

(l) $x^3 \sin x \sec^2 x + x^3 \cos x \tan x + 3x^2 \sin x \tan x$

(m) $-3x^{-2} = -3/x^2$

(n) $D\frac{1}{3}x^{-1} = -\frac{1}{3}x^{-2} = -1/3x^2$

4. (a) $f'(r) = 5r^4$, $f''(r) = 5 \cdot 4r^3, \cdots$, $f^{(5)}(r) = 5 \cdot 4 \cdot 3 \cdot 2 \cdot 1 = 5!$

(b) $f^{(4)}(r) = 4 \cdot 3 \cdot 2 \cdot 1$, $f^{(5)}(r) = 0$

(c) $f'(r) = r^4 \cdot \frac{1}{r} + 4r^3 \ln r = r^3 + 4r^3 \ln r$,

$f''(r) = 3r^2 + 4r^3 \cdot \frac{1}{r} + 12r^2 \ln r$
$= 7r^2 + 12r^2 \ln r$,

$f'''(r) = 14r + 12r^2 \cdot \frac{1}{r} + 24r \ln r$
$= 26r + 24r \ln r$,

$f^{(4)}(r) = 26 + 24r\left(\frac{1}{r}\right) + 24 \ln r$
$= 50 + 24 \ln r$,

$f^{(5)}(r) = \frac{24}{r}$

5. $f(-2) = 48 + 4 = 52$, $f'(x) = 12x^3 - 2$, $f''(x) = 36x^2$, $f'(-2) = -96 - 2 = -98$, $f''(-2) = 144$

6. (a) (product rule) $xe^x + e^x$

(b) $xe^x + e^x + e^x = xe^x + 2e^x$

(c) $xe^x + e^x + 2e^x = xe^x + 3e^x$

(d) Guess $xe^x + ne^x$.

7. (a) $\dfrac{(6x + x^2)3 - (1 + 3x)(6 + 2x)}{(6x + x^2)^2}$ (quot rule)

$= -\dfrac{3x^2 + 2x + 6}{(6x + x^2)^2}$

(b) $\dfrac{x \cos x - \sin x}{x^2}$

(c) $\dfrac{(1 + 3e^x)(xe^x + e^x) - xe^x \cdot 3e^x}{(1 + 3e^x)^2}$

$= \dfrac{xe^x + e^x + 3e^{2x}}{(1 + 3e^x)^2}$

8. $D \sec x = D\dfrac{1}{\cos x} = \dfrac{\cos x \cdot 0 - (-\sin x)}{\cos^2 x} = \dfrac{\sin x}{\cos^2 x}$

$= \dfrac{1}{\cos x}\dfrac{\sin x}{\cos x} = \sec x \tan x$

9. (a) $f(x) = \begin{cases} \sin x & \text{if } 0 \leq x \leq \pi \\ -\sin x & \text{if } \pi \leq x \leq 2\pi \\ \sin x & \text{if } 2\pi \leq x \leq 3\pi, \text{ etc.} \end{cases}$

so $f'(x) = \begin{cases} \cos x & \text{if } 0 < x < \pi \\ -\cos x & \text{if } \pi < x < 2\pi \\ \cos x & \text{if } 2\pi < x < 3\pi, \text{ etc.} \end{cases}$

(b) $f'(x)$ is $6x^2$ if $x \leq 2$, and is 0 if $x > 2$

10. $y' = 6x^2 + 6$, slope at $(1, 8)$ is $y'|_{x=1} = 12$. Tangent line is $y - 8 = 12(x - 1)$, $y = 12x - 4$.

11. $y' = -4x^3$, $y'|_{x=2} = -32$. Slope of perpendicular line is $\frac{1}{32}$. Perpendicular line is $y + 11 = \frac{1}{32}(x - 2)$.

12. (a) $y' = \cos x$, $y'' = -\sin x$, negative if $0 < x < \pi$, positive if $\pi < x < 2\pi$, etc. So sine curve is concave down on $[0, \pi]$, concave up on $[\pi, 2\pi]$ as in Fig. 8, Sect. 1.3.

(b) $y' = 3x^2$, $y'' = 6x$, pos if $x > 0$, neg if $x < 0$. So x^3 is concave down for $x < 0$, concave up for $x > 0$ as in Fig. 12, Section 1.2.

13. vel $= 2t - 9t^2$, acc $= 2 - 18t$. At time $t = 2$, vel $= -32$, acc $= -34$. Speed is 32 (mph). Car is speeding up by 34 (mph per hour).

14. $f'(x) = 2x + a$. Tangent line at $(3,4)$ has slope 2 so $f'(3) = 2$, $2 \cdot 3 + a = 2$, $a = -4$. Graph of f passes through $(3,4)$ so $f(3) = 4$, $4 = 9 - 12 + b$, $b = 7$.

15. $y' = -6x - 4$, slope is 0 when $x = -\frac{2}{3}$. If $x = -\frac{2}{3}$ then $y = \frac{10}{3}$. (See fig.)

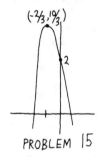

PROBLEM 15

16. The graph of f starts out as a parabola and then switches to a line at $x = 4$. Need the heights x^2 and $ax + b$ to agree at $x = 4$ for continuity. Otherwise the graph of f will jump. (See left-hand fig.) So need $16 = 4a + b$. Also $f'(x)$ is $2x$ if $x \le 4$ and is a if $x > 4$. Need slopes $2x$ and a to agree at $x = 4$ for smoothness. Otherwise f will have a cusp. (See center fig.) So need $8 = a$; then $b = 16 - 32 = -16$. Then graph of f is continuous and smooth. (See right-hand fig.)

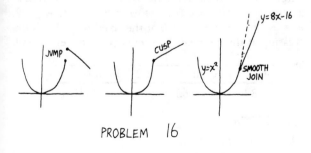

PROBLEM 16

17. $y' = x \cos x + \sin x$,
$y'' = x \cdot -\sin x + \cos x + \cos x = -x \sin x + 2 \cos x$.
Then $y'' + y = -x \sin x + 2 \cos x + x \sin x = 2 \cos x$
18. $T' = 3t^2 - 15$, $T'' = 6t$. If $t = 3$ then $T = -18$, $T' = 12$, $T'' = 18$, so temp is $-18°$ (cold), temp is rising

by 12° per hour (hopeful) and the 12° per hour rate is itself increasing by 18°/hour per hour (still better).

19. Graph of $y = -x^2 + 8x$ is a parabola, $y' = -2x + 8$, slope is 0 if $x = 4$. If $x = 4$ then $y = 16$ so turning point is $(4, 16)$. Graph of $y = 16$ is a horizontal line. The two pieces join continuously since each has height 16 at $x = 4$. They also join smoothly (no cusp) since each piece has slope 0 at $x = 4$. Graph of $y = x^2 - 20x + 100$ is a parabola, $y' = 2x - 20$, slope is 0 if $x = 10$. If $x = 10$ then $y = 0$ so turning point is $(10, 0)$. The line and the second parabola join without a jump since each has height 16 at $x = 6$. They do not join smoothly since parabola slope is -8 and line slope is 0 at $x = 6$. (See fig.)

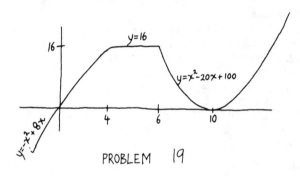

PROBLEM 19

20. If $f(t) = 12t - t^3$ then $f'(t) = 12 - 3t^2$, $f''(t) = -6t$. $f'(t) = 0$ if $t = \pm 2$, $f'(t)$ is neg in $(-\infty, -2)$, pos in $(-2, 2)$, neg in $(2, \infty)$. $f''(t)$ is pos if $t < 0$, neg if $t > 0$.

time interval	sign of f'	sign of f''	particle
$(-\infty, -2)$	neg	pos	moves left, decelerates
$(-2, 0)$	pos	pos	moves right, accelerates
$(0, 2)$	pos	neg	moves right, decelerates
$(0, \infty)$	neg	neg	moves left, accelerates

Key values are $f(-\infty) = \infty$, $f(-2) = -16$, $f(0) = 0$, $f(2) = 16$, $f(\infty) = -\infty$. (See fig.)

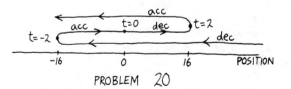

PROBLEM 20

Section 3.6 (page 80)

1. $6e^{6x}$
2. $2 \cos 2x$
3. $-e^{-x}$
4. $-e^x$
5. $\dfrac{1}{\sqrt{1 - (3 - x)^2}} \cdot -1$
6. $-10 \sin 5x$
7. (product rule and chain rule)
$x^2 \cdot \cos 5x \cdot 5 + 2x \sin 5x = 5x^2 \cos 5x + 2x \sin 5x$
8. $5x \cdot e^{2x} \cdot 2 + e^{2x} \cdot 5 = 10xe^{2x} + 5e^{2x}$
9. $D(2 + \sin x)^{-1} = -(2 + \sin x)^{-2} \cdot \cos x$
$= -\cos x/(2 + \sin x)^2$
10. $e^x \cos e^x$
11. $e^{-x} \cdot -\sin 4x \cdot 4 + \cos 4x \cdot e^{-x} \cdot -1$
$= -4e^{-x} \sin 4x - e^{-x} \cos 4x$
12. $x^3 \cdot 6(2x + 5)^5 \cdot 2 + 3x^2(2x + 5)^6$
$= 12x^3(2x + 5)^5 + 3x^2(2x + 5)^6$
13. $-2 \sin 5x \cdot 5 = -10 \sin 5x$
14. $\dfrac{1}{5 - x} \cdot -1 = -\dfrac{1}{5 - x}$
15. $\dfrac{1}{\cos x} \cdot -\sin x = -\tan x$
16. $2e^{5 + 2x}$
17. $\frac{1}{2}(3 + x^2)^{-1/2} \cdot 2x = x/\sqrt{3 + x^2}$
18. $\dfrac{1}{1 + (\frac{1}{2}x)^2} \cdot \frac{1}{2} = \dfrac{1}{2 + \frac{1}{2}x^2}$
19. $D4 \sec 5x = 4 \sec 5x \tan 5x \cdot 5 = 20 \sec 5x \tan 5x$
20. $\pi \cos \pi x$
21. $D(\cos x)^3 = 3 \cos^2 x \cdot D \cos x = -3 \cos^2 x \sin x$
22. $\cos \dfrac{1}{x} \cdot D_x \dfrac{1}{x} = -\dfrac{1}{x^2} \cos \dfrac{1}{x}$
23. $e^{\sqrt{x}} D_x \sqrt{x} = e^{\sqrt{x}}/2\sqrt{x}$
24. $e^{1/x} D \dfrac{1}{x} = -e^{1/x}/x^2$
25. $3(\tan^{-1}x)^2/(1 + x^2)$
26. $3(x^2 + 4)^2 \cdot 2x = 6x(x^2 + 4)^2$
27. $\cos x^4 \cdot 4x^3 = 4x^3 \cos x^4$
28. $D_x(\cos x)^4 = 4(\cos x)^3 \cdot D \cos x = -4 \sin x \cos^3 x$
29. $D(x^2 + 4x)^{-1/2} = -\frac{1}{2}(x^2 + 4x)^{-3/2} \cdot (2x + 4)$
30. $\ln x^3 = 3 \ln x$ so derivative is $3/x$ (alternatively, derivative is $(1/x^3) \cdot 3x^2 = 3/x$).
31. $3(\ln x)^2 \cdot D \ln x = 3(\ln x)^2/x$
32. $D(\ln x)^{-1} = -(\ln x)^{-2} D \ln x = -1/x(\ln x)^2$
33. $D(\sin x)^2 = 2 \sin x \cdot D \sin x = 2 \sin x \cos x$
34. $x \cdot -\sin 2x \cdot 2 + \cos 2x = -2x \sin 2x + \cos 2x$
35. $-\sin(3 - x) \cdot -1 = \sin(3 - x)$
36. $-e^x \csc^2 e^x$
37. (product and chain)
$4x^3e^{8x} \cos 4x + 8x^3e^{8x} \sin 4x + 3x^2e^{8x} \sin 4x$
38. $x \cdot \dfrac{1}{2x + 1} \cdot 2 + \ln(2x + 1) = \dfrac{2x}{2x + 1} + \ln(2x + 1)$
39. $6(3x + 4)^5 \cdot 3 = 18(3x + 4)^5$

40. $D(\sec 3x^4)^3 = 3(\sec 3x^4)^2 D \sec 3x^4$
$= 3 \sec^2 3x^4 \sec 3x^4 \tan 3x^4 \cdot 12x^3$
41. $-6(4 - x)^5$
42. $\frac{7}{2}$
43. $(3/\sqrt{1 - (\frac{1}{2}x)^2}) \cdot \frac{1}{2} = 3/2\sqrt{1 - \frac{1}{4}x^2}$
44. $(1/\sin e^x) \cdot D \sin e^x = (e^x \cos e^x)/(\sin e^x) = e^x \cot e^x$
45. $D(\cos 4x)^3 = 3(\cos 4x)^2 \cdot D \cos 4x$
$= -12 \cos^2 4x \sin 4x$
46. $e^x/x + e^x \ln x$
47. $D(e^x + 1)^{-1} = -(e^x + 1)^{-2}e^x = -e^x/(e^x + 1)^2$
48. $-4 \csc 4x \cot 4x$
49. $(4/\ln x) \cdot D \ln x = 4/(x \ln x)$
50. $D(\ln x)^{1/2} = \frac{1}{2}(\ln x)^{-1/2} \cdot (1/x) = 1/(2x\sqrt{\ln x})$
51. $\ln\sqrt{x} = \frac{1}{2} \ln x$ so derivative is $1/2x$. Alternatively, derivative is $(1/\sqrt{x}) \cdot \frac{1}{2}x^{-1/2} = 1/2x$.
52. $\ln 3$ is a *constant* so the problem is of the form cx^2 where c is the *number* $\ln 3$. Derivative is $2x \ln 3$.
53. $|x| = -x$ if $x < 0$ and $|x| = x$ if $x > 0$ so
$\ln|x| = \begin{cases} \ln(-x) & \text{if } x < 0 \\ \ln x & \text{if } x > 0 \end{cases}$ and
$D \ln|x| = \begin{cases} (1/-x) \cdot -1 & \text{if } x < 0 \\ 1/x & \text{if } x > 0 \end{cases} = \dfrac{1}{x}$ in both cases.
54. (quotient and chain)
$\dfrac{\sqrt{2x + 3} \cdot 4 - 4x \cdot \frac{1}{2}(2x + 3)^{-1/2} \cdot 2}{2x + 3} = \dfrac{4x + 12}{(2x + 3)^{3/2}}$
55. $\cos \dfrac{x^2 + 2}{x + 1} \cdot D \dfrac{x^2 + 2}{x + 1}$
$= \cos \dfrac{x^2 + 2}{x + 1} \cdot \dfrac{(x + 1)2x - (x^2 + 2)}{(x + 1)^2}$
$= \dfrac{x^2 + 2x - 2}{(x + 1)^2} \cos \dfrac{x^2 + 2}{x + 1}$
56. $\dfrac{1}{2}\left(\dfrac{2 - x}{3x + 4}\right)^{-1/2} \cdot \dfrac{(3x + 4) \cdot -1 - (2 - x) \cdot 3}{(3x + 4)^2}$
$= \dfrac{-5}{(3x + 4)^2} \sqrt{\dfrac{3x + 4}{2 - x}}$
57. Remember that by the chain rule, $D_t v^2(t) = 2v(t)v'(t)$. Then by the product rule $D_t \frac{1}{2} m(t)v^2(t) = \frac{1}{2} m(t) \cdot 2v(t)v'(t) + v^2(t) \cdot \frac{1}{2}m'(t)$. Set $m = 5$, $v = 3$, $m' = 2$, $v' = -1$ to get $(KE)' = -6$ at the fixed time. KE is decreasing at this moment by 6 energy units per second.
58. $\underbrace{\dfrac{1}{\ln \ln \cdots \ln 2x}}_{638 \text{ logs}} \cdot \underbrace{\dfrac{1}{\ln \ln \cdots \ln 2x}}_{637 \text{ logs}} \cdot \underbrace{\dfrac{1}{\ln \ln \cdots \ln 2x}}_{636 \text{ logs}} \cdots$
$\dfrac{1}{\ln \ln 2x} \cdot \dfrac{1}{\ln 2x} \cdot \dfrac{1}{2x} \cdot 2$
59. (a) $-\csc^2 f(x) \cdot f'(x)$
(b) $xf'(x) + f(x)$
(c) $3(f(x))^2 f'(x)$
(d) $f'(x)/f(x)$
(e) $e^{f(x)} f'(x)$
60. $D_x \text{ star } 3x = e^{3x}([3x]^3 + 3) \cdot 3$
61. $w' = 3e^{\sec 2\theta} \cdot \sec 2\theta \tan 2\theta \cdot 2$
$= 6e^{\sec 2\theta} \sec 2\theta \tan 2\theta,$

$w'' = 6e^{\sec 2\theta} \sec 2\theta \sec^2 2\theta \cdot 2$
$\quad + 6e^{\sec 2\theta} \sec 2\theta \tan 2\theta \cdot 2 \tan 2\theta$
$\quad + 6e^{\sec 2\theta} \cdot \sec 2\theta \tan 2\theta \cdot 2 \cdot \sec 2\theta \tan 2\theta$
$\quad = 12 \sec 2\theta e^{\sec 2\theta}(\sec^2 2\theta + \tan^2 2\theta + \sec 2\theta \tan^2 2\theta)$

62. $y' = 4(2 - x)^3 \cdot -1$. If $x = 3$ then $y' = 4$. Tangent line has slope 4 and equation $y - 1 = 4(x - 3)$. Perpendicular line has slope $-\frac{1}{4}$, equation $y - 1 = -\frac{1}{4}(x - 3)$.

63. First derivative $= -(2 + 3x)^{-2} \cdot 3$, second derivative $= 2(2 + 3x)^{-3} \cdot 3 \cdot 3$, third derivative $=$
$-3 \cdot 2(2 + 3x)^{-4} \cdot 3^3, \cdots$, 99th deriv $= \dfrac{-99! \times 3^{99}}{(2 + 3x)^{100}}$,

100th derivative $= \dfrac{100! \times 3^{100}}{(2 + 3x)^{101}}$

64. (See fig.) $y = \sqrt{100 - x^2}$, $y' = \frac{1}{2}(100 - x^2)^{-1/2} \cdot -2x$ $= -x/\sqrt{100 - x^2}$. If $x = 1$ then $y' = -1/\sqrt{99}$; if $x = 9$ then $y' = -9/\sqrt{19}$. As the ladder begins to slide, x is small and y is decreasing slowly, by $1/\sqrt{99}$ ft/sec. When the ladder is nearly horizontal and $x = 9$, y is decreasing more rapidly, by $9/\sqrt{19}$ ft/sec.

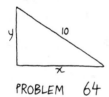

PROBLEM 64

Section 3.7 (page 84)

1. (a) $y' = x \cos y \cdot y' + \sin y$,
$\quad y' = \sin y/(1 - x \cos y)$
(b) $1 + y'$
$\quad = y \sec^2 y \cdot y' + \tan y \cdot y' + x \sec^2 x + \tan x$,
$\quad y'(1 - y \sec^2 y - \tan y) = x \sec^2 x + \tan x - 1$,
$\quad y' = \dfrac{x \sec^2 x + \tan x - 1}{1 - y \sec^2 y - \tan y}$

2. $y' = -\sin(x^2 + y^2) \cdot (2x + 2yy')$,
$y'(1 + 2y \sin(x^2 + y^2)) = -2x \sin(x^2 + y^2)$,
$\dfrac{dy}{dx} = \dfrac{-2x \sin(x^2 + y^2)}{1 + 2y \sin(x^2 + y^2)}$,

$\dfrac{dx}{dy} = \dfrac{1}{dy/dx} = \dfrac{1 + 2y \sin(x^2 + y^2)}{-2x \sin(x^2 + y^2)}$

3. (a) *Method 1:* $y = -\sqrt{1 - x^2}$ (choose the negative square root because point $(\frac{1}{2}, -\frac{1}{2}\sqrt{3})$ is on the lower half of the circle, where y values are neg). Then $y' = -\frac{1}{2}(1 - x^2)^{-1/2} \cdot -2x = x/\sqrt{1 - x^2}$. If $x = \frac{1}{2}$ then $y' = 1/\sqrt{3}$. Tangent line is $y + \frac{1}{2}\sqrt{3} = (1/\sqrt{3})(x - \frac{1}{2})$, $y\sqrt{3} - x = -2$.
Method 2: $2x + 2yy' = 0$, $y' = -x/y$. If $x = \frac{1}{2}$, $y = -\frac{1}{2}\sqrt{3}$ then $y' = 1/\sqrt{3}$ as above, etc.

(b) *Method 1:* $\sqrt{y} = 3 - \sqrt{x}$, $y = (3 - \sqrt{x})^2$, $y' = 2(3 - \sqrt{x}) \cdot -1/2\sqrt{x}$. If $x = 1$ then $y' = -2$. Tangent line is $y - 4 = -2(x - 1)$, $2x + y = 6$. *Method 2:* $1/2\sqrt{x} + y'/2\sqrt{y} = 0$, $y' = -\sqrt{y}/\sqrt{x}$. If $x = 1$, $y = 4$ then $y' = -2$ as above, etc.

4. $\dfrac{1}{y} y' = -(xy' + y)$, $y' = \dfrac{-y^2}{1 + xy}$. Want y' when $x = 0$ but first need y. If $x = 0$ then $\ln y = 1$, $y = e$. Then $y' = f'(0) = -e^2$.

5. For ellipse, $8x + 18yy' = 0$, $y' = -4x/9y$. For hyp, $2x - 2yy' = 0$, $y' = x/y$. Product of slopes is $-4x^2/9y^2$. Solve system $4x^2 + 9y^2 = 72$, $x^2 - y^2 = 5$ to get points of intersection $(-3, 2)$, $(-3, -2)$, $(3, 2)$, $(3, -2)$. For these values of x and y the product of slopes is -1, so curves are perpendicular.

6. $e^{xy} \cdot (xy' + y) = y'$, $y' = \dfrac{ye^{xy}}{1 - xe^{xy}} = \dfrac{y^2}{1 - xy}$

since $e^{xy} = y$. So $(1 - xy)y' = (1 - xy)\dfrac{y^2}{1 - xy} = y^2$,

as desired.

7. (a) $y = x^3 \sin x \cdot (x^2 + 4)^{-1}$,
$\quad y' = x^3 \sin x \cdot -(x^2 + 4)^{-2} \cdot 2x$
$\quad\quad + x^3 \cos x \cdot (x^2 + 4)^{-1}$
$\quad\quad + 3x^2 \sin x \cdot (x^2 + 4)^{-1}$
$\quad = \dfrac{-2x^4 \sin x}{(x^2 + 4)^2} + \dfrac{x^3 \cos x}{x^2 + 4} + \dfrac{3x^2 \sin x}{x^2 + 4}$

(b) $\ln y = 3 \ln x + \ln \sin x - \ln(x^2 + 4)$,
$\dfrac{1}{y}y' = \dfrac{3}{x} + \dfrac{1}{\sin x} \cdot \cos x - \dfrac{2x}{x^2 + 4}$,

$y' = y\left(\dfrac{3}{x} + \cot x - \dfrac{2x}{x^2 + 4}\right)$. Replace y by

$x^3 \sin(x^2 + 4)^{-1}$ to get same answer as (2).

8. (a) If $y = 2^x$ then $\ln y = x \ln 2$, $(1/y)y' = \ln 2$, $y' = y \ln 2 = 2^x \ln 2$.

(b) $y = x^x$, $\ln y = x \ln x$, $y'/y = x(1/x) + \ln x$, $y' = y + y \ln x = x^x + x^x \ln x$

(c) $y = x^{\sin x}$, $\ln y = \sin x \ln x$,
$(1/y)y' = (\sin x)(1/x) + \cos x \ln x$,

$y' = y\left(\dfrac{\sin x}{x} + \cos x \ln x\right)$

$\quad = x^{\sin x}\left(\dfrac{\sin x}{x} + \cos x \ln x\right)$

(d) $3x^2$
(e) $4(2x + 3)^3 \cdot 2 = 8(2x + 3)^3$
(f) $y = 4^{2x+3}$, $\ln y = (2x + 3) \ln 4$, $(1/y)y' = 2 \ln 4$, $y' = 2y \ln 4 = 2 \cdot 4^{2x+3} \ln 4$
(g) e^x
(h) $y = (2x + 3)^{4x}$, $\ln y = 4x \ln(2x + 3)$,
$\dfrac{1}{y}y' = 4x \cdot \dfrac{1}{2x + 3} \cdot 2 + 4 \ln(2x + 3)$,

$y' = (2x + 3)^{4x}\left(\dfrac{8x}{2x + 3} + 4 \ln(2x + 3)\right)$

Section 3.8 (page 90)

1. (a) $-3 \cos x + C$
   (b) $-\frac{1}{3} \cos 3x + C$
   (c) $u^5/5 + C$
   (d) $\pi \sec(x/\pi) + C$
   (e) $\int t^{-3} dt = t^{-2}/-2 + C = -1/2t^2 + C$
   (f) $\ln|x| + C$
   (g) $x^{-4}/-4 + C = -1/4x^4 + C$
   (h) $\frac{2}{3}x^{3/2} + C$
   (i) $\int x^{-1/2} dx = x^{1/2}/\frac{1}{2} + C = 2\sqrt{x} + C$
   (j) $x^9/9 + C$
   (k) $\frac{1}{2}\int x^{-2} dx = \frac{1}{2}x^{-1}/-1 + C = -1/2x + C$
   (l) $\int 4x^{-2} dx = 4x^{-1}/-1 + C = -4/x + C$

2. $f(x) = -\cos x + \frac{1}{3}x^3 + C$. Set $x = 0, f(x) = 10$ to get $10 = -1 + C, C = 11, f(x) = -\cos x + \frac{1}{3}x^3 + 11$

3. $f''(x) = 5x + A, f'(x) = \dfrac{5x^2}{2} + Ax + B$,

   $f(x) = \frac{5}{6}x^3 + \frac{1}{2}Ax^2 + Bx + C$

4. If s is position at time t then $s'(t) = 7 - t^2$, $s = 7t - \frac{1}{3}t^3 + C$. Set $t = 3, s = 4$ to get $4 = 21 - 9 + C, C = -8, s = 7t - \frac{1}{3}t^3 - 8$. If $t = 6$ then $s = -38$.

5. $y = x^2 + 3x + C, -2 = 1 + 3 + C, C = -6, y = x^2 + 3x - 6$

6. No because $D \ln \sin x = (1/\sin x)$ TIMES $D \sin x = (1/\sin x) \cos x$.

7. (a) No, since $D \sin^2 x = 2 \sin x \cos x$.
   (b) No, since $D \sin 2x = \cos 2x$ TIMES 2. (Correct answer is $\frac{1}{2} \sin 2x + C$.)
   (c) Yes
   (d) No, since $D \sin x^2 = \cos x^2$ TIMES $2x$.

8. Let y be height at time t. Then $y'' = -32, y' = -32t + C$. Given $y' = 40$ when $t = 0$, so $40 = 0 + C, C = 40$. Then $y = -16t^2 + 40t + K$. Given $y = 24$ when $t = 0$, so $24 = K, y = -16t^2 + 40t + 24$. Stone reaches peak when velocity is 0 (turns from positive to neg), $-32t + 40 = 0, t = \frac{5}{4}, y = 49$. Stone hits ground when $y = 0, -16t^2 + 40t + 24 = 0, 2t^2 - 5t - 3 = 0, (2t + 1)(t - 3) = 0, t = -\frac{1}{2}$ (ignore) or $t = 3$ (answer).

9. $-3 \ln|3 - x|$
10. $\frac{1}{2} \ln|2x + 5|$
11. Can't do.
12. $5 \ln|x|$
13. $\frac{1}{5}\int (1/x) dx = \frac{1}{5} \ln|x|$
14. $\ln|2 + x|$
15. Can't do.
16. $(7/\pi) \sin \pi x$
17. Can't do.
18. $\frac{1}{5}(\frac{1}{3}x^3 + 3x^2)$
19. Can't do.

20. $5 \int (3x + 6)^{-2} dx = \dfrac{5(3x + 6)^{-1}}{-1} \cdot \dfrac{1}{3} = \dfrac{-5}{3(3x + 6)}$

21. $4\dfrac{(2 + \frac{1}{4}x)^{3/2}}{3/2} = \dfrac{8}{3}(2 + \frac{1}{4}x)^{3/2}$

22. $2 \tan^{-1}x$
23. Can't do.

24. $-6 \cos(x/6)$
25. $\dfrac{3}{2\pi} \sin \dfrac{2\pi x}{3}$

26. $x^7/42$
27. $-3e^{-x}$
28. Can't do.
29. $\frac{1}{3}x^3 + \frac{1}{2}x^2 + \ln|x| - 1/x - 1/2x^2$

30. $\frac{1}{2}e^{2x}$
31. πx
32. $\frac{1}{3}(3x + 4)^5/5$
33. $2x^{-2}/-2 = -1/x^2$
34. $x^4/8$
35. $\frac{1}{2}\int x^{-3} dx = \frac{1}{2}(x^{-2}/-2) = -1/4x^2$
36. $\frac{1}{5}(x^{-2}/-2) + C = -1/10x^2 + C$
37. $t^{1/2}/\frac{1}{2} + C = 2\sqrt{t} + C$
38. $\frac{3}{4}x^4 + C$
39. $\frac{1}{3}(x^{-2}/-2) + C = -1/6x^2 + C$
40. $\ln|x| + C$
41. Can't do.
42. $2x - x^3 + C$
43. $\int (2 - 3x)^{-3} dx = -\frac{1}{3} \cdot (2 - 3x)^{-2}/-2 + C = 1/6(2 - 3x)^2 + C$
44. $\frac{1}{5}x^5 + 5x + C$
45. $x + C$
46. $-\frac{1}{3} \cos 3u + C$
47. Can't do.
48. $-\frac{1}{2}e^{-2x} + C$
49. Can't do.
50. $-1/x^3 + C$
51. $-\ln|1 - v| + C$
52. $2 \cdot \frac{1}{4} \ln|3 + 4x| + C = \frac{1}{2} \ln|3 + 4x| + C$
53. Can't do.
54. $\frac{4}{5}e^{5x} + C$
55. $-(3 - x)^{3/2}/\frac{3}{2} = -\frac{2}{3}(3 - x)^{3/2}$
56. $\frac{1}{2}[(5t^2/2) + 3t] + C = \frac{5}{4}t^2 + \frac{3}{2}t + C$
57. $\frac{1}{2}[(5x^4/4) + 3x] = \frac{5}{8}x^4 + \frac{3}{2}x + C$
58. Can't do.
59. $\frac{1}{2}(2x + 3)^6/6 + C = \frac{1}{12}(2x + 3)^6 + C$

Chapter 3 Review Problems (page 92)

1. (a) $f'(t)$ is the rate in gallons/hour flowing in instantaneously at time t. With the given data, 20 gal/hour flow in at time 3 (unhappy) but the figure 20 is in the process of decreasing by 1 gal/hour per hour (happy).
   (b) Want $f'(t) = 0$ so that the rate of flow into the flood plain is 0. Negative values would be nice but unrealistic since water will not spurt back through the hole from the land.

2. $2 \cos(2x + 3\pi)$
3. $x \cos x + \sin x$
4. $\dfrac{1}{1 + (x^2)^2} \cdot 2x = \dfrac{2x}{1 + x^4}$
5. $D(2 - x)^{-1} = -(2 - x)^{-2} \cdot -1 = 1/(2 - x)^2$
6. $\dfrac{1}{2 - x} \cdot -1$
7. $-1/x^2$
8. $D\frac{1}{4}x^{-2} = \frac{1}{4} \cdot -2x^{-3} = -1/2x^3$
9. $-2e^{-2x}$
10. $-e^x$
11. $3 \sec^2 3x$
12. $-3/x^2$
13. $x^2 \cdot 7(2 - 3x)^6 \cdot -3 + 2x(2 - 3x)^7 = -21x^2(2 - 3x)^6 + 2x(2 - 3x)^7$
14. $x/\sqrt{1 - x^2} + \sin^{-1}x$
15. $\frac{4}{5} \cos 4x$

16. $3xe^x \sec x \tan x + 3xe^x \sec x + 3e^x \sec x$

17. $-\frac{1}{6}\sin x$

18. Let $y = 4^x$. Then $\ln y = x \ln 4$, $(1/y)y' = \ln 4$, $y' = y \ln 4 = 4^x \ln 4$

19. $4x^3$

20. $-e^{8-x}$

21. $3(8-x)^2 \cdot -1$

22. Let $y = (8-x)^x$. Then $\ln y = x \ln(8-x)$, $(1/y)y' = x \cdot 1/(8-x) \cdot -1 + \ln(8-x)$,
$$y' = (8-x)^x\left(\frac{-x}{8-x} + \ln(8-x)\right).$$

23. $4\left(\dfrac{2x+3}{5}\right)^3 \cdot \dfrac{2}{5} = \dfrac{8}{5}\left(\dfrac{2x+3}{5}\right)^3$

24. $\frac{1}{2}(2x+5)^{-1/2} \cdot 2 = 1/\sqrt{2x+5}$

25. $2 \cdot -(3+2x)^{-2} \cdot 2 = -4/(3+2x)^2$

26. $e^{\sqrt{x}} \cdot \frac{1}{2}x^{-1/2} = e^{\sqrt{x}}/2\sqrt{x}$

27. $-\frac{2}{7}$

28. $\dfrac{(2x+3) \cdot 1 - x \cdot 2}{(2x+3)^2} = \dfrac{3}{(2x+3)^2}$

29. (product and chain)
$$x \cdot \cos\frac{1}{x} \cdot -\frac{1}{x^2} + \sin\frac{1}{x} = -\frac{1}{x}\cos\frac{1}{x} + \sin\frac{1}{x}$$

30. $(xe^x - e^x)/x^2$

31. $\frac{1}{4}$

32. $\frac{2}{3} \cdot -x^{-2} = -2/3x^2$

33. $D(\cos 2x)^3 = 3(\cos 2x)^2 \cdot -\sin 2x \cdot 2$
$$= -6\cos^2 2x \sin 2x$$

34. $3\cos e^{2x} \cdot e^{2x} \cdot 2 = 6e^{2x}\cos e^{2x}$

35. $D(7x^3 + 2x - 5)^{-1} = -(7x^3 + 2x - 5)^{-2}(21x^2 + 2)$
$$= -(21x^2 + 2)/(7x^3 + 2x - 5)^2$$

36. $\dfrac{(5x-4) \cdot 2 - (2x+3) \cdot 5}{(5x-4)^2} = \dfrac{-23}{(5x-4)^2}$

37. $s' = 2t - 6t^2$, $s'' = 2 - 12t$. If $t = 2$ then $s = -11$, $s' = -20$, $s'' = -22$. Particle is at position -11, moving left at 20 meters/sec, speeding up (since acceleration and velocity have same sign) by 22 meters/sec per sec.

38. Graph rises until $x = 3$, falls until $x = 5$, then rises again, concave down until $x = 4$, then concave up. (See fig.)

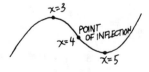

x=3

x=4 POINT OF INFLECTION

x=5

PROBLEM 38

39. $f(x) = \frac{1}{4}x^4 - x^2 + C$, $f(2) = -2$ so $-2 = 4 - 4 + C$, $C = -2$, $f(x) = \frac{1}{4}x^4 - x^2 - 2$

40. $f'(t) = 3t^2 + 6t = 3t(t+2)$, $f''(t) = 6t + 6$; f' is zero if $t = 0, -2$; never is disc; f' is pos in $(-\infty, -2)$, neg in $(-2, 0)$, pos in $(0, \infty)$; f'' is zero when $t = -1$; never disc; f'' is neg if $t < -1$, pos if $t > -1$.

time interval	f'	f''	particle
$(-\infty, -2)$	pos	neg	moves right, slows down
$(-2, -1)$	neg	neg	moves left, speeds up
$(-1, 0)$	neg	pos	moves left, slows down
$(0, \infty)$	pos	pos	moves right, speeds up

$f(-\infty) = -\infty$, $f(-2) = 5$, $f(-1) = 3$, $f(0) = 1$, $f(\infty) = \infty$. (See fig.)

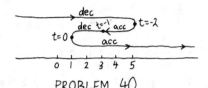

PROBLEM 40

41. $\cos \underbrace{\sin\sin\cdots\sin}_{824 \text{ sines}} 2x \cdot \cos \underbrace{\sin\sin\cdots\sin}_{823 \text{ sines}} 2x \cdot$
$\cdots \cos\sin 2x \cdot \cos 2x \cdot 2$

42. (a) $y' = x^2 + \frac{1}{2}$, always positive. So y increases. Alternatively, $\frac{1}{3}x^3$ and $\frac{1}{2}x$ are both increasing functions, by inspection. So their sum is increasing.

(b) If $x = \frac{1}{2}$ then $y' = \frac{3}{4}$. So as x goes up, y goes up $\frac{3}{4}$ as fast. So y is increasing slower than x at this instant.

43. Second derivative of e^x is e^x which is always positive. So graph of e^x is concave up.

44. Diff implicitly (a) $xy' + y + 3x \cdot 2yy' + 3y^2 = -1$,
$y' = (-1 - 3y^2 - y)/(x + 6xy)$

(b) $\cos x + y' \cos y = 0$, $y' = -\cos x/\cos y$

45. (a) $\dfrac{(x^2 + 3x) \cdot 5 - (5x + 2)(2x + 3)}{(x^2 + 3x)^2}$
$$= -\frac{5x^2 + 4x + 6}{(x^2 + 3x)^2}$$

(b) (product rule)
$x^3(1/x) + 3x^2 \ln x = x^2 + 3x^2 \ln x$

(c) Let $y = (\ln t)^{2t}$, then $\ln y = 2t \ln \ln t$,
$$\frac{1}{y}y' = 2t \cdot \frac{1}{\ln t} \cdot \frac{1}{t} + 2\ln\ln t,$$
$$y = (\ln t)^{2t}\left[\frac{2}{\ln t} + 2\ln\ln t\right].$$

(d) $3x - 6$ is positive if $x > 2$, negative if $x < 2$ so
$$|3x - 6| = \begin{cases} 3x - 6 & \text{if } x \geq 2 \\ -(3x - 6) & \text{if } x < 2 \end{cases}$$
$$\text{derivative is } \begin{cases} 3 & \text{if } x > 2 \\ -3 & \text{if } x < 2 \end{cases}$$
(cusp at $x = 2$; see fig).

2

PROBLEM 45(d)

(e) $1 + e^x$

(f) (product rule) $te^t + e^t$

(g) $\dfrac{\sqrt{3x + 4} \cdot 2 - 2x \cdot \frac{1}{2}(3x + 4)^{-1/2} \cdot 3}{3x + 4}$

$= \dfrac{3x + 8}{(3x + 4)^{3/2}}$

(h) $e^{|x|}$ is e^{-x} if $x < 0$ and is e^x if $x \geq 0$ so $De^{|x|}$ is $-e^{-x}$ if $x < 0$ and is e^x if $x > 0$.

46. (a) (product rule)

$y' = 3x \cos x + 3 \sin x$,

$y'' = -3x \sin x + 3 \cos x + 3 \cos x$
$\quad = -3x \sin x + 6 \cos x$,

$y''' = -3x \cos x - 3 \sin x - 6 \sin x$
$\quad\; = -3x \cos x - 9 \sin x$

(b) $1 - \ln x$ is pos if $0 < x < e$; neg if $x > e$. So

$y = \begin{cases} 1 - \ln x & \text{if } 0 < x \leq e \\ -(1 - \ln x) & \text{if } x > e \end{cases}$

$y' = \begin{cases} -1/x & \text{if } 0 < x < e \\ 1/x & \text{if } x > e \end{cases}$

$y'' = \begin{cases} 1/x^2 & \text{if } 0 < x < e \\ -1/x^2 & \text{if } x > e \end{cases}$

There is a cusp at $x = e$. When y' changes formulas at $x = e$, the two formulas, $-1/x$ and $1/x$, do not agree; i.e., left-hand slope and right-hand slope disagree at $x = e$. For confirmation, consider graph of $|1 - \ln x|$. First reflect $\ln x$ to get $-\ln x$, then translate up to get $1 - \ln x$, and finally reflect all points that are below the x-axis. (See fig.)

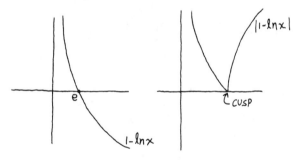

PROBLEM 46(b)

(c) $y' = x^4 \cdot -\sin x^2 \cdot 2x + 4x^3 \cos x^2$
$\quad = -2x^5 \sin x^2 + 4x^3 \cos x^2$,

$y'' = -2x^5 \cos x^2 \cdot 2x - 10x^4 \sin x^2$
$\quad\quad + 4x^3 \cdot -\sin x^2 \cdot 2x + 12x^2 \cos x^2$
$\quad = (12x^2 - 4x^6) \cos x^2 - (10x^4 + 8x^3) \sin x^2$

(d) $\ln y = x \ln 5$, $(1/y)y' = \ln 5$, $y' = y \ln 5 = 5^x \ln 5$, $y'' = (\ln 5)D 5^x$, and we just got $D5^x = 5^x \ln 5$ so $y'' = 5^x(\ln 5)^2$

47. 1st derivative $= -\frac{1}{2}(2 + 5x)^{-3/2} \cdot 5$,

2nd derivative $= \frac{1}{2} \cdot \frac{3}{2}(2 + 5x)^{-5/2} \cdot 5^2$,

3rd derivative $= -\frac{1}{2}\frac{3}{2}\frac{5}{2}(2 + 5x)^{-7/2} \cdot 5^3, \cdots$,

19th derivative $= -\dfrac{3 \cdot 5 \cdot 7 \cdots 37}{2^{19}} 5^{19}(2 + 5x)^{-39/2}$,

20th derivative $= \dfrac{3 \cdot 5 \cdot 7 \cdots 39}{2^{20}} 5^{20}(2 + 5x)^{-41/2}$

48. $y = 1/x$, $y' = -1/x^2$. A typical point Q on the graph has coordinates $(a, 1/a)$. Tangent line at Q has slope $-1/a^2$, equation $y - (1/a) = -(1/a^2)(x - a)$, $a^2 y + x = 2a$. Tan hits x-axis at $C = (2a, 0)$, hits y-axis at $A = (0, 2/a)$. $\overline{BC} = 2a$, $\overline{BA} = 2/a$ so area of triangle $= \frac{1}{2}\overline{BC}\,\overline{BA} = \frac{1}{2} \cdot 2a \cdot 2/a = 2$. All such triangles have same area, namely area 2.

49. By product rule,

$(fg)'' = fg'' + f'g' + f'g' + f''g$
$\quad\quad = fg'' + 2f'g' + f''g$,

$(fg)''' = fg''' + f'g'' + 2(f'g'' + f''g') + f''g' + f'''g$
$\quad\quad = fg''' + 3f'g'' + 3f''g' + f'''g$ Similarly,

$(fg)^{(4)} = fg^{(4)} + 4f'g''' + 6f''g'' + 4f'''g' + f^{(4)}g$.

Same pattern as binomial expansion for $(x + y)^n$ (Appendix A4).

Guess $(fg)^{(n)} = fg^{(n)} + nf'g^{(n-1)} + \dfrac{n(n-1)}{2!} f''g^{(n-2)}$

$+ \cdots + nf^{(n-1)}g' + f^{(n)}g$.

50. (a) f' negative because reversed image appears when $f(x)$ goes *down* as x goes *up*.

(a') f' is positive.

(b) $|f'(x)| > 1$ so that $f(x)$ goes up or down *faster* than x goes up.

(b') $|f'(x)| < 1$

(c) $f'(x)$ *not* constant so that $f(x)$ doesn't change steadily as x goes up.

(c') $f'(x)$ is constant.

51. $\frac{1}{7} \ln|x| + C$

52. $\frac{1}{7} \int x^{-2} \, dx = -1/7x + C$

53. $\dfrac{(4x - 2)^{-2}}{-2} \cdot \dfrac{1}{4} + C = \dfrac{-1}{8(4x - 2)^2} + C$

54. $2x^2 + 2x + C$

55. $\frac{1}{5}e^{5x} + C$

56. $-(2/\pi) \cos \frac{1}{2}\pi x + C$

57. Can't do.

58. Can't do.

59. $-\ln|3 - t| + C$

60. $\int (3 - t)^{-1/2} \, dt = \dfrac{(3 - t)^{1/2}}{1/2} \cdot -1 + C$
$\quad\quad = -2\sqrt{3 - t} + C$

61. $\dfrac{(1 + 2x)^{3/2}}{3/2} \cdot \dfrac{1}{2} + C = \dfrac{1}{3}(1 + 2x)^{3/2} + C$

62. Can't do.

63. (a) $5x^4$

(b) $x^6/6 + C$

(c) $-4x^{-5} + C = -4/x^5 + C$

(d) $x^{-3}/-3 + C = -1/3x^3 + C$

4/THE DERIVATIVE   PART II

Section 4.1 (page 97)

1. (a) $f'(x) = 3x^2 - 6x - 24$; $f'(x) = 0$ if
$3(x^2 - 2x - 8) = 0$, $(x - 4)(x + 2) = 0$,
$x = 4, -2$, the candidates.

   (i) f' is pos on $(-\infty, -2)$, neg on $(-2, 4)$, pos on
$(4, \infty)$ so f has max at $x = -2$, min at $x = 4$.
   (ii) $f''(x) = 6x - 6$, $f''(-2) = -18$, neg, so f has
max at $x = -2$; $f''(4) = 18$, pos, so f has
min at $x = 4$.

   (b) $f'(x) = 4x^3 - 2x$, $f'(x) = 0$ iff $2x(2x^2 - 1) = 0$,
$x = 0, \pm\sqrt{\frac{1}{2}}$

   (i) f' is neg in $(-\infty, -\sqrt{\frac{1}{2}})$, pos in $(-\sqrt{\frac{1}{2}}, 0)$
(test say $x = -.01$), neg in $(0, \sqrt{\frac{1}{2}})$, pos in
$(\sqrt{\frac{1}{2}}, \infty)$; f has min at $x = -\sqrt{\frac{1}{2}}$, max at $x = $
0, min at $x = \sqrt{\frac{1}{2}}$.
   (ii) $f'(x) = 12x^2 - 2$, $f''(-\sqrt{\frac{1}{2}}) = 4$, pos, min at
$x = -\sqrt{\frac{1}{2}}$, $f''(0) = -2$, neg, max at $x = 0$,
$f''(\sqrt{\frac{1}{2}}) = 4$, pos, min at $x = \sqrt{\frac{1}{2}}$.

   (c) $f''(x) = 5x^4 + 1$, never 0. No candidates, no extrema (f' is always positive so f is an increasing
function with no relative extrema).
   (d) $f'(x) = (xe^x - e^x)/x^2$, $f' = 0$ if $xe^x - e^x = 0$,
$e^x(x - 1) = 0$, $x = 1$

   (i) f' is neg if $x < 1$, pos if $x > 1$, f has min at
$x = 1$.
   (ii) Using quotient and product rule, $f''(x) = $
$(x^2e^x - 2xe^x + 2e^x)/x^3$, $f''(1) = e$, pos, so f
has min at $x = 1$.

   (e) $f'(x) = x \cdot (1/x) + \ln x = 1 + \ln x$, $f' = 0$ if
$\ln x = -1$, $x = e^{-1} = 1/e$

   (i) f and f' are defined only for $x > 0$. Consider interval $(0, 1/e)$. Test number from
interval, say $.0001$; $f'(.0001) = 1 + $
$\ln .0001 = 1 - $ large neg = neg. So f' is
neg in interval; f' is pos in $(1/e, \infty)$ so f has
min at $x = 1/e$.
   (ii) $f''(x) = 1/x$, $f''(1/e) = e$, pos, so min at $1/e$.

2. (a) Min at 2 by first derivative test.
   (b) $x = 2$ is a candidate; no further conclusion.
   (c) No conclusion.
   (d) No rel extremum at $x = 3$ since slope is not 0.

   (e) Min at $x = 2$ by second derivative test.
   (f) $x = 2$ is a candidate; no further conclusion.
   (g) $x = 7$ is a candidate; no further conclusion.

3. No. In diagram, the rel min at x_0 is higher than the
rel max at x_1.

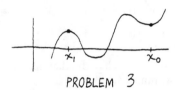

PROBLEM 3

4. For each function, the derivative is 0 when $x = 0$ and
the second derivative is also 0 when $x = 0$, an inconclusive second derivative test. By inspection, x^3 does not
have a rel extremum at $x = 0$, x^4 has min at $x = 0$, $-x^4$
has max at $x = 0$. So inconclusive situation can go any of
three ways.

5.

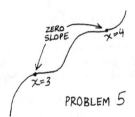

ZERO
SLOPE

$x = 4$

$x = 3$

PROBLEM 5

Section 4.2 (page 104)

1. (a) $f'(x) = 3x^2 + 2x - 5$; $f'(x) = 0$ if
$(3x + 5)(x - 1) = 0$, $x = -\frac{5}{3}, 1$.

   (i) Candidates are end values $f(-\infty) = -\infty$,
$f(\infty) = \infty$, and critical values $f(-\frac{5}{3})$, $f(1)$.
Max is ∞, min is $-\infty$.
   (ii) Candidates are $f(0) = -5$, $f(1) = -8$,
$f(2) = -3$. Max is -3, min is -8.
   (iii) Candidates are $f(-1) = 0, f(0) = -5$. Max
is 0, min is -5.

   (b) $f'(x) = (xe^x - e^x)/x^2$; $f'(x) = 0$ if $xe^x - e^x = 0$,
$e^x(x - 1) = 0$, $x = 1$.

   Note that f has an infinite disc at $x = 0$.

(i) Candidates are $f(0+) = \lim_{x \to 0+} (e^x/x) = 1/0+ = \infty$, $f(0-) = \lim_{x \to 0-} (e^x/x) = 1/0- = -\infty$ (and end values $f(-2), f(2)$ and critical value $f(1)$). Max is ∞, min is $-\infty$.

(ii) Candidates are $f(0+) = \infty$, $f(1) = e$, $f(2) = e^2/2$. Max is ∞, min is e.

(iii) Candidates are $f(-\infty) = e^{-\infty}/-\infty = 0/-\infty = 0$, $f(0-) = -\infty$. Min is $-\infty$, max is 0.

(c) $f'(x) = (-x^2 + 4x - 3)/(x^2 - 3)^2$ (quotient rule); $f' = 0$ if $-x^2 + 4x - 3 = 0$, $x = 3, 1$. Inf disc at $x = \pm\sqrt{3}$.

(i) Candidates are $x = 0, 5, 1, 3, \sqrt{3}$.
$$\lim_{x \to \sqrt{3}+} f(x) = (\sqrt{3} - 2)/0+ = -\infty,$$
$$\lim_{x \to \sqrt{3}-} f(x) = (\sqrt{3} - 2)/0- = \infty.$$
No need to look further. Max is $f(\sqrt{3} -) = \infty$, min is $f(\sqrt{3} +) = -\infty$.

(ii) Candidates are $f(2) = 0, f(3) = \frac{1}{6}, f(5) = \frac{3}{22}$. Max is $\frac{1}{6}$, min is 0.

(d) $f'(x) = 3x^2 + 2x - 1 = (3x - 1)(x + 1)$, zero if $x = \frac{1}{3}, -1$. Candidates are $f(0) = 3, f(\frac{1}{3}) = \frac{76}{27}$, $f(4) = 79$. Max is 79, min is $\frac{76}{27}$.

2. f is decreasing, graph falls to the left. Max is $f(3)$, min is $f(4)$.

3. By inspection, max is ∞ when $x = \pm\infty$, min is $\sqrt{2}$ when $x = 0$.

4. Let $f(x)$ be the revenue with x passengers. The number of passengers over 200 is $x - 200$ so each ticket is reduced from \$300 by $x - 200$ dollars. So

$$f(x) = \text{number of passengers} \times \text{ticket price}$$
$$= x(300 - [x - 200]) = 500x - x^2$$
where $200 \le x \le 350$.

$f'(x)$ is $500 - 2x$ and is 0 when $x = 250$. Candidates are $f(200) = 60,000, f(250) = 62,500$ and $f(350) = 52,500$. Max revenue is with 250 passengers, min is with 350 passengers.

5. Let $\overline{CD} = x$. Then $\overline{AB} = 100 + x$ and $200 - (100 + 2x)$ feet of wire remain for sides $\overline{AE}$ and $\overline{BC}$. So $\overline{BC} = \frac{1}{2}(200 - (100 + 2x)) = 50 - x$ and area $A(x) = (100 + x)(50 - x) = 5000 - 50x - x^2$ for $0 \le x \le 50$. $A'(x) = -50 - 2x$; $A'(x) = 0$ if $x = -25$; not in $[0, 50]$ so ignore. $A(50) = 0$ ("garden" dimensions are 150×0), $A(0) = 5000$. Max area is 5000 using wall as entire side, min area is 0 as garden collapses to a segment.

6. Let $x = \overline{BC}$ (this is not the only way to begin). Then
$$\overline{AB} = 150 - x, \frac{\overline{EB}}{150 - x} = \frac{100}{150}, \overline{EB} = -\tfrac{2}{3}x + 100, \text{ area}$$
$A(x) = \overline{BC} \times \overline{EB} = x(-\tfrac{2}{3}x + 100) = -\tfrac{2}{3}x^2 + 100x$ for $0 \le x \le 150$; $A'(x) = -\tfrac{4}{3}x + 100$. Critical number is $x = 75$. By inspection, the ends $x = 0, 150$ produce a

collapsed house with area 0. So max area has dimensions $x = 75$ by $\overline{EB} = 50$.

7. Let $p(x)$ be profit when farmer sells after x more days. Then $p(x) = (100 + 1.2x)(12 - \frac{1}{40}x) = -.03x^2 + 11.9x + 1200$ where $x \ge 0$, or better still $0 \le x \le 480$ since after 480 days the 12¢ figure is down to 0. Then $p'(x) = -.06x + 11.9$. Critical x is $11.9/.06$ or approximately 198.3; $p(198.3)$ is about \$23.80, $p(0) = \$12$, $p(480) = \$0$. Sell after 198 days.

8. Slope $s(x) = -3x^2 - 10x - 13$. We want max and min values of s (*not* f). Then $s'(x) = f''(x) = -6x - 10$. Critical value is $x = -\frac{5}{3}$, not in interval. Candidates are $s(0) = -13, s(1) = -26$. Max slope on f graph in $[0, 1]$ is -13, min slope is -26.

9. (See figs.) At 5 AM, car B has reached the A road and the cars are 75 miles apart. From then on, the distance between them is increasing so min must occur at or before 5 AM. At t hours after midnight, B has gone $20t$ miles, A has gone $15t$ miles so
$$s(t) = \sqrt{(100 - 20t)^2 + 225t^2}, \qquad 0 \le t \le 5.$$
For convenience, can work with $R(t) = (100 - 20t)^2 + 225t^2$ instead of $s(t)$; $R'(t) = -40(100 - 20t) + 450t$, 0 when $t = \frac{16}{5}$. Candidates are $s(\frac{16}{5}) = 60, s(0) = 100$, $s(5) = 75$. Cars are closest when $t = 3.2$, i.e., at 3:12 AM.

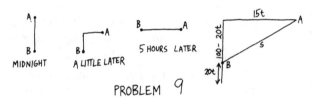

PROBLEM 9

10. Let (x, y) be a typical point on the ellipse. (See fig.) Distance s from point to $(1, 0)$ is $\sqrt{(x - 1)^2 + y^2} = \sqrt{(x - 1)^2 + \frac{1}{9}(36 - 4x^2)}$ where $-3 \le x \le 3$. For convenience, one can work with $R(x) = (x - 1)^2 + \frac{1}{9}(36 - 4x^2)$ instead of s; $R'(x) = 2(x - 1) - \frac{8}{9}x$, zero if $x = \frac{9}{5}$. Candidates are $x = \frac{9}{5}$ and ends $x = \pm 3$ so candidate points are $A = (3, 0)$, $B = (-3, 0)$, $C = (\frac{9}{5}, \frac{8}{5})$, $D = (\frac{9}{5}, -\frac{8}{5})$. $\overline{AQ} = 2, \overline{BQ} = 4, \overline{CQ} = \frac{1}{5}\sqrt{80}, \overline{DQ} = \frac{1}{5}\sqrt{80}$. Points C and D are closest, B is furthest.

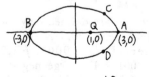

PROBLEM 10

11. Let r be the fixed radius, let x be the height of the inscribed rectangle. (See fig.) Area $A(x) = \text{base} \times \text{height} = 2x\sqrt{r^2 - x^2}$ where $0 \le x \le r$.

$$A'(x) = 2x\tfrac{1}{2}(r^2 - x^2)^{-1/2} \cdot -2x + 2\sqrt{r^2 - x^2}$$
$$= 2\sqrt{r^2 - x^2} - 2x^2/\sqrt{r^2 - x^2}.$$

A' is 0 when $2(r^2 - x^2) - 2x^2 = 0$, $x = \pm r/\sqrt{2}$. Ignore the neg x. Candidates are ends $x = 0$, r when rectangle collapses to a segment with area 0, and $x = r/\sqrt{2}$. Rectangle with min area is the degenerate case of a segment, rectangle with max area has dimensions $r/\sqrt{2}$ by $2r/\sqrt{2}$.

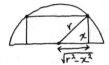

PROBLEM 11

12. Let speed of truck be s. Trip takes $600/s$ hours and total cost C = gas and oil + driver
= 600 miles $\times$ $(5 + \frac{1}{10}s)$ cents per mile
  + $600/s$ hours $\times$ 360 cents per hour
where $30 \le s \le 80$; $C'(s) = 60 + 360 \cdot (-600/s^2)$, zero if $s = \pm 60$. Ignore $s = -60$. Candidates are $s = 30, 60, 80$. Corresponding costs are $120, \$102, \$105$. Best speed is 60 mph, worst is 30 mph.

13. Let x be the square's share and $16 - x$ the circle's piece. Square has perimeter x, side $x/4$, area $x^2/16$. Circle has circum $16 - x$, radius $(16 - x)/2\pi$, area $\pi(16 - x)^2/4\pi^2 = (16 - x)^2/4\pi$. Total area is

$A(x) = x^2/16 + (16 - x)^2/4\pi$ where $0 \le x \le 16$.
$A'(x) = \frac{x}{8} + \frac{1}{4\pi} \cdot 2(16 - x) \cdot -1 = \frac{x}{8} - \frac{16 - x}{2\pi}$.

$A'(x) = 0$ when $x = 64/(\pi + 4)$. Candidates are $A(64/(\pi + 4)) = 64/(\pi + 4)$, $A(0) = 64/\pi$, $A(16) = 16$. Max is $A(0)$ so for max area, use whole wire for circle.

14. Let one dimension of garden be x. (See fig.) To keep area fixed at A, other dimension is A/x. Perimeter p is $2A/x + 2x$, $x \ge 0$. When $x = 0$, have long thin garden with huge perimeter. When $x = \infty$, have tall skinny garden with huge perimeter. So endpoints produce max perimeter. Expect min perimeter at a critical point. $p'(x) = -2A/x^2 + 2$, zero when $x = \sqrt{A}$, i.e., best garden is square. Min perimeter itself is $p(\sqrt{A}) = 4\sqrt{A}$.

PROBLEM 14

15. Let x be the price charged. Then
number of vacancies is $\frac{1}{2}(x - 50)$,
number of rentals is $100 - \frac{1}{2}(x - 50) = -\frac{1}{2}x + 125$.
Income $I(x) = x \times$ no. of rentals
$= x(-\frac{1}{2}x + 125) = 125x - \frac{1}{2}x^2$ where $50 \le x \le 250$.
(Once the price reaches $250, all rooms are vacant.)
$I'(x) = 125 - x$. Critical x is 125. Candidates are $x = 50, 125, 250$. Corresponding incomes are $5000, \$7812.50, \$0$. Charge $125 a night (have $62\frac{1}{2}$ vacancies, but max profit).

Section 4.3 (page 109)

1. (a) $0/0 = \lim_{x\to 1}(3x^2 - 5)/(2x - 3)$ (L'Hôpital)
       $= -2/-1 = 2$
   (b) $\frac{4}{2} = 2$
   (c) $\lim_{x\to\infty}(x^3/x^2)$ (highest power rule) $= \lim x = \infty$
2. (a) ∞ since x^2 has higher order of magnitude.
   (b) $0/0 = \lim \dfrac{1/(x - 1)}{1}$ (L'Hôpital) $= 1$
   (c) $-\infty/0+ = -\infty$ (*not* indeterminate)
   (d) 0; e^x has higher order of magnitude.
   (e) $0/0 = \lim \dfrac{\cos x - 1}{-\sin x} = 0/0 = \lim \dfrac{-\sin x}{-\cos x}$
       $= \dfrac{0}{-1} = 0$
   (f) $0/(1 + 0) = 0$
   (g) $\dfrac{-\infty}{e^\infty} = \dfrac{-\infty}{\infty} = \lim \dfrac{1/x}{e^{1/x} \cdot -1/x^2}$ (L'Hôpital)
       $= \lim \dfrac{-x}{e^{1/x}}$ (algebra) $= \dfrac{0}{\infty} = 0$
   (h) $-\infty/0+ = -\infty$
   (i) 0 since $3x$ has higher order of magnitude.

3. $\lim_{x\to\infty} \dfrac{(\ln x)^{27}}{x} = \dfrac{\infty}{\infty} = \lim \dfrac{27(\ln x)^{26} \cdot \frac{1}{x}}{1}$ (L'Hôpital)

$= 27 \lim \dfrac{(\ln x)^{26}}{x}$ (algebra) $= \dfrac{\infty}{\infty} = 27 \lim \dfrac{26(\ln x)^{25} \cdot \frac{1}{x}}{1}$

(L'Hôpital) $= 27 \cdot 26 \lim \dfrac{(\ln x)^{25}}{x} = \dfrac{\infty}{\infty}$. Keep using

L'Hôpital to eventually get $27! \lim \dfrac{\ln x}{x} = \dfrac{\infty}{\infty} =$

$27! \lim \dfrac{1/x}{1} = 0$. So x is faster than $(\ln x)^{27}$.

4. (a) *Method 1:* Let $u = 3x$. If $x \to 0$ then $u \to 0$ and limit becomes

$$\lim_{u\to 0} \dfrac{\sin u}{\frac{2}{3}u} = \dfrac{3}{2} \lim_{u\to 0} \dfrac{\sin u}{u} = \dfrac{3}{2} \times 1 = \dfrac{3}{2}.$$

Method 2: (L'Hôpital) $\lim(3 \cos 3x)/2 = \frac{3}{2}$.

   (b) *Method 1:* $\lim \dfrac{\sin x}{x} \cdot \sin x = 1 \times 0 = 0$.

   Method 2: (L'Hôpital) $\lim \dfrac{2 \sin x \cos x}{1} = 0$.

5. First application is OK since original problem is of the form $0/0$. But $\lim_{x\to 1} \dfrac{8x - 2}{6x - 4} = \dfrac{6}{2}$. It is not an indeterminate quotient so L'Hôpital's rule can't be used a second time. No need for a special rule anyway. Answer is $\frac{6}{2} = 3$.

6. (a) Same order of mag since each is a multiple of the other.

(b) $\lim_{x\to\infty} e^{5x}/e^{3x} = \infty/\infty = \lim 5e^{5x}/3e^{3x} = \infty/\infty$, getting nowhere. Instead use algebra to get $\lim_{x\to\infty} e^{5x}/e^{3x} = \lim e^{2x} = \infty$. So e^{5x} has higher order of magnitude.

(c) *Method 1:* $\lim_{x\to\infty} \dfrac{\ln 3x}{\ln 4x} = \dfrac{\infty}{\infty}$

$$= \lim \frac{\dfrac{1}{3x}\cdot 3}{\dfrac{1}{4x}\cdot 4} \text{ (L'Hôpital)} = \lim 1 = 1.$$

Same order of magnitude.

Method 2: $\ln 3x = \ln 3 + \ln x$, $\ln 4x = \ln 4 + \ln x$. Each has same order of mag as $\ln x$.

7. We already know that $\lim_{x\to 0}(\sin x)/x = 1$. Draw the hyperbola $y = 1/x$ and its reflection in the x-axis to serve as the envelope. (See fig.)

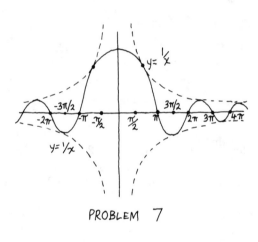

PROBLEM 7

Section 4.4 (page 112)

1. (a) $\infty \times e^{-\infty} = \infty \times 0$; $\lim x/e^x = 0$ since e^x has higher order of magnitude.

(b) $0e^0 = 0 \times 1 = 0$

(c) $(-\infty)e^\infty = -\infty \times \infty = -\infty$

2. (a) $1 - \ln 1 = 1 - 0 = 1$

(b) $0 - (-\infty) = \infty$

(c) $\infty - \infty$. Answer is ∞ since x^2 has higher order of magnitude.

3. (a) $-\infty$ since e^x has higher order of magnitude.

(b) $-\infty - 0 = -\infty$

4. $\lim_{x\to 0+} xe^{1/x} = 0 \times e^\infty = 0 \times \infty$. Let $u = 1/x$. Then $u \to \infty$, problem becomes $\lim_{u\to\infty} e^u/u = \infty$ since e^u has higher order of magnitude. Also $\lim_{x\to 0-} xe^{1/x} = 0 \times e^{-\infty} = 0 \times 0 = 0$. For the purpose of the graph, can be more precise to get $0- \times 0+ = 0-$. (See fig.)

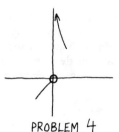

PROBLEM 4

5. (a) $0 \times -\infty$; $\lim \dfrac{\ln x}{\cot x} = \dfrac{-\infty}{\infty} = \lim \dfrac{1/x}{-\csc^2 x}$

$= \lim \dfrac{-\sin^2 x}{x} = 0/0 = \cdots = 0$. (See Prob. 4b, Section 4.3.)

(b) $1 \times -\infty = -\infty$

(c) $\infty \times \sin 0 = \infty \times 0$. Let $u = 1/x$ to get $\lim_{u\to 0+}(\sin u)/u^2 = \frac{0}{0} = \lim(\cos u)/2u$ (L'Hôpital) $= 1/0+ = \infty$

(d) ∞^0 (indet). Let $y = x^{1/x}$. Then $\ln y = (1/x)\ln x$, $\lim_{x\to\infty} \ln y = \lim_{x\to\infty}(\ln x)/x = 0$ ($\ln x$ has lower order of magnitude). Answer is $e^0 = 1$.

(e) $(0+)^\infty = 0$ (not indet). If you don't see the answer 0 immediately, let $y = x^{1/x}$; then $\ln y = (1/x)\ln x$, $\lim_{x\to 0+} \ln y = \lim_{x\to 0+}(\ln x)/x = -\infty/0+ = -\infty$. Answer is $e^{-\infty} = 0$, as before (no indeterminancy in this approach either).

(f) 1^∞ (indet). Let $y = (1 + x)^{1/x}$; then $\ln y = (1/x)\ln(1 + x)$,

$\lim_{x\to 0+} \ln y = \lim_{x\to 0+} \dfrac{\ln(1 + x)}{x}$

$= \dfrac{0}{0} = \lim \dfrac{\dfrac{1}{1 + x}}{1}$ (L'Hôpital) $= 1$.

Answer is $e^1 = e$.

(g) $\infty^\infty = \infty$

(h) $\infty \times (e^0 - 1) = \infty \times 0$. Let $u = 1/x$. Then $u \to 0+$, and $\lim_{u\to 0+}(e^u - 1)/u = \frac{0}{0}$ $= \lim_{u\to 0+} e^u/1$ (L'Hôpital) $= 1$.

(i) $(0+)^2 = 0$

(j) 1^∞ (indet). Let $y = (e^x + 4x)^{2/x}$; then $\ln y = (2/x)\ln(e^x + 4x)$,

$\lim_{x\to 0+} \ln y = \lim_{x\to 0+} \dfrac{2\ln(e^x + 4x)}{x} = \dfrac{0}{0}$

$= \lim \dfrac{\dfrac{2}{e^x + 4x}\cdot(e^x + 4)}{1}$ (L'Hôpital)

$= \lim_{x\to 0+} \dfrac{2(e^x + 4)}{e^x + 4x} = 10$. Answer is e^{10}.

Section 4.5 (page 115)

Each problem in this section has a diagram.

1. Parabola, opening down. Turning point when derivative is 0; $f'(x) = -2x + 4$, zero if $x = 2$.

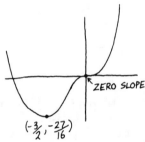

PROBLEM 1

2. $f'(x) = 4x^3 + 6x^2$, zero if $2x^2(2x + 3) = 0$, $x = 0$, $-\frac{3}{2}$; $f''(x) = 12x^2 + 12x$, $f''(0) = 0$ (2nd derivative test is inconclusive), $f''(-\frac{3}{2}) = 9$, pos, rel min. On $(-\frac{3}{2}, 0)$, f' is pos and on $(0, \infty)$, f' is pos so no rel extrema at $x = 0$; $\lim_{x \to \infty} f(x) = \infty$ and $\lim_{x \to -\infty} f(x) = \infty$ (x^4 term dominates).

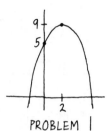

ZERO SLOPE

$(-\frac{3}{2}, -\frac{27}{16})$

PROBLEM 2

3. f is defined for $x \geq 0$ only; $f'(x) = \frac{3}{2}x^{1/2}$, zero only if $x = 0$, otherwise positive so f increases; $\lim_{x \to \infty} f(x) = \infty$. $f''(x) = \frac{3}{4}x^{-1/2}$, positive for all $x > 0$ so concave up.

4. $f(x) = \sqrt[3]{x^2}$. By inspection, graph of f falls until $x = 0$ and then rises; $f'(x) = \frac{2}{3}x^{-1/3} = 2/3\sqrt[3]{x}$, neg if $x < 0$, $-\infty$ if $x = 0-$, ∞ if $x = 0+$, pos if $x > 0$. So again, graph of f falls and then rises. At origin, left-hand slope is $-\infty$, right-hand slope is ∞; $f(\infty) = \infty$, $f(-\infty) = \infty$.

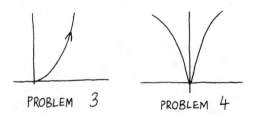

PROBLEM 3   PROBLEM 4

5. $f'(x) = 4x^3 + 3x^2 + 10x$, zero if $x(4x^2 + 3x + 10) = 0$. Equation $4x^2 + 3x + 10 = 0$ has no real roots so only sol is $x = 0$; f' is neg in $(-\infty, 0)$ and pos in $(0, \infty)$ so rel min at $x = 0$; $f(\infty) = \infty, f(-\infty) = \infty$ (x^4 dominates)

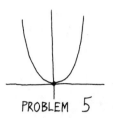

PROBLEM 5

6. Exponential curve; $f(\infty) = 2e^{-\infty} = 0$, $f(-\infty) = 2e^{\infty} = \infty$, $f(0) = 2$.

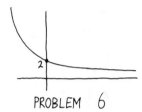

PROBLEM 6

7. Sine curve, period π. Plot a few points to pinpoint location.

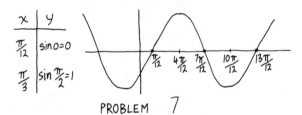

x	y
$\frac{\pi}{12}$	$\sin 0 = 0$
$\frac{\pi}{3}$	$\sin \frac{\pi}{2} = 1$

PROBLEM 7

8. f is defined for $-\sqrt{2} \leq x \leq \sqrt{2}$. $f(-\sqrt{2}) = 0$, $f(\sqrt{2}) = 0$. Use product rule, and simplify, to get $f'(x) = (2 - 2x^2)/\sqrt{2 - x^2}$, zero if $x = \pm 1$; f' is neg in $(-\sqrt{2}, -1)$, pos in $(-1, 1)$, neg in $(1, \sqrt{2})$. Graph falls, rises, falls.

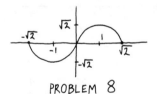

PROBLEM 8

9. $\lim_{x \to 0+} = 1/0+ = \infty$, $\lim_{x \to 0-} = 1/0- = -\infty$. Otherwise, fit cosine curve in envelope $y = \pm 1/x$, and reflect cosine when $1/x$ is neg, i.e., when $x < 0$ (see fig.).

10. $f'(x) = e^{-1/x} \cdot 1/x^2$,
$$f''(x) = e^{-1/x} \cdot \frac{-2}{x^3} + \frac{1}{x^2} \cdot e^{-1/x} \cdot \frac{1}{x^2} = e^{-1/x}\left(\frac{-2}{x^3} + \frac{1}{x^4}\right).$$
$f'(x) > 0$ for $x \neq 0$, graph of f rises on $(-\infty, 0)$ and rises in $(0, \infty)$; $f(0-) = e^{\infty} = \infty$, $f(0+) = e^{-\infty} = 0$, $f(\infty) = e^0 = 1$, $f(-\infty) = e^0 = 1$; $f''(x) = 0$ if $x = \frac{1}{2}$. In $(-\infty, 0)$, f'' pos and graph concave up; in $(0, \frac{1}{2})$, f'' is pos and graph is concave up; f'' is neg in $(\frac{1}{2}, \infty)$ and graph is concave down.

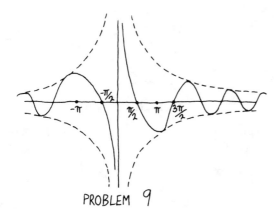

PROBLEM 9

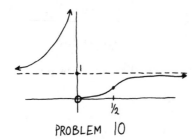

PROBLEM 10

11. $f'(x) = xe^x + e^x = e^x(x + 1)$, zero if $x = -1$; $f''(x) = e^x + e^x(x + 1) = e^x(x + 2)$; $f''(-1)$ is pos so rel min at $x = -1$; $f(\infty) = \infty$, $f(-\infty) = -\infty \times 0 = \lim_{x \to -\infty}(x/e^{-x}) = -\infty/\infty = \lim_{x \to -\infty}(1/-e^{-x})$ (L'Hôpital) $= \lim_{x \to -\infty}(-e^x) = 0-$.

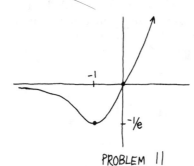

PROBLEM 11

12. $f'(x) = e^{-x}(2x - x^2)$, zero if $x = 0, 2$; pos if x in $(0, 2)$; neg if $x > 2$ or $x < 0$, so relative max at $x = 2$; $f(\infty) = \infty \times 0 = \lim(x^2/e^x) = 0$ since e^x has higher order of magnitude. $f(-\infty) = \infty \times \infty = \infty$.

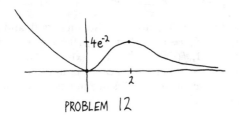

PROBLEM 12

13. $f'(x) = 1 + \ln x$, zero if $x = 1/e$; $f''(x) = 1/x$; $f''(1/e)$ is pos, rel min; $f(\infty) = \infty$, $f(0+) = 0 \times -\infty$
$$= \lim_{x \to 0+} \frac{\ln x}{1/x} = \frac{-\infty}{\infty} = \lim_{x \to 0+} \frac{1/x}{-1/x^2} \text{ (L'Hôpital)}$$
$$= \lim(-x) = 0.$$

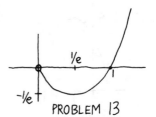

PROBLEM 13

14. *Method 1:* Draw $y = x$, $y = \ln x$ and subtract heights. *Method 2:* $f'(x) = 1 - 1/x$; zero if $x = 1$; $f''(x) = 1/x^2$, $f''(1)$ is pos so rel min; $f(\infty) = \infty$ (x term dominates), $f(0+) = 0 - (-\infty) = \infty$.

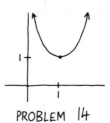

PROBLEM 14

15. $f'(x) = 2/(x + 1)^2$, pos for $x > -1$ and for $x < -1$. Curve rises on $(-\infty, -1)$, rises on $(-1, \infty)$; $f(\infty) = \lim(x/x) = 1$ (highest power rule), $f(-\infty) = 1$, $\lim_{x \to (-1)+} f(x) = -2/0+ = -\infty$, $\lim_{x \to (-1)-} f(x) = -2/0- = \infty$

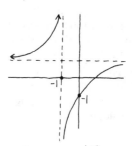

PROBLEM 15

16.

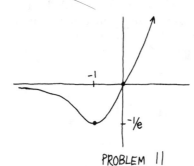

PROBLEM 16

17. Exponential curve.
$f(\infty) = -4, f(-\infty) = -\infty, f(0) = -5$

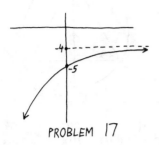

PROBLEM 17

18.

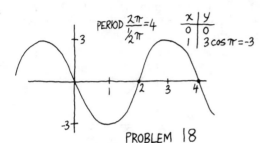

PERIOD $\dfrac{2\pi}{\frac{1}{2}\pi} = 4$

x	y
0	0
1	$3\cos\pi = -3$

PROBLEM 18

19. $f'(x) = e^x(x - 5)/x^6$, zero if $x = 5$; $f'(x)$ is neg on $(-\infty, 0)$, neg on $(0, 5)$, pos on $(5, \infty)$, rel min at $x = 5$. $f(\infty) = \infty$ since e^x has higher order of mag; $f(-\infty) = 0/-\infty = 0$; $f(0+) = 1/0+ = \infty, f(0-) = 1/0- = -\infty$.

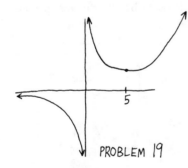

PROBLEM 19

20. $f'(x) = -2xe^{-x^2}$, zero if $x = 0$, pos if $x < 0$, neg if $x > 0$, rel max at $x = 0$; $f(\infty) = e^{-\infty} = 0, f(-\infty) = 0$.

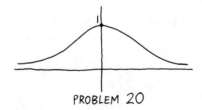

PROBLEM 20

21. *Method 1:* Draw line $y = x$ and hyperbola $y = 1/x$ and add heights.
Method 2: $f'(x) = 1 - 1/x^2$, zero if $x = \pm 1$. $f''(x) = 2/x^3$; $f''(1)$ is pos, $f''(-1)$ is neg. So rel min at point $(1, 2)$ and rel

max at $(-1, -2)$; $f(\infty) = \infty + 0 = \infty$, $f(-\infty) = -\infty + 0 = -\infty$, $f(0+) = 0 + \infty = \infty$, $f(0-) = 0 + (-\infty) = -\infty$. Results agree with method 1.

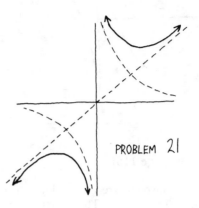

PROBLEM 21

22. $f(\infty) = 0, f(-\infty) = 0$. By inspection, f has max value of 4, when $x = 0$.

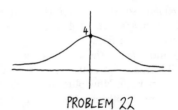

PROBLEM 22

23. (a) f is defined for $x > 0$; $f(0+) = -\infty/0+ = -\infty$, $f(\infty) = 0$ since x has higher order of magnitude than $\ln x$; $f'(x) = (1 - \ln x)/x^2$, zero if $x = e$, pos if $0 < x < e$, neg if $x > e$, rel max at $x = e$.

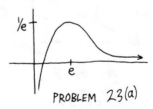

PROBLEM 23(a)

(b) The graph of $(\ln|x|)/x$ agrees with the graph of $(\ln x)/x$ for $x > 0$; i.e., the right half of $(\ln|x|)/x$ is the same as the graph in (a). But now there is a left half as well. Consider say $x = -5$. Then $\dfrac{\ln|x|}{x} = \dfrac{\ln 5}{-5} = -\dfrac{\ln 5}{5}$ which is the opposite of the height in the part (a) graph at $x = 5$. So left half of graph in (b) is obtained by reflecting the graph of part (a) as indicated.

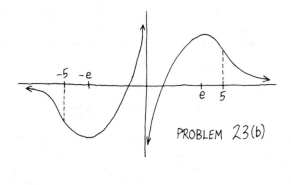

PROBLEM 23(b)

Section 4.6 (page 118)

1. Functions involved are volume $V(t)$ and radius $r(t)$, related by $V(t) = \frac{4}{3}\pi[r(t)]^3$. Then $V'(t) = 4\pi r^2 r'$. Given $V' = -10$ (*neg* because melting means *decreasing* volume). When $r = 2$, we have $-10 = 4\pi \cdot 4 \cdot r'$, $r' = -5/8\pi$. At this instant, radius is decreasing by $5/8\pi$ feet per sec.

2. $A(t) = b(t)h(t)$ so $A'(t) = b(t)h'(t) + h(t)b'(t)$ (product rule). If $b = 6, h = 8, b' = 4, h' = -3$ then $A' = 14$. Area is growing by 14 ft²/sec.

3. Let $s(t)$ and $x(t)$ be the distances indicated in the diagram. Then $s^2 = 8100 + x^2$ so (diff w.r.t. t) $2ss' = 2xx'$, $s' = xx'/s$. Now plug in specific data for instant when runner is 30 feet down the line. Set $x = 30$, $s = 30\sqrt{10}$, $x'(t) = 25$ (positive since x is increasing) to get $s' = (30)(25)/30\sqrt{10} = 25/\sqrt{10}$. Runner moves away from home plate at rate of $25/\sqrt{10}$ ft/sec.

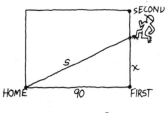

PROBLEM 3

4. Let $h(t)$ be the water level at time t, $V(t)$ the corresponding volume of water. $V(t) = \pi r^2 h(t) = 16\pi h(t)$, $V'(t) = 16\pi h'(t)$. So $h'(t) = V'(t)/16\pi = 8/16\pi$. At every instant of time, the height is increasing by $1/2\pi$ ft per min.

5. If PQR is equilateral, center of circle is A, radius is r (see fig.), then $\overline{AP} = r$, PBA is a 30°, 60°, 90° triangle, $\overline{AB} = \frac{1}{2}r$, $\overline{PB} = \frac{1}{2}r\sqrt{3}$, altitude $\overline{RB} = r + \frac{1}{2}r = \frac{3}{2}r$, base $\overline{PQ} = r\sqrt{3}$, area $A(t) = \frac{1}{2}bh = \frac{3}{4}\sqrt{3}[r(t)]^2$, $A'(t) = \frac{3}{2}\sqrt{3}\,rr' = \frac{9}{2}\sqrt{3}\,r$ since we are given $r' = 3$. If $r = 4$, then $A' = 18\sqrt{3}$. Triangle's area is increasing at this instant by $18\sqrt{3}$ ft²/sec.

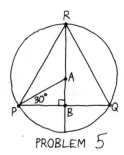

PROBLEM 5

6. In figure, $x(t) = 5\tan\theta(t)$, $x'(t) = 5\sec^2\theta(t) \cdot \theta'(t) = 10\pi\sec^2\theta$ since we are given $\theta'(t) = 2\pi$ radians/min. If $x = 12$ then $\overline{BC} = 13$, $\sec\theta = \frac{13}{5}$, $x' = 338\pi/5$ miles per min, speed of the spot of light.

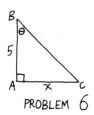

PROBLEM 6

7. In figure, if $h(t)$ is height of water at time t then radius of the water-cone is $h/4$ by similar triangles, $V = \frac{1}{3}\pi r^2 h = \frac{1}{48}\pi h^3$, $V'(t) = \frac{1}{16}\pi h^2(t)h'(t)$, $h'(t) = 48/\pi h^2$ since $V' = 3$ is given. If $h = 2$ then $h' = 12/\pi$. Water level is rising by $12/\pi$ meters per min.

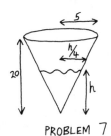

PROBLEM 7

8. In figure, by similar triangles

$$\frac{y}{15} = \frac{y-x}{6}, \quad 6y = 15y - 15x, \quad y = \frac{5}{3}x, \quad y'(t) = \frac{5}{3}x'(t) = 5$$

since $x' = 3$ by hypothesis. Since y is increasing by 5 ft/sec, speed of shadow's head is 5 ft/sec.

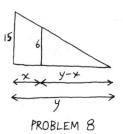

PROBLEM 8

9. In figure, let $h(t)$ be water level at time t. By similar triangles, the corresponding radius of the water-cone is $\frac{1}{2}h$. Then

$$V(t) = \frac{1}{3}\pi r^2 h = \frac{1}{12}\pi [h(t)]^3, \qquad V'(t) = \frac{1}{4}\pi h^2 h'.$$

(a) $V' = -10$ (*negative* because leak makes water volume *decrease*). If $h = 3$ then
$h' = (4)(-10)/9\pi = -40/9\pi$.
Water level is dropping by $40/9\pi$ cm per min.

(b) If $h = 6$ and $h' = -2$ then $V' = -18\pi$. Volume is decreasing by 18π cubic cm/min, leak is 18π cubic cm/min.

(c) If $h = 2$ then $r = 1$, exposed area $= \pi r^2 = \pi$, $V' = -\sqrt{\pi}$ (*negative* because evaporation makes V *decrease*), $h' = 4V'/\pi h^2 = -1/\sqrt{\pi}$. Water level is dropping by $1/\sqrt{\pi}$ cm per min.

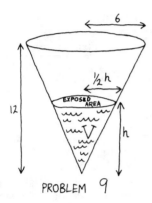

PROBLEM 9

10. $A(t) = \pi[r(t)]^2$, $A'(t) = 2\pi r r' = 4\pi r$ since $r' = 2$. If $r = 5$ then $A' = 20\pi$, disturbed area is growing by 20π square m/sec.

11. Let r be inner radius, R the outer radius. Then
$A(t) = \pi[R(t)]^2 - \pi[r(t)]^2$,
$A'(t) = 2\pi R R' - 2\pi r r' = 4\pi R - 8\pi r$ since
$r' = 4$, $R' = 2$. If $r = 5$, $R = 9$ then $A' = 36\pi - 40\pi = -4\pi$. Area is decreasing at this instant by 4π m²/sec.

12. In figure, $A(t) = \frac{1}{2}x(t)y(t)$, $A'(t) = \frac{1}{2}xy' + \frac{1}{2}x'y$ (prod rule) $= 3x - 2y$ since $y' = 6$ (*positive* since y is *increasing*), $x' = -4$ (*negative* since x is *decreasing*). If $y = 12$, $x = 10$ then $A = 30 - 24 = 6$. Area is increasing by 6 square meters/sec.

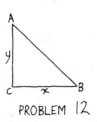

PROBLEM 12

13. Let $r(t)$ be radius of ice-coated sphere. Then $V(t) = \frac{4}{3}\pi[r(t)]^3$, $V'(t) = 4\pi r^2 r'$. Surface area is $4\pi r^2$, and melting rate of ice is proportional to surface area. So $V' = -k(4\pi r^2)$ for some positive constant k (V' is *negative* because ice is *melting*), $r' = V'/4\pi r^2 = -k(4\pi r^2)/4\pi r^2 = -k$. Radius is decreasing at the *constant* rate of k volume units per time unit. So thickness of ice is decreasing at a *constant* rate.

14. In figure, $x^2(t) = y^2(t) - 900$, $2xx' = 2yy'$, $x' = yy'/x = -2y/x$ since $y' = -2$. If $y = 50$ then $x = 40$ and $x' = -\frac{5}{2}$ (appropriately *neg* since fish is moving toward the dock and x is *decreasing*). Speed of fish is $\frac{5}{2}$ m/sec. If $y = 31$ then $x = \sqrt{61}$, $x' = -62/\sqrt{61}$. Fish speed is now $62/\sqrt{61}$ m per sec (faster than before).

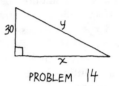

PROBLEM 14

15. Remember that if y is a function of t then derivative of $1/y$ with respect to t is $(-1/y^2) \cdot y'$ by the chain rule. Differentiate w.r.t. t to get $-\dfrac{1}{R^2}R' = -\dfrac{1}{R_1^2}R_1' - \dfrac{1}{R_2^2}R_2'$,

$R' = R^2\left(\dfrac{2}{R_1^2} + \dfrac{-3}{R_2^2}\right)$ since $R_1' = 2$, $R_2' = -3$. If $R_1 = 10$, $R_2 = 20$ then $1/R = \frac{1}{10} + \frac{1}{20} = \frac{3}{20}$, $R = \frac{20}{3}$,

$R' = \dfrac{400}{9}\left(\dfrac{2}{100} - \dfrac{3}{400}\right) = \dfrac{5}{9}$; R is increasing at the moment by $\frac{5}{9}$ ohms/min.

Section 4.7 (page 122)

These problems were solved using a TI-55 calculator.

1. Let $f(x) = x^2 - 39$. Then $f'(x) = 2x$,

new x = old $x - \dfrac{(\text{old})^2 - 39}{2 \times \text{old}}$. If old $x = 6$ then

new $x = 6 - \dfrac{36 - 39}{12} = 6.25$. If old $x = 6.25$ then

new $x = 6.25 - \dfrac{(6.25)^2 - 39}{12.5} = 6.245$. If old $x = 6.245$

then new $x = 6.244998$. The last two approximations agree on two decimal places. Take $\sqrt{39}$ to be 6.24. To check on accuracy, $f(6.24) < 0$, while $f(6.245) > 0$ so there is a root between 6.24 and 6.245. The decimal places 6.24 are correct.

2. We want to solve equ $x^3 = 173$. Let $f(x) = x^3 - 173$.

Then $f'(x) = 3x^2$, new x = old $x - \dfrac{(\text{old})^3 - 173}{3(\text{old})^2}$.

Starting with old $x = 5.5$ as first guess (since 173 is between 5^3 and 6^3) we get successive approximations 5.5730028, 5.5720548, 5.5720547. Have agreement in six places, so stop and take 5.572054 as the approximation. To check accuracy, note that $f(5.572054) < 0$, $f(5.5720548) > 0$ so there is a root between 5.572054 and 5.5720548. Our six places are correct.

3. Figure shows two solutions to $e^x = 3 - x^2$, namely x_0 and x_1. Let $f(x) = e^x + x^2 - 3$. Then $f'(x) = e^x + 2x$, new $x = $ old $- \dfrac{e^{\text{old}} + (\text{old})^2 - 3}{e^{\text{old}} + 2(\text{old})}$. Starting with old $x = 1$ as first guess, successive approximations are .8477662, .8345815, .8344869. Take .834 as an approximate solution. As accuracy check, $f(.834) < 0$. $f(.8345) > 0$ so there is a solution between .834 and .8345. The three places .834 are accurate.

Starting with old $x = -2$, successive approximations are $-1.7062267, -1.6775167, -1.6772327$. Approximate solution is -1.677. As accuracy check, $f(-1.677) < 0$, $f(-1.6773) > 0$. Solution is between -1.6773 and -1.677. Our 3 places are accurate.

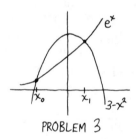

PROBLEM 3

4. (a) The graphs of $y = x$ and $y = \tan x$ (see fig.) do not intersect in interval $(0, \pi/2)$ so there is no solution.

   (b) See intersection at a point where $x = x_0$, the desired solution. To find x_0 make a first guess near

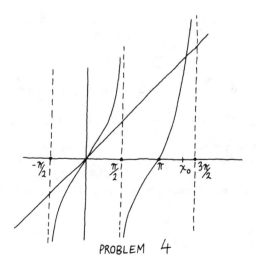

PROBLEM 4

$3\pi/2$, say old $x = 4.5$. Let $f(x) = x - \tan x$. Then $f'(x) = 1 - \sec^2 x$ and new $x = $

$$\text{old} - \frac{\text{old} - \tan(\text{old})}{1 - \sec^2(\text{old})} = \text{old} - \frac{\text{old} - \tan(\text{old})}{-\tan^2(\text{old})}.$$

Next approximations after 4.5 are 4.4936139, 4.4934097. Have 3 place agreement so take solution to be approximately 4.493. To test accuracy, note that $f(4.493) > 0$ and $f(4.4935) < 0$ so root lies between 4.493 and 4.4935. Our three places are accurate.

Section 4.8 (page 127)

1. (a) $dx/2\sqrt{x}$

   (b) $-\sin x\,dx$

   (c) $x^5\,d(\sin x) + \sin x\,d(x^5)$
      $= x^5 \cos x\,dx + 5x^4 \sin x\,dx$

   (d) $\dfrac{x\,d(\sin x) - \sin x\,d(x)}{x^2} = \dfrac{x \cos x\,dx - \sin x\,dx}{x^2}$

   (e) $\cos x^5\,d(x^5) = 5x^4 \cos x^5\,dx$

   (f) 0

2. $6x^2\,dx$

3. $df = dx$

4. (a) $d(x^3 + x^2) = (3x^2 + 2x)\,dx$. Set $x = 3$, $dx = -.0001$ to get $d(x^3 + x^2) = -.0033$.

   (b) $d(x^{1/4}) = \frac{1}{4}x^{-3/4}\,dx$. Set $x = 16$, $dx = .1$ to get $d(x^{1/4}) = \frac{1}{4}(16)^{-3/4} \times .1 = (\frac{1}{4})(\frac{1}{8})(.1) = 1/320$.

5. (a) By trigonometry (see fig.), $\overline{AE} = r\sqrt{3}$, $\overline{CD} = \overline{AD} = 2r$, base $= 2r\sqrt{3}$, height $= 3r$, $A = \frac{1}{2}bh = 3r^2\sqrt{3}$. If r changes by dr then $dA = 6r\sqrt{3}\,dr$, the area of the triangular shell.

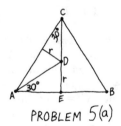

PROBLEM 5(a)

   (b) $V = \frac{1}{3}\pi r^2 h$. If h stays fixed and r changes by dr then $dV = \frac{2}{3}\pi r h\,dr$, the shell volume.

Section 4.9 (page 134)

1. (a) $\cos y\,dy = -x\,dx$,
      $\sin y = -\frac{1}{2}x^2 + C$ (implicit solution)

(b) $y\,dy = -dx/x^3, \frac{1}{2}y^2 = 1/2x^2 + C,$
$y = \pm\sqrt{1/x^2 + D}$

(c) $y^4\,dy = -x^2\,dx, \frac{1}{5}y^5 = -\frac{1}{3}x^3 + C,$
$y = \sqrt[5]{-\frac{5}{3}x^3 + D}$

(d) $dy/y = dx/(2x + 3), \ln Ky = \frac{1}{2}\ln(2x + 3) = \ln\sqrt{2x + 3}, Ky = \sqrt{2x + 3}, y = A\sqrt{2x + 3}$

(e) $e^{-y}\,dy = dx/x^2, -e^{-y} = -1/x + C, e^{-y} = 1/x + D, -y = \ln(1/x + D), y = -\ln(1/x + D)$

(f) $y\,dy = (5x + 3)\,dx, \frac{1}{2}y^2 = \frac{5}{2}x^2 + 3x + C,$
$y = \pm\sqrt{5x^2 + 6x + D}$

2. (a) $dy/y = x\,dx, \ln Ky = \frac{1}{2}x^2, Ky = e^{x^2/2}, y = Ae^{x^2/2},$
$3 = Ae^{1/2}, A = 3e^{-1/2},$
$y = 3e^{-1/2}e^{x^2/2} = 3e^{(x^2-1)/2}$

(b) $y\,dy = (3 - 5x)\,dx, \frac{1}{2}y^2 = 3x - \frac{5}{2}x^2 + C.$ Set $x = 2, y = 4$ to get $C = 12.$ Then $\frac{1}{2}y^2 = 3x - \frac{5}{2}x^2 + 12, y = \sqrt{6x - 5x^2 + 24}.$ (Choose the *positive* square root since y is *positive* when $x = 2$.)

(c) $e^y\,dy = 3x\,dx, e^y = \frac{3}{2}x^2 + C.$ Set $x = 0, y = 2$ to get $C = e^2.$ Then $e^y = \frac{3}{2}x^2 + e^2,$
$y = \ln(\frac{3}{2}x^2 + e^2).$

(d) $dy/y^4 = \cos x\,dx, -1/3y^3 = \sin x + C.$ Set $x = 0, y = 2$ to get $C = -1/24, y = -1/\sqrt[3]{3\sin x - \frac{1}{8}}.$

3. (a) $dy/y = 2\,dx/x, \ln Ky = 2\ln x = \ln x^2, Ky = x^2, y = Ax^2.$ (See fig.)

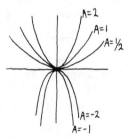

PROBLEM 3(a)

(b) $3 = 4A, A = 3/4, y = \frac{3}{4}x^2$

4. (a) Differentiate w.r.t. x; $2x + 4yy' = 0, y' = -x/2y.$ For orthog family, $y' = 2y/x, dy/y = 2\,dx/x, \ln Ky = 2\ln x = \ln x^2, Ky = x^2, y = Ax^2,$ a family of parabolas. (See fig.)

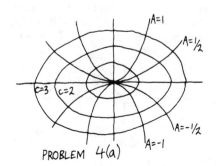

PROBLEM 4(a)

(b) Isolate the C first so that it differentiates away: $ye^{3x} = C.$ Then differentiate w.r.t. x to get $y \cdot 3e^{3x} + y'e^{3x} = 0, y' = -3y.$ For orthog family, $y' = 1/3y, y\,dy = \frac{1}{3}dx, \frac{1}{2}y^2 = \frac{1}{3}x + C,$
$x = \frac{3}{2}y^2 + D,$
a family of parabolas. (See fig.)

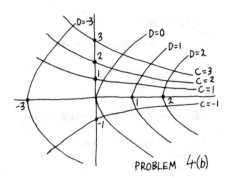

PROBLEM 4(b)

(c) Using differential notation for variety, $4x\,dx - 2y\,dy = 0.$ For orthog family $4x\,dy + 2y\,dx = 0,$
$dy/y = -dx/2x, \ln Ky = -\frac{1}{2}\ln x = \ln x^{-1/2}, Ky = 1/\sqrt{x}, x = A/y^2.$ The original is a family of hyperbolas, all with asymptotes $y = \pm x\sqrt{2}.$ For the graph of the orthog family note that $\lim_{y\to\infty} 1/y^2 = 0, \lim_{y\to-\infty} 1/y^2 = 0, \lim_{y\to 0} 1/y^2 = 1/0+ = \infty.$ (See fig.)

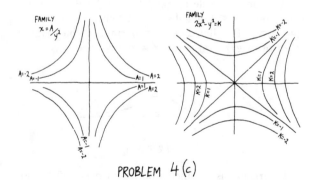

PROBLEM 4(c)

5. (a) $y'(t) = -\frac{1}{10}y(t), dy/y = -dt/10, \ln Ky = -\frac{1}{10}t, Ky = e^{-t/10}, y = Ae^{-t/10}$

(b) Set $t = 0, y = 75$ to get $A = 75.$ Solution is $y = 75e^{-t/10}.$

(c) Note that in the general solution in part (a), if $t = 0$ then $y = A,$ so the constant A represents the initial amount. For half-life set $y = \frac{1}{2}A$ and solve for t; $\frac{1}{2}A = Ae^{-t/10}, \frac{1}{2} = e^{-t/10}, \ln\frac{1}{2} = -t/10, t = 10\ln 2,$ for any initial amount $A.$

6. $m'(t) = \frac{1}{2}m(t), dm/m = \frac{1}{2}dt, \ln Km = \frac{1}{2}t, Km = e^{t/2}, m = Ae^{t/2}.$ Set $t = 0, m = 2$ to get $A = 2, m(t) = 2e^{t/2}.$ Then if $t = 3$ we have $m = 2e^{3/2}.$

7. $m\dfrac{dv}{dt} = mg - cv$, $\dfrac{dv}{cv - mg} = -\dfrac{dt}{m}$,

$\dfrac{1}{c}\ln K(cv - mg) = -\dfrac{t}{m}$, $K(cv - mg) = e^{-ct/m}$,

$cv - mg = Ae^{-ct/m}$, $v = \dfrac{mg}{c} + \dfrac{A}{c}e^{-ct/m}$. Set $t = 0$, $v = 0$

to get $A = -mg$. Solution is $v = \dfrac{mg}{c}(1 - e^{-ct/m})$.

Set $t = \infty$ to get steady state velocity mg/c.

Chapter 4 Review Problems (page 134)

1. $PV = kT$, $V = kT/P$. Differentiate w.r.t. time t; $V'(t) = k\left(\dfrac{PT' - TP'}{P^2}\right)$. If $T = 20$, $V = 10$ then $P = 20k/10 = 2k$. With $P' = -2$, $T' = 3$ we have

$V' = k\left(\dfrac{6k + 40}{4k^2}\right) = \dfrac{3k + 20}{2k}$, positive because $k > 0$.

V is increasing by $(3k + 20)/2k$ volume units per sec.

2. (a) $\dfrac{\infty}{\infty} = \lim \dfrac{\frac{1}{\ln x}\cdot\frac{1}{x}}{\frac{1}{x}}$ (L'Hôpital) $= \lim \dfrac{1}{\ln x} = \dfrac{1}{\infty} = 0$

(b) $\ln 0+/\ln 1+ = -\infty/0+ = -\infty$

3. Use product rule to get $f'(x) = e^{-x}(1 - x)$, zero if $x = 1$, pos if $x < 1$, neg if $x > 1$. Rel max at $x = 1$. Also, $f(\infty) = \infty \times 0 = \lim_{x\to\infty} x/e^x = 0$ since e^x has higher order of mag, $f(-\infty) = -\infty \times \infty = -\infty$. (See fig.)

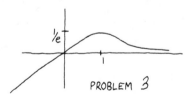

PROBLEM 3

4. Clearly min is $-\infty$ since products such as -90×100, -990×1000, etc. can get unboundedly low. To search for max product let the numbers be x and $10 - x$. Then product $p = x(10 - x) = -x^2 + 10x$; $p'(x) = -2x + 10$, zero if $x = 5$. Get max product from factors 5 and 5.

5. $x\,d(e^{2x}) + e^{2x}\,d(x) = 2xe^{2x}\,dx + e^{2x}\,dx$

6. (a) $\ln x^2 = 2\ln x$, a multiple of $\ln x$. Same order of magnitude.

(b) $\lim_{x\to\infty} e^{x^2}/e^x = \infty/\infty = 2xe^{x^2}/e^x$ (L'Hôpital). Still ∞/∞ and getting more complicated. Instead, $\lim_{x\to\infty} e^{x^2}/e^x = \lim_{x\to\infty} e^{x^2-x} = e^\infty$ (since x^2 has higher order of mag than x) $= \infty$. So e^{x^2} has higher order of magnitude.

7. $V(t) = [e(t)]^3$, $V'(t) = 3e^2e'$. If $e = 3$, $e' = 2$ then $V' = 54$. Volume is increasing at the moment by 54 cubic meters/sec.

8. (a)

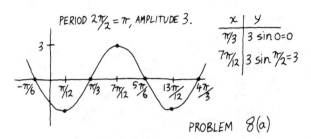

PERIOD $2\pi/2 = \pi$, AMPLITUDE 3.

x	y
$\pi/3$	$3\sin 0 = 0$
$7\pi/12$	$3\sin \pi/2 = 3$

PROBLEM 8(a)

(b) Exponential curve with $f(\infty) = 2 + 0 = 2$, $f(-\infty) = 2 + \infty = \infty$, $f(0) = 2 + 5 = 7$. (See fig.)

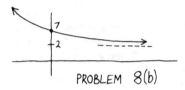

PROBLEM 8(b)

9. (a) $1 \times -\infty = -\infty$

(b) indeterminate 0^0. Let $y = x^{\tan x}$. Then $\ln y = \tan x \ln x$, $\lim_{x\to 0+} \ln y = \lim_{x\to 0+} \tan x \ln x = 0 \times -\infty$ (indet) $= \cdots = 0$ (see Prob. 5(a), Section 4.4). Final answer is $e^0 = 1$.

10. Let diagonal be d, one side x. (See fig.) Other side is $\sqrt{d^2 - x^2}$; area $A(x) = x\sqrt{d^2 - x^2}$ where $0 \le x \le d$. If $x = 0$ or d, rectangle collapses to segment with area 0, a minimum. Maximum will be at critical point. $A'(x) = x \cdot \frac{1}{2}(d^2 - x^2)^{-1/2} \cdot -2x + \sqrt{d^2 - x^2}$ $= (d^2 - 2x^2)/\sqrt{d^2 - x^2}$, zero if $x = \pm d/\sqrt{2}$. Ignore neg value. Rectangle with max area has sides $x = d/\sqrt{2}$ and $\sqrt{d^2 - x^2} = \sqrt{d^2 - \frac{1}{2}d^2} = d/\sqrt{2}$, a square.

PROBLEM 10

11. By quotient rule, $y' = -2(x^2 - 1)/(x^2 + 1)^2$, zero if $x = \pm 1$, neg in $(-\infty, -1)$, pos in $(-1, 1)$, neg in $(1, \infty)$. Min at $x = -1$, max at $x = 1$; $\lim_{x\to\infty} y = 0$ (x^2 has higher order of mag). Similarly $\lim_{x\to-\infty} y = 0$. (See fig.)

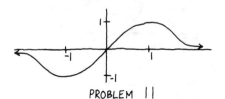

PROBLEM 11

12. (a) $f'(x) = 4\sin^3 x \cos x$, zero if $x = \cdots, -\pi/2, 0, \pi/2, \pi, 3\pi/2, \cdots$; f' is neg in $(-\pi/2, 0)$, pos in

$(0, \pi/2)$, neg in $(\pi/2, \pi)$, pos in $(\pi, 3\pi/2), \cdots$. By first derivative test, min at $x = 0$, max at $\pi/2$, min at π, etc.
$f''(x) = 4 \sin^3 x \cdot - \sin x + \cos x \cdot 12 \sin^2 x \cos x$
$= -4 \sin^4 x + 12 \cos^2 x \sin^2 x$.
$f''(0) = 0$, $f''(\pi/2) = -4$, $f''(\pi) = 0, \cdots$. Second deriv test is inconclusive about critical numbers $0, \pi, \cdots$, shows max at $\pi/2, 3\pi/2$, etc.

By inspection, $\sin^4 x \geq 0$ and has min when $\sin^4 x = 0$, $\sin x = 0$, $x = 0, \pi, 2\pi, \cdots$. Sines are between -1 and 1 so $\sin^4 x$ has max value of 1 when $\sin x = \pm 1$, namely at $x = \pm\pi/2, \pm3\pi/2$, etc.

(b) $f'(x) = 2(x + 2)$, zero if $x = -2$; f' is neg if $x < -2$, pos if $x > -2$. Rel min at $x = -2$ by first deriv test; $f''(x) = 2$, $f''(-2) = 2$, pos, so rel min at $x = -2$ by second derivative test.

By inspection, $(x + 2)^2$ is always ≥ 0 and is smallest when it *is* 0, namely when $x = -2$. So $(x + 2)^2 + 1$ has min value of 1 when $x = -2$.

13. $dy/y = dt/t^2$, $\ln Ky = -1/t$, $Ky = e^{-1/t}$, $y = Ae^{-1/t}$. Set $t = \infty$, $y = 2$ to get $2 = Ae^0$, $A = 2$, $y = 2e^{-1/t}$.

14. Let $\overline{AD} = x$. Then $100 - 5x$ is left for $\overline{AB}$ and $\overline{DC}$, $\overline{AB} = \frac{1}{2}(100 - 5x)$,
area $A = x \times \frac{1}{2}(100 - 5x) = 50x - \frac{5}{2}x^2$
where $0 \leq x \leq 20$. If $x = 0$ or 20, plot collapses and has zero area, min. Will find max area at critical number. $A'(x) = 50 - 5x$, zero if $x = 10$. Other dimension is 25. For max area make outer rectangle 10×25, (with 3 fences of length 10 to subdivide into the four smaller rectangles).

15. f is defined on $(0, 1)$; $f'(x) =$
$x \cdot \dfrac{1}{x} + \ln x + (1 - x) \cdot \dfrac{1}{1 - x} \cdot -1 + \ln(1 - x) \cdot -1$
$= \ln x - \ln(1 - x)$.
Zero if $\ln x = \ln(1 - x)$, $x = 1 - x$, $x = \frac{1}{2}$. Candidates are $f(0+), f(1-), f(\frac{1}{2})$; $\lim_{x \to 0+} x \ln x$ is indet form $0 \times \infty$; ans is 0 (see § 4.4, (1)–(3)). So $f(0+) = 0 + 1 \times 0 = 0$; $\lim_{x \to 1-}(1 - x)\ln(1 - x)$ is indeterminate $0 \times \infty$. Let $u = 1 - x$ to get $\lim_{u \to 0+} u \ln u$ which is 0, as above. So $f(1-) = 1 \times 0 + 0 = 0$; $f(\frac{1}{2}) = \frac{1}{2}\ln\frac{1}{2} + \frac{1}{2}\ln\frac{1}{2} = \ln\frac{1}{2} = -\ln 2$, neg since $\ln 2$ is pos. Min is $-\ln 2$, max is 0.

16. (a) $f'(x) = 3x^2 - 4x + 3$. f' never jumps and is never 0 (the equation $3x^2 - 4x + 3 = 0$ has no real roots since $b^2 - 4ac$ is neg). So f' has only one sign. $f'(0)$ is pos so f' is positive for all x, and f is an increasing function.

(b) $f(-\infty) = -\infty$, $f(\infty) = \infty$, f increases, so graph of f crosses the x-axis only once; equation has only one root.

(c) $f(1)$ is neg, $f(2)$ is pos; root lies between $x = 1$ and $x = 2$. One sensible starting x is 1.5.

(d) If first guess is $x = 1.5$ then next approximations are 1.6666667, 1.6507937, 1.6506292. Choose approx solution 1.650. For check on accuracy, $f(1.650) < 0$, $f(1.6507) > 0$. Root lies between 1.650 and 1.6507. Newton's method produces three accurate places.

5/THE INTEGRAL   PART I

Section 5.2 (page 145)
In Problems 1–3, each part has a diagram.

1. (a) $\int_{-1}^{4} 6\,dx =$ area under graph $= 5 \times 6 = 30$
(b) $\int_{-1}^{3} x\,dx =$ area II $-$ area I $= \frac{9}{2} - \frac{1}{2} = 4$
(c) $\int_{-2}^{2} x^3\,dx =$ II $-$ I $= 0$

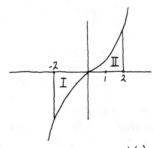

PROBLEM 1(c)

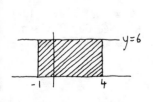

PROBLEM 1(a)

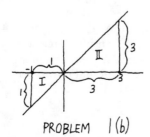

PROBLEM 1(b)

2. (a) $\int_{1}^{5} \ln x\,dx$
(b) $-\int_{1/2}^{1} \ln x\,dx$
(c) $\int_{1}^{7} \ln x\,dx - \int_{1/3}^{1} \ln x\,dx$

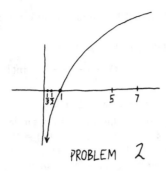

PROBLEM 2

3. (a) $\int_0^3 x^2\,dx = $ II, $\int_{-1}^3 x^2\,dx = $ I + II (larger)

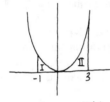

PROBLEM 3(a)

(b) $\int_0^3 x^3\,dx = A$, $\int_{-1}^3 x^3\,dx = A - B$ (smaller)
(c) $\int_{-1}^0 x^3\,dx = -B$, $\int_{-2}^0 x^3\,dx = -(C + B)$ (smaller)

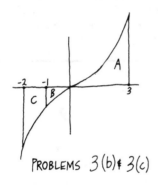

PROBLEMS 3(b) & 3(c)

4. (a) More area below than above. Negative.
   (b) $\cos^2 x$ is always ≥ 0, graph lies above x-axis. Integral must be positive.
5. (a) True. (See Fig. 5 where graph lies below x-axis, integral = $-$area = neg.)
   (b) False. See diagram in problem 1(c) where $\int_{-2}^1 x^3\,dx < 0$ (more area below than above) but x^3 is not < 0 throughout $[-2, 1]$.
   (c) True. If graph of f lies below graph of g then "f area" is $\leq$ "g area". Alternatively if $f(x) \leq g(x)$ then $\Sigma f(x)\,dx \leq \Sigma g(x)\,dx$.
6. (a) Area under $\cos^2 x$ for $0 \leq x \leq 2\pi$ equals area under $\sin^2 x$ so integrals are equal. (See fig.)

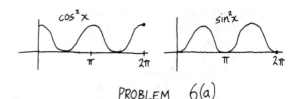

PROBLEM 6(a)

(b) $\int_0^{2\pi} \sin^2 x\,dx = \int_0^{2\pi}(1 - \cos^2 x)\,dx$
   $= \int_0^{2\pi} 1\,dx - \int_0^{2\pi} \cos^2 x\,dx$. But
   $\int_0^{2\pi} dx = $ area of rect (ht 1, base 2π) $= 2\pi$,
   and $\int_0^{2\pi} \cos^2 x\,dx = \int_0^{2\pi} \sin^2 x\,dx$ by (a).
   So $\int_0^{2\pi} \sin^2 x\,dx = 2\pi - \int_0^{2\pi} \sin^2 x\,dx$,
   $2\int_0^{2\pi} \sin^2 x\,dx = 2\pi$, $\int_0^{2\pi} \sin^2 x\,dx = \pi$.
7. $A_1 = $ area under graph of f between $x = a$ and $x = b$.
   (a) (See fig.) Will get same area if graph of f and interval $[a, b]$ are both translated similarly. In A_2, interval is translated right 3, graph doesn't move; in A_3, interval moves right 3, graph moves left 3; in A_4, interval and graph both move right 3. So $A_1 = A_4$.

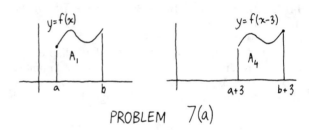

PROBLEM 7(a)

(b) (See fig.) In A_5, both graph and interval have contracted horizontally by a factor of 2. New area is half the old area, so $A_5 = \frac{1}{2}A_1$.

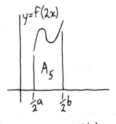

PROBLEM 7(b)

8. (a) 10 (change in dummy variable doesn't change value)
   (b) $\int_a^b 4x^3\,dx = 10$ so $4\int_a^b x^3\,dx = 10$, $\int_a^b x^3\,dx = 10/4$
9. Consider circle $x^2 + y^2 = R^2$. (See fig.) Top semicircular area lies under graph of $y = \sqrt{R^2 - x^2}$ between $x = -R$ and $x = R$. Semicircle area $= \int_{-R}^R \sqrt{R^2 - x^2}\,dx$, circle area $= 2\int_{-R}^R \sqrt{R^2 - x^2}\,dx$.

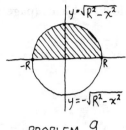

PROBLEM 9

Section 5.3 (page 150)

1. $(2x^3 - \frac{3}{2}x^2 + 2x)|_{-1}^{2} = 14 - (-\frac{11}{2}) = \frac{39}{2}$
2. $(3t - \frac{1}{2}t^2)|_{1}^{3} = \frac{9}{2} - \frac{5}{2} = 2$
3. $(\frac{1}{2}x^6 - \frac{2}{3}x^3)|_{0}^{2} = \frac{80}{3} - 0 = \frac{80}{3}$
4. $-\frac{1}{2}\cos 2x |_{\pi/3}^{\pi/2} = \frac{1}{2} - \frac{1}{4} = \frac{1}{4}$
5. $\tan^{-1} x |_{0}^{1} = \frac{1}{4}\pi - 0 = \frac{1}{4}\pi$
6. $-(1/\pi)\cos \pi x |_{0}^{1/2} = 0 - (-1/\pi) = 1/\pi$
7. $\ln x |_{1}^{5} = \ln 5 - \ln 1 = \ln 5$
8. $-1/12x^2 |_{2}^{3} = -\frac{1}{108} - (-\frac{1}{48}) = \frac{5}{432}$
9. $2x^{3/2} |_{1}^{5} = 2\sqrt{5}^3 - 2 = 10\sqrt{5} - 2$
10. $-\frac{2}{3}(10 - x)^{3/2} |_{1}^{9} = -\frac{2}{3} - (-\frac{2}{3}\sqrt{9^3}) = -\frac{2}{3} + \frac{54}{3} = \frac{52}{3}$
11. $\frac{1}{2}\ln(2x + 1)|_{3}^{4} = \frac{1}{2}\ln 9 - \frac{1}{2}\ln 7$
12. $4(5 - -2) = 28$
13. $\tan x |_{0}^{\pi/4} = 1 - 0 = 1$
14. $\int_{2}^{5} 1\, dx = 1(5 - 2) = 3$
15. $\int_{-1}^{2} (x^6 + 4x^3 + 4)\, dx = (\frac{1}{7}x^7 + x^4 + 4x)|_{-1}^{2}$
$= \frac{296}{7} - (-\frac{22}{7}) = \frac{318}{7}$
16. $\frac{1}{4} \cdot \frac{1}{4}(\frac{1}{2}x + 7)^4 \cdot 2 |_{2}^{4} = \frac{1}{8}(9^4 - 8^4) = \frac{2465}{8}$
17. $\frac{1}{8}\left(\dfrac{x + 3}{5}\right)^8 \cdot 5 |_{-1}^{1} = \frac{5}{8}[(\frac{4}{5})^8 - (\frac{2}{5})^8]$
18. $\frac{1}{3}\ln|x| |_{-5}^{-4} = \frac{1}{3}\ln 4 - \frac{1}{3}\ln 5$
19. $(10/\pi)\sin \frac{1}{2}\pi x |_{-3}^{0} = 0 - 10/\pi = -10/\pi$
20. $-1/4(2x - 9)^2 |_{2}^{3} = -\frac{1}{4}[\frac{1}{9} - \frac{1}{25}] = \frac{-4}{225}$
21. $\frac{1}{3}e^{3x} |_{0}^{1} = \frac{1}{3}e^3 - \frac{1}{3}$
22. (See fig.)
    (a) Base $\overline{AC} = 6$, height $= 2$, area $= \frac{1}{2}bh = 6$.
    (b) Line AB has equation $y = \frac{1}{2}x$, line BC has equation $y = -x + 6$.

    Area $= \int_{0}^{4} \frac{1}{2}x\, dx + \int_{4}^{6}(-x + 6)\, dx$

    $= \frac{1}{4}x^2 |_{0}^{4} + (-\frac{1}{2}x^2 + 6x)|_{4}^{6} = 4 + 2 = 6$.

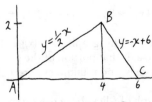

PROBLEM 22

---

23. $\dfrac{\int_{0}^{\pi} \sin x\, dx}{\pi - 0} = \dfrac{-\cos x |_{0}^{\pi}}{\pi} = 2/\pi$
24. (a) $\frac{1}{4}x^4 + C$
    (b) $\frac{1}{4}x^4 |_{1}^{2} = 15/4$
25. $\int_{2}^{3} 5\, dx + \int_{3}^{4} 0\, dx + \int_{4}^{6} x^3\, dx$
$= 5(3 - 2) + 0 + \frac{1}{4}x^4 |_{4}^{6} = 5 + 260 = 265$
26. $4 - x$ is positive if $x < 4$, negative if $x > 4$. So $|4 - x|$ is $4 - x$ if $x < 4$ and is $-(4 - x)$ if $x > 4$. So $\int_{3}^{10}|4 - x|\, dx = \int_{3}^{4}(4 - x)\, dx + \int_{4}^{10}(x - 4)\, dx$
$= (4x - \frac{1}{2}x^2)|_{3}^{4} + (\frac{1}{2}x^2 - 4x)|_{4}^{10} = \frac{1}{2} + 18 = 37/2$.
27. (a) Graph crosses the x-axis at $x = 0, 2, 4$. Area $=$
    $\int_{0}^{2}(x^3 - 6x^2 + 8x)\, dx - \int_{2}^{4}(x^3 - 6x^2 + 8x)\, dx$
    $= (\frac{1}{4}x^4 - 2x^3 + 4x^2)|_{0}^{2} - (\frac{1}{4}x^4 - 2x^3 + 4x^2)|_{2}^{4}$
    $= 4 - (-4) = 8$.

    (b) (See fig.)
    Method 1: Given area $= \int_{0}^{5}\sqrt{9 - x}\, dx - 10$
$= -\frac{2}{3}(9 - x)^{3/2} |_{0}^{5} - 10 = \frac{38}{3} - 10 = \frac{8}{3}$
    Method 2: Point B has coordinates $x = 0, y = 3$. Turn sideways so that x-axis is vertical, positive y-axis is to your left. Region lies under graph of $x = 9 - y^2$, area $= \int_{y=2}^{3}(9 - y^2)\, dy = (9y - \frac{1}{3}y^3)|_{2}^{3} = 18 - \frac{46}{3} = \frac{8}{3}$.

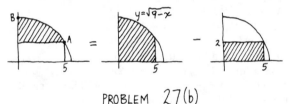

PROBLEM 27(b)

Section 5.4 (page 155)

1. (a) $h = \frac{1}{4}$; $x_0 = 0$, $y_0 = f(x_0) = 1$; $x_1 = \frac{1}{4}$, $y_1 = f(x_1) = 1.0019512$; $x_2 = \frac{1}{2}$, $y_2 = f(x_2) = 1.0307764$; $x_3 = \frac{3}{4}$, $y_3 = f(x_3) = 1.1473475$; $x_4 = 1$, $y_4 = f(x_4) = 1.4142136$; $\frac{1}{3}h(y_0 + 4y_1 + 2y_2 + 4y_3 + y_4) = 1.0894134$

   (b) $h = \frac{1}{6}$; $x_0 = 0$, $y_0 = f(x_0) = 1$; $x_1 = \frac{1}{6}$, $y_1 = f(x_1) = .027399$; $x_2 = \frac{1}{3}$, $y_2 = f(x_2) = .1053605$; $x_3 = \frac{1}{2}$, $y_3 = f(x_3) = .2231436$; $x_4 = \frac{2}{3}$, $y_4 = f(x_4) = .3677248$; $x_5 = \frac{5}{6}$, $y_5 = f(x_5) = .5273549$; $x_6 = 1$, $y_6 = f(x_6) = .6931472$; $\frac{1}{3}h(y_0 + 4y_1 + 2y_2 + 4y_3 + 2y_4 + 4y_5 + y_6) = .2639393$

   (c) $h = \frac{1}{8}$; $x_0 = 1$, $y_0 = f(x_0) = .5$; $x_1 = \frac{9}{8}$, $y_1 = f(x_1) = .4125705$; $x_2 = \frac{5}{4}$, $y_2 = f(x_2) = .3386243$; $x_3 = \frac{11}{8}$, $y_3 = f(x_3) = .2778079$; $x_4 = \frac{3}{2}$, $y_4 = f(x_4) = .2285714$; $x_5 = \frac{13}{8}$, $y_5 = f(x_5) = .1889996$; $x_6 = \frac{7}{4}$, $y_6 = f(x_6) = .1572482$; $x_7 = \frac{15}{8}$,

$y_7 = f(x_7) = .1317211$; $x_8 = 2$, $y_8 = f(x_8) = .1111111$; $\frac{1}{3}h(y_0 + 4y_1 + 2y_2 + 4y_3 + 2y_4 + 4y_5 + 2y_6 + 4y_7 + y_8) = .2543498$

(d) $h = \frac{1}{6}$; $x_0 = 0$, $y_0 = f(x_0) = 1$; $x_1 = \frac{1}{6}$, $y_1 = f(x_1) = .9992287$; $x_2 = \frac{1}{3}$, $y_2 = f(x_2) = .9877302$; $x_3 = \frac{1}{2}$, $y_3 = f(x_3) = .9394131$; $x_4 = \frac{2}{3}$, $y_4 = f(x_4) = .8207548$; $x_5 = \frac{5}{6}$, $y_5 = f(x_5) = .6173908$; $x_6 = 1$, $y_6 = f(x_6) = .3678794$; $\frac{1}{3}h(y_0 + 4y_1 + 2y_2 + 4y_3 + 2y_4 + 4y_5 + y_6) = .8449433$

2. $h = \frac{1}{4}$; $x_0 = 1$, $y_0 = f(x_0) = 1$; $x_1 = \frac{5}{4}$, $y_1 = f(x_1) = .64$; $x_2 = \frac{3}{2}$, $y_2 = f(x_2) = \frac{4}{9}$; $x_3 = \frac{7}{4}$, $y_3 = f(x_3) = .3265306$; $x_4 = 2$, $y_4 = f(x_4) = .25$; $\frac{1}{3}h(y_0 + 4y_1 + 2y_2 + 4y_3 + y_4) = .5004176$. The exact answer is $-(1/x)|_1^2 = -\frac{1}{2} + 1 = .5$.

Section 5.6 (page 160)

1. $-1/4x^4|_3^\infty = 0 + 1/4(81) = 1/324$

2. $\frac{5}{6}x^{6/5}|_2^\infty = \infty - \frac{5}{6}\cdot 2\sqrt[5]{2} = \infty$

3. $-1/2x^2|_{-\infty}^{-2} = -\frac{1}{8} + 0 = -\frac{1}{8}$

4. $-1/x|_{-1}^{0^-} = -(1/0-) - 1 = \infty$

5. $\ln x|_{0+}^2 = \ln 2 - (-\infty) = \infty$

6. (Integrand blows up at $x = 0$.)
$\int_{-2}^{0^-} 1/x^3\,dx + \int_{0+}^3 1/x^3\,dx = -1/2x^2|_{-2}^{0^-} - 1/2x^2|_{0+}^3$
$= -(1/0+) + \frac{1}{8} - \frac{1}{18} + \frac{1}{0+} = -\infty + \infty$. The integral diverges, and there is no "answer".

7. $\tan^{-1} x|_{-\infty}^0 = 0 - (-\pi/2) = \pi/2$

8. $\frac{1}{2}e^{4x}|_{-\infty}^0 = \frac{1}{2} - 0 = \frac{1}{2}$

9. (Integrand blows up at $x = 4$.)
$\int_2^{4^-} + \int_{4+}^5 = -\frac{3}{2}(4 - x)^{2/3}|_2^{4^-} - \frac{3}{2}(4 - x)^{2/3}|_{4+}^5$
$= \frac{3}{2}\sqrt[3]{4} - \frac{3}{2}$

10. ($1/x^2$ blows up at $x = 0$.)
$\int_{-2}^{0^-} + \int_{0+}^3 = -1/x|_{-2}^{0^-} - 1/x|_{0+}^3 = \infty + \infty = \infty$

11. (Improper because $1/x$ blows up at $x = 0$ *and* because interval is infinite.)
$\ln x|_{0+}^\infty = \infty - \infty$. The integral diverges.

12. $-\cos x|_0^\infty = -\cos\infty + 1$ but $\lim_{x\to\infty}\cos x$ doesn't exist since $\cos x$ oscillates between -1 and 1 as $x \to \infty$. The integral diverges.

13. $e^{-|x|}$ is e^{-x} if $x \geq 0$ and is e^x if $x < 0$.
$\int_{-\infty}^0 e^x\,dx + \int_0^\infty e^{-x}\,dx = e^x|_{-\infty}^0 + -e^{-x}|_0^\infty$
$= 1 - 0 - 0 + 1 = 2$

14. Improper because $\tan x$ blows up at $x = \pi/2$.
$-\ln\cos x|_0^{(\pi/2)^-} = -\ln(0+) + \ln 1 = \infty + 0 = \infty$

15. $\lim_{x\to\infty} x/(x^2 + 1) = \lim x/x^2$ (highest power rule) $= \lim 1/x = 0$. Similarly for $x \to -\infty$.
$\tan^{-1}\infty = \pi/2$, $\tan^{-1}(-\infty) = -\pi/2$.
$F(\infty) - F(-\infty) = \frac{1}{2}[0 + \pi/2 - (0 - \pi/2)] = \pi/2$

16. (Improper since $\ln x$ blows up at $x = 0$).
$\lim_{x\to0+} x\ln x$ is $0 \times \infty$ and turns out to be 0 (see (1)–(3), Section 4.4). So $(x\ln x - x)|_{0+}^1 = -1$.

Chapter 5 Review Problems (page 161)

1. (a) $\frac{1}{7}x^7|_{-1}^1 = \frac{1}{7} - (-\frac{1}{7}) = \frac{2}{7}$

(b) (Improper, x^6 blows up at $x = 0$.)
$\int_{-1}^{0^-} + \int_{0+}^1 = -1/5x^5|_{-1}^{0^-} - 1/5x^5|_{0+}^1 = \infty + \infty$
$= \infty$

(c) $-1/5x^5|_1^\infty = 0 + \frac{1}{5} = \frac{1}{5}$

(d) $(\frac{1}{3}x^3 + 3x)|_1^2 = \frac{26}{3} - \frac{10}{3} = \frac{16}{3}$

(e) $\frac{1}{3}\cdot\frac{2}{3}(3x + 4)^{3/2}|_1^2 = \frac{2}{9}(10\sqrt{10} - 7\sqrt{7})$

(f) $-\frac{1}{3}e^{-3x}|_2^\infty = \frac{1}{3}e^{-6}$

(g) $2\sin\frac{1}{2}x|_0^\pi = 0$

(h) $3(7 - 4) = 9$

(i) $e^{-|x|}$ is e^{-x} if $x \geq 0$ and e^x if $x < 0$;
$\int_{-1}^0 e^x\,dx + \int_0^3 e^{-x}\,dx = e^x|_{-1}^0 - e^{-x}|_0^3$
$= 1 - e^{-1} - e^{-3} + 1 = 2 - 1/e - 1/e^3$

(j) $\frac{1}{4}\cdot\frac{1}{6}(2x + 5)^6\cdot\frac{1}{2}|_{-1}^0 = (1/48)[5^6 - 3^6]$
$= 14896/48 = 310\frac{1}{3}$

(k) $-4\ln(2 - x)|_0^1 = 4\ln 2$

(l) $17 - 15 = 2$

2. (a) I + II + III − IV − V
$= \frac{3}{2} + 1 + 2 - 2 - 12 = -\frac{19}{2}$ (See fig.)

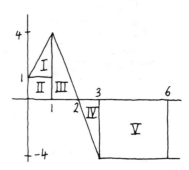

PROBLEM 2

(b) Line AB has equation $y = 3x + 1$, line BC has equation $y = -4x + 8$.
$\int_0^6 f(x)\,dx$
$= \int_0^1(3x + 1)\,dx + \int_1^3(-4x + 8)\,dx + \int_3^6(-4)\,dx$
$= (\frac{3}{2}x^2 + x)|_0^1 + (-2x^2 + 8x)|_1^3 - 4(3)$
$= \frac{5}{2} + 0 - 12 = -19/2$

3. $h = \frac{1}{6}$; $x_0 = 0$, $y_0 = f(x_0) = 1$; $x_1 = \frac{1}{6}$, $y_1 = f(x_1) = 1.0068733$; $x_2 = \frac{1}{3}$, $y_2 = f(x_2) = 1.0266901$; $x_3 = \frac{1}{2}$, $y_3 = f(x_3) = 1.0573713$; $x_4 = \frac{2}{3}$, $y_4 = f(x_4) = 1.0962894$; $x_5 = \frac{5}{6}$, $y_5 = f(x_5) = 1.1409243$; $x_6 = 1$, $y_6 = f(x_6) = 1.1892071$.
So $\frac{1}{3}h(y_0 + 4y_1 + 2y_2 + 4y_3 + 2y_4 + 4y_5 + y_6) = 1.069769$

4. See diagrams. The graph of $|f(x)|$ is found by reflecting in the x-axis that portion of the graph of f which lies below the x-axis, and retaining the rest. (See Problem 4, Sect. 1.7.)

Case 1: Graph of f lies above the x-axis, i.e., $f(x) \geq 0$. Then $|f(x)| = f(x)$. Both I and II equal area A, so I = II.

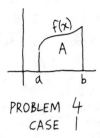

PROBLEM 4
CASE 1

Case 2: Graph lies below x-axis. Then I = C, II = B, and I = II.

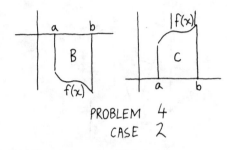

PROBLEM 4
CASE 2

Case 3: Graph crosses x-axis.
Then I = D + F = D + E, II = |D − E|; I is larger.
In general can conclude that in any case, II ≤ I.

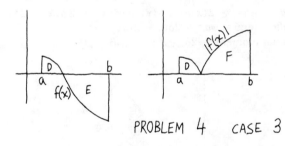

PROBLEM 4   CASE 3

5. $(\int_1^e 1/x \, dx)/(e - 1)$. Numerator is $\ln x \big|_1^e = 1$. Answer is $1/(e - 1)$.

6. Graph crosses the x-axis at $x = 1, -2, 3$.
Area $= -\int_{-1}^{1} [-(x^3 - 2x^2 - 5x + 6)] \, dx$
$\qquad + \int_1^3 [-(x^3 - 2x^2 - 5x + 6)] \, dx$
$= (\tfrac{1}{4}x^4 - \tfrac{2}{3}x^3 - \tfrac{5}{2}x^2 + 6x)\big|_{-1}^1 - (\tfrac{1}{4}x^4 - \tfrac{2}{3}x^3 - \tfrac{5}{2}x^2 + 6x)\big|_1^3$
$= \tfrac{32}{3} + \tfrac{16}{3} = 16$

7. (a) The graph of f is symmetric w.r.t. the origin so that as much area lies above as below. (See fig.) Answer is 0.

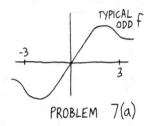

PROBLEM 7(a)

(b) Graph of f is symmetric w.r.t. the y-axis. (See fig.) $\int_{-3}^3 f(x) \, dx = 2\int_0^3 f(x) \, dx$.

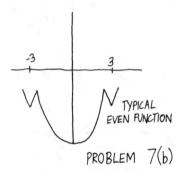

PROBLEM 7(b)

6/THE INTEGRAL   PART II

Section 6.1 (page 170)

Each problem in this section has a diagram, except for 9, 15, 19.

1. The indicated piece of wire has almost constant density x^3, length dx so mass dm = density × length = $x^3 \, dx$. Total mass = $\int_0^8 x^3 \, dx$.

2. Consider the time interval [3, 5]. During the dt hours around hour t, speed is almost constant at t^2. Distance

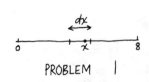

PROBLEM 1

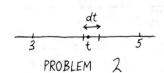

PROBLEM 2

traveled in the dt hours is $d$distance = speed × time = $t^2 dt$. Total distance is $\int_3^5 t^2 \, dt$.

3. Indicated strip of wall has area $7 \, dx$, height x, $d$cost = .01 $x^2 \times 7 \, dx$, total cost = $\int_0^6 .07 \, x^2 \, dx$.

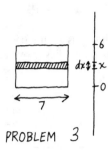

PROBLEM 3

4. Consider slab at position x with thickness dx. Slab radius is $\sqrt{R^2 - x^2}$. Treat slab as a cylinder with height dx, radius $\sqrt{R^2 - x^2}$, $dV = \pi r^2 h = \pi (R^2 - x^2) \, dx$, $V = \int_{-R}^{R} \pi (R^2 - x^2) \, dx = \pi (R^2 x - \frac{1}{3} x^3)\big|_{-R}^{R} = \frac{4}{3} \pi R^3$.

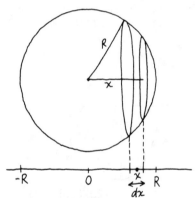

PROBLEM 4

5. Indicated strip has width dx, distance to tracks is $x + 6$; $\dfrac{\text{strip height}}{7 - x} = \dfrac{5}{7}$, strip height $= \frac{5}{7}(7 - x)$.
Strip area is $dA = \frac{5}{7}(7 - x) \, dx$,
$d$price = area × distance to tracks
$= \frac{5}{7}(7 - x) \, dx \times (x + 6)$.
Total price = $\int_0^7 \frac{5}{7}(7 - x)(x + 6) \, dx$
$= \frac{5}{7} \int_0^7 (-x^2 + x + 42) \, dx = 137.1$

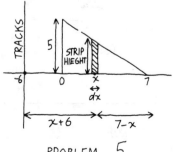

PROBLEM 5

6. Consider slab x feet up, with thickness dx.
$\dfrac{\text{slab radius}}{x} = \frac{5}{20}$, slab radius = $x/4$. Slab is a cylinder with radius $x/4$, height dx, vol $dV = \pi r^2 h = \pi (x^2/16) \, dx$, $d$weight = density × vol = $\frac{1}{8} \pi x^2 \, dx$. Slab must be moved up $20 - x$ feet,
$d$cost = weight × distance moved = $\frac{1}{8} \pi x^2 \, dx \times (20 - x)$,
total cost $= \int_0^{20} d$cost $= \int_0^{20} \frac{1}{8} \pi x^2 (20 - x) \, dx$
$= \frac{1}{8} \pi \int_0^{20} (-x^3 + 20x^2) \, dx$.

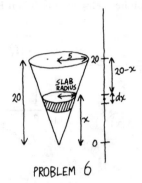

PROBLEM 6

7. Indicated strip has width dx, distance x from pole. $\dfrac{\text{strip height}}{x - 2} = \dfrac{6}{3}$, strip height = $2(x - 2)$. Strip area $dA = bh = 2(x - 2) \, dx$, $d$mass $= 2\delta (x - 2) \, dx$, $d$moment $= md^2 = 2\delta (x - 2) \, dx \times x^2$, total moment = $2\delta \int_2^5 x^2 (x - 2) \, dx$.

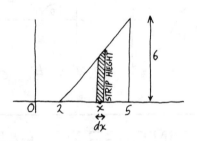

PROBLEM 7

8. If the temp is t degrees then the specific heat is t^3 and the number of calories dc required to raise the temp by dt degrees is $dc = t^3 \, dt$. Total calories = $\int_{54}^{61} t^3 \, dt$.

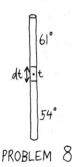

PROBLEM 8

9. (a) The time interval $[2, 14]$ was subdivided, and dx represents a small amount of time around time x.

(b) Since $\Sigma f(x)\, dx$ is the total number of words, $f(x)\, dx$ is the number of words (i.e., $d$words) typed during dx minutes around time x.

(c) Since $f(x)\, dx$ is words and dx is number of minutes, $f(x)$ is words/min, the instantaneous typing rate *at* at time x. If $f(3.2) = 25$, the secretary is typing at the rate of 25 words/min at time 3.2.

10. (a) Slab in Fig. 12(ii) has height dx, radius x^2, volume $dV = \pi r^2 h = \pi x^4\, dx$. Total volume of solid is $\int_0^2 \pi x^4\, dx$.

(b) Revolve a strip to get a small slab with height dy, radius $\sqrt{y}$, volume $dV = \pi r^2 h = \pi y\, dy$. Total volume $= \int_0^4 \pi y\, dy$.

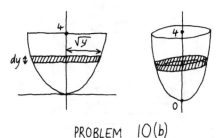

PROBLEM 10(b)

11. Consider slab x feet above square base, with thickness dx. Then $\dfrac{\text{slab edge}}{a} = \dfrac{h - x}{h}$, slab edge $a(h - x)/h$.

Slab volume = (area of base) × height = (edge)$^2\, dx$
$= a^2(h - x)^2/h^2 \cdot dx$,
total volume $= (a^2/h^2) \int_0^h (h - x)^2\, dx$
$= (a^2/h^2) \cdot -\frac{1}{3}(h - x)^3 \big|_0^h = \frac{1}{3}a^2 h = \frac{1}{3}$ base × height.

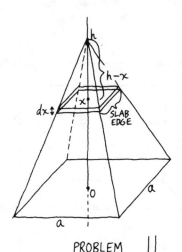

PROBLEM 11

12. Convenient to let P be the 0 point on the wire. The small piece of wire around position x with length dx is roughly distance $|x|$ from P (*not* plain x) so

$d$charge = density × length = $e^{-|x|}\, dx$ and
total charge $= \int_{-\infty}^{\infty} e^{-|x|}\, dx = 2$ (Section 5.6, Prob. 13).

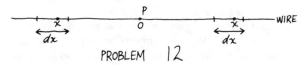

PROBLEM 12

13. The points in the indicated shell are all roughly distance x from the center. Density in shell is x^2. Shell has radius x, thickness dx, $dA = 2\pi x\, dx$ (Section 4.8, (8)), $d$mass = density × area = $2\pi x^3\, dx$, total mass $= \int_0^6 2\pi x^3\, dx$.

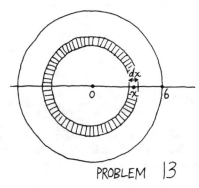

PROBLEM 13

14. The indicated cylindrical shell has radius x, thickness dx, height $2\sqrt{R^2 - x^2}$, $dV = 2\pi x \times 2\sqrt{R^2 - x^2}\, dx$ (Section 4.8, (9)), $d$mass $= 4\pi\delta x\sqrt{R^2 - x^2}\, dx$. Shell is (almost) all at distance x from the pole so by (5), $d$moment $= 4\pi\delta x^3\sqrt{R^2 - x^2}\, dx$,
total moment $= 4\pi\delta \int_0^R x^3\sqrt{R^2 - x^2}\, dx$.

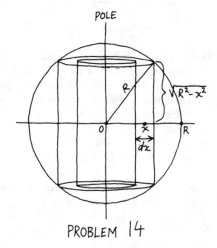

PROBLEM 14

15. (a) Pipeline was divided into pieces, dx is the length of a typical piece located around mile x.

(b) The *sum* of $g(x)\, dx$'s is total cost, $g(x)\, dx$ is the $d$cost of the typical piece of length dx around mile x.

(c) dx is miles, $g(x)\,dx$ is dollars so $g(x)$ is \$/mile, the instantaneous cost per mile of the pipeline at mile x. If $g(4) = 17000$ then at mile marker 4, the pipeline costs \$17000 *per mile.*

16. Look at a small piece of the rod around position x with length dx. All of it (almost) travels in a circle of radius x. Circumference of circle is $2\pi x$ so the piece travels at speed $v = 2\pi x$ feet per sec. Piece has mass $3\,dx$ (use density $\times$ length),
denergy $= \frac{1}{2}mv^2 = \frac{1}{2} \times 3\,dx\,(2\pi x)^2 = 6\pi^2 x^2\,dx$,
total energy $= 6\pi^2 \int_0^{10} x^2\,dx$.

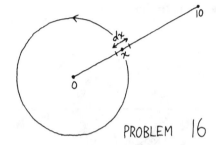

PROBLEM 16

17. Divide the "sector" into subsectors. The indicated radial strip has angle $d\theta$, is located roughly at angle θ, has radius $\cos\theta$ and area $dA = \frac{1}{2}d\theta\,\cos^2\theta$.
Total area $= \frac{1}{2}\int_0^{\pi/4}\cos^2\theta\,d\theta$.

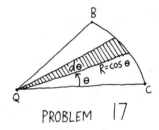

PROBLEM 17

18. (a) Divide solid cylinder into cylindrical shells (significance of the shell is that its points are (almost) all the same distance from the axis). Typical shell has radius x, thickness dx, volume $dV = 2\pi xh\,dx$ (Section 4.8, (9)); distance to axis is x so density is x,
dmass $=$ density $\times$ vol $= 2\pi x^2 h\,dx$.
Total mass $= 2\pi h\int_0^R x^2\,dx$.

(b) Divide solid cylinder into cylindrical slabs (because slab is (almost) at a constant distance from base). Typical slab x feet above base has height dx, radius R, volume $dV = \pi R^2\,dx$, density x, dmass $= \pi R^2 x\,dx$. Total mass $= \pi R^2\int_0^h x\,dx$.

19. (a) Machine is dead when $225 - t^2 = 0$ (no earnings), $t = 15$.

(b) Consider the time interval $[0, 15]$. During the dt years around year t, machine earns at the rate of $225 - t^2$ dollars per year so dearnings $= (225 - t^2)\,dt$, total earnings $= \int_0^{15}(225 - t^2)\,dt$.

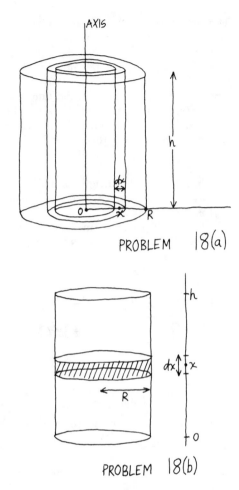

PROBLEM 18(a)

PROBLEM 18(b)

20. The indicated slab is distance x above earth, dimensions are 5 by 6 by dx, volume $dV = 30\,dx$, density is δ, mass $dm = 30\delta\,dx$, $dw = 30\delta\,dx/(2 + x^2)$.
Total weight $= \int_{4000}^{4020} 30\delta/(2 + x^2)\,dx$.

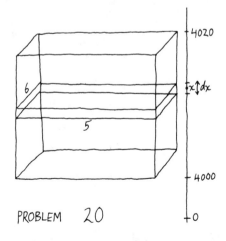

PROBLEM 20

21. All the points in the circular ring are (almost) at distance x from the pump. Ring has radius x, thickness

dx, area $dA = 2\pi x\, dx$ (Section 4.8, (8)),
$d$cost $= 2\pi x\, dx \times x^3$. Total cost $= 2\pi \int_0^R x^4\, dx$.

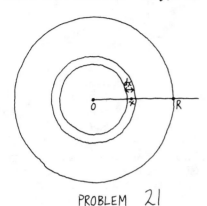

PROBLEM 21

22. Indicated slab is at distance $d = 12 - x$ from collector, dimensions are 9 by 10 by dx, volume $dV = 90\, dx$, $d$heat $= 90\, dx/(12 - x + 1)$.
Total heat $= \int_0^{12} 90/(13 - x)\, dx$.

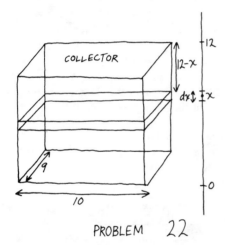

PROBLEM 22

23. Indicated cylindrical slab has radius 3, height dx, (almost) all water in it fell distance $9 - x$, $dV = 9\pi\, dx$, $d$splash $= 9\pi\, dx \times (9 - x)$.
Total splash $= 9\pi \int_0^5 (9 - x)\, dx$.

24. If the moving charge is at position x, so that distance from A is x, then $d$effort required to move dx feet closer is $(1/x^2)\, dx$ (number of feet $\times$ effort per foot).

(a) Total effort to move from C to B is
$\int_2^5 1/x^2\, dx = -1/x\, |_2^5 = 3/10$.

(b) $\int_0^5 1/x^2\, dx = -1/x\, |_{0+}^5 = -\frac{1}{5} + \infty = \infty$.

25. (a) During a small time interval of dt hours around time t, the snow falls at the rate of $R(t)$ flakes per hour. So $d$flakes $= R(t)\, dt$,
total flakes $= \int_0^{10} R(t)\, dt$.

(b) $R(t)\, dt$ flakes *land* in the dt hours around time t but in the remaining $10 - t$ hours until time t,

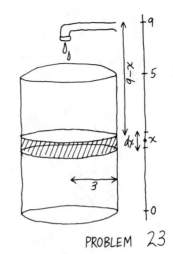

PROBLEM 23

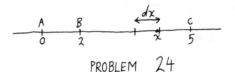

PROBLEM 24

only $\dfrac{1}{(10 - t) + 1}$-th of them will *last*. So
d(lasting flakes) $= R(t)\, dt/(11 - t)$,
total lasting flakes $= \int_0^{10} R(t)/(11 - t)\, dt$.

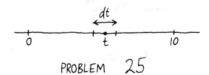

PROBLEM 25

26. The indicated onion shell has thickness dx, radius x, surface area $4\pi x^2$. Current flowing through it travels distance $L = dx$, $dR = dx/4\pi x^2$.
Total $R = (1/4\pi) \int_0^{10} 1/x^2\, dx$.

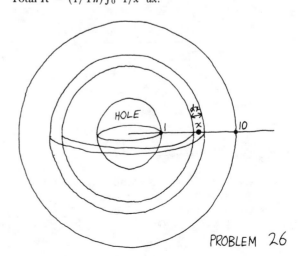

PROBLEM 26

Section 6.3 (page 180)

1. (See figs.)
(a) $u(x) = 3x$, $l(x) = x^2$.
Area $= \int_0^3 (3x - x^2)\,dx = (\frac{3}{2}x^2 - \frac{1}{3}x^3)\,|_0^3 = \frac{9}{2}$.

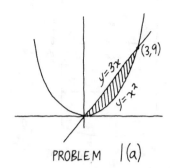

PROBLEM 1(a)

(b) Graph of $x = y^2$ is a parabola. Upper part is $y = \sqrt{x}$ so $u(x) = \sqrt{x}$, $l(x) = x^2$.
Area $= \int_0^1 (\sqrt{x} - x^2)\,dx = (\frac{2}{3}x^{3/2} - \frac{1}{3}x^3)\,|_0^1 = \frac{1}{3}$.

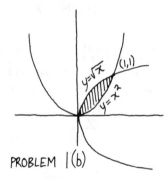

PROBLEM 1(b)

(c) Line AB has equation $y = -4x - 12$, $u(x) = 8/x$, $l(x) = -4x - 12$.
Area $= \int_{-2}^{-1} (8/x - (-4x - 12))\,dx$
$= (8 \ln |x| + 2x^2 + 12x)\,|_{-2}^{-1} = 6 - 8 \ln 2$.

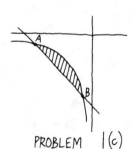

PROBLEM 1(c)

(d) Need points where parabola crosses x-axis. $x^2 - 4x + 3 = 0$, $(x - 3)(x - 1) = 0$, $x = 3, 1$.
Area $= -\int_1^3 (x^2 - 4x + 3)\,dx$
$= -(\frac{1}{3}x^3 - 2x^2 + 3x)\,|_1^3 = \frac{4}{3}$.

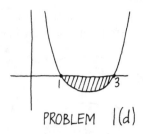

PROBLEM 1(d)

2. (a) $A = (\pi/4, \frac{1}{2}\sqrt{2})$, $B = (5\pi/4, -\frac{1}{2}\sqrt{2})$. To the left of A, $u(x) = \cos x$, $l(x) = \sin x$; situation reverses between A and B. Area $=$
$\int_0^{\pi/4} (\cos x - \sin x)\,dx + \int_{\pi/4}^{3\pi/4} (\sin x - \cos x)\,dx$
$= (\sin x + \cos x)\,|_0^{\pi/4} + (-\cos x - \sin x)\,|_{\pi/4}^{5\pi/4}$
$= (\sqrt{2} - 1) + 2\sqrt{2} = 3\sqrt{2} - 1$
(b) $A = (-2, 4)$, $B = (\frac{1}{4}, 4)$, $C = (1, 1)$. Area $=$
$\int_{-2}^{1/4} (4 - x^2)\,dx + \int_{1/4}^1 (1/x - x^2)\,dx$
$= (4x - \frac{1}{3}x^3)\,|_{-2}^{1/4} + (\ln x - \frac{1}{3}x^3)\,|_{1/4}^1$
$= (1 - \frac{1}{192} + 8 - \frac{8}{3}) + (-\frac{1}{3} - \ln \frac{1}{4} + \frac{1}{192})$
$= 6 + \ln 4$

3. (a) $dy = e^x\,dx$, $ds = \sqrt{dx^2 + (e^x\,dx)^2} = \sqrt{1 + e^{2x}}\,dx$,
$s = \int_{x=0}^{x=1} \sqrt{1 + e^{2x}}\,dx$
(b) $dx = 3y^2\,dy$, $ds = \sqrt{9y^4\,dy^2 + dy^2} = \sqrt{1 + 9y^4}\,dy$, $s = \int_{y=0}^{y=4} \sqrt{1 + 9y^4}\,dy$
(c) $y = 1/x$, $dy = -(1/x^2)\,dx$, $ds = \sqrt{1 + 1/x^4}\,dx$,
$s = \int_{x=1}^2 \sqrt{1 + 1/x^4}\,dx$
(d) $dx = 2\,dt$, $dy = 2t\,dt$, $ds = \sqrt{(2\,dt)^2 + (2t\,dt)^2} = 2\sqrt{1 + t^2}\,dt$. Point $(3, 1)$ corresponds to $t = 1$, point $(9, 16)$ to $t = 4$ so $s = 2\int_{t=1}^{t=4} \sqrt{1 + t^2}\,dt$

4. Line AB has eq $y = m(x - x_1) + y_1$ where $m = (y_2 - y_1)/(x_2 - x_1)$. Then $dy = m\,dx$, $ds = \sqrt{dx^2 + m^2\,dx^2} = \sqrt{1 + m^2}\,dx$. Say $x_2 > x_1$. Then $s = \int_{x_1}^{x_2} \sqrt{1 + m^2}\,dx$. Note that $\sqrt{1 + m^2}$ is a *constant*, so $s = \sqrt{1 + m^2}\,(x_2 - x_1)$
$= \sqrt{1 + \dfrac{(y_2 - y_1)^2}{(x_2 - x_1)^2}} \cdot (x_2 - x_1)$
$= \sqrt{\dfrac{(x_2 - x_1)^2 + (y_2 - y_1)^2}{(x_2 - x_1)^2}} \cdot (x_2 - x_1)$
$= \sqrt{(x_2 - x_1)^2 + (y_2 - y_1)^2}$, standard distance formula.

Section 6.5 (page 187)

1. $(\frac{1}{2}t^2 + 5t)\,|_{t=2}^{t=x} = \frac{1}{2}x^2 + 5x - 12$
2. (a) The piece of wire indicated in the diagram is (almost all) at distance x from A, charge density

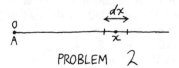

PROBLEM 2

in piece is e^{-x}, $d\text{charge} = e^{-x}\,dx$.
Total charge $= \int_0^\infty e^{-x}\,dx$.
(b) $\int_0^x e^{-t}\,dt$

3. (a) During the time interval of duration dx hours, roughly x hours after 3 PM, the drainfall is $x^3\,dx$. Total rainfall $= \int_0^2 x^3\,dx$.
(b) $\int_0^x t^3\,dt$

4. (See figs.) *Case 1*: $0 \le x \le 1$. By similar triangles, $2/1 = \overline{ED}/(1-x)$, $\overline{ED} = 2 - 2x$. Then $\overline{AB} = 2 - (2-2x) = 2x$.
$I(x) = $ area $ACDE = ABE + BCDE$
$= x^2 + x(2-2x) = 2x - x^2$.

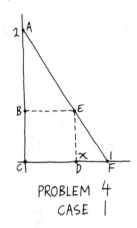

PROBLEM 4
CASE 1

Case 2: $1 \le x \le 3$. $I(x) = ACF$ (can use $I(1)$ from case 1) $+ FGHJ = 1 + 4(x-1) = 4x - 3$.

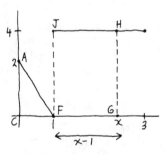

PROBLEM 4
CASE 2

Case 3: $x \ge 3$. $I(x) = ACF + FGHJ$ (can use $I(3)$ from case 2 for this sum) $+ HLM + GHLK$
$= 9 + \frac{1}{2}(x-3)^2 + 4(x-3) = \frac{1}{2}x^2 + x + \frac{3}{2}$.

All in all, $I(x) = \begin{cases} 2x - x^2 & \text{if } 0 \le x \le 1 \\ 4x - 3 & \text{if } 1 \le x \le 3 \\ \frac{1}{2}x^2 + x + \frac{3}{2} & \text{if } x \ge 3 \end{cases}$

5. (See figs.) $I(1) = 0$, $I(2) = \int_1^2 f(t)\,dt = $ area A, $I(3) = \int_1^3 f(t)\,dt = $ area B, $I(10) = $ area C, $I(0) = \int_1^0 f(t)\,dt = $

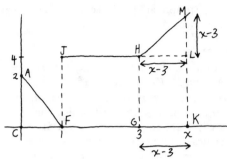

PROBLEM 4
CASE 3

$-D$, $I(-1) = E - D$, $I(-5) = E - D - F$, etc. Can get rough graph by plotting points. Can also use the fact that $I'(x) = f(x)$; graph of I has slope 0 at $x = 0, -1$, slope 3 at $x = 2$, slope $\frac{1}{2}$ for large x, slope ∞ at $-\infty$.

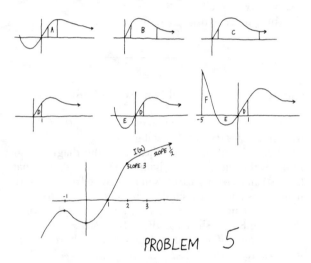

PROBLEM 5

6. (a) $I(\frac{1}{2}) = \int_0^{1/2} f(t)\,dt = \int_0^{1/2} dt = \frac{1}{2}$
(b) $I(2) = \int_0^2 f(t)\,dt = \int_0^1 dt + \int_1^2 \frac{1}{t}\,dt$
$= 1 + \ln t \big|_1^2 = 1 + \ln 2$
(c) If $0 \le x \le 1$, then $I(x) = \int_0^x f(t)\,dt = \int_0^x dt = x$. If $x > 1$ then $I(x) = \int_0^x f(t)\,dt$
$= \int_0^1 dt + \int_1^x (1/t)\,dt = 1 + \ln t \big|_1^x = 1 + \ln x$. So $I(x)$ is x if $0 \le x \le 1$ and $1 + \ln x$ if $x > 1$.

7. (a) $J(7)$ includes the "extra" amount $\int_{1/2}^1 \ln t\,dt$. That amount is negative, namely $-A$. (See fig.) So $J(7)$ is smaller; $J(7) = I(7) - A$.

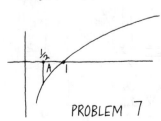

PROBLEM 7

(b) The graphs are parallel curves. The $J(x)$ graph is lower by A.

8. (a) $(2/\sqrt{\pi})e^{-x^2}$

(b) e^{-x}/x

(c) Continue from (b). $D(e^{-x}/x) = (-xe^{-x} - e^{-x})/x^2$

9. $I'(x) = \sin x^2$, $I''(x) = 2x \cos x^2$

10. $\mathrm{Si}'(x) = (\sin x)/x$; zero if $x = \cdots, -\pi, \pi, 2\pi, 3\pi, \cdots$. (If $x = 0$ then $\mathrm{Si}'(x) = (\sin 0)/0$ which for all practical purposes is 1 (Example 2, Section 4.3). So $x = 0$ is not a critical number.) $\mathrm{Si}''(x) = (x \cos x - \sin x)/x^2$. $\mathrm{Si}''(-\pi) = 1/\pi$, positive, $\mathrm{Si}\,x$ has rel min at $x = -\pi$ by 2nd derivative test; $\mathrm{Si}''(\pi) = -1/\pi$, negative, rel max at $x = \pi$; $\mathrm{Si}''(2\pi)$ is positive, rel min at $x = 2\pi$, etc.

11. $f(x) = \mathrm{Si}(x^3)$ so $f'(x) = 3x^2\,\mathrm{Si}'(x^3)$. But $\mathrm{Si}'(x) = (\sin x)/x$ so $\mathrm{Si}'(x^3) = (\sin x^3)/x^3$ and $f'(x) = (3 \sin x^3)/x$.

12. Numerator $\to 0$ (since Si becomes $\int_0^0$) so lim is $0/0$. Use L'Hôpital; $\lim \mathrm{Si}'(x)/1 = \lim (\sin x)/x = 0/0 = \lim(\cos x)/1 = 1$

13. (a) $(\frac{1}{2}x^2 - 5x)\big|_4^2 = -8 - (-12) = 4$

(b) $\frac{1}{2}\ln(2x + 5)\big|_2^0 = \frac{1}{2}\ln 5 - \frac{1}{2}\ln 9$.

Chapter 6 Review Problems (page 188)

1. In dt days around day t, colony grows at $f(t)$ cm^3/day so $d\text{growth} = f(t)\,dt$ cm^3. Total growth $= \int_3^7 f(t)\,dt$.

2. The cylindrical shell indicated in the diagram has radius x, thickness dx. The significance of the shell is that (almost) all its points are the same distance from the axis, namely distance x. By similar triangles,

$(h - \text{shell height})/x = h/R$, shell height $= h(R - x)/R$. By Sect. 4.8, (9), shell vol $dV = 2\pi x \cdot h(R - x)/R \cdot dx$, $d\text{mass} = \delta\,dV$, $d\text{moment} = d\text{mass} \times d^2$ (where $d = x$) $= 2\pi h \delta/R \cdot x^3(R - x)\,dx$.

Total moment $= \int_0^R d\text{moment}$
$= (2\pi h \delta/R)\int_0^R (Rx^3 - x^4)\,dx$
$= (2\pi h \delta/R)(\frac{1}{4}Rx^4 - \frac{1}{5}x^5)\big|_0^R = \frac{1}{10}\pi hR^4\delta$.

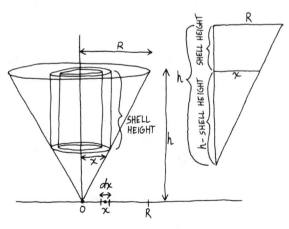

PROBLEM 2

3. The indicated strip (see fig.) has length 4, width dx, and most importantly is (almost) all at depth $11 - x$. Area is $4\,dx$, $d\text{reading} = \text{depth} \times \text{area} = 4(11 - x)\,dx$. Total reading $= \int_0^5 d\text{reading} = 4\int_0^5(11 - x)\,dx$ $= 4(11x - \frac{1}{2}x^2)\big|_0^5 = 170$.

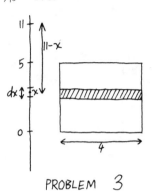

PROBLEM 3

4. Line AB (see fig.) has equation $y = 2x - 4$, $u(x) = 2x - 4$, $l(x) = \sin \pi x$, area $= \int_{3/2}^2 (2x - 4 - \sin \pi x)$ $dx = (x^2 - 4x + \frac{1}{\pi}\cos \pi x)\big|_{3/2}^2 = 1/\pi - 1/4$

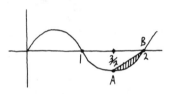

PROBLEM 4

5. Region is triangle ABC where $B = (4, 8)$. (See fig.)

(a) ABC has height 8, base 12, area 48.

(b) Area $= \mathrm{I} + \mathrm{II} = \int_0^4 2x\,dx + \int_4^{12}(12 - x)\,dx$ $= x^2\big|_0^4 + (12x - \frac{1}{2}x^2)\big|_4^{12} = 16 + 32 = 48$.

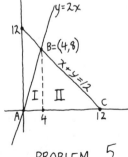

PROBLEM 5

6. (a) In the dt years around age t the sheep produces wool at the rate of $100 - t$ pounds per year so $d\text{wool} = (100 - t)\,dt$, total wool $= \int_2^4(100 - t)\,dt$.

(b) $\int_2^t(100 - x)\,dx$ (use dummy variable *other* than t)

7. (a) $I(x) = \int_2^x(2t + 3)\,dt = (t^2 + 3t)\big|_2^x$ $= x^2 + 3x - 10$

(b) If $x \le 7$ then
$$I(x) = \int_2^x f(t)\,dt = \int_2^x 3t^2\,dt = t^3\,|_2^x = x^3 - 8.$$
If $x > 7$ then
$$I(x) = \int_2^x f(t)\,dt = \int_2^7 3t^2\,dt + \int_7^x 5\,dt$$
$$= I(7) + 5(x - 7) = (7^3 - 8) + 5(x - 7)$$
$$= 5x + 300.$$
All in all, $I(x)$ is $x^3 - 8$ if $x \le 7$ and is $5x + 300$ if $x > 7$.

(c) Can find formula for f and proceed as in (b) or can use geometry (see fig.); $I(2) = \int_2^2 f(t)\,dt = 0$. If $2 \le x \le 4$ then $I(x) = \int_0^x f(t)\,dt = $ area ABC. Base $\overline{AC} = x - 2$; by similar triangles, height $\overline{BC} = 4(x - 2)$, $I(x) = \frac{1}{2}\overline{AC} \cdot \overline{BC} = 2(x - 2)^2$.

If $4 \le x \le 6$ then $I(x) = $ area $ADEF = AGD - EFG = 16 - 2(6 - x)^2$.

If $x \ge 6$ then $I(x) = $ area $AGD = 16$.

If $x \le 2$, $I(x) = 0$ since there is no area under the graph of f to the left of $x = 2$. All in all,

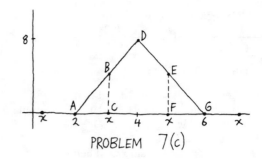

PROBLEM 7(c)

$$I(x) = \begin{cases} 0 & \text{if } x \le 2 \\ 2(x - 2)^2 & \text{if } 2 \le x \le 4 \\ 16 - 2(6 - x)^2 & \text{if } 4 \le x \le 6 \\ 16 & \text{if } x \le 6. \end{cases}$$

8. $I'(x) = e^{x^2}$ (*not* e^{t^2}), $I''(x) = D_x e^{x^2} = 2x e^{x^2}$

7/ANTIDIFFERENTIATION

Section 7.2 (page 195)

1. Let $u = x^2$, $du = 2x\,dx$. Then
$\int xe^{x^2}\,dx = \frac{1}{2}\int e^u\,du = \frac{1}{2}e^u + C = \frac{1}{2}e^{x^2} + C.$

2. Let $u = 3x^2 + 7$, $du = 6x\,dx$. Then
$\frac{1}{6}\int \sqrt{u}\,du = \frac{1}{6}u^{3/2}/\frac{3}{2} + C = \frac{1}{9}(3x^2 + 7)^{3/2} + C.$

3. By inspection or let $u = 3 + 5x$, $du = 5\,dx$. Then
$\frac{1}{5}\int \sqrt{u}\,du = \frac{1}{5}u^{3/2}/\frac{3}{2} + C = \frac{2}{15}(3 + 5x)^{3/2} + C.$

4. Let $u = 3 + 7x$, $du = 7\,dx$. Then
$\frac{1}{7}\int 1/\sqrt{u}\,du = \frac{2}{7}\sqrt{u} + C = \frac{2}{7}\sqrt{3 + 7x} + C.$

5. Let $u = \tan x$, $du = \sec^2 x\,dx$. Then
$\int u^{14}\,du = \frac{1}{15}u^{15} + C = \frac{1}{15}\tan^{15}x + C.$

6. Let $u = x + 1$, $du = dx$, $x = u - 1$. Then
$$\int \frac{u - 2}{u^5}\,du = \int (u^{-4} - 2u^{-5})\,du = \frac{u^{-3}}{-3} - 2\frac{u^{-4}}{-4} + C$$
$$= -\frac{1}{3(x + 1)^3} + \frac{1}{2(x + 1)^4} + C.$$

7. Let $u = 1 + 2\sec\theta$, $du = 2\sec\theta\tan\theta\,d\theta$. Then
$\int = \frac{1}{2}\int 1/\sqrt{u}\,du = \frac{1}{2}u^{1/2}/\frac{1}{2} + C = \sqrt{1 + 2\sec\theta} + C.$

8. Let $u = \ln x$, $du = 1/x\,dx$. Then
$\int 1/u\,du = \ln|u| + C = \ln|\ln x| + C.$

9. Let $u = x^2$, $du = 2x\,dx$. Then $\int x^3 \sin u\,du/2x = \frac{1}{2}\int x^2 \sin u\,du = \frac{1}{2}\int u \sin u\,du = \frac{1}{2}(\sin u - u \cos u) + C$ (formula 48) $= \frac{1}{2}(\sin x^2 - x^2 \cos x^2) + C.$

10. By inspection or let $u = 1 + 3x$, $du = 3\,dx$. Then
$\frac{1}{3}\int u^7\,du = \frac{1}{24}u^8 + C = \frac{1}{24}(1 + 3x)^8 + C.$

11. By inspection or let $u = 2 - 3x$, $du = -3\,dx$. Then
$-\frac{1}{3}\int 1/u\,du = -\frac{1}{3}\ln|u| + C = -\frac{1}{3}\ln|2 - 3x| + C.$

12. By inspection or let $u = 2 - x$, $du = -dx$. Then
$$-\int \frac{1}{u^3}\,du = \frac{1}{2u^2} + C = \frac{1}{2(2 - x)^2} + C.$$

13. By inspection or let $u = \frac{1}{2}\theta - 1$, $du = \frac{1}{2}d\theta$. Then
$2\int \cos u\,du = 2 \sin u + C = 2 \sin(\frac{1}{2}\theta - 1) + C.$

14. Let $u = -x$, $du = -dx$. Then $\int (-u)e^u \cdot -du$
$= \int ue^u\,du = e^u(u - 1) + C$ (formula 61)
$= e^{-x}(-x - 1) + C.$

15. Let $u = \cos x$, $du = -\sin x\,dx$. Then
$-\int u^3\,du = -\frac{1}{4}u^4 + C = -\frac{1}{4}\cos^4 x + C.$

16. By inspection or let $u = -x$, $du = -dx$. Then
$-\int e^u\,du = -e^u + C = -e^{-x} + C.$

17. Let $u = 3x$, $du = 3\,dx$. Then $\int \frac{1}{3}\sin u \cdot \frac{1}{3}du$
$= \frac{1}{9}\int u \sin u\,du = \frac{1}{9}(\sin u - u \cos u) + C$ (formula 48)
$= \frac{1}{9}(\sin 3x - 3x \cos 3x) + C.$

18. Let $u = \pi x$, $du = \pi\,dx$. Then $(1/\pi)\int \sin^2 u\,du$
$= (1/\pi)\frac{1}{2}(u - \sin u \cos u) + C$ (formula 39)
$= (1/2\pi)(\pi x - \sin \pi x \cos \pi x) + C.$

19. $3\int x \sin x\,dx = 3(\sin x - x \cos x) + C$ (formula 48).

20. Let $u = 3x$, $du = 3\,dx$.
$\int (\frac{1}{3}u)^2 \cos u \cdot \frac{1}{3}du = \frac{1}{27}\int u^2 \cos u\,du$ (use 51)
$= \frac{1}{27}[(u^2 - 2)\sin u + 2u \cos u] + C$
$= \frac{1}{27}(9x^2 - 2)\sin 3x + \frac{6}{27}x \cos 3x + C.$

21. Let $u = 2x + 3$, $du = 2\,dx$. Then
$\frac{1}{2}\int \ln u\,du = \frac{1}{2}(u \ln u - u) + C$ (formula 62)
$= \frac{1}{2}(2x + 3)\ln(2x + 3) - \frac{1}{2}(2x + 3) + C.$

22. $\int \sec x\,dx = \ln|\sec x + \tan x| + C$ (formula 33)

23. NO! The first step is OK, but the second step im-

plicitly lets $u = \sqrt{3}\,x$ in which case $du = \sqrt{3}\,dx$. Then
$$\int \frac{1}{1 + (\sqrt{3}\,x)^2}\,dx = \int \frac{1}{1 + u^2}\frac{du}{\sqrt{3}} = \frac{1}{\sqrt{3}}\tan^{-1}u + C$$
$$= \frac{1}{\sqrt{3}}\tan^{-1}\sqrt{3}\,x + C.$$

24. (a) Let $u = 3x$, $du = 3\,dx$. Then $\frac{1}{3}\int \tan^{-1}u\,du$
$\qquad = \frac{1}{3}(u\,\tan^{-1}u - \frac{1}{2}\ln(1 + u^2)) + C$ (formula 59)
$\qquad = x\,\tan^{-1}3x - \frac{1}{6}\ln(1 + 9x^2) + C.$
$\quad$ (b) Not possible yet.

25. Let $u = \cos x$, $du = -\sin x\,dx$. Then
$-\int 1/u\,du = -\ln|u| + C = -\ln|\cos x| + C.$

26. $\int \sec x\,dx = \int \dfrac{\sec x(\sec x + \tan x)}{\sec x + \tan x}\,dx$
$= \int \dfrac{\sec^2 x + \sec x \tan x}{\sec x + \tan x}\,dx.$

Now let $u = \sec x + \tan x$, $du = (\sec x \tan x + \sec^2 x)\,dx$.
Then $\int du/u = \ln|u| + C = \ln|\sec x + \tan x| + C.$

27. $\int \sin^2 x\,dx = \frac{1}{2}\int(1 - \cos 2x)\,dx = \frac{1}{2}\int dx -$
$\frac{1}{2}\int \cos 2x\,dx = \frac{1}{2}x - \frac{1}{2}\cdot\frac{1}{2}\sin 2x + C = \frac{1}{2}x - \frac{1}{4}\sin 2x + C.$

Section 7.3 (page 198)

1. $2 + 6x - x^2 = -(x^2 - 6x - 2)$
$= -(x^2 - 6x + 9 - 9 - 2) = -([x - 3]^2 - 11)$
$= 11 - (x - 3)^2.$ Then to do
$\int 1/\sqrt{11 - (x - 3)^2}\,dx$ let $u = x - 3$, $du = dx$, use (19)
to get $\displaystyle\int \frac{du}{\sqrt{11 - u^2}} = \sin^{-1}\frac{u}{\sqrt{11}} + C = \sin^{-1}\frac{x - 3}{\sqrt{11}} + C.$

2. $x + 2x^2 = 2(x^2 + \frac{1}{2}x + \frac{1}{16} - \frac{1}{16}) = 2([x + \frac{1}{4}]^2 - \frac{1}{16}).$
Let $u = x + \frac{1}{4}$, $du = dx$.
$$\frac{1}{\sqrt{2}}\int \frac{dx}{\sqrt{(x + \frac{1}{4})^2 - \frac{1}{16}}} = \frac{1}{\sqrt{2}}\int \frac{du}{\sqrt{u^2 - \frac{1}{16}}}$$
$= \dfrac{1}{\sqrt{2}}\ln\left|x + \frac{1}{4} + \sqrt{(x + \frac{1}{4})^2 - \frac{1}{16}}\right| + C$ (formula 27).

3. $\dfrac{1}{\sqrt{3}}\displaystyle\int \frac{dx}{\sqrt{x^2 - \frac{5}{3}}} = \dfrac{1}{\sqrt{3}}\ln\left|x + \sqrt{x^2 - \frac{5}{3}}\right| + C$
(formula 27).

4. $\displaystyle\int\left(x^2 - 4 + \frac{2x + 16}{x^2 + 4}\right)dx$ (long division)
$= \dfrac{x^3}{3} - 4x + 2\displaystyle\int \frac{x}{x^2 + 4}\,dx + 16\int \frac{1}{x^2 + 4}\,dx$
$= \frac{1}{3}x^3 - 4x + \ln(x^2 + 4) + 8\tan^{-1}\frac{1}{2}x + C.$
(Sub $u = x^2 + 4$, $du = 2x\,dx$ in first integral; use 1(b)
or 16 for second.)

5. $\int x\sqrt{(x + 1)^2 - 1}\,dx.$ Let $u = x + 1$, $du = dx$;
$\int(u - 1)\sqrt{u^2 - 1}\,du$
$= \int u\sqrt{u^2 - 1}\,du - \int\sqrt{u^2 - 1}\,du$
$= \frac{1}{3}(u^2 - 1)^{3/2} - \frac{1}{2}u\sqrt{u^2 - 1} + \frac{1}{2}\ln|u + \sqrt{u^2 - 1}| + C$
$= \frac{1}{3}(x^2 + 2x)^{3/2} - \frac{1}{2}(x + 1)\sqrt{x^2 + 2x}$
$\quad + \frac{1}{2}\ln|x + 1 + \sqrt{x^2 + 2x}| + C.$
(Sub $v = u^2 - 1$, $dv = 2u\,du$ in first integral;
use 28 in second.)

6. (a) $\displaystyle\int\left(\frac{1}{2} - \frac{3}{2x + 6}\right)dx$ (long division)
$\quad = \frac{1}{2}x - 3\cdot\frac{1}{2}\ln|2x + 6| + C$ (by inspection).
$\quad$ (b) Use formula 5 with $a = 6$, $b = 2$ to get
$\quad \frac{3}{2} + \frac{1}{2}x - \frac{3}{2}\ln|2x + 6| + C$
$\quad = \frac{1}{2}x - \frac{3}{2}\ln|2x - 6| + K.$
$\quad$ (c) Let $u = 2x + 6$, $du = 2\,dx$. Then
$$\int \frac{\frac{1}{2}(u - 6)}{u}\frac{du}{2} = \frac{1}{4}\int\left(1 - \frac{6}{u}\right)du$$
$\quad = \frac{1}{4}u - \frac{3}{2}\ln|u| + C$
$\quad = \frac{1}{4}(2x + 6) - \frac{3}{2}\ln|2x + 6| + C$ as in (b).

7. $\displaystyle\int\left(1 - \frac{1}{x^2 + 1}\right)dx$ (long division) $= x - \tan^{-1}x + C.$

Section 7.4 (page 202)

1. (a) $\dfrac{A}{x} + \dfrac{B}{x^2} + \dfrac{C}{x^3} + \dfrac{D}{x + 1} + \dfrac{E}{2x + 3}$
$\quad$ (b) $x^2 + 2x - 2$ factors into $(x - [-1 + \sqrt{3}])\cdot$
$\quad (x - [-1 - \sqrt{3}])$ and $x^2 - 2x + 2$ doesn't
$\quad$ factor since $b^2 - 4ac < 0$, so decomp is
$$\frac{A}{x - (-1 + \sqrt{3})} + \frac{B}{x - (-1 - \sqrt{3})}$$
$$+ \frac{Cx + D}{x^2 - 2x + 2}.$$

2. (a) $\dfrac{12}{(x - \sqrt{3})(x + \sqrt{3})} = \dfrac{A}{x + \sqrt{3}} + \dfrac{B}{x - \sqrt{3}}$,
$\quad 12 = A(x - \sqrt{3}) + B(x + \sqrt{3}).$
$\quad$ If $x = \sqrt{3}$ then
$\quad 12 = 2\sqrt{3}B$, $B = 6/\sqrt{3} = 2\sqrt{3}.$
$\quad$ If $x = -\sqrt{3}$ then $12 = -2\sqrt{3}A$, $A = -2\sqrt{3}.$
$\quad$ Decomp is $\dfrac{-2\sqrt{3}}{x + \sqrt{3}} + \dfrac{2\sqrt{3}}{x - \sqrt{3}}.$
$\quad$ (b) $\dfrac{1}{(x - 4)(2x + 3)} = \dfrac{A}{x - 4} + \dfrac{B}{2x + 3}$,
$\quad 1 = A(2x + 3) + B(x - 4).$
$\quad$ If $x = 4$ then $1 = 11A$, $A = 1/11.$
$\quad$ If $x = -3/2$ then $1 = -\frac{11}{2}B$, $B = -2/11.$
$\quad$ Decomp is $\dfrac{1/11}{x - 4} - \dfrac{2/11}{2x + 3}.$
$\quad$ (c) $\dfrac{5x}{(x^2 + 1)(x - 2)} = \dfrac{Ax + B}{x^2 + 1} + \dfrac{C}{x - 2}$,
$\quad 5x = (Ax + B)(x - 2) + C(x^2 + 1).$
$\quad$ If $x = 2$ then $10 = 5C$, $C = 2.$
$\quad$ Equate x^2 coeffs to get $0 = A + C$, $A = -2.$
$\quad$ Equate x coeffs to get $5 = B - 2A$, $B = 1.$
$\quad$ Decomp is $\dfrac{-2x + 1}{x^2 + 1} + \dfrac{2}{x - 2}.$
$\quad$ (d) $\dfrac{2x + 3}{(x - 2)^2} = \dfrac{A}{x - 2} + \dfrac{B}{(x - 2)^2}$,
$\quad 2x + 3 = A(x - 2) + B.$
$\quad$ If $x = 2$ then $7 = B.$

Equate x coefficients to get $2 = A$.

So decomp is $\dfrac{2}{x - 2} + \dfrac{7}{(x - 2)^2}$.

3. (a) $\dfrac{3}{(2 - x)(x + 1)} = \dfrac{A}{2 - x} + \dfrac{B}{x + 1}$,

$3 = A(x + 1) + B(2 - x)$.

If $x = 2$ then $3 = 3A$, $A = 1$.

If $x = -1$ then $3 = 3B$, $B = 1$. Decomp is

$\dfrac{1}{2 - x} + \dfrac{1}{1 + x}$. Antidiff by inspection to get

answer $-\ln|2 - x| + \ln|x + 1| + C$.

(b) For $\displaystyle\int \dfrac{3\,dx}{-x^2 + x + 2}$ use formula 1(a) with $a =$

-1, $b = 1$, $c = 2$ to get $\ln\left|\dfrac{-2x + 1 - 3}{-2x + 1 + 3}\right| + C$

$= \ln\left|\dfrac{-x - 1}{2 - x}\right| + C = \ln\dfrac{|x + 1|}{|2 - x|} + C$

$= \ln|x + 1| - \ln|2 - x| + C$.

4. (a) Can write as $2\displaystyle\int \dfrac{x\,dx}{x^2 - 4x + 4} +$

$3\displaystyle\int \dfrac{dx}{x^2 - 4x + 4}$ and use formulas 2 and 1(c). Or

can decompose (see problem 2(d)) into $\dfrac{2}{x - 2} +$

$\dfrac{7}{(x - 2)^2}$. Then antidiff by inspection to get an-

swer $2\ln|x - 2| - 7/(x - 2) + C$.

(b) $\dfrac{8x}{(x^2 - 1)(x^2 + 1)} = \dfrac{8x}{(x + 1)(x - 1)(x^2 + 1)}$

$= \dfrac{A}{x + 1} + \dfrac{B}{x - 1} + \dfrac{Cx + D}{x^2 + 1}$,

$8x = A(x - 1)(x^2 + 1) + B(x + 1)(x^2 + 1)$
$\qquad + (Cx + D)(x^2 - 1)$.

If $x = 1$ then $8 = 4B$, $B = 2$. If $x = -1$ then

$-8 = -4A$, $A = 2$. Equate x^3 coeffs to get $0 = A + B + C$, $C = -4$. Equate x^2 coeffs to get $0 = -A + B + D$, $D = 0$. So decomp is

$\dfrac{2}{x + 1} + \dfrac{2}{x - 1} - \dfrac{4x}{x^2 + 1}$. Antidiff the first frac-

tions by inspection, substitute $u = x^2 + 1$ for the third. Answer is

$2\ln|x + 1| + 2\ln|x - 1| - 2\ln(x^2 + 1) + C$.

(c) Either use formula 10 or decompose to

$\dfrac{-2/9}{x} - \dfrac{1/3}{x^2} + \dfrac{4/9}{2x - 3}$. Antidiff by inspection

to get $-\frac{2}{9}\ln|x| + 1/3x + \frac{2}{9}\ln|2x - 3| + C$.

5. $\dfrac{1}{x(a + bx)^2} = \dfrac{A}{x} + \dfrac{B}{a + bx} + \dfrac{C}{(a + bx)^2}$,

$1 = A(a + bx)^2 + Bx(a + bx) + Cx$.

If $x = 0$ then $1 = a^2A$ so $A = 1/a^2$.

If $x = -a/b$ then $1 = -(a/b)C$, $C = -b/a$.

Equate x^2 coeffs; $0 = b^2A + bB$, $B = -b/a^2$. Get

$\dfrac{1}{a^2}\displaystyle\int \dfrac{1}{x}\,dx - \dfrac{b}{a^2}\displaystyle\int \dfrac{1}{a + bx}\,dx - \dfrac{b}{a}\displaystyle\int \dfrac{1}{(a + bx)^2}\,dx$. Anti-

diff by inspection or substitute $u = a + bx$ to get

$\dfrac{1}{a^2}\ln|x| - \dfrac{1}{a^2}\ln|a + bx| + \dfrac{1}{a}\dfrac{1}{a + bx} + C$

$= -\dfrac{1}{a^2}(\ln|a + bx| - \ln|x|) + \dfrac{1}{a(a + bx)} + C$

$= -\dfrac{1}{a^2}\ln\left|\dfrac{a + bx}{x}\right| + \dfrac{1}{a(a + bx)} + C$.

6. $x^2 + 5x + 4\,\overline{)\,x^2}$

$\,\dfrac{x^2 + 5x + 4}{-5x - 4}$

so $\dfrac{x^2}{x^2 + 5x + 4} = 1 - \dfrac{5x + 4}{x^2 + 5x + 4}$ and

$\displaystyle\int \dfrac{x^2}{x^2 + 5x + 4}\,dx$

$= \displaystyle\int dx - 5\displaystyle\int \dfrac{x\,dx}{x^2 + 5x + 4} - 4\displaystyle\int \dfrac{dx}{x^2 + 5x + 4}$.

First integral is x. For second integral use formula 2 with $a = 1$, $b = 5$, $c = 4$, or factor denom and use decomposition. Second is

$-5 \cdot \frac{1}{2}\ln|x^2 + 5x + 4| + \dfrac{5 \cdot 5}{2}\displaystyle\int \dfrac{1}{x^2 + 5x + 4}\,dx$.

Add in first and third to get

$x - \frac{5}{2}\ln|x^2 + 5x + 4| + \frac{17}{2}\displaystyle\int \dfrac{dx}{x^2 + 5x + 4}$

$\qquad$ (formula 1(a) or decompose)

$= x - \frac{5}{2}\ln|x^2 + 5x + 4| + \frac{17}{2} \cdot \frac{1}{3}\ln\left|\dfrac{2x + 5 - 3}{2x + 5 + 3}\right|$

$= x - \frac{5}{2}\ln|x^2 + 5x + 4| + \frac{17}{6}\ln\left|\dfrac{x + 1}{x + 4}\right|$

$= x - \frac{5}{2}\ln|x + 4||x + 1| + \frac{17}{6}\ln\left|\dfrac{x + 1}{x + 4}\right|$

$= x - \frac{5}{2}\ln|x + 4| - \frac{5}{2}\ln|x + 1| + \frac{17}{6}\ln|x + 1|$
$\quad - \frac{17}{6}\ln|x + 4|$

$= x + \frac{1}{3}\ln|x + 1| - \frac{16}{3}\ln|x + 4|$.

Section 7.5 (page 204)

1. (a) Let $u = x$, $dv = e^x\,dx$. Then $du = dx$, $v = e^x$ and $\displaystyle\int xe^x\,dx = xe^x - \int e^x\,dx = xe^x - e^x + C$.

(b) Let $u = \tan^{-1}x$, $dv = dx$. Then
$du = dx/(1 + x^2)$, $v = x$ and

$\displaystyle\int \tan^{-1}x\,dx = x\tan^{-1}x - \int \dfrac{x}{1 + x^2}\,dx$

$= x\tan^{-1}x - \frac{1}{2}\ln(1 + x^2) + C$
$\qquad$ (use formula 2 or sub $u = 1 + x^2$).

(c) Let $u = \sin^{-1}x$, $dv = dx$. Then
$du = dx/\sqrt{1 - x^2}$, $v = x$ and

$\displaystyle\int \sin^{-1}x\,dx = x\sin^{-1}x - \int \dfrac{x}{\sqrt{1 - x^2}}\,dx$

(now sub $u = 1 - x^2$)
$$= x \sin^{-1}x + \sqrt{1 - x^2} + C.$$

(d) Let $u = \ln x$, $dv = dx$. Then $du = dx/x$, $v = x$,
$\int \ln x\, dx = x \ln x - \int dx = x \ln x - x + C.$

2. (a) Let $u = \cos(\ln x)$, $dv = dx$. Then

$$du = -\sin(\ln x) \cdot (1/x)\, dx, \quad v = x,$$
$$\int \cos(\ln x)\, dx = x \cos(\ln x) + \int \sin(\ln x)\, dx.$$

Now let $u = \sin(\ln x)$, $dv = dx$. Then $du = \cos(\ln x) \cdot (1/x)\, dx$, $v = x$ and
$\int \cos(\ln x)\, dx =$
$x \cos(\ln x) + x \sin(\ln x) - \int \cos(\ln x)\, dx.$
Collect terms to get
$2 \int \cos(\ln x)\, dx = x \cos(\ln x) + x \sin(\ln x)$
so $\int \cos(\ln x)\, dx = \frac{1}{2}x[\cos(\ln x) + \sin(\ln x)] + C.$

(b) Let $u = x^2$, $dv = e^x\, dx$. Then $du = 2x\, dx$, $v = e^x$,
$\int x^2 e^x\, dx = x^2 e^x - 2 \int x e^x\, dx.$ Now either use formula 61 or use parts again with $u = x$, $dv = e^x\, dx$. Then $du = dx$, $v = e^x$ and
$\int x^2 e^x\, dx = x^2 e^x - 2(x e^x - \int e^x\, dx)$
$= x^2 e^x - 2x e^x + 2e^x + C.$

(c) Let $u = \tan^{-1}x$, $dv = x\, dx$. Then
$du = dx/(1 + x^2)$, $v = \frac{1}{2}x^2$,

$$\int x \tan^{-1}x\, dx = \frac{1}{2}x^2 \tan^{-1}x - \frac{1}{2} \int \frac{x^2}{1 + x^2}\, dx$$

$$= \frac{1}{2}x^2 \tan^{-1}x - \frac{1}{2} \int \left(1 - \frac{1}{1 + x^2}\right) dx \text{ (long div)}$$

$$= \frac{1}{2}x^2 \tan^{-1}x - \frac{1}{2}x + \frac{1}{2} \tan^{-1}x + C.$$

3. Let $u = \sec x$, $dv = \sec^2 x\, dx$. Then
$du = \sec x \tan x\, dx$, $v = \tan x$,
$\int \sec^3 x\, dx = \sec x \tan x - \int \sec x \tan^2 x\, dx$
$= \sec x \tan x - \int \sec x \cdot (\sec^2 x - 1)\, dx$ (trig identity)
$= \sec x \tan x - \int \sec^3 x\, dx + \int \sec x\, dx$
$= \sec x \tan x - \int \sec^3 x\, dx + \ln|\sec x + \tan x|.$
So $2 \int \sec^3 x\, dx = \sec x \tan x + \ln|\sec x + \tan x|$,
$\int \sec^3 x\, dx = \frac{1}{2}(\sec x \tan x + \ln|\sec x + \tan x|) + C.$

4. Let $u = x$, $dv = xe^{-x^2}\, dx$. Then $du = dx$, $v = \int xe^{-x^2}\, dx = -\frac{1}{2}e^{-x^2}$ (sub $u = -x^2$, $du = -2x\, dx$). Then
$\int x^2 e^{-x^2}\, dx = -\frac{1}{2}xe^{-x^2} + \frac{1}{2} \int e^{-x^2}\, dx = -\frac{1}{2}xe^{-x^2} + \frac{1}{2}Q(x) + C.$

Section 7.6 (page 206)

1. Let $u = x^n$, $dv = e^x\, dx$. Then $du = nx^{n-1}\, dx$, $v = e^x$,
$\int x^n e^x\, dx = x^n e^x - n \int x^{n-1} e^x\, dx.$
2. $\int (\sec^2 x - 1) \tan^{n-2}x\, dx = \int \sec^2 x \tan^{n-2}x\, dx - \int \tan^{n-2}x\, dx.$ For first integral, sub $u = \tan x$,
$du = \sec^2 x\, dx$ to get $(\tan^{n-1}x)/(n - 1) - \int \tan^{n-2}x\, dx.$
3. Let $u = (\ln x)^n$, $dv = dx$. Then $du = n(\ln x)^{n-1} \cdot \frac{1}{x}\, dx$,
$v = x$, $\int (\ln x)^n\, dx = x(\ln x)^n - n \int (\ln x)^{n-1}\, dx$. Using it,
$\int (\ln x)^3\, dx = x(\ln x)^3 - 3 \int (\ln x)^2\, dx$
$= x(\ln x)^3 - 3[x(\ln x)^2 - 2 \int \ln x\, dx]$
$= x(\ln x)^3 - 3x(\ln x)^2 + 6(x \ln x - x) + C.$
4. First use 52(a) to get $\int \sin^m x \cos^n x\, dx =$

$-\dfrac{\sin^{m-1}x \cos^{n+1}x}{m + n} + \dfrac{m - 1}{m + n} \int \sin^{m-2}x \cos^n x\, dx$. Continue

with 52(b) on the last integral. Note that 52(b) uses the letter m as the sine exponent but our integral has $m - 2$ as the sine exponent so 52(b) has to be used with its letter m replaced by $m - 2$ to get

$$\int \sin^m x \cos^n x\, dx = -\frac{\sin^{m-1}x \cos^{n+1}x}{m + n} + \frac{m - 1}{m + n} \cdot$$
$$\left[\frac{\sin^{m-1}x \cos^{n-1}x}{m - 2 + n} + \frac{n - 1}{m - 2 + n} \int \sin^{m-2}x \cos^{n-2}x\, dx\right].$$

5. New integral is missing the numerator x and is *not* of the same form as the original.

6. (a) Let $u = \sin x$, $du = \cos x\, dx$.
$\int \sin x \cos x\, dx = \int u\, du = \frac{1}{2}u^2 + C$
$= \frac{1}{2} \sin^2 x + C.$
(Can also use $u = \cos x$, $du = -\sin x\, dx$.)

(b) Let $u = \cos x$, $du = -\sin x\, dx$.
$-\int u^{12}\, du = -\frac{1}{13}u^{13} + C = -\frac{1}{13} \cos^{13}x + C.$

(c) Use 52(c) with $m = 0$, $n = -5$ to get
$\dfrac{-\sin x \cos^{-4}x}{-4} + \dfrac{-3}{-4} \int \sec^3 x\, dx.$
Now use 43 to get $\frac{1}{4} \sin x \sec^4 x +$
$\frac{3}{4}[\frac{1}{2} \sec x \tan x + \frac{1}{2} \ln|\sec x + \tan x|] + C.$

(d) Use 53; $\int \tan^4 x\, dx = \frac{1}{3} \tan^3 x - \int \tan^2 x\, dx$
$= \frac{1}{3} \tan^3 x - (\tan x - x) + C$ (by 41).

(e) Let $u = \sin x$, $du = \cos x\, dx$.
$\int du/u^2 = -1/u + C = -1/\sin x + C.$

(f) $-\dfrac{\sin^3 x \cos^{-2}x}{-2} - \dfrac{1}{2} \int \dfrac{\sin^2 x}{\cos x}\, dx$ (by 52(c))
$= \frac{1}{2} \sin^3 x \sec^2 x - \frac{1}{2}(-\sin x + \int \sec x\, dx)$ (52(a))
$= \frac{1}{2} \sin^3 x \sec^2 x - \frac{1}{2}(-\sin x + \ln|\sec x + \tan x|)$
$+ C$ (by 33).

(g) Can use 52(b) once or can use
$\int \sin^4 x(1 - \sin^2 x) \cos x\, dx = \int (u^4 - u^6)\, du$
(let $u = \sin x$, $du = \cos x\, dx$)
$= \frac{1}{5}u^5 - \frac{1}{7}u^7 + C = \frac{1}{5} \sin^5 x - \frac{1}{7} \sin^7 x + C.$

(h) Let $u = 3x$, $du = 3\, dx$. Then $\frac{1}{3} \int \sin^4 u\, du$
$= \frac{1}{3}(-\frac{1}{4} \sin^3 u \cos u + \frac{3}{4} \int \sin^2 u\, du)$ (by 52(a))
$= -\frac{1}{12} \sin^3 u \cos u + \frac{1}{4} \cdot \frac{1}{2}(u - \sin u \cos u) + C$
$= -\frac{1}{12} \sin^3 3x \cos 3x + \frac{1}{8}(3x - \sin 3x \cos 3x)$
$+ C.$

7. $\dfrac{-\sin^2 x \cos^{99}x}{101} - \dfrac{2}{101(99)} \cos^{99}x$
$= \dfrac{-\cos^{99}x}{99}\left[\dfrac{99}{101} \sin^2 x + \dfrac{2}{101}\right]$ (use $\sin^2 x = 1 - \cos^2 x$)
$= -\dfrac{\cos^{99}x}{99}\left[1 - \dfrac{99}{101} \cos^2 x\right] = \dfrac{-\cos^{99}x}{99} + \dfrac{\cos^{101}x}{101}.$

Section 7.7 (page 209)

In each problem, use the indicated diagram.

1. (a) $\sqrt{x^2 - a^2} = a \tan u$, $x = a \sec u$,
$dx = a \sec u \tan u\, du$.

$\displaystyle\int \frac{1}{a \tan u} a \sec u \tan u\, du$

$= \int \sec u\, du = \ln|\sec u + \tan u| + C$ (by 33)

$= \ln\left|\dfrac{x}{a} + \dfrac{\sqrt{x^2 - a^2}}{a}\right| + C$ (keep going with

algebra to make answer look like the tables)

$= \ln\dfrac{|x + \sqrt{x^2 - a^2}|}{a} + C$

$= \ln|x + \sqrt{x^2 - a^2}| - \ln a + C$

$= \ln|x + \sqrt{x^2 - a^2}| + K$.

PROBLEM $1(a)$

(b) $\sqrt{a^2 - x^2} = a \cos u$, $x = a \sin u$,
$dx = a \cos u\, du$.

$\int a \cos u \cdot a \cos u\, du$

$= a^2 \int \cos^2 u\, du = \frac{1}{2}a^2(u + \sin u \cos u) + C$

$= \frac{1}{2}a^2\left(\arcsin \dfrac{x}{a} + \dfrac{x}{a}\dfrac{\sqrt{a^2 - x^2}}{a}\right) + C$

$= \frac{1}{2}a^2 \arcsin x/a + \frac{1}{2}x\sqrt{a^2 - x^2} + C$.

PROBLEM $1(b)$

(c) $\sqrt{a^2 + x^2} = a \sec u$, $x = a \tan u$,
$dx = a \sec^2 u\, du$.

$\displaystyle\int \frac{a \sec u}{a \tan u} a \sec^2 u\, du = a \int \frac{du}{\cos^2 u \sin u}$

$\qquad$ (now use 52(c) with $m = -1$, $n = -2$)

$= a(\sec u + \int \csc u\, du)$

$= a \sec u - a \ln|\csc u + \cot u| + C$ (by 34)

$= \sqrt{a^2 + x^2} - a \ln\left|\dfrac{\sqrt{a^2 + x^2}}{x} + \dfrac{a}{x}\right| + C$.

PROBLEM $1(c)$

2. $\sqrt{3 - x^2} = \sqrt{3} \cos u$, $x = \sqrt{3} \sin u$,
$dx = \sqrt{3} \cos u\, du$.

$\displaystyle\int \frac{\sqrt{3} \cos u}{3 \sin^2 u} \sqrt{3} \cos u\, du = \int \cot^2 u\, du$

$= -\cot u - u + C$ (by 42)

$= -\dfrac{\sqrt{3 - x^2}}{x} - \sin^{-1}\dfrac{x}{\sqrt{3}} + C$.

PROBLEM 2

3. $\sqrt{x^2 - 5} = \sqrt{5} \tan u$, $x = \sqrt{5} \sec u$,

$dx = \sqrt{5} \sec u \tan u\, du$. Then $\displaystyle\int \frac{\sqrt{5} \sec u \tan u\, du}{5 \sec^2 u \sqrt{5} \tan u} =$

$\frac{1}{5}\int \cos u\, du = \frac{1}{5} \sin u + C = \frac{1}{5}\dfrac{\sqrt{x^2 - 5}}{x} + C$.

PROBLEM 3

4. $\sqrt{7 + x^2} = \sqrt{7} \sec u$, $(7 + x^2)^2 = (\sqrt{7} \sec u)^4 = $
$49 \sec^4 u$, $x = \sqrt{7} \tan u$, $dx = \sqrt{7} \sec^2 u\, du$;

$\displaystyle\int \frac{\sqrt{7} \sec^2 u\, du}{49 \sec^4 u} = \frac{1}{7\sqrt{7}} \int \cos^2 u\, du$

$= \dfrac{1}{7\sqrt{7}} \frac{1}{2}(u + \sin u \cos u) + C$

$= \dfrac{1}{14\sqrt{7}}\left(\tan^{-1}\dfrac{x}{\sqrt{7}} + \dfrac{x\sqrt{7}}{7 + x^2}\right) + C$.

PROBLEM 4

5. $x = a \tan u$, $dx = a \sec^2 u\, du$, $\sqrt{a^2 + x^2} = a \sec u$;

$\displaystyle\int \frac{a \sec^2 u\, du}{(a \sec u)^3} = \frac{1}{a^2} \int \cos u\, du = \frac{1}{a^2} \sin u + C$

$= \dfrac{x}{a^2\sqrt{a^2 + x^2}} + C$.

PROBLEM 5

Section 7.8 (page 210)

1. Sub $u = \sqrt{x}$
2. Sub $u = 1 - x^2$
3. Formula 16 with $a = \sqrt{3}$
4. $\sqrt{2x + 3} + C$ by inspection (or sub $u = 2x + 3$)
5. Sub $u = x - 1$ or parts with $u = x$, $dv = (x - 1)^{20}\,dx$
6. $\int e^{-x}\,dx = -e^{-x} + C$ by inspection (or sub $u = -x$)
7. Long division and then inspection
8. Sub $u = 4 - x^2$
9. $x^2 + 9x + C$ by inspection
10. Formula 19
11. Formula 9 or partial fractions
12. Sub $u = 3x$. Then formula 31.
13. Formula 11 or partial fractions
14. Factor out the 2 (or sub $u = \sqrt{2}x$). Then formula 21. Or trig sub.
15. Partial fractions
16. Long division and then inspection
17. Formula 52(b) once, then 42. Or use $\cos^4 x = (1 - \sin^2 x)^2$.
18. $-(1/\pi) \cos \pi x + C$ by inspection (or sub $u = \pi x$)
19. $-\dfrac{1}{24(3x + 1)^8} + C$ by inspection (or sub $u = 3x + 1$)
20. Sub $u = 9 + 4x^3$
21. $\int \dfrac{\sin^3 x}{\cos x}\,dx$. Use 52(a) once, then 31. Or use $\dfrac{\sin^3 x}{\cos x} = \dfrac{\sin x \cdot (1 - \cos^2 x)}{\cos x} = \tan x - \sin x \cos x$.
22. Tables 22
23. $\frac{1}{16}(9 + 4x)^4$ by inspection (or sub $u = 9 + 4x$)
24. $\int \sec^2 x\,dx$. Formula 35.
25. Trig sub (see fig.)

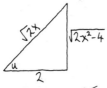

PROBLEM 25

26. Use 52(a) twice or use $\sin^4 x = (1 - \cos^2 x)^2$. Then sub $u = \cos x$.
27. $\int (1 - 4/x^2)\,dx = x + 4/x + C$
28. Formula 43
29. $\frac{1}{2} \ln|2x + 1| + C$ by inspection (or sub $u = 2x + 1$)
30. Sub $u = x^2$
31. Use 52(c) once. Then 35.
32. Use 50 (or integrate by parts twice).
33. Factor out the 3 (or sub $u = \sqrt{3}x$) and use formula 23.
34. $\frac{1}{3}e^{3x} + C$ by inspection (or sub $u = 3x$)

35. Long division (or formula 5)
$\int \frac{1}{2}dx - \frac{3}{2} \int \dfrac{dx}{2x + 3} = \frac{1}{2}x - \frac{3}{4} \ln|2x + 3| + C$
36. $-\frac{1}{5} \cos 5x + C$ by inspection (or sub $u = 5x$)
37. Sub $u = 2 - r^2$
38. Use 42.
39. 52(a) once. Or $\sin x \cdot (1 - \cos^2 x)$ and sub $u = \cos x$.
40. $2x + C$
41. Sub $u = 2x$. Then 33.
42. $\int (2 + 3/x)\,dx = 2x + 3 \ln|x| + C$
43. Use 46.
44. $-\frac{1}{2}\pi \cos(2x/\pi)$ by inspection
45. Sub $u = \sin 2x$ (or sub $u = \cos 2x$).
46. $\frac{1}{5} \ln|5x - 2| + C$ by inspection
47. Multiply out numerator and then use long division.
48. Sub $u = x^2 + 7$
49. Sub $u = \cos x$
50. Parts with $u = \sin^{-1}x$, $dv = x\,dx$. Get $\frac{1}{2}x^2 \sin^{-1}x - \frac{1}{2} \int \dfrac{x^2\,dx}{\sqrt{1 - x^2}}$. Then trig sub (or big table).
51. Sub $u = x^2$
52. Partial fractions
53. Use 61 (or integration by parts)
54. 52(a) or $\dfrac{1 - \cos^2 x}{\cos x} = \sec x - \cos x$
55. Factor out 2 (or sub $u = \sqrt{2}x$). Then 27.
56. Sub $u = 3 - 2x$. Then 31.
57. $-\frac{2}{3}(3 - x)^{3/2} + C$ by inspection
58. $-\frac{3}{5}(2 - \frac{1}{3}x)^5 + C$ by inspection
59. Sub $u = e^x$, $du = e^x\,dx$. Get $\int \dfrac{1/u}{u + 1/u}\,du = \int \dfrac{1}{u^2 + 1}\,du$. Then formula 16.
60. $\int \sin x \cos^3 x\,dx$. Sub $u = \cos x$.
61. Factor out 3 (or sub $u = \sqrt{3}x$). Then formula 20.
62. Complete square. Then formula 24.
63. Use 1(b).
64. Sub $u = 4x + 5$. Then 62.
65. $2e^{\theta/2} + C$ by inspection (or sub $u = \frac{1}{2}\theta$)
66. $\int \dfrac{dx}{1 + x^2}$ (use 16) $+ 2 \int \dfrac{x\,dx}{1 + x^2}$ (sub $u = 1 + x^2$)
67. Trig sub (see fig.)

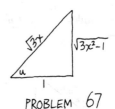

PROBLEM 67

68. 52(a) once. Then 39 (or 52(a) twice).
69. 52(b) once or use $\cos^2 x = 1 - \sin^2 x$. Then sub $u = \sin x$.

70. $\int \tan 2x \, dx$. Sub $u = 2x$ and use 31.

71. $(1 + e^x)^2 = 1 + 2e^x + 2^{2x}$

antideriv $= x + 2e^x + \dfrac{1}{2} e^{2x}$

72. 52(a) twice or $\sin x \cdot (1 - \cos^2 x)^2$ and sub $u = \cos x$.

73. Complete square. Then 19.

74. Sub $u = \cos 2x$. Get $\int \dfrac{-\frac{1}{2} \, du}{9 - u^2}$. Then 17 or partial fractions.

75. Partial fractions

76. Sub $u = 2 + 3x$. Or parts with $u = x$, $dv = (2 + 3x)^4 \, dx$. Or multiply out whole integrand to get a polynomial.

77. Formula 64

78. $\displaystyle\int \dfrac{x \, dx}{2x^2 + x - 1} + \int \dfrac{4 \, dx}{2x^2 + x - 1}$

Use 2, 1(a) on first; use 1(a) on second.

79. Parts twice starting with $u = (\ln x)^3$, $dv = dx$, $du = 3(\ln x)^2 \cdot dx/x$, $v = x$.

80. Sub $u = 3x$. Then 41.

81. Parts; $u = x^3$, $dv = \sin x \, dx$. Then 51.

82. Sub $u = 2 + \cos x$.

83. Formula 40

84. Can start with 52(b). For another method, $\int \cos^3 x \, dx = \int \cos x (1 - \sin^2 x) \, dx$ $= \int \cos x \, dx - \int \cos x \, \sin^2 x \, dx$. Sub $u = \sin x$ for second integral.

85. Sub $u = \cos x$.

Section 7.9 (page 214)

1. (a) $du = 3 \, dx$, $\int_2^5 \sin^5 x \, dx = \frac{1}{3} \int_6^{15} \sin^5 \frac{1}{3} u \, du$

(b) $du = \frac{1}{x} dx$, $\int_1^{e^3} \sin(\ln x) \, dx = \int_0^3 \sin u \cdot x \, du$
$= \int_0^3 e^u \sin u \, du$

(c) $x = 2 \csc u$, $dx = -2 \csc u \cot u \, du$, $\sqrt{x^2 - 4} = 2 \cot u$. If $x = 2$ then $u = \pi/2$ (degenerate triangle). If $x = 4$ then $u = \pi/6$. (See fig.)
$\displaystyle\int_2^4 \dfrac{\sqrt{x^2 - 4}}{x^2} \, dx = \int_{\pi/2}^{\pi/6} \dfrac{2 \cot u}{4 \csc^2 u} \cdot -2 \csc u \cot u \, du$
$\displaystyle = -\int_{\pi/2}^{\pi/6} \dfrac{\cos^2 u}{\sin u} \, du$

PROBLEM 1(c)

2. (a) Let $u = 3x^2 - 1$, $du = 6x \, dx$.
$\frac{1}{6} \int_{11}^{47} u^{10} \, du = \frac{1}{6} \cdot \frac{1}{11} u^{11} \big|_{11}^{47} = \frac{1}{66}(47^{11} - 11^{11})$

(b) Parts with $u = e^{-x}$, $dv = \cos x \, dx$.
$\int_0^\infty e^{-x} \cos x \, dx = e^{-x} \sin x \big|_0^\infty + \int_0^\infty e^{-x} \sin x \, dx$
$= \int_0^\infty e^{-x} \sin x \, dx$
$= -e^{-x} \cos x \big|_0^\infty - \int_0^\infty e^{-x} \cos x \, dx$ (parts again).
So $2 \int_0^\infty e^{-x} \cos x \, dx = 1$, $\int_0^\infty e^{-x} \cos x \, dx = \frac{1}{2}$.

(c) Let $u = \ln x$, $du = \frac{1}{x} dx$; $\int_0^1 u^5 \, du = \frac{1}{6} u^6 \big|_0^1 = \frac{1}{6}$.

(d) $-\frac{1}{4} \sin x \cos^3 x \big|_{\pi/2}^\pi + \frac{1}{4} \int_{\pi/2}^\pi \cos^2 x \, dx$ (by 52(a))
$= \frac{1}{4} \cdot \frac{1}{2}(x + \sin x \cos x) \big|_{\pi/2}^\pi$ (by (40)) $= \pi/16$

(e) Let $u = x^3$, $du = 3x^2 \, dx$.
$\int_{-\infty}^2 x^2 e^{x^3} \, dx = \frac{1}{3} \int_{-\infty}^8 e^u \, du = \frac{1}{3} e^8$

(f) Let $u = x^2 + 4$, $du = 2x \, dx$.
$\frac{1}{2} \int_4^8 \sqrt{u} \, du = \frac{1}{3} u^{3/2} \big|_4^8 = 8\sqrt{8}/3 - 8/3$

3. (a) Let $u = 1 - x$, $du = -dx$.
$\int_0^1 x^m (1 - x)^n \, dx = -\int_1^0 (1 - u)^m u^n \, du$ (sub)
$= \int_0^1 (1 - u)^m u^n \, du$ (since $\int_b^a f(x) \, dx$
$= -\int_a^b f(x) \, dx) = \int_0^1 (1 - x)^m x^n \, dx$ (change dummy variable from u to x)

(b) Let $u = x + 20$, $du = dx$.
$\int_0^{10} (x + 20)^2 \, dx = \int_{20}^{30} u^2 \, du$ (sub) $= \int_{20}^{30} x^2 \, dx$ (change dummy variable)

(c) Let $u = \frac{1}{2} x$, $du = \frac{1}{2} dx$.
$\int_{2a}^{2b} \sqrt{\sin \frac{1}{2} x} \, dx = 2 \int_a^b \sqrt{\sin u} \, du = 2 \int_a^b \sqrt{\sin x} \, dx$ (change dummy variable)

4. $u = \ln \ln x$, $dv = x \, dx$, $du = \dfrac{1}{\ln x} \cdot \dfrac{1}{x} \, dx$, $v = \frac{1}{2} x^2$,

$\int_2^3 x \ln \ln x \, dx = \frac{1}{2} x^2 \ln \ln x \big|_2^3 - \frac{1}{2} \int_2^3 \dfrac{x \, dx}{\ln x}$
$= \frac{9}{2} \ln \ln 3 - 2 \ln \ln 2 - \frac{1}{2} k$.

Chapter 7 Review Problems (page 214)

1. (a) Formula 2 with $a = 1$, $b = 0$, $c = 1$. Get
$\frac{1}{2} \ln(x^2 + 1) + C$.

(b) Let $u = x^2 + 1$, $du = 2x \, dx$. Then
$\frac{1}{2} \int du/u = \frac{1}{2} \ln|u| + C = \frac{1}{2} \ln(x^2 + 1) + C$.

(c) (See fig.) $\tan u = x$, $dx = \sec^2 u \, du$,
$\sqrt{x^2 + 1} = \sec u$;
$\displaystyle\int \dfrac{x}{x^2 + 1} \, dx = \int \dfrac{\tan u \, \sec^2 u \, du}{\sec^2 u} = \int \tan u \, du$
$= \ln|\sec u| + C$ (by formula 31)
$= \ln \sqrt{x^2 + 1} + C = \ln(x^2 + 1)^{1/2} + C$
$= \frac{1}{2} \ln(x^2 + 1) + C$.

PROBLEM 1(c)

(d) Let $u = x$, $dv = \dfrac{1}{x^2 + 1}\,dx$. Then $du = dx$,

$v = \tan^{-1}x$ and

$\displaystyle\int \frac{x}{x^2 + 1}\,dx = x\tan^{-1}x - \int \tan^{-1}x\,dx$ (use 59)

$= x\tan^{-1}x - (x\tan^{-1}x - \tfrac{1}{2}\ln(1 + x^2)) + C$

$= \tfrac{1}{2}\ln(x^2 + 1) + C.$

2. (a) Let $u = x + 2$ (could also sub $u = x - 4$), $du = dx$. Then $x = u - 2$. Get

$\displaystyle\int \frac{1}{u(u - 2 - 4)}\,du = \int \frac{1}{u(u - 6)}\,du$

$= -\dfrac{1}{-6}\ln\left|\dfrac{u - 6}{u}\right| + C = \dfrac{1}{6}\ln\left|\dfrac{x - 4}{x + 2}\right| + C.$

(b) $\displaystyle\int \frac{1}{x^2 - 2x - 8}\,dx = \int \frac{1}{(x - 1)^2 - 9}\,dx$

(now sub $u = x - 1$, $du = dx$)

$= \dfrac{1}{6}\ln\left|\dfrac{x - 1 - 3}{x - 1 + 3}\right| + C = \dfrac{1}{6}\ln\left|\dfrac{x - 4}{x + 2}\right| + C.$

(c) $\displaystyle\int \frac{1}{x^2 - 2x - 8}\,dx = \dfrac{1}{6}\ln\left|\dfrac{2x - 2 - 6}{2x - 2 + 6}\right| + C$

$= \dfrac{1}{6}\ln\left|\dfrac{x - 4}{x + 2}\right| + C.$ (Use 1(a).)

(d) $\dfrac{1}{(x + 2)(x - 4)}$ decomposes to $\dfrac{-1/6}{x + 2} + \dfrac{1/6}{x - 4}$ so

integral is $-\tfrac{1}{6}\ln|x + 2| + \tfrac{1}{6}\ln|x - 4| + C$ which

is $\dfrac{1}{6}\ln\left|\dfrac{x - 4}{x + 2}\right| + C.$

3. (a) Formula 64 (or parts twice)

(b) $\tfrac{1}{3}\ln|3x + 4| + C$ by inspection

(c) Formula 19

(d) Sub $u = 1 + 2x^3$

(e) Sub $u = 3x$. Then formula 41.

(f) $-\tfrac{1}{6}e^{-6x} + C$ by inspection

(g) $\tfrac{1}{5}\ln|x| + C$ by inspection

(h) Long division and then inspection (or formula 5)

(i) $\int (1 + 3/x)\,dx = x + 3\ln|x| + C$

(j) Long division and then inspection

(k) Formula 13 or sub $u = 3x + 4$

(l) Complete square. Then formula 27.

(m) Partial fractions or formula 10

(n) Formula 1(b)

4. (a) By 45 with $a = 5$, $b = 3$.

$\tfrac{1}{4}\sin 2x - \tfrac{1}{16}\sin 8x + C$

(b) $\tfrac{1}{2}\int (\cos 2x - \cos 8x)\,dx$

$= \tfrac{1}{4}\sin 2x - \tfrac{1}{16}\sin 8x + C$

(c) Let $u = \sin 3x$, $dv = \sin 5x\,dx$ (or vice versa).

Then $du = 3\cos 3x\,dx$, $v = -\tfrac{1}{5}\cos 5x$.

$\int \sin 3x\,\sin 5x\,dx$

$= -\tfrac{1}{5}\sin 3x\,\cos 5x + \tfrac{3}{5}\int \cos 3x\,\cos 5x\,dx.$

Now let $u = \cos 3x$, $dv = \cos 5x\,dx$,

$du = -3\sin 3x\,dx$, $v = \tfrac{1}{5}\sin 5x$. Then

$\int \sin 3x\,\sin 5x\,dx = -\tfrac{1}{5}\sin 3x\,\cos 5x +$

$\tfrac{3}{5}(\tfrac{1}{5}\cos 3x\,\sin 5x + \tfrac{3}{5}\int \sin 3x\,\sin 5x\,dx).$

Collect terms to get $\tfrac{16}{25}\int \sin 3x\,\sin 5x\,dx$

$= -\tfrac{1}{5}\sin 3x\,\cos 5x + \tfrac{3}{25}\cos 3x\,\sin 5x.$

So $\int \sin 3x\,\sin 5x\,dx$

$= -\tfrac{5}{16}\sin 3x\,\cos 5x + \tfrac{3}{16}\cos 3x\,\sin 5x + C$

$= \tfrac{1}{4}\sin 2x - \tfrac{1}{16}\sin 8x + C$ using the identities

for $\cos x\,\sin y$ and $\sin x\,\cos y$ in Sect. 1.3.

5. Let $u = \tan x$, $dv = e^x\,dx$. Then $du = \sec^2x\,dx$, $v = e^x$

and $\int_0^{\pi/3} e^x \tan x\,dx = e^x \tan x\,\big|_0^{\pi/3} - \int_0^{\pi/3} e^x \sec^2x\,dx$

$= e^{\pi/3}\sqrt{3} - Q.$

6. Let $u = 2 + x^2$, $du = 2x\,dx$.

$\int_0^1 x(2 + x^2)^5\,dx = \tfrac{1}{2}\int_2^3 u^5\,du = \tfrac{1}{12}u^6\big|_2^3 = 665/12$

7. (a) $-\cos x + C$

(b) $\tfrac{1}{2}(x - \sin x\,\cos x) + C$ by 39

(c) One method is to use 52(a). As another method

$\int \sin^3 x\,dx = \int \sin x(1 - \cos^2 x)\,dx$

$= \int \sin x\,dx - \int \sin x\,\cos^2 x\,dx.$

First integral is $-\cos x$. For second,

sub $u = \cos x$, $du = -\sin x\,dx$ to get

$\int \sin x\,\cos^2 x\,dx = \int (-u^2)\,du = -\tfrac{1}{3}u^3 =$

$-\tfrac{1}{3}\cos^3 x.$ Final answer is $-\cos x + \tfrac{1}{3}\cos^3 x + C.$

(d) Let $u = \sin x$, $du = \cos x\,dx$ (or could sub

$u = \cos x$). Get $\int u\,du = \tfrac{1}{2}u^2 + C = \tfrac{1}{2}\sin^2 x + C$

(e) Let $u = \sin x$, $du = \cos x\,dx$. Get

$\int u^2\,du = \tfrac{1}{3}u^3 + C = \tfrac{1}{3}\sin^3 x + C.$

(f) Begin with 52(a) (or 52(b)) to get

$-\tfrac{1}{4}\sin x\,\cos^3 x + \tfrac{1}{4}\int \cos^2 x\,dx$ (now use 40)

$= -\tfrac{1}{4}\sin x\,\cos^3 x + \tfrac{1}{4}\cdot\tfrac{1}{2}(x + \sin x\,\cos x) + C.$

(g) $\dfrac{x^{-1}}{-1} + C = -1/x + C$

(h) $\ln|x| + C$

(i) $\dfrac{x^{1/2}}{1/2} + C = \tfrac{1}{2}\sqrt{x} + C$

8/SERIES

Section 8.1 (page 219)

1. (a) $(-1)^3 \frac{1}{7} + (-1)^4 \frac{1}{9} + (-1)^5 \frac{1}{11}$, i.e., $-\frac{1}{7} + \frac{1}{9} - \frac{1}{11}$
   (b) $a_1 + 4a_2 + 9a_3$

2. (a) Partial sums are $1, -1, 2, -2, 3, -3, \cdots$. They oscillate wildly so series diverges (not to ∞ or $-\infty$ but plain diverges)
   (b) Partial sums are $\frac{1}{2}, 1, 1\frac{1}{2}, 2, 2\frac{1}{2}, 3, \cdots$ which $\to \infty$. Series diverges to ∞.

3. (a) $S_n \to \infty$ as $n \to \infty$. Series diverges (to ∞). Since partial sums are $1, 2, 3, 4, 5, \cdots$, series itself is $1 + 1 + 1 + 1 + 1 + \cdots$.
   (b) $S_n \to 1$ as $n \to \infty$ so series converges to 1. Since partial sums are $1, 1, 1, 1, 1, \cdots$, series is $1 + 0 + 0 + 0 + 0 + 0 + \cdots$.

4. Series is $(1 - \frac{1}{2}) + (\frac{1}{2} - \frac{1}{3}) + (\frac{1}{3} - \frac{1}{4}) + (\frac{1}{4} - \frac{1}{5}) + \cdots$. Partial sums are $S_1 = 1 - \frac{1}{2}$, $S_2 = 1 - \frac{1}{3}$ (the $\frac{1}{2}$'s cancel in the sum), $S_3 = 1 - \frac{1}{4}$ (other terms cancel in the sum), etc. Limit is 1; series converges to 1.

5. $1 + 2 + 3 + 4 + \cdots$ and $1 + 1 + 1 + \cdots$ both diverge (to ∞) and so does the sum series $2 + 3 + 4 + 5 + \cdots$. On the other hand, $1 + 1 + 1 + \cdots$ and $(-1) + (-1) + (-1) + (-1) + \cdots$ both diverge but the sum series $0 + 0 + 0 + \cdots$ converges (to 0).

6. S_{99} adds up the first 99 a's and S_{100} adds up the first 100 a's. So $S_{100} - S_{99} = a_{100}$.

Section 8.2 (page 221)

1. $r = -\frac{1}{6}, a = -1$. Series converges since $-1 < r < 1$. Sum is $-1/(1 - -\frac{1}{6}) = -\frac{6}{7}$.

2. $r = -\frac{1}{4}$, $a = \frac{1}{4}$. Converges to $\frac{1}{4}/(1 - -\frac{1}{4}) = \frac{1}{5}$.

3. $r = \frac{3}{2} > 1$. Series diverges (to ∞).

4. $r = 3 > 1$. Diverges (to ∞). (Even without geom series test, series obviously diverges since partial sums $\to \infty$.)

5. Series is $1/4^3 + 1/4^4 + 1/4^5 + \cdots$; $a = 1/4^3$, $r = 1/4$. Converges to $\dfrac{1/4^3}{1 - 1/4} = 1/48$.

6. $a = \frac{1}{4}$, $r = (\frac{2}{3})^2$. Converges to $\frac{1}{4}/(1 - (\frac{2}{3})^2) = \frac{9}{20}$

7. $a = .1$, $r = .1$. Converges to $.1/(1 - .1) = \frac{1}{9}$ (i.e., the repeating decimal $.1111111...$ is the fraction $\frac{1}{9}$)

8. Series is $\sin^2\theta + \sin^4\theta + \sin^6\theta + \cdots$; $a = \sin^2\theta$, $r = \sin^2\theta$ which is between -1 and 1 provided $\theta \neq \pi/2$,

$3\pi/2, 5\pi/2$, etc. In that case series converges to $\sin^2\theta/(1 - \sin^2\theta) = \sin^2\theta/\cos^2\theta = \tan^2\theta$.

9. Series is $1/\pi + 1/\pi^3 + 1/\pi^5 + \cdots$; $a = 1/\pi$, $r = 1/\pi^2$. Converges to $\dfrac{1/\pi}{1 - 1/\pi^2} = \pi/(\pi^2 - 1)$.

Section 8.3 (page 227)

1. (use nth term test)
   (a) F (c) F
   (b) T (d) T

2. (a) $n!/4^n \to \infty$ as $n \to \infty$ since $n!$ is listed in (4) as having higher order of magnitude. Series diverges by nth term test.
   (b) $n^2/4^n \to 0$ as $n \to \infty$ since 4^n has higher order of magnitude. No conclusion from nth term.

3. (a) $\Sigma 1/3^n$, geometric, $r = \frac{1}{3}$, converges.
   (b) $\Sigma 1/n^2$, p-series, $p = 2$, converges.
   (c) $-\Sigma 1/\sqrt{n}$, p-series, $p = \frac{1}{2}$, diverges.
   (d) $\Sigma \sqrt{n}$, terms do not $\to 0$, diverge by nth term test.
   (e) $3 \Sigma 1/n!$, standard convergent series.
   (f) $\Sigma 1/n^3$ p-series, $p = 3$, convergent.
   (g) Every other term (i.e., subseries) of $\Sigma 1/n^{3/2}$, p-series, $p = 3/2$, convergent.
   (h) (Drop 5 and 6) $\Sigma 1/n$, diverges (harmonic).
   (i) $\Sigma_{n=3}^{\infty} n/(n + 1)$, terms $\to 1$ not 0, diverges by nth term test.
   (j) $1/2^n n! < 1/n!$ so $\Sigma 1/2^n n!$ conv by comparison with convergent series $\Sigma 1/n!$ (or by comparison with $\Sigma 1/2^n$).
   (k) $1/n2^n < 1/2^n$ so $\Sigma 1/n2^n$ converges by comparison with convergent geom series $\Sigma 1/2^n$.
   (l) $\Sigma 1/n^5$, p-series, $p = 5$, converges.
   (m) $\Sigma 1/5^n$, geom, $r = 1/5$, converges.
   (n) Converges by comparison with $1/e^3 + 1/e^4 + \cdots$ (conv geom, $r = 1/e$).
   (o) $\Sigma 1/n^2$, p-series, $p = 2$, converges.
   (p) Many methods. Terms are less than those of $\frac{1}{8} + \frac{1}{80} + \frac{1}{800} + \cdots$, conv geom. $(r = 1/10)$; given series converges by comparison.
   (q) Converges; subseries of the convergent series $\Sigma 1/n!$.

4. (a) Σa_n converges so $a_n \to 0$, $1/a_n$ does not $\to 0$, $\Sigma 1/a_n$ diverges by nth term test.

(b) $a_n/n! < a_n$; if Σa_n converges so does $\Sigma a_n/n!$ by comparison.

(c) Can't tell. If Σa_n is $\Sigma 1/n!$ then $\Sigma n! a_n$ is $1 + 1 + 1 + 1 + \cdots$ which diverges. But if Σa_n is $\Sigma 1/(n!)^2$ then $\Sigma n! a_n$ is $\Sigma 1/n!$ which converges.

(d) Σa_n converges so $a_n \to 0$. But then $\cos a_n \to 1$ so $\Sigma \cos a_n$ diverges by nth term test.

Section 8.4 (page 231)

1. Acts like $\frac{1}{2} \Sigma 1/n^2$, convergent p-series. So original converges also.

2. $\dfrac{a_{n+1}}{a_n} = \dfrac{[2(n+1)]! \, (3n)!}{[3(n+1)]! \, (2n)!} = \dfrac{(2n+2)! \, (3n)!}{(3n+3)! \, (2n)!}$
$= \dfrac{(2n+2)(2n+1)}{(3n+3)(3n+2)(3n+1)}$.
Limit as $n \to \infty$ is 0 (denominator contains n^3, numerator only contains n^2); series converges by ratio test.

3. p-series, $p = \frac{1}{3}$, diverges.

4. Same as $\Sigma 1/3^n$, geometric series, $r = \frac{1}{3}$, converges.

5. $n!/10^n$ does not approach 0 as $n \to \infty$; series diverges by nth term test.

6. $\Sigma 1/\sqrt[n]{2}$, $\sqrt[n]{2} \to 1$ as $n \to \infty$ so $1/\sqrt[n]{2}$ does not $\to 0$, series diverges by nth term test.

7. Acts like $\Sigma 1/n^3$ (p-series, $p = 3$), original converges.

8. $-\Sigma (n-1)/n^2$, acts like $-\Sigma 1/n$ (div harmonic) so original diverges.

9. $\dfrac{a_{n+1}}{a_n} = \dfrac{(n+1)^2 (\frac{3}{4})^{n+1}}{n^2 (\frac{3}{4})^n} = \dfrac{3}{4} \left(\dfrac{n+1}{n} \right)^2$. Limit as $n \to \infty$ is $\frac{3}{4}$. Series converges by ratio test.

10. $\dfrac{a_{n+1}}{a_n} = \dfrac{10^{n+1}}{(n+1)!} \dfrac{n!}{10^n} = \dfrac{10}{n+1}$. Limit is 0 as $n \to \infty$. Series converges by ratio test.

11. $\Sigma 1/\sqrt{n}$ diverges ($p = \frac{1}{2}$). Terms of $\Sigma \ln n/\sqrt{n}$ are even larger so it too diverges.

12. $(n-1)/n$ does not $\to 0$ as $n \to \infty$, series diverges by nth term test.

13. Acts like $2 \Sigma 1/n$, diverges.

14. $\dfrac{a_{n+1}}{a_n} = \dfrac{(n+1)^2}{5^{n+1}} \dfrac{5^n}{n^2} = \dfrac{1}{5} \left(\dfrac{n+1}{n} \right)^2$. Limit is $\frac{1}{5}$. Series converges by ratio test.

15. Geometric, $r = .1$, converges.

16. Ratio test, $a_n = \dfrac{1 \cdot 3 \cdot 5 \cdots (2n-1)}{3 \cdot 6 \cdot 9 \cdots 3n}$, $a_{n+1} = \dfrac{1 \cdot 3 \cdots (2n-1)(2n+1)}{3 \cdot 6 \cdots (3n)(3n+3)}$, $a_{n+1}/a_n = (2n+1)/(3n+3)$, limit as $n \to \infty$ is $\frac{2}{3}$, series converges.

17. Geometric, $r = 1/5$, converges.

18. Ratio test, $a_n = \dfrac{(n+1)!}{1 \cdot 3 \cdots (2n+1)}$, $a_{n+1} = \dfrac{(n+2)!}{1 \cdot 3 \cdots (2n+1)(2n+3)}$, $a_{n+1}/a_n = (n+2)/(2n+3)$. Limit as $n \to \infty$ is $\frac{1}{2}$. Series converges.

19. Geom, $r = e/3$ between -1 and 1, converges.

20. Geom, $r = e/2 > 1$, diverges.

21. Every other term of $\Sigma 1/n^2$, converges.

22. $\Sigma 1/n(n+1)$ acts like $\Sigma 1/n^2$, converges.

23. Acts like $\Sigma n/n\sqrt{n} = \Sigma 1/\sqrt{n}$ which diverges; original diverges.

24. $n/(n-1) \to 1$ not 0 as $n \to \infty$. Diverges by nth term test.

25. $\dfrac{a_{n+1}}{a_n} = \dfrac{[(n+1)!]^2}{[2(n+1)]!} \dfrac{(2n)!}{(n!)^2} = \dfrac{[(n+1)!]^2}{(2n+2)!} \dfrac{(2n)!}{(n!)^2}$
$= \dfrac{(n+1)(n+1)}{(2n+2)(2n+1)}$.
Limit is $\frac{1}{4}$ (since numerator contains n^2 and denominator contains $4n^2$). Converges by ratio test.

26. $\dfrac{a_{n+1}}{a_n} = \dfrac{\sqrt{n+1}}{3^{n+1}} \dfrac{3^n}{\sqrt{n}} = \dfrac{1}{3} \sqrt{\dfrac{n+1}{n}}$. Limit is $\frac{1}{3}$. Series converges by ratio test.

27. Ratio test, $a_n = \dfrac{2 \cdot 4 \cdots (2n+2)}{(2n+3)!}$,
$a_{n+1} = \dfrac{2 \cdot 4 \cdots (2n+2)(2n+4)}{(2n+5)!}$,
$\dfrac{a_{n+1}}{a_n} = \dfrac{2n+4}{(2n+5)(2n+4)} = \dfrac{1}{2n+5}$.
Limit as $n \to \infty$ is 0. Series converges.

28. Acts like $\Sigma \sqrt{n}/n = \Sigma 1/\sqrt{n}$, diverges.

29. $\Sigma (\frac{3}{4})^n$, geom, $r = \frac{3}{4}$, converges.

30. $\Sigma (n+1)/n^2$, acts like $\Sigma 1/n$, diverges.

31. $\Sigma 1/(4n-2)$ (equivalently $\Sigma 1/(4n+2)$), acts like $\frac{1}{4} \Sigma 1/n$, diverges.

32. Terms smaller than those of conv series $\Sigma 1/4^n$; converges by comparison.

33. $\ln n < \sqrt{n}$ (since $\ln n$ has lower order of magnitude), $(\ln n)/n^2 < \sqrt{n}/n^2 = 1/n^{3/2}$; $\Sigma 1/n^{3/2}$ converges ($p = \frac{3}{2}$); original series converges by comparison.

34. Terms smaller than those of $\Sigma 1/n^2$, converges by comparison.

35. $\dfrac{1}{2 \ln 2} + \dfrac{1}{3 \ln 3} + \cdots = A_1 + A_2 + A_3 + \cdots$
$\geq \displaystyle\int_2^\infty \dfrac{1}{x \ln x} \, dx$ (let $u = \ln x$, $du = \dfrac{1}{x} dx$)
$= \int_{\ln 2}^\infty 1/u \, du = \ln u \big|_{\ln 2}^\infty = \infty$. Series diverges. See fig.

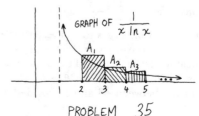

GRAPH OF $\dfrac{1}{x \ln x}$

PROBLEM 35

36. (a) $1 + \frac{1}{3} + \frac{1}{5} + \cdots = \sum_{n=1}^{\infty} 1/(2n-1)$ which acts like $\frac{1}{2}\sum 1/n$, diverges; $\frac{1}{2} + \frac{1}{4} + \frac{1}{6} + \cdots = \sum 1/2n = \frac{1}{2}\sum 1/n$, diverges.

(b) There are millions, e.g., $\frac{1}{2} + \frac{1}{4} + \frac{1}{8} + \frac{1}{16} + \cdots$ (geom, $r = \frac{1}{2}$), $1 + \frac{1}{4} + \frac{1}{9} + \frac{1}{16} + \frac{1}{25} + \cdots (p = 2)$.

37. (a) If $\sum a_n$ is $\sum 1/n^2$ then it converges but $\sum na_n$ is $\sum 1/n$ which diverges. On the other hand, if $\sum a_n$ is $\sum 1/n^3$ then it converges and $\sum na_n$ is $\sum 1/n^2$ which converges also.

(b) Suppose $\sum a_n$ converges by ratio test, i.e., $\lim_{n\to\infty} a_{n+1}/a_n < 1$. Then ratio test for $\sum na_n$ uses $\lim_{n\to\infty} \dfrac{(n+1)a_{n+1}}{na_n} = \lim_{n\to\infty}\dfrac{n+1}{n}\lim_{n\to\infty}\dfrac{a_{n+1}}{a_n} = 1 \times (\text{less than } 1)$. So limit is less than 1, new series also converges by ratio test.

Section 8.5 (page 236)

1. $\sum (-1)^{n+1}/\sqrt{n}$ converges by alternating series test since $1/\sqrt{n} \downarrow 0$. Error using S_{24} is less than $1/\sqrt{25} = \frac{1}{5}$. Approximation is under since last term, $1/\sqrt{24}$, was subtracted.

2. (a) $1/n! \downarrow 0$, series converges by alt series test. First term less than .001 is $1/7!$. So use
$1 - 1/2! + 1/3! - 1/4! + 1/5! - 1/6! = \frac{455}{720}$
as approximation.

(b) $1/n^n \downarrow 0$ so series converges by alt series test. First term less than .001 is $1/5^5$ so use $1/4^4$ as the approximation (i.e., just add one term).

3. (a) False (see warning 1).

(b) True by nth term test.

4. (a) $n^2/n! \downarrow 0$, series converges by alt series test.

(b) $n!/n^2$ does not $\to 0$, series diverges by nth term test.

(c) $1/n \ln n \downarrow 0$, series converges by alt series test.

(d) $2n/(n^2 + 4) \downarrow 0$, series converges by alt series test.

(e) Geometric series, $r = -.1$, converges (can also apply alt series test).

(f) Terms $\to 1$, not 0, diverges by nth term test.

(g) $\sqrt{n} - 1/n \downarrow 0$, converges by alt series test.

5. (a) True. If $\sum b_n$ converges then $b_n \to 0$. Eventually $0 < b_n < 1$. Then $b_n^2 < b_n$ so $\sum b_n^2$ converges by comparison.

(b) False. If $b_n = 1/\sqrt{n}$ then $\sum (-1)^{n+1}b_n$ converges by alt series test, but $\sum b_n^2$ is $\sum 1/n$ and diverges.

6. (a) All converge since in each case $a_n \downarrow 0$.

(b) For the alternating version of Table 1, the dividing line between conditional conv and absolute conv is where the dividing line is in Table 1 itself between convergence and divergence, i.e., $\sum (-1)^{n+1}1/n^p$ is conditionally convergent if $p \le 1$ and is abs convergent if $p > 1$.

7. (a) $n/(1 + n^2) \downarrow 0$ so series conv. But series of abs values is $\sum n/(1 + n^2)$, acts like $\sum 1/n$, diverges. Original series is conditionally convergent.

(b) Series of abs values is $\sum (n + 2)/(n^3 + 3)$, acts like $\sum n/n^3 = \sum 1/n^2$, converges. Original is abs convergent.

8. (a) $\sum a_n$ can't be absolutely convergent but we can't tell if it is conditionally convergent or divergent.

(b) $\sum a_n$ converges, and furthermore is abs convergent.

9. (a) $\sum |a_n|$ diverges.

(b) No conclusion (Fig. 2 shows that convergent series may have $\sum |a_n|$ converge or div.

10. (a) $1/n! \downarrow 0$ so series converges by alt series test. Series of abs values is $\sum 1/n!$, converges. Original converges by (9).

(b) $1/\sqrt{n} \downarrow 0$, series converges by alt series test. Series of abs values is $\sum 1/\sqrt{n}$, diverges. No information about original.

11. (a) Convergent geom series has r between -1 and 1. Series of abs values has r between 0 and 1 (For example if r is $-\frac{2}{3}$ then series of abs values has $r = \frac{2}{3}$. If r is $\frac{1}{7}$ then series of abs values has $r = \frac{1}{7}$). Series of abs values converges. So original is abs convergent.

(b) p-series are positive series. Series of abs values is same as original. Original is abs convergent.

Section 8.6 (page 240)

1. $\dfrac{|x^{n+1}\text{ term}|}{|x^n\text{ term}|} = \dfrac{|(n+2)x^{n+1}|}{|(n+1)x^n|} = \dfrac{n+2}{n+1}|x|$. Lim as $n \to \infty$ is $|x|$. Interval of convergence is $|x| < 1$, $-1 < x < 1$.

2. $\dfrac{|x^{n+1}\text{ term}|}{|x^n\text{ term}|} = \left|\dfrac{x^{n+1}}{3^{n+1}(n+1)^2}\right|\left|\dfrac{3^n n^2}{x^n}\right|$
$= \dfrac{1}{3}\left(\dfrac{n}{n+1}\right)^2 |x|$.

Limit as $n \to \infty$ is $\frac{1}{3}|x|$. Interval of convergence is $\frac{1}{3}|x| < 1$, $|x| < 3$, $-3 < x < 3$.

3. $\dfrac{|(n+1)!\, x^{n+1}|}{|n!\, x^n|} = (n+1)|x|$. Limit as $n \to \infty$ is ∞ (except when $x = 0$). Series converges only when $x = 0$, i.e., radius of convergence is 0.

4. $\left|\dfrac{x^{n+1}}{(n+1)!}\right|\left|\dfrac{n!}{x^n}\right| = \dfrac{1}{n+1}|x|$. Limit as $n \to \infty$ is 0. Series converges for all x, i.e., interval of convergence is $(-\infty, \infty)$.

5. $\sum_{n=0}^{\infty}(-1)^n x^{2n+1}$, $\dfrac{|x^{2n+3}\text{ term}|}{|x^{2n+1}\text{ term}|} = \dfrac{|x^{2n+3}|}{|x^{2n+1}|} = |x^2| = x^2$. Limit as $n \to \infty$ is x^2. Interval is $x^2 < 1$, $-1 < x < 1$.

6. $\sum_{n=1}^{\infty} 2^{2n}x^{2n+1}$, $\dfrac{|x^{2n+3}\text{ term}|}{|x^{2n+1}\text{ term}|} = \dfrac{|2^{2n+2}x^{2n+3}|}{|2^{2n}x^{2n+1}|} = 2^2|x^2| =$

$4x^2$. Limit as $n \to \infty$ is $4x^2$. Interval is $4x^2 < 1$, $x^2 < \frac{1}{4}$, $-\frac{1}{2} < x < \frac{1}{2}$.

7. $\sum 3^n x^n/n$, $\left|\dfrac{3^{n+1}x^{n+1}}{n+1}\right|\left|\dfrac{n}{3^n x^n}\right| = 3\dfrac{n}{n+1}|x|$. Limit as $n \to \infty$ is $3|x|$. Interval is $3|x| < 1$, $|x| < 1/3$, $-\frac{1}{3} < x < \frac{1}{3}$.

Section 8.7 (page 246)

1. (a) Binomial series with $q = 1/3$, $(1 + x)^{1/3} =$
$$1 + \tfrac{1}{3}x + \frac{\frac{1}{3}(-2/3)}{2!}x^2 + \frac{1/3(-\frac{2}{3})(-\frac{5}{3})}{3!}x^3 + \cdots$$
for $-1 < x < 1$
$$= 1 + \tfrac{1}{3}x - \frac{2}{3^2 2!}x^2 + \frac{2 \cdot 5}{3^3 3!}x^3 - \frac{2 \cdot 5 \cdot 8}{3^4 4!}x^4 + \cdots$$
for $-1 < x < 1$.

(b) Multiply series in equation (1) by x; get
$x + x^2 + x^3 + x^4 + \cdots$ for $-1 < x < 1$.

(c) Binomial series with $q = -3$, $(1 + x)^{-3} =$
$$1 - 3x + \frac{(-3)(-4)}{2!}x^2 + \frac{(-3)(-4)(-5)}{3!}x^3 + \cdots$$
$$= 1 - 3x + \frac{3 \cdot 4}{2!}x^2 - \frac{3 \cdot 4 \cdot 5}{3!}x^3 + \cdots$$
for $-1 < x < 1$.

(d) $\dfrac{1}{2 - 3x} = \dfrac{1}{2(1 - \frac{3}{2}x)} = \dfrac{1}{2}\dfrac{1}{1 - \frac{3}{2}x}$. Use series for $1/(1 - x)$ with x replaced by $\frac{3}{2}x$. Get
$\frac{1}{2}(1 + \frac{3}{2}x + (\frac{3}{2}x)^2 + (\frac{3}{2}x)^3 + \cdots)$
for $-1 < \frac{3}{2}x < 1$
$$= \frac{1}{2} + \frac{3}{2^2}x + \frac{3^2}{2^3}x^2 + \frac{3^3}{2^4}x^3 + \cdots$$
for $-\frac{2}{3} < x < \frac{2}{3}$.

(e) Factor, then binomial series. $\dfrac{1}{(3 + x)^6} =$
$\dfrac{1}{3^6(1 + \frac{1}{3}x)^6} = \dfrac{1}{3^6}(1 + \frac{1}{3}x)^{-6} = \dfrac{1}{3^6}(1 - 6(\frac{1}{3}x) + \dfrac{(-6)(-7)}{2!}(\frac{1}{3}x)^2 + \dfrac{(-6)(-7)(-8)}{3!}(\frac{1}{3}x)^3 + \cdots)$ for
$-1 < \frac{1}{3}x < 1$. Answer is $\dfrac{1}{3^6} - \dfrac{6}{3^7}x + \dfrac{6 \cdot 7}{3^8 2!}x^2 - \dfrac{6 \cdot 7 \cdot 8}{3^9 3!}x^3 + \cdots$ for $-3 < x < 3$.

(f) *Method 1:* $\dfrac{x}{(1 - x)(1 - 3x)} = \dfrac{-\frac{1}{2}}{1 - x} + \dfrac{\frac{1}{2}}{1 - 3x}$
by partial fraction decomposition. Get series for first fraction directly from (1). Get series for second fraction from (1) by replacing x by $3x$. Get $-\frac{1}{2}(1 + x + x^2 + x^3 + \cdots) + \frac{1}{2}(1 + 3x + 9x^2 + 27x^3 + \cdots) = \sum_{n=1}^{\infty}\frac{1}{2}(3^n - 1)x^n$. First series converges for $-1 < x < 1$, second series for $-1 < 3x < 1$, i.e., $-\frac{1}{3} < x < \frac{1}{3}$. Sum series converges on smaller interval $-\frac{1}{3} < x < \frac{1}{3}$.

Method 2: Multiply the series for $1/(1 - x)$ and for $1/(1 - 3x)$ and then multiply by x.

(g) $\dfrac{1}{x - 2} = \dfrac{1}{-2 + x} = \dfrac{1}{-2(1 - \frac{1}{2}x)} = -\dfrac{1}{2}\dfrac{1}{1 - \frac{1}{2}x}$.
Replace x by $\frac{1}{2}x$ in the $1/(1 - x)$ series. Get
$-\frac{1}{2}(1 + \frac{1}{2}x + (\frac{1}{2}x)^2 + (\frac{1}{2}x)^3 + \cdots)$, $-1 < \frac{1}{2}x < 1$,
$-\frac{1}{2} - \dfrac{1}{2^2}x - \dfrac{1}{2^3}x^2 - \cdots$ for $-2 < x < 2$,
$\sum_{n=0}^{\infty} -x^n/2^{n+1}$ for $-2 < x < 2$.

(h) $\ln(2 + x) = \ln 2[1 + \frac{1}{2}x] = \ln 2 + \ln(1 + \frac{1}{2}x)$.
Use (10), get $\ln 2 + \frac{1}{2}x - \frac{1}{2}(\frac{1}{2}x)^2 + \frac{1}{3}(\frac{1}{2}x)^3 - \frac{1}{4}(\frac{1}{2}x)^4 + \cdots$ for $-1 < \frac{1}{2}x < 1$, $\ln 2 + \dfrac{x}{2} - \dfrac{1}{2^2 \cdot 2}x^2 + \dfrac{1}{2^3 \cdot 3}x^3 - \dfrac{1}{2^4 \cdot 4}x^4 + \cdots$ for $-2 < x < 2$ (the $\ln 2$ term is the constant term in the series).

2. Binomial series, $q = \frac{1}{2}$, x replaced by $-3x^2$, $\sqrt{1 - 3x^2}$
$$= 1 + \tfrac{1}{2}(-3x^2) + \frac{\frac{1}{2}(-\frac{1}{2})}{2!}(-3x^2)^2 + \frac{\frac{1}{2}(-\frac{1}{2})(-\frac{3}{2})}{3!}(-3x^2)^3$$
$+ \cdots$ for $-1 < -3x^2 < 1$
$$= 1 - \tfrac{3}{2}x^2 - \frac{3^2}{2^2 2!}x^4 - \frac{3^3 \cdot 1 \cdot 3}{2^3 3!}x^6 - \frac{3^4 \cdot 1 \cdot 3 \cdot 5}{2^4 4!}x^8$$
$+ \cdots$ for $x^2 < \frac{1}{3}$, $-\sqrt{\frac{1}{3}} < x < \sqrt{\frac{1}{3}}$.
The x^{34} term is $-\dfrac{3^{17} \cdot 1 \cdot 3 \cdot 5 \cdots 31}{2^{17} 17!}x^{34}$. Series is
$$1 - \sum_{n=1}^{\infty} \frac{3^n \cdot 1 \cdot 3 \cdots (2n - 3)}{2^n n!}x^{2n}.$$

3. (a) $(1 + -x^2)^{-1} = 1 + (-1)(-x^2)$
$+ \dfrac{(-1)(-2)}{2!}(-x^2)^2 + \dfrac{(-1)(-2)(-3)}{3!}(-x^2)^3$
$+ \cdots$ for $-1 < -x^2 < 1$
$= 1 + x^2 + x^4 + x^6 + \cdots$ for $-1 < x < 1$.

(b) Replace x by x^2 in (1), $1/(1 - x^2)$
$= 1 + x^2 + (x^2)^2 + (x^2)^3 + \cdots$ for $-1 < x^2 < 1$
$= 1 + x^2 + x^4 + x^6 + \cdots$ for $-1 < x < 1$

(c) $\dfrac{1}{1 - x^2} = \dfrac{1}{1 - x}\dfrac{1}{1 + x} = (1 + x + x^2 + x^3 + \cdots)(1 - x + x^2 - x^3 + \cdots)$ for $-1 < x < 1$
$= 1 + x^2 + x^4 + x^6 + x^8 + \cdots$ for $-1 < x < 1$.
(use (1) twice, once with x replaced by $-x$)

(d) $\dfrac{1}{1 - x^2} = \dfrac{1}{(1 + x)(1 - x)} = \dfrac{\frac{1}{2}}{1 + x} + \dfrac{\frac{1}{2}}{1 - x}$
$= \frac{1}{2}(1 - x + x^2 - x^3 + \cdots)$
$+ \frac{1}{2}(1 + x + x^2 + x^3 + \cdots)$
$= 1 + x^2 + x^4 + x^6 + \cdots$ for $-1 < x < 1$.

(e)
$$1 - x^2 \overline{\smash{\big)}\,1} \quad \begin{array}{l} 1 + x^2 + x^4 + \cdots \end{array}$$
$$\underline{1 - x^2}$$
$$x^2$$
$$\underline{x^2 - x^4}$$
$$x^4 \text{ etc.}$$
So $1/(1 - x^2) = 1 + x^2 + x^4 + \cdots$. Series was not obtained from a previously known series, its

interval of convergence is not simply "inherited." With this method, must use ratio test to find interval $(-1, 1)$.

4. (a) $(1 + -x)^{-2} = 1 - 2(-x) + \dfrac{(-2)(-3)}{2!}(-x)^2 +$

$\dfrac{(-2)(-3)(-4)}{3!}(-x)^3 + \cdots$ for $-1 < -x < 1$.

$= 1 + 2x + 3x^2 + 4x^3 + \cdots$ for $-1 < x < 1$.

(b) $\dfrac{1}{1 - x}\dfrac{1}{1 - x} = (1 + x + x^2 + x^3 + \cdots)(1 + x$

$+ x^2 + x^3 + \cdots)$ for $-1 < x < 1$

$= 1 + 2x + 3x^2 + 4x^3 + \cdots$ for $-1 < x < 1$.

5. (a) First find series for $1/(1 + x^2)$ by replacing x by $-x^2$ in (1), get $1 - x^2 + x^4 - x^6 + \cdots$ for $-1 < -x^2 < 1$, $-1 < x < 1$. Then antidiff to get $\tan^{-1}x = C + x - x^3/3 + x^5/5 - x^7/7 + \cdots$ for $-1 < x < 1$. Set $x = 0$ to get $C = 0$.

(b) $\tan^{-1}x^2 = x^2 - x^6/3 + x^{10}/5 - x^{14}/7 + \cdots$ for $-1 < x^2 < 1$, $-1 < x < 1$ by part (a). So

$\int_0^{1/2} \tan^{-1}x^2 \, dx = \dfrac{x^3}{3}\Big|_0^{1/2} - \dfrac{x^7}{3 \cdot 7}\Big|_0^{1/2} + \dfrac{x^{11}}{5 \cdot 11}\Big|_0^{1/2} -$

$\cdots = \dfrac{1}{24} - \dfrac{1}{(21)(128)}$ + third term which is less

than $.0001 - \cdots$. Use sum of first two terms, $\frac{111}{2688}$, as the approx. Underestimate, last term used was subtracted.

6. Can get $x + 2x^2 + 3x^3 + 4x^4 + \cdots$ by differentiating $1 + x + x^2 + x^3 + \cdots$ and then multiplying by x. $1/(1 - x) = 1 + x + x^2 + x^3 + \cdots$ so (diff and then multiply by x) $x/(1 - x)^2 = x + 2x^2 + 3x^3 + 4x^4 + \cdots$. Answer is $x/(1 - x)^2$.

7. (a) $f(x) = \displaystyle\sum_{n=1}^{\infty} \dfrac{1}{n \, 2^{n-1}} x^n$

(b) $f'(x) = 1 + \dfrac{x}{2} + \dfrac{x^2}{2^2} + \dfrac{x^3}{2^3} + \cdots$

(c) Series for $f'(x)$ is geom, $r = \frac{1}{2}x$. If $-1 < \frac{1}{2}x < 1$ it converges to $1/(1 - \frac{1}{2}x)$, i.e., $f'(x)$ is the function $1/(1 - \frac{1}{2}x)$; $f(x)$ is an antiderivative so $f(x) = -2\ln(1 - \frac{1}{2}x) + C$. To find C, set $x = 0$, get $0 = 0 + C$, $C = 0$. So $f(x)$ is $-2\ln(1 - \frac{1}{2}x)$.

8. (a) $4(1 + \frac{3}{16})^{1/2} = 4(1 + \frac{1}{2}(\frac{3}{16}) + \frac{1}{2}(-\frac{1}{2})(\frac{3}{16})^2/2! + \cdots)$
$= 4 + 4(.09375) - 4(.0043945) + \cdots$ Alternating series, fourth term is less than $.01$. Use sum of first three terms as approximation.

(b) binomial series converges for $-1 < x < 1$ so can't use it with $x = 18$. In part (a) we used it with $x = \frac{3}{16}$ which *is* between -1 and 1.

Section 8.8 (page 251)

1. $f(x) = (1 + x)^q$, $f'(x) = q(1 + x)^{q-1}$,
$f''(x) = (q - 1)q(1 + x)^{q-2}$,
$f'''(x) = (q - 2)(q - 1)q(1 + x)^{q-3}, \cdots$,

$f(0) = 1, f'(0) = q, f''(0) = q(q - 1)$,
$f'''(0) = q(q - 1)(q - 2), \cdots$. The Maclaurin series is
$1 + qx + \dfrac{q(q - 1)}{2!}x^2 + \dfrac{q(q - 1)(q - 2)}{3!}x^3 + \cdots$ agreeing with binomial series.

2. Let $f(x) = \dfrac{1}{1 - x}$. Then $f'(x) = \dfrac{1}{(1 - x)^2}, f''(x) = \dfrac{2}{(1 - x)^3}, f'''(x) = \dfrac{3 \cdot 2}{(1 - x)^4}, f^{(4)}(x) = \dfrac{4!}{(1 - x)^5}, \cdots$,
$f(0) = 1, f'(0) = 1, f''(0) = 2, f'''(0) = 3!$,
$f^{(4)}(0) = 4!, \cdots$. Maclaurin series is $1 + x + \dfrac{2}{2!}x^2 + \dfrac{3!}{3!}x^3 + \cdots = 1 + x + x^2 + x^3 + x^4 + \cdots$ again. If $f(x) = \ln(1 + x)$ then $f'(x) = \dfrac{1}{1 + x}, f''(x) = -\dfrac{1}{(1 + x)^2}$,
$f'''(x) = \dfrac{2}{(1 + x)^3}, f^{(4)}(x) = -\dfrac{3 \cdot 2}{(1 + x)^4}, f^{(5)}(x) = \dfrac{4!}{(1 + x)^5}, \cdots, f(0) = 0, f'(0) = 1, f''(0) = -1, f'''(0) = 2$,
$f^{(4)}(0) = -3!, f^{(5)}(0) = 4!, \cdots$. Maclaurin series is $x - \dfrac{1}{2!}x^2 + \dfrac{2}{3!}x^3 - \dfrac{3!}{4!}x^4 + \cdots$
$= x - \frac{1}{2}x^2 + \frac{1}{3}x^3 - \frac{1}{4}x^4 + \cdots$. In each case the interval of conv would have to be found using a ratio test.

3. (a) *Method 1:* If $f(x) = \frac{1}{2}(e^x - e^{-x})$ then $f'(x) = \frac{1}{2}(e^x + e^{-x})$, $f''(x) = \frac{1}{2}(e^x - e^{-x}), \cdots, f(0) = 0$, $f'(0) = 1, f''(0) = 0, f'''(0) = 1 \cdots$. Maclaurin series is $x + x^3/3! + x^5/5! + \cdots$. Need ratio test to find interval of convergence $(-\infty, \infty)$.

Method 2: $e^x = 1 + x + x^2/2! + x^3/3! + \cdots$ for all x, so $e^{-x} = 1 + (-x) + (-x)^2/2! + (-x)^3/3! + \cdots = 1 - x + x^2/2! - x^3/3! + \cdots$ for all x. Subtract series, $\frac{1}{2}(e^x - e^{-x})$
$= \frac{1}{2}(2x + 2x^3/3! + 2x^5/5! + \cdots)$
$= x + x^3/3! + x^5/5! + \cdots$ for all x.

(b) *Method 1:* Let $f(x) = \dfrac{1}{3 - 2x}$. Then $f'(x) = \dfrac{2}{(3 - 2x)^2}, f''(x) = \dfrac{2^2 \cdot 2}{(3 - 2x)^3}, f'''(x) = \dfrac{2^3 \cdot 3!}{(3 - 2x)^4}$,
$f^{(4)}(x) = \dfrac{2^4 \cdot 4!}{(3 - 2x)^5}, \cdots, f(0) = \frac{1}{3}, f'(0) = 2/3^2$,
$f''(0) = 2^2 \cdot 2/3^3, f'''(0) = 2^3 \cdot 3!/3^4, f^{(4)}(0) = 2^4 \cdot 4!/3^5, \cdots$. Maclaurin series is $\dfrac{1}{3} + \dfrac{2}{3^2}x + \dfrac{2^2 \cdot 2}{3^3}\dfrac{x^2}{2!} + \dfrac{2^3 \cdot 3!}{3^4}\dfrac{x^3}{3!} + \cdots$ which cancels to
$\dfrac{1}{3} + \dfrac{2}{3^2}x + \dfrac{2^2}{3^3}x^2 + \dfrac{2^3}{3^4}x^3 + \cdots$. Need ratio test to find interval of convergence $(-\frac{3}{2}, \frac{3}{2})$.

Method 2: $\dfrac{1}{3 - 2x} = \dfrac{1}{3}\dfrac{1}{1 - \frac{2}{3}x} =$
$\frac{1}{3}(1 + \frac{2}{3}x + (\frac{2}{3}x)^2 + (\frac{2}{3}x)^3 + \cdots)$

for $-1 < \frac{2}{3}x < 1$. Simplifies
to same answer as method 1.

4. $\sin x = \sum_{n=0}^{\infty}(-1)^n x^{2n+1}/(2n+1)!$,
$\cos x = \sum_{n=0}^{\infty}(-1)^n x^{2n}/(2n)!$

5. (a) By (4), $\cos 3x = 1 - (3x)^2/2! + (3x)^4/4! - (3x)^6/6! + \cdots$ for $-\infty < 3x < \infty$

$$= 1 - \frac{3^2}{2!}x^2 + \frac{3^4}{4!}x^4 - \frac{3^6}{6!}x^6 + \cdots \text{ for all } x.$$

(b) By (3), $x^3 \sin x = x^3(x - x^3/3! + x^5/5! - \cdots)$
$= x^4 - x^6/3! + x^8/5! - \cdots$ for all x.

(c) By (5), $1 + (4x) + (4x)^2/2! + (4x)^3/3! + \cdots$ for
all x, $1 + 4x + \dfrac{4^2}{2!}x^2 + \dfrac{4^3}{3!}x^3 + \cdots$ for all x.

6. $\sin^2 x = \frac{1}{2}(1 - \cos 2x)$

$$= \frac{1}{2}\left(1 - \left[1 - \frac{(2x)^2}{2!} + \frac{(2x)^4}{4!} - \frac{(2x)^6}{6!} + \cdots\right]\right)$$

$$= \frac{2}{2!}x^2 - \frac{2^3}{4!}x^4 + \frac{2^5}{6!}x^6 - \cdots \text{ for all } x.$$

7. $f''(x) = g'(x) = f(x)$, $f'''(x) = f'(x) = g(x)$,
$f^{(4)}(x) = g'(x) = f(x), \cdots, f(0) = 1, f'(0) = g(0) = 0$,
$f''(0) = f(0) = 1, f'''(0) = g(0) = 0, f^{(4)}(0) = 1, \cdots$. Maclaurin series is $1 + x^2/2! + x^4/4! + x^6/6! + \cdots$. Use
ratio test to get interval.

$$\frac{|x^{2n+2} \text{ term}|}{|x^{2n} \text{ term}|} = \frac{|x^{2n+2}|}{(2n+2)!} \cdot \frac{(2n)!}{|x^{2n}|} = \frac{|x|^2}{(2n+2)(2n+1)}.$$

Limit as $n \to \infty$ is 0, series converges
for all x.

8. $\sin(-x) = (-x) - (-x)^3/3! + (-x)^5/5! - \cdots$
$= -x + x^3/3! - x^5/5! + \cdots = -\sin x$

9. Derivative of e^x = derivative of $(1 + x + x^2/2! + x^3/3! + \cdots) = 0 + 1 + 2x/2! + 3x^2/3! + 4x^3/4! + \cdots$
$= 1 + x + x^2/2! + x^3/3! + \cdots$ (old series back again).

10. $\sin 1 = 1 - 1/3! + 1/5! - 1/7! + \cdots$. The first
term under .0001 is $1/9!$ so use sum of the first four
terms as the approximation, underestimate since $1/7!$
was subtracted.

11. (a) Use e^x series with x replaced by $-x^2$.
$\int_0^1 (1 - x^2 + x^4/2! - x^6/3! + \cdots)\,dx$

$$= x\Big|_0^1 - \frac{x^3}{3}\Big|_0^1 + \frac{x^5}{5 \cdot 2!}\Big|_0^1 - \cdots$$

$$= 1 - \frac{1}{3} + \frac{1}{10} - \frac{1}{42} + \cdots.$$

Alternating series, fourth term is the first one
less than .1, so use $1 - \frac{1}{3} + \frac{1}{10}$ as the
(over)approximation.

(b) Binomial series with $q = -4$, x replaced by x^2.
$\int_0^{1/3}\left(1 - 4x^2 + \dfrac{(-4)(-5)}{2!}x^4 + \dfrac{(-4)(-5)(-6)}{3!}x^6 + \cdots\right)dx = \frac{1}{3} - \frac{4}{3}(\frac{1}{3})^3 + \frac{20}{10}(\frac{1}{3})^5 - \cdots$. Series alternates and first term less than .01 is the third one;
use sum of first two terms as the (under)estimate.

12. Use (10), Section 8.7 to get $\ln(1 + x^2) = x^2 - (x^2)^2/2 + (x^2)^3/3 - \cdots = x^2 - x^4/2 + x^6/3 - x^8/4 + \cdots$. By

(4), $1 - \cos x = x^2/2! - x^4/4! + x^6/6! - \cdots$ so

$$\frac{\ln(1 + x^2)}{1 - \cos x} = \frac{x^2 - x^4/2 + x^6/3 - x^8/4 + \cdots}{x^2/2! - x^4/4! + x^6/6! - \cdots} \text{ (now can-}$$

cel) $= \dfrac{1 - x^2/2 + x^4/3 - \cdots}{1/2! - x^2/4! + x^4/6! - \cdots}$. Limit as $x \to 0$ is
$1/(1/2!) = 2$.

(b) $\displaystyle\lim_{x \to 0} \frac{x - x^3/3! + x^5/5! - x^7/7! + \cdots}{x}$
$= \lim_{x \to 0}(1 - x^2/3! + x^4/5! - \cdots) = 1$

13. By (5), $e^x = \sum_{n=0}^{\infty} x^n/n!$ for all x. Set $x = 1$ to get $e = \sum_{n=0}^{\infty} 1/n!$; i.e., answer is e.

Section 8.10 (page 256)

1. $\left|\dfrac{(x-4)^{n+1}}{(n+1)3^{n+1}}\right|\left|\dfrac{n3^n}{(x-4)^n}\right| = \dfrac{1}{3}\dfrac{n}{n+1}|x-4|$. Limit
as $n \to \infty$ is $\frac{1}{3}|x - 4|$. Interval is $\frac{1}{3}|x - 4| < 1$,
$-3 < x - 4 < 3$, $1 < x < 7$ (centered about 4).

2. (a) Function blows up at $x = -8$, can't expand
around -8, can't do powers of $x - -8$, i.e.,
powers of $x + 8$.

(b) $\ln x$ blows up at $x = 0$ and is totally undefined
for $x < 0$. So can't expand around 0 or any
negative number, i.e., can't expand in powers of
x, $x + 1$, $x + 2$, $x + \frac{1}{2}$, $x + \pi$, etc.

3. (a) *Method 1:* $\ln x = \ln([x - 1] + 1)$ (let $u = x - 1$)
$= \ln(1 + u) = u - u^2/2 + u^3/3 - u^4/4 + \cdots$
for $-1 < u < 1$ by eq. (10), Section, 8.7;
$= (x - 1) - \frac{1}{2}(x - 1)^2 + \frac{1}{3}(x - 1)^3 - \frac{1}{4}(x - 1)^4 + \cdots$ for $-1 < x - 1 < 1$, $0 < x < 2$.

Method 2: If $f(x) = \ln x$ then $f'(x) = 1/x$,
$f''(x) = -1/x^2$, $f'''(x) = 2/x^3$, $f^{(4)}(x) = -3!/x^4$,
$\cdots, f(1) = 0, f'(1) = 1, f''(1) = -1, f'''(1) = 2$,
$f^{(4)}(1) = -3!, f^{(5)}(1) = 4!, \cdots$. Use (1) to get
$(x - 1) - \dfrac{1}{2!}(x - 1)^2 + \dfrac{2}{3!}(x - 1)^3 - \dfrac{3!}{4!}(x - 1)^4 + \cdots$ which cancels to the answer from method
1. Need ratio test to get interval $(0, 2)$.

(b) $\sin x = \sin([x - \pi] + \pi)$ (let $u = x - \pi$)
$= \sin(u + \pi) = \sin u \cos \pi + \cos u \sin \pi$
$= -\sin u$ (trig) $= -(u - u^3/3! + u^5/5! - \cdots)$
for all u (by (3) Section 8.8)
$= -(x - \pi) + (x - \pi)^3/3! - (x - \pi)^5/5! + \cdots$
for all $x - \pi$, i.e., for all x.

(c) $e^x = e^{(x-1)+1}$ (let $u = x - 1$) $= e^{u+1} = ee^u$
$= e(1 + u + u^2/2! + u^3/3! + \cdots)$ for all u
(by (5), Section 8.8)

$$= e + e(x - 1) + \frac{e}{2!}(x - 1)^2 + \frac{e}{3!}(x - 1)^3$$
$+ \cdots$ for all $x - 1$, i.e., for all x.

(d) $\dfrac{1}{-6 - x} = \dfrac{1}{-6 - [(x + 1) - 1]}$

(let $u = x + 1$)

$= \dfrac{1}{-6 - (u - 1)}$

$= \dfrac{1}{-5 - u} = -\dfrac{1}{5}\dfrac{1}{1 + \frac{1}{5}u}$.

Now use series for $1/(1 - x)$ with x replaced by $-\frac{1}{5}u$ to get
$-\frac{1}{5}(1 + (-\frac{1}{5}u) + (-\frac{1}{5}u)^2 + (-\frac{1}{5}u)^3 + \cdots)$
for $-1 < -\frac{1}{5}u < 1$
$= -\frac{1}{5} + u/5^2 - u^2/5^3 + u^3/5^4 - \cdots$
for $-5 < u < 5$
$= -\frac{1}{5} + (x + 1)/5^2 - (x + 1)^2/5^3 + (x + 1)^3/5^4$
$- \cdots$ for $-5 < x + 1 < 5$, $-6 < x < 4$.

(e) $\dfrac{1}{x} = \dfrac{1}{(x + 2) - 2}$ (let $u = x + 2$) $= \dfrac{1}{u - 2}$

$= -\dfrac{1}{2}\dfrac{1}{1 - \frac{1}{2}u}$. Use $1/(1 - x)$ series. Get

$-\frac{1}{2}(1 + \frac{1}{2}u + (\frac{1}{2}u)^2 + (\frac{1}{2}u)^3 + \cdots)$
for $-1 < \frac{1}{2}u < 1$, $-2 < u < 2$
$= -\frac{1}{2} - (x + 2)/2^2 - (x + 2)^2/2^3 - (x + 2)^3/2^4$
$- \cdots$ for $-2 < x + 2 < 2$, $-4 < x < 0$.

(f) $\sqrt{x} = \sqrt{(x - 9) + 9}$ (let $u = x - 9$)
$= \sqrt{u + 9} = 3(1 + \frac{1}{9}u)^{1/2}$ (use binomial series)
$= 3(1 + \frac{1}{2}(\frac{1}{9}u) + \dfrac{(\frac{1}{2})(-\frac{1}{2})}{2!}(\frac{1}{9}u)^2$

$+ \dfrac{(\frac{1}{2})(-\frac{1}{2})(-\frac{3}{2})}{3!}(\frac{1}{9}u)^3 + \cdots)$

for $-1 < \frac{1}{9}u < 1$, i.e., $-9 < u < 9$

$= 3 + \dfrac{1}{2 \cdot 3}(x - 9) - \dfrac{1}{3^3 2^2 2!}(x - 9)^2$

$+ \dfrac{3}{3^5 2^3 3!}(x - 9)^3 - \dfrac{3 \cdot 5}{3^7 2^4 4!}(x - 9)^4 + \cdots$

for $-9 < x - 9 < 9$, $0 < x < 18$.

Coefficient of $(x - 9)^{50}$ is $-\dfrac{3 \cdot 5 \cdot 7 \cdots 97}{3^{99} 2^{50} 50!}$.

(g) *Method 1:* $\dfrac{1}{(x + 8)^5} = \dfrac{1}{([x - 1] + 1 + 8)^5}$

$= \dfrac{1}{(u + 9)^5}$ (let $u = x - 1$) $= \dfrac{1}{9^5}(1 + \frac{1}{9}u)^{-5}$

(use binomial series)

$= \dfrac{1}{9^5}(1 - 5(\frac{1}{9}u) + \dfrac{(-5)(-6)}{2!}(\frac{1}{9}u)^2$

$+ \dfrac{(-5)(-6)(-7)}{3!}(\frac{1}{9}u)^3 + \cdots)$

for $-1 < \frac{1}{9}u < 1$, $-9 < u < 9$

$= \dfrac{1}{9^5} - \dfrac{5}{9^6}(x - 1) + \dfrac{5 \cdot 6}{9^7 2!}(x - 1)^2 - \cdots$

for $-9 < x - 1 < 9$, $-8 < x < 10$.
Coefficient of $(x - 1)^{19}$ is $-\dfrac{5 \cdot 6 \cdot 7 \cdots 23}{9^{24} 19!}$.

Method 2: $f(x) = \dfrac{1}{(x + 8)^5}$, $f'(x) = \dfrac{-5}{(x + 8)^6}$,

$f''(x) = \dfrac{5 \cdot 6}{(x + 8)^7}$, $f'''(x) = \dfrac{-5 \cdot 6 \cdot 7}{(x + 8)^8}$, $\cdots$, $f(1) = $
$1/9^5$, $f'(1) = -5/9^6$, $f''(1) = 5 \cdot 6/9^7$,
$f'''(1) = -5 \cdot 6 \cdot 7/9^8, \cdots$. Series is $\dfrac{1}{9^5} - $

$\dfrac{5}{9^6}(x - 1) + \dfrac{5 \cdot 6}{9^7 2!}(x - 1)^2 - \dfrac{5 \cdot 6 \cdot 7}{9^8 3!}(x - 1)^3$

$+ \cdots$. Must use ratio test to get interval of convergence $(-8, 10)$.

(h) $\cos 2x = \cos 2([x + \frac{1}{2}\pi] - \frac{1}{2}\pi)$
$= \cos(2[x + \frac{1}{2}\pi] - \pi)$
$= \cos 2[x + \frac{1}{2}\pi] \cos \pi + \sin 2[x + \frac{1}{2}\pi] \sin \pi$
$= -\cos 2[x + \frac{1}{2}\pi]$ (let $u = 2[x + \frac{1}{2}\pi]$)
$= -(1 - u^2/2! + u^4/4! - u^6/6! + u^8/8! - \cdots)$
for all u (standard cosine series). So

$\cos 2x = -1 + \dfrac{2^2}{2!}(x + \frac{1}{2}\pi)^2 - \dfrac{2^4}{4!}(x + \frac{1}{2}\pi)^4$

$+ \dfrac{2^6}{6!}(x + \frac{1}{2}\pi)^6 - \cdots$

for all $2(x + \frac{1}{2}\pi)$, i.e., for all x.

(i) $\ln 3x = \ln 3([x - 2] + 2) = \ln(3[x - 2] + 6)$
$= \ln 6(\frac{1}{2}[x - 2] + 1) = \ln 6 + \ln(1 + \frac{1}{2}[x - 2])$
(let $u = \frac{1}{2}[x - 2]$)
$= \ln 6 + u - u^2/2 + u^3/3 - u^4/4 + \cdots$
for $-1 < u < 1$ (use standard $\ln(1 + x)$ series).

So $\ln 3x = \ln 6 + \dfrac{1}{2}(x - 2) - \dfrac{1}{2^2 \cdot 2}(x - 2)^2$

$+ \dfrac{1}{2^3 \cdot 3}(x - 2)^3 - \dfrac{1}{2^4 \cdot 4}(x - 2)^4 + \cdots$

for $-1 < \frac{1}{2}(x - 2) < 1$, $0 < x < 4$.

(j) $\dfrac{1}{1 + 2x} = \dfrac{1}{1 + 2([x + 4] - 4)}$

$= \dfrac{1}{1 + 2[x + 4] - 8} = \dfrac{1}{-7 + 2[x + 4]}$

$= \dfrac{1}{-7(1 - \frac{2}{7}(x + 4))} = -\dfrac{1}{7}\dfrac{1}{1 - \frac{2}{7}(x + 4)}$

(let $u = \frac{2}{7}(x + 4)$)

$= -\frac{1}{7}(1 + u + u^2 + u^3 + u^4 + \cdots)$
for $-1 < u < 1$ (use standard $1/(1 - x)$ series).

So $\dfrac{1}{1 + 2x} = -\dfrac{1}{7} - \dfrac{2}{7^2}(x + 4) - \dfrac{2^2}{7^3}(x + 4)^2$

$- \dfrac{2^3}{7^4}(x + 4)^3 - \cdots$

for $-1 < \frac{2}{7}(x + 4) < 1$, $-\frac{15}{2} < x < -\frac{1}{2}$.

Chapter 8 Review Problems (page 257)

1. (a) Geom, $r = \frac{1}{7}$, converges.
    (b) p-series, $p = 7$, converges.
    (c) Geom, $r = 2$, diverges. Also series is $\Sigma\, 2^n$ which diverges obviously by nth term test.
    (d) p-series, $p = \frac{1}{2}$, diverges.
    (e) $-\Sigma\, 1/n$, harmonic series, diverges.
    (f) $3n/(n^2 + n) \downarrow 0$, converges by alternating series test.
    (g) Acts like $3\Sigma\, 1/n$, diverges.
    (h) $\dfrac{6^{n+1}}{n!} \dfrac{(n-1)!}{6^n} = \dfrac{6}{n}$. Limit as $n \to \infty$ is 0. Converges by ratio test.
    (i) It is an alternating series. It diverges by nth term test since terms approach 1, not 0.
    (j) Converges by comparison with say $\frac{1}{7} + \frac{1}{70} + \frac{1}{700} + \cdots$ which is convergent geom $(r = \frac{1}{10})$.
    (k) $a_2/a_1 = 7/16$, $a_3/a_2 = 9/32$, $a_{n+1}/a_n = (2n+3)/2^{n+3}$. Lim as $n \to \infty$ is 0. Series converges by ratio test.
    (l) $\dfrac{(n+1)^2}{5^{n+1}} \dfrac{5^n}{n^2} = \dfrac{1}{5}\left(\dfrac{n+1}{n}\right)^2$. Limit as $n \to \infty$ is $\frac{1}{5}$. Converges by ratio test.
    (m) Partial sums are $1, -2, -1, -4, -3, -6, \cdots$. Series diverges to $-\infty$.

2. Geometric, $a = (\frac{1}{9})^5$, $r = (\frac{1}{9})^2$,
    sum is $\dfrac{(\frac{1}{9})^5}{1 - (\frac{1}{9})^2} = \dfrac{1}{9^5 - 9^3}$.

3. $2^8/8!$ is first term less than .01. So use the sum of the first eight terms $1 - 2 + 2^2/2! - \cdots - 2^7/7!$ as the approx (underestimate) to get error $< .01$.

4. (a) $1/(\ln n)^2 \downarrow 0$ so series converges by alternating series test. But series of absolute values is $\Sigma\, 1/(\ln n)^2$ which diverges. So original series is conditionally convergent.
    (b) Converges absolutely since series is positive already and is convergent geom $(r = \frac{1}{3})$.

5. (a) Converges. Series of absolute values is $\Sigma\, a_n$ which converges by hypothesis. So $\Sigma(-1)^{n+1}a_n$ is absolutely convergent, so converges.
    (b) May converge or may diverge. For example if $\Sigma\, a_n$ is $\Sigma\, 1/n^3$ then $\Sigma\, n^2 a_n$ is $\Sigma\, 1/n$ which diverges. But if $\Sigma\, a_n$ is $\Sigma\, 1/n^4$ then $\Sigma\, n^2 a_n$ is $\Sigma\, 1/n^2$ which converges.

6. Must have $e^{a_n} \to 0$ by nth term test so $a_n \to -\infty$. But then $\Sigma\, a_n$ is adding many huge negative numbers and series diverges to $-\infty$.

7. (a) Let $\Sigma\, a_n$ and $\Sigma\, b_n$ both be $1/\sqrt{2} - 1/\sqrt{3} + 1/\sqrt{4} - \cdots$. Each converges by alternating series test. But $\Sigma\, a_n b_n$ is $\frac{1}{2} + \frac{1}{3} + \frac{1}{4} + \cdots$ which diverges.
    (b) If $\Sigma\, b_n$ converges then $b_n \to 0$ so eventually b_n is between 0 and 1. So $a_n b_n < a_n$ eventually. But

$\Sigma\, a_n$ converges by hypothesis so $\Sigma\, a_n b_n$ converges by comparison.

8. Series is $\Sigma_{n=4}^{\infty}\, x^{n+1}/4^n$ so $\left|\dfrac{x^{n+2}}{4^{n+1}}\right|\left|\dfrac{4^n}{x^{n+1}}\right| = \dfrac{|x|}{4}$. Limit as $n \to \infty$ is $\frac{1}{4}|x|$ so interval of convergence is $\frac{1}{4}|x| < 1$, $|x| < 4, -4 < x < 4$.

9. (a) $\dfrac{1}{3-x} = \dfrac{1}{3}\left(\dfrac{1}{1 - \frac{1}{3}x}\right)$
    (now use series for $1/(1-x)$)
    $= \frac{1}{3}(1 + \frac{1}{3}x + (\frac{1}{3}x)^2 + (\frac{1}{3}x)^3 + \cdots)$
    for $-1 < \frac{1}{3}x < 1$, $-3 < x < 3$.
    $= \dfrac{1}{3} + \dfrac{1}{3^2}x + \dfrac{1}{3^3}x^2 + \dfrac{1}{3^4}x^3 + \cdots$, $-3 < x < 3$.

    (b) $\dfrac{1}{(x-1)(1-2x)} = \dfrac{-1}{x-1} + \dfrac{-2}{1-2x}$
    (partial fraction decomposition)
    $= \dfrac{1}{1-x} - 2\left(\dfrac{1}{1-2x}\right)$
    $= 1 + x + x^2 + x^3 + \cdots$
    $\quad - 2(1 + [2x] + [2x]^2 + [2x]^3 + \cdots)$
    using series for $1/(1-x)$. First expansion holds for $-1 < x < 1$, second for $-1 < 2x < 1$, $-\frac{1}{2} < x < \frac{1}{2}$. So sum series has interval of conv $-\frac{1}{2} < x < \frac{1}{2}$ (the smaller) and answer is
    $-1 + (1 - 2^2)x + (1 - 2^3)x^2 + (1 - 2^4)x^3 + \cdots$ for $-\frac{1}{2} < x < \frac{1}{2}$, i.e., series is $\Sigma_{n=0}^{\infty}\, (1 - 2^{n+1})x^n$.
    Series can also be found by multiplying series instead of adding but that method doesn't produce as clear a pattern.

    (c) (Binomial series) $(1+x)^{-6}$
    $= 1 - 6x + \dfrac{(-6)(-7)}{2!}x^2 + \dfrac{(-6)(-7)(-8)}{3!}x^3$
    $\quad + \cdots$ for $-1 < x < 1$

    (d) Use series for $1/(1-x)$.
    $1/(1 + x^6) = 1 + (-x^6) + (-x^6)^2 + (-x^6)^3$
    $\quad + \cdots$ for $-1 < -x^6 < 1$, $-1 < x < 1$.
    Get $1 - x^6 + x^{12} - x^{18} + \cdots$ for $-1 < x < 1$.

10. *Method 1:* $f(x) = x^2 e^x$, $f'(x) = x^2 e^x + 2xe^x$ (product rule), $f''(x) = x^2 e^x + 2xe^x + 2xe^x + 2e^x = x^2 e^x + 4xe^x + 2e^x$, $f(0) = 0$, $f'(0) = 0$, $f''(0) = 2$. Series begins with $0 + 0x + (2/2!)x^2$, i.e., $0 + 0x + x^2$.
    Method 2: $e^x = 1 + x + x^2/2! + x^3/3! + \cdots$ so $x^2 e^x$ begins $x^2 + x^3 + \cdots$, actually with $0 + 0x + x^2$.

11. $\lim_{x\to 0} \dfrac{1 - (1 - x^2/2! + x^4/4! - \cdots)}{x^2}$
    $= \lim_{x\to 0}(1/2! - x^2/4! + x^4/6! - \cdots) = \frac{1}{2}$.

12. (a) $\cos x = \cos([x - \frac{1}{4}\pi] + \frac{1}{4}\pi)$ (let $u = x - \frac{1}{4}\pi$)
    $= \cos(u + \frac{1}{4}\pi) = \cos u \cos \frac{1}{4}\pi - \sin u \sin \frac{1}{4}\pi$
    $= \frac{1}{2}\sqrt{2} \cos u - \frac{1}{2}\sqrt{2} \sin u$
    $= \frac{1}{2}\sqrt{2}(1 - u^2/2! + u^4/4! - u^6/6! + \cdots) -$
    $\frac{1}{2}\sqrt{2}(u - u^3/3! + u^5/5! - u^7/7! + \cdots)$, all u,

so $\cos x = \frac{1}{2}\sqrt{2} - \frac{1}{2}\sqrt{2}\,(x - \frac{1}{4}\pi) -$
$\dfrac{\sqrt{2}}{2\cdot 2!}(x - \frac{1}{4}\pi)^2 + \dfrac{\sqrt{2}}{2\cdot 3!}(x - \frac{1}{4}\pi)^3 +$
$\dfrac{\sqrt{2}}{2\cdot 4!}(x - \frac{1}{4}\pi)^4 - \dfrac{\sqrt{2}}{2\cdot 5!}(x - \frac{1}{4}\pi)^5 - \cdots$
for all $x - \frac{1}{4}\pi$, i.e., for all x.

(b) $\sqrt[3]{x} = \sqrt[3]{(x - 8)} + 8$ (now let $u = x - 8$)
$= 2(1 + \frac{1}{8}u)^{1/3}$ (use binomial series)
$= 2\left(1 + \frac{1}{3}(\frac{1}{8}u) + \dfrac{(\frac{1}{3})(-\frac{2}{3})}{2!}(\frac{1}{8}u)^2 + \cdots\right)$
for $-1 < \frac{1}{8}u < 1$, $-8 < u < 8$
$= 2 + \dfrac{2}{3\cdot 8}(x - 8) - \dfrac{2\cdot 2}{3^2\, 2!\, 8^2}(x - 8)^2 +$
$\dfrac{2\cdot 2\cdot 5}{3^3\, 3!\, 8^3}(x - 8)^3 - \dfrac{2\cdot 2\cdot 5\cdot 8}{3^4\, 4!\, 8^4}(x - 8)^4 + \cdots$
for $-8 < x - 8 < 8$, $0 < x < 16$.

13. $x^3 e^{-x^3} = x^3(1 + (-x^3) + \frac{1}{2!}(-x^3)^2 + \frac{1}{3!}(-x^3)^3 + \cdots)$
for all $-x^3$, i.e., for all x so

$\int_0^1 (x^3 - x^6 + x^9/2! - x^{12}/3! + \cdots)\,dx$
$= \dfrac{x^4}{4}\Big|_0^1 - \dfrac{x^7}{7}\Big|_0^1 + \dfrac{x^{10}}{2!\cdot 10}\Big|_0^1$
$= \dfrac{1}{4} - \dfrac{1}{7} + \dfrac{1}{2!\,10} - \dfrac{1}{3!\,13} + \dfrac{1}{4!\,16} - \dfrac{1}{5!\,19} + \cdots.$
First term less than .001 is $1/(5!\cdot 19)$ so add five terms to get (over) estimate.

14. $(1 + -x^2)^{-1/2} = 1 - \dfrac{1}{2}(-x^2) + \dfrac{(-\frac{1}{2})(-\frac{3}{2})}{2!}(-x^2)^2 +$
$\cdots$ (binomial series) $= 1 + \dfrac{1}{2}x^2 + \dfrac{3}{2^2\,2!}x^4 + \dfrac{3\cdot 5}{2^3\,3!}x^6 +$
$\dfrac{3\cdot 5\cdot 7}{2^4\,4!}x^8 + \cdots.$ Antidifferentiate to get
$\sin^{-1}x = C + x + \dfrac{1}{2\cdot 3}x^3 + \dfrac{3}{2^2\,2!\,5}x^5 + \dfrac{3\cdot 5}{2^3\,3!\,7}x^7 +$
$\dfrac{3\cdot 5\cdot 7}{2^4\,4!\,9}x^9 + \cdots.$ Set $x = 0$ to find C;
$0 = C + 0 + 0 + 0 + \cdots, C = 0.$

9/VECTORS

Section 9.1 (page 262)

1. (See fig.)

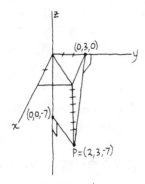

PROBLEM 1

(a) Use (*) in Section 9.1. $\sqrt{1 + 4 + 81} = \sqrt{86}$.
(b) Use (*), Section 9.1 with P and $(0,0,0)$.
$\sqrt{4 + 9 + 49} = \sqrt{62}$.
(c) P is distance 7 below x, y plane; answer is 7.
(d) P is distance 2 in front of y, z plane; answer is 2.
(e) Foot of perp from P to z-axis has coords $(0, 0, -7)$. Answer is distance between P and foot: $\sqrt{4 + 9} = \sqrt{13}$.

(f) Foot of perp from P to y-axis has coords $(0, 3, 0)$. Answer is distance between P and foot: $\sqrt{4 + 49} = \sqrt{53}$.
(g) Foot of perp from (x, y, z) to x-axis is $(x, 0, 0)$. Distance between P and foot is $\sqrt{y^2 + z^2}$.
(h) (See fig.) Foot of perp from (x, y, z) to z-axis is $(0, 0, z)$. Distance between P and foot is $\sqrt{x^2 + y^2}$.

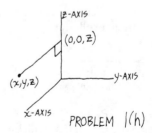

PROBLEM 1(h)

2. $\overrightarrow{AF} = (-4, 0, 3)$, $\overrightarrow{HB} = (4, 5, 0)$, $\overrightarrow{HE} = (4, 5, 3)$
3. $(0, 2)$
4. (See fig.) Line is $y = -\frac{2}{3}x - \frac{4}{3}$ and has slope $-\frac{2}{3}$, i.e., goes over 3, down 2. Arrows parallel to line are $(3, -2)$, $(-3, 2)$, $(6, -4)$, $(-6, 4)$, etc.

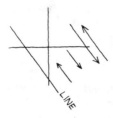

PROBLEM 4

5. $\overrightarrow{AB} = B - A = (-3, -3)$
6. head − tail $= (3, 1, 6)$,
head $= (3, 1, 6) + (1, 0, 4) = (4, 1, 10)$
7. $\vec{u} = (3\cos 120°, 3\sin 120°) = (-\frac{3}{2}, \frac{3}{2}\sqrt{3})$

Section 9.2 (page 269)

1. (a) $\overrightarrow{DB}$
    (b) $\overrightarrow{DB}$
    (c) $\overrightarrow{AB} + \overrightarrow{CB} = \overrightarrow{AB} + \overrightarrow{DA} = \overrightarrow{DB}$
    (d) $\overrightarrow{AB} - \overrightarrow{CB} = \overrightarrow{AB} + \overrightarrow{BC} = \overrightarrow{AC}$
    (e) $\vec{0}$ (not the *number* 0 but the *vector* $\vec{0}$)

2. Need $\overrightarrow{AB} = \overrightarrow{DC}$ so need $C - D = B - A = (-1, -2, -3)$, $D = (5, 5, 5) - (-1, -2, -3) = (6, 7, 8)$.

3. $\overrightarrow{AB} = (1, 4, -4)$, $\overrightarrow{CD} = (-2, -8, 8)$, $\overrightarrow{CD} = -2\overrightarrow{AB}$ so lines are parallel.

4. $\overrightarrow{AB} = (3, 6, -4)$, $\overrightarrow{QP} = (10, y, z - 2)$. Need $\overrightarrow{QP}$ a multiple of $\overrightarrow{AB}$ so need $10/3 = y/6$, $y = 20$ and $10/3 = (z - 2)/(-4)$, $z = -34/3$.

5. A, B, C are collinear if and only if vectors $\overrightarrow{AB}$ and $\overrightarrow{AC}$ are parallel (see fig). $\overrightarrow{AB} = (-1, -6, 4)$ and

PROBLEM 5

$\overrightarrow{AC} = (-4, -3, -3)$. Vectors are not multiples so not parallel. Points are not collinear.

6. (See fig.)

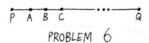

PROBLEM 6

$\overrightarrow{PA} = \frac{1}{10}\overrightarrow{PQ}$ so $A - P = \frac{1}{10}(Q - P)$, $A = \frac{1}{10}(Q + 9P)$. Similarly $B = \frac{1}{10}(2Q + 8P)$, $C = \frac{1}{10}(3Q + 7P)$.
7. (See fig.) $\vec{w} = -\vec{v} + 2\vec{u}$

PROBLEM 7

8. *Method 1:* $\overrightarrow{AE} + \overrightarrow{BF} + \overrightarrow{CD}$
$= E - A + F - B + D - C$
$= \frac{1}{2}(B + C) - A + \frac{1}{2}(A + C) - B + \frac{1}{2}(A + B) - C = \vec{0}$
  Method 2: $\overrightarrow{AE} + \overrightarrow{BF} + \overrightarrow{CD}$
$= \overrightarrow{AB} + \overrightarrow{BE} + \overrightarrow{BC} + \overrightarrow{CF} + \overrightarrow{CA} + \overrightarrow{AD}$
$= (\overrightarrow{AB} + \overrightarrow{BC} + \overrightarrow{CA}) + (\overrightarrow{BE} + \overrightarrow{CF} + \overrightarrow{AD})$
$= \vec{0} + \frac{1}{2}\overrightarrow{BC} + \frac{1}{2}\overrightarrow{CA} + \frac{1}{2}\overrightarrow{AB} = \frac{1}{2}(\overrightarrow{BC} + \overrightarrow{CA} + \overrightarrow{AB})$
$= \frac{1}{2}\vec{0} = \vec{0}$

9. (a) $\|\vec{u}\| = \sqrt{9 + 1 + 25} = \sqrt{35}$
    (b) $\|\vec{u}\| = \sqrt{\pi^2 + \pi^2 + \pi^2 + \pi^2 + \pi^2} = \pi\sqrt{5}$
10. $\vec{u}/\|\vec{u}\| = (2/\sqrt{104}, -6/\sqrt{104}, 8/\sqrt{104})$
11. $\vec{v} = -5\vec{u}_{\text{normalized}} = -5\vec{u}/\|\vec{u}\|$
12. $\overrightarrow{BC} = 12\overrightarrow{BA}_{\text{normalized}} = 12(0, -1/\sqrt{17}, -4/\sqrt{17})$,
$C - B = \left(0, \dfrac{-12}{\sqrt{17}}, \dfrac{-48}{\sqrt{17}}\right)$,
$C = \left(1, 2 - \dfrac{12}{\sqrt{17}}, 6 - \dfrac{48}{\sqrt{17}}\right)$
(See fig.)

PROBLEM 12

13. $\|\vec{u}\| = \sqrt{38}$, so $\|217\vec{u}\| = 217\sqrt{38}$
14. Use Section 9.1, (3), with $r = 1$ to get $\vec{u} = (\cos\theta, \sin\theta)$.
15. (See fig.) Foot of perp from $A = (4, 5, 6)$ to y-axis is $B = (0, 5, 0)$; $\vec{u}$ points like $\overrightarrow{AB}$ but has length 3 so $\vec{u} = 3\overrightarrow{AB}_{\text{normalized}} = 3(-4, 0, -6)_{\text{normalized}}$
$= 3(-4/\sqrt{52}, 0, -6/\sqrt{52}) = (-12/\sqrt{52}, 0, -18/\sqrt{52})$.

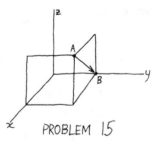

PROBLEM 15

16. (See fig.) If origin is named B then $\vec{u}$ points like $\overrightarrow{AB} = (-5, -6, -7)$ and $\|\vec{u}\| = 1/(25 + 36 + 49) = 1/110$ so $\vec{u} = \frac{1}{110}\overrightarrow{AB}_{\text{normalized}}$
$= \left(\dfrac{-5}{110\sqrt{110}}, \dfrac{-6}{110\sqrt{110}}, \dfrac{-7}{110\sqrt{110}}\right)$.

PROBLEM 16

17. $\vec{u} - 2\vec{v} = 5\vec{j} - 3\vec{k}$, $\|\vec{u}\| = \sqrt{4 + 9 + 1} = \sqrt{14}$,
$\vec{u}_{\text{normalized}} = (2/\sqrt{14})\vec{i} + (3/\sqrt{14})\vec{j} - (1/\sqrt{14})\vec{k}$
18. Norm of $r^3\vec{r}$ is r^3 times the norm of $\vec{r}$ so it is $r^3 r = r^4$.
19. (See fig.) $\vec{u}/\|\vec{u}\|$ and $\vec{v}/\|\vec{v}\|$ are unit vectors in the di-

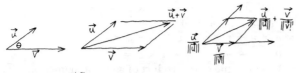

PROBLEM 19

rections of $\vec{u}$ and $\vec{v}$. They have the same length so their sum is the diagonal of a *rhombus*. On the other hand, $\vec{u} + \vec{v}$ is the diagonal of a parallelogram, not necessarily a rhombus. Angle bisection occurs only in the rhombus case.

Section 9.3 (page 275)

1. $\vec{u} \cdot \vec{v} = 5 + 12 - 15 = 2$. Angle is acute since $\vec{u} \cdot \vec{v} > 0$.

2. $\vec{u} \cdot \vec{v} = \|\vec{u}\| \|\vec{v}\| \cos 180° = 30(-1) = -30$.

3. $\cos A = \dfrac{\overrightarrow{AB} \cdot \overrightarrow{AC}}{\|\overrightarrow{AB}\| \|\overrightarrow{AC}\|}$, $\overrightarrow{AB} = (1, -3, 9)$,

$\overrightarrow{AC} = (3, -1, 5)$ so $\cos A = \dfrac{51}{\sqrt{91}\sqrt{35}}$, about .904 so A is

approximately $25°$.

4. (a) θ_1 is determined by $\vec{u}$ and $\vec{i}$ so
$$\cos \theta_1 = \frac{\vec{u} \cdot \vec{i}}{\|\vec{u}\| \|\vec{i}\|} = \frac{u_1}{\|\vec{u}\|}.$$
Similarly $\cos \theta_2 = u_2/\|\vec{u}\|$ and $\cos \theta_3 = u_3/\|\vec{u}\|$.

(b) $(\cos \theta_1, \cos \theta_2, \cos \theta_3) = (u_1/\|\vec{u}\|, u_2/\|\vec{u}\|, u_3/\|\vec{u}\|)$ (by part (a)) $= \vec{u}/\|\vec{u}\| = \vec{u}_{\text{normalized}}$.

5. (a) Slope $AB = \frac{5}{3}$, slope $CD = -\frac{3}{5}$. Product of slopes is -1 so lines are perpendicular.

(b) $\overrightarrow{AB} = (3,5)$, $\overrightarrow{CD} = (5,-3)$,
$\overrightarrow{AB} \cdot \overrightarrow{CD} = 15 - 15 = 0$ so lines are perp.

6. You start walking in direction of vector $\overrightarrow{AB} = (6,5)$. After a left turn you are walking in the perp direction $\vec{u} = (-5,6)$. $\overrightarrow{BC}$ points like $\vec{u}$ and has length 7 so $\overrightarrow{BC} = 7\vec{u}/\|\vec{u}\|$, $C - B = (-35/\sqrt{61}, 42/\sqrt{61})$,
$$C = \left(8 - \frac{35}{\sqrt{61}}, 9 + \frac{42}{\sqrt{61}}\right).$$

7. (See fig.) First line has parallel vector $\vec{u} = (-2, 7)$. Second line has parallel vector $\vec{v} = (1, 4)$,
$$\cos \theta = \frac{\vec{u} \cdot \vec{v}}{\|\vec{u}\| \|\vec{v}\|} = \frac{26}{\sqrt{53}\sqrt{17}}.$$

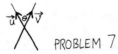

PROBLEM 7

8. Let $\vec{u} \cdot \vec{u}$ be denoted by a and $\vec{v} \cdot \vec{u}$ by b. To show $a\vec{v} - b\vec{u}$ perp to $\vec{u}$, dot them. We have $\vec{u} \cdot (a\vec{v} - b\vec{u})$ $= a(\vec{u} \cdot \vec{v}) - b(\vec{u} \cdot \vec{u}) = (\vec{u} \cdot \vec{u})(\vec{u} \cdot \vec{v}) - (\vec{v} \cdot \vec{u})(\vec{u} \cdot \vec{u})$ $= 0$ so $\vec{u}$ perp to $a\vec{u} - b\vec{v}$.

9. $\|-6\vec{u}\| = 6\|\vec{u}\| = 18$, $\vec{u} \cdot 3\vec{u} = 3(\vec{u} \cdot \vec{u}) = 3\|\vec{u}\|^2 = 27$,
$\|\vec{u} - \vec{v}\| = \sqrt{(\vec{u} - \vec{v}) \cdot (\vec{u} - \vec{v})}$
$= \sqrt{\vec{u} \cdot \vec{u} - 2\vec{u} \cdot \vec{v} + \vec{v} \cdot \vec{v}} = \sqrt{\|\vec{u}\|^2 - 2\vec{u} \cdot \vec{v} + \|\vec{v}\|^2}$
$= \sqrt{9 - 10 + 4} = \sqrt{3}$

10. (a) $\vec{u} \cdot \vec{v} = -33$ so $|\vec{u} \cdot \vec{v}| = $ abs value of $-33 = 33$
(b) Can't take the norm of the scalar $\vec{u} \cdot \vec{v}$.
(c) $\|\vec{v}\| = \sqrt{16 + 9 + 1 + 16} = \sqrt{42}$ so $\|\vec{v}\|\vec{u} = (5\sqrt{42}, 2\sqrt{42}, 3\sqrt{42}, -4\sqrt{42})$
(d) Can't divide by a vector.
(e) $\|\vec{u}\| = \sqrt{54}$ so $2/\|\vec{u}\|$ is $2/\sqrt{54}$.
(f) $\vec{u} \cdot \vec{v} = -33$ so $(\vec{u} \cdot \vec{v})\vec{v} = (132, -99, 33, -132)$
(g) Can't dot the scalar $\vec{u} \cdot \vec{v}$ with the vector $\vec{v}$.

11. (See figs.)

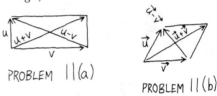

PROBLEM 11(a)

PROBLEM 11(b)

(a) (i) Draw $\vec{u}$ and $\vec{v}$ perp since $\vec{u} \cdot \vec{v} = 0$. In same picture can draw $\vec{u} + \vec{v}$ and $\vec{u} - \vec{v}$. Then $\|\vec{u} + \vec{v}\|$ and $\|\vec{u} - \vec{v}\|$ are (lengths of) diagonals of a rectangle. Diagonals of a rectangle have same length so $\|\vec{u} + \vec{v}\| = \|\vec{u} - \vec{v}\|$.

(ii) $\|\vec{u} + \vec{v}\| = \sqrt{(\vec{u} + \vec{v}) \cdot (\vec{u} + \vec{v})}$
$= \sqrt{\vec{u} \cdot \vec{u} - 2\vec{u} \cdot \vec{v} + \vec{v} \cdot \vec{v}}$
$= \sqrt{\vec{u} \cdot \vec{u} + \vec{v} \cdot \vec{v}}$,
$\|\vec{u} - \vec{v}\| = \sqrt{(\vec{u} - \vec{v}) \cdot (\vec{u} - \vec{v})}$
$= \sqrt{\vec{u} \cdot \vec{u} - 2\vec{u} \cdot \vec{v} + \vec{v} \cdot \vec{v}}$
$= \sqrt{\vec{u} \cdot \vec{u} + \vec{v} \cdot \vec{v}}$,
so $\|\vec{u} + \vec{v}\| = \|\vec{u} - \vec{v}\|$

(b) (i) Draw $\vec{u}, \vec{v}, \vec{u} + \vec{v}, \vec{u} - \vec{v}$. Parallelogram is a rhombus because $\|\vec{u}\| = \|\vec{v}\|$. Diagonals of rhombus are perp so $\vec{u} + \vec{v}$ perp to $\vec{u} - \vec{v}$.

(ii) If $\|\vec{u}\| = \|\vec{v}\|$ then $(\vec{u} + \vec{v}) \cdot (\vec{u} - \vec{v})$
$= \vec{u} \cdot \vec{u} - \vec{v} \cdot \vec{v} = \|\vec{u}\|^2 - \|\vec{v}\|^2 = 0$
so $\vec{u} + \vec{v}$ and $\vec{u} - \vec{v}$ perp.

12. (a) $(\vec{u} \cdot \vec{v})/\|\vec{v}\| = -7/\sqrt{11}$
(b) $(\vec{v} \cdot \vec{u})/\|\vec{u}\| = -7/\sqrt{29}$

13. With rays DA, DC, and DH as axes, we have $\overrightarrow{FH} = (-2, -10, 0)$, $\overrightarrow{AG} = (-2, 10, 7)$ so $(\overrightarrow{FH} \cdot \overrightarrow{AG})/\|\overrightarrow{AG}\| = -96/\sqrt{153}$. *Length* of the projection is $96/\sqrt{153}$.

14. (See figs.) (a) *Geometric method* (more sensible). Projecting onto the line determined by $4\vec{v}$ is same as projecting on the line determined by $\vec{v}$; answer is 6 again.
Algebraic method (overkill). Given $\vec{u} \cdot \vec{v}/\|\vec{v}\| = 6$.
Then $\dfrac{\vec{u} \cdot 4\vec{v}}{\|4\vec{v}\|} = \dfrac{4(\vec{u} \cdot \vec{v})}{4\|\vec{v}\|} = \dfrac{\vec{u} \cdot \vec{v}}{\|\vec{v}\|} = 6$
by (10), Section 9.3; (15) Section 9.2.

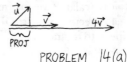

PROBLEM 14(a)

(b) Angle between $\vec{u}$ and $-\vec{v}$ is obtuse; component of $\vec{u}$ in direction of $-\vec{v}$ (i.e., *signed* projection) is -6.

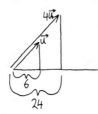

PROBLEM 14(b)

(c) If $\vec{u}$ is quadrupled, its projection is 4 times as long as before. Answer is 24.

PROBLEM 14(c)

15. (See fig.) $p = \|\vec{u}\|\cos 60° = 6(\tfrac{1}{2}) = 3$; component of $\vec{u}$ in direction of $\vec{v}$ is -3.

PROBLEM 15

16. Component of $\vec{w}$ in track direction is $(\vec{w}\cdot\vec{t})/\|\vec{t}\| = 6/\sqrt{5} > 2$. Disqualified!

17. $(\vec{f}\cdot\vec{v})/\|\vec{v}\| = 46/\sqrt{29} < 10$. Not enough force.

18. Component is max in direction of $\vec{u}$ itself. Max value is $\|\vec{u}\|$.

19. $\dfrac{\vec{v}\cdot\vec{q}}{\vec{q}\cdot\vec{q}}\vec{q} = \tfrac{4}{29}(5\vec{i}-2\vec{j}) = \tfrac{20}{29}\vec{i} - \tfrac{8}{29}\vec{j}$

20. $\vec{p}$ since it has the longer projection onto $\vec{u}$.

Section 9.4 (page 281)

1. $\vec{u}\times\vec{v}$ is into the page, $\vec{p}\times\vec{q}$ is in plane of the page pointing east, $\vec{s}\times\vec{t} = \vec{0}$.

2. $\vec{u} = \vec{0}$ or $\vec{v} = \vec{0}$ or $\vec{u}$ and $\vec{v}$ are nonzero parallel perp vectors. The latter is impossible so either $\vec{u} = \vec{0}$ or $\vec{v} = \vec{0}$.

3. $\vec{u}$ and $\vec{u}_{normalized}$ are parallel; cross product of parallel vectors is $\vec{0}$.

4. If $\vec{a}\times\vec{x} = \vec{b}$ then $\vec{b}$ is perp to $\vec{a}$ and to $\vec{x}$ But if $\vec{a}\cdot\vec{b}\neq 0$ then $\vec{b}$ is not perp to $\vec{a}$. So there is no such $\vec{x}$.

5. (a) $(\vec{u}\cdot\vec{v})\times\vec{w}$ is scalar × vector; makes no sense. $\vec{u}\cdot(\vec{v}\times\vec{w})$ is vector · vector; OK.

(b) $\vec{v}\times\vec{u}$ is perp to $\vec{u}$ (and to $\vec{v}$) so $\vec{u}\cdot(\vec{v}\times\vec{u}) = 0$ (the *number* 0, not $\vec{0}$).

(c) $\vec{v}\times\vec{v}$ is $\vec{0}$ so $\vec{u}\cdot(\vec{v}\times\vec{v}) = \vec{u}\cdot\vec{0} = 0$.

6. *Method 1:* $(\vec{u}+\vec{v})\times(\vec{u}+\vec{v})$
$= \vec{u}\times\vec{u} + \vec{v}\times\vec{v} + \vec{v}\times\vec{u} + \vec{u}\times\vec{v}$
$= \vec{0} + \vec{0} + -(\vec{u}\times\vec{v}) + \vec{u}\times\vec{v} = \vec{0}$.

Method 2: the cross product of a vector, namely $\vec{u}+\vec{v}$, with itself is $\vec{0}$.

7. $\vec{u}\times\vec{v}$ and $\vec{p}\times\vec{q}$ are perp to the floors of the building so they are parallel. So $(\vec{u}\times\vec{v})\times(\vec{p}\times\vec{q}) = \vec{0}$.

8. $3\vec{u}\times(4\vec{u}+5\vec{v}) = 12(\vec{u}\times\vec{u}) + 15(\vec{u}\times\vec{v})$
$= \vec{0} + 15(\vec{u}\times\vec{v}) = 15(\vec{u}\times\vec{v})$

9. (See fig.) $\vec{v}\times\vec{w}$ is perp to $\vec{v},\vec{w}$ plane; $\vec{u}\times(\vec{v}\times\vec{w})$ is perp to $\vec{v}\times\vec{w}$ (and to $\vec{u}$). All perps to $\vec{v}\times\vec{w}$ land back in the $\vec{v},\vec{w}$ plane. So $\vec{u}\times(\vec{v}\times\vec{w})$ is in the plane.

PROBLEM 9

10. (a) $(-11,-12,27)$
(b) $-17\vec{i} + 13\vec{j} + \vec{k}$
(c) $(0,0,21)$
(d) $(4,-2,11)$

11. $\vec{w}\times\vec{v} = (-13,11,-3)$,
$\vec{v}\times\vec{w} = -(\vec{w}\times\vec{v}) = (13,-11,3)$

12. $\cos\theta = \dfrac{\vec{u}\cdot\vec{v}}{\|\vec{u}\|\,\|\vec{v}\|} = \dfrac{5}{\sqrt{14}\sqrt{30}}$,

$\sin\theta = \dfrac{\|\vec{u}\times\vec{v}\|}{\|\vec{u}\|\,\|\vec{v}\|} = \dfrac{\|(9,-5,17)\|}{\|\vec{u}\|\,\|\vec{v}\|} = \dfrac{\sqrt{395}}{\sqrt{14}\sqrt{30}}$,

$\cos^2\theta + \sin^2\theta = \dfrac{25}{(14)(30)} + \dfrac{395}{(14)(30)} = 1$

13. (a) Anything that dots with $\vec{u}$ to give 0 is OK, e.g., $\vec{i}+2\vec{j}$, $6\vec{i}+3\vec{j}-3\vec{k}$, etc.

(b) Only $\vec{u}\times\vec{v} = (-3,27,11)$ and multiples of $\vec{u}\times\vec{v}$ will do.

14. $\overrightarrow{AB} = (4,-6,3)$, $\overrightarrow{AC} = (-1,-6,7)$, $\overrightarrow{AB}\times\overrightarrow{AC} = (-24,-31,-30)$, parallelogram area $= \|\overrightarrow{AB}\times\overrightarrow{AC}\|$, triangle area $= \tfrac{1}{2}\sqrt{(24)^2 + (31)^2 + (30)^2}$.

Section 9.5 (page 284)

1. $(1,2,3)\cdot(1,4,-3) = 0$

2. $\vec{u}\cdot\vec{v}\times\vec{w} = -16$, volume is 16.

3. $\overrightarrow{AB}\cdot\overrightarrow{AC}\times\overrightarrow{AD} = (1,2,3)\cdot(1,-1,2)\times(1,-4,-3)$
$= (1,2,3)\cdot(11,5,-3) = 12 \neq 0$; vectors not coplanar. So points A, B, C, D not coplanar (see fig.).

PROBLEM 3

4. $\vec{v} \times \vec{w}$ points down, makes an obtuse angle with $\vec{u}$; $\vec{u} \cdot \vec{v} \times \vec{w}$ is negative.

5. If vectors are placed with a common initial point, they are coplanar. So $\vec{u} \cdot \vec{v} \times \vec{w} = 0$

6. (a) -5 (cyclic perm)
  (b) 5 (noncyclic perm)
  (c) 5
  (d) $60(\vec{q} \cdot \vec{p} \times \vec{r}) = -300$
  (e) *Method 1:* $\vec{q}, \vec{p}, \vec{q}$ are coplanar (since there are only two distinct vectors) so scalar triple product is 0.
  Method 2: $\vec{q} \cdot \vec{p} \times \vec{q} = \vec{p} \cdot \vec{q} \times \vec{q}$ (cyclic perm) $= \vec{p} \cdot \vec{0} = 0$.
  Method 3: $\vec{p} \times \vec{q}$ is perp to $\vec{q}$ so dot product $\vec{q} \cdot (\vec{p} \times \vec{q})$ is 0.
  (f) *Method 1:* same as (e).
  Method 2: $\vec{q} \cdot \vec{r} \times \vec{r} = \vec{q} \cdot \vec{0} = 0$.

Section 9.6 (page 289)

1. (See figs.)
  (a) $x = y^2 + 5$

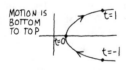

PROBLEM 1(a)

  (b) $x = y^2 + 5$

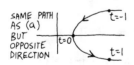

PROBLEM 1(b)

  (c) $y = 4 - 2(x - 2) = 8 - 2x$

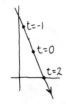

PROBLEM 1(c)

(d) $y = 8 - 2x$ again, but particle travels on only half the line and then doubles back.

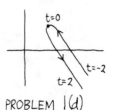

PROBLEM 1(d)

(e) $y = 2x$. Graph of $y = 2x$ is a line but particle travels on only part of the line.

t	$\vec{r}(t)$
-100	$(e^{-100}, 2e^{-100})$
-1	$(e^{-1}, 2e^{-1})$
0	$(1, 2)$
10	$(e^{10}, 2e^{10})$
etc.	

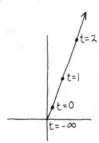

PROBLEM 1(e)

2. (See figs.) (a) circle, radius 4, one rev per 6π sec, ccl

PROBLEM 2(a)

  (b) "Circle with radius $1/t$", a spiral

$$\lim_{t \to 0+} (\cos t)/t = 1/0+ = \infty,$$
$$\lim_{t \to 0+} (\sin t)/t = 0/0 = \lim_{t \to 0+} (\cos t)/1 (\text{L'Hôpital})$$
$$= 1.$$

So at time $t = 0$, particle comes in from $(\infty, 1)$ and at time $t = \pi/2$ hits the y-axis and starts spiraling in toward the origin.

PROBLEM 2(b)

(c) Particle circles *and* rises since $z = t$.

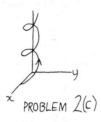

PROBLEM 2(c)

3. (a) One possibility is $x = 3 \sin t$, $y = 3 \cos t$. Another is $x = 3 \cos t$, $y = -3 \sin t$.

   (b) $x = 2 + 3 \cos t$, $y = 7 + 3 \sin t$

   (c) $x = 3 \cos 2\pi t$, $y = 3 \sin 2\pi t$

4. (a) If $x = -3$ then $2 - t = -3$, $t = 5$ but $t = 5$ does not produce $z = 4$; point is not on the path.

   (b) If $x = 3$ then $2 - t = 3$, $t = -1$. If $t = -1$ then $y = 4$, $z = -6$; point is on the path, reached at time $t = -1$.

5. The particles travel on the same route but second particle gets everywhere 5 seconds later than the first. For example at time $t = 0$, first particle is at the origin but second particle doesn't reach the origin until $t = 5$.

6. Distance from particle to origin is always 7. (a) Travels on circle with center at origin, radius 7. (b) Travels on surface of a sphere centered at origin, radius 7.

7. $\vec{v} = 3t^2 \vec{i} + 2\vec{j} - \sin t \, \vec{k}$

8. $\vec{v} = (-t \sin t + \cos t)\vec{i} + (t \cos t + \sin t)\vec{j}$ (product rule). At time $t = \pi$, $\vec{v} = -\vec{i} - \pi\vec{j}$. Path is a "circle with radius t", i.e., a spiral. (See fig.)

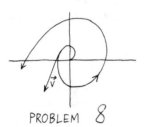

PROBLEM 8

9. $\vec{v} = -6 \sin t\vec{i} + 6 \cos t\vec{j}$

   (a) $\|\vec{v}\| = \sqrt{36 \sin^2 t + 36 \cos^2 t}$
$= \sqrt{36(\sin^2 t + \cos^2 t)} = 6$. Circle has radius 6, circum 12π. A speed of one rev per 2π seconds is 12π feet per 2π sec or 6 ft/sec, same as $\|\vec{v}\|$.

   (b) $\vec{v} = -6\vec{i}$ which does point ccl. (See fig.)

PROBLEM 9(b)

10. $\vec{v} = (1, 2t)$; speed is $\|\vec{v}\| = \sqrt{1 + 4t^2}$. As t goes from $-\infty$ to 0, speed decreases; as t goes from 0 to ∞, speed increases. So particle decelerates, then accelerates.

11. Particle has polar coords $r = 1$, $\theta = t^2$. Moves on circle with radius 1; $\vec{v} = (-2t \sin t^2, 2t \cos t^2)$, $\|\vec{v}\| = \sqrt{4t^2([\sin t^2]^2 + \cos t^2]^2)} = \sqrt{4t^2} = 2|t|$. (Note that $\sqrt{t^2}$ is $|t|$, not t.) As t goes from $-\infty$ to 0, speed $2|t|$ decreases (from ∞ to 0). As t goes from 0 to ∞, speed increases (from 0 to ∞).

12. $\vec{v} = 3\vec{i} - 2\vec{j} + 4\vec{k}$.

   (a) $\vec{v}$ is constant so particle has constant direction; path is a line.

   (b) $\|\vec{v}\| = \sqrt{29}$.

   (c) Want to change $(3, -2, 4)$ to keep same direction but get norm 2. Use $\vec{v} = 2(3, -2, 4)_{\text{normalized}} = (6/\sqrt{29}, -4/\sqrt{29}, 8/\sqrt{29})$; $\vec{r} =$
$$\left(-1 + \frac{6}{\sqrt{29}}t\right)\vec{i} + \left(1 - \frac{4}{\sqrt{29}}t\right)\vec{j} + \frac{8}{\sqrt{29}}t\vec{k}.$$

13. (a) Particle sits at the origin and refuses to move.

   (b) Zero velocity, zero speed. Particle doesn't move (but it isn't necessarily sitting at the origin).

14. (a) Antidiff to get $x = t^2 + C_1$, $y = \frac{5}{3}t^3 + C_2$, $z = 6t + C_3$. Need $x = 1$, $y = 4$, $z = 6$ when $t = 3$ so need $1 = 9 + C_1$, $C_1 = -8$. Similarly $C_2 = -41$, $C_3 = -12$. So
$\vec{r} = (t^2 - 8)\vec{i} + (\frac{5}{3}t^3 - 41)\vec{j} + (6t - 12)\vec{k}$.

   (b) If $t = 2$ then $\vec{v} = 4\vec{i} + 20\vec{j} + 6\vec{k}$. Unit tangent is $\vec{v}_{\text{normalized}}$
$= (4/\sqrt{452})\vec{i} + (20/\sqrt{452})\vec{j} + (6/\sqrt{452})\vec{k}$.

15. (a) $(\frac{1}{3}x)^2 + (\frac{1}{2}y)^2 = \cos^2 t + \sin^2 t = 1$ so path is along ellipse $x^2/9 + y^2/4 = 1$.

   (b) $\vec{v} = (-3 \sin t, 2 \cos t)$,
speed $= \|\vec{v}\| = \sqrt{9 \sin^2 t + 4 \cos^2 t}$
$= \sqrt{9 \sin^2 t + 4(1 - \sin^2 t)}$
$= \sqrt{5 \sin^2 t + 4}$. Speed is min when $\sin^2 t = 0$, $t = 0, \pi, 2\pi$, etc. Min value of speed is 2. Speed is max when $\sin^2 t = 1$, $t = \pi/2, 3\pi/2$, etc. Max value of speed is 3.

Section 9.7 (page 293)

1. $xy = 1$; path is a hyperbola (see fig.)

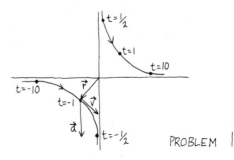

PROBLEM 1

$\vec{v} = \vec{i} - (1/t^2)\vec{j}$, $\vec{a} = (2/t^3)\vec{j}$.
If $t = -1$ then $\vec{r} = -\vec{i} - j$, $\vec{v} = \vec{i} - j$, $\vec{a} = -2\vec{j}$,
$a_{tan} = (\vec{a} \cdot \vec{v})/\|\vec{v}\| = 2/\sqrt{2} = \sqrt{2}$. Particle is speeding up by $\sqrt{2}$ meters/sec per sec; $\|\vec{a}\| = 2$ so if mass is m then $2m$ pounds of force act (in direction of $\vec{a}$, i.e., down).

2. $xy = 1$ but only one branch is traversed since $x > 0$, $y > 0$ (see fig.); $\vec{v} = e^t\vec{i} - e^{-t}\vec{j}$, $\vec{a} = e^t\vec{i} + e^{-t}\vec{j}$.
If $t = -1$ then $\vec{r} = (1/e)\vec{i} + e\vec{j}$, $\vec{v} = (1/e)\vec{i} - e\vec{j}$,
$\vec{a} = (1/e)\vec{i} + e\vec{j}$, a_{tan} is negative at $t = -1$ since
$\vec{a} \cdot \vec{v} = 1/e^2 - e^2 < 0$; particle is slowing down.

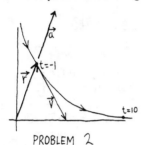

PROBLEM 2

3. $\vec{r}'' = \vec{a} = \vec{f}/m$,
$$\vec{r} \times \vec{r}'' = \vec{r} \times \left(\frac{\vec{f}}{m}\right) = \left(\frac{1}{m}\right)(\vec{r} \times \vec{f}).$$

$\vec{r}$ points from the origin to the particle. By hypothesis, $\vec{f}$ points from the particle toward the origin (see fig.). So $\vec{f}$ and $\vec{r}$ are parallel (opposite directions) and their cross prod is $\vec{0}$.

ORIGIN

PROBLEM 3

4. $\vec{f} = -mg\vec{j}$ and $\vec{f} = m\vec{a}$ so $\vec{a} = -g\vec{j}$, i.e., $x'' = 0$, $y'' = -g$. Antidiff to get $x' = C_1$, $y' = -gt + C_2$. If $t = 0$ then $\vec{v} = 4\vec{i} + 2\vec{j}$ so $C_1 = 4$, $C_2 = 2$. Antidiff again to get $x = 4t + K_1$, $y = -\frac{1}{2}gt^2 + 2t + K_2$. If $t = 0$ then $x = 1$, $y = 2$ so need $K_1 = 1$, $K_2 = 2$. So
$\vec{r} = (4t + 1)\vec{i} + (-\frac{1}{2}gt^2 + 2t + 2)\vec{j}$.

5. (a) ds/dt is the rate of change of distance traveled w.r.t. time so it is the car's speed, i.e., $ds/dt = \|\vec{v}\|$.
$$\frac{d^2s}{dt^2} = \frac{d(ds/dt)}{dt} = \frac{d(\text{speed})}{dt} = \text{rate of change of}$$
speed = car's acc. So $d^2s/dt^2 = a_{tan}$.

(b) $\dfrac{d\vec{r}}{ds} = \dfrac{d\vec{r}/dt}{ds/dt} = \dfrac{\vec{v}}{\|\vec{v}\|} = \vec{v}_{normalized}$
= unit tangent vector

6. (a) $\vec{v} = (-5 \sin t, 5 \cos t)$, $\|\vec{v}\| =$
$\sqrt{25 \sin^2 t + 25 \cos^2 t} = \sqrt{25(\sin^2 t + \cos^2 t)} = 5$

(b) $\vec{a} = (-5 \cos t, -5 \sin t)$,
$a_{tan} = 0$ because $\vec{a} \cdot \vec{v} = 0$.

(c) $\vec{r} = (-5 \cos t, -5 \sin t)$ is radial direction;
$a_{rad} =$ comp in radial direction

$= (\vec{a} \cdot -\vec{r})/\|-\vec{r}\| = \frac{1}{5}(25 \cos^2 t + 25 \sin^2 t) = 5$;
$\|v\|^2/r = (5)^2/5 = 5$ also.

7. (a) Force $m\vec{a}$ acting on particle never makes it change direction so $\vec{a}$ parallel to $\vec{v}$.

(b) No change in speed, so $a_{tan} = 0$ so $\vec{a}$ perp to $\vec{v}$.

8. (a) $\vec{v} = (-3t^2, t^3 + 1)$, $\vec{a} = (-6t, 3t^2)$. Particle is at point $(3, 5/4)$, moving in direction of
$\vec{v} = -3\vec{i} + 2\vec{j}$ at speed $\|\vec{v}\| = \sqrt{13}$ ft/sec.
$\vec{a} = -6\vec{i} + 3\vec{j}$ and $(\vec{a} \cdot \vec{v})/\|\vec{v}\| = 24/\sqrt{13}$;
speeding up by $24/\sqrt{13}$ ft/sec per sec. If mass is m then amount of force is $m\|\vec{a}\| = m\sqrt{45}$ lbs in direction of $\vec{a}$.

Chapter 9 Review Problems (page 294)

1. (a) $-8 + 15 - 2 = 5$
   (b) $\sqrt{14}$
   (c) $(-11, 0, 22)$
   (d) $\dfrac{\vec{u} \cdot \vec{v}}{\|\vec{u}\|\|\vec{v}\|} = \dfrac{5}{\sqrt{14}\sqrt{45}}$
   (e) $(\vec{u} \cdot \vec{v})/\|\vec{v}\| = 5/\sqrt{45}$
   (f) $\dfrac{\vec{u} \cdot \vec{v}}{\vec{v} \cdot \vec{v}}\vec{v} = \frac{5}{45}\vec{v} = \frac{1}{9}\vec{v} = (-\frac{4}{9}, \frac{5}{9}, -\frac{2}{9})$
   (g) $\vec{v}/\|v\| = (-4/\sqrt{45})\vec{i} + (5/\sqrt{45})\vec{j} - (2/\sqrt{45})\vec{k}$
   (h) $6\vec{u}/\|u\| = (12/\sqrt{14})\vec{i} + (18/\sqrt{14})\vec{j} - (6/\sqrt{14})\vec{k}$

2. $\vec{AB} = (2, -3, 1)$, $\vec{AC} = (4, -6, 3)$. Vectors are not multiples, so not parallel; C is not on the line (see fig.).

PROBLEM 2

3. (a) $\vec{v} \times \vec{u} = -(\vec{u} \times \vec{v}) = \vec{j} - 5\vec{k}$
   (b) $\|\vec{v} \times \vec{u}\| = \|\vec{u} \times \vec{v}\| = 6$

4. (a) Meaningless to even write $(\vec{u} \cdot \vec{v}) \cdot \vec{w}$ since scalar $\vec{u} \cdot \vec{v}$ can't dot vector $\vec{w}$.
   (b) False. Make up almost any vectors $\vec{u}, \vec{v}, \vec{w}$ to get a counterexample; e.g., $\vec{i} \times (\vec{i} \times \vec{j}) =$
   $\vec{i} \times \vec{k} = -\vec{j}$ but $(\vec{i} \times \vec{i}) \times \vec{j} = \vec{0} \times \vec{j} = \vec{0}$.

5. $\vec{PQ} = (2, 4, z)$, $\vec{AB} = (6, -2, 2)$. Need $\vec{PQ} \cdot \vec{AB} = 0$
so $12 - 8 + 2z = 0$, $z = -2$.

6. $\vec{u} \times \vec{v} = \left(0, 0, \begin{vmatrix} u_1 & u_2 \\ v_1 & v_2 \end{vmatrix}\right)$. Area of parallelogram is $\|\vec{u} \times \vec{v}\|$ which in this case is absolute value of third component $\begin{vmatrix} u_1 & u_2 \\ v_1 & v_2 \end{vmatrix}$.

7. (a) True by Pythagorean theorem (see fig.)

PROBLEM 7(a)

(b) If $\vec{u} \cdot \vec{v} = 0$ then $\|\vec{u} + \vec{v}\|^2 =$
$(\vec{u} + \vec{v}) \cdot (\vec{u} + \vec{v}) = \vec{u} \cdot \vec{u} + 2\vec{u} \cdot \vec{v} + \vec{v} \cdot \vec{v}$
$= \|\vec{u}\|^2 + 0 + \|\vec{v}\|^2 = \|\vec{u}\|^2 + \|\vec{v}\|^2$.

8. $\vec{u} \times \vec{v}$ is perp to $\vec{u}$ and to $\vec{v}$ by definition of cross product; $\|\vec{u} \times \vec{v}\| = \|\vec{u}\|\|\vec{v}\| \sin 90° = (1)(1)(1) = 1$ so $\vec{u} \times \vec{v}$ is a unit vector

9. (See fig.) $\|\vec{u} + \vec{v}\| \leq \|\vec{u}\| + \|\vec{v}\|$ because third side of a triangle is shorter than sum of other sides. In special case that $\vec{u}$ and $\vec{v}$ point the same way then $\|\vec{u} + \vec{v}\|$ equals $\|\vec{u}\| + \|\vec{v}\|$.

PROBLEM 9

10. (a) $\|\vec{u} + \vec{v}\|^2 + \|\vec{u} - \vec{v}\|^2$
$= (\vec{u} + \vec{v}) \cdot (\vec{u} + \vec{v}) + (\vec{u} - \vec{v}) \cdot (\vec{u} - \vec{v})$
$= \vec{u} \cdot \vec{u} + 2\vec{u} \cdot \vec{v} + \vec{v} \cdot \vec{v} + \vec{u} \cdot \vec{u} - 2\vec{u} \cdot \vec{v} + \vec{v} \cdot \vec{v}$
$= 2\vec{u} \cdot \vec{u} + 2\vec{v} \cdot \vec{v} = 2\|\vec{u}\|^2 + 2\|\vec{v}\|^2$

(b) Sum of squares of four sides of a parallelogram equals sum of squares of the diagonals (see fig.).

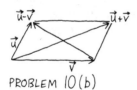

PROBLEM 10(b)

11. Component of wind in plane direction is $(\vec{u} \cdot \vec{v})/\|\vec{v}\| = -8/\sqrt{42}$. So it is a head wind slowing the plane down by $8/\sqrt{42}$ mph per hour.

12. $\|\vec{u} \times \vec{v}\|$ is the area of the parallelogram determined by $\vec{u}$ and $\vec{v}$. That area will be the product of the two sides if and only if $\vec{u}$ and $\vec{v}$ are perp. Alternatively $\|\vec{u} \times \vec{v}\| =$

$\|\vec{u}\|\|\vec{v}\| \sin \theta$ so $\|\vec{u} \times \vec{v}\| = \|\vec{u}\|\|\vec{v}\|$ if and only if $\sin \theta = 1$, $\theta = 90°$, $\vec{u}$ perp to $\vec{v}$.

13. Spirals in x direction (see fig.).

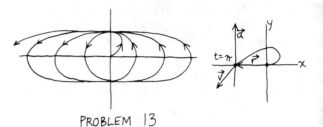

PROBLEM 13

$\vec{v} = (-t \sin t + \cos t)\vec{i} + \cos t \, \vec{j}$,
$\vec{a} = (-t \cos t - \sin t - \sin t)\vec{i} - \sin t \, \vec{j}$
$= (-t \cos t - 2 \sin t)\vec{i} - \sin t \, \vec{j}$.
If $t = \pi$ then $\vec{r} = -\pi\vec{i}$, $\vec{v} = -\vec{i} - \vec{j}$, $\vec{a} = \pi\vec{i}$, speed $= \|\vec{v}\| = \sqrt{2}$, $a_{tan} = (\vec{a} \cdot \vec{v})/\|\vec{v}\| = -\pi/\sqrt{2}$; slowing down by $\pi/\sqrt{2}$ meters/sec per sec.

14. (a) $\vec{v}$ perp to $\vec{r}$ so particle circles the origin. (See fig.)

PROBLEM 14(a)

(b) $\vec{v}$ parallel to $\vec{r}$. Particle moves on a line through the origin. (See fig.)

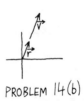

PROBLEM 14(b)

(c) $\vec{v}$ parallel to $\vec{a}$. Force never makes particle change direction. Path is a line.

(d) $\vec{v}$ perp to $\vec{a}$ so $a_{tan} = 0$. Speed never changes (shape of path unknown).

10/ TOPICS IN THREE-DIMENSIONAL ANALYTIC GEOMETRY

Section 10.1 (page 297)

1. Distance from center to origin is $\sqrt{50}$. Equation is $(x - 4)^2 + (y + 3)^2 + (z - 5)^2 = 50$.

2. $x^2 + y^2 + z^2 + z + \frac{1}{4} = \frac{1}{4}$, $x^2 + y^2 + (z + \frac{1}{2})^2 = \frac{1}{4}$. Sphere, center $(0, 0, -\frac{1}{2})$, radius $\frac{1}{2}$.

3. Set $x = 0$, $y = 0$, $z = 0$ in left side. Result is > 2; origin is outside. Equivalently, distance from origin to center $(-2, 3, 2)$ is $\sqrt{17}$, larger than the radius $\sqrt{2}$; origin is outside.

4. Distance from $(3, 5, 6)$ to x, y plane is z-coordinate 6. Equation is $(x - 3)^2 + (y - 5)^2 + (z - 6)^2 = 36$.

Section 10.2 (page 301)

1. (a) Normal is $(5, 3, 1)$. Plane is
$5(x - 5) + 3(y - 5) + z - 4 = 0$,
$5x + 3y + z = 44$.
   (b) Intercepts are $a = 2$, $b = 5$, $c = 7$. Plane is $x/2 + y/5 + z/7 = 1$.
   (c) Perpendicular to x-axis means parallel to y, z plane; equation is $x = 3$.
   (d) $z = 5$
   (e) Normal is $2\vec{i} + 9\vec{j} - 6\vec{k}$. Plane is
$2(x - 3) + 9(y - \pi) - 6(z - 7) = 0$,
$2x + 9y - 6z = 9\pi - 36$.

2. Normal to plane ABC is
$\overrightarrow{AB} \times \overrightarrow{AC} = (1, -4, 2) \times (3, 1, 5) = (-22, 1, 13)$.
Plane is $-22(x - 1) + (y - 3) + 13(z + 2) = 0$,
$-22x + y + 13z = -45$. Test point D in the equation;
$-22(1) + 2 + 13(3) = ? -45$, NO. Points not coplanar.

3. Write equation as $3x - 4y + 2z - 6 = 0$. Distance is $\dfrac{|3(2) - 4(3) + 2(-4) - 6|}{\sqrt{9 + 16 + 4}} = \dfrac{20}{\sqrt{29}}$.

4. Write equation as $3x - y + 4 = 0$. Distance is $\dfrac{|3(0) - 0 + 4|}{\sqrt{9 + 1}} = \dfrac{4}{\sqrt{10}}$.

5. Radius of sphere is distance from center to plane, $\dfrac{|2(1) + 3 - (-1) - 4|}{\sqrt{4 + 1 + 1}} = \dfrac{2}{\sqrt{6}}$. Sphere is
$(x - 1)^2 + (y - 3)^2 + (z + 1)^2 = \frac{4}{6}$.

6. Planes have common normal $\vec{n} = 2\vec{i} - \vec{j} + 3\vec{k}$ so they are parallel. (See fig.) To find distance between them pick any point in one plane and find distance from it to other plane. Point $(0, 0, 2)$ is in first plane. Its distance to second plane is $\dfrac{|2(0) - 0 + 3(2) - 8|}{\sqrt{4 + 1 + 9}} = \dfrac{2}{\sqrt{14}}$.

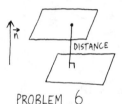

PROBLEM 6

7. Plane is $\dfrac{x}{a} + \dfrac{y}{b} + \dfrac{z}{c} - 1 = 0$ so

$D = \dfrac{|0 + 0 + 0 - 1|}{\sqrt{\left(\dfrac{1}{a}\right)^2 + \left(\dfrac{1}{b}\right)^2 + \left(\dfrac{1}{c}\right)^2}}$ by distance formula.

Square both sides and rearrange to get

$D^2 = \dfrac{1}{\dfrac{1}{a^2} + \dfrac{1}{b^2} + \dfrac{1}{c^2}}$, $\dfrac{1}{a^2} + \dfrac{1}{b^2} + \dfrac{1}{c^2} = \dfrac{1}{D^2}$.

8. A normal to the plane is (see fig.) $\overrightarrow{AB} \times \overrightarrow{CD} = (1, 1, -1) \times (3, 13, 3) = (16, -6, 10)$. Use point A (or B) to get equation $16(x + 1) - 6(y - 2) + 10(z - 4) = 0$, $16x - 6y + 10z = 12$.

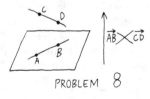

PROBLEM 8

Section 10.3 (page 306)

1. (a) Line has same parallel vector as given line, namely $(1, -2, 5)$. Equations are $x = 1 + t$, $y = 2 - 2t$, $z = 3 + 5t$.
   (b) Plane's normal is $(3, -4, 6)$. This is parallel to

line so line is $x = 1 + 3t$, $y = 4 - 4t$, $z = 5 + 6t$.

(c) *Method 1:* Parallel vector is $\vec{i} = (1, 0, 0)$; line is $x = 2 + t$, $y = 3$, $z = 4$.

Method 2: Points on the line satisfy $y = 3$, $z = 4$, $x = $ anything; equations are $x = t$, $y = 3$, $z = 4$.

(d) Line is parallel to the y-axis. Has $x = 2$, $z = 4$, $y = $ anything; equations are $x = 2$, $y = t$ ($y = 3 + t$ is OK too), $z = 4$.

(e) Parallel vector is $7\vec{i} - \vec{j} + 16\vec{k}$. Line is $x = 7t$, $y = -t$, $z = 16t$.

(f) (See fig.) Planes have normals $\vec{u} = (2, -1, 1)$, $\vec{v} = (3, 1, 4)$; $\vec{u} \times \vec{v} = (-5, -5, 5)$ is parallel to the line. With the simpler parallel $(1, 1, -1)$, the line is $x = 1 + t$, $y = 5 + t$, $z = 7 - t$.

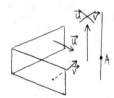

PROBLEM 1(f)

2. Parallel vector is $\overrightarrow{AB} = (14, -5)$. With point A, get $x = -1 + 14t$, $y = 3 - 5t$. (With B get $x = 13 + 14t$, $y = -2 - 5t$.)

3. (a) $y = 0$ (b) $x = 0$, $y = 0$, $z = t$

4. (See fig.) Line contains point $P = (6, 2, 7)$, has parallel vector $\vec{u} = (4, -1, 8)$. So $\vec{u}$ and $\overrightarrow{AP} = (3, 3, 5)$ are both parallel to plane and $\vec{u} \times \overrightarrow{AP} = (-29, 4, 15)$ is normal to plane. Plane is $-29(x - 3) + 4(y + 1) + 15(z - 2) = 0$, $-29x + 4y + 15z = -61$.

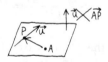

PROBLEM 4

5. (a) Point is $t = 2$ on first line and $t = -1$ on second line so it's on both. Lines aren't the same since their parallel vectors $\vec{u} = (1, -4, 5)$, $\vec{v} = (2, -1, -6)$ are not parallel to each other.

(b) (See fig.) Normal to plane is $\vec{u} \times \vec{v} = (29, 16, 7)$. With points $(2, 3, 6)$ from first line, plane is $29(x - 2) + 16(y - 3) + 7(z - 6) = 0$, $29x + 16y + 7z = 148$.

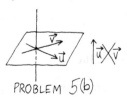

PROBLEM 5(b)

(c) $\vec{u} \times \vec{v}$ is parallel to line; line is $x = 4 + 29t$, $y = -5 + 16t$, $z = 16 + 7t$.

6. Set $x = 0$ to get $t = 2$. Then $y = 11$, $z = -1$ so intersection is $(0, 11, -1)$.

7. (See fig.) Common parallel vector $\vec{u} = (-3, 1, 2)$; lines are parallel. Not same line since $A = (2, 5, 4)$ is on first line but not on second (it takes $t = -3$ to get $x = 2$ but $t = -3$ does not produce $y = 5$, $z = 4$). To get plane's equation, need a normal. Point $B = (-7, 6, 0)$ and $A = (2, 5, 4)$ are in plane so $\overrightarrow{AB}$ is parallel to plane. Normal is $\vec{u} \times \overrightarrow{AB} = (-3, 1, 2) \times (-9, 1, -4) = (-6, -30, 6)$. Use simpler normal $(1, 5, -1)$ and point A to get $x - 2 + 5(y - 5) - (z - 4) = 0$, $x + 5y - z = 23$.

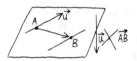

PROBLEM 7

8. $\overrightarrow{AB} = (3, 2, 2)$. Line AB is $x = 1 + 3t$, $y = 3 + 2t$, $z = -2 + 2t$. Point C isn't on the line since to get $x = 3$ we need $t = \frac{2}{3}$ but $t = \frac{2}{3}$ doesn't give $y = 3$, $z = 5$.

9. By (*), line's parallel vector (a, b, c) and the plane's normal vector (A, B, C) are perp, so line and plane are parallel. By (**), point (x_0, y_0, z_0), which we know is on line, satisfies equation of plane so line and plane have (x_0, y_0, z_0) in common. All in all, the line must lie *in* the plane.

10. (a) Parallel vectors are $(-6, 1, 3)$, $(12, -2, -6)$. They are parallel to each other. Point $(1, 2, 4)$ is on first line (when $t = 0$) but not on second (need $t = \frac{1}{2}$ to get $x = 1$ but $t = \frac{1}{2}$ doesn't give $z = 4$). Lines are parallel (but different).

(b) Parallels are $(-1, 2, -3)$, $(1, -4, 6)$; not parallel to each other. Lines are not parallel or coincident. Solve $2 - t = s$, $3 + 2t = 5 - 4s$, $5 - 3t = -1 + 6s$. Solution to first two equations is $s = -1$, $t = 3$ which doesn't satisfy third. Lines are skew.

(c) Parallel vectors are $(-1, 1, 2)$, $(-1, 2, 1)$; not parallel to each other. Solve $2 - t = 3 - s$, $3 + t = 4 + 2s$, $5 + 2t = 1 + s$. Solution to first two equations is $s = -2$, $t = -3$. Works in third equation too. Lines intersect, at point $x = 5$, $y = 0$, $z = -1$.

(d) Parallels are $(3, -1, 1)$, $(-3, 1, -1)$ which are parallel. Point $(2, 5, 3)$ is on first line (when $t = 0$) and also on second (when $t = -2$). Lines are coincident.

11. (a) $2(1 + 2t) + 6(3 - t) + 2 + 2t = 8$, $0 = -14$. No solutions. Line is parallel to but not contained in plane.

(b) $2(1 + 2t) + 6(3 - t) + 2 + 2t = 22$, $0 = 0$. All values of t are solutions. Line lies in plane.

12. Normals are $\vec{u} = (2, 1, 3)$, $\vec{v} = (1, -1, 1)$. Parallel to line is $\vec{u} \times \vec{v} = (4, 1, -3)$. To get point, set $z = 0$ and solve $2x + y = 5$, $x - y = 4$. Get $x = 3$, $y = -1$. Point on line is $(3, -1, 0)$. Line is $x = 3 + 4t$, $y = -1 + t$, $z = -3t$.

13. Parallel to line is $\vec{u} = (1, 2, 8) \times (1, -1, 2) = (12, 6, -3)$. Normal to 3rd plane is $\vec{v} = (3, -2, 8)$. Since $\vec{u} \cdot \vec{v} = 0$, vectors are perpendicular so line is parallel to plane. To show line is not *in* plane, find a point on the line. Set $z = 0$ and solve $x + 2y - 20 = 0$, $x - y - 8 = 0$ to get $x = 12$, $y = 4$. So $(12, 4, 0)$ is on line. But it doesn't satisfy equation of 3rd plane $(36 - 8 + 0 \neq 5)$. So line is parallel to but not contained in plane.

14. (See fig.) Plane has normal $\vec{u} = (2, -3, 1)$. So line has parallel $\vec{u}$. Line is $x = -1 + 2t$, $y = 1 - 3t$, $z = 1 + t$. P is intersection of line and plane. Solve $2(-1 + 2t) - 3(1 - 3t) + 1 + t + 1 = 0$, get $t = \frac{3}{14}$, $P = (-1, +\frac{6}{14}, 1 - \frac{9}{14}, 1 + \frac{3}{14}) = (-\frac{8}{14}, \frac{5}{14}, \frac{17}{14})$.

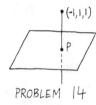

PROBLEM 14

15. (See fig.) Parallel to line is $\vec{u} = (1, -1, 4)$. So $\vec{u}$ is plane's normal. Plane is $x - 4 - (y + 1) + 4(z - 4) = 0$, $x - y + 4z = 21$; P is intersection of plane with line L. Solve $1 + t - (2 - t) + 4(3 + 4t) = 21$; $t = \frac{5}{9}$. $P = (1 + \frac{5}{9}, 2 - \frac{5}{9}, 3 + \frac{20}{9}) = (\frac{14}{9}, \frac{13}{9}, \frac{47}{9})$. Distance from Q to line L is $\overline{PQ} = \sqrt{(4 - \frac{14}{9})^2 + (-1 - \frac{13}{9})^2 + (4 - \frac{47}{9})^2}$.

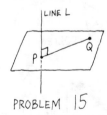

PROBLEM 15

16. Parallels are $\vec{u} = (-1, 1, 2)$, $\vec{v} = (-1, 2, 7)$; not parallel. Solve $2 - t = 3 - s$, $3 + t = 4 + 2s$, $5 + 2t = 1 + 7s$. Solution to first two equations is $s = -2$, $t = -3$; doesn't work in third. Lines skew. Consider plane containing second line and parallel to first. (See fig.) Normal is $\vec{u} \times \vec{v} = (3, 5, -1)$. Point is $(3, 4, 1)$. Plane is $3(x - 3) + 5(y - 4) - (z - 1) = 0$, $3x + 5y - z = 28$. Distance between lines is distance from any point on first

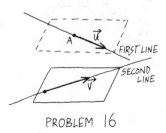

PROBLEM 16

line to the plane. Use $A = (2, 3, 5)$. Distance to plane is $\frac{|3(2) + 5(3) - 5 - 28|}{\sqrt{9 + 25 + 1}} = \frac{12}{\sqrt{35}}$.

17. (a) (See fig.) $\overrightarrow{AB} = (-1, 4, -2)$, $\overrightarrow{CD} = (3, -12, 6)$; parallel. But $\overrightarrow{AB}$ and $\overrightarrow{AC} = (1, 5, 4)$ are not parallel. Lines parallel (but not coincident).

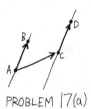

PROBLEM 17(a)

(b) (See fig.) Plane has normal $\overrightarrow{AB}$; equation is $-(x - 3) + 4(y - 0) - 2(z - 2) = 0$, $-x + 4y - 2z = -7$. Find point of intersection P of line CD and plane. Line CD is $x = 4 - t$, $y = 5 + 4t$, $z = 6 - 2t$. Solve $-(4 - t) + 4(5 + 4t) - 2(6 - 2t) = -7$; $t = -\frac{11}{21}$. $P = (4 + \frac{11}{21}, 5 - \frac{44}{21}, 6 + \frac{22}{21}) = (\frac{95}{21}, \frac{61}{21}, \frac{148}{21})$. Distance between the parallel lines is $\overline{AP} = \sqrt{(3 - \frac{95}{21})^2 + (\frac{61}{21})^2 + (2 - \frac{148}{21})^2}$.

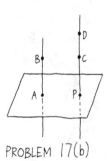

PROBLEM 17(b)

Section 10.4 (page 312)

1.

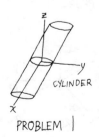

PROBLEM 1

2.

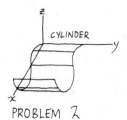

PROBLEM 2

3.

"PLANE" CYLINDER
(AND A PLAIN PLANE
SINCE IT IS OF
THE FORM
$ax+by+cz+d=0$)

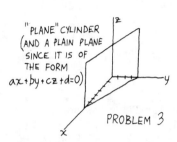

PROBLEM 3

4.

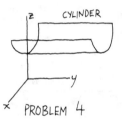

PROBLEM 4

5.

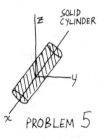

PROBLEM 5

6.

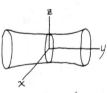

PROBLEM 6

7.

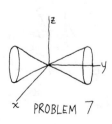

PROBLEM 7

8.

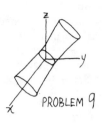

PROBLEM 8

9.

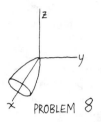

PROBLEM 9

10.

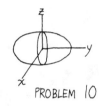

PROBLEM 10

11.

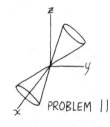

PROBLEM 11

12.

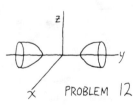

PROBLEM 12

13.

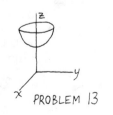

PROBLEM 13

14.

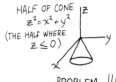

HALF OF CONE
$z^2 = x^2 + y^2$
(THE HALF WHERE
$z \leq 0$)

PROBLEM 14

15.

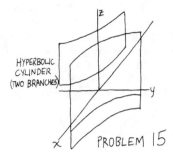

HYPERBOLIC
CYLINDER
(TWO BRANCHES)

PROBLEM 15

16. Hyperbolic paraboloid, saddle shaped; hard to draw.
17. Need $a = b$ to get circles rather than ellipses. If $z = 6$ get $x^2/a^2 + y^2/a^2 = 36$, $x^2 + y^2 = 36a^2$. Need $36a^2 = 9$, $a = \frac{1}{2}$, equation is $4x^2 + 4y^2 = z^2$.

Section 10.5 (page 317)

1. (a) (See fig.) Point is in y, z plane; $\theta = 90°$. Distance to z-axis is 2 so $r = 2$. Given that $z = 3$.

PROBLEM 1(a)

(b) (See fig.) Distance to origin is 5 so $\rho = 5$. Point is in x, y plane so $\phi = 90°$. Point is around 90° so $\theta = 90°$.

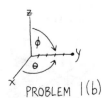

PROBLEM 1(b)

2. (See figs.)
(a) $z = 5$, $r = \sqrt{9 + 4} = \sqrt{13}$, $\tan \theta = \frac{2}{3}$; θ is in quad I, approx 34°

PROBLEM 2(a)

(b) $z = 5$, $r = \sqrt{13}$, $\tan \theta = \frac{-2}{-3} = \frac{2}{3}$; θ is in quad III, approx 214°

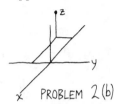

PROBLEM 2(b)

3. (a) $x = 2 \cos 150° = 2(-\frac{1}{2}\sqrt{3}) = -\sqrt{3}$, $y = 2 \sin 150° = 2(\frac{1}{2}) = 1$, $z = 7$
(b) $x = 2 \sin 30° \cos 120° = 2(\frac{1}{2})(-\frac{1}{2}) = -\frac{1}{2}$, $y = 2 \sin 30° \sin 120° = \frac{1}{2}\sqrt{3}$, $z = 2 \cos 30° = \sqrt{3}$

4. (See figs.)
(a) Region between two cylindrical surfaces plus the surfaces themselves.

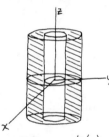

PROBLEM 4(a)

(b) Circle of intersection of plane $z = 2$ and cylinder $r = 3$.

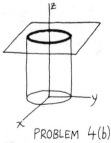

PROBLEM 4(b)

(c) Ray of intersection of half plane $\theta = \pi/3$ and plane $z = 7$.

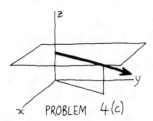

PROBLEM 4(c)

(d) Line of intersection of cylinder $r = 3$ and half plane $\theta = \pi/3$.

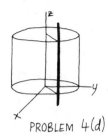

PROBLEM 4(d)

5. (a) (See fig.) Circular cone. In cylindrical coordinates, $r^2 = 4z^2$. In spherical, $\rho^2 \sin^2\phi(\cos^2\theta + \sin^2\theta) = 4\rho^2 \cos^2\phi$, $\tan^2\phi = 4$, $\tan \phi = \pm 2$, equ is $\phi = 63°$ or $117°$ (approx). Can see this geometrically since circular cone(s) should be described entirely in terms of cone angle(s).

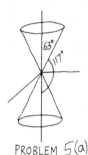

PROBLEM 5(a)

(b) Sphere. In cylindrical, $r^2 + z^2 = 10$. In spherical, $\rho = \sqrt{10}$.

6. (a) x, y plane is $z = 0$, z-axis is $r = 0$.
(b) x, y plane is $\phi = 90°$, z-axis is $\phi = 0°$ or $180°$.

7. (a) $\sqrt{x^2 + y^2 + z^2}$ in rect, ρ in spherical; convert the rect answer using $r^2 = x^2 + y^2$ to get $\sqrt{r^2 + x^2}$ in cyl (can also see this in triangle ABP in Fig. 5 where hypot AP is $\sqrt{r^2 + x^2}$).
(b) r in cyl, convert r to get $\sqrt{x^2 + y^2}$ in rect. Use triangle ABP in Fig. 5 to convert r to $\rho \sin \phi$ in spherical.

Chapter 10 Review Problems (page 317)

1. L_1 has parallel $\vec{u} = (-1, 1, 3)$;
L_2 has parallel $\vec{v} = (2, -1, 1)$.
(a) $x = 2 - t, y = 3 + t, z = -3 + 3t$
(b) $\vec{u} \times \vec{v} = (4, 7, -1)$ is parallel to line. Eqs are $x = 2 + 4t, y = 3 + 7t, z = -3 - t$.
(c) Plane has normal $\vec{u}$; Equation is
$-(x - 2) + y - 3 + 3(z + 3) = 0$,
$-x + y + 3z = -8$.

(d) $\vec{u} \times \vec{v} = (4, 7, -1)$ is normal to plane. Equation is $4(x - 2) + 7(y - 3) - (z + 3) = 0$, $4x + 7y - z = 32$.

2. (a) $t + 5(3 + 2t) - (2 - t) = 1$, $t = -1$. Intersection is $x = -1, y = 1, z = 3$.
(b) Parallels are $(1, -2, 3)$ and $(-1, 1, -2)$, not multiples so lines not parallel. Solve $2 + t = 2 - s$, $1 - 2t = 3 + s$, $3 + 3t = 1 - 2s$. Sol to first two eqs is $t = -2$, $s = 2$. Satisfies third equ. Lines intersect at $x = 0, y = 5, z = -3$.
(c) Vector parallel to line is $(3, -1, 1) \times (1, 1, -6)$ $= (5, 19, 4)$. To get point, set $z = 0$ and solve $3x - y = 5, x + y = 3$, get $x = 2, y = 1$. Line is $x = 2 + 5t, y = 1 + 19t, z = 4t$.

3. $\overrightarrow{AB} = (-3, -4, -4)$. Line is $x = 9 - 3t, y = 8 - 4t$, $z = 7 - 4t$.

4. $\dfrac{|5(0) + 2(0) - 6(0) - 8|}{\sqrt{25 + 4 + 36}} = \dfrac{8}{\sqrt{65}}$

5.

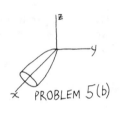

CYLINDER   PROBLEM 5(a)

PROBLEM 5(b)

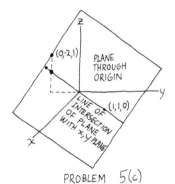

PROBLEM 5(c)

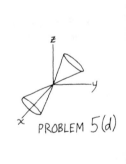

PROBLEM 5(d)

6. $(-1, 1, \sqrt{3})$ satisfies sphere's equ so is on sphere. (See fig.) Plane has normal vector $(-1, 1, \sqrt{3})$, passes through $(-1, 1, \sqrt{3})$; equ is $-(x + 1) + y - 1 + \sqrt{3}\,(z - \sqrt{3}) = 0$, $-x + y + \sqrt{3}\,z = 5$.

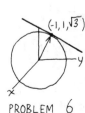

PROBLEM 6

7. (See fig.) Point lies in y, z plane. $x = 0, y = 2, z = 5$
$\rho = \sqrt{25 + 4} = \sqrt{29}$, $\theta = 90°$, $\phi = \tan^{-1}\frac{2}{5}$

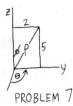

PROBLEM 7

PROBLEM 8(c)

8. (a) $x^2 + y^2 = 4$
(b) $r = 2$
(c) (See fig.) Convert from r to $\rho \sin \phi$ to get
$\rho \sin \phi = 2$, $\rho = 2 \csc \phi$.

9. $2 + 4t + 3 - 2t - 2(1 + t) = 3$, $0 = 0$. All t's work, L lies in plane p_1. (See fig.) L has parallel $\vec{u} = (4, -2, 1)$; p_1 has normal $\vec{v} = (1, 1, -2)$. Both $\vec{u}$ and $\vec{v}$ are parallel to p_2 so $\vec{u} \times \vec{v} = (3, 9, 6)$ is normal to p_2. A

PROBLEM 9

point in p_2 is $(2, 3, 1)$ from line L. Equ of p_2 is $3(x - 2) + 9(y - 3) + 6(z - 1) = 0$, $x + 3y + 2z = 13$.

11/PARTIAL DERIVATIVES

Section 11.1 (page 324)

1. (See figs.) (a) Graph is circular paraboloid $z = x^2 + y^2$. No neg level sets. Zero level set is $x^2 + y^2 = 0$, the origin. Pos level sets are circles.

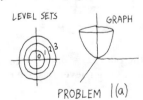

PROBLEM 1(a)

(b) Graph is hyperbolic paraboloid $z = x^2 - y^2$.

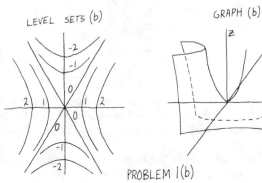

PROBLEM 1(b)

The 5 level set is $x^2 - y^2 = 5$, a hyperbola; -5 level set is $x^2 - y^2 = -5$, hyperbola; 0 level set is $x^2 - y^2 = 0$, the pair of lines $y = \pm x$ (the asymptotes for the hyperbolas).
(c) Graph is plane $z = x + y$. Level sets are the lines $x + y = C$.

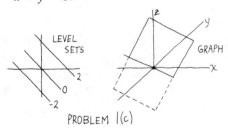

PROBLEM 1(c)

2. (See figs.) (a) Like Problem 1(a) but spheres.

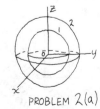

PROBLEM 2(a)

(b) $x^2 + y^2 - z^2 = 2$ is a hyperboloid of one sheet; $x^2 + y^2 - z^2 = -2$ is a hyperboloid of two

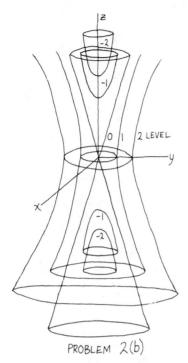

PROBLEM 2(b)

sheets; $x^2 + y^2 - z^2 = 0$ is a cone, etc. The cone is an asymptotic surface for all the hyperboloids, the two-piece ones are inside, the one piece ones outside.

(c) $1/(x + y) = 100$ is the line $x + y = \frac{1}{100}$; $1/(x + y) = -\frac{1}{7}$ is the line $x + y = -7$. Can't have $1/(x + y) = 0$, etc.

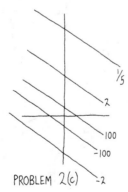

PROBLEM 2(c)

(d) Points where $y \geq x$ are those on and to the left

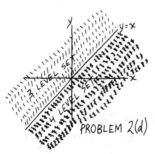

PROBLEM 2(d)

of line $y = x$; this is the 3 level set. Remaining points are 4 level. Other levels are empty.

3. $f(2,6) = 12$ so point has f value 12. It is on the 12 level set, the hyperbola $xy = 12$. (See fig.)

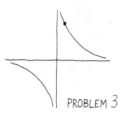

PROBLEM 3

4. (a) Graph of $f(x) = \frac{1}{3}(6 - 2x)$;
6 level set of $g(x,y) = 2x + 3y$;
0 level set of $h(x,y) = 2x + 3y - 6$, etc.

(b) Graph of $f(x,y) = x^2 + 2y^2$;
0 level set of $h(x,y,z) = z - x^2 - 2y^2$;
0 level set of $k(x,y,z) = x^2 + 2y^2 - z$, etc.

(c) Not the graph of a function since no unique solution for z. It is the 0 level set of $f(x,y,z) = z^2 + 2y^2 - x$.

(d) Equ is $x^2 + y^2 = 4$. Not the graph of a function (no z to even try to solve for). It is 4 level set of $f(x,y,z) = x^2 + y^2$.

5. (See fig.) 6 level set is set of points at distance 6 from x-axis, a pair of lines; 0 level set is x-axis itself. No neg level sets since distance is never neg.

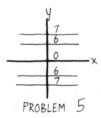

PROBLEM 5

6. Graph is the plane $z = 6$. (See fig.) The 6 level set is entire x, y plane. Other level sets are empty.

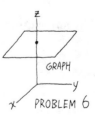

PROBLEM 6

7. (See figs.) (a) f is given as a function of 3 variables (even though $y^2 - x$ has no z in it). The 6 level set is surface $y^2 - x = 6$, a parabolic cylinder, etc.

(b) Level sets are the lines $y = C$ in 2-space.

(c) Level sets are planes $y = C$ in 3-space.

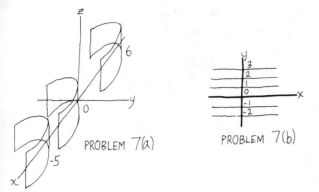

PROBLEM 7(a)

PROBLEM 7(b)

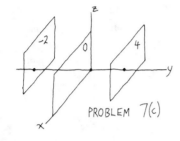

PROBLEM 7(c)

8. The 6 level is $2/\sqrt{\cdots} = 6$ or
$\sqrt{(x+2)^2 + (y-1)^2 + (z-3)^2} = \frac{1}{3}$,
$(x+2)^2 + (y-1)^2 + (z-3)^2 = \frac{1}{9}$ (sphere). There are
no 0 or neg levels. In general, if $C > 0$ the C level set is
a sphere with center $(-2, 1, 3)$ and radius $2/C$.

9. (a) Of points A, B, C, the surface is highest above A,
i.e., z value is largest for A. So $f(3, 2)$ is largest.
   (b) (See fig.) Cross section where $z = 3$ is a single
   line. Cross section where $z = 4$ is a pair of lines.
   No cross section if $z < 3$.

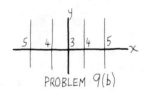

PROBLEM 9(b)

10.

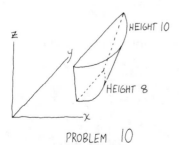

PROBLEM 10

11. NO because then Q has more than one f value (e.g.,
more than one temperature) and a function can't assign
two values to the same input.

Section 11.2 (page 329)

1. (a) $\partial z/\partial x = 2x + 6x^2 y^2$, $\partial z/\partial y = 4x^3 y$
   (b) $\partial z/\partial x = e^{-y}$, $\partial z/\partial y = -xe^{-y}$
   (c) $\dfrac{\partial z}{\partial x} = \dfrac{(x + y - x)}{(x + y)^2} = \dfrac{y}{(x + y)^2}$ (quotient rule)

   $\dfrac{\partial z}{\partial y} = -x(x + y)^{-2} = \dfrac{-x}{(x + y)^2}$
   (d) $\partial z/\partial x = x \cdot 4(2x + 5y)^3 \cdot 2 + (2x + 5y)^4$
   $= 8x(2x + 5y)^3 + (2x + 5y)^4$, (product rule)
   $\partial z/\partial y = 4x(2x + 5y)^3 \cdot 5 = 20x(2x + 5y)^3$
   (e) $\partial z/\partial x = -3yx^{-2} = -3y/x^2$, $\partial z/\partial y = 3/x$

2. (a) $\partial f/\partial y = 3/(2x + 3y)$,
   $\partial^2 f/\partial y^2 = -3(2x + 3y)^{-2} \cdot 3 = -9/(2x + 3y)^2$,
   $\partial^2 f/\partial x\, \partial y = -3(2x + 3y)^{-2} \cdot 2 = -6/(2x + 3y)^2$
   (b) $\dfrac{\partial f}{\partial y} = \dfrac{1}{1 + (y/x)^2} \cdot \dfrac{1}{x} = \dfrac{x}{x^2 + y^2}$,

   $\dfrac{\partial^2 f}{\partial y^2} = -x(x^2 + y^2)^{-2} \cdot 2y = \dfrac{-2xy}{(x^2 + y^2)^2}$,

   $\dfrac{\partial^2 f}{\partial x\, \partial y} = \dfrac{x^2 + y^2 - x \cdot 2x}{(x^2 + y^2)^2}$ (quotient rule)

   $= (y^2 - x^2)/(x^2 + y^2)^2$
   (c) $\dfrac{\partial f}{\partial y} = \dfrac{x - y - (x + y) \cdot -1}{(x - y)^2} = \dfrac{2x}{(x - y)^2}$,

   $\dfrac{\partial^2 f}{\partial y^2} = 2x \cdot -2(x - y)^{-3} \cdot -1 = \dfrac{4x}{(x - y)^3}$,

   $\dfrac{\partial^2 f}{\partial x\, \partial y} = \dfrac{(x - y)^2 \cdot 2 - 2x \cdot 2(x - y)}{(x - y)^4}$

   $= (-2x - 2y)/(x - y)^3$

3. (a) g_{cba} (diff first w.r.t. c, then b then a)
   (b) u_{txx} (first w.r.t. t, then x, x.)

4. (a) $\dfrac{\partial f}{\partial x} = z \cos \dfrac{x}{y} \cdot \dfrac{1}{y} = \dfrac{z}{y} \cos \dfrac{x}{y}$
   (b) $\dfrac{\partial f}{\partial y} = z \cos \dfrac{x}{y} \cdot -xy^{-2} = \dfrac{-xz}{y^2} \cos \dfrac{x}{y}$
   (c) $\partial f/\partial z = \sin(x/y)$
   (d) $\dfrac{\partial^2 f}{\partial x\, \partial z} = \dfrac{1}{y} \cos \dfrac{x}{y}$

5. (a) $\partial/\partial y = -e^x \sin y$, $\partial^2/\partial y^2 = -e^x \cos y$,
   $\partial^3/\partial y^3 = e^x \sin y$
   (b) Successive derivatives w.r.t. x just keep producing $e^x \cos y$.
   (c) $\partial/\partial x = \sin y$, $\partial^2/\partial x^2 = 0$, $\partial^3/\partial x^3 = 0$
   (d) $\partial/\partial y = x \cos y$, $\partial^2/\partial y^2 = -x \sin y$,
   $\partial^3/\partial y^3 = -x \cos y$
   (e) $\partial(-x \sin y)/\partial x = -\sin y$

6. (a) $\partial x/\partial \rho = \sin \phi \cos \theta$, $\partial^2 x/\partial \rho^2 = 0$
   (b) $\partial x/\partial \theta = -\rho \sin \phi \sin \theta$,
   $\partial^2 x/\partial \phi\, \partial \theta = -\rho \cos \phi \sin \theta$

7. $\partial f/\partial x$ is probably negative because if the price of a
camera goes up, sales usually go down; $\partial f/\partial y$ is probably
neg because if film goes up, camera sales usually go
down.

8. $\partial f/\partial a$ is negative because if prices on airline A go up, number of passengers on A goes down; $\partial f/\partial b$ is positive because your passengers increase if competitor's price goes up.

9. Consider employees jogging $\frac{1}{2}$ mile and cycling 1 mile. Since both partials are positive, increasing jogging or cycling alone boosts company profits. Furthermore increasing the jogging alone boosts them more than increasing cycling alone.

10. $\partial temp/\partial x = 8(2x - 3y)^3$, $\partial temp/\partial y = -12(2x - 3y)^3$; if $x = 4$, $y = 3$ get -8 and 12. Eastbound particle feels temp dropping by 8°/foot. Northbound particle feels temp rising by 12°/foot. Southbound particle feels temp dropping by 12°/foot.

11. Particle moving east through A feels temp rising so $\partial f/\partial x$ is pos. Particle moving north through A feels no temp change instantaneously so $\partial f/\partial y = 0$ at A.

12. (See figs.) (a) Particle moving on graph through B with x fixed and y increasing isn't going up or down so $\partial f/\partial y = 0$, i.e., slope on indicated curve through B is 0.

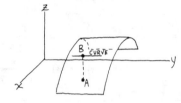

PROBLEM 12(a)

(b) Particle moving on graph through B with y fixed and x increasing (i.e., forward) is descending so $\partial f/\partial x$ is negative, i.e., slope at B on indicated curve is neg.

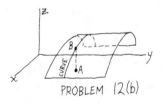

PROBLEM 12(b)

13. Note that P lies *behind* the y, z plane. A particle moving through P with y, z fixed and x increasing (i.e., moving forward, but behind the y, z plane) moves from outer to inner cylinders, from lower to higher levels; feels temp increasing. So $\partial f/\partial x > 0$. Particle moving through P with x and z fixed and y increasing is moving to the right, from inner to outer cyls, from higher to lower levels; feels temp decreasing. So $\partial f/\partial y < 0$. Particle moving through P with z increasing is moving up a cyl, not changing levels; feels no change in temp. So $\partial f/\partial z = 0$.

14. $\partial g/\partial x$ is larger because for a small eastward step, g changes more than f.

Section 11.3 (page 334)

All problems have diagrams.

1. $\dfrac{\partial w}{\partial s} = \dfrac{\partial w}{\partial x}\dfrac{\partial x}{\partial s} + \dfrac{\partial w}{\partial y}\dfrac{\partial y}{\partial s} + \dfrac{\partial w}{\partial z}\dfrac{\partial z}{\partial s}$

PROBLEM 1

2. $\dfrac{du}{dt} = \dfrac{\partial u}{\partial x}\dfrac{\partial x}{\partial a}\dfrac{da}{dt} + \dfrac{\partial u}{\partial y}\dfrac{\partial y}{\partial a}\dfrac{da}{dt} + \dfrac{\partial u}{\partial z}\dfrac{\partial z}{\partial a}\dfrac{da}{dt} + \dfrac{\partial u}{\partial x}\dfrac{\partial x}{\partial b}\dfrac{db}{dt}$

$\qquad\qquad + \dfrac{\partial u}{\partial y}\dfrac{\partial y}{\partial b}\dfrac{db}{dt} + \dfrac{\partial u}{\partial z}\dfrac{\partial z}{\partial b}\dfrac{db}{dt}$

PROBLEM 2

3. $\dfrac{\partial p}{\partial y} = \dfrac{dp}{dt}\dfrac{\partial t}{\partial y}$

PROBLEM 3

4. (a) *Directly* $w = \sin(t^3 \ln t)$ so
$dw/dt = \cos(t^3 \ln t) \cdot (t^3 \cdot 1/t + 3t^2 \ln t)$
$= t^2 \cos(t^3 \ln t) + 3t^2 \ln t \cos(t^3 \ln t)$.
With chain rule
$$\dfrac{dw}{dt} = \dfrac{\partial w}{\partial x}\dfrac{dx}{dt} + \dfrac{\partial w}{\partial y}\dfrac{dy}{dt}$$
$= y \cos xy \cdot 1/t + x \cos xy \cdot 3t^2$
$= t^2 \cos(t^3 \ln t) + 3t^2 \ln t \cos(t^3 \ln t)$.

PROBLEM 4(a)

(b) *Directly* $w = \dfrac{1}{x \sin y}$ so

PROBLEM 4(b)

$$\frac{\partial w}{\partial x} = \frac{1}{\sin y} \cdot -x^{-2} = -\frac{1}{x^2 \sin y}.$$

With chain rule

$$\frac{\partial w}{\partial x} = \frac{dw}{du}\frac{\partial u}{\partial x} = -\frac{1}{u^2}\sin y = -\frac{1}{x^2 \sin y}.$$

5. (a) $\dfrac{dw}{dt} = \dfrac{\partial w}{\partial x}\dfrac{dx}{dt} + \dfrac{\partial w}{\partial y}\dfrac{dy}{dt}$

PROBLEM 5(a)

(b) At time 3 the traveling particle feels temp dropping by 2° per *second*.

6. $\dfrac{\partial z}{\partial u} = \dfrac{\partial z}{\partial x}\dfrac{dx}{dt}\dfrac{\partial t}{\partial u} + \dfrac{\partial z}{\partial y}\dfrac{dy}{dt}\dfrac{\partial t}{\partial u}$

$$= \cos t \frac{\partial z}{\partial x}\frac{\partial t}{\partial u} + 6t^2\frac{\partial z}{\partial y}\frac{\partial t}{\partial u}$$

PROBLEM 6

7. $\dfrac{\partial u}{\partial x} = \dfrac{du}{d\rho}\dfrac{\partial \rho}{\partial x} = \dfrac{x}{\sqrt{x^2+y^2+z^2}}\dfrac{du}{d\rho} = \dfrac{x}{\rho}\dfrac{du}{d\rho}.$ Similarly

$\dfrac{\partial u}{\partial y} = \dfrac{y}{\rho}\dfrac{du}{d\rho}, \dfrac{\partial u}{\partial z} = \dfrac{z}{\rho}\dfrac{du}{d\rho}.$ Then $\left(\dfrac{\partial u}{\partial x}\right)^2 + \left(\dfrac{\partial u}{\partial y}\right)^2 + \left(\dfrac{\partial u}{\partial z}\right)^2$

$= \left(\dfrac{x^2}{\rho^2}+\dfrac{y^2}{\rho^2}+\dfrac{z^2}{\rho^2}\right)\left(\dfrac{du}{d\rho}\right)^2 = \dfrac{x^2+y^2+z^2}{\rho^2}\left(\dfrac{du}{d\rho}\right)^2 = \left(\dfrac{du}{d\rho}\right)^2.$

PROBLEM 7

8. $u = u(p,q)$ where $p = (y-x)/xy,\ q = (z-x)/xz.$

Then $u_x = \dfrac{\partial u}{\partial x} = \dfrac{\partial u}{\partial p}\dfrac{\partial p}{\partial x} + \dfrac{\partial u}{\partial q}\dfrac{\partial q}{\partial x}$

$= \dfrac{\partial u}{\partial p}\dfrac{xy(-1)-(y-x)y}{(xy)^2} + \dfrac{\partial u}{\partial q}\dfrac{xz(-1)-(z-x)z}{(xz)^2}$

PROBLEM 8

$= -\dfrac{1}{x^2}\dfrac{\partial u}{\partial p} - \dfrac{1}{x^2}\dfrac{\partial u}{\partial q}.$ Similarly

$u_y = \dfrac{\partial u}{\partial p}\dfrac{\partial p}{\partial y} = \dfrac{\partial u}{\partial p}\dfrac{xy-(y-x)x}{x^2y^2} = \dfrac{1}{y^2}\dfrac{\partial u}{\partial p}, u_z = \dfrac{1}{z^2}\dfrac{\partial u}{\partial q}.$

Then $x^2u_x + y^2u_y + z^2u_z$ does equal 0 (everything cancels out as you add).

9. The three functions are functions of x and y in a special way, namely they are functions of the combination $x^2 + y^2$. In general let $z = z(t)$ where $t = x^2 + y^2$.

Then $\dfrac{\partial z}{\partial x} = \dfrac{dz}{dt}\dfrac{\partial t}{\partial x} = 2x\dfrac{dz}{dt}, \dfrac{\partial z}{\partial y} = 2y\dfrac{dz}{dt},$ and $y\dfrac{\partial z}{\partial x}$ does

equal $x\dfrac{\partial z}{\partial y}$ (both equal $2xy\dfrac{dz}{dt}$).

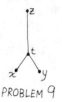

PROBLEM 9

10. By product rule $u_x = x^2\,\partial w/\partial x + 2xw$ where $w = w(p,q),\ p = y/x,\ q = z/x$ so

$u_x = x^2\left(\dfrac{\partial w}{\partial p}\dfrac{\partial p}{\partial x} + \dfrac{\partial w}{\partial q}\dfrac{\partial q}{\partial x}\right) + 2xw$

$= x^2\left(\dfrac{\partial w}{\partial p}\cdot -\dfrac{y}{x^2} + \dfrac{\partial w}{\partial q}\cdot -\dfrac{z}{x^2}\right) + 2xw$

$= -y\dfrac{\partial w}{\partial p} - z\dfrac{\partial w}{\partial q} + 2xw$

$u_y = x^2\dfrac{\partial w}{\partial y} = x^2\dfrac{\partial w}{\partial p}\dfrac{\partial p}{\partial y} = x^2\dfrac{\partial w}{\partial p}\cdot\dfrac{1}{x} = x\dfrac{\partial w}{\partial p},$

$u_y = x^2\dfrac{\partial w}{\partial y} = x^2\dfrac{\partial w}{\partial p}\dfrac{\partial p}{\partial y} = x^2\dfrac{\partial w}{\partial p}\cdot\dfrac{1}{x} = x\dfrac{\partial w}{\partial p},\ u_z = x\dfrac{\partial w}{\partial q}.$

Then $xu_x + yu_y + zu_z$

$= -xy\dfrac{\partial w}{\partial p} - xz\dfrac{\partial w}{\partial q} + 2x^2w + xy\dfrac{\partial w}{\partial p} + xz\dfrac{\partial w}{\partial q}$

$= 2x^2w = 2u$

PROBLEM 10

Section 11.4 (page 337)

All problems have diagrams.

1. $\dfrac{\partial p}{\partial u} = \dfrac{\partial p}{\partial a}\dfrac{\partial a}{\partial u} + \dfrac{\partial p}{\partial b}\dfrac{\partial b}{\partial u} = 3\dfrac{\partial p}{\partial a} + 5\dfrac{\partial p}{\partial b}$,

$\dfrac{\partial^2 p}{\partial u^2} = 3\dfrac{\partial}{\partial u}\left(\dfrac{\partial p}{\partial a}\right) + 5\dfrac{\partial}{\partial u}\left(\dfrac{\partial p}{\partial b}\right)$

$= 3\left(\dfrac{\partial^2 p}{\partial a^2}\dfrac{\partial a}{\partial u} + \dfrac{\partial^2 p}{\partial b\,\partial a}\dfrac{\partial b}{\partial u}\right) + 5\left(\dfrac{\partial^2 p}{\partial a\,\partial b}\dfrac{\partial a}{\partial u} + \dfrac{\partial^2 p}{\partial b^2}\dfrac{\partial b}{\partial u}\right)$

$= 9\dfrac{\partial^2 p}{\partial a^2} + 30\dfrac{\partial^2 p}{\partial a\,\partial b} + 25\dfrac{\partial^2 p}{\partial b^2}$

PROBLEM 1

2. $\dfrac{dz}{dt} = \dfrac{\partial z}{\partial x}\dfrac{dx}{dt} + \dfrac{\partial z}{\partial y}\dfrac{dy}{dt} = 3\dfrac{\partial z}{\partial x} + 4\dfrac{\partial z}{\partial y}$. Use product rule to diff again w.r.t. t since $\partial z/\partial x$, $\partial z/\partial y$, dx/dt, dy/dt are functions of t. Get $\dfrac{d^2 z}{dt^2} = 3\dfrac{\partial}{\partial t}\left(\dfrac{\partial z}{\partial x}\right) + 4\dfrac{\partial}{\partial t}\left(\dfrac{\partial z}{\partial y}\right)$

$= 3\left(3\dfrac{\partial^2 z}{\partial x^2} + 4\dfrac{\partial^2 z}{\partial y\,\partial x}\right) + 4\left(3\dfrac{\partial^2 z}{\partial x\,\partial y} + 4\dfrac{\partial^2 z}{\partial y^2}\right)$

$= 9\dfrac{\partial^2 z}{\partial x^2} + 16\dfrac{\partial^2 z}{\partial y^2} + 24\dfrac{\partial^2 z}{\partial x\,\partial y}$

PROBLEM 2

3. $\dfrac{\partial u}{\partial a} = \dfrac{\partial u}{\partial x}\dfrac{\partial x}{\partial a} + \dfrac{\partial u}{\partial y}\dfrac{\partial y}{\partial a} = 2\dfrac{\partial u}{\partial x} + 2ab\dfrac{\partial u}{\partial y}$. Need product rule to diff $2ab\dfrac{\partial u}{\partial y}$ w.r.t. b.

$\dfrac{\partial^2 u}{\partial b\,\partial a} = 2\dfrac{\partial}{\partial b}\left(\dfrac{\partial u}{\partial x}\right) + 2ab\dfrac{\partial}{\partial b}\left(\dfrac{\partial u}{\partial y}\right) + 2a\dfrac{\partial u}{\partial y}$

$= 2\left[\dfrac{\partial^2 u}{\partial x^2}\dfrac{\partial x}{\partial b} + \dfrac{\partial^2 u}{\partial y\,\partial x}\dfrac{\partial y}{\partial b}\right]$

$+ 2ab\left(\dfrac{\partial^2 u}{\partial x\,\partial y}\dfrac{\partial x}{\partial b} + \dfrac{\partial^2 u}{\partial y^2}\dfrac{\partial y}{\partial b}\right) + 2a\dfrac{\partial u}{\partial y}$

PROBLEM 3

$= 6\dfrac{\partial^2 u}{\partial x^2} + 2a^3 b\dfrac{\partial^2 u}{\partial y^2} + \dfrac{\partial^2 u}{\partial x\,\partial y}(2a^2 + 6ab) + 2a\dfrac{\partial u}{\partial y}$.

4. $\dfrac{dw}{dt} = \dfrac{\partial w}{\partial x}\dfrac{dx}{dt} + \dfrac{\partial w}{\partial y}\dfrac{dy}{dt} = 3t^2\dfrac{\partial w}{\partial x} + 2t\dfrac{\partial w}{\partial y}$. Need product rule to diff again w.r.t. t since everything on the right-hand side is a function of t.

$\dfrac{d^2 w}{dt^2} = 3t^2\dfrac{\partial}{\partial t}\left(\dfrac{\partial w}{\partial x}\right) + 6t\dfrac{\partial w}{\partial x} + 2t\dfrac{\partial}{\partial t}\left(\dfrac{\partial w}{\partial y}\right) + 2\left(\dfrac{\partial w}{\partial y}\right)$

$= 3t^2\left(\dfrac{\partial^2 w}{\partial x^2}\dfrac{dx}{dt} + \dfrac{\partial^2 w}{\partial y\,\partial x}\dfrac{dy}{dt}\right) + 6t\dfrac{\partial w}{\partial x}$

$+ 2t\left(\dfrac{\partial^2 w}{\partial x\,\partial y}\dfrac{dx}{dt} + \dfrac{\partial^2 w}{\partial y^2}\dfrac{dy}{dt}\right) + 2\dfrac{\partial w}{\partial y}$.

Use $dx/dt = 3t^2$, $dy/dt = 2t$ to get answer

$9t^4\dfrac{\partial^2 w}{\partial x^2} + 4t^2\dfrac{\partial^2 w}{\partial y^2} + 12t^3\dfrac{\partial^2 w}{\partial x\,\partial y} + 6t\dfrac{\partial w}{\partial x} + 2\dfrac{\partial w}{\partial y}$.

PROBLEM 4

5. (a) $\dfrac{\partial r}{\partial x} = \dfrac{x}{\sqrt{x^2 + y^2}} = \dfrac{x}{r}$,

$\dfrac{\partial \theta}{\partial x} = \dfrac{1}{1 + \left(\dfrac{y}{x}\right)^2}\left(-\dfrac{y}{x^2}\right) = -\dfrac{y}{r^2}$

(b) $\dfrac{\partial^2 r}{\partial x^2} = \dfrac{r - x\dfrac{\partial r}{\partial x}}{r^2}$ (quotient rule) $= \dfrac{1}{r} - \dfrac{x^2}{r^3} = \dfrac{y^2}{r^3}$,

$\dfrac{\partial^2 \theta}{\partial x^2} = -y\left(-\dfrac{2}{r^3}\right)\dfrac{\partial r}{\partial x} = \dfrac{2xy}{r^4}$

(c) $\dfrac{\partial v}{\partial x} = \dfrac{\partial v}{\partial r}\dfrac{\partial r}{\partial x} + \dfrac{\partial v}{\partial \theta}\dfrac{\partial \theta}{\partial x}$. Now use product and chain rule.

$\dfrac{\partial^2 v}{\partial x^2} = \dfrac{\partial v}{\partial r}\dfrac{\partial^2 r}{\partial x^2} + \left[\dfrac{\partial\left(\dfrac{\partial v}{\partial r}\right)}{\partial r}\dfrac{\partial r}{\partial x} + \dfrac{\partial\left(\dfrac{\partial v}{\partial r}\right)}{\partial \theta}\dfrac{\partial \theta}{\partial x}\right]\dfrac{\partial r}{\partial x}$

$+ \dfrac{\partial v}{\partial \theta}\dfrac{\partial^2 \theta}{\partial x^2} + \left[\dfrac{\partial\left(\dfrac{\partial v}{\partial \theta}\right)}{\partial r}\dfrac{\partial r}{\partial x} + \dfrac{\partial\left(\dfrac{\partial v}{\partial \theta}\right)}{\partial \theta}\dfrac{\partial \theta}{\partial x}\right]\dfrac{\partial \theta}{\partial x}$

$= \dfrac{\partial v}{\partial r}\dfrac{\partial^2 r}{\partial x^2} + \left[\dfrac{\partial^2 v}{\partial r^2}\dfrac{\partial r}{\partial x} + \dfrac{\partial^2 v}{\partial \theta\,\partial r}\dfrac{\partial \theta}{\partial x}\right]\dfrac{\partial r}{\partial x} + \dfrac{\partial v}{\partial \theta}\dfrac{\partial^2 \theta}{\partial x^2}$

$+ \left[\dfrac{\partial^2 v}{\partial r\,\partial \theta}\dfrac{\partial r}{\partial x} + \dfrac{\partial^2 v}{\partial \theta^2}\dfrac{\partial \theta}{\partial x}\right]\dfrac{\partial \theta}{\partial x}$

$= \dfrac{\partial v}{\partial r}\dfrac{\partial^2 r}{\partial x^2} + \dfrac{\partial^2 v}{\partial r^2}\left(\dfrac{\partial r}{\partial x}\right)^2 + 2\dfrac{\partial^2 v}{\partial r\,\partial \theta}\dfrac{\partial \theta}{\partial x}\dfrac{\partial r}{\partial x}$

$+ \dfrac{\partial v}{\partial \theta}\dfrac{\partial^2 \theta}{\partial x^2} + \dfrac{\partial^2 v}{\partial \theta^2}\left(\dfrac{\partial \theta}{\partial x}\right)^2$.

So (*) $\quad \dfrac{\partial^2 v}{\partial x^2} = \dfrac{x^2}{r^2}\dfrac{\partial^2 v}{\partial r^2} - 2\dfrac{xy}{r^3}\dfrac{\partial^2 v}{\partial r\,\partial\theta} + \dfrac{y^2}{r^4}\dfrac{\partial^2 v}{\partial\theta^2} + \dfrac{y^2}{r^3}\dfrac{\partial v}{\partial r}$

$\qquad\qquad + 2\dfrac{xy}{r^4}\dfrac{\partial v}{\partial\theta}$.

Remark: Similarly, (**) $\dfrac{\partial^2 v}{\partial y^2} = \dfrac{y^2}{r^2}\dfrac{\partial^2 v}{\partial r^2} + 2\dfrac{xy}{r^3}\dfrac{\partial^2 v}{\partial r\,\partial\theta} +$

$\dfrac{x^2}{r^4}\dfrac{\partial^2 v}{\partial\theta^2} + \dfrac{x^2}{r^3}\dfrac{\partial v}{\partial r} - 2\dfrac{xy}{r^4}\dfrac{\partial v}{\partial\theta}$. Then from (*) and (**) we

have $\dfrac{\partial^2 v}{\partial x^2} + \dfrac{\partial^2 v}{\partial y^2}$ (Laplacian in rect coords) equal to

$\dfrac{\partial^2 v}{\partial r^2} + \dfrac{1}{r}\dfrac{\partial v}{\partial r} + \dfrac{1}{r^2}\dfrac{\partial^2 v}{\partial\theta^2}$ (Laplacian in polar coords).

$$v, \dfrac{\partial v}{\partial r}, \dfrac{\partial v}{\partial\theta}$$

PROBLEM 5(c)

6. $\dfrac{dv}{dt} = \dfrac{\partial v}{\partial x}\dfrac{dx}{dt} + \dfrac{\partial v}{\partial y}\dfrac{dy}{dt}$. Now need product rule.

$\dfrac{d^2 v}{dt^2} = \dfrac{\partial v}{\partial x}\dfrac{d^2 x}{dt^2} + \dfrac{\partial}{\partial t}\left(\dfrac{\partial v}{\partial x}\right)\dfrac{dx}{dt} + \dfrac{\partial v}{\partial y}\dfrac{d^2 y}{dt^2} + \dfrac{\partial}{\partial t}\left(\dfrac{\partial v}{\partial y}\right)\dfrac{dy}{dt}$

$= \dfrac{\partial v}{\partial x}\dfrac{d^2 x}{dt^2} + \left(\dfrac{\partial^2 v}{\partial x^2}\dfrac{dx}{dt} + \dfrac{\partial^2 v}{\partial y\,\partial x}\dfrac{dy}{dt}\right)\dfrac{dx}{dt} + \dfrac{\partial v}{\partial y}\dfrac{d^2 y}{dt^2}$

$+ \left(\dfrac{\partial^2 v}{\partial x\,\partial y}\dfrac{dx}{dt} + \dfrac{\partial^2 v}{\partial y^2}\dfrac{dy}{dt}\right)\dfrac{dy}{dt}$

$= \dfrac{\partial v}{\partial x}\dfrac{d^2 x}{dt^2} + \dfrac{\partial v}{\partial y}\dfrac{d^2 y}{dt^2} + \dfrac{\partial^2 v}{\partial x^2}\left(\dfrac{dx}{dt}\right)^2$

$+ \dfrac{\partial^2 v}{\partial y^2}\left(\dfrac{dy}{dt}\right)^2 + 2\dfrac{\partial^2 v}{\partial x\,\partial y}\dfrac{dx}{dt}\dfrac{dy}{dt}$.

$$v\,\dfrac{\partial v}{\partial x}, \dfrac{\partial v}{\partial y}$$

PROBLEM 6

7. $w_x = \dfrac{\partial w}{\partial p}\dfrac{\partial p}{\partial x} + \dfrac{\partial w}{\partial q}\dfrac{\partial q}{\partial x} = \dfrac{\partial w}{\partial p} + \dfrac{\partial w}{\partial q}$,

$w_{xx} = \dfrac{\partial}{\partial x}\left(\dfrac{\partial w}{\partial p}\right) + \dfrac{\partial}{\partial x}\left(\dfrac{\partial w}{\partial q}\right)$

$\quad = \dfrac{\partial^2 w}{\partial p^2}\dfrac{\partial p}{\partial x} + \dfrac{\partial^2 w}{\partial q\,\partial p}\dfrac{\partial q}{\partial x} + \dfrac{\partial^2 w}{\partial p\,\partial q}\dfrac{\partial p}{\partial x} + \dfrac{\partial^2 w}{\partial q^2}\dfrac{\partial q}{\partial x}$

$\quad = \dfrac{\partial^2 w}{\partial p^2} + 2\dfrac{\partial^2 w}{\partial p\,\partial q} + \dfrac{\partial^2 w}{\partial q^2}$,

$w_t = \dfrac{\partial w}{\partial p}\dfrac{\partial p}{\partial t} + \dfrac{\partial w}{\partial q}\dfrac{\partial q}{\partial t} = -c\dfrac{\partial w}{\partial p} + c\dfrac{\partial w}{\partial q}$,

$w_{tt} = -c\dfrac{\partial}{\partial t}\left(\dfrac{\partial w}{\partial p}\right) + c\dfrac{\partial}{\partial t}\left(\dfrac{\partial w}{\partial q}\right)$

$= -c\left(\dfrac{\partial^2 w}{\partial p^2}\dfrac{\partial p}{\partial t} + \dfrac{\partial^2 w}{\partial q\,\partial p}\dfrac{\partial q}{\partial t}\right) + c\left(\dfrac{\partial^2 w}{\partial p\,\partial q}\dfrac{\partial p}{\partial t} + \dfrac{\partial^2 w}{\partial q^2}\dfrac{\partial q}{\partial t}\right)$

$= c^2\dfrac{\partial^2 w}{\partial p^2} - 2c^2\dfrac{\partial^2 w}{\partial p\,\partial q} + c^2\dfrac{\partial^2 w}{\partial q^2}$.

Then $c^2 w_{xx} - w_{tt}$ does cancel to $4c^2 w_{pq}$ as desired.

$$w, \dfrac{\partial w}{\partial p}, \dfrac{\partial w}{\partial q}$$

PROBLEM 7

Section 11.5 (page 346)

1. (a) (See fig.) *Critical:* $\partial f/\partial x = 3y - 4x$, $\partial f/\partial y = 3x + 2$. Sol to $3y - 4x = 0$, $3x + 2 = 0$ is $x = -2/3$, $y = -8/9$. Not in region; ignore.

Lower boundary: Here $y = 0$ so $f = -2x^2 + 8$ for $0 \le x \le \sqrt{3}$; $f'(x) = -4x$, zero when $x = 0$. Ends are $x = 0$, $\sqrt{3}$. Candidates are $(0,0)$ and $(\sqrt{3}, 0)$.

Left boundary: Here $x = 0$ so $f = 2y + 8$ for $0 \le y \le 3$; $f'(y) = 2$, never 0; only candidates are ends where $y = 0, 3$, i.e., points $(0,0)$, $(0,3)$.

Boundary $y = 3 - x^2$: Substitute for y to get $f = 3x(3 - x^2) - 2x^2 + 2(3 - x^2) + 8$ $= -3x^3 - 4x^2 + 9x + 14$, $0 \le x \le \sqrt{3}$. So $f'(x) = -9x^2 - 8x + 9$. Critical x's are approx $.65$ and -1.5 (not in interval). Candidates are $.65$ and ends $0, \sqrt{3}$, i.e., points $(.65, 2.58)$, $(0,3)$, $(\sqrt{3}, 0)$.

For final decision find $f(0,0) = 8$, $f(0,3) = 14$, $f(\sqrt{3}, 0) = 2$, $f(.65, 2.58) = 17.3$ approx. Min value is 2, max value is approx 17.

PROBLEM 1(a)

(b) Same criticals as in (a). Again, not in region. On boundary $x + y = 2$,
$f = 3x(2 - x) - 2x^2 + 2(2 - x) + 8$
$= -5x^2 + 4x + 12$ for $0 \le x \le 2$. Then $f'(x) = -10x + 4$ so $x = \frac{4}{10}$ is critical. Ends are $x = 0$, $x = 2$; candidates are $(\frac{4}{10}, \frac{16}{10})$, $(0,2)$, $(2,0)$. On lower boundary y is 0, $f = -2x^2 + 8$ for $0 \le x \le 2$; $f'(x) = -4x$, zero when $x = 0$. Candidates are $(0,0)$ and $(2,0)$. On left bdry x is 0 and $f = 2y + 8$ for $0 \le y \le 2$; $f'(y) = 2$, never 0.

Only candidates are $y = 0, 2$, i.e., points $(0, 0)$ and $(0, 2)$ again.

For final decision find $f(0, 0) = 8$, $f(0, 2) = 12$, $f(2, 0) = 0$, $f(\frac{4}{10}, \frac{16}{10}) = \frac{64}{5}$. Min value is 0, max value is $\frac{64}{5}$.

(c) (See fig.) $\partial f/\partial x = 2$, never 0; no criticals. On bdry $y = x^2$, $f = 2x - 3x^2$ for $-2 \le x \le 2$; $f'(x) = 2 - 6x$, $x = \frac{1}{3}$ is critical. Ends are $-2, 2$. Candidates are $(\frac{1}{3}, \frac{1}{9})$, $(\pm 2, 4)$.

On boundary $y = 4$, $f = 2x - 12$ where $-2 \le x \le 2$; $f'(x) = 2$ never 0. No criticals. Candidates are ends $x = \pm 2$, points $(\pm 2, 4)$.

Then $f(\frac{1}{3}, \frac{1}{9}) = \frac{1}{3}$ (MAX), $f(-2, 4) = -16$(MIN), $f(2, 4) = -8$ (LOSER).

PROBLEM 1(c)

(d) (See fig.) $\partial f/\partial x = 2x + 2xy$, $\partial f/\partial y = x^2 - 1$. Solve $2x + 2xy = 0$, $x^2 - 1 = 0$ to get criticals $x = 1, y = -1$; $x = -1, y = -1$. On boundary, $f = 4 - 2y^2 + (4 - 2y^2)y - y = 4 + 3y - 2y^2 - 2y^3$ for $-\sqrt{2} \le y \le \sqrt{2}$. Then $f'(y) = 3 - 4y - 6y^2$, zero if $y = -1.1, .5$ approx. Ends are $y = \pm\sqrt{2}$. Candidates are $(1, -1)$, $(-1, -1)$, $(0, \pm\sqrt{2})$, $(\pm\sqrt{1.58}, -1.1)$, $(\pm\sqrt{3.5}, .5)$. Max value (at $\pm\sqrt{3.5}, .5$) is 4.75; min value (at $(0, \sqrt{2})$) is $-\sqrt{2} = -1.414$.

PROBLEM 1(d)

2. (a) (See fig.) $\partial f/\partial x = 2x + 3y + 10$, $\partial f/\partial y = 2y + 3x$. Solve $2x + 3y + 10 = 0$, $3x + 2y = 0$ to get critical point $x = 4, y = -6$. Not in region, ignore.

On top boundary $y = 3$; $f = x^2 + 9 + 19x$, $0 \le x \le 5$; $f' = 2x + 19$, zero if $x = -\frac{19}{2}$, not in interval. Candidates are ends $x = 0, 5$, points $(0, 3)$ and $(5, 3)$.

On right boundary $y = x - 2$; $f = 5x^2 + 4$, $-1 \le x \le 5$; $f'(x) = 10x$, zero if $x = 0$. Candidates are $x = 0$ and ends $x = -1, 5$, i.e., points $(0, -2)$, $(-1, -3)$, $(5, 3)$.

On left boundary $y = 6x + 3$; $f = 19x^2 + (6x + 3)^2 + 19x$, $-1 \le x \le 0$; $f'(x) = 38x + 12(6x + 3) + 19$.

Critical x is $-\frac{1}{2}$, ends are $-1, 0$.
Candidates are $(-\frac{1}{2}, 0)$, $(-1, -3)$, $(0, 3)$.

For final decision find $f(-\frac{1}{2}, 0) = -\frac{19}{4}$, $f(-1, -3) = 9$, $f(5, 3) = 129$, $f(0, -2) = 4$, $f(0, 3) = 9$. Max at $(5, 3)$, min at $(-\frac{1}{2}, 0)$.

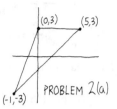

PROBLEM 2(a)

(b) $\partial f/\partial x = 2x + 1$, $\partial f/\partial y = 4y$. Solve $2x + 1 = 0$, $4y = 0$ to get critical point $(-\frac{1}{2}, 0)$. On bdry, $y^2 = 1 - x^2$, $f = -x^2 + x + 2$, $-1 \le x \le 1$; $f'(x) = -2x + 1$, zero if $x = \frac{1}{2}$. Ends in the sub-problem are $x = \pm 1$. Then $f(-\frac{1}{2}, 0) = -\frac{1}{4}$, $f(\frac{1}{2}, \pm\frac{1}{2}\sqrt{3}) = \frac{9}{4}$, $f(1, 0) = 2$, $f(-1, 0) = 0$. Max is at points $(\frac{1}{2}, \pm\frac{1}{2}\sqrt{3})$, min at $(-\frac{1}{2}, 0)$

3. (a) If (x, y) approaches bdry at ∞ then $f \to -\infty$ because of terms $-x^2 - y^2$. Min is $-\infty$. Max will be at a critical point. $\partial f/\partial x = -2 - 2x$, $\partial f/\partial y = 6 - 2y$. They are 0 when $x = -1$, $y = 3$. Have $f(-1, 3) = 10$, max value.

(b) If $x = 0$ and $y \to \infty$ (i.e., point approaches bdry at infinity along positive y-axis) then $f \to \infty$. If $x = 0$ and $y \to -\infty$ (approach bdry at infinity along negative y-axis) then $f \to -\infty$. Max is ∞, min is $-\infty$.

(c) On any path to bdry at ∞, $f \to \infty$. For crit points, $\partial f/\partial x = 2x - y + 2$, $\partial f/\partial y = -x + 2y + 2$. They are 0 when $x = -2, y = -2$. Min value is $f(-2, -2) = -8$, max is ∞.

(d) Walk up the y-axis, get $f \to -\infty$. Walk east on the x-axis, get $f \to \infty$. Max is ∞, min is $-\infty$.

4. Typical point on plane is $(x, y, 14 - 3x - 2y)$. Its distance to origin is $s(x, y) = \sqrt{x^2 + y^2 + (14 - 3x - 2y)^2}$, the function to be minimized over all (x, y), i.e., the region is the entire x, y plane. On the boundary at ∞, s is ∞, the max (i.e., points (x, y, z) on given plane are far from origin if $x \to \infty$ or $y \to \infty$.) Expect min s at a critical point. As a shortcut, minimize the radicand
$r(x, y) = x^2 + y^2 + (14 - 3x - 2y)^2$ over all (x, y).
This minimizes s too and is simpler.
$\partial r/\partial x = 2x + 2(14 - 3x - 2y) \cdot -3$,
$\partial r/\partial y = 2y + 2(14 - 3x - 2y) \cdot -2$. Partials are 0 when $x = 3$, $y = 2$. Corresponding z from equ of plane is $z = 1$. Point on plane nearest origin is $(3, 2, 1)$.

5. (a) $\dfrac{|2(1) - 2(3) + 0 + 10|}{\sqrt{4 + 4 + 1}} = 2$

(b) Typical point on plane is $(x, y, -2x + 2y - 10)$. Distance to $(1, 3, 0)$ is $f(x, y) = \sqrt{(x - 1)^2 + (y - 3)^2 + (-2x + 2y - 10)^2}$. Get max of ∞ when x and/or $y \to \infty$. Need critical

point for min. Let
$r = (x - 1)^2 + (y - 3)^2 + (-2x + 2y - 10)^2$.
$\partial r / \partial x = 2(x - 1) - 4(-2x + 2y - 10)$,
$\partial r / \partial y = 2(y - 3) + 4(-2x + 2y - 10)$. Partials
are 0 when $x = -\frac{1}{3}$, $y = \frac{13}{3}$. Corresponding z
from equ of plane is $z = -\frac{2}{3}$. Distance from
$(-\frac{1}{3}, \frac{13}{3}, -\frac{2}{3})$ to $(1, 3, 0)$ is 2, the min.

6. A typical point on the first line is
$(2 + t, 3 - t, 4 + 2t)$ and a typical point on second
line is $(1 + s, 2 + s, 7 + 3s)$ (must use different parameter letters). Distance between them is
$f = \sqrt{(1 + t - s)^2 + (1 - t - s)^2 + (-3 + 2t - 3s)^2}$.
Let $r = (1 + t - s)^2 + (1 - t - s)^2 + (-3 + 2t - 3s)^2$,
all s, t. Can work with r instead of f.
$\partial r / \partial s = -2(1 + t - s) - 2(1 - t - s)$
$\qquad\qquad\qquad\qquad - 6(-3 + 2t - 3s)$,
$\partial r / \partial t = 2(1 + t - s) - 2(1 - t - s)$
$\qquad\qquad\qquad\qquad + 4(-3 + 2t - 3s)$.
Partials are 0 when $s = -\frac{1}{5}, t = \frac{4}{5}$. Max value of f (namely
∞) comes when $s \to \pm\infty$, $t \to \pm\infty$. Min must be at
$s = -\frac{1}{5}, t = \frac{4}{5}$. Plug into equations of line to get points of
closest contact $(\frac{14}{5}, \frac{11}{5}, \frac{28}{5})$ and $(\frac{4}{5}, \frac{9}{5}, \frac{32}{5})$.

7. Let dimensions of base be x and y. Volume is 256 so
height is $256/xy$. Surface area is
$$A(x,y) = xy + 2x\frac{256}{xy} + 2y\frac{256}{xy} = xy + \frac{512}{y} + \frac{512}{x}$$
for $x \geq 0, y \geq 0$ (see fig.).

PROBLEM 7(a),(b)

(a) If x and or $y \to \infty$ (i.e., on bdry at ∞), $A \to \infty$.
Low tank, large base. If $x \to 0+$ or $y \to 0+$ then
$A \to \infty$. Tall tank, small base. Both are worst
options. Between extremes is best tank.

(b) Part (a) showed that boundaries give max A.
Find critical point for min.
$\partial A / \partial x = y - 512/x^2$, $\partial A / \partial y = x - 512/y^2$.
Partials are 0 when $x^2 y = 512$, $y^2 x = 512$,
$x^2 y = y^2 x$, $x = y$, $x^3 = 512$, $x = 8$, $y = 8$.
Dimensions of best tank are $8 \times 8 \times \frac{256}{64}$.

(c) New region. (See fig.) Critical point $(8, 8)$ from
(b) isn't in region. Must get min from boundaries. Again, x-axis and y-axis give $A = \infty$. On
line $x = 4$ we have $A = 4y + 512/y + 512/4$,
$0 < y \leq 20$. Then $A'(y) = 4 - 512/y^2$, zero if
$y = \sqrt{128}$. End is $y = 20$.

PROBLEM 7(c)

On line $y = 20$, $A = 20x + \frac{512}{20} + 512/x$,
$0 < x \leq 4$; $A'(x) = 20 - 512/x^2$ never 0
if $0 < x \leq 4$. Only candidate is end $x = 4$.
$A(4, \sqrt{128}) = 4\sqrt{128} + \frac{512}{\sqrt{128}} + \frac{512}{4}$,
$A(4, 20) = 80 + \frac{512}{20} + \frac{512}{4}$. First is smaller; best
tank has dimensions 4 by $\sqrt{128}$ by $256/4\sqrt{128}$.

Section 11.6 (page 353)

1. ∇ temp $= (y^2 + 6, 2xy)$. At point $(1, 2)$, $\nabla t = (10, 4)$.
(a) $SW = (-1, -1)$ so D_{SW} temp $= \dfrac{\nabla t \cdot SW}{\|SW\|} = \dfrac{-14}{\sqrt{2}}$
(temp is dropping by $14/\sqrt{2}$ degrees per meter).

(b) $\overrightarrow{PQ} = (2, -6)$, $\dfrac{\nabla t \cdot \overrightarrow{PQ}}{\|\overrightarrow{PQ}\|} = \dfrac{-4}{\sqrt{40}}$.

(c) Direction is south, i.e., $-\vec{j}$, $\nabla t \cdot (-\vec{j}) = -4$.

(d) $\|\nabla$ temp$\| = \sqrt{116}$

(e) WNW makes angle of 157.5° with x-axis (see fig.);
a unit WNW vector is $(\cos 157.5°, \sin 157.5°)$,
D_{WNW} temp $= \nabla t \cdot WNW$
$\qquad\qquad = 10 \cos 157.5° + 4 \sin 157.5°$.

PROBLEM 1(e)

2. ∇ temp $= (2xy, x^2)$. At point A, ∇ temp is $(12, 4)$.
$NE = (1, 1)$, $(\nabla$ temp $\cdot NE)/\|NE\| = 16/\sqrt{2}$. Both runners feel temp increase by $16/\sqrt{2}$ degrees per meter.

3. (See fig.) $\nabla f = (y, x - 2y, 1)$ and at A, $\nabla f = (2, 1, 1)$.

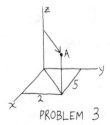

PROBLEM 3

(a) Away from z-axis is vector $\vec{u} = 5\vec{i} + 2\vec{j}$,
$(\nabla f \cdot \vec{u})/\|\vec{u}\| = 12/\sqrt{29}$.

(b) Any arrow perp to ∇f such as $(-1, 2, 0)$,
$(5, -8, -2)$, etc.

(c) At A, $\nabla f = (2, 1, 1)$. Particle at A moves in direction of $\overrightarrow{AB} = (1, 2, 1)$ so feels temp changing by
$(\nabla f \cdot \overrightarrow{AB})/\|\overrightarrow{AB}\| = 5/\sqrt{6}$ degrees per meter. At
B, $\nabla f = (4, -2, 1)$. Particle at B moves in direction of $\overrightarrow{BA} = (-1, -2, -1)$ so feels temp changing by $(\nabla f \cdot \overrightarrow{BA})/\|\overrightarrow{BA}\| = -1/\sqrt{6}$. When they

meet midway at $(\frac{11}{2}, 3, \frac{3}{2})$, $\nabla f = (3, -\frac{1}{2}, 1)$. First particle has direction $\vec{AB}$ so feels temp *increasing* by $(\nabla f \cdot \vec{AB})/\|\vec{AB}\| = 3/\sqrt{6}°$ per m. The other particle feels temp *dropping by* $3/\sqrt{6}°$ per m.

(d) Direction of $-\nabla f = -2\vec{i} - \vec{j} - \vec{k}$. In that direction, pressure decreases by $\|\nabla f\| = \sqrt{6}$ units per meter.

4. The direction of motion is tangent $\vec{u}$. (See fig.) Makes obtuse angle with ∇f so component of ∇f in $\vec{u}$ direction is neg. Particle feels temp dropping.

PROBLEM 4

5. ∇f is a positive multiple of $3\vec{i} + 2\vec{j}$ so $\nabla f = c(3\vec{i} + 2\vec{j})$ where $c > 0$. Also $\|\nabla f\| = 2$ so $c\sqrt{13} = 2$, $c = 2/\sqrt{13}$ and $\nabla f = (6/\sqrt{13}, 4/\sqrt{13})$. $D_{\text{north}}f = \nabla f \cdot \vec{j} = 4/\sqrt{13}$ degrees per m.

6. Direction toward $(1, 1)$ is $\vec{u} = -\vec{j}$ and direction toward $(7, 10)$ is $\vec{v} = 6\vec{i} + 8\vec{j}$. Let ∇f at A be (a, b). Given $(\nabla f \cdot \vec{u})/\|\vec{u}\| = 2$ so $-b = 2$, $b = -2$. Also $(\nabla f \cdot \vec{v})/\|\vec{v}\| = -4$ so $\frac{1}{10}(6a + 8b) = -4$, $a = -4$. Then $\nabla f = (-4, -2)$ at A. This is direction in which temp rises maximally. Max rate is $\|\nabla f\| = \sqrt{20}$ deg per m.

7. (a) $f(-1, 2) = 2$ so P is on 2-level set $x^2 y = 2$, $y = 2/x^2$. (See fig.) $\nabla f = (2xy, x^2)$ which is $(-4, 1)$ at P.

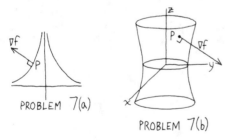

PROBLEM 7(a)

PROBLEM 7(b)

(b) $f(1, 2, 1) = 12$ so P is on 12-level set $x^2 + 2y^2 - z^2 + 4 = 12$, $x^2 + 2y^2 - z^2 = 8$. (See fig.) $\nabla f = (2x, 4y, -2z)$ which is $(2, 8, -2)$ at P.

8. The gradients are perp to level sets and point to higher levels. (See fig.) The function f is changing more per meter in direction of ∇f than g is changing in the

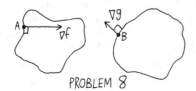

PROBLEM 8

direction of ∇g because the f levels jump by 10 among level sets in Fig. 13 while g jumps by only 1 for similarly spaced level sets. So ∇f is longer than ∇g.

9. At a rel max/min, partials are 0. So $\nabla f = \vec{0}$.

10. Distance is never neg, no neg level sets. Line L is 0 level set. The 2 level set (points at distance 2 from line) is cylinder with axis L and radius 2, etc. (See fig.) Gradients are perp to level sets and point to higher levels, so point out of each cylinder. Each ∇f has length 1 because if you walk away from L in the direction of ∇f (i.e., perpendicularly away from L), f increases by 1 foot for each foot you walk (since f *is* distance to L).

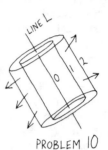

PROBLEM 10

11. (a) $1 \times 2 \times 6$ does equal 12.

(b) $z = 12/xy$, $\nabla z = (-12/x^2 y, -12/xy^2)$, $\nabla z|_{x=1, y=2} = -6\vec{i} - 3\vec{j}$ (more west than south). The path in direction $-6\vec{i} - 3\vec{j}$ is steepest. Slope on path at P is $\|\nabla z\| = \sqrt{45}$.

(c) $\vec{SE} = \vec{i} - \vec{j}$, $\partial z/\partial SE = (\nabla z \cdot \vec{SE})/\|\vec{SE}\| = -3/\sqrt{2}$. Path descends. Slope is $-3/\sqrt{2}$.

(d) $\nabla(xyz) = (yz, xz, xy)$; at P it is $12\vec{i} + 6\vec{j} + 2\vec{k}$, a normal to the surface at P. Tangent plane is $12(x - 1) + 6(y - 2) + 2(z - 6) = 0$, $6x + 3y + z = 18$. Normal line is $x = 1 + 12t$, $y = 2 + 6t$, $z = 6 + 2t$.

12. (a) Graph has equation $z = 3x^2 - 2y^2$. P is on graph since $1 = 3 - 2$.

(b) $\nabla z = 6x\vec{i} - 4y\vec{j}$, $\nabla z|_{x=1, y=-1} = 6\vec{i} + 4\vec{j}$ (more east than north). Steepest path is in direction of $6\vec{i} + 4\vec{j}$. Slope at P on steepest path is $\|\nabla z\| = \sqrt{52}$.

(c) $\partial z/\partial SE = (\nabla z \cdot \vec{SE})/\|\vec{SE}\| = 2/\sqrt{2}$. So path ascends; slope is $2/\sqrt{2}$.

(d) Surface is $z - 3x^2 + 2y^2 = 0$, $\nabla(z - 3x^2 + 2y^2) = (-6x, 4y, 1)$, $\nabla = (-6, -4, 1)$ at P. This is normal vector to surface. Tangent plane is $-6(x - 1) - 4(y + 1) + (z - 1) = 0$, $6x + 4y - z = 1$. Normal line is $x = 1 - 6t$, $y = -1 - 4t$, $z = 1 + t$.

13. (See fig.) (a) $\nabla f = (4x, 2y)$, $\nabla f|_{x=1, y=3} = 4\vec{i} + 6\vec{j}$. It is perp to level set of f through point $(1, 3)$. Since $f(1, 3) = 11$, it's the 11 level set, ellipse $2x^2 + y^2 = 11$.

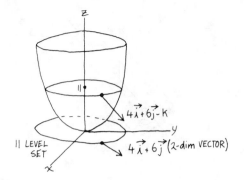

PROBLEM 13

(b) $\nabla(2x^2 + y^2 - z) = (4x, 2y, -1)$ which is
$4\vec{i} + 6\vec{j} - \vec{k}$ if $x = 1, y = 3$. It is perp to graph
of f, $z = 2x^2 + y^2$, at point $(1, 3, 11)$.

14. (a) $\nabla(x^2 + 2y^2 + 3z^2) = (2x, 4y, 6z)$ which is
$(2, 4, 12)$ at P. This is perp to the earth. From a
rough sketch can tell it is OUT not in.

(b) ∇ temp $= \nabla(2xz + y^2 + 6) = (2z, 2y, 2x)$ which is
$(4, 2, 2)$ at P. Takeoff direction is $\vec{n} = (2, 4, 12)$.
Then ∂temp$/\partial\vec{n} = (\nabla$temp $\cdot \vec{n})/\|\vec{n}\| =$
$40/\sqrt{164}$ degrees per meter.

(c) Choose any direction perp to ∇ temp such as
$(1, -1, -1)$, $(-3, 2, 4)$, $(0, -2, 2)$, etc.

(d) Nonburrowing directions must make acute angles with the OUTWARD normal $\vec{n}$.
But $(1, -1, -1) \cdot \vec{n}$ is negative so $(1, -1, -1)$
burrows. Change it to $(-1, 1, 1)$. Fortunately,
$(-3, 2, 4) \cdot \vec{n}$ is positive so $(-3, 2, 4)$ is OK, etc.

(e) On northern hemisphere (where P is),
$z = \sqrt{\frac{1}{3}(15 - x^2 - 2y^2)}$. Then $\nabla z =$
$(-\frac{1}{3}x/\sqrt{\ }, -\frac{2}{3}y/\sqrt{\ })$, $\nabla z|_{x=1,y=1} = -\frac{1}{6}\vec{i} - \frac{1}{3}\vec{j}$.
Mother's slope $= \partial z/\partial NE$
$$= (\nabla z \cdot NE)/\|NE\| = -1/2\sqrt{2}.$$

Section 11.7 (page 361)

1. (a) $d(y/x)$ is found immediately by quotient rule.

(b) $d(x^2 + y^2)^{-1} = -(x^2 + y^2)^{-2} d(x^2 + y^2)$ (chain)
$$= -(x^2 + y^2)^{-2}(2x\,dx + 2y\,dy)$$
$$= \frac{-2x\,dx - 2y\,dy}{(x^2 + y^2)^2}$$

(c) $d(\pm\sqrt{x^2 + y^2}) = \pm\frac{1}{2}(x^2 + y^2)^{-1/2} d(x^2 + y^2)$
(chain rule) $= \dfrac{2x\,dx + 2y\,dy}{\pm 2\sqrt{x^2 + y^2}} = \dfrac{x\,dx + y\,dy}{\pm\sqrt{x^2 + y^2}}$

(d) $\dfrac{1}{x^2 + y^2} d(x^2 + y^2) = \dfrac{2x\,dx + 2y\,dy}{x^2 + y^2}$

(e) $\dfrac{1}{1 + \left(\frac{y}{x}\right)^2} d\left(\frac{y}{x}\right) = \dfrac{1}{1 + \left(\frac{y}{x}\right)^2} \dfrac{x\,dy - y\,dx}{x^2}$
$$= \dfrac{-y\,dx + x\,dy}{x^2 + y^2}$$

2. $x = r \cos\theta$ so by (1), $dx = \cos\theta\,dr - r\sin\theta\,d\theta$.
$y = r\sin\theta$ so $dy = \sin\theta\,dr + r\cos\theta\,d\theta$.

3. (a) $p = 2xy$, $q = y$, $\partial q/\partial x = 0$, $\partial p/\partial y = 2x$; not
equal so form not exact.

(b) $\partial q/\partial x = \partial p/\partial y = 3x^2$; exact. Antidiff p w.r.t. x
to get terms $\frac{1}{4}x^4 + x^3y$. Diff this temporary an-
swer w.r.t. y to get x^3. Compare with q and see we
must tack on $\frac{1}{4}y^4$ to get final answer
$f(x, y) = \frac{1}{4}x^4 + x^3y + \frac{1}{4}y^4 + C$.

(c) $f(x, y) = -y/x + 5y + C$

4. Need $\partial q/\partial x = \partial p/\partial y$, $\partial q/\partial x = 3xy^2$, $q = \frac{3}{2}x^2y^2 + $ any
$f(y)$, e.g., q could be $\frac{3}{2}x^2y^2 + y^3 \sin y + 7$.

5. (a) $d(2x^3 + xy^2 + y^3) = 0$,
implicit sol is $2x^3 + xy^2 + y^3 = C$.

(b) $d(x^3 + xy) = 0$, implicit sol is $x^3 + xy = C$, explicit sol is $y = (C - x^3)/x$.

(c) $(x - y\cos x)\,dx - (y + \sin x)\,dy = 0$,
$d(\frac{1}{2}x^2 - y\sin x - \frac{1}{2}y^2) = 0$,
implicit sol is $\frac{1}{2}x^2 - y\sin x - \frac{1}{2}y^2 = C$.

(d) $e^{xy}\,dx - dy = 0$, not exact since $\partial q/\partial x = 0$,
$\partial p/\partial y = xe^{xy}$ (not equal).

(e) $(2r\cos\theta - 1)\,dr - r^2\sin\theta\,d\theta = 0$,
$d(r^2\cos\theta - r) = 0$, implicit sol $r^2\cos\theta - r = C$.

(f) Not exact.

(g) Can move everything to left side of equation or
better still, each side is an exact differential as it
stands, namely $d(\sin x \cos y) = d(\frac{1}{4}x^4)$. Implicit sol
is $\sin x \cos y = \frac{1}{4}x^4 + C$.

(h) $(ye^{-x} - \sin x)\,dx - (e^{-x} + 2y)\,dy = 0$,
$d(-ye^{-x} + \cos x - y^2) = 0$,
implicit sol is $-ye^{-x} + \cos x - y^2 = C$.

6. (a) $d(x^2y + \frac{1}{2}y^2) = 0$, implicit sol is $x^2y + \frac{1}{2}y^2 = C$. If
$x = 1, y = 4$ then $C = 12$ so implicit particular
sol is $x^2y + \frac{1}{2}y^2 = 12$.

(b) $d(-\cos(2x + 3y)) = 0$, implicit solution is
$-\cos(2x + 3y) = C$. If $x = 0$, $y = \frac{1}{2}\pi$ then
$C = -\cos\frac{3}{2}\pi$, $C = 0$. So particular sol (still
implicit) is $\cos(2x + 3y) = 0$.

(c) Each side is exact, $d\ln(x + y) = d(x)$, so
$\ln(x + y) = x + C$. If $x = 0, y = 1$ then $C = 0$;
implicit sol is $\ln(x + y) = x$. Then $x + y = e^x$,
explicit sol is $y = e^x - x$.

7. (a) $d(\frac{1}{3}x^3 + 2x + \frac{3}{2}y^2) = 0$, $\frac{1}{3}x^3 + 2x + \frac{3}{2}y^2 = K$

(b) $(x^2 + 2)\,dx = -3y\,dy$, $\frac{1}{3}x^3 + 2x = -\frac{3}{2}y^2 + K$

8. (a) See (22). Use integrating factor $\dfrac{1}{x^2 + y^2}$. Then
$$dx = \dfrac{x\,dy - y\,dx}{x^2 + y^2}, \quad d(x) = d\left(\tan^{-1}\frac{y}{x}\right),$$
$$x = \tan^{-1}\frac{y}{x} + K.$$

(b) See (17). Use integrating factor $1/y^2$,
$$\frac{(y\,dx - x\,dy)}{y^2} = dx, \quad \frac{x}{y} = x + K, \quad y = \frac{x}{(x + K)}.$$

(c) See (20). Use integrating factor $\dfrac{1}{\sqrt{x^2 + y^2}}$,
$$dy = \frac{x\,dx + y\,dy}{\sqrt{x^2 + y^2}}, \quad y = \sqrt{x^2 + y^2} + K.$$

(d) See (21). Use integrating factor $\dfrac{2}{x^2 + y^2}$.
$$\frac{2x\,dx + 2y\,dy}{x^2 + y^2} = 2dy, \quad \ln(x^2 + y^2) = 2y + K.$$

Chapter 11 Review Problems (page 361)

1. (a) (See fig.) The 6 level set for example is the line
$2x + 3y = 6$. The graph is the plane
$z = 2x + 3y$ (which passes through the origin).

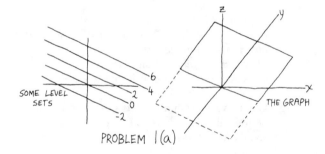

PROBLEM 1(a)

(b) (See fig.). The 3 level set for example is
$\sqrt{x^2 + y^2} = 3$, a circle with radius 3. There are
no neg level sets. Graph is $z = \sqrt{x^2 + y^2}$, top
half of cone $z^2 = x^2 + y^2$.

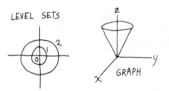

PROBLEM 1(b)

2. (a) (See fig.) The 6 level set is $y^2 + z^2 = 6$, a cylinder. There are no neg level sets. The 0 level set
is $y^2 + z^2 = 0$, the x-axis where $y = z = 0$.

(b) (See fig.) The -3 level set is $5 - x^2 - 2y^2$
$- 3z^2 = -3$, ellipsoid $x^2 + 2y^2 + 3z^2 = 8$.
Highest level possible is 5, a point ellipsoid.

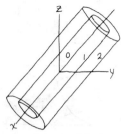

PROBLEM 2

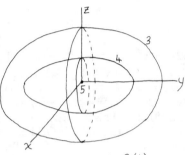

PROBLEM 2(b)

3. (a) y
   (b) 0

4. (See fig.) $\dfrac{\partial z}{\partial r} = \dfrac{\partial z}{\partial x}\dfrac{\partial x}{\partial r} + \dfrac{\partial z}{\partial y}\dfrac{\partial y}{\partial r} = \dfrac{\partial z}{\partial x}\cos\theta + \dfrac{\partial z}{\partial y}\sin\theta,$

$$\frac{\partial z}{\partial\theta} = \frac{\partial z}{\partial x}\frac{\partial x}{\partial\theta} + \frac{\partial z}{\partial y}\frac{\partial y}{\partial\theta}$$
$$= -\frac{\partial z}{\partial x}r\sin\theta + \frac{\partial z}{\partial y}r\cos\theta,$$

$$\left(\frac{\partial z}{\partial r}\right)^2 + \frac{1}{r^2}\left(\frac{\partial z}{\partial\theta}\right)^2$$
$$= \left(\frac{\partial z}{\partial x}\right)^2\cos^2\theta + \left(\frac{\partial z}{\partial y}\right)^2\sin^2\theta$$
$$+ 2\frac{\partial z}{\partial x}\frac{\partial z}{\partial y}\cos\theta\sin\theta$$
$$+ \frac{1}{r^2}\left(\left(\frac{\partial z}{\partial x}\right)^2 r^2\sin^2\theta + \left(\frac{\partial z}{\partial y}\right)^2 r^2\cos^2\theta\right.$$
$$\left. - 2r^2\frac{\partial z}{\partial x}\frac{\partial z}{\partial y}\cos\theta\sin\theta\right)$$
$$= \left(\frac{\partial z}{\partial x}\right)^2(\cos^2\theta + \sin^2\theta) + \left(\frac{\partial z}{\partial y}\right)^2(\sin^2\theta + \cos^2\theta)$$
(by canceling and factoring)
$$= \left(\frac{\partial z}{\partial x}\right)^2 + \left(\frac{\partial z}{\partial y}\right)^2$$

PROBLEM 4

PROBLEM 5

5. (See fig.) $\dfrac{\partial u}{\partial c} = \dfrac{\partial u}{\partial y}\dfrac{\partial y}{\partial c} + \dfrac{\partial u}{\partial z}\dfrac{\partial z}{\partial c} = 4\dfrac{\partial u}{\partial y} + a\dfrac{\partial u}{\partial z}$,

$\dfrac{\partial^2 u}{\partial a\,\partial c} = 4\dfrac{\partial}{\partial a}\left(\dfrac{\partial u}{\partial y}\right) + a\dfrac{\partial}{\partial a}\left(\dfrac{\partial u}{\partial z}\right) + \dfrac{\partial u}{\partial z}$ (prod rule)

$= 4\left(\dfrac{\partial^2 u}{\partial x\,\partial y}\cdot 2 + \dfrac{\partial^2 u}{\partial z\,\partial y}\cdot c\right)$

$\quad + a\left(\dfrac{\partial^2 u}{\partial x\,\partial z}\cdot 2 + \dfrac{\partial^2 u}{\partial z^2}\cdot c\right) + \dfrac{\partial u}{\partial z}$

$= 8\dfrac{\partial^2 u}{\partial x\,\partial y} + 4c\dfrac{\partial^2 u}{\partial z\,\partial y} + 2a\dfrac{\partial^2 u}{\partial x\,\partial z} + ac\dfrac{\partial^2 u}{\partial z^2} + \dfrac{\partial u}{\partial z}$

6. Let $f(x,y) = xy + 2y^2 - 12y$. Then $\partial f/\partial x = y$, $\partial f/\partial y = x + 4y - 12$. Partials are 0 when $y = 0, x = 12$ (critical point); NOT IN REGION, ignore.

On bdry $y = x$, $f = y^2 + 2y^2 - 12y = 3y^2 - 12y$ where $0 \le y \le 4$. Then $f'(y) = 6y - 12$, zero if $y = 2$. Ends are $y = 0, 4$. Candidates are $(2,2), (0,0), (4,4)$.

On bdry $x = 4$, $f = 2y^2 - 8y$ where $0 \le y \le 4$. Then $f'(y) = 4y - 8$; zero if $y = 2$. Ends are $y = 0, 4$. Candidates are $(4,2), (4,0), (4,4)$.

On bdry $y = 0$, f is always 0.
To make final decision find $f(2,2) = -12$, $f(4,4) = 0$, $f(4,2) = -8$, $f(\text{lower bdry}) = 0$.
Max is 0, min is -12.

7. $\nabla f = (6x, 8y)$. At any point, max directional deriv is in direction of ∇f and has max value $\|\nabla f\|$. So problem wants to know at what point on circle $\|\nabla f\|$ is largest. $\|\nabla f\| = \sqrt{36x^2 + 64y^2}$ and y^2 is $1 - x^2$ on circle, so on circle, $\|\nabla f\| = \sqrt{36x^2 + 64(1 - x^2)} = \sqrt{64 - 28x^2}$ for $-1 \le x \le 1$. By inspection, max occurs at $x = 0$ (value is 8) which corresponds to points $(0, \pm 1)$ on circle. At $(0, 1)$, $\nabla f = (0, 8)$, at $(0, -1)$, $\nabla f = (0, -8)$.

Answer is that max dir deriv on circle has value 8. It occurs at $(0, 1)$ in north direction and at $(0, -1)$ in south direction (Note that the circle is not and never was intended to be a level set of f.)

8. $\nabla f = (2xyz, x^2z, x^2y)$, $\nabla f|_P = (12, 3, 2)$, $\overrightarrow{QP} = (1, 3, 2)$, $\overrightarrow{PR} = (1, -5, 2)$. Arriving at P from Q, dir deriv is

$\dfrac{\partial f}{\partial \overrightarrow{QP}} = \dfrac{\nabla f|_P \cdot \overrightarrow{QP}}{\|\overrightarrow{QP}\|} = \dfrac{25}{\sqrt{14}}$. Departing P for R, dir deriv

is $\dfrac{\partial f}{\partial \overrightarrow{PR}} = \dfrac{\nabla f \cdot \overrightarrow{PR}}{\|\overrightarrow{PR}\|} = \dfrac{1}{\sqrt{30}}$.

9. (a) $T(-2, 2) = 4 - 2 = 2$ so point is on the 2 level set which is $x^2 - y = 2$, $y = x^2 - 2$, parabola. (See fig.) $\nabla T = (2x, -1)$,
$\nabla T|_{x=-2, y=2} = -4\vec{i} - \vec{j}$, a perp vector.

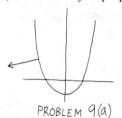

PROBLEM 9(a)

(b) (i) Graph is $z = x^2 - y$. Point Q satisfies equation so is on graph.
(ii) Surface is $x^2 - y - z = 0$, $\nabla(x^2 - y - z) = (2x, -1, -1)$. At Q get $\nabla = (-4, -1, -1)$ which is perp to surface at Q.
(iii) $\nabla z = (2x, -1)$, $\nabla z|_{x=-2, y=2} = (-4, -1)$.
$\overrightarrow{NW} = (-1, 1)$. Slope at Q on $\overrightarrow{NW}$ path is $\partial z/\partial NW = (\nabla z \cdot NW)/\|NW\| = 3/\sqrt{2}$. Slope is $3/\sqrt{2}$.
(iv) Path in direction of $\nabla z = -4\vec{i} - \vec{j}$ (more west, some south) rises most steeply.

10. $dS = 2\pi(r\,dh + h\,dr) + 4\pi r\,dr$ (product rule)
$= (2\pi h + 4\pi r)\,dr + 2\pi r\,dh$.

11. $(2x - y)\,dx - x\,dy = 0$, $d(x^2 - xy) = 0$, $x^2 - xy = C$. If $x = 1$, $y = 2$ then $C = -1$. So sol is $x^2 - xy = -1$, explicit solution is $y = (x^2 + 1)/x$, $y = x + 1/x$.

12/MULTIPLE INTEGRALS

Section 12.1 (page 369)

1. $\int_R 5\,dA = 5\int_R dA = 5 \times$ area of R
$= 5 \times 4\pi = 20\pi$

2. (See figs.) (a) Consider hemisphere which is top half of $x^2 + y^2 + z^2 = 9$. It is under graph of $z = \sqrt{9 - x^2 - y^2}$ and over the circular region R with radius 3. So
hemisphere volume $= \int_R \sqrt{9 - x^2 - y^2}\,dA$.

PROBLEM 2(a)

PROBLEM 2(b)

(b) Indicated cyclinder is under plane $z = 5$ and over circular region R with radius 2 so cyl volume $= \int_R 5\,dA$.

3. (a) Roof of solid is plane $z = 5$. The floor is the region R_1. Solid is box in fig.

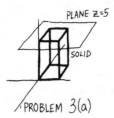

PROBLEM 3(a)

(b) Roof of solid is paraboloid $z = x^2 + y^2$. Floor is region R_2. Solid lies *under* roof and over floor as indicated in fig.

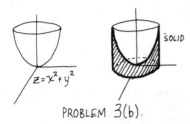

PROBLEM 3(b).

4. (a) Each $xy\,dA$ in III is positive (since $x < 0, y < 0$) so $\Sigma xy\,dA$ is pos, $\int_{III} xy\,dA$ is positive. Alternatively, integral is positive because graph of $z = xy$ is a surface in space *above* region III in x, y plane.

(b) (See fig.) For each subregion in I where $xy\,dA$ is positive there is a corresponding subregion in II where $xy\,dA$ has the negative value (because $x < 0$). So $\Sigma xy\,dA = 0$ and $\int_{I \text{ and } II} xy\,dA = 0$.

PROBLEM 4(b)

(c) Same answer as (b).

(d) Negative because each $xy^2\,dA$ is neg.

(e) Pos. In fact the integral is the area of region IV.

5. Not necessarily. If $f(x, y)$ is neg in the extension then the extra $f(x, y)\,dA$'s being added make the sum smaller, not larger.

6. The sum of 0 dA's is 0. Integral is 0.

7. (a) True. If $f(x, y) > g(x, y)$ then for any subregion, $f(x, y)\,dA > g(x, y)\,dA$ and $\Sigma f(x, y)\,dA > \Sigma g(x, y)\,dA$. Alternatively, graph of f is higher than graph of g so more volume (above − below) is caught by the f graph.

(b) False. If f is larger on the smaller region R_2 it is possible for the integral on R_2 to be larger. In particular, it will happen if f is positive on the smaller region R_2 and negative on the larger region R_1.

(c) False. Even if x and y are positive, it may be that $f(x, y)$ is negative at all or some points (e.g., $f(x, y) = -xy$, $f(x, y) = x + y - 100$) in which case the integral can be neg.

(d) True, since the integrals are the areas of R_1 and R_2.

(e) True. If $f(x, y) > 0$ then $f(x, y)\,dA > 0$ and $\Sigma f(x, y)\,dA > 0$.

8. Rearrange (3) to get $\int_R x\,dA =$ (av value of x in R) × area of $R = 4 \times 9\pi = 36\pi$.

9. No. $\int x\,dA$ on the entire region is 0 (since $\Sigma x\,dA$ on the right half cancels $\Sigma x\,dA$ on the left half) but $4 \int x\,dA$ on region I is positive.

Section 12.2 (page 376)

Each problem has a diagram.

1. (a) $\int_{x=0}^{3/2} \int_{y=2x}^{y=3} x^3\,dy\,dx$ and $\int_{y=0}^{3} \int_{x=0}^{x=y/2} x^3\,dx\,dy$.
With second version, inner $= \frac{1}{4}x^4 |_{x=0}^{x=y/2} = \frac{1}{64}y^4$,
outer $= \int_0^3 \frac{1}{64}y^4\,dy = \frac{1}{65} \cdot \frac{1}{5}y^5 |_0^3 = 243/320$.

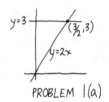

PROBLEM 1(a)

(b) $\int_{x=0}^{x=1} \int_{y=x^3}^{y=x} 3dy\,dx$ and $\int_{y=0}^{1} \int_{x=y}^{x=\sqrt[3]{y}} 3\,dx\,dy$.
With first version, inner $= 3y |_{y=x^3}^{y=x} = 3(x - x^3)$,
outer $= 3(\frac{1}{2}x^2 - \frac{1}{4}x^4) |_0^1 = \frac{3}{4}$.

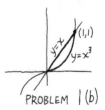

PROBLEM 1(b)

(c) $\int_{x=-\sqrt{10}}^{\sqrt{10}} \int_{y=x^2}^{y=10} 2xy\,dy\,dx$; and $\int_{y=0}^{10} \int_{x=-\sqrt{y}}^{x=\sqrt{y}} 2xy\,dx\,dy$
$= \int_{y=0}^{10} (x^2y |_{x=-\sqrt{y}}^{\sqrt{y}})\,dy = \int_0^{10} 0\,dy = 0$. (The answer 0 is predictable since $2xy\,dA$'s in the left half of the region are the negatives of $2xy\,dA$'s in the right half, so sum is 0.)

PROBLEM 1(c)

2. (a) $\int_{x=0}^{x=4}\int_{y=x^2/8}^{y=\sqrt{x}} f(x,y)\,dy\,dx$ and $\int_{y=0}^{y=2}\int_{x=y^2}^{x=\sqrt{8y}} f(x,y)\,dx\,dy$

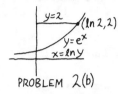

PROBLEM $2(a)$

(b) $\int_{x=0}^{x=\ln 2}\int_{y=e^x}^{y=2} f(x,y)\,dy\,dx$ and $\int_{y=1}^{y=2}\int_{x=0}^{x=\ln y} f(x,y)\,dx\,dy$

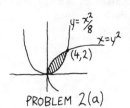

PROBLEM $2(b)$

(c) Note that for $dy\,dx$ order of integration, upper boundary is two curves so use $\int_{\text{left part}} + \int_{\text{right part}}$
$= \int_{x=0}^{x=2}\int_{y=x/3}^{y=2x} f(x,y)\,dy\,dx + \int_{x=2}^{x=3}\int_{y=x/3}^{-3x+10} f(x,y)\,dy\,dx$.
Note that for order of integration $dx\,dy$, right boundary is two curves so use $\int_{\text{top}} + \int_{\text{bottom}}$
$= \int_{y=1}^{4}\int_{x=y/2}^{x=(10-y)/3} f(x,y)\,dx\,dy$
$+ \int_{y=0}^{y=1}\int_{x=y/2}^{x=3y} f(x,y)\,dx\,dy$.

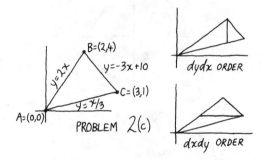

PROBLEM $2(c)$

(d) $\int_{y=0}^{y=5}\int_{x=2}^{x=7} f(x,y)\,dx\,dy$ and $\int_{x=2}^{x=7}\int_{y=0}^{y=5} f(x,y)\,dy\,dx$

PROBLEM $2(d)$

(e) Solve $4y = x^2$, $x - 2y + 4 = 0$ to get points of intersection. One set-up is
$\int_{x=-2}^{x=4}\int_{y=x^2/4}^{y=2+x/2} f(x,y)\,dy\,dx$. To use the other order of integration note that left boundary consists of two curves so divide up region to get
$\int_I + \int_{II} = \int_{y=0}^{y=1}\int_{x=-2\sqrt{y}}^{x=2\sqrt{y}} f(x,y)\,dx\,dy$
$+ \int_{y=1}^{y=4}\int_{x=2y-4}^{x=2\sqrt{y}} f(x,y)\,dx\,dy$.

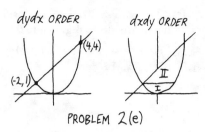

PROBLEM $2(e)$

(f) $\int_{x=2}^{x=3}\int_{y=0}^{y=1/x} f(x,y)\,dy\,dx$. For other order of integration note that right boundary is two curves so divide up region:
$\int_I + \int_{II} = \int_{y=0}^{y=1/3}\int_{x=2}^{x=3} f(x,y)\,dx\,dy$
$+ \int_{y=1/3}^{y=1/2}\int_{x=2}^{x=1/y} f(x,y)\,dx\,dy$.

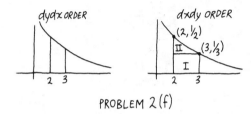

PROBLEM $2(f)$

(g) "Top" of region is boundary at infinity:
$\int_{x=-\infty}^{x=\infty}\int_{y=x^2}^{y=\infty} f(x,y)\,dy\,dx$ and
$\int_{y=0}^{y=\infty}\int_{x=-\sqrt{y}}^{x=\sqrt{y}} f(x,y)\,dx\,dy$.

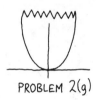

PROBLEM $2(g)$

(h) Region is enclosed by $xy = 1$ and boundary at infinity. Each boundary consists of two curves; e.g., the left boundary is partly lower branch of hyperbola and partly boundary at infinity. For one method, $\int_{\text{lower half}} + \int_{\text{upper half}}$
$= \int_{y=-\infty}^{y=0}\int_{x=1/y}^{x=\infty} f(x,y)\,dx\,dy$
$+ \int_{y=0}^{y=\infty}\int_{x=-\infty}^{x=1/y} f(x,y)\,dx\,dy$.

Also $\int_{\text{left half}} + \int_{\text{right half}}$
$= \int_{x=-\infty}^{x=0}\int_{y=1/x}^{y=\infty} f(x,y)\,dy\,dx$
$+ \int_{x=0}^{x=\infty}\int_{y=-\infty}^{y=1/x} f(x,y)\,dy\,dx$.

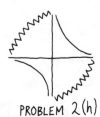

PROBLEM $2(h)$

(i) $\int_{y=0}^{y=2} \int_{x=y}^{x=y+3} f(x,y)\, dx\, dy$. To use other order of integration must divide up region into three parts:

$\int_{x=0}^{2} \int_{y=0}^{y=x} f(x,y)\, dy\, dx + \int_{x=2}^{3} \int_{y=0}^{y=2} f(x,y)\, dy\, dx$
$+ \int_{x=3}^{x=5} \int_{y=x-3}^{y=2} f(x,y)\, dy\, dx.$

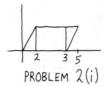

PROBLEM 2(i)

(j) $\int_{x=-\sqrt{2}}^{\sqrt{2}} \int_{y=-\sqrt{4-2x^2}}^{y=\sqrt{4-2x^2}} f(x,y)\, dy\, dx$ and
$\int_{y=-2}^{y=2} \int_{x=-\sqrt{2-y^2/2}}^{x=\sqrt{2-y^2/2}} f(x,y)\, dx\, dy$

PROBLEM 2(j)

3. (a) Lower and upper boundaries each consist of two curves so divide up:

$\int_{\text{left part}} + \int_{\text{right part}}$
$= \int_{x=-1}^{x=0} \int_{y=-x-1}^{y=x+1} f\, dy\, dx + \int_{x=0}^{x=1} \int_{y=x-1}^{y=-x+1} f\, dy\, dx.$

PROBLEM 3(a)

(b) It depends on f. If $f(x,y)$ takes on the same values in quads II, III, IV as it does in I then yes. But usually no. For example if $f(x,y) = x$ then NO since $\int_{\text{whole}} = 0$ but $4\int_{\text{quad I part}}$ is positive.

4. (a) Left boundary is line $x = \frac{1}{2}y$, right boundary is line $x = 1$, extreme y's are 0 and 1. For reverse order, divide up region since upper boundary is two curves:

$\int_{x=0}^{x=1/2} \int_{y=0}^{y=2x} f(x,y)\, dy\, dx + \int_{x=1/2}^{x=1} \int_{y=0}^{y=1} f(x,y)\, dy\, dx.$

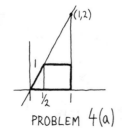

PROBLEM 4(a)

(b) Lower bdry is line $y = -x$, upper is parabola $y = x^2$. Extremes are $x = 0$, 1. For reverse order, divide up region:

$\int_{\text{upper part}} + \int_{\text{lower part}}$
$= \int_{y=0}^{y=1} \int_{x=\sqrt{y}}^{x=1} f(x,y)\, dx\, dy + \int_{y=-1}^{y=0} \int_{x=-y}^{x=1} f(x,y)\, dx\, dy.$

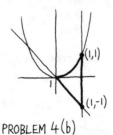

PROBLEM 4(b)

(c) Lower boundary is $y = -\sqrt{1-x^2}$, lower half of circle $x^2 + y^2 = 1$, upper boundary is parabola $y = 1 - x^2$. For other order of integration divide up region:

$\int_{\text{top}} + \int_{\text{bottom}} = \int_{y=0}^{y=1} \int_{x=0}^{x=\sqrt{1-y}} f(x,y)\, dx\, dy$
$+ \int_{y=-1}^{0} \int_{x=0}^{x=\sqrt{1-y^2}} f(x,y)\, dx\, dy.$

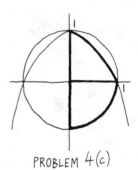

PROBLEM 4(c)

(d) $\int_{y=0}^{\infty} \int_{x=0}^{x=y/2} f(x,y)\, dx\, dy$

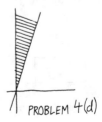

PROBLEM 4(d)

(e) Lower boundary is $y = \sin^{-1}x$, upper is $y = 2$. Note that the two boundaries don't actually meet since $\sin^{-1}x$ doesn't get that high. Extremes are $x = 0$ and $x = 1$. For other order divide up region since right boundary is two curves:

$\int_{\text{I}} + \int_{\text{II}} = \int_{y=\pi/2}^{y=2} \int_{x=0}^{x=1} f(x,y)\, dx\, dy$
$+ \int_{y=0}^{y=\pi/2} \int_{x=0}^{x=\sin y} f(x,y)\, dx\, dy.$

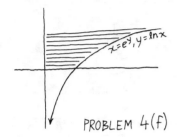

PROBLEM 4(e)

(f) Left boundary is y-axis, right boundary is $x = e^y$. For other order, divide region since lower boundary is two curves:

$$\int_{x=0}^{x=1} \int_{y=0}^{y=\infty} f(x,y)\, dy\, dx \;+\; \int_{x=1}^{x=\infty} \int_{y=\ln x}^{y=\infty} f(x,y)\, dy\, dx\,.$$

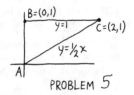

PROBLEM 4(f)

5. $\int_{x=0}^{2} \int_{y=x/2}^{y=1} e^{y^2}\, dy\, dx$ and $\int_{y=0}^{1} \int_{x=0}^{x=2y} e^{y^2}\, dx\, dy$. With first set-up we have to antidiff e^{y^2} w.r.t. y; hard. For second set-up,

inner integral $= \int_{x=0}^{x=2y} e^{y^2}\, dx = xe^{y^2}\big|_{x=0}^{x=2y} = 2ye^{y^2}$

outer integral $= \int_{y=0}^{y=1} 2ye^{y^2}\, dy = e^{y^2}\big|_{0}^{1} = e - 1$

(get antiderivative of $2ye^{y^2}$ by inspection or substitute $u = y^2$, $du = 2y\, dy$).

PROBLEM 5

Section 12.3 (page 381)

1. (a) $\int_{\theta=-\pi/2}^{\pi/2} \int_{r=0}^{4} r \cos\theta\, r\, dr\, d\theta$. Inner $= \frac{1}{3}r^3 \cos\theta\big|_{r=0}^{4}$
$= \frac{64}{3}\cos\theta$. Outer $= \frac{64}{3} \sin\theta\big|_{-\pi/2}^{\pi/2} = \frac{128}{3}$.

(b) $\int_{\theta=0}^{\pi/2} \int_{r=0}^{4} (r\cos\theta)(r\sin\theta)\, r\, dr\, d\theta$
$= \int_{\theta=0}^{\pi/2} \int_{r=0}^{4} r^3 \cos\theta \sin\theta\, dr\, d\theta$
$= \int_{\theta=0}^{\pi/2} \frac{1}{4}r^4 \cos\theta \sin\theta\big|_{r=0}^{4}\, d\theta$
$= \int_{0}^{\pi/2} 64 \cos\theta \sin\theta\, d\theta = 64 \cdot \frac{1}{2} \sin^2\theta\big|_{0}^{\pi/2} = 32$

(c) $\int_{\theta=0}^{2\pi} \int_{r=0}^{4} \dfrac{1}{1+r^2} r\, dr\, d\theta$. Antideriv of $r/(1+r^2)$ is $\frac{1}{2}\ln(1+r^2)$ (sub $u = 1+r^2$, $du = 2r\, dr$) so get $\int_{\theta=0}^{2\pi} (\frac{1}{2} \ln 17 - \frac{1}{2} \ln 1)\, d\theta$
$= \frac{1}{2} \ln 17 \int_0^{2\pi} d\theta = \frac{1}{2} \ln 17 \times 2\pi = \pi \ln 17$.

2. $\int_{\theta=0}^{\pi/2} \int_{r=0}^{\infty} e^{-r^2} r\, dr\, d\theta = \int_{\theta=0}^{\pi/2} (-\frac{1}{2}e^{-r^2}\big|_0^{\infty})\, d\theta = \int_0^{\pi/2} \frac{1}{2}\, d\theta$
$= \frac{1}{2} \times \frac{1}{2}\pi = \frac{1}{4}\pi$

3. Lower boundary of region is x-axis, upper boundary is top half of circle $x^2 + y^2 = 9$, extreme x's are 3, -3 (see fig.); $\int_{\theta=0}^{\pi} \int_{r=0}^{3} \ln(1 + r^2) r\, dr\, d\theta$. Let $u = 1 + r^2$, $du = 2r\, dr$. So inner integral is $\frac{1}{2} \int_1^{10} \ln u\, du$ (now use integral tables) $= \frac{1}{2}(u \ln u - u)\big|_1^{10} = \frac{1}{2}(10 \ln 10 - 9)$. Outer $= \frac{1}{2}\pi(10 \ln 10 - 9)$.

PROBLEM 3

4. (a) $\int_{\theta=0}^{2\pi} \int_{r=3}^{4}$

(b) $\int_{\theta=0}^{2\pi} \int_{r=2}^{\infty}$

(c) Inner boundary is $r = 0$, outer is line $x + y = 2$ which in polar coords is $r \cos\theta + r \sin\theta = 2$, $r = 2/(\cos\theta + \sin\theta)$; get $\int_{\theta=0}^{\pi/2} \int_{r=0}^{2/(\cos\theta + \sin\theta)}$.

(d) Inner boundary is $r = 0$. Outer is circle $(x - 2)^2 + y^2 = 4$, $x^2 + y^2 = 4x$, $r^2 = 4r \cos\theta$, $r = 4 \cos\theta$. Extreme θ's are $-\pi/2$ and $\pi/2$ so get $\int_{\theta=-\pi/2}^{\pi/2} \int_{r=0}^{r=4\cos\theta}$.

(e) $\int_{\theta=0}^{\pi} \int_{r=5}^{\infty}$

(f) Inner boundary is $r = 0$. Outer boundary is two curves, line AB and line BC. (See fig.) Divide up region. Line AB has equ $y = 3$, $r \sin\theta = 3$, $r = 3 \csc\theta$. Line BC has equ $x = 6$, $r \cos\theta = 6$, $r = 6 \sec\theta$.
$\int_{\text{square}} = \int_{\text{I}} + \int_{\text{II}} = \int_{\theta=0}^{\theta_0} \int_{r=0}^{6\sec\theta} + \int_{\theta=\theta_0}^{\pi/2} \int_{r=0}^{3\csc\theta}$
where θ_0 is in diagram, can be called $\tan^{-1}\frac{1}{2}$.

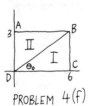

PROBLEM 4(f)

Section 12.4 (page 386)

All problems except 1(d), 2(c), and 4 have diagrams.

1. (a) *Method (B):* Upper curve is line and parabola; divide up region. For left part, $u(x) = \sqrt{x + 1}$, $l(x) = -\sqrt{x + 1}$. For right part, $u(x) = 1 - x$, $l(x) = -\sqrt{x + 1}$.
Area $= \int_{-1}^{0} (\sqrt{x + 1} - -\sqrt{x + 1})\, dx$
$\quad + \int_0^3 (1 - x - -\sqrt{x + 1})\, dx$
$= 2\int_{-1}^{0} \sqrt{x + 1}\, dx + \int_0^3 (1 - x + \sqrt{x + 1})\, dx$
$= 2 \cdot \frac{2}{3}(x + 1)^{3/2}\big|_{-1}^{0} + (x - \frac{1}{2}x^2 + \frac{2}{3}(x + 1)^{3/2})\big|_0^3$
$= \frac{9}{2}$.

Method (C): Area $= \int_{\text{region}} dA$
$= \int_{y=-2}^{y=1} \int_{x=y^2-1}^{1-y} dx\, dy = \cdots = \frac{9}{2}$.

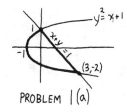

PROBLEM 1 (a)

(b) To get points of intersection solve $x(5 - x) = 4$ to get $x = 4, 1$.

Method (B): $u(x) = 5 - x$, $l(x) = 4/x$.
Area $= \int_1^4 (5 - x - 4/x)\, dx$
$= (5x - \frac{1}{2}x^2 - 4 \ln x)|_1^4 = \frac{15}{2} - 4 \ln 4$.

Method (C): Area $= \int_{\text{region}} dA$
$= \int_{x=1}^4 \int_{y=4/x}^{y=5-x} dy\, dx = \int_1^4 (5 - x - 4/x)\, dx$, as above.

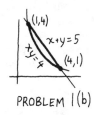

PROBLEM 1 (b)

(c) *Method (B):* $u(x) = 4$, $l(x) = x^2$.
Area $= \int_{-2}^2 (4 - x^2)\, dx = (4x - \frac{1}{3}x^3)|_{-2}^2 = 32/3$.
Method (C): Area $= \int_{\text{region}} dA$
$= \int_{x=-2}^{x=2} \int_{y=x^2}^{y=4} dy\, dx = \int_{-2}^2 (4 - x^2)\, dx$, as above.

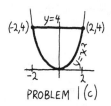

PROBLEM 1 (c)

(d) By (C), area $= \int_{\text{region}} dA = \int_{\theta=0}^{2\pi} \int_{r=0}^{\theta} r\, dr\, d\theta$.
Inner $= \frac{1}{2} r^2 |_{r=0}^{\theta} = \frac{1}{2}\theta^2$;
outer $= \int_0^{2\pi} \frac{1}{2}\theta^2\, d\theta = \frac{1}{6}\theta^3 |_0^{2\pi} = \frac{4}{3}\pi^3$.

2. (a) Use (A'). Solid lies under graph of $f(x,y) = 6 - \frac{1}{2}x - 3y$ and above circular region R with radius 2; volume $= \int_R (6 - \frac{1}{2}x - 3y)\, dA$
$= \int_{\theta=0}^{2\pi} \int_{r=0}^{2} (6 - \frac{1}{2}r \cos \theta - 3r \sin \theta) r\, dr\, d\theta$;
inner $= (3r^2 - \frac{1}{6}r^3 \cos \theta - r^3 \sin \theta)|_0^2$
$= 12 - \frac{4}{3} \cos \theta - 8 \sin \theta$;
outer $= \int_{\theta=0}^{2\pi} 12\, d\theta - \frac{4}{3}\int_0^{2\pi} \cos \theta\, d\theta$
$\qquad - 8\int_0^{2\pi} \sin \theta\, d\theta = 24\pi + 0 + 0 = 24\pi$.

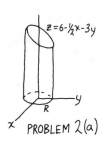

PROBLEM 2 (a)

(b) *Method (A'):* Find vol of top hemispherical region and then double. Sphere is $x^2 + y^2 + z^2 = R^2$. Hemispherical region lies under graph of $f(x,y) = \sqrt{R^2 - x^2 - y^2}$ and over circular region of radius R in x, y plane so
hemisphere vol $= \int_{\text{circ region}} \sqrt{R^2 - x^2 - y^2}\, dA$
$= \int_{\theta=0}^{2\pi} \int_{r=0}^{R} \sqrt{R^2 - r^2}\, r\, dr\, d\theta$. (To get antideriv of $r\sqrt{R^2 - r^2}$, let $u = R^2 - r^2$, $du = -2r\, dr$.)
Inner $= -\frac{1}{3}(R^2 - r^2)^{3/2}|_{r=0}^{R} = \frac{1}{3}R^3$;
outer $= \int_{\theta=0}^{2\pi} \frac{1}{3}R^3\, d\theta = 2\pi \cdot \frac{1}{3}R^3$;
sphere vol $= 2 \cdot 2\pi \cdot \frac{1}{3}R^3 = \frac{4}{3}\pi R^3$.

PROBLEM 2 (b)

Method (B'): $u(x,y) = \sqrt{R^2 - x^2 - y^2}$,
$l(x,y) = -\sqrt{R^2 - x^2 - y^2}$,
sphere vol $= \int_{\text{proj in x,y plane}} (u(x,y) - l(x,y))\, dA$
$= \int_{\text{circ region 2}} \sqrt{R^2 - x^2 - y^2}\, dA$, etc.

(c) Find vol of top half and double. Use (A'). Solid lies under sphere $x^2 + y^2 + z^2 = 36$, i.e., under graph of $f(x,y) = \sqrt{36 - x^2 - y^2}$, and above circ region R of radius 3.
Volume $= 2\int_R \sqrt{36 - x^2 - y^2}\, dA$
$= 2\int_{\theta=0}^{2\pi} \int_{r=0}^{3} \sqrt{36 - r^2}\, r\, dr\, d\theta$;
inner $= -\frac{1}{3}(36 - r^2)^{3/2}|_0^3 = \frac{1}{3}(216 - 27^{3/2})$;
outer $= 2 \cdot 2\pi \cdot \frac{1}{3}(216 - 27^{3/2})$.

(d) Use (B'). Upper surface is plane $z = h$, lower surface is the cone. Projection in x, y plane is circular with radius R. Will use polar coords so need cone's equ in terms of z, r, θ. By similar triangles $z/h = r/R$ so $z = rh/R$.
Volume $= \int_{\theta=0}^{2\pi} \int_{r=0}^{R} (h - rh/R) r\, dr\, d\theta$
$= \int_{\theta=0}^{2\pi} (\frac{1}{2}hr^2 - \frac{1}{3}hr^3/R)|_{r=0}^{R}\, d\theta$
$= \int_{\theta=0}^{2\pi} \frac{1}{6}hR^2\, d\theta = 2\pi \cdot \frac{1}{6}hR^2 = \frac{1}{3}\pi R^2 h$.

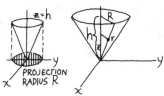

PROBLEM 2 (d)

(e) Use (A'). Solid lies under graph of $f(x,y) = h$ and above circ region with radius R.
Volume $= \int_{\text{circ region}} h\, dA = \int_{\theta=0}^{2\pi} \int_{r=0}^{R} hr\, dr\, d\theta$
$= \int_{\theta=0}^{2\pi} \frac{1}{2}hr^2 |_{r=0}^{R}\, d\theta = \int_{\theta=0}^{2\pi} \frac{1}{2}hR^2\, d\theta = 2\pi \cdot \frac{1}{2}hR^2$
$= \pi R^2 h$.

PROBLEM 2(e)

3. (a) Plane intersects parabola in circle $2x^2 + 2y^2 = 12$ so projection is circle $x^2 + y^2 = 6$. Use (B'). Volume $= \int_{\text{proj}} [12 - (2x^2 + 2y^2)]\, dA$
$= \int_{\theta=0}^{2\pi} \int_{r=0}^{\sqrt{6}} (12 - 2r^2)r\, dr\, d\theta$.

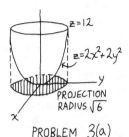

PROBLEM 3(a)

(b) Use (B'). To get projection use
$8 - 3x^2 - 3y^2 = x^2 + y^2$, i.e., $x^2 + y^2 = 2$.
Volume $= \int_{\text{proj}} (8 - 3x^2 - 3y^2 - [x^2 + y^2])\, dA$
$= \int_{\theta=0}^{2\pi} \int_{r=0}^{\sqrt{2}} (8 - 4r^2)r\, dr\, d\theta$.

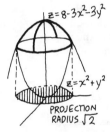

PROBLEM 3(b)

(c) Can use (B'). Upper surface is sphere $x^2 + y^2 + z^2 = 25$ so $u(x, y) = \sqrt{25 - x^2 - y^2}$. Lower surface is plane $z = \sqrt{21}$. Projection is circ region with radius 2.
Volume $= \int_{\text{proj}} (\sqrt{25 - x^2 - y^2} - \sqrt{21})\, dA$
$= \int_{\theta=0}^{2\pi} \int_{r=0}^{2} (\sqrt{25 - r^2} - \sqrt{21})r\, dr\, d\theta$.
Or can find volume of a "half apple core" and subtract vol of cylinder with radius 2 and height $\sqrt{21}$ (which is essentially what the preceding double integral does anyway).

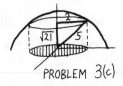

PROBLEM 3(c)

4. Use (A'). Half-solid lies under cylinder $x^2 + z^2 = 9$, i.e., under graph of $f(x, y) = \sqrt{9 - x^2}$, and above a circ region R of radius 3 in x, y plane.
$\frac{1}{2}$vol $= \int_R \sqrt{9 - x^2}\, dA = \int_{x=-3}^{x=3} \int_{y=-\sqrt{9-x^2}}^{\sqrt{9-x^2}} \sqrt{9 - x^2}\, dy\, dx$;
inner $= \sqrt{9 - x^2}(\sqrt{9 - x^2} - -\sqrt{9 - x^2}) = 2(9 - x^2)$;
outer $= 2 \int_{-3}^{3} (9 - x^2)\, dx = \cdots = 72$;
total volume $= 2 \times 72 = 144$.

5. See (A'). Integral is vol of solid with floor R and roof $z = \sqrt{x^2 + y^2}$ (top half of cone $z^2 = x^2 + y^2$). (Note that integral is not the volume *inside* cone since that volume lies above the cone, not under it.)

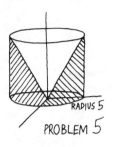

PROBLEM 5

Section 12.5 (page 390)

All problems have diagrams.

1. (a) Distance from (x, y) to longer side is x.
Av value of x in region $= \dfrac{\int_{\text{region}} x\, dA}{\text{area(which is 25)}}$;
$\int_{\text{region}} x\, dA = \int_{x=0}^{5} \int_{y=0}^{10-2x} x\, dy\, dx$
$= \int_0^5 x(10 - 2x)\, dx = (5x^2 - \frac{2}{3}x^3)\,|_0^5 = \frac{125}{3}$
so average value $= \frac{1}{25} \times \frac{125}{3} = \frac{5}{3}$.

(b) Distances to legs are x and y so
$d$mass $=$ mass density $\times$ area $= xy\, dA$;
total mass
$= \int_{\text{region}} xy\, dA = \int_{x=0}^{5} \int_{y=0}^{10-2x} xy\, dy\, dx$
$= \int_0^5 \frac{1}{2}xy^2\,|_{y=0}^{10-2x}\, dx = \int_0^5 \frac{1}{2}x(10 - 2x)^2\, dx$
$= \int_0^5 \frac{1}{2}(100x - 40x^2 + 4x^3)\, dx = \cdots = 625/6$.

PROBLEM 1

2. *Method 1:* Consider sand in strip at distance $150 - x$ from desert. Area is $500\, dx$, sand density in strip is $1/(150 - x)$ so

$$d\text{sand} = \text{density} \times \text{area} = \frac{500\, dx}{150 - x},$$

$$\text{total sand} = \int_0^{100} \frac{500\,dx}{150 - x} = -500 \ln(150 - x)\Big|_0^{100}$$
$$= -500(\ln 50 - \ln 150) = 500 \ln 3.$$

Method 2: Consider small subregion containing point (x, y) with area dA. Its distance to desert is $150 - y$ so $d\text{sand} = dA/(150 - y)$,

$$\text{total sand} = \int_{\text{town}} \frac{1}{150 - y}\, dA = \int_{y=0}^{100} \int_{x=0}^{500} \frac{1}{150 - y}\, dx\, dy$$
$$= \int_{y=0}^{100} \frac{500}{150 - y}\, dy = \cdots = 500 \ln 3 \text{ as above}.$$

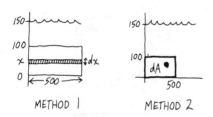

METHOD 1   METHOD 2

PROBLEM 2

3. *Method 1:* Consider small subregion with area dA containing point with polar coords r, θ. Then
$d\text{water} = \text{density} \times \text{area} = r^3 dA$,
total water $= \int_{\theta=0}^{\pi} \int_{r=0}^{6} r^3 r\, dr\, d\theta$
$= \int_{\theta=0}^{\pi} \frac{1}{5} r^5 \big|_{r=0}^{6}\, d\theta = 6^5 \pi/5.$

Method 2: Consider semi-ring of radius r and thickness dr. By (8), Section 4.8, its area is $\pi r\, dr$; its density is r^3 so $d\text{water} = r^3 \pi r\, dr = \pi r^4\, dr$;
total water $= \int_0^6 \pi r^4\, dr = 6^5 \pi/5.$

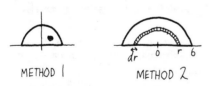

METHOD 1   METHOD 2

PROBLEM 3

4. (a) *Method 1:* Consider small subregion of area dA containing point r, θ. Then $d\text{cost} = r\, dA$,
total cost $= \int_{\text{land}} r\, dA = \int_{\theta=0}^{2\pi} \int_{r=0}^{2} r r\, dr\, d\theta$
$= \int_0^{2\pi} \int_{r=0}^{2} r^2\, dr\, d\theta.$

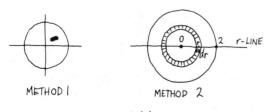

METHOD 1   METHOD 2

PROBLEM 4(a)

Method 2: Consider circular ring of radius r, thickness dr. Then density in ring is r, area is $2\pi r\, dr$, $d\text{cost} = r \cdot 2\pi r\, dr = 2\pi r^2\, dr$, total cost $= \int_{r=0}^{2} 2\pi r^2\, dr.$

(b) Consider small subregion with area dA containing point (x, y): $d\text{cost} = \sqrt{x^2 + y^2}\, dA$;
cost $= \int_{\text{land}} \sqrt{x^2 + y^2}\, dy\, dx$
$= \int_{x=1}^{x=5} \int_{y=0}^{-3(x-5)/4} \sqrt{x^2 + y^2}\, dy\, dx.$

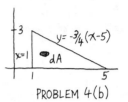

PROBLEM 4(b)

5. (a) The small number of people (i.e., $d$people) living in a little subdivision containing point (x, y) and with area dA.

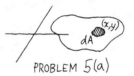

PROBLEM 5(a)

(b) $f(x, y)$ is people per unit area, i.e., population density. If $f(2, 3) = 8$ it does NOT mean that 8 people live (all squashed together) at point $(2, 3)$. It means that the pop density at $(2, 3)$ is 8 people per (say) square mile.

6. Consider small subregion containing point (x, y) with area dA. Density is $f(x, y)$ people per square mile so number of people in small region is $f(x, y)\, dA$ and

$$d\text{disease} = \frac{f(x, y)\, dA}{\sqrt{(x - 8)^2 + y^2}},$$

total disease $= \int_{\text{region}} \frac{f(x, y)\, dA}{\sqrt{(x - 8)^2 + y^2}}$

$$= \int_{y=0}^{4} \int_{x=y/2}^{x=8-3y/2} \frac{f(x, y)}{\sqrt{(x - 8)^2 + y^2}}\, dx\, dy.$$

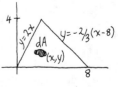

PROBLEM 6

7. *Method 1:* Consider a strip. By similar triangles, $\dfrac{\text{strip height}}{5 - x} = \dfrac{5}{5}$; strip height $= 5 - x$,

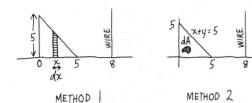

METHOD 1   METHOD 2

PROBLEM 7

area of strip $= (5 - x)\,dx$; $d$heat $= \dfrac{(5 - x)\,dx}{8 - x}$;

total heat $= \displaystyle\int_0^5 \frac{5 - x}{8 - x}\,dx = \int_0^5 \left(1 - \frac{3}{8 - x}\right)dx$

(divide out) $= (x + 3\ln(8 - x))\big|_0^5$

$= 5 + 3\ln 3 - 3\ln 8$.

   Method 2: Consider a small subregion with area dA, containing point (x, y).

$d$heat $= \dfrac{dA}{8 - x}$, heat $= \displaystyle\int_{\text{region}} \frac{dA}{8 - x}$

$= \displaystyle\int_{x=0}^{x=5}\int_{y=0}^{y=5-x} \frac{1}{8 - x}\,dy\,dx = \int_0^5 \frac{5 - x}{8 - x}\,dx$ etc. (as above).

8. Consider a small subregion containing point (x, y) with area dA. Distance to tracks is $x + 2$, distance to TD is $7 - y$; $d$cost $= \dfrac{x + 2}{7 - y}\,dA$;

total cost $= \displaystyle\int_{\text{land}} \frac{x + 2}{7 - y}\,dA = \int_{x=0}^{x=6}\int_{y=0}^{(6-x)/2} \frac{x + 2}{7 - y}\,dy\,dx.$

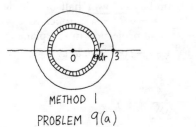

PROBLEM 8

9. (a) *Method 1:* Consider a circular ring with radius r, thickness dr. Area is $2\pi r\,dr$; d(energy) $= 2\pi r\,dr/r = 2\pi\,dr$; total energy $= \int_0^3 2\pi\,dr = 6\pi$.

   Method 2: Consider a small region with area dA containing point r, θ; d(energy) $= dA/r$,

energy $= \displaystyle\int_{\text{region}}(1/r)\,dA = \int_{\theta=0}^{2\pi}\int_{r=0}^{3}(1/r)\,r\,dr\,d\theta$

$= 6\pi$.

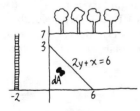

METHOD 1
PROBLEM 9(a)

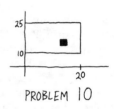

METHOD 2
PROBLEM 9(a)

(b) Like part (a), but dist to heat source is $\sqrt{25 + r^2}$ not r so heat $= \displaystyle\int_0^3 \frac{2\pi r\,dr}{\sqrt{25 + r^2}}$, also

$\displaystyle\int_{\theta=0}^{2\pi}\int_{r=0}^{3} \frac{1}{\sqrt{25 + r^2}}\,r\,dr\,d\theta.$

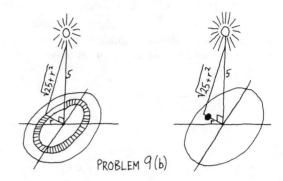

PROBLEM 9(b)

10. Consider a small subregion containing (x, y) with area dA. Height above ground is y, distance to ladder is x, $d$cost $= xy^2\,dA$, cost $= \int_{x=0}^{20}\int_{y=10}^{25} xy^2\,dy\,dx$

$= \int_0^{20} \frac{1}{3}xy^3\big|_{y=10}^{25}\,dx = \frac{1}{3}(25^3 - 1000)\cdot\frac{1}{2}x^2\big|_0^{20}$

$= \frac{200}{3}(25^3 - 1000)$.

PROBLEM 10

Section 12.6 (page 397)

All problems have diagrams except 2(a), 3(d), and (5).

1. (a) Lower boundary is plane $z = 0$, upper is plane $z = 1$. Use x, y projection.

$\int x^2\,dV = \int_{x=0}^{1}\int_{y=0}^{y=1-x}\int_{z=0}^{2} x^2\,dz\,dy\,dx$.

Left boundary is plane $y = 0$, right is $x + y = 1$. Use x, z projection.

$\int x^2\,dV = \int_{x=0}^{1}\int_{z=0}^{2}\int_{y=0}^{1-x} x^2\,dy\,dz\,dx$.

Rear boundary is plane $x = 0$, forward is plane $x + y = 1$. Use y, z projection

$\int x^2\,dV = \int_{z=0}^{2}\int_{y=0}^{1}\int_{x=0}^{1-y} x^2\,dx\,dy\,dz$.

We used version $dy\,dz\,dx$ and got answer $1/6$.

PROBLEM 1(a)

(b) Lower boundary is plane $z = 0$, upper is $z = 5$. Use x, y projection.
$\int x^2 z\, dV = \int_{\theta=0}^{2\pi} \int_{r=0}^{2} \int_{z=0}^{5} r^2 \cos^2\theta \cdot z \cdot r\, dz\, dr\, d\theta$.
Left and right boundaries are the cylinder. Use x, z projection.
$\int x^2 z\, dV = \int_{z=0}^{5} \int_{x=-2}^{2} \int_{y=-\sqrt{4-x^2}}^{y=\sqrt{4-x^2}} x^2 z\, dy\, dx\, dz$. Also
$\int x^2 z\, dV = \int_{z=0}^{5} \int_{y=-2}^{2} \int_{x=-\sqrt{4-y^2}}^{x=\sqrt{4-y^2}} x^2 z\, dx\, dy\, dz$.

Use version $dz\, dr\, d\theta$ (antiderivative of $\cos^2\theta$ is $\frac{1}{2}\theta + \frac{1}{4}\sin 2\theta$ from tables) to get answer 50π.

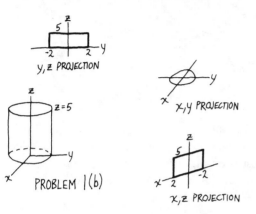

y, z PROJECTION

x, y PROJECTION

x, z PROJECTION

PROBLEM 1(b)

2. (a) Lower boundary is plane $z = 0$, upper is plane ABC which has equ $x/2 + y/3 + z/4 = 1$, $z = \frac{1}{3}(12 - 6x - 4y)$. Use projection in x, y plane. Line BC is $x/2 + y/3 = 1$, $3x + 2y = 6$.
$\int_{x=0}^{2} \int_{y=0}^{(6-3x)/2} \int_{z=0}^{z=(12-6x-4y)/3} f(x, y, z)\, dz\, dy\, dx$.
Also, rear boundary is plane $x = 0$, forward is plane ABC. Project into y, z plane where line AB has equation $y/3 + z/4 = 1$.
$\int_{y=0}^{3} \int_{z=0}^{z=(12-4y)/3} \int_{x=0}^{x=(12-4y-3z)/6} f(x, y, z)\, dx\, dz\, dy$.

(b) Can use polar coords and z. Lower boundary is lower half of sphere where
$z = -\sqrt{R^2 - x^2 - y^2} = -\sqrt{R^2 - r^2}$;
upper boundary is $z = \sqrt{R^2 - r^2}$. Projection in x, y plane is circular with radius R.
$\int_{\theta=0}^{2\pi} \int_{r=0}^{R} \int_{z=-\sqrt{R^2-r^2}}^{z=\sqrt{R^2-r^2}} f(r\cos\theta, r\sin\theta, z)\, r\, dz\, dr\, d\theta$.

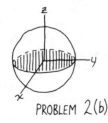

PROBLEM 2(b)

(c) See Example 2.
$\int_{\theta=0}^{2\pi} \int_{r=0}^{R} \int_{z=rh/R}^{z=h} f(r\cos\theta, r\sin\theta, z)\, r\, dz\, dr\, d\theta$.
Can also project into y, z plane. Cone has equation
$z = \frac{rh}{R} = \frac{h}{R}\sqrt{x^2 + y^2}$,

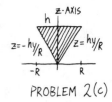

PROBLEM 2(c)

$z^2 = \frac{h^2}{R^2}(x^2 + y^2)$.
Rear boundary is $x = -\sqrt{R^2 z^2/h^2 - y^2}$,
forward boundary is $x = \sqrt{R^2 z^2/h^2 - y^2}$.
$\int_{z=0}^{h} \int_{y=-Rz/h}^{y=Rz/h} \int_{x=-\sqrt{R^2z^2/h^2-y^2}}^{x=\sqrt{R^2z^2/h^2-y^2}} f(x, y, z)\, dx\, dy\, dz$.

(d) Lower boundary is plane $z = 0$, upper is plane $z = h$. Project into x, y plane.
$\int_{\theta=0}^{\pi/2} \int_{r=0}^{R} \int_{z=0}^{h} f(r\cos\theta, r\sin\theta, z)\, r\, dz\, dr\, d\theta$.
Also, rear boundary is plane $x = 0$, forward boundary is
$x = \sqrt{R^2 - y^2}$. Project into y, z plane.
$\int_{z=0}^{h} \int_{y=0}^{R} \int_{x=0}^{x=\sqrt{R^2-y^2}} f(x, y, z)\, dx\, dy\, dz$.

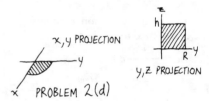

x, y PROJECTION

y, z PROJECTION

PROBLEM 2(d)

(e) Upper boundary is not clear (two surfaces are involved). Instead, rear boundary is plane $x = 0$, forward is cylinder $x^2 + z^2 = 9$, $x = \sqrt{9 - z^2}$. Can use polar coords for projection in y, z plane so that $y = r\cos\theta$, $z = r\sin\theta$.
$\int_{\theta=0}^{\pi/2} \int_{r=0}^{3} \int_{x=0}^{x=\sqrt{9-r^2\sin^2\theta}} f(x, r\cos\theta, r\sin\theta)\, r\, dx\, dr\, d\theta$.

PROBLEM 2(e)

(f) Left boundary is plane $y = 0$, right boundary is plane $y = h$. Use polar coords for circular projection in x, z plane so that $x = r\cos\theta$, $z = r\sin\theta$.
$\int_{\theta=0}^{2\pi} \int_{r=0}^{R} \int_{y=0}^{h} f(r\cos\theta, y, r\sin\theta)\, r\, dy\, dr\, d\theta$.
Also, lower boundary is lower half of cylinder $x^2 + z^2 = R^2$, $z = -\sqrt{R^2 - x^2}$; upper is $z = \sqrt{R^2 - x^2}$. Use projection in x, y plane:
$\int_{y=0}^{h} \int_{x=-R}^{R} \int_{z=-\sqrt{R^2-x^2}}^{\sqrt{R^2-x^2}} f(x, y, z)\, dz\, dx\, dy$.

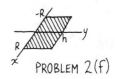

PROBLEM 2(f)

(g) See Example 3 of Section 12.4. Lower boundary is $z = 2x^2 + y^2$, upper is $z = 4 - y^2$. Can use polar coords for projection.
$$\int_{\theta=0}^{2\pi} \int_{r=0}^{\sqrt{2}} \int_{z=2r^2\cos^2\theta+r^2\sin^2\theta}^{z=4-r^2\sin^2\theta} f(r\cos\theta, r\sin\theta, z)\, r\, dz\, dr\, d\theta.$$
Also, rear boundary and forward boundary are both $z = 2x^2 + y^2$. Use proj in y, z plane.
$$\int_{y=-\sqrt{2}}^{\sqrt{2}} \int_{z=y^2}^{z=4-y^2} \int_{x=-\sqrt{(z-y^2)/2}}^{\sqrt{(z-y^2)/2}} f(x, y, z)\, dx\, dz\, dy$$

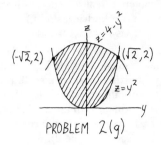

PROBLEM 2(g)

(h) Lower boundary is $z = x^2$, upper is $z = 5$. Use x, y projection; $\int_{x=-\sqrt{5}}^{\sqrt{5}} \int_{y=-2}^{3} \int_{z=x^2}^{z=5} f(x, y, z)\, dz\, dy\, dx$.
Also, rear boundary and forward boundary are $z = x^2$. Use y, z projection.
$$\int_{y=-2}^{3} \int_{z=0}^{5} \int_{x=-\sqrt{z}}^{x=\sqrt{z}} f(x, y, z)\, dx\, dz\, dy.$$
Also, left boundary is plane $y = -2$, right is plane $y = 3$. Use x, z projection.
$$\int_{x=-\sqrt{5}}^{\sqrt{5}} \int_{z=x^2}^{z=5} \int_{y=-2}^{y=3} f(x, y, z)\, dy\, dz\, dx.$$

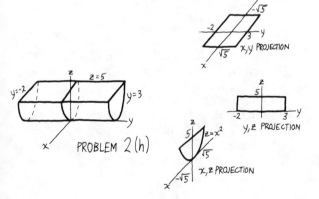

PROBLEM 2(h)

(i) Lower and upper boundaries are the ellipsoid. Projection in x, y plane is bounded by ellipse $x^2 + 2y^2 = 12$.
$$\int_{y=-\sqrt{6}}^{\sqrt{6}} \int_{x=-\sqrt{12-2y^2}}^{\sqrt{12-2y^2}} \int_{z=-\sqrt{(12-x^2-2y^2)/3}}^{\sqrt{(12-x^2-2y^2)/3}} f(x, y, z)\, dz\, dx\, dy.$$

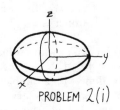

PROBLEM 2(i)

(j) Lower boundary is plane $z = 0$, upper is plane $z = x$. Project into x, y plane.
$$\int_{y=0}^{2} \int_{x=0}^{1} \int_{z=0}^{z=x} f(x, y, z)\, dz\, dx\, dy.$$
Also, rear boundary is plane $x = z$, forward is plane $x = 1$.
$$\int_{z=0}^{1} \int_{y=0}^{2} \int_{x=z}^{1} f(x, y, z)\, dx\, dy\, dz.$$
Also, left boundary is plane $y = 0$, right is plane $y = 2$.
$$\int_{x=0}^{1} \int_{z=0}^{z=x} \int_{y=0}^{y=2} f(x, y, z)\, dy\, dz\, dx.$$

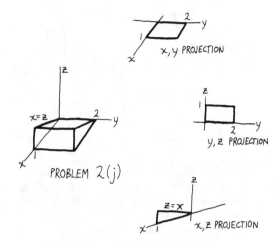

PROBLEM 2(j)

(k) Lower boundary is plane $z = 0$, upper is plane $y + z = 5$. Can use polar coords for x, y projection.
$$\int_{\theta=0}^{2\pi} \int_{r=0}^{1} \int_{z=0}^{z=5-r\sin\theta} f(r\cos\theta, r\sin\theta, z)\, r\, dz\, dr\, d\theta.$$
Also, rear and forward boundaries are the cyl $x^2 + y^2 = 1$.
$$\int_{y=-1}^{1} \int_{z=0}^{z=5-y} \int_{x=-\sqrt{1-y^2}}^{\sqrt{1-y^2}} f(x, y, z)\, dx\, dz\, dy.$$

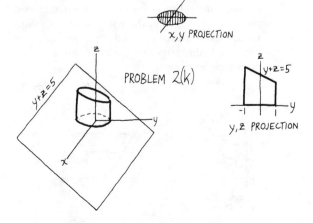

PROBLEM 2(k)

(l) Lower boundary is plane $z = 0$, upper is plane $z = x$. Can use polar coords for projection in x, y plane
$$\int_{\theta=0}^{\pi/2} \int_{r=0}^{1} \int_{z=0}^{r\cos\theta} f(r\cos\theta, r\sin\theta, z)\, r\, dz\, dr\, d\theta.$$

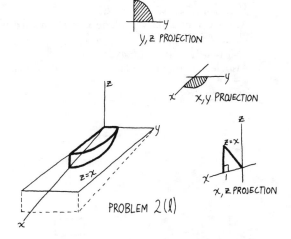

PROBLEM 2(ℓ)

Also, left boundary is plane $y = 0$, right is cyl $x^2 + y^2 = 1$;

$\int_{x=0}^{1} \int_{z=0}^{x} \int_{y=0}^{y=\sqrt{1-x^2}} f(x, y, z) \, dy \, dz \, dx.$

Rear boundary is plane $z = x$, forward is cyl. Projection in y, z plane is not obvious. Plane and cyl intersect to give $z^2 = 1 - y^2$. So projection is bounded by circle $y^2 + z^2 = 1$. Can use polar coords with $y = r \cos \theta$, $z = r \sin \theta$.

$\int_{\theta=0}^{\pi/2} \int_{r=0}^{1} \int_{x=z=r\sin\theta}^{x=\sqrt{1-y^2}=\sqrt{1-r^2\cos^2\theta}} f(x, r \cos \theta, r \sin \theta)$
$r \, dx \, dr \, d\theta.$

(m) Let cone height be H. Then $\dfrac{H}{R_1} = \dfrac{H-h}{R_2}$,

$H = \dfrac{R_1 h}{R_1 - R_2}$. See Example 2. Equation of cone

is $z = rH/R_1$ in cyl coords and $z^2 = \dfrac{H^2}{R_1^2}(x^2 + y^2)$

in rect coords. Lower boundary is *two* surfaces, plane $z = H - h$ and the cone, so with this approach must divide up solid. Instead, rear and forward boundaries are the cone surface. Use projection in y, z plane.

$\int_{z=H-h}^{z=H} \int_{y=-R_1 z/H}^{y=R_1 z/H} \int_{x=-\sqrt{R_1^2 z^2/H^2 - y^2}}^{x=\sqrt{R_1^2 z^2/H^2 - y^2}} f(x, y, z) \, dx \, dy \, dz.$

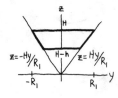

PROBLEM 2(m)

3. (a) Use cyl coords. Consider a small subregion with volume dV, at point r, θ, z.

(i) Density in subregion is r, $d\text{mass} = r \, dV$,

$\text{mass} = \int_{\text{solid cyl}} r \, dV$
$= \int_{\theta=0}^{2\pi} \int_{r=0}^{R} \int_{z=0}^{h} rr \, dz \, dr \, d\theta = \frac{2}{3} \pi R^3 h.$

PROBLEM 3(a)

(Problem can also be done with a single integral and cylindrical shells.)

(ii) Density is z, $d\text{mass} = z \, dV$.

$\text{mass} = \int_{\theta=0}^{2\pi} \int_{r=0}^{R} \int_{z=0}^{h} zr \, dz \, dr \, d\theta = \frac{1}{2} \pi h^2 R^2.$
(Can also use a single integral and slabs.)

(b) Use cyl coords. Consider a small piece with volume dV containing point r, θ, z. Distance to line of revolution is r, mass is $\delta \, dV$, $d\text{moment} = \delta \, dV \cdot r^2$. Lower boundary is cone surface where $z = rh/R$ (see Example 2), upper is $z = h$.
total moment
$= \int_{\text{solid}} \delta r^2 \, dV = \int_{\theta=0}^{2\pi} \int_{r=0}^{R} \int_{z=rh/R}^{z=h} \delta r^2 r \, dz \, dr \, d\theta$
$= \frac{1}{10} \pi h R^4 \delta.$
(Problem can also be done with a single integral using cyl shells.)

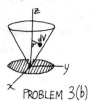

PROBLEM 3(b)

(c) Consider a little blob of liquid of volume dV at point r, θ, z. It has weight $2 \, dV$ and must move up $20 - z$ feet so
$d\text{work} = (20 - z) \cdot 2 \, dV = (40 - 2z) \, dV$,
total work $= \int_{\text{half-cyl}} (40 - 2z) \, dV$
$= \int_{\theta=0}^{2\pi} \int_{r=0}^{5} \int_{z=0}^{10} (40 - 2z) r \, dz \, dr \, d\theta = 7500\pi.$
(Problem can also be done with a single integral using slabs of water.)

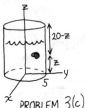

PROBLEM 3(c)

(d) Volume is $\int_{\text{solid}} dV$. Lower boundary is $z = 2x^2 + 2y^2$, upper is $z = 12$. When they intersect, $2x^2 + 2y^2 = 12$, $x^2 + y^2 = 6$; x, y projection is circular with radius $\sqrt{6}$. Use polar coords; vol $= \int_{\theta=0}^{2\pi} \int_{r=0}^{\sqrt{6}} \int_{z=2r^2}^{z=12} r \, dz \, dr \, d\theta$. (Problem can also be done with a double integral.)

4. Vol $= \int_{\text{solid}} dV$. Lower boundary is two surfaces, plane $z = 0$ and hyperboloid. Don't use this version.

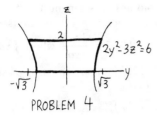

PROBLEM 4

Rear and forward boundaries are the hyperboloid. Use y, z projection. Then volume is

$$\int_{\text{solid}} dV = \int_{z=0}^{2} \int_{y=-\sqrt{(6+3z^2)/2}}^{\sqrt{(6+3z^2)/2}} \int_{x=-\sqrt{6+3z^2-2y^2}}^{x=\sqrt{6+3z^2-2y^2}} dx\, dy\, dz.$$

5. Not necessarily. It depends on the f. If the values of f in the missing hemisphere match the values of f in the hemisphere used, then OK. But otherwise NO. In other words, the sum of 100 terms of the form $f(x, y, z)\, dV$ is not the same as twice the sum of the first fifty terms unless the last fifty match the first fifty. For example, with sphere centered at origin, $\int_{\text{solid sphere}} z\, dV$ is 0 but $2 \int_{\text{top hemi}} z\, dV$ is not 0.

---

3. Let origin be center of sphere. Density is $1/\rho^2$ so mass $= \int (1/\rho^2)\, dV$

$= \int_{\theta=0}^{2\pi} \int_{\phi=0}^{\pi} \int_{\rho=0}^{R} (1/\rho^2)\rho^2 \sin\phi\, d\rho\, d\phi\, d\theta = 4\pi R.$

4. (a) Like Fig. 9 but with "radius" $\rho_0 = \infty$

(b) $\int_{\theta=0}^{2\pi} \int_{\phi=0}^{\pi/2} \int_{\rho=0}^{\infty}$

(c) $\int_{\theta=0}^{\pi} \int_{\phi=0}^{\pi} \int_{\rho=0}^{\infty}.$

5. (See fig.)

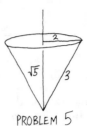

PROBLEM 5

Volume $= \int dV = \int_{\theta=0}^{2\pi} \int_{\phi=0}^{\phi_0} \int_{\rho=0}^{3} \rho^2 \sin\phi\, d\rho\, d\phi\, d\theta.$

Inner $= \frac{1}{3}\rho^3 \big|_0^3 = 9$; middle $= -9\cos\phi \big|_{\phi=0}^{\phi_0} = 9(1 - \cos\phi_0) = 9(1 - \frac{1}{3}\sqrt{5})$; outer $= 18\pi(1 - \frac{1}{3}\sqrt{5}).$

6. (See fig.) Inner boundary is $\rho = 0$; outer is plane $z = h$, $\rho\cos\phi = h$, $\rho = h\sec\phi$. For any point, distance to z-axis is $r = \rho\sin\phi.$

Moment $= \delta \int_{\theta=0}^{2\pi} \int_{\phi=0}^{\phi_0} \int_{\rho=0}^{\rho=h\sec\phi} (\rho\sin\phi)^2 \rho^2 \sin\phi\, d\rho\, d\phi\, d\theta.$

Section 12.7 (page 404)

1. Vol $= \int_{\text{sphere}} dV = \int_{\theta=0}^{2\pi} \int_{\phi=0}^{\pi} \int_{\rho=0}^{R} \rho^2 \sin\phi\, d\rho\, d\phi\, d\theta.$
Inner $= \frac{1}{3}\rho^3 \big|_0^R = \frac{1}{3}R^3$; middle $= -\frac{1}{3}R^3\cos\phi \big|_0^\pi = \frac{2}{3}R^3$; outer $= \frac{2}{3}R^3 \cdot 2\pi = \frac{4}{3}\pi R^3.$

2. (See fig.) Vol $= \int_{\text{cone}} dV.$ Inner boundary is $\rho = 0$, outer is plane $z = h$, $\rho\cos\phi = h$, $\rho = h/\cos\phi$ so vol $= \int_{\theta=0}^{2\pi} \int_{\phi=0}^{\phi_0} \int_{\rho=0}^{h/\cos\phi} \rho^2 \sin\phi\, d\rho\, d\phi\, d\theta.$

Inner $= \frac{1}{3}\rho^3 \big|_{\rho=0}^{h} = \dfrac{h^3}{3\cos^3\phi}$; middle $= \frac{1}{3}h^3 \int_{\phi=0}^{\phi_0} \dfrac{\sin\phi}{\cos^3\phi}\, d\phi.$

To get antideriv, substitute $u = \cos\phi$, $du = -\sin\phi\, d\phi.$

Get $\frac{1}{3}h^3 \dfrac{1}{2\cos^2\phi} \Big|_{\phi=0}^{\phi_0} = \frac{1}{6}h^3\left(\dfrac{1}{\cos^2\phi_0} - 1\right).$

From right triangle, $\cos\phi_0 = \dfrac{h}{\sqrt{h^2 + R^2}}$ so

middle $= \frac{1}{6}h^3\left(\dfrac{h^2 + R^2}{h^2} - 1\right) = \frac{1}{6}R^2 h.$

Outer $= \frac{1}{6}R^2 h \cdot 2\pi = \frac{1}{3}\pi R^2 h.$

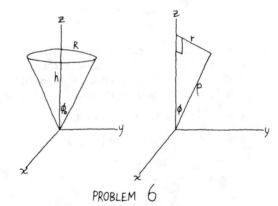

PROBLEM 6

7. (See fig.) Inner boundary is $z = 3$, $\rho\cos\phi = 3$,

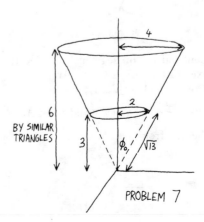

BY SIMILAR TRIANGLES

PROBLEM 7

PROBLEM 2

$\rho = 3/\cos\phi$; outer is $\rho = 6/\cos\phi$. Mass of a small piece is $(1/\rho)\,dV$ so total mass is

$\int (1/\rho)\,dV = \int_{\theta=0}^{2\pi}\int_{\phi=0}^{\phi_0}\int_{\rho=3/\cos\phi}^{6/\cos\phi}(1/\rho)\rho^2\sin\phi\,d\rho\,d\phi\,d\theta$;

inner = $27/2\cos^2\phi$; antideriv of $\dfrac{\sin\phi}{\cos^2\phi}$ is $\dfrac{1}{\cos\phi}$ so

middle = $\dfrac{27}{2}\left(\dfrac{1}{\cos\phi_0} - 1\right) = \dfrac{27}{2}(\tfrac{1}{3}\sqrt{13} - 1)$;

outer = $27\pi(\tfrac{1}{3}\sqrt{13} - 1)$.

Section 12.8 (page 408)

1. By physical considerations, the centroid lies on segment AB. (See fig.) Insert axes so that AB is the z-axis. Hemisphere volume = $\frac{2}{3}\pi R^3$, $\int z\,dV$

$= \int_{\theta=0}^{2\pi}\int_{\phi=0}^{\pi/2}\int_{\rho=0}^{R}\rho\cos\phi\,\rho^2\sin\phi\,d\rho\,d\phi\,d\theta = \frac{1}{4}\pi R^4$.

So $\bar z = \dfrac{\int z\,dV}{\text{vol}} = \frac{3}{8}R$. The centroid is on AB, $\frac{3}{8}$ths of the way from A to B.

PROBLEM 1

2. (a) Centroid is on segment CD. (See fig.) Insert axes so that CD is z-axis. Cone vol is $\frac{1}{3}\pi R^2 h$.
$\int z\,dV = \int_{\theta=0}^{2\pi}\int_{r=0}^{R}\int_{z=-rh/R}^{h}zr\,dz\,dr\,d\theta = \frac{1}{4}\pi h^2 R^2$
(See Example 2, Section 12.6 for limits.)
$\bar z = \dfrac{\int z\,dV}{\text{vol}} = \frac{3}{4}h$. Centroid is on CD,
$\frac{3}{4}$ths of the way from C to D.

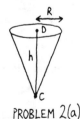

PROBLEM 2(a)

(b) Centroid lies on segment AB. (See fig.) Insert axes so that AB is y-axis. Area of semicircle is $\frac{1}{2}\pi R^2$; $\int y\,dA = \int_{\theta=0}^{\pi}\int_{r=0}^{R} r\sin\theta\,r\,dr\,d\theta = \frac{2}{3}R^3$,
$\bar y = \dfrac{\int y\,dA}{\text{area}} = \dfrac{4R}{3\pi}$. Centroid is on AB, $\frac{4}{3\pi}$-ths of the way from A to B.

PROBLEM 2(b)

3. (a) Centroid lies on segment AB. (See fig.) Use AB as z-axis. Then $\bar x = \bar y = 0$, and (see Example 2, Section 12.7) $\bar z = \dfrac{\int z\,dV}{\text{vol}}$ where $\int z\,dV$

$= \int_{\theta=0}^{2\pi}\int_{\phi=0}^{\phi_0}\int_{\rho=4\sqrt{2}\sec\phi}^{\rho=6}\rho\cos\phi\,\rho^2\sin\phi\,d\rho\,d\phi\,d\theta$,
vol = $\int dV$
$= \int_{\theta=0}^{2\pi}\int_{\phi=0}^{\phi_0}\int_{\rho=4\sqrt{2}\sec\phi}^{\rho=6}\rho^2\sin\phi\,d\rho\,d\phi\,d\theta$.

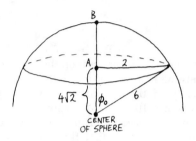

PROBLEM 3(a)

(b) (See fig.) $\bar x = \dfrac{\int x\,dA}{\text{area}} = \frac{1}{6}\int_{x=0}^{3}\int_{y=0}^{4-4x/3} x\,dy\,dx$,

$\bar y = \dfrac{\int y\,dA}{\text{area}} = \frac{1}{6}\int_{x=0}^{3}\int_{y=0}^{4-4x/3} y\,dy\,dx$

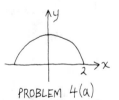

PROBLEM 3(b)

4. (a) With indicated axes (see fig.), $\bar x = 0$,
$\bar y = \dfrac{\int y \times \text{density}\,dA}{\int \text{density}\,dA}$
$= \dfrac{\int_{\theta=0}^{\pi}\int_{r=0}^{2}(r\sin\theta)(r^2)r\,dr\,d\theta}{\int_{\theta=0}^{\pi}\int_{r=0}^{2}r^2 r\,dr\,d\theta}$

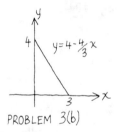

PROBLEM 4(a)

(b) With indicated axes, (see fig.) $\bar x = \bar y = 0$,
$\bar z = \dfrac{\int z \times \text{density}\,dV}{\int \text{density}\,dV}$. For each integral, limits are $\int_{\theta=0}^{2\pi}\int_{r=0}^{R}\int_{z=0}^{h}$ and $dV = r\,dz\,dr\,d\theta$. At an arbitrary point P in the solid cylinder, with cylindrical coords r, θ, z, the distance to the top of cylinder is $h - z$ and distance to z-axis is r. So in (i) density is $h - z$, in (ii) density is r.

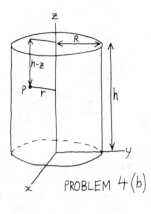

PROBLEM 4(b)

___

Chapter 12 Review Problems (page 408)

1. (a) (See fig.) Left and right boundaries are the circle; $\int_{y=-\sqrt{2}}^{y=1} \int_{x=-\sqrt{2-y^2}}^{x=\sqrt{2-y^2}} x^3 y \, dx \, dy$. Lower boundary is circle, upper is partly circle and partly line, divide up region (into three parts);

$$\int_{x=-\sqrt{2}}^{x=-1} \int_{y=-\sqrt{2-x^2}}^{\sqrt{2-x^2}} + \int_{x=-1}^{1} \int_{y=-\sqrt{2-x^2}}^{1}$$
$$+ \int_{x=1}^{\sqrt{2}} \int_{y=-\sqrt{2-x^2}}^{\sqrt{2-x^2}}$$

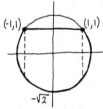

PROBLEM 1(a)

(b) (See fig.) Outer boundary is line and circle. Divide up region. Line is
$y = 1$, $r \sin \theta = 1$, $r = \csc \theta$;
$\int_{\theta=\pi/4}^{3\pi/4} \int_{r=0}^{\csc\theta} (r \cos \theta)^3 r \sin \theta \, r \, dr \, d\theta$
$+ \int_{\theta=3\pi/4}^{9\pi/4} \int_{r=0}^{\sqrt{2}} (r \cos \theta)^3 r \sin \theta \, r \, dr \, d\theta$

Warning: For the second integral you can use $\int_{\theta=-5\pi/4}^{\pi/4}$ but NOT $\int_{3\pi/4}^{\pi/4}$.

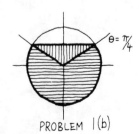

PROBLEM 1(b)

(c) (See fig.) $x^3 y \, dA$'s are positive in I and III and take on precisely the negatives of those values in II and IV so the sum of $x^3 y \, dA$'s is 0.

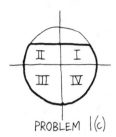

PROBLEM 1(c)

2. Not necessarily. It depends on the values of f. If values of f are higher in R_1 then $\int_{R_1} f(x,y) \, dA$ will be larger.

3. Area $= \int_a^b (u(x) - l(x)) \, dx = \int_{-1}^2 (x + 2 - x^2) \, dx$. Also area $= \int_{\text{region}} dA = \int_{x=-1}^2 \int_{y=x^2}^{y=x+2} dy \, dx$. (See fig.)

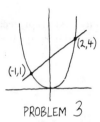

PROBLEM 3

4. (See fig.) $\int e^{-x^2-y^2} dA = \int_{\theta=0}^{2\pi} \int_{r=3}^{\infty} e^{-r^2} r \, dr \, d\theta$
$= \int_{\theta=0}^{2\pi} -\frac{1}{2} e^{-r^2} \big|_{r=3}^{\infty} \, d\theta = \int_{\theta=0}^{2\pi} \frac{1}{2} e^{-9} \, d\theta = \frac{1}{2} e^{-9} \cdot 2\pi$
$= \pi/e^9$.

PROBLEM 4

5. (See fig.) Consider a small piece of land with area dA containing point (x, y). Then $d\text{cost} = xy \, dA$,
$\text{cost} = \int_{\text{land}} xy \, dA = \int_{\theta=0}^{\pi/2} \int_{r=0}^{2} (r \cos \theta)(r \sin \theta) r \, dr \, d\theta$.

PROBLEM 5

6. (See fig.) Lower boundary is $z = 0$, upper is cyl $z = 16 - y^2$. Use x, y projection; $\int_{x=0}^3 \int_{y=-4}^4 \int_{z=0}^{z=16-y^2}$.
Rear is plane $x = 0$, forward is plane $x = 3$.
Use y, z projection; $\int_{y=-4}^4 \int_{z=0}^{16-y^2} \int_{x=0}^3$.
Left and right boundaries are the cyl. Use x, z proj.
$\int_{x=0}^3 \int_{z=0}^{16} \int_{y=-\sqrt{16-z}}^{\sqrt{16-z}}$.

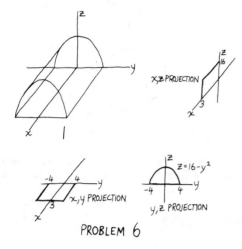

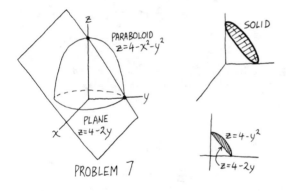

PROBLEM 6

7. (See fig.) The only obvious projection is in y, z plane. Rear and forward boundaries are the paraboloid.
$$\int_{y=0}^{2} \int_{z=4-2y}^{z=4-y^2} \int_{x=-\sqrt{4-y^2-z}}^{\sqrt{4-y^2-z}}.$$

PROBLEM 7

8. $3 \int dA = 3 \times$ area of region $= 3 \cdot 81\pi = 243\pi$.

9. (See fig.) (a) Region lies under $z = 10 - 3x^2 - 3y^2$ and over circular floor.
$$\text{vol} = \int_{\text{circ region}} (10 - 3x^2 - 3y^2)\, dA$$
$$= \int_{\theta=0}^{2\pi} \int_{r=0}^{2} (10 - 3r^2) r\, dr\, d\theta.$$
(b) $\text{vol} = \int_{\text{solid}} dV$
$$= \int_{\theta=0}^{2\pi} \int_{r=0}^{2} \int_{z=0}^{z=10-3x^2-3y^2=10-3r^2} r\, dz\, dr\, d\theta.$$

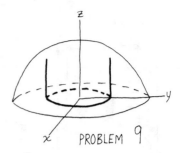

PROBLEM 9

10. (See fig.) Consider small piece of solid with vol dV at point with cyl coords r, θ, z and spher coords ρ, ϕ, θ. Its distance to z-axis (around which solid revolves) is r. Mass of small piece is $\delta\, dV$ so $d\text{moment} = r^2 \delta\, dV$. In cyl coords, lower and upper bdries are the sphere where $z = \pm\sqrt{R^2 - x^2 - y^2} = \pm\sqrt{R^2 - r^2}$. Projection in x, y plane is circular with radius R. So
$$\text{moment} = \int_{\text{solid}} r^2 \delta\, dV$$
$$= \delta \int_{\theta=0}^{2\pi} \int_{r=0}^{R} \int_{z=-\sqrt{R^2-r^2}}^{\sqrt{R^2-r^2}} r^2 r\, dz\, dr\, d\theta.$$
In spher coords, distance from small piece to z-axis is $\rho \sin \phi$ (see right triangle), $d\text{moment} = \rho^2 \sin^2\phi \delta\, dV$,
$$\text{moment} = \delta \int_{\theta=0}^{2\pi} \int_{\phi=0}^{\pi} \int_{\rho=0}^{R} \rho^2 \sin^2\phi\, \rho^2 \sin\phi\, d\rho\, d\phi\, d\theta.$$
The integration is easier in spherical coords. Use tables for $\int \sin^3\phi\, d\phi$ or use $\sin^3\phi = \sin\phi(1 - \cos^2\phi) = \sin\phi - \sin\phi \cos^2\phi$. To find $\int \sin\phi \cos^2\phi$ let $u = \cos\phi$, $du = -\sin\phi\, d\phi$. All in all,
$$\int \sin^3\phi\, d\phi = -\cos\phi + \tfrac{1}{3}\cos^3\phi.$$
Final answer is $\frac{8}{15}\pi R^5 \delta$.

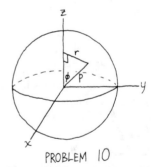

PROBLEM 10

APPENDIX

Solutions Section A1 (page 410)

1. (a) $m = \frac{1}{3}$, dist $= \sqrt{36 + 4} = \sqrt{40}$.
   (b) $m = 0$, dist is 2.
   (c) $m = \frac{1}{2}$, dist $= \sqrt{4 + 1} = \sqrt{5}$.

2. Slope of $AB = -2$, slope of $CD = \dfrac{y - 1}{-8}$.

   (a) Need $(-2)\left(\dfrac{y - 1}{-8}\right) = -1, y = -3$.

   (b) Need $\dfrac{y - 1}{-8} = -2, y = 17$.

Section A2 (page 411)

1. (a) $m = -7/6$, $y - 5 = -\frac{7}{6}(x - 1)$, $7x + 6y = 37$
   (b) Given line is $y = 3x + 7/2$, $m = 3$. Parallel line is $y - 7 = 3(x - 1)$, $3x - y + 4 = 0$.
   (c) Given line has $m = 3$. Perp line has $m = -1/3$, $y - 8 = -\frac{1}{3}(x + 2)$.

2. (a) $y = \frac{2}{5}x + \frac{8}{5}$, $m = \frac{2}{5}$. (b) $y = -\frac{4}{3}x + 4$, $m = -\frac{4}{3}$.

3. (See figs.) (a) If $x = 0$ then $y = -4/3$. If $y = 0$ then $x = 4$.

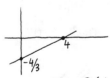

PROBLEM 3(a)

(b) $m = 2$, y-intercept 1

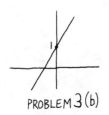

PROBLEM 3(b)

Appendix A3 (page 412)

1.

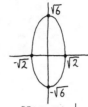

PROBLEM 1

2.

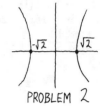

PROBLEM 2

3.

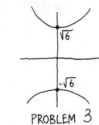

PROBLEM 3

4.

PROBLEM 4

5.

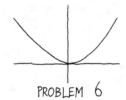

PROBLEM 5

6.

PROBLEM 6

7.

PROBLEM 7

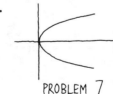

Section A4 (page 413)

1. Next two lines in Pascal's triangle are
$1, 6, 15, 20, 15, 6, 1$ and $1, 7, 21, 35, 35, 21, 7, 1$;
$(x + y)^7 = x^7 + 7x^6y + 21x^5y^2 + 35x^4y^3$
$$+ 35x^3y^4 + 21x^2y^5 + 7xy^6 + y^7.$$

2. $(2p)^4 = 4(2p)^3q + 6(2p)^2q^2 + 4(2p)q^3 + (2p)q^4$
$= 16p^4 + 32p^3q + 24p^2q^2 + 8pq^3 + 2pq^4$

3. $\dfrac{14 \times 13 \times 12 \times 11 \times \cdots \times 4 \text{ (eleven factors)}}{11!}$

$= \dfrac{14 \times 13 \times 12}{1 \times 2 \times 3} = 364$. Alternatively, can use coeff of

$x^{11}y^3$ to get $\dfrac{14 \times 13 \times 12}{3!}$ directly.

4. Use Pascal line $1, 6, 15, 20, 15, 6, 1$;
$1^6 + 6(1)^5(-x) + 15(1)^4(-x)^2 + 20(1)^3(-x)^3 +$
$15(1)^2(-x)^4 + 6(1)(-x)^5 + (-x)^6$
$= 1 - 6x + 15x^2 - 20x^3 + 15x^4 - 6x^5 + x^6$.

5. $\dfrac{11 \times 10}{1 \times 2} = 55$

Section A5 (page 415)

1. $10 - (-12) = 22$

2. (a) $-4\begin{vmatrix} 2 & 3 \\ -3 & 7 \end{vmatrix} + 5\begin{vmatrix} 10 & 3 \\ 1 & 7 \end{vmatrix} - -6\begin{vmatrix} 10 & 2 \\ 1 & -3 \end{vmatrix}$

$= -4(23) + 5(67) + 6(-32) = 51$

(b) $3\begin{vmatrix} 4 & 5 \\ 1 & -3 \end{vmatrix} - -6\begin{vmatrix} 10 & 2 \\ 1 & -3 \end{vmatrix} + 7\begin{vmatrix} 10 & 2 \\ 4 & 5 \end{vmatrix}$

$= 3(-17) + 6(-32) + 7(42) = 51$

3. $3\begin{vmatrix} 0 & 3 & 1 \\ 2 & 1 & 3 \\ 1 & 1 & 2 \end{vmatrix} - -1\begin{vmatrix} 1 & -1 & 4 \\ 0 & 3 & 1 \\ 2 & 1 & 3 \end{vmatrix}$

$= 3(-2) + (-18) = -24$

Appendix A6 (page 418)

1. (a) $x = 3\cos 60° = 3/2$, $y = 3\sin 60° = \frac{3}{2}\sqrt{3}$
(b) By inspection (see fig.), $x = 0, y = 2$ (it's OK but overkill to use the formulas).

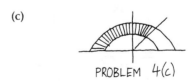

PROBLEM 1(b)

(c) $x = 3\cos(-\pi/4) = \frac{3}{2}\sqrt{2}$.
$y = 3\sin(-\pi/4) = -\frac{3}{2}\sqrt{2}$.
(d) $x = 2, y = 0$ by inspection. (See fig.)

PROBLEM 1(d)

2. (a) $r = \sqrt{4 + 16} = \sqrt{20}$, $\tan^{-1}y/x = \tan^{-1}2 = 63°$ approx. Fig. shows that θ is in quad I so θ is 63°.

PROBLEM 2(a)

(b) $r = \sqrt{20}$, $\tan^{-1}y/x = 63°$ again. This time $\theta = 63° + 180° = 243°$ because fig. shows θ in the third quad.

PROBLEM 2(b)

(c) $r = \sqrt{18} = 3\sqrt{2}$, $\theta = 225°$ by inspection (the ray bisects quad III).
(d) (See fig.) $r = 3$, $\theta = 3\pi/2$ (or $-\pi/2$) by inspection.

PROBLEM 2(d)

(e) $r = \sqrt{20}$, $\tan^{-1}y/x = \tan^{-1}(-2) = -63°$. Point is in quad II so θ is not $-63°$ but is $-63° + 180° = 117°$.

3. $r\cos\theta = 3$

4. (a)

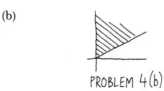

PROBLEM 4(a)

(b)

PROBLEM 4(b)

(c)

PROBLEM 4(c)

5. (a) The rect coords are $x_1 = r_1 \cos \theta_1$, $y_1 = r_1 \sin \theta_1$ and $x_2 = r_2 \cos \theta_2$, $y_2 = r_2 \sin \theta_2$.
By distance formula for rect coords, distance is

$$\sqrt{(r_1 \cos \theta_1 - r_2 \cos \theta_2)^2 + (r_1 \sin \theta_1 - r_2 \sin \theta_2)^2}$$

$$= \sqrt{r_1^2 \cos^2\theta_1 - 2r_1r_2 \cos \theta_1 \cos \theta_2 + r_2^2 \cos^2\theta_2 + r_1^2 \sin^2\theta_1 - 2r_1r_2 \sin \theta_1 \sin \theta_2 + r_2^2 \sin^2\theta_2}$$

$$= \sqrt{r_1^2 \underbrace{(\cos^2\theta_1 + \sin^2\theta_1)}_{1} + r_2^2 \underbrace{(\cos^2\theta_2 + \sin^2\theta_2)}_{1} - 2r_1r_2 \underbrace{(\cos \theta_1 \cos \theta_2 + \sin \theta_1 \sin \theta_2)}_{\cos(\theta_2 - \theta_1)}}$$

(b) (See fig. which assumes $\theta_2 > \theta_1$.) By law of cosines,
distance$^2 = r_1^2 + r_2^2 - 2r_1r_2 \cos(\theta_2 - \theta_1)$.

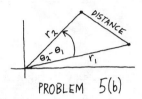

PROBLEM 5(b)

6. (See figs.) (a)

θ	0	$\pi/6$	$\pi/4$	$\pi/3$	$\pi/2$	120°	135°	150°	180°	210°	235°	240°	270°	300°	315°	330°	360°
$\sin \theta$	0	$\frac{1}{2}$	$\frac{1}{2}\sqrt{2}$	$\frac{1}{2}\sqrt{3}$	1	$\frac{1}{2}\sqrt{3}$	$\frac{1}{2}\sqrt{2}$	$\frac{1}{2}$	0	$-\frac{1}{2}$	$-\frac{1}{2}\sqrt{2}$	$-\frac{1}{2}\sqrt{3}$	-1	$-\frac{1}{2}\sqrt{3}$	$-\frac{1}{2}\sqrt{2}$	$-\frac{1}{2}$	0
r	2	1	$2-\sqrt{2}$	$2-\sqrt{3}$	0	$2-\sqrt{3}$	$2-\sqrt{2}$	1	2	3	$2+\sqrt{2}$	$2+\sqrt{3}$	4	$2+\sqrt{3}$	$2+\sqrt{2}$	3	2

Using $\theta < 0$ or $\theta > 360°$ doesn't produce any new points. For instance $\theta = -30°$, $r = 3$
is just point $\theta = 330°$, $r = 3$ again. And point $\theta = 390°$, $r = 1$ is just point $\theta = \pi/6$,
$r = 1$ again.

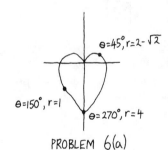

PROBLEM 6(a)

(b)

θ	0	10°	15°	20°	30°		90°	100°	105°	110°	120°	130°	140°
r	2	$\sqrt{3}$	$\sqrt{2}$	1	0	neg, no points	0	1	$\sqrt{2}$	$\sqrt{3}$	2	$\sqrt{3}$	1

150°		210°	220°	225°	230°	240°	250°	255°	260°
0	neg, no points	0	1	$\sqrt{2}$	$\sqrt{3}$	2	$\sqrt{3}$	$\sqrt{2}$	1

270°		330°	340°	345°	350°	360°
0	neg, no points	0	1	$\sqrt{2}$	$\sqrt{3}$	2

Using $\theta < 0$ and $\theta > 360°$ produces no new points in this problem.

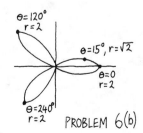

PROBLEM 6(b)

(c)

θ	$0° \to 45°$	$45° \to 90°$	$90° \to 180°$	$180° \to 270°$	$270° \to 360°$
r^2	$0 \to 4$	$4 \to 0$	neg	$0 \to 4 \to 0$	neg
r	$0 \to 2$	$2 \to 0$	impossible	$0 \to 2 \to 0$	impossible

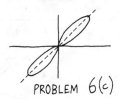

PROBLEM 6(c)

(d)

θ	$0° \to 180°$	$180° \to 360°$	$360° \to 720°$
r	$0 \to 4$	$4 \to 0$	neg, no points

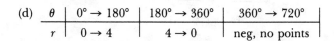

PROBLEM 6(d)

(e)

θ	0	$\pi/4$	$\pi/2$	π	$3\pi/2$	2π
r	0	$\pi/4$(approx. $\frac{3}{4}$)	$\pi/2$(approx. $\frac{3}{2}$)	π	$3\pi/2$	2π

θ	$5\pi/2$	3π	$\cdots$	negative
r	$5\pi/2$(about $7\frac{1}{2}$)	3π		neg, no points

Note that $\theta > 2\pi$ DOES produce new points here while $\theta < 0$ produces negative r so produces non-points.

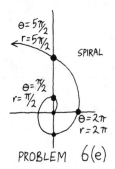

PROBLEM 6(e)

Abbreviations Used in the Solutions

abs conv	absolutely convergent	inf disc	infinite discontinuity
abs value	absolute value	lim	limit
alt	alternating	mag	magnitude
antideriv	antiderivative	max	maximum
antidiff	antidifferentiate	min	minimum
av	average	neg	negative
ccl	counterclockwise	1-1	one-to-one
cond conv	conditionally convergent	orthog	orthogonal
conv	converges	perp	perpendicular
coord	coordinate	pos	positive
cyl	cylinder, cylindrical	prod	product
dec	decreasing	proj	projection
denom	denominator	quot	quotient
deriv	derivative	rect	rectangular
det	determinant	rel	relative
diff	differentiate	sol	solution
dist	distance	sub	substitute, substitution
div	diverges	tan	tangent
eq, equ	equation	temp	temperature
iff	if and only if	vol	volume
inc	increasing	w.r.t.	with respect to
indet	indeterminate		

LIST OF SYMBOLS

antiderivative $\int$, 85

change Δ, 53

derivative f', $f'(x)$, $\dfrac{df}{dx}$, $\dfrac{d}{dx}f(x)$, $D_x f$, Df, y', $\dfrac{dy}{dx}$, 57

differentials dx, dy, 123

directional derivative $D_{\vec{u}}f$, $\dfrac{\partial f}{\partial \vec{u}}$, 346

double integral $\iint$, 373

factorial $n!$, 224, 413

infinity ∞, 3, 42

integral $\int_a^b$, 140

interval $[a, b]$, (a, b), $(-\infty, b]$ etc., 3

limit $\lim_{x \to a}$, 41

partial derivative ∂, 325

substitution $\big|_a^b$, 146

sum Σ, 138

triple integral $\iiint$, 393

INDEX

BIOGRAPHIES

Carol Ash has an M.A. in mathematics from the University of California, Berkeley, 1963. She is an instructor in the Department of Mathematics at the University of Illinois at Urbana-Champaign, teaching courses in calculus, computer calculus, advanced calculus, differential equations, linear algebra, discrete mathematics, and engineering mathematics. She has published several articles on the teaching of mathematics.

Robert Ash received his Ph.D. degree in electrical engineering in 1960 from Columbia University, and subsequently became a mathematician. After teaching at Columbia and at the University of California, Berkeley, he arrived at the University of Illinois at Urbana-Champaign in 1963, where he is now a professor of mathematics. He has taught and written in many areas, and is the author of textbooks on information theory, probability, complex variables, real analysis and stochastic processes.

FIG. 30026

RESULT OF $(x^3 + \vec{u} + \vec{v})$

(Continued from inside front cover)

Forms Involving $\sqrt{a^2 \pm u^2}$

19. $\displaystyle\int \frac{du}{\sqrt{a^2 - u^2}} = \sin^{-1}\frac{u}{a} + C$

20. $\displaystyle\int \sqrt{a^2 - u^2}\, du = \frac{u}{2}\sqrt{a^2 - u^2} + \frac{a^2}{2}\sin^{-1}\frac{u}{a} + C$

21. $\displaystyle\int \frac{\sqrt{a^2 - u^2}}{u}\, du = \sqrt{a^2 - u^2} - a\,\ln\left|\frac{a + \sqrt{a^2 - u^2}}{u}\right| + C$

22. $\displaystyle\int \frac{du}{u\sqrt{a^2 - u^2}} = -\frac{1}{a}\ln\left|\frac{a + \sqrt{a^2 - u^2}}{u}\right| + C$

23. $\displaystyle\int \frac{du}{\sqrt{a^2 + u^2}} = \ln(u + \sqrt{a^2 + u^2}) + C$

24. $\displaystyle\int \sqrt{a^2 + u^2}\, du = \tfrac{1}{2}u\sqrt{a^2 + u^2} + \tfrac{1}{2}a^2\ln(u + \sqrt{a^2 + u^2}) + C$

25. $\displaystyle\int \frac{\sqrt{a^2 + u^2}}{u}\, du = \sqrt{a^2 + u^2} - a\,\ln\left|\frac{a + \sqrt{a^2 + u^2}}{u}\right| + C$

26. $\displaystyle\int \frac{du}{u\sqrt{a^2 + u^2}} = -\frac{1}{a}\ln\left|\frac{\sqrt{a^2 + u^2} + a}{u}\right| + C$

Forms Involving $\sqrt{u^2 - a^2}$

27. $\displaystyle\int \frac{du}{\sqrt{u^2 - a^2}} = \ln|u + \sqrt{u^2 - a^2}| + C$

28. $\displaystyle\int \sqrt{u^2 - a^2}\, du = \tfrac{1}{2}u\sqrt{u^2 - a^2} - \tfrac{1}{2}a^2\ln|u + \sqrt{u^2 - a^2}| + C$

29. $\displaystyle\int \frac{\sqrt{u^2 - a^2}}{u}\, du = \sqrt{u^2 - a^2} - a\,\cos^{-1}\frac{a}{u} + C$

30. $\displaystyle\int \frac{du}{u\sqrt{u^2 - a^2}} = \frac{1}{a}\cos^{-1}\frac{a}{u} + C$

Trigonometric Forms

31. $\displaystyle\int \tan u\, du = -\ln|\cos u| + C = \ln|\sec u| + C$

32. $\displaystyle\int \cot u\, du = \ln|\sin u| + C$

33. $\displaystyle\int \sec u\, du = \ln|\sec u + \tan u| + C$

34. $\displaystyle\int \csc u\, du = -\ln|\csc u + \cot u| + C = \ln|\csc u - \cot u| + K$

35. $\displaystyle\int \sec^2 u\, du = \tan u + C$

36. $\displaystyle\int \csc^2 u\, du = -\cot u + C$

37. $\displaystyle\int \sec u \tan u\, du = \sec u + C$

38. $\displaystyle\int \csc u \cot u\, du = -\csc u + C$